Surveying Computations

John Walker · Joseph Awange

Surveying Computations

A Necessity for Surveyors and Engineers

Springer

John Walker
Perth, WA, Australia

Joseph Awange
Department of Land Surveying
and Geo-Informatics
The Hong Kong Polytechnic University
Kowloon, Hong Kong

ISBN 978-3-031-94381-2 ISBN 978-3-031-94382-9 (eBook)
https://doi.org/10.1007/978-3-031-94382-9

This Springer imprint is published by the registered company Springer Nature Switzerland AG
The registered company address is: Gewerbestrasse 11, 6330 Cham, Switzerland

SURVEY COMPUTATIONS: A necessity for surveyors and engineers.

Preface

Efficient computational skills are required for engineering, cadastral and mining surveys works. More often pre-survey calculations are made in the office from existing survey or engineering plans to be used to find existing survey marks needed for further tasks such as construction on one hand, and for planning/ tender submission on the other hand.

Following pre-survey calculations, field survey calculations are undertaken after connection to survey marks and used to set out new/existing boundaries, engineering marks, etc. These field calculations cannot be done as pre-survey calculations in the office since field information is needed before decisions on where to put marks are made, e.g., check calculations needed to prove field measurements and use of observations in field book such as calculations in the field book calculation sheet(s).

Finally, post-survey calculations are needed to convert data into the appropriate format for the final plan/presentation. In the sequence pre-survey, field-survey and post-survey computational patterns above, a surveyor/engineer should set out calculations clearly and neatly including sufficient information so that others (and oneself) can see what has been done, i.e., should be 'repeatable'.

Setting out calculations should be done as follows: Draw a diagram – approx. to scale – always use a ruler! show known information, state the aim of the calculation / what are the unknowns? Show all stages of the calculation – it should be clear on how the problem is solved, and make the answer/results clear – these may also be shown on the diagram.

Assuming no computational surveying knowledge, this book starts by introducing basic computational skills essential for undergraduate students in part 1. In this part, comprising 10 chapters, the book starts by looking at general formulars, coordinate systems and conversions. It then moves to angular computation introducing the necessary trigonometry employed, delves into intersection, resection and traversing procedures used to establish horizontal controls before looking at area calculations needed for earthworks and simple horizontal curve computations needed for setting out circular road/rail alignments. The book ends by looking at subdivisions, necessary for property/cadastral surveying. Chapter 18 consolidated the various formulæ implemented in part 1. Chapter 18 also examines the application of computational skills using an algebraic calculator.

Part 2 of the book presents intermediate undergraduate concepts by examining significant figures in calculations and error propagation of observations. It then extends the simple horizontal curves to transition/composite incorporating superelevation and vertical curves. Geological strikes and dips as well as coordinate transformations are also presented. A detailed examination of the transverse Mercator projection is accompanied by examples covering a variety of different projection parameters.

Finally, part 3 considers a miscellany of survey topics such as survey target design, least squares adjustments of plane intersections, reenforcing modern EDM positioning methods and alignment of survey lines are presented. An examination of transverse Mercator zone to zone transformation methods is included for interest.

The 22 chapters each contain worked examples throughout the chapter as well as chapter exercises and answers to reinforce the particular subject.

There are over 500 illustrations, 90 linked equations and 70 tables in this book

Historical aspects of land measurement.

The Surveyor and Boundaries.

The definition of real property boundaries of parcels of land is a task that has been entrusted to the SURVEYOR since ancient times. The annual flooding of the Nile River in Egypt required a surveyor to re-establish the property boundaries after each inundation.

In more modern history, near the start of the 16th century, in the English feudal system of land tenure, the surveyor (from the French *sur*= over, *voir* = see) was an overseer, an executive officer, of a landed nobleman. His job was to inspect and note the boundaries of the nobleman's lands. To this day parcels of land are described by "metes and bounds", that is to say; by the physical relationship between points and lines on the ground forming the boundary of the property.

Its most familiar form to us is the corner peg of a block of land. The definition of the parcel of land is tied to the physical "peg in the ground." The surveyor is charged with describing the physical boundary of a property in words and with a plan.

The change from "overseer" to measurer and mapmaker began in the 1530s as feudal lords tried to maximise returns from their holdings. They became more interested in their holdings in terms of being able to extract rents from their tenants so that they could pay their own dues to the Crown, which OWNED the land.

The feudal system started to break down in England after King Henry VIII, starting in 1538, dissolved and sold some four hundred Catholic monasteries as a way of making money for the Crown. The rising wealth of the merchant class meant that they were able to buy (cheap) land and become landed gentry. This important change in land tenure meant that land was valued for its unit of area, rather than its unit of production under the feudal system. The problem became one of being able to measure plots of land to determine their areas.

Prior to this time, from William the Conqueror's great survey of England noted in the Doomsday Book of 1086, the area of the various estates was described in terms such as "virgates"; an amount of land that could support a person, and "hides", which could support a family. This meant that these descriptors varied in size, depending on the fertility of the land.

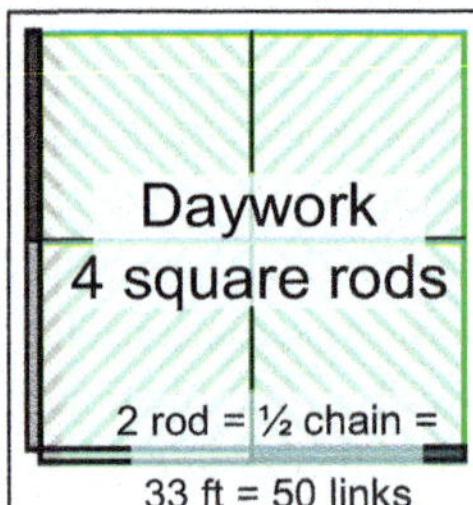

Fig 1. Daywork area

A new measurement was needed that would be constant and repeatable. So land was measured by a fixed length called a ROD, defined as 16½ feet (5½ yards). The foot had been defined as a fixed physical measure by the 12 century and incorporated nationally in the 13th century Magna Carta. [1] (Skinner, 1952) describes the derivation of the *ulna* (yard) from "*The Statute of Measuring Land, 33 Edward I, Stat. 6 1305*".

By 1601 Queen Elizabeth I had established the yardstick of 36 inches. ".. foure graines of barley make a finger, foure fingers a hand, foure hands a foot", three feet a yard and 1760 yards make a mile. 1 statute mile = 5280 feet.

Other research [3] shows multiples resulting in the yard: 1 barleycorn = 4 poppyseeds, 1 inch = 3 barleycorns, 1 hand = 4 inches, 3 hands (12 inches) = 1 foot (ft), 3 feet = 1 yard.

The hand is used to measure the height of a horse to the withers (1 hand ≈ 0.1m).

In typography, 12 points = line, 6 lines = 1 inch = 72 points per inch (ppi). 1 point ≈ 0.353mm. This means that 10-point type has an upper X height of ≈ 3.5mm).

Measuring AREA, Gunter's contribution.

Enter Edmund Gunter, born in Wales in 1581. He had a passion for mathematics, especially ratios and proportions. By 1607 he had invented the "chain" with which to measure property. It was a clever distillation of current practice with the rod and the adaptation of the new "decimal" mathematics. Simon Stevin, a Flemish engineer, first published an account of decimals in 1586.

Gunter made the chain 4 rods long. This was 22 yards or 66 feet. It also meant that an acre was now defined as 1 chain × 10 chains, 66 × 660 feet or 43,560 square feet (4,840 sq yards). To subdivide this chain into manageable decimal units, Gunter incorporated 100 LINKS in the chain. Each 10 links had a brass marker. Thus the link had a value of 0.66 ft or 7.92 inches. (1 link = 0.201168m).

In Australia the Gunter's chain was replaced by the surveyor's steel band (generally 5 chain, 100m long) by about 1860. The band was made of crinoline wire, a flat wire use in lady's underwear. The wire was either 1/8″, 1/12″ or 1/16″ (about 3mm, 2mm or 1.5mm) wide with stamped brass markers every 10 links (2m). A 15 link (3m) leader with 1/100 link (2mm) subdivisions was used for fine measurement.

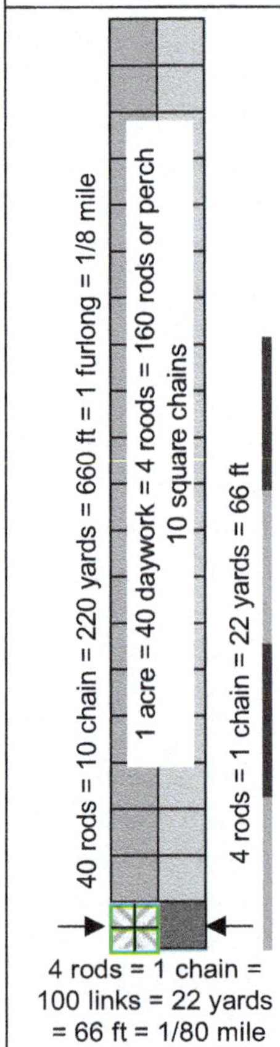

Fig 2. The acre

Measuring imperial area.

It was considered that a man could cultivate or tend a "daywork area" of 4 square rods: i.e. 2 rods × 2 rods or 33ft × 33ft. (Fig 1. Dayworks area).

A team of oxen could work an area of 40 dayworks per day. This became the ACRE, the equivalent of 160 square rods, just called rods because of the implied areal quality. An acre was thus 4 rods x 40 rods. (Fig 2. The acre).

A subdivision of the acre that is seen on old plans is the rood. This is the ¼ acre, it is an area of 40 (square) rods. So, a cadastral plan will show an area, for example, as 2 a. 3 r. 27 p.

Now 2 acres, 3 roods and 27 perch = 2 + ¾ + 27/160 acres

2 a. 3 r. 27 p = 2 +0.75 +0.16875 = 2.91875 acres = 29.1875 sq chains.

Or 2 x 160 + 3 x 40 + 27 = 320 + 120 +27 = 467 (sq) rods.

Fig 3 shows the breakdown of the area of 2 acres, 3 roods (3×¼ acre) and 27 perch or (square) rods.

Remembering that an acre was an area 1 chain wide x 10 chains long we then have the measure of the FURLONG, a furrow *long*, which is a 10-chain distance that would be ploughed by a team of oxen. Long runs were favoured as being more efficient: – it takes a long time to turn an ox team around.

Thus, if the 2 a. 3 r. 27 p. field was 1 chain wide, it would be 29 chain and 19 links long (29.1875). Or it can be estimated to 29 ch. and 18¾ links.

And there are 8 furlongs in an English (statute) mile. The Melbourne Cup was run over 2 miles or 16 furlongs. It is now run over a rather mundane 3200 metres.

A mile is 80 chains long. A SQUARE mile is 80 × 80 = 6400 square chains = 6400/10 = 640 ACRES. Square miles are the basis of the sections in US plats.

Metrication of the chain.

Many road reserves were 1 chain wide. 1 chain = 66 ft = (66 x 0.3048) = 20.1168m. We now use a 20m road reserve: very much the equivalent of the one chain reserve. 5 links are about a metre long (1.006m), Many old Reference Marks (RMs) are offset at 5 links. A mile of 1 chain road reserve = 8 acres, ≈ 3.2Ha.
1Km of 20m reserve = 2Ha.

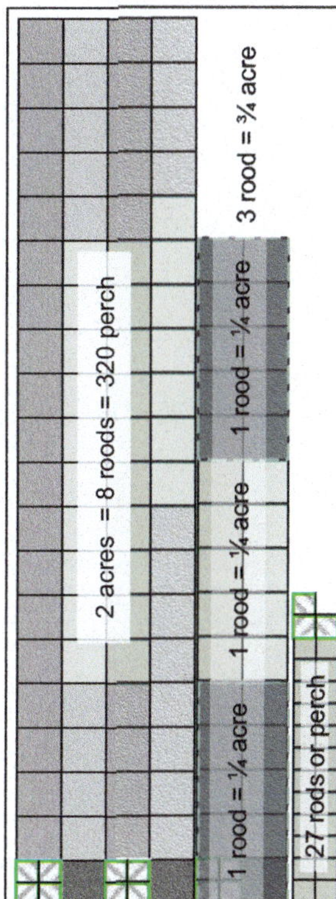

Fig 3. 2a 3r 27p area

Comparison with metric units

However, it gets better. If we accept the concept of the English "daywork" area, what are the metric equivalents? A rod = 16½ feet ≈ 5m (5.029m). A familiar rod equivalent is the levelling staff, normally extended to 5m.

Thus, a daywork in metric measure is (5 + 5) x (5 + 5) m2 or 100m^2 = 1 ARE (France) - and a square metre thus becomes a centiare. Within sensible limits (≈ 1%) the *daywork* and the *are* have a great similarity. 1 daywork = 101.2m^2.

The Hectare = 100 Are = 10,000 sq metres. And a square kilometre = 100 Hectares. There is no "nice" conversion between acres and hectares, just a nasty multiplier of 0.404 685 642 24. However, 1 acre ≈ 4 047m^2.

There are similarities between the Acre, the French *journal*, the north German and Dutch *morgen*, south German and Swiss *juchart* or *tagwerk*, Austrian *joch*, and Czech *jitro*. These are all measures of a day's ploughing with oxen.

Death of the chain?

Not likely. The settlement of North America, from the beginning of the 17th century, saw the chain cross the Atlantic. After independence, America was founded and funded by the sale of land [2] Linklater, 2002. That imperial measure, the acre, and the square mile sections that were stamped across the land by the Public Lands Survey System (PLSS), has taken such a hold. It is inconceivable that the hectare will ever invade. Thomas Jefferson's continued efforts prior to and during his presidency (1801 – 1809) to metricate America always fell to the supremacy of the Gunter's chain and the foot which defined the area of the land. The USA may metricate one day, but the land will always be expressed in acres. Long live the chain!

Be aware of the difference between the International foot (0.3048m) and US Survey foot (1200/3937m), especially when using mapping programs originating in the US. (The PLSS defined the metre as 39.37 inches).

Further Reading.

[1] "The English Yard and Pound Weight", Skinner, F.G., Bulletin of the British Society for the History of Science, Vol.1 No. 7 (May, 1952). Pp179-187. Cambridge University Press.

[2] "Measuring America. How the United States was Shaped by the Greatest Land Sale in History", Linklater, Andro, Harper Collins Publishers, London, 2002. ISBN 0 00 710887 7

[3] http://en.wipedia.org/wiki/English_unit (December 2007)

Examples to try:

As you study this course it is as well to see how other people solve problems.
https://www.thecivilengineer.org/education/professional-examinations-preparation/calculation-examples

If it's not checked it's wrong.

Table of Contents

If it's not checked it's wrong.

PART 2: Intermediate undergraduate concepts

PART 3: A miscellany of survey topics

About the Authors

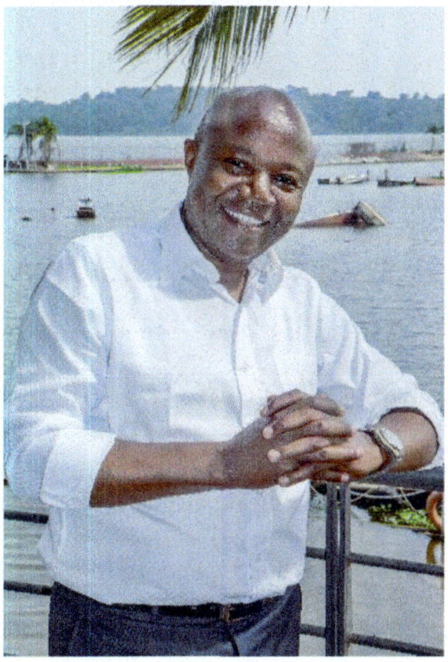

Prof Joseph Awange's Consistently ranked among the world's top 2% most-cited scientists since 2020 (Stanford), Joseph Awange is a Professor of Environmental Geoinformatics at The Hong Kong Polytechnic University's LSGI Department, having previously worked at Curtin University, Australia.

Prof. Awange began his academic journey with BSc and MSc degrees in Surveying from the University of Nairobi, Kenya. His exceptional promise was recognized through a DAAD merit scholarship, enabling him to earn a second MSc and a PhD in Geodesy at Stuttgart University, Germany, under the supervision of the late, world-renowned Prof. Erik W. Grafarend. His postdoctoral training was completed at Kyoto University (Japan) under the prestigious Japanese Society for the Promotion of Science (JSPS) Fellowship.

His distinguished research career has been honoured with several international awards, including multiple Alexander von Humboldt Fellowships, a 2015 JSPS Fellowship, and recognition from the Brazil Frontier of Science. Beyond his research, Prof. Awange holds a Postgraduate Diploma in Environmental Impact Assessment from Murdoch University, Australia. He actively contributes to the global scientific community as the Acquisition Manager for Earth Sciences, Geography, and Environment at Springer Nature, and as an Associate Editor for the top-ranked journal *Remote Sensing of Environment* and *Science of Remote Sensing*. His scholarly output includes more than 26 books and over 260 high-impact journal articles, and he has cultivated the next generation of scientists by mentoring more than 50 BSc (Honours), MSc, PhD, and postdoctoral researchers globally.

If it's not checked it's wrong.

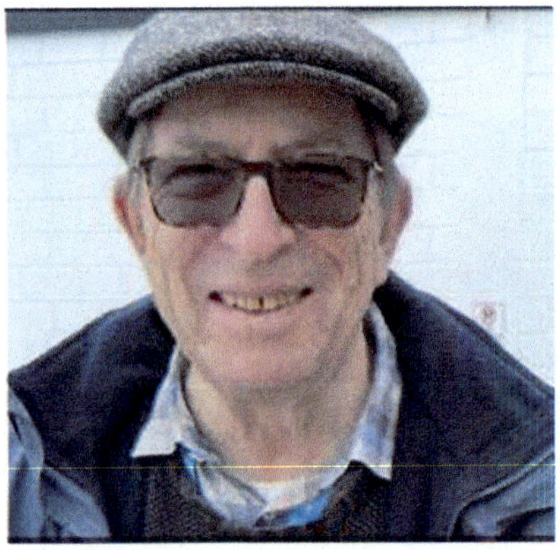

John Walker is an octogenarian who has been active in the practical and academic world of surveying, as well as working in agriculture, aviation and electronics.

He started out as a wool classer, shearing shed manager and company manager in Queensland.

He gained a commercial pilot licence and has flown over 2200 hours in mainly single engine aircraft. The time included aerial mustering of livestock and glider towing.

Time was spent in the Pilbara region of Western Australia managing a pastoral property and flying aerial mustering operations in the Pilbara and Ashburton areas of Western Australia.

He then studied Surveying and Mapping at the Western Australian Institute of Technology (now Curtin University) and worked as a engineering surveyor on earthworks projects.

Leaving surveying, he joined the Australian Civil Aviation Authority and spent 17 years as an Air Traffic Controller, rising to a Senior Operations Controller and Search and Rescue Mission Controller (SOC/SARMC) covering the Western Australian Flight Information Region (FIR). During that time, prior to the introduction of GPS, he pioneered the calculation and introduction of direct air routes for aircraft equipped with self-contained navigation systems (INS and Omega low frequency radio navigation stations). As GPS was introduced, he incorporated its positioning capabilities into Search and Rescue (SAR) pattern flying and participated in several search and rescue efforts in the Indian Ocean. His surveying and aviation background allowed him to develop SAR planning software on early PCs as they were introduced into SAR operations. As an air traffic controller he also participated in the design of the controller/machine interface of the Eurocat 2000 air traffic control system introduced as The Australian Advanced Air Traffic SYSTEM (TAAATS).

During his employment as an ATC he completed his degree and postgraduate diploma in Surveying and Mapping at Curtin. Upon his retirement from ATC he became involved in the design and manufacture of electronic equipment, concentrating mainly on vehicle tracking systems. His employer was awarded a contract to provide the real time tracking and prediction software and hardware for the Perth Central Area Transit System (CAT). The initial system, introduced in 1996, involved 16 buses and 65 bus stops. The bus stop had voice and visual predictions of bus arrival times, while the buses had voice and visual displays of "The next stop is…". Vehicle tracking was by polling the bus positions for Differential Global Positioning Systems (DGPS) enhanced position. Involvement in other systems included the Wastetrack system on waste disposal trucks and tracking transcontinental heavy haulage trucks using Global Positioning Systems (GPS) and Global System for Mobile Communication (GSM) telephones.

He returned to Curtin University as a casual academic and has taught in an Air Transport Pilot degree embracing aircraft performance, navigation, flight planning and crew resource management. In Surveying he has lectured and tutored in the undergraduate areas of survey computations, cartography, engineering surveying and astronomy. Practical sessions were designed for a civil engineering surveying course, these exercises emphasised the theoretic basis of undergraduate surveying.

With Dr Awange, he is the co-author of Springer textbooks:

Walker, John and Joseph Awange, Surveying for Civil and Mine Engineers, First Edition, 2018, Springer International Publishing AG. ISBN 978-3-319-53128-1

Walker, John and Joseph Awange, Surveying for Civil and Mine Engineers, Second Edition, 2020, Springer Nature Switzerland. ISBN 978-3-030-45802-7

He participated in the comparison of CAT bus real time single frequency DGPS tracking compared with the (then) emerging dual frequency GPS receivers which required extensive post processing.

Barry Pantall, Mike Stewart, Maria Tsakiri, John Walker. CATS on the Prowl, Improving Public Transport in Perth, GPS World, Advanstar Communications Inc., USA, April 1999.

If it's not checked it's wrong.

PART 1: Basic undergraduate concepts

If it's not checked it's wrong.

Chapter 1 GENERAL FORMULA, COORDINATE SYSTEMS, CONVERSIONS, CALCULATIONS

Structures and engineering works are normally established on the physical earth (known as the plane system) whereas other surveying/geodesist structures such as the global navigation satellite system (GNSS) to which global positioning system (GPS) belong orbit some 20000 km above the surface of the Earth. In both cases, engineering and surveying/geodesy, it is essential that the computations be done in the proper system. For example, the engineers will undertake their computations on a plane defined by two perpendicular axes (x and y) whereas the geodesist or surveyor will use an ellipsoid of revolution (imagine an egg rotated) to get the positions from the satellites. This Chapter starts by introducing general formulae before presenting these two coordinate systems and presents the methods of converting from one system to another. Knowledge and skills from this fundamental chapter is a necessity for the subsequent chapters as most if not all computational surveys for engineering and surveying work will be undertaken in either of these two systems.

The aim of the first half of this book is to introduce the student to the fundamentals of survey computations and, in particular, the accuracies involved in arithmetic computations and the proper use of mathematical results. The person performing the calculations is called the computer. A basic electronic calculator with hardwired trigonometric functions is required for this part of the course.

The second part of the course, using these fundamental methods, introduces the student to the solution of some basic survey calculations, and to solve particular survey problems.

1.1 Overview

1.1.1 Setting out calculations.

There is a variety of opinions on this matter. Generally, setting out should be done in the following form.

(a) Diagrams should be drawn to scale where possible
(b) Information given should be stated or drawn on the diagram
(c) The aim of the exercise should be stated
(d) The computation should be set out in a neat logical manner with a minimum amount of explanatory matter. With some problems it may be convenient to draw a line about 80mm from the right-hand side of the page and perform calculations to the right of this line. The left-hand side should then contain the diagram, formula used, results etc.
(e) All stages in the computation process should be fully documented. Note that this does not mean covering reams of paper with computations, making it difficult for others to follow.

It should be noted that many computations in surveying are best set out in tabular form. This is the case with many of the calculations encountered in this course. While it is recommended that you use the setting out methods given here, as they have been developed from experience, you should feel free to use any format provided it is neat and logical.

Setting out in tabular form means that numbers should be vertically aligned, making sure decimal numbers are **decimal aligned** under each other.

1.1.2 Checking Calculations.

I.J "Jack" Good (1916 - 2009) had a maxim: *"If it's not checked it's wrong."*
(Copeland, 2010)(Colossus, The Secret of Bletchley Park's Codebreaking Computers, B. Jack Copeland et al, p215. Oxford University Press, 2010. Jack Good was a WWII codebreaker and became a famous mathematician and statistician.)

The checking of calculations is an integral part of survey computations. There are two accepted methods of carrying out checking procedures:

(a) An independent calculation of the problem by a second computer is performed and the results compared,
OR
(b) Another calculation of the problem by the same computer but using a different mathematical process.

There are some calculations which will provide built-in checks, but many others will not be able to be checked without a considerable amount of work. Where calculations are unchecked, the computer should be aware of this and treat the results with special care.

Another source of error arises when students use incorrect data in calculations. All transcribed data must be double checked.

J. Walker and J. Awange, *Surveying Computations*,
https://doi.org/10.1007/978-3-031-94382-9_1

1.2 Significance of calculation.

You will be introduced to the significance of figures early in this course. What you must ensure is that the results of calculations reflect the significance of the data used in the calculation. See Chapter 11, page 169, § 11.1-§ 11.4.

1.3 Angles in plane triangles

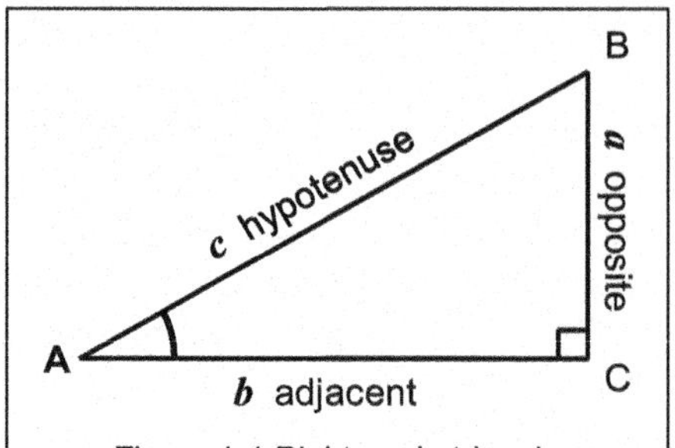

Figure 1-1 Right angle triangle

1.3.1 General formula. ABC is a right-angle triangle where angle C = 90° (Figure 1-1).

Considering the angle **A**

$$\sin(A) = \frac{a \ \ opposite}{c \ \ hypotenuse}. \quad \mathrm{cosec}(A) = \frac{c \ \ hypotenuse}{a \ \ opposite}$$

$$\cos(A) = \frac{b \ \ adjacent}{c \ \ hypotenuse}. \quad \sec(A) = \frac{c \ \ hypotenuse}{b \ \ adjacent}$$

$$\tan(A) = \frac{a \ opposite}{b \ adjacent}. \quad \mathrm{cosec}(A) = \frac{b \ adjacent}{a \ opposite}$$

The general mnemonic for remembering the relationship of angles and sides is:

Sin: opp/hypot; Cos: adj/hypot; Tan: opp/adj

SOH **CAH** **TOA**

1.3.2 General triangle, no right angles.

A number of rules are developed which govern the relationship between the angular functions and the triangle's sides. The rules are cyclical. See Figure 1-2

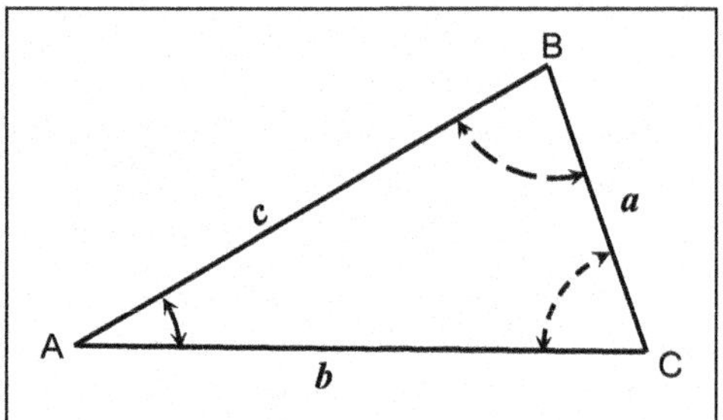

Figure 1-2 General triangle

sin *rule*
$$\frac{a}{\sin A} = \frac{b}{\sin B} = \frac{c}{\sin C} \quad (1)$$

Angle from sides
$$\sin\left(\frac{A}{2}\right) = \sqrt{\frac{(s-b)(s-c)}{bc}} \quad (2)$$

$$\cos\left(\frac{A}{2}\right) = \sqrt{\frac{s(s-a)}{bc}} \quad (3)$$

$$\tan\left(\frac{A}{2}\right) = \sqrt{\frac{(s-b)(s-c)}{s(s-a)}} \quad (4)$$

cos *rule*
$$\cos A = \frac{b^2 + c^2 - a^2}{2bc} \quad (5)$$

tan *rule*
$$\tan\left(\frac{(A-B)}{2}\right) = \frac{(a-b)}{(a+b)}\cot\left(\frac{C}{2}\right) \quad (6)$$

AREA
$$Area = A = \sqrt{s(s-a)(s-b)(s-c)} \quad (7)$$

$$Area = A = \frac{1}{2} bc \sin A \quad (8)$$

$$Area = A = \frac{1}{2}\frac{a^2 \sin B \sin C}{\sin A} \quad (9)$$

Semiperimeter *where* $s = \frac{a+b+c}{2}$

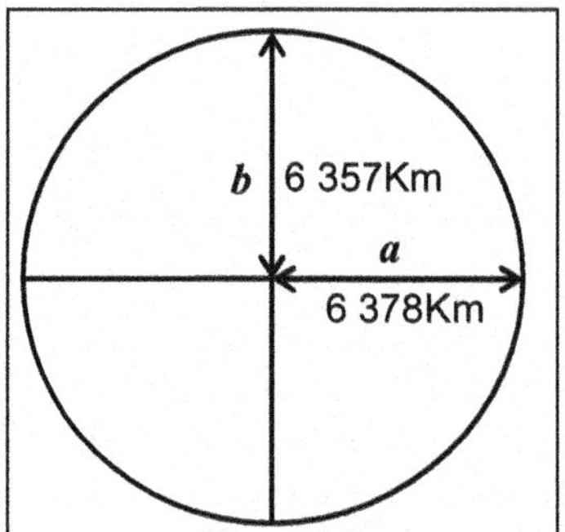

Figure 1-3

1.4 Coordinate Systems.

Because the earth's surface (topographical surface) is irregular, it cannot be used as a reference system for defining points. The figure which best fits this surface is an ellipsoid (a reference ellipsoid).

The semi-axes of the reference ellipsoid, ***a*** and ***b***, are about 6 378Km (equatorial) and 6 357Km (polar) (See Figure 1-3) due to gravitational changes. This is a *flattening* of about 1:300 (*a*/(*a*–*b*)).

The standard Geodetic Reference SYSTEM, adopted in 1980, (GRS80), has a coordinate system based on the mass centre of the earth with values:

GRS80 a = 6 378 137.0, b = 6 356 752.314 140m, f = 1/298.257 222 101.

A small change by the United States Defense Mapping Agency created the World Geodetic System 1984 (WGS84), which is used by the Global Positioning System (GPS) satellite navigation system.

Parameters are:

WGS84 a = 6 378 137.0, b = 6 356 752.314 245m, f = 1/298.257 223 563.

The difference is 0.105mm (0.000105m) in the semi-polar axis and is inconsequential in all but the most esoteric survey computations.

A point in this system is defined by its ellipsoidal latitude (Φ - phi) and ellipsoidal longitude (λ - lambda).

1.4.1 Rectangular Cartesian* Coordinates.

Rectangular coordinates in the plane are the basic values used in survey work. The rectangular coordinate 3- dimension frame can be expressed as 3 orthogonal vectors, as illustrated, or as a circular vector of magnitude and direction. The circular vector is examined in a later chapter regarding coordinate conversion.

The system is defined within a local area in the vicinity of **O** (See Figure 1-4).

OZ = zenith direction of gravity = plumb line vector (See Figure 1-5)

OY = North direction

OX = East direction

Point A in this system is referenced as (X,Y,Z) or (E,N,H)

E = Easting = OX

N = Northing = OY

H = OZ = Height with respect to a datum

* Cartesian: named after Rene Descartes (1596-1650), from his work in philosophy and analytical geometry.

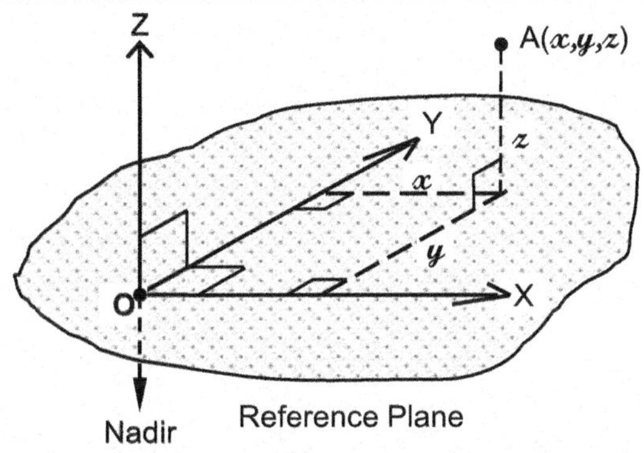

Figure 1-4 The (X, Y, Z) system

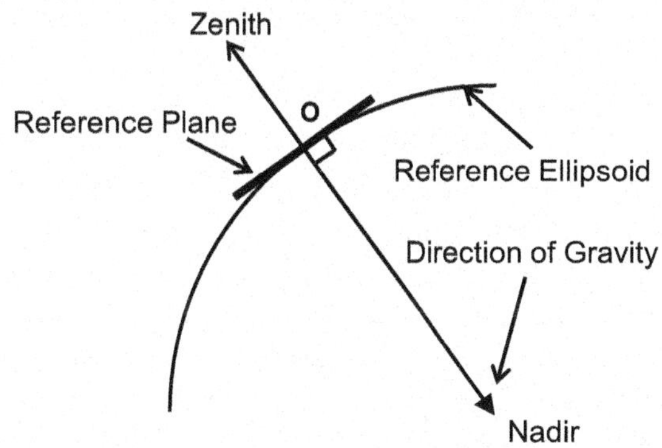

Figure 1-5 On the reference ellipsoid

1.4.2 Spheroidal or Ellipsoidal Coordinates of point P.

As surveyors you will work either on a plane surface, an almost spherical surface called the ellipsoid or the actual (local) geoidal surface of the earth.

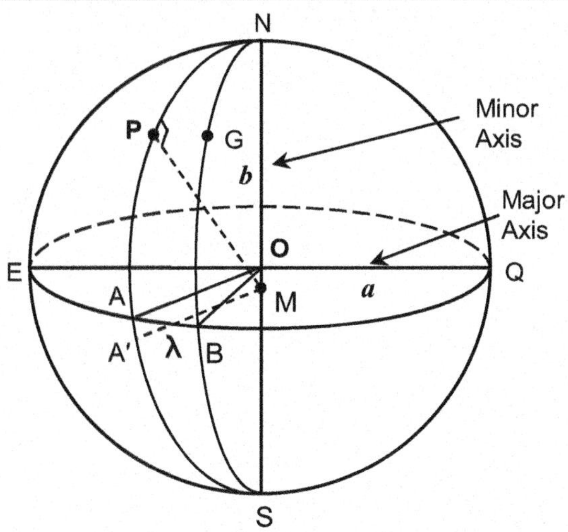

Figure 1-6 Reference ellipsoid

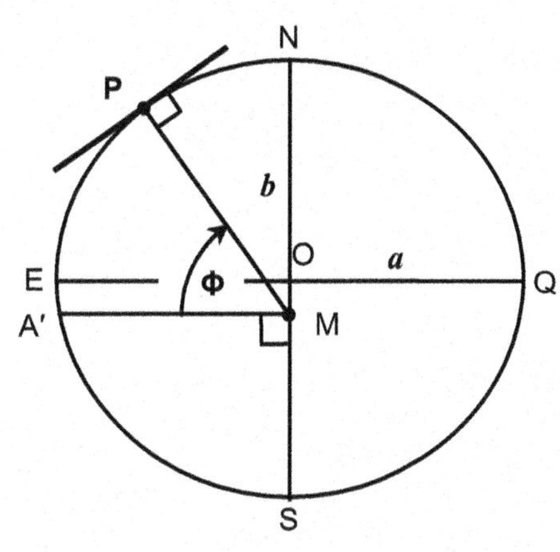

Figure 1-7 Reference ellipsoidal meridian plane

Point O is the centre of the ellipsoid from which the major and minor radii (a and b) are measured.

Point M (Figure 1-6) is the ellipsoidal centre where the ellipsoidal normal from a place P intercepts the minor (NS) axis of the ellipse.

EABQ = Equator in Figure 1-6 and EOQ = Equator in Figure 1-7 of the ellipsoid.

PM = Ellipsoidal Normal in the meridian plane is shown in

NGBS = ICRS, Greenwich or nominated reference meridian

Ellipsoidal latitude = $\Phi = \angle PMA'$ (Figure 1-7) A′ M perpendicular to minor axis NS

Ellipsoidal longitude = $\lambda = \angle AOB$, (Figure 1-6) i.e. angle between the reference meridian and meridian of P.

1.5 Electronic Calculators.

Electronic calculators have transformed many of the methods and procedures previously used by surveyors, both in the office and in the field. Electronic calculators have been widely used since the introduction of the Hewlett Packard HP-35, the "Electronic Slide-rule", in 1972. This series of calculators is now discontinued.

Computations are now performed in a fraction of the time previously taken to complete hand computations. The time saved must be used to check the work. The main problems presented by electronic calculators are those made by the user: mis-keying input values, transposition of digits; misreading answers, either missing decimal point, incorrect transcription of the answer; acceptance of "wrong" answers; lack of knowledge of the calculator's functions and limitations.

The success to using electronic calculators is the development of a sound methodology in problem solving. The clear and logical set out of a problem means that the calculations can be made quickly and systematically. The intermediate answers allow progressive checking and build a confidence in the result. Don't "search" for the answers using the calculator in a haphazard manner.

A recommendation will be made as to the most suitable calculator available at the time. As models change and improve no recommendation can be made for today which will be good for tomorrow. The recommendation should be made during the Semester 1 orientation week each year.

1.5.1 Alternate calculators

The use of the different calculators for survey calculations is covered as an addendum covering Survey Calculations on HP 10s+, HP300s+ and similar, § 18.16, page 298. These addendums overcome the limitations of the User Manuals as applied to surveying problems.

In some calculators it is possible to combine several operators in a single formula. The calculator performs the operations in the order shown according to the **precedence** of the operator. If a formula contains operators with the same precedence – for example, if a formula contains both a multiplication and division operator – then the formula evaluates the operators from left to right. If in doubt, use parentheses.

Order of precedence. Brackets, indices, multiplication, division, addition, subtraction (BIMDAS):

( ) parentheses are used to contain the order of operation

^ exponentiation (raised to a power, 5^2 = 5^2, 5^0.5 = $5^{½} = \sqrt{5}$)

– negation (as in –5) is a problem in some calculators. $-5^2 = -25$. Use $(-5)^2 = 25$ (Pythagoras)

* and / multiplication and division (× and ÷)

+ and - addition and subtraction.

1.6 Systems of Measurement Conversions.

Conversion between units of:

Feet/inches

Feet/decimal feet

Links/chains

Metres (SI Units)

Exact conversions.

12 inches = 1 foot

3 feet = 1 yard

66 feet = 22 yards = 1 chain

100 links = 1 chain

1 link = 7.92 inches

1 inch = 25.4mm

1 International foot = 12 inches = 0.3048m

1 U.S. survey foot = 1200/3937 = 0.304800609601…m (they accepted 1m = 39.37feet)
Links × 0.201168 = metres
Feet x 0.3048 = metres

Conversion used in cadastral surveying.

5 links = 1.00584m = 1.006m
1 chain = 20.1168 m = 20.117m
10 chain = 1 furlong = 201.168m
8 furlongs = 1 mile (statute) = 1760 yards = 5280 feet = 1609.344m

Imperial area convention and conversion. Acres, rood and perches.

1 rod, pole or perch = ¼ chain = 5½ yards = 16½ feet = 5.0292m (approx length 5m survey staff)
1 square rod, pole or perch = 1 perch (of area)
1 rood = 40 perches (area) (= 4 rods x 10 rods = 22 x 55 yards)
= 1210 sq yards = 1011.7141m^2
1 acre = 4 roods = 160 perches = 4840 sq yards= 43560 sq feet = **4046.8564 224m^2**

Metric area convention and conversion. Hectare = 10,000m^2.

1 centiare = 1m^2
1 are = 100 centiare = 100m^2
1 hectare = 100 are = **10,000m^2**
1 ha = (10000/4046.8564224) = 2.471054 acres
= 2 acres 1 rood 35 perches (2a.1r.35p)

Other interesting Conversions.

1 nautical mile = 1852 metres (previously 1 minute arc of latitude)
1 rod = 5½ yards = 16½ feet = 5.0292m. About 5m
4 square rods = 1 daywork = 11 x 11 yards = 121 sq yards
1 daywork = 1089 sq feet = 101.171m^2. About 100m^2.
1 are = 100m^2. About 1 daywork.
1 furlong = 10 chain = 1000 links = 660 feet = 220 yards = 201.168m

1.7 Workshop Exercise 1.1.

Verify the following results, correct any errors

a) $17.54+26.21-14.37$ = 29.38
b) $3.74\text{x}(-237.15)$ = 887
c) $(16.37+(20.21\times4.44))\div12.17$ = 8.35
d) $0.666328\times237.14\div258.91$ = 0.61030
e) $0.666328\times237.14\div0.610301$ = 258.91
f) $\sqrt{(17.54)^2+(18.92)^2}$ = 25.80
g) $\sqrt{17.54^2+18.92^2-2\times17.54\times18.92\times0.317985}$ = 20
h) $(473.21^2+87.94^2-442.84^2)\div(2\times473.21\times87.94)$ = 0.4
i) $\sqrt{2571.418^2+2195.773^2}$ = 3381.362
j) $\sqrt{-25743.418^2+3381.362^2-2\times25743.418\times3381.362\times0.760469}$ = 2200
k) $(2571.418^2+3381.362^2-2195.773^2)\div(2\times2571.418\times3381.362)$ = 0.8
l) $0.649374\times2571.418\div2195.773$ = 0.760466
m) $\sqrt{-3^2+-4^2}$ = 5

1.8 Workshop Exercise 1.2.

Use the memory of your calculator to hold the conversion factor, feet to metres = 0.3048.
Verify the following results:
Convert feet to metres

a) 0.3048×2375.144 = 723.9
b) 0.3048×6378160.000 = 1944000.168
c) 0.3048×0.717 = 0.219
d) $0.3048\times(-141561.662)$ = −43150.995

e) $0.3048 \times 151.070 = 46.05$
f) $0.3048 \times 66.212 = 20.18$

1.9 Laboratory Exercise 1.1

Part I: Conversions between units (38 points; 1 point per answer)

a.) Convert these horizontal distances between metric (metre) and imperial (feet) system

Use the following relations/conversions:
1 foot = 12 inches = 0.3048 m; 1 inch = 25.4 mm; 1 link = 7.92 inches

Inches	Feet	Metres	Links
5.00			
	537.52		
		4129.57	
			2.345

b.) Convert these angular values between sexagesimals and radian system

Use the following relations/conversions:
$1° = 60'$; $1' = 60''$; $1' = 1°/60 = 0.016666667°$; $1'' = 1°/3{,}600 = 0.000277778°$
1 rad = $360°/2\pi = 180°/\pi = 57.2957759°$ (conversion factor)
1 rad = $360\times60'/2\pi = 180\times60'/\pi = 3437.746771'$ (conversion factor)
1 rad = $360\times3{,}600'/2\pi = 180\times3{,}600''/\pi = 206{,}264.8063''$ (conversion factor)

Deg. Min. Sec.	Decimal Degree	Radians
93° 18' 26"		
	192.8155556°	
		4.993784538

c.) Convert these areas between metric (metre) and imperial (feet) systems

Use the following relations/conversions:
1 chain = 100 links = 66 feet = 4 rods; 1 rod = 1 perch = 16½ feet =5.0292 m;
1 sq rod = 1 sq perch or 1 perch of area = 5.0292×5.0292 m^2
1 rood = 40 sq rods or 40 perches; 1 acre = 4 roods
decimal acres = acres + roods/4 + perches/160

square metre	sq rod or sq perch	rood	acre	acre rood perches
4,567.0				
	2.479			
		1.247		
			5.294	
				7ac 3r 35p

If it's not checked it's wrong.

Part II: Circular Curves (14 points; 1 point per answer)

a.) Provide the relation between length of arc, radius and angle given in radians

See text book or lecture notes.

Relation:

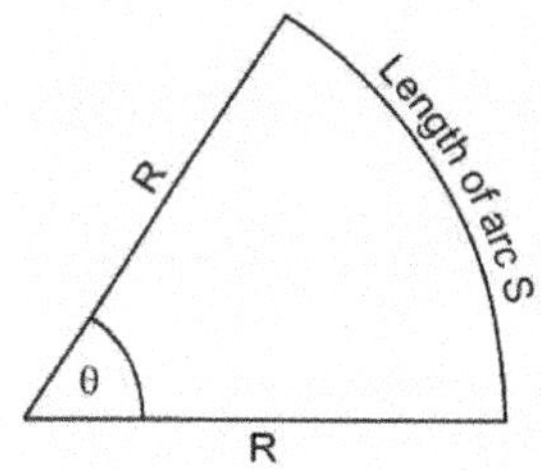

b.) Derive the length of arcs for the following radius and angle combinations

Use the above relation.

Angle θ [DMS]	Angle θ [rad]	R [m]	S [m]
93° 18' 26"		264.34	
93° 18' 26"		5,284.26	
357° 59' 16.47"		264.34	
357° 59' 16.47"		5,284.26	

c.) A satellite is on a circular orbit at an altitude of 20,183 km (e.g. GPS satellite) above the Earth surface. What distance does the satellite travel when covering an angular distance of 18° 32′ 18.76″? Assume a mean Earth radius of 6,378,000 m.

Distance travelled: ________________ km

d.) For the same level of accuracy provide the accuracies for angle and distance measurements assuming distances of 500 m.

Use the angle-distance relationship.

Angle accuracy	Distance accuracy
± 15″	
± 3.5′	
	± 3 mm
	± 0.05 m

Part III: Directions, Angles and Bearings (14 points; 1 point per answer)

e.) Add the following angles together

If required reduce the result to a value between 0° and 360°

93° 17′ 25.00″ + 193° 48′ 16.00″ = ________________

324° 07′ 58.00″ + 193° 56′ 56.00″ = ________________

5° 07′ 55.63″ + 356° 58′ 16.47″ = ________________

f.) Provide all combinations of clockwise angles based on the directions given figure)

∠ AOB = ____________________

∠ AOC = ____________________

∠ BOC = ____________________

∠ BOA = ____________________

∠ COA = ____________________

∠ COB = ____________________

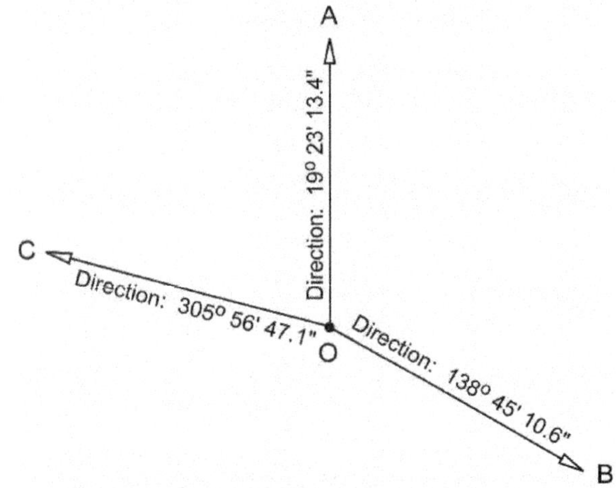

g.) Provide the forward and back bearings of all lines in the given figure

Pay attention to the definition of the angles.

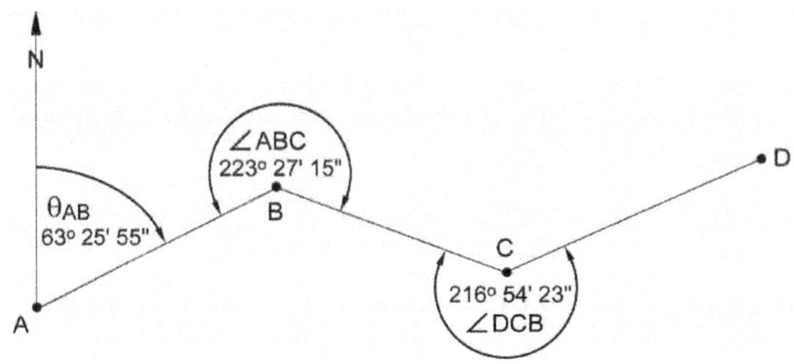

Forward Bearing		**Back Bearing**	
θ_{AB} =	63° 25′ 55″	θ_{BA} =	
θ_{BC} =		θ_{CB} =	
θ_{CD} =		θ_{DC} =	

1.10 Greek Symbols used in Calculations.

The Greek alphabet, and the names of its letters, is used to symbolize many of the constants and variables used in survey calculations. The following Table 1-1 shows the upper- and lower-case symbols. They can be generated using the **Symbol** font in Microsoft programs. However, they can also be found as symbols in various other **font sets** of your writing programs (Inset > Symbol). The "safest" fonts are for serif, Times New Roman (TNR), for sans serif, use Arial (at a slightly lower font size). The Lucida font set handles most characters. Fixed pitch fonts best used are Courier New and Lucida Console. The previous Calibri font was popular but displayed a nasty degree symbol as º (note the underscore) rather than ° in TNR, Arial and the others.

The new (2023) default MS Aptos font (in its many variations) does not use the º symbol. (ALT+0186 = º)

The lowercase epsilon normally looks like a reversed 3 (ε, Symbol e). Most texts favour the lunate epsilon (ϵ). It is a bit hard to find and is easiest inserted by typing 03f5 and then pressing the ALT key and the X key (03f5 Alt+X). Other alternates (and their HEX values in Table 1-1) are ς (ending small sigma), ϖ (Greek pi symbol), ϒ (upsilon with hook) and ϑ (Greek small theta). The HTML mnemonic codes are included for reference.

Most symbols can be found in normal word processing software. However, some of the more obscure lower-case characters are hard to find. The HTML codes are used as escape codes in LaTex documents. The ASCII codes for the more common Greek characters are quicker to input as the ALT+ code sequence instead of using the INSERT>SYMBOL menu. MS Word and PowerPoint play nicely. Excel requires the use of the Symbol font.

Table 1-1 Table of Greek symbols

Name	Upper	Lower	Letter	HEXIDECIMAL		ACSII		HTML code	
	Upper	Lower		UPPER	lower	UPPER	lower	UPPER	lower
alpha	Α	α	A a	0391	03B1	0913	0945	Α	α
beta	Β	β	B b	0392	03B2	0914	0946	Β	β
gamma	Γ	γ	G g	0393	03B3	0915	0947	Γ	γ
delta	Δ	δ	D d	0394	03B4	0915	0948	Δ	δ
epsilon	Ε	ε	E e	0395	03B5	0917	0949	&Epsilo;n	ε
lunate epsilon		ϵ			03F5		1013		ϵ
back epsilon		϶			03F6		1014		϶
zeta	Ζ	ζ	Z z	0396	03B6	0917	0950	Ζ	ζ
eta	Η	η	H h	0397	03B7	0918	0951	Η	η
theta	Θ	θ	Q q	0398	03B8	0920	0952	Θ	θ
	ϴ	ϑ		03F4	03D1	1012	0977		ϑ
iota	Ι	ι	I i	0399	03B9	0921	0953	Ι	ι
kappa	Κ	κ	K k	039A	03BA	0922	0954	Κ	κ
lambda	Λ	λ	L l	039B	03BB	0923	0955	Λ	λ
mu	Μ	μ	M m	039C	03BC	0924	0956	Μ	μ
nu	Ν	ν	N n	039D	03BD	0925	0957	Ν	ν
xi	Ξ	ξ	X x	039E	03BE	0926	0958	Ξ	ξ
omicron	Ο	ο	O o	039F	03BF	0927	0959	Ο	ο
pi	Π	π	P p	03A0	03C0	0928	0960	Π	π
		ϖ			03D6		0982		ϖ
rho	Ρ	ρ	R r	03A1	03C1	0929	0961	Ρ	ρ
sigma	Σ	σ	S s	03A3	03C3	0931	0963	Σ	σ
sigma final		ς			03C2		0962		ς
tau	Τ	τ	T t	03A4	03C4	0932	0964	Τ	τ
upsilon	Υ	υ	U u	03A5	03C5	0933	0965	Υ	υ
upsilon hook		ϒ			03D2		0978		ϒ
phi	Φ	ϕ	F f	03A6	03D5	0934	0981	Φ	
small phi		φ			03C6		0968		φ
chi	Χ	χ	C c	03A7	03C7	0935	0967	Χ	χ
psi	Ψ	ψ	Y y	03A8	03C8	0936	0968	Ψ	ψ
omega	Ω	ω	W w	03A9	03C9	0937	0969	Ω	ω

The ALT codes (ALT+nnnn) **must** be entered from the numeric keypad. This can be a problem on some laptops, which may not have a number pad. The HEX codes can use the numeric row of the main keyboard or the numeric pad (nnnn ALT+X). Some symbols don't appear to have matching HTML entities.

1.11 Keycodes for some common math entities

Table 1-2 is a list of common characters used in this book. Most will work with the conventional fonts however a few use either Cambria Math or Segoe UI Symbol fonts. For non-numeric keypad keyboards, mainly on laptops, the use of the HEX code escape sequence may be easier than using the Insert>Symbol method. First type the hex code, then the ALT key and x (ALT+x). the alpha (A-F) keys are not case sensitive.

Table 1-2 Math symbols

	Entity	ASCII	HEX	Character		Entity	ASCII	HEX	Character
←	←	8592	2190	Leftwards arrow	≠	≠	8800	2260	Not equal to
↑	↑	8593	2191	Upwards arrow	≡	≡	8801	2261	Identical to
→	→	8594	2192	Rightwards arrow	≤	≤	8804	2264	Less-than or equal to
↓	↓	8595	2193	Downwards arrow	≥	≥	8805	2265	Greater-than or equal to
↔	↔	8596	2194	Left right arrow	⊕	⊕	8853	2295	Direct sum
↵	↵	8629	21B5	Carriage return	⊗	⊗	8855	2297	Vector product
⇐	⇐	8656	21D0	Leftwards double arrow	∀	&floral;	8704	2200	For all
⇑	⇑	8657	21D1	Upwards double arrow	∃	∃	8707	2203	There exists
⇒	⇒	8658	21D2	Rightwards double arrow	∈	∈	8712	2208	Element of
⇓	⇓	8659	21D3	Downwards double arrow	∉	∉	8713	2209	Not element of
⇔	⇔	8660	21D4	Left right double arrow	∋	∋	8715	220B	Contains as member
∫	∫	8747	222B	Integral	∏	∏	8719	220F	N-ary product=product sign
∂	∂	8706	2202	Partial differential	⊂	⊂	8834	2282	Subset of
∅	∅	8709	2205	Empty-null-diameter	⊃	⊃	8835	2283	Superset of
·	⋅	8901	22C5	Dot operator	⊄	⊄	8836	2284	Not subset of
∇	∇	8711	2207	Nabla-Del operator	⊆	⊆	8838	2286	Subset of or equal to
∑	∑	8721	2211	N-ary summation	∅	∅	8709	2205	Null, empty
+	+	0043	002B	Plus sign	⊿	⊿	8895	22BF	Right triangle
–	−	0150	0096	Minus sign	§	§	0167	00A7	Section sign
×	×	0215	00D7	Times sign	º	º	0186	00BA	Degree
÷	÷	0247	00F7	Divided by	¹	¹	0185	00B9	Superscript 1
−	−	8722	2212	Minus sign (alternate)	²	²	0178	00B2	Superscript 2
±	±	0177	00B1	Plus-minus suign	³	³	0179	00B3	Superscript 3
∓	∓	8723	2213	Minus-plus sign	µ	µ	0181	00B5	Micro sign
∠	∠	8736	2220	Angle	‧		8231	2027	Hyphen point
∡	∡	8737	2221	Measured angle	°		0176	00B0	Degree sign
∢	∢	8738	2222	Spherical angle	∴	∴	8756	2234	Therefore
∗	&lowest;	8727	2217	Asterisk operator	∵	∵	8757	2235	Because
√	√	8730	221A	Square root = radical sign	◠		9686	25D6	Upper half circle
∛		8731	221B	Cube root	Æ	Æ	0198	00C6	Upper AE ligature
½	½	0189	00BD	Fraction half	æ	æ	0230	00E6	Lower ae ligature
¼	¼	0188	00BC	Fraction quarter	Ø	Ø	0216	00D8	Uppercase O slash
¾	¾	0190	00BE	Fraction three quarters	¶	&pilcrow;	0182	00B6	Pilcrow, paragraph
⅓	⅓	8531	2153	Fraction one third	∧	∧	8743	2227	AND
∝	∝	8733	221D	Proportional to	∨	∨	8744	2228	OR
∞	∞	8734	221E	Infinity	∩	∩	8745	2229	Intersection
⊥	⊥	8869	22A5	Orthogonal to	∪	∪	8746	222A	Union
∥	∥	8741	2225	Parallel to	☉		9737	2609	Sun
∦	∦	8742	2226	Not parallel to	☼		9788	263C	Sun with rays
∼	∼	8764	223C	Similar to	☽		9789	263D	Moon first quarter
≅	≅	8773	2245	Approx equal to	○		11096	2B58	Full moon
≈	≈	8776	2248	Almost equal to - asymptotic	℄		8452	2104	Centreline

1.12 References for Chapter 1

Abbott, P and H. Neill. Trigonometry. A complete introduction. John Murray Learning, London 2018. ISBN978 1 473 67849 1

Chapter 2 ANGLES AND BEARINGS

For elementary school children, the concept of distance is easier to comprehend (i.e., from one point to another). However, ask the same child what an angle is, and one soon realizes that it is not easy to define. Over several years of teaching both undergraduate engineering and surveying students, we have noticed that even at this level, students still have difficulties in differentiating angles, directions and bearings, essential tools needed for computations. This chapter, therefore, assumes no previous knowledge and proceeds to introduce angles and bearings. The relationship between them is highlighted and how to calculate forward and back bearing presented. Following this chapter, the students should understand that angles are measurable quantities that must be converted into bearing (i.e., known directions measured from the north) for them to be useful in computing coordinates (positions).

2.1 Angles and Bearings.

Angles and bearings are the basis on which a surveyor provides direction for a line in space, called a vector. A vector has direction and magnitude but is commonly called a line with a bearing and distance.

Angles are formed by determining the difference in the direction of two vectors.

Angles are measured directly with a theodolite, or they may be calculated from known lines.

2.1.1 Angle systems.

Angles can be measured in any of four main systems.

2.1.1.1 Sexagesimal angles.

Sexagesimal System – (base 60) in degree (°), minutes (′) and seconds (″)

360°	=	1 circle
60′	=	1 degree
60″	=	1 minute
3600″	=	1 degree

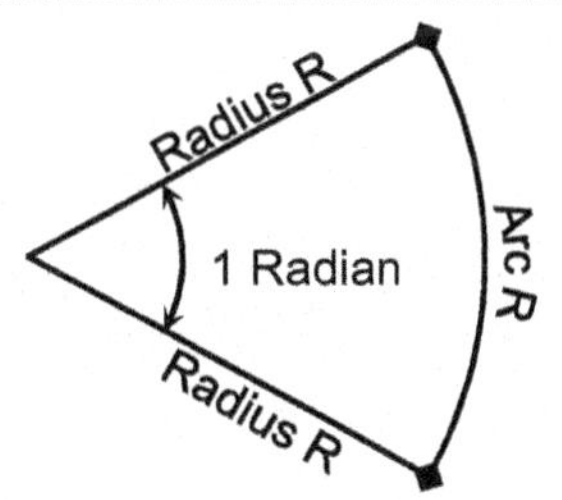

Figure 2-1 1 RADIAN

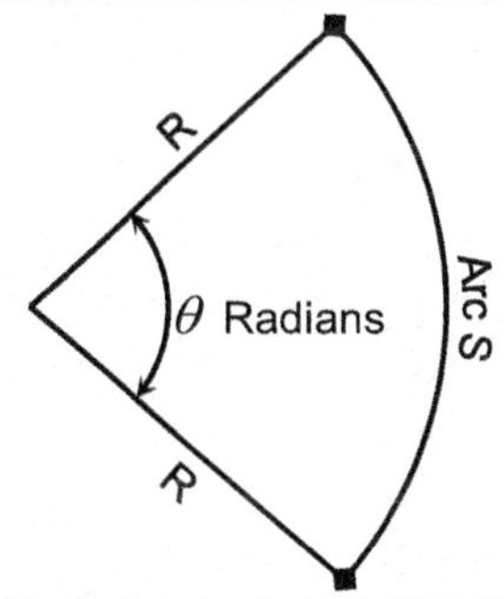

Figure 2-2 Angle in RADIANS

2.1.1.2 Circular or RADIAN System – in **radians**

One RADIAN is the angle subtended at the centre of a circle by an arc equal in length to the radius. (Figure 2-1)

The circular measure of any angle **θ** is given by: (Figure 2-2)

$$\theta = \frac{S}{R} \quad \text{and} \quad \text{Arc } S = R\theta, \ (\theta \text{ in radians})$$

Radian conversion to degrees, minutes and seconds

Now $\theta^{Radians} = \frac{Arc}{R} = \frac{\pi R}{R} = \pi$

i.e., $\theta^{Radians} = \pi^{Radians} = 180^{Degrees}$

or $180° = \pi^{Radians} = 3.1415927\ldots^{R}$ (Figure 2-3)

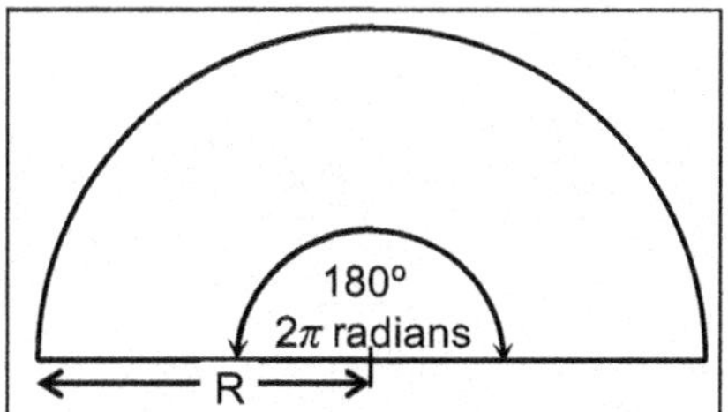

Figure 2-3 PI (π) radians

1°	=	$0.017\,4533^{R}$	1^{R}	=	57.295 7759°
1′	=	$0.000\,2909^{R}$	1^{R}	=	3437.7467′
1″	=	$0.000\,0048^{R}$	1^{R}	=	206264.806247″ (common conversion **1 radian** = **206 265**″)

The book will use the Sexagesimal (base 60) and circular (Radian) system.

The symbol ∠ (Alt+8763) is used for the word ***angle***.

In Figure 2-4 the angle of 25° 27′ 03″ is represented as ∠AOB.

When writing angles there is a golden rule: use leading zeros, especially for minutes and seconds. This avoids ambiguity.

Generally, when set out neatly especially in column form, there is no need for unit symbols of °, ′ or ″. Insert a space between units. The notation for 25° 27′ 03″ as, e.g., 25 27 03 is acceptable. However, 25° 27′ 3″ as, e.g., 25 27 3 is **not** acceptable as it may be ambiguous and lead to errors in interpretation.

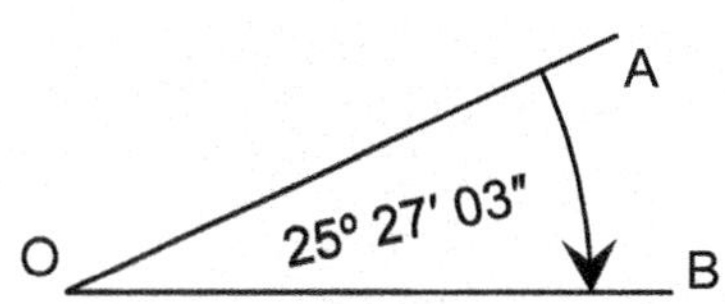

Figure 2-4 Angular notation

If it's not checked it's wrong.

J. Walker and J. Awange, *Surveying Computations*,
https://doi.org/10.1007/978-3-031-94382-9_2

2.2 Bearing.

With reference to Figure 2-5 the bearing of the point B is the clockwise angle FROM some reference line (in this case OA) TO the point B. i.e., the bearing FROM **O** TO **B** is 115°05′50″

The reference line OA is usually the north point; either magnetic north, true north, grid north or assumed north.

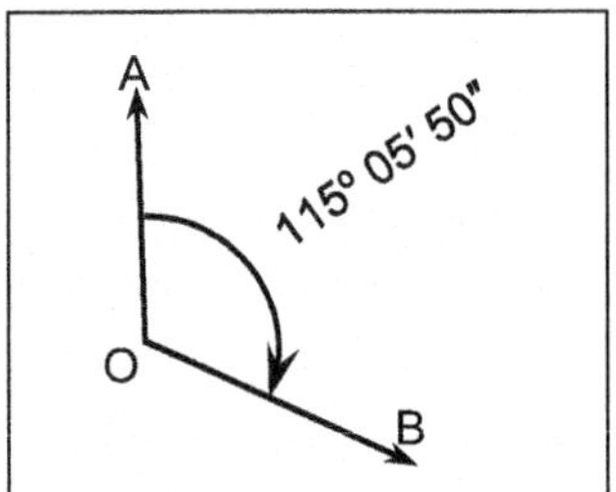

Figure 2-5 The BEARING line

* Magnetic north refers to the direction taken up by a freely floating and properly balanced compass needle in the magnetic meridian.
* True north is the meridian line joining the geographic north pole and the observer. It may be located by astronomic observations.
* Grid north refers to the north orientation of a map projection.
* There is another reference or north line – known as "assumed" north. It is any convenient direction used to orient a local survey or facilitate calculations.

2.2.1 Bearing Notation

We can refer to the bearing from A to B as $\overrightarrow{AB}$ (note the overscore arrow), Brg AB or θ_{AB} (theta AB)

Methods of designating bearings

Bearings are designated in either of two systems of notation:

(a) The whole circle system.

(b) The quadrantal system.

2.2.1.1 Whole bearing system (WCB).

In the whole circle system bearings originate from north, which is marked 0° or 360°, and are measured clockwise from the meridian, through east or 90°, south or 180° and west or 270°.

2.2.1.2 Quadrantal system.

In the quadrantal system bearings are measured in four quadrants, increasing from 0° to 90° from north to east, south to east, south to west and north to west.

If in Figure 2-6, O is a survey station and NOS the meridian through it, the bearings of the lines from O are :

Line	Whole circle Bearing (WCB)	Quadrantal bearing
OA	a	N a° E
OB	e	S b° E
OC	f	S c° W
OD	g	N d° W

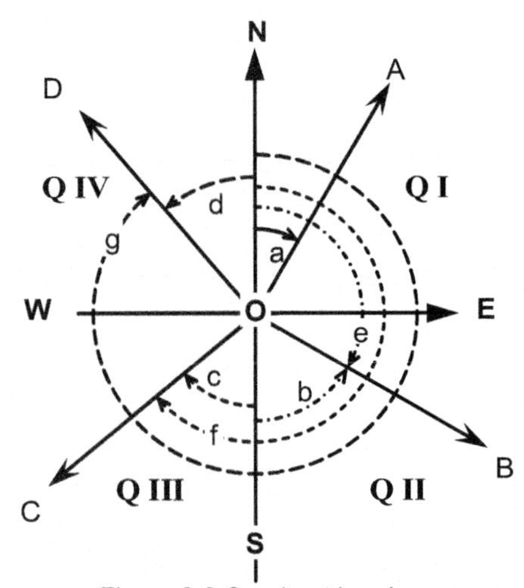

Figure 2-6 Quadrant bearings

where a, b, c, d, e, f and g are angles (a, b, c, d are less than 90°) expressed in degrees, minutes and seconds.

Note. Quadrantal bearings are never reckoned from the east – west line. The letter which precedes the number must be either N or S.

2.2.2 Workshop Exercise 2.1

(i) Reduce the following whole circle bearings to quadrantal bearings:

(a) 237°10′

(b) 136°12′18″

(c) 76°14′

(d) 260°48′19″

(ii) write the following quadrantal bearings as whole circle bearings.

(a) S 34°12′ W

(b) N 27°19′35″ E

(c) N 72°35′12″ W

(d) S 86°15′ E

The Quadrantal bearing is now seldom used and no further reference to it will be given in this text.

2.3 Back Bearings

Bearing $\overrightarrow{OA}$ indicates the bearing of the line from O to A. (See Figure 2-6).

The bearing from A to O is termed the "back" or "reverse" bearing $\overrightarrow{AO}$

In whole circle reckoning, the back bearing of a line is obtained from the given (forward) bearing by adding 180° when the forward bearing is less than 180°, and by subtracting 180° when the forward bearing is more than 180°.

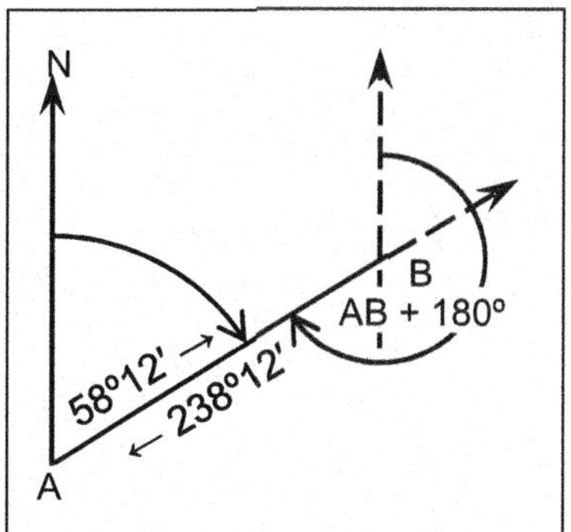

Figure 2-7 Back bearing

2.3.1 Example of Back Bearing Calculations

(a) $\overrightarrow{AB} = 58°12'$

$\overrightarrow{BA} = 58°12' + 180°$ See Figure 2-7.

$= 238°12'$

2.3.2 Workshop Exercise 2.2

Write down the back or reverse bearings corresponding to the following forward bearings :

(a) 16°14′42″	(d) 328°16′54″
(b) 234°17′28″	(e) 93°56′27″
(c) 196°36′13″	(f) 291°06′21″

2.4 Angles Between Bearings.

In Figure 2-8 the angle AOB is the difference between the bearings of the lines $\overrightarrow{OB}$ and $\overrightarrow{OA}$

$\angle AOB = \angle NOB - \angle NOA \quad = \overrightarrow{OB} - \overrightarrow{OA}$

The bearing you're going **from** minus the bearing you're going **to**.

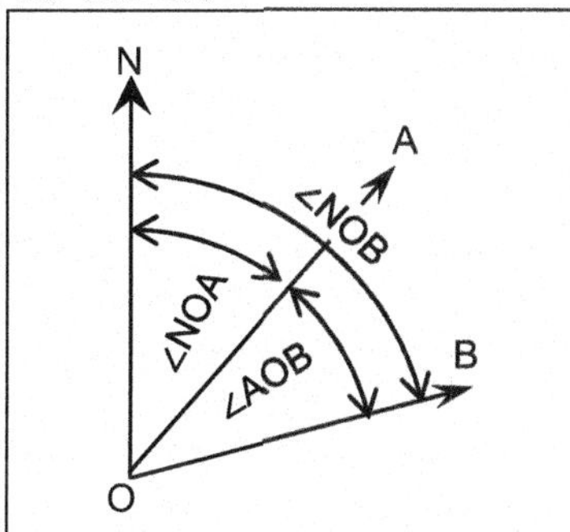

Figure 2-8 Angles between bearings

2.4.1 Examples of Angles Between Bearings.

(See Figure 2-9 (a), (b)).

(a) $\overrightarrow{OB}$=50°12′, OC=1323°45′

$\angle BOC = \overline{OC} - \overline{OB}$

$= 132°45' - 50°12'$

$= 82°33'$

(b) $\overrightarrow{OB}$=50°12′, $\overrightarrow{EO}$=103°44′

$\angle EOB = \overrightarrow{OB} - \overline{OE}$, reciprocal of $\overline{EO}$

(need the direction from O to E)

$= \overline{OB} - (180° + \overline{EO})$

$= 50°12' - 283°44'$

(this will be negative and nasty)

(add 360° to make subtraction easier)

$= 360° + 52°12 - 283°44'$

$= 410°12' - 283°44'$

$= 126°28'$

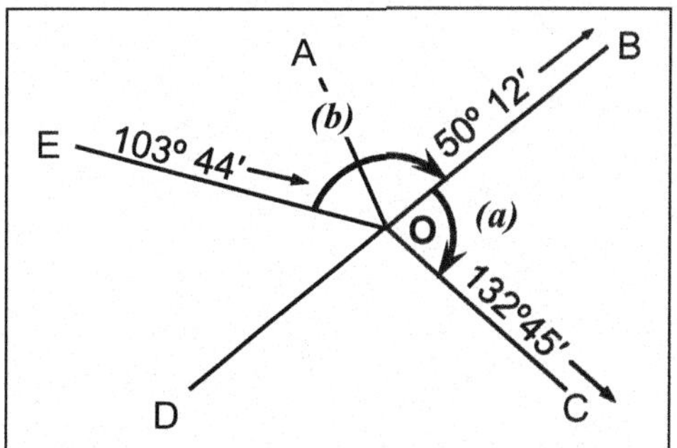

Figure 2-9 Derived angles from bearings

The method of calculation of an angle between two bearings is as follows:

1. Imagine yourself standing at the intersection of the two lines.
2. Take the bearing on your right (i.e., the more clockwise of the two) and subtract the bearing on your left.
3. If the result is negative, simply add 360 to the result, or if the RH bearing is less than the LH bearing, add 360 to the RH bearing and then subtract the LH bearing.

HINT: Use a protractor to draw the two bearings and then check measure to ensure you have the required angle.

2.4.2 Interior angles from bearings.

In any polygon the interior angles can be calculated from the bearings of the sides. The reason for the calculation of the angles is to the condition that the sum equals the theoretical sum of (n-1)×180 where n = number of angles. If the condition is not met, then there is an *angular misclose*. See Figure 2-10.

2.4.2.1 Worked Exercise 2.3

Line	FWD Brg		Line	BACK Brg	
AB	60° 38′ 32″	+180	BA	240° 38′ 32″	
BC	158° 44′ 58″	+180	CB	338° 44′ 58″	
CA	109° 44′ 55″	+180	AC	469° 41 ′55″	– 360° = 109° 41′ 55″

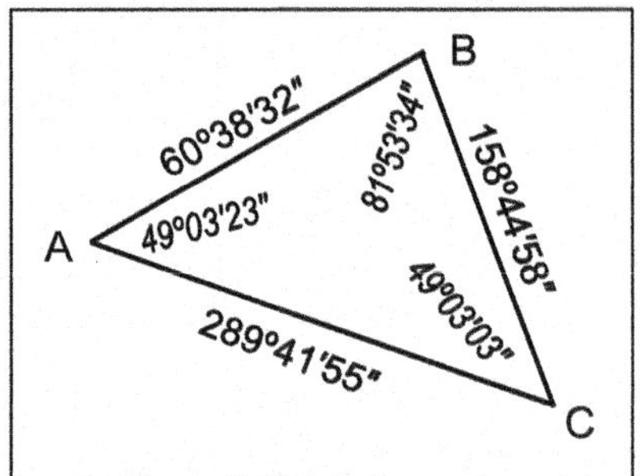

Figure 2-10 Interior angles

Solution:

Angle	Line	FWD Brg	Angle	Line	BACK Brg
∠A	AC	109° 41′ 55″	∠B	BA	240° 38′ 32″
	– AB	60° 38′ 32″		– BC	158° 44′ 58″
		49° 03′ 23″			81° 53′ 34″
∠C	CB	338° 44′ 58″	Check	∠A	49° 03′ 23″
	– CA	289° 41′ 55″		∠B	81° 53′ 34″
		49° 03′ 03″		∠C	49° 03′ 03″
CHECK				SUM	180° 00′ 00″

2.4.3 Bearings from interior angles by the deflection angle.

If an initial bearing is known, the other bearings can be calculated using the forward bearing and the corresponding deflection angle. The deflection angle is 180° – interior angle.

2.4.3.1 Worked Exercise 2.4

Figure 2-11 shows an initial bearing, AB and the internal angles of the traverse. The corresponding deflection angles are tabled. The deflection angles are clockwise and thus positive.

Line	FWD Brg		
AB	60° 38′ 32″		
Interior angle		Deflection angle	
∠A	49° 03′ 23″	180 – ∠A =	130° 56′ 37″
∠B	81°53′ 34″	180 – ∠B =	98° 06′ 26″
∠C	49° 03′ 03″	180 – ∠C =	130° 56′ 57″

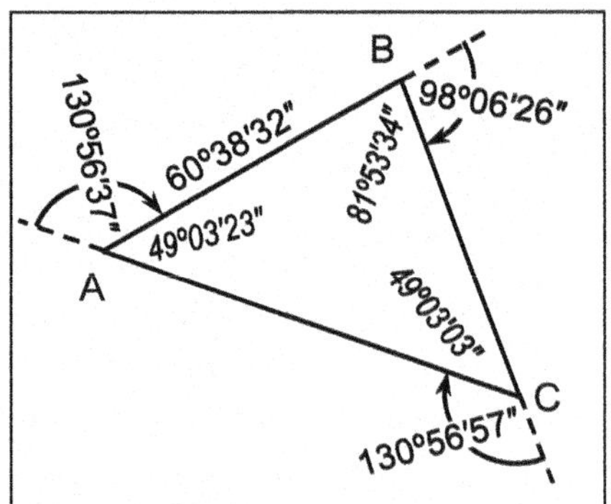

Figure 2-11 Station deflection angles

Solution:

BC = AB + deflection ∠B			CA = BC + deflection ∠C	
BC =	AB	60° 38′ 32″	CA = BA	158° 44′ 58″
	+ ∠B	98° 06′ 26″	+ ∠C	130° 56′ 57″
		158° 44′ 58″		289° 41′ 55″
AB	CA	289° 41′ 55″		
	+ ∠A	130° 56′ 37″		
		420° 38′ 32″	– 360 = 60° 38′ 32″ CHECK Brg AB	

2.5 Calculating Traverse Bearings. Clockwise and anticlockwise angles.

Because of terrain difficulties, it is not an easy matter to measure from one point to another in a straight line. Therefore, points are joined, in practice, by a series of straight lines whose distances are measured and between which the angles are also measured (see Figure 2-12). This is called a traverse. The angles can be expressed as clockwise (positive) or anticlockwise (negative). The calculated bearing is reduced to a value between 0 and 360° (MODULO 360).

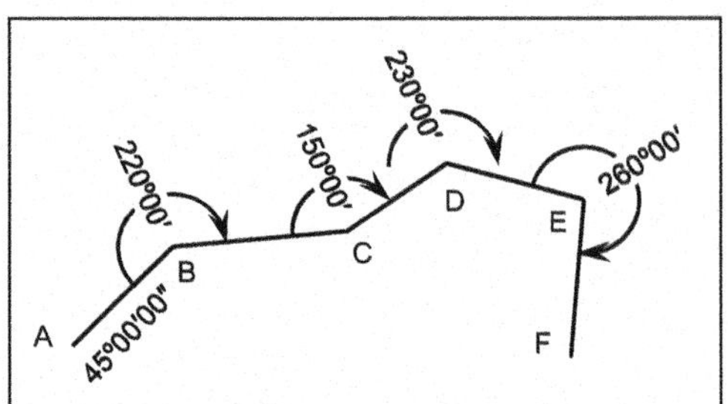

Figure 2-12 Traverse bearings

2.5.1 Calculating Traverse Bearings. Clockwise angles.

In Figure 2-12 we can compute the bearing of each line in the traverse as a function of the back bearing of the previous line and the clockwise angle between the two lines.

Now that we are accustomed to the overscore arrow for bearing we can dispense with it.

Now $BA = AB + 180°$

$= 45°00'00'' + 180°00'00'' = 225°00'00''$

Thus $BC = BA + \angle B$

$= 225°00'00'' + 220°00'00'' = 445°00'00'' - 360°$

$BC = \underline{85°00'00''}$

$CB = 265°00'00''\ (BC + 180)$

$CD = CB + \angle C$

$= 265°00'00'' + 150°00'00'' = 415°00'00'' - 360°$

$CD = \underline{55°00'00''}$

$DC = 235°00'00''\quad (CD + 180)$

$DE = DC + \angle D$

$= 235°00'00'' + 230°00'00'' = 465°00'00'' - 360°$

$DE = \underline{105°00'00''}$

$ED = 285°00'00''\,(DE + 180)$

$EF = ED + \angle E$

$= 285°00'00'' + 260°00'00'' = 545°00'00'' - 360°$

$EF = \underline{185°00'00''}$

Check: $EF = AB + \angle B + \angle C + \angle D + \angle E - (4 \times 180)$

$= 45°00'00'' + 220°00'00'' + 150°00'00'' + 230°00'00'' + 260°00'00''$

$- (4 \times 180°) \quad = 905°00'00'' - 720°$

$= \underline{185°00'00''}$

Final bearing = first **inbound** bearing + sum of angles – (n × 180°)
where n = number of positive angles in the traverse.

2.5.2 Calculating Traverse Bearings. Mixed angles

In Figure 2-13 we can compute the bearing of each line in the traverse as a function of the back bearing of the previous line and the clockwise (positive) or anticlockwise (negative) angle between the two lines.

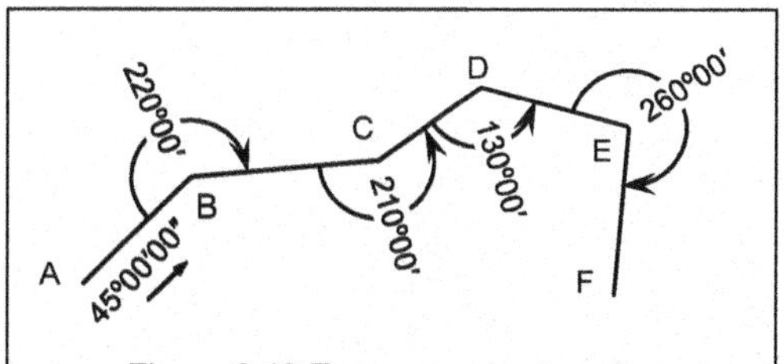

Figure 2-13 Traverse, mixed angles

Now $BA = AB + 180°$

$= 45°00'00'' + 180°00'00'' = 225°00'00''$

Thus $BC = BA + \angle B$

$= 225°00'00'' + 220°00'00''$

$= 445°00'00'' - 360°$

$BC = \underline{85°00'00''}$

$CB = 265°00'00''\ (BC + 180)$

$CD = CB + \angle C$

$= 265°00'00'' - 210°00'00''$

$CD = \underline{55°00'00''}$

$DC = 235°00'00''\quad (CD + 180)$

$DE = DC + \angle D$

$= 235°00'00'' - 130°00'00''$

$DE = \underline{105°00'00''}$

$ED = 285°00'00''\,(DE + 180)$

$EF = ED + \angle E$

$= 285°00'00'' + 260°00'00''$

$= 545°00'00'' - 360°$

$EF = \underline{185°00'00''}$

Check: $EF = AB + \angle B - \angle C - \angle D + \angle E - (0 \times 180)$

$= 45°00'00'' + 220°00'00'' - 210°00'00'' - 130°00'00'' + 260°00'00''$

$- ((2+ + 2-) \times 180°) \;= 185°00'00'' - 0°$

$= \underline{185°00'00''}$

If it's not checked it's wrong.

Final bearing = first **inbound** bearing + sum of angles – (n × 180°)
where n = number of positive angles in the traverse.
In this second case of two clockwise (+) and 2 anticlockwise (–) angles, n = 0.

2.5.2.1 Worked Example 2.1

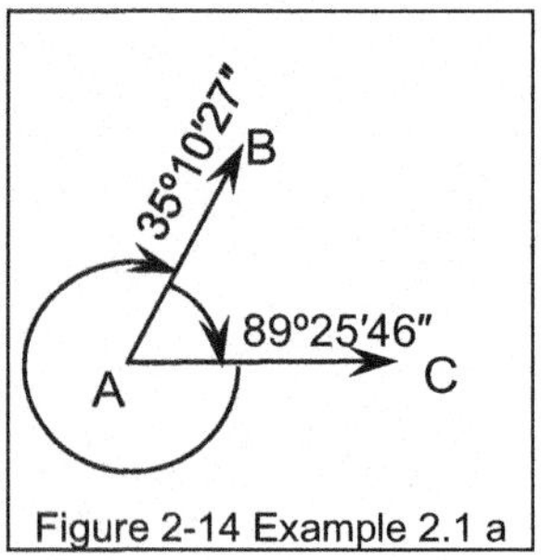

Figure 2-14 Example 2.1 a

The figures show two bearings, AB and AC:

1) In Figure 2-14. Find the clockwise angles ∠BAC and ∠CAB

∠BAC = AC – AB = 89°25′46″ – 35°10′27″ = 54°15′19″

∠CAB = AB – AC = 35°10′27″ – 89°25′46″ = –54°15′49″

(Gives∠CAB a negative answer, convert to positive by the addition of 360°. Easier to add 360 first.)

∠CAB = 360 + AB – AC = 360 + 35°10′27″ (395°10′27″) – 89°25′46″

∠CAB = 305°44′41″

Check ∠BAC + ∠CAB = 54°15′19″ + 305°44′41″ = 360°

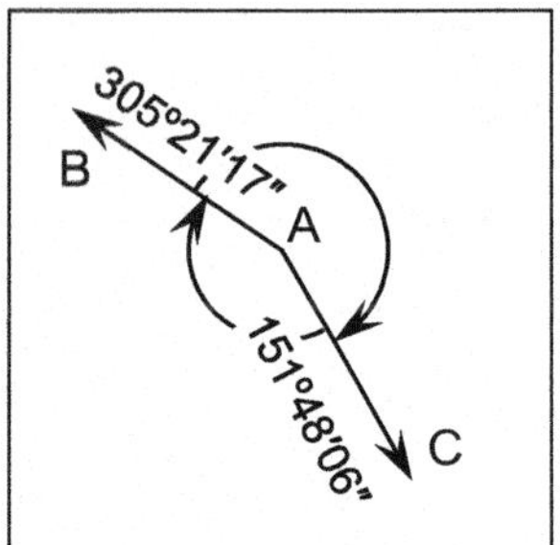

Figure 2-15 Example 2.1 b

2) In Figure 2-15. Find the clockwise angles ∠BAC and ∠CAB

∠BAC = AC – AB = 151°48′06″ – 305°21′17″ = –153°33′11″

(Gives a negative answer, should convert to positive by the addition of 360°. Easier to add 360 first.)

∠BAC = 360 + 151°48′06″ (511°48′06″) – 305°21′17″ = 206°26′49″

∠CAB = AB – AC = 305°21′17″ – 151°48′06″ = 153°33′11″

Check ∠BAC + ∠CAB = 206°26′49″ + 153°33′11″ = 360°

2.5.2.2 Worked Example 2.2

Figure 2-16 shows an initial bearing, AB and a number of observed angles, ∠B and ∠E are negative. Find the final bearing EF

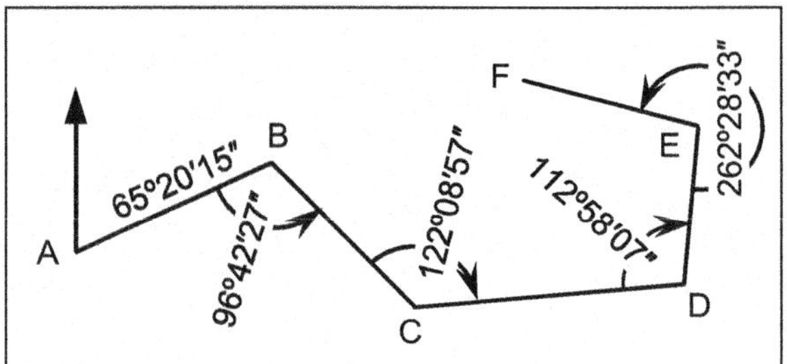

Figure 2-16 Worked Example 2.2

	° ′ ″	
Brg AB	65 20 15	+ 180=
Rev BA	245 20 15	
–∠B	– 96 42 27	
Brg BC	148 37 48	+ 180=
Rev CB	328 37 48	
+∠C	+122 08 57	
Brg CD	450 46 45	– 360=
Brg CD	90 46 45	+ 180=
Rev DC	270 46 45	
+∠D	+112 58 07	
Brg DE	383 44 52	– 360=
Brg DE	23 44 52	+ 180=
Rev ED	203 44 52	+ 360= (add 360 to avoid negative answer because of negative angle)
	563 44 52	
–∠F	– 262 28 33	
Brg EF	**301 16 19**	

Check: Fwd AB – ∠B + ∠C + ∠D – ∠E – (0 × 180)

65 20 15 – 96 42 27 + 122 08 57 +112 58 07 – 262 28 33 – 0

= –58 43 41 + 360 = 301 16 19. (Add 360° for positive bearing

This is better laid out for a check as: AB + 360 + C + D – B – E

where it can be seen that the sum of the negative angles is greater than the sum of the positives.

Old school method with negatives. I borrow from bottom line, pay back and increment.
New school method borrows from top line and decrements values

	65 20 15+ 360	425 20 15	**Old school.** Add 360 to avoid negatives.
	+C	+122 08 57	
1		547 29 12	Seconds: 57 + 15 = 72 = 1′ 12“ write down 12 and carry 1 to minutes. 20 + 8 + 1 = 29
	+D	+112 58 07	
2		660 27 19	Seconds: 12 + 7 = 20. Minutes: 29 + 58 = 87 = 1° 27′. Write 27 and carry 1°. 547 + 112 + 1 = 660
3	-B	– 96 42 27	–ve sec. result. Take 1′ = 60″ from lower line minutes and add. 19 + 60 – 27 = 52
4		563 44 52	–ve min. result. Take 1° = 60′ from lower line degrees and add. 27 + 60 – (42 + 1) (borrow) = 44
5			Degree sum: 660 – (96+1)(borrow) = 563
	-E	–262 28 33	
		301 16 19	Check. Normal subtraction.

If it's not checked it's wrong.

	65 20 15 + 360	425 20 15	**New school**. Add 360 to avoid negatives.
	+C	+122 08 57	
1		547 29 12	Seconds: 57 + 15 = 72 = 1' 12" write down 12 and carry 1 to minutes. 20 + 8 + 1 = 29
	+D	+112 58 07	
2		660 27 19	Seconds: 12 + 7 = 20. Minutes: 29 + 58 = 87 = 1° 27'. Write 27 and carry 1°. 547 + 112 + 1 = 660
3	-B	− 96 42 27	−ve sec. result. Take 1' = 60" from (top line) minutes and add. 19 + 60 − 27 = 52. Dec. minutes
4		563 44 52	−ve min. result. Take 1° = 60' from top degrees and add. (27-1) (decrement) + 60 − 42 = 44
5			Degree sum: 660-1(decrement) − 96 = 563
	-E	−262 28 33	
		301 16 19	Check. Normal subtraction.

2.6 Alternate angular systems

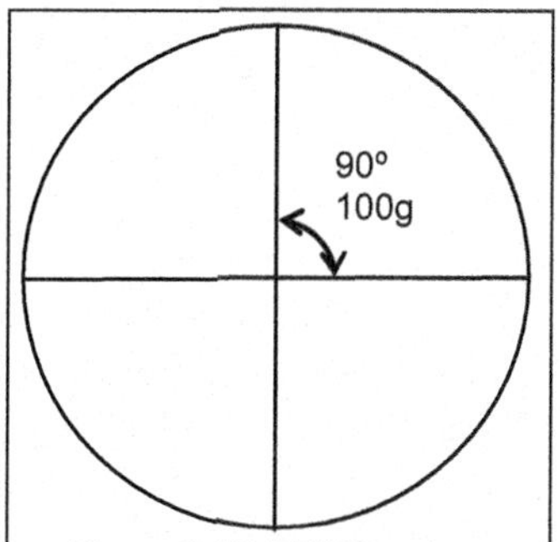

Figure 2-17 GRAD system

2.6.1 Grad System – Centesimal (base 100): grad/gon.
(France, Germany for topographic survey work). (Figure 2-17).

1 quadrant = 90° = 100 grads
1 grad = 1g = 100c (centigrads)
1 centigrad = 1c = 0.01g
1 centi centigrad = 1cc= 0.01c = 0.0001g
90° = 100g
1° = 1.1111g

Since the original determination of the metre was to be 1/10 000 000 quadrant of the Earth's circumference it can be seen that 1 grad ≈ 100 000m or 100Km of latitude. (1 degree ≈ 111.1Km)

2.6.2 Mil System – milliarc radians (1/1000 radian)

The **mil** is a special measure used in the armed forces for rifle and artillery aiming and fire correction. Introduced in First World War.

Two units of mil exist, based on radian measure:

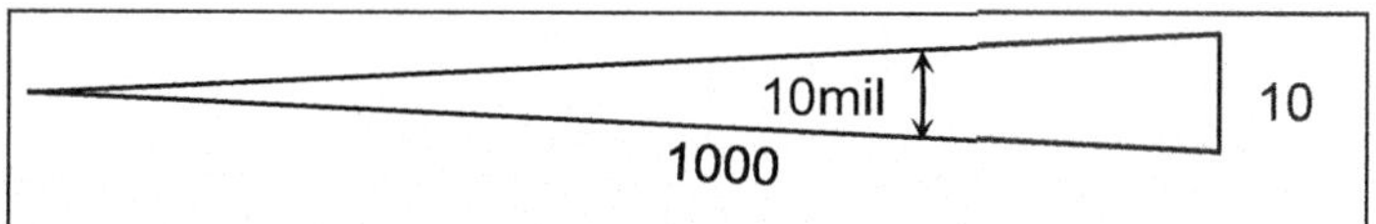

Figure 2-18 10 mils (milliarc radians)

1. 1 mil = 0.001 Radian. 2000·π mils = 6283 mils in a circle. (2 π radians)
2. The mil has been rationalized to 6400 mils in a circle:
   1° =17.78 mils (6400/360)
   1 mil =0.05625° =3.375′ = 202.5″

The US Army and NATO armies use 1 mil = 1/6400 circle. The US Marine Corp use 1 mil = 1/6283 for telescope reticule marks.

The stadia wires in levels and theodolites are 10 milliradians (0.01^R) apart (see Figure 2-18). This provides the *stadia constant* of 100. Some military theodolites were graduated in **mils**; with subdivisions to: Wild T16 MIL to 0.2 mil = 40″ and Wild T2 MIL to 0.002mil = 0.4″.

2.7 Chapter Exercise 2.3

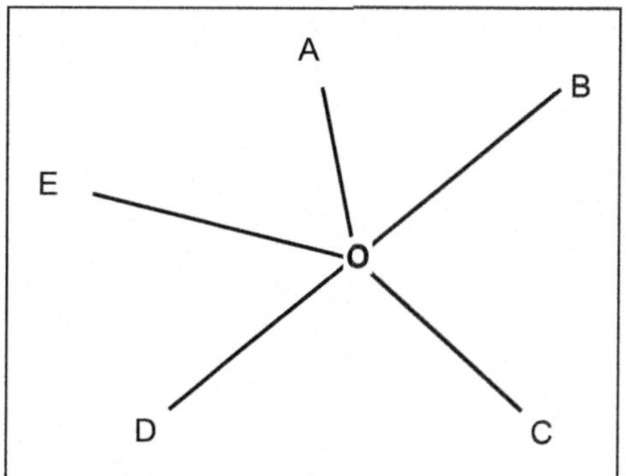

Figure 2-19 Chapter exercise 2.3

In Figure 2-19 let the bearings be as follows:

draw and label the figure with **forward** bearings

$\overrightarrow{OA}$ = 358°14′16″
$\overrightarrow{OB}$ = 44°33′56″
$\overrightarrow{CO}$ = 307°09′41″ OC = ?
$\overrightarrow{DO}$ = 12°59′23″ OD = ?
$\overrightarrow{OE}$ = 281°01′50″

Calculate the clockwise values of the following angles, from the start bearing to the end bearing, where:

∠AOC = $\overrightarrow{OA}$ clockwise to $\overrightarrow{OC}$

(a) ∠AOC ____________________ (d) ∠DOE ____________________
(b) ∠BOE ____________________ (e) ∠BOC ____________________
(c) ∠COD ____________________ (f) ∠EOB ____________________

2.7.1 Answers to Workshop Exercises (Refer § 2.2.2 and § 2.3.2).

2.1	(i)	(a)	S 57°10′W	(ii)	(a)	214°12′
		(b)	S 43°47′42″E		(b)	27°19′35″
		(c)	N 76°14′E		(c)	287°24′48″
		(d)	S 80°48′19″W		(d)	93°45′

2.2	(a)	196°14′42″	2.3	(a)	∠AOC 128°55′25″
	(b)	54°17′28″		(b)	∠BOE 236°27′54″
	(c)	16°36′13″		(c)	∠COD 65°49′42″
	(d)	148°16′54″		(d)	∠DOE 88°02′27″
	(e)	273°56′27″		(e)	∠BOC 82°35′45″
	(f)	111°06′21″		(f)	∠EOB 123°32′06″

2.8 References for Chapter 2

Clark, D and J. Clendinning. Plane and Geodetic Surveying for Engineers, Volume One, Plane Surveying. Fourth Edition, Constable & Company Ltd, London 1948.

Bomford, G. Geodesy 4th ed. Oxford University Press 1980. ISBN 0-19-851946-X

Chapter 3 SOLUTION OF RIGHT ANGLED AND GENERAL TRIANGLES

In engineering setting out of structures, it is often the practice to break the ground area under operation into regular geometrical shapes/figures such as triangles and rectangles as well as irregular figures. In this chapter, once the student has grasped the concepts of angles in Chapter 2, they are now placed into a regular geometrical figure, a triangle. First, we start with a right-angle triangle in which one of the angles is 90° to build trigonometric relationships between the sides of the triangle and the respective angles. This enables the computation of missing angles and sides of the triangle as well as areas of the triangles. Once the student has mastered these skills, a general triangle and its related formulae are presented.

3.1 Trigonometric functions.

The trigonometric functions used by a practising surveyor are as follows: included are math equation editors; Mathtype, MATLAB, and spreadsheet function abbreviations. Your calculator is set to work in degrees. But MATLAB and Excel work in radians, so you must always convert to/from radians in these programs.

	Name	Abbreviation	Math	Excel
(a)	Sine	Sin	sin	SIN()
(b)	Cosine	Cos	cos	COS()
(c)	Tangent	Tan	tan	TAN()
(d)	Cotangent	Cot	cot	1/TAN()
(e)	Secant	Sec	sec	1/COS()
(f)	Cosecant	Cosec	csc	1/SIN()
(g)	Arc Sine	Asin	arcsin	ASIN()
(h)	Arc Cosine	Acos	arccos	ACOS()
(i)	Arc Tangent	Atan	arctan	ATAN()

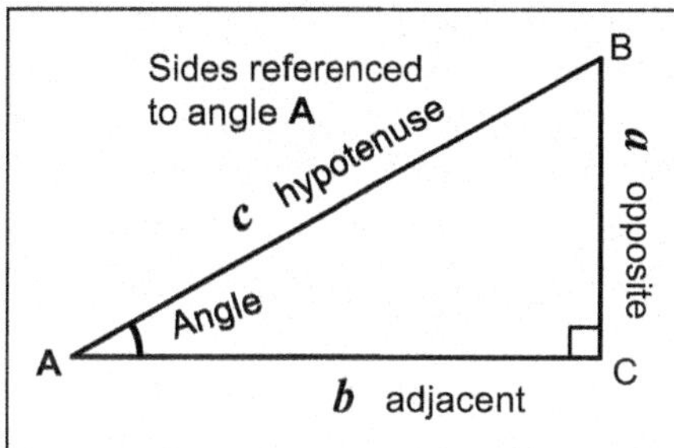

Figure 3-1 Triangle components

In Figure 3-1 each of the above 6 functions (a – f) of the right-angled triangle (angle C = 90°) can be defined as follows and memorized from the mnemonic **SOH CAH TOA**:

Note the NAME of the *inverted* Sine, Cosine and Tangent function.

$$\sin A = \frac{a}{c} \Rightarrow \frac{opposite}{hypotenuse}$$ **SOH** (sin = **o**pposite over **h**ypotenuse)

$$\cos A = \frac{b}{c} \Rightarrow \frac{adjacent}{hypotenuse}$$ **CAH** (cos = **a**djacent over **h**ypotenuse)

$$\tan A = \frac{a}{b} \Rightarrow \frac{opposite}{adjacent}$$ **TOA** (tan = **o**pposite over **a**djacent)

$$\cot A = \frac{b}{a} \Rightarrow \frac{adjacent}{opposite} = \frac{1}{\tan A}$$ inverse tangent

$$\sec A = \frac{c}{b} \Rightarrow \frac{hypotenuse}{adjacent} = \frac{1}{\cos A}$$ inverse cosine

$$\text{cosec}A = \frac{c}{a} \Rightarrow \frac{hypotenuse}{opposite} = \frac{1}{\sin A}$$ inverse sine

3.1.1 Define the tangent

An important function simile is the definition that the *tangent* of an angle is *sine* divided by the *cosine* of that angle.

$$\text{Show } \tan A = \frac{\sin A}{\cos A} = \frac{\frac{a}{c}}{\frac{b}{c}} = \frac{a}{c} \times \frac{c}{c} = \frac{a}{b}, \text{ and } \frac{opposite}{adjacent} = \tan A$$

3.1.2 Two other functions

versine$A$ (CD) = (1 – cos$A$) = $\left(1 - \frac{b}{c}\right)$. and external secant$A$ (DE) = (sec$A$ – 1) = $\left(\frac{c}{b} - 1\right)$ in Figure 3-2.

3.1.3 Inverse trigonometric functions. Refer to Figure 3-1.

Often shown as: ASIN, ACOS, ATAN (MS Excel) or $\sin^{-1}$, $\cos^{-1}$, $\tan^{-1}$ (HP10s+).

J. Walker and J. Awange, *Surveying Computations*,
https://doi.org/10.1007/978-3-031-94382-9_3

$\arcsin\left(\frac{a}{c}\right) = \angle A$ the angle whose sine is $\left(\frac{a}{c}\right)$

$\arccos\left(\frac{b}{c}\right) = \angle A$ the angle whose cosine is $\left(\frac{b}{c}\right)$

$\arctan\left(\frac{a}{b}\right) = \angle A$ the angle whose tangent is $\left(\frac{a}{b}\right)$

ALWAYS draw the diagram and use the correct names of the sides opposite the required angle. Angles are upper case letters; sides are lower case letters.
Figure 3-2 shows the various trigonometric functions on the unit circle.

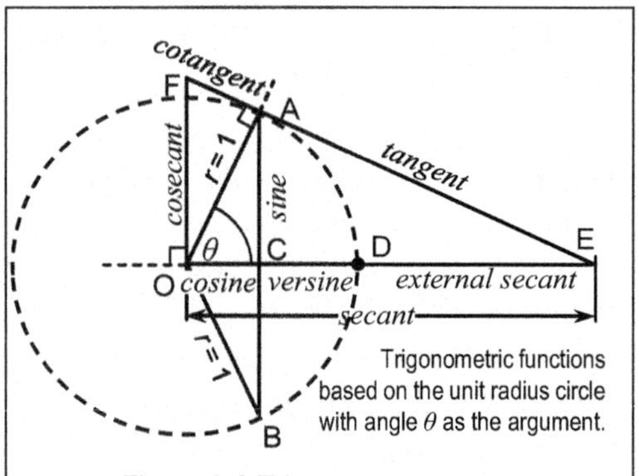

Figure 3-2 Triangle components

3.2 Right Angle Triangles.

In survey computations involving triangle of any sort, we attempt to calculate all missing values rather than deduce the missing values.

For example, let us suppose two sides of a right-angled triangle are stated. It is possible to calculate one angle and deduce the remaining angle by subtracting the calculated angle from 90°.

It is better surveying practice to calculate both angles and check that the sum of the two angles is 90°. Similarly, if we are calculating distances, it is better to calculate all distances in the triangle and verify that:

$c^2 = a^2 + b^2$.

3.2.1 Worked Example 3.1

Case: Two Sides
find third side and two angles.

Given: Figure 3-3:

a = 542.792
b = 695.883

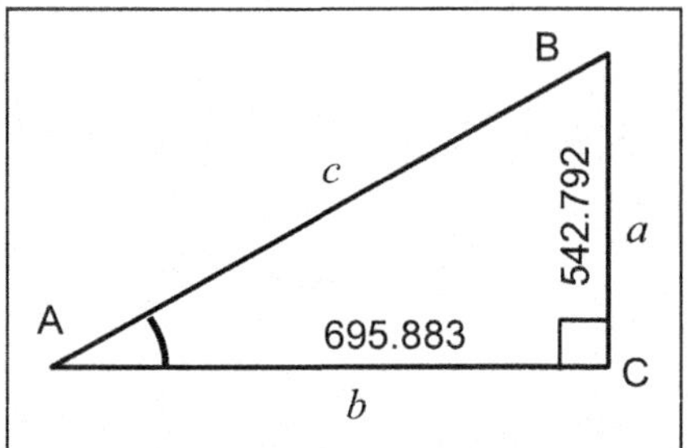

Figure 3-3 Solve for 2 sides

Solution:(a)

$$\angle A = \arctan\left(\frac{a}{b}\right) = \arctan\left(\frac{542.792}{695.883}\right)$$
$$= \arctan(0.780005) = 37.954398^\circ$$
$$\angle A = 37^\circ 57' 16''$$

(b) Find side AB = side c (Angle A calculated, side b = 695.883)

$$\cos A = \frac{adjacent}{hypotenuse} = \frac{b}{c}\text{ , thus } c = \frac{b}{\cos A} = \frac{695.883}{0.788501}$$

c = 882.540

Check: $c = \sqrt{a^2 + b^2}$

c = 882.540 ✓

(c) $\angle B = \arcsin\left(\frac{b}{c}\right) = \arcsin\left(\frac{695.883}{882.540}\right) = \arcsin(0.788500)$

$\angle B = 52°02'44''$

Check $\angle A + \angle B = 52°02'44'' + 37°57'16'' = 90°00'00'' = \angle C$ ✓

3.2.2 Worked Example 3.2

Case: One Side and One Angle,
find other two sides and other angle.

Given: Figure 3-4.

$\angle A$ = 48°15′21″
Side a = 637.912m

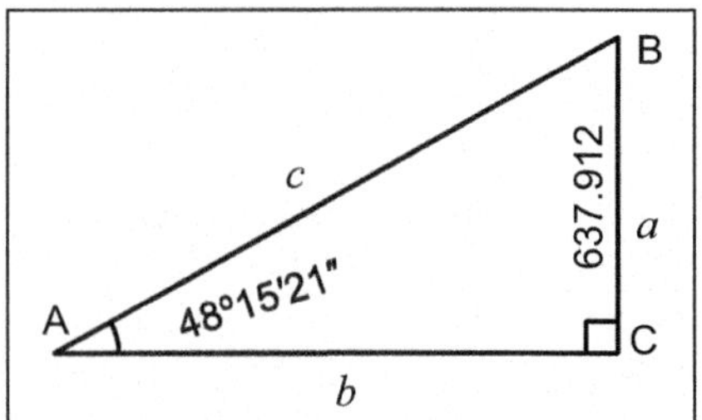

Figure 3-4 Side and angle

If it's not checked it's wrong.

Solution:

(a) $\sin A = \dfrac{opposite}{hypotenuse} = \dfrac{a}{c}$ thus side $c = \dfrac{a}{\sin A} = \dfrac{637.912}{0.746125}$

$c \quad = 854.966\text{m}$

(b) $\tan A = \dfrac{opposite}{adjacent} = \dfrac{a}{b}$ thus side $b = \dfrac{a}{\tan A} = \dfrac{637.912}{1.120635}$

$b \quad = 569.241\text{m}$

Check $a = \sqrt{c^2 - b^2}$. Check the two calculated sides equal the original data.

$a = 637.912$ Checks ✓

(c) $\angle B = \arcsin\left(\dfrac{b}{c}\right) = \arcsin\left(\dfrac{5692.411}{8549.658}\right) = \arcsin(0.665806)$

$\angle B \quad = 41°44'39''$

Check $\angle A + \angle B = 41°44'39'' + 48°15'21'' = 90°00'00'' = \angle C$ ✓

You must calculate the second angle by trigonometry. Do not find the second angle as (90 – angle 1). This method does not guard against errors in calculating the first angle.

3.2.3 Chapter Exercise 3.1

Using Figure 3-5, not drawn to scale

compute and check the missing elements:

(1) $\angle A = 37°18'51''$ $c = 6378.115$
(2) $a = 6782.719$ $c = 13141.159$
(3) $a = 6888.919$ $c = 6888.895$
(4) $\angle A = 85°17'48''$ $c = 937.444$
(5) $\angle A = 0°14'22''$ $a = 21.514$

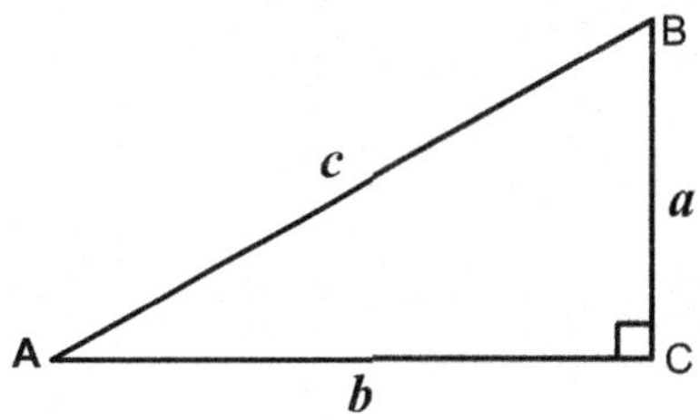

Figure 3-5 Right triangle components

3.2.3.1 Answers to Chapter Exercise 3.1

1)	$\angle B = 52°41'09''$	$b = 5072.666$	$c = 3866.318$
2)	$\angle A = 31°04'26''$	$\angle C = 58°55'34''$	$b = 11255.433$
3)	$\angle A = 45°00'00''$	$\angle B = 45°00'00''$	$c = 9742.386$
4)	$\angle B = 4°42'12''$	$b = 76.867$	$a = 934.287$
5)	$\angle B = 89°45'38''$	$b = 5147.976$	$c = 5148.021$

3.3 General Triangle.

We now introduce the concept of calculating all missing elements in a triangle and checking that they fulfill the appropriate geometric conditions. When solving for the missing elements of any triangle, the above concept of checking must also be applied rigorously.

There are five general triangle problems solved by using one of the applicable trigonometric formulae. The combinations of data are as follows:

Case 1 Given 3 sides
find the three angles.

Case 2 Given 2 sides and an angle opposite one of the sides
find the other side and two angles (two possible solutions).

Case 3 Given 2 sides and an included angle,
find the other side and two angles.

Case 4 Given 1 side and two angles,
find the other angle (implied) and two sides.

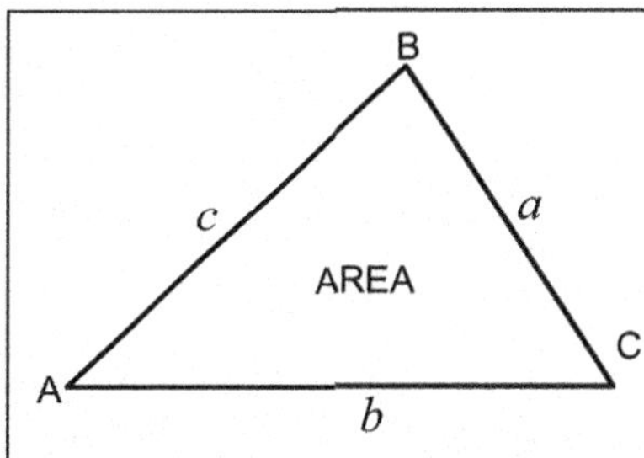

Figure 3-6 General triangle components

A plane triangle with only measured angles is not solvable.

Case 5 Given the AREA of a triangle the following solutions are available:

Case 5.1 Given AREA and the 3 angles,
find the three sides.

Case 5.2 Given AREA and 2 angles,
find the other angle and the 3 sides.

Case 5.3 Given AREA and 2 sides,

find the three angles and one side.

Case 5.4 Given AREA, 1 side and an adjacent angle, find the other two angles and two sides.

Case 5.5 Given AREA, 1 side and the opposite angle (ambiguous case), find the other two angles and 2 sides.

A plane triangle with no sides measured is not solvable.

The following formulae are used for solving the general triangle ABC as depicted in Figure 3-6.

3.4 Formulae for triangle solutions.

3.4.1 Basic rules

Sine Rule (sides)	a.	$\frac{a}{\sin A}=\frac{b}{\sin B}=\frac{c}{\sin C}=\frac{a+b+c}{\sin A+\sin B+\sin C}$	Eqn 3-1
Sine Rule (angles)	b.	$\frac{\sin A}{a}=\frac{\sin B}{b}=\frac{\sin C}{c}$	
Cosine rule (angles)	a.	$\cos A=\frac{b^2+c^2-a^2}{2bc}$	Eqn 3-2
Cosine rule (sides)	b.	$a=\sqrt{b^2+c^2-2bc\cos A}$	
(quadratic form)	c.	$a=b\cos C\pm\sqrt{c^2-b^2\sin^2 C}$	

Compound angles, see § 3.8.2, Compound identities:, page 35 for the listing of compound (A+B) identities

3.4.2 Half angle formula on the plane triangle.

Tangent Rule (angles) TAN HALF angles	a) $\tan\left(\frac{A-B}{2}\right)=\left(\frac{a-b}{a+b}\right)\cot\left(\frac{C}{2}\right)$ b) $\tan\left(\frac{A+B}{2}\right)=\left(\frac{a+b}{a-b}\right)\tan\left(\frac{A-B}{2}\right)$	Eqn 3-3
Perimeter Rule semi-perimeter	$S=\frac{a+b+c}{2}$	Eqn 3-4
Sin rule (half angle)	$\sin\left(\frac{A}{2}\right)=\sqrt{\frac{(S-b)(S-c)}{bc}}$	Eqn 3-5
Cos rule (half angle)	$\cos\left(\frac{A}{2}\right)=\sqrt{\frac{S(S-a)}{bc}}$	Eqn 3-6
Tan rule (half angle)	$\tan\left(\frac{A}{2}\right)=\sqrt{\frac{(S-b)(S-c)}{S(S-a)}}$	Eqn 3-7
Heron's Rule for area	$\text{Area}=\sqrt{S(S-a)(S-b)(S-c)}$	Eqn 3-8
Area rule for angles	a. $Area=\frac{1}{2}bc\sin A$ b. $Area=\frac{a^2\sin B\sin C}{2\sin A}$	Eqn 3-9
Area rule for sides	$a=\sin A\sqrt{\frac{2\,Area}{\sin A\sin B\sin C}}$	Eqn 3-10

The Sin Rule and the Cos Rule are the two formulae generally used, but the Tan Rule is useful in some circumstances.

3.4.3 Cyclic rules

Rules are cyclical in nature, with the unknown substituting: e.g.;

Cos Rule: Angle: $\cos A=\frac{b^2+c^2-a^2}{2bc}$ (Eqn 3-2.a1)

$\cos B=\frac{a^2+c^2-b^2}{2ac}$ (Eqn 3-2.a2)

$$\cos C = \frac{a^2 + b^2 - c^2}{2ab} \quad \text{(Eqn 3-2.a3)}$$

Cos Rule: Side:

$$a = \sqrt{b^2 + c^2 - 2bc\cos A} \quad \text{(Eqn 3-2.b1)}$$

$$b = \sqrt{a^2 + c^2 - 2ac\cos B} \quad \text{(Eqn 3-2.b2)}$$

$$c = \sqrt{a^2 + b^2 - 2ab\cos C} \quad \text{(Eqn 3-2.b3)}$$

Quadratic Cos Rule:

$$a = b\cos C \pm \sqrt{c^2 - b^2 \sin^2 C} \quad \text{(Eqn 3-2.c1)}$$

$$b = c\cos A \pm \sqrt{a^2 - c^2 \sin^2 A} \quad \text{(Eqn 3-2.c2)}$$

$$c = a\cos B \pm \sqrt{b^2 - a^2 \sin^2 B} \quad \text{(Eqn 3-2.c3)}$$

3.5 Rules for Triangle Solution.

3.5.1 Case 1. Three sides (SSS)

Given three sides (see Figure 3-7)

(a) Solve for the angle opposite the **largest** side by cos rule (angle) [1]

(b) Solve for remaining 2 angles by sin rule or continue with cos rule

(c) Check ∠A + ∠B + ∠C = 180°

[1] directly solves angles greater than 90° (resolves ambiguity).

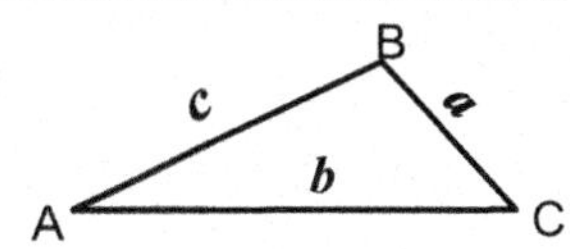

Figure 3-7 SSS

3.5.2 Case 2. Two sides, angle opposite (SSA$_O$)

Given two sides and an angle opposite one of the sides, two solutions (ambiguous)

There are 3 possible situations to this problem:

1. side $a > c \sin B$; 2 solutions (see Figure 3-8)
   opposite side intersects unknown side in two places, the ambiguous solution.
2. side $a = c \sin B$; 1 solution (see Figure 3-9)
   opposite side perpendicular to unknown side.
3. side $a < c \sin B$; 0 solution (see Figure 3-10)
   opposite side does not intersect base, unknown side too short. Check data.

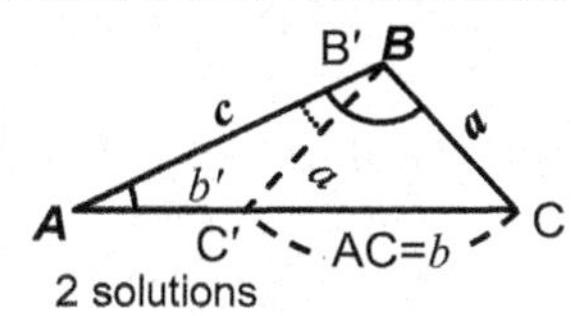

Figure 3-8 SSA$_2$

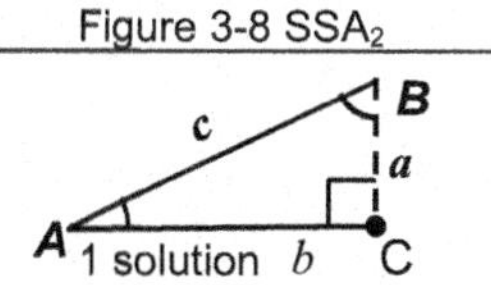

Figure 3-9 SSA$_1$

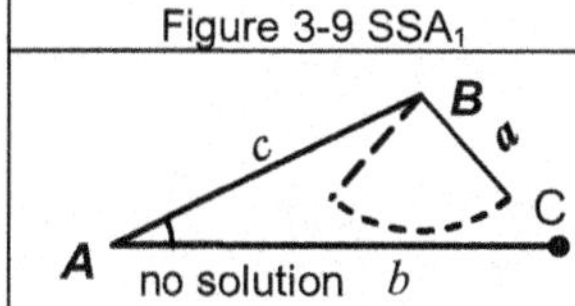

Figure 3-10 SSA$_0$

There are two solutions methods for case 2:

1. (a) Solve other opposite angle by sin rule
   (b) Solve remaining supplementary angle
   (c) Solve remaining side using sin rule
   (d) Check third angle using cos rule
   (e) Resolve ambiguity by additional data, or present two solutions
2. (a) Solve for 2 solutions of remaining side by quadratic cos rule (Eqn 3-2c).
   (b) Solve other opposite angles *A* and *B* by cyclic cos rule for angles using both solutions of side *a*,

3.5.3 Case 3. Two sides, included angle (SA$_I$S)

Given two sides and an included angle (see Figure 3-11).

There are two solutions methods for case 3:

1. (a) Solve side opposite the included angle by Cos Rule for sides (Eqn 3-2b).
   (b) Use this side to solve for other angles by Sin Rule (Eqn 3-1b), OR
2. (a) Solve two angles DIRECTLY from tan half rule (Eqn 3-3a, b),
   (b) opposite side from sin rule (Eqn 3-1a).

Check ∠A +∠B + ∠C = 180°

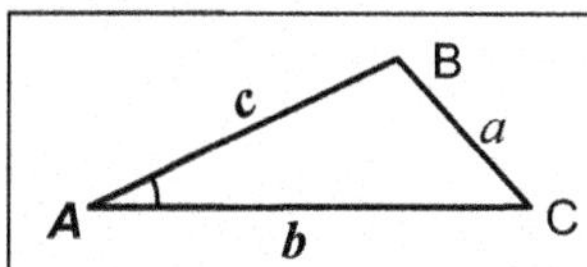

Figure 3-11 SAS

If it's not checked it's wrong.

3.5.4 Case 4. One side, two angles (SAA)

Given one side and two angles (see Figure 3-12)

(a) Third angle is the supplement of the sum of the other two ($C = 180 - (A + B)$)

(b) Solve for each of the other sides by sin rule (Eqn 3-1a).

Check third angle by cos rule (Eqn 3-2a).

Check ∠A + ∠B + ∠C = 180° (Check your subtractions).

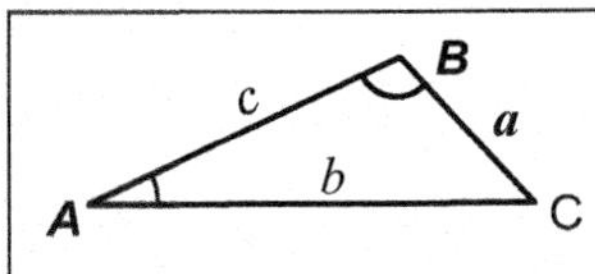

Figure 3-12 SAA

3.6 Examples of Triangle Solutions.

The following Worked Examples will illustrate the applications of the above rules for triangle solution.

3.6.1 Worked Example 3.3.

Case 1. Three sides, find the three missing angles.

Given: Figure 3-13

a = BC = 945.875

b = AC = 1000.000

c = AB = 1046.514

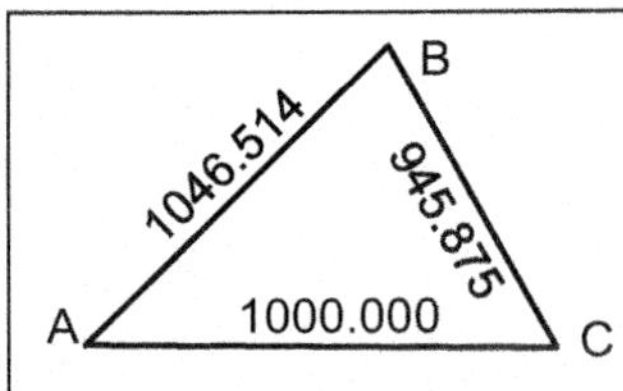

Figure 3-13 Case 1

Solution:

Step (a): Find the angle opposite the largest side by COS rule (angle) (Eqn 3-2a)

$$\angle C = \arccos\left(C = \left(\frac{a^2+b^2-c^2}{2ab}\right)\right) = \arccos\left(\frac{1000.000^2 + 945.875^2 - 1046.514^2}{2\times 1000.000 \times 945.875}\right)$$

$$= \arccos(0.422618)$$

$$\angle C = 65°00'00''$$

Step (b) (1) Solve for first of remaining 2 angles by Sin Rule (Eqn 3-1b)

$$\frac{\sin B}{b} = \frac{\sin C}{c} \text{ then } \sin B = \frac{b \sin C}{c}$$

$$\angle B = \arcsin\left(\frac{a \sin C}{c}\right) = \arcsin\left(\frac{1000.000 \times (0.906308)}{1046.514}\right) = \arcsin(0.866026)$$

$$\angle B = 60°00'00''$$

Step (b) (2) Solve for remaining angle by Sin Rule (Eqn 3-1b)

$$\frac{\sin A}{a} = \frac{\sin C}{c} \text{ then } \sin A = \frac{a \sin C}{c}$$

$$\angle A = \arcsin\left(\frac{a \sin C}{c}\right) = \arcsin\left(\frac{945.875 \times (0.906308)}{1046.514}\right) = \arcsin(0.819152)$$

$$\angle A = 55°00'00''$$

Step (c) Check

$$\angle A + \angle B + \angle C = 180°$$

Note:

The fraction $\left(\frac{\sin C}{c}\right) = \left(\frac{0.906308}{1046.514}\right) = 0.000866025$ is a constant and can be entered into the calculator memory as it is used in calculations for angles A and B.

3.6.2 Worked Example 3.4.

Case 2. Two sides, an opposite angle (AMBIGUOUS case). find the missing angles and distances.

Given: Figure 3-14. Diagram NOT to scale.

$c =$ 991.461

$a =$ 70.023

$\angle A =$ 4°

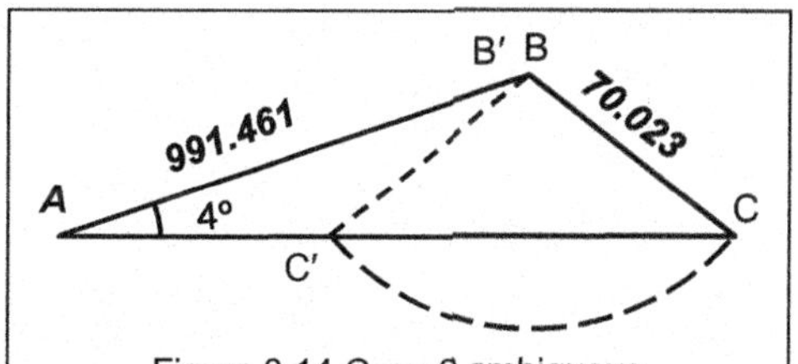

Figure 3-14 Case 2 ambiguous

Solution:

The solution is ambiguous. Side BC could also occupy the dotted position at C′. You will need extra information to determine the location of C, and hence the distance AC = b and the angle B. This data may come from field notes, it is especially important when angle B is close to 90°.

Step (a) Solve the other opposite angle.

by SIN Rule for ANGLES (Eqn 3-1b)

$$\frac{\sin A}{a} = \frac{\sin C}{c} \quad \text{then} \quad \sin C = \frac{c \sin A}{a}$$

$$\angle C = \arcsin\left(\frac{c \sin A}{a}\right) = \arcsin\left(\frac{991.461 \times (0.069756)}{70.023}\right)$$

$= \arcsin(0.987687)$

$\angle C$ is POSITIVE in quadrant I and II.

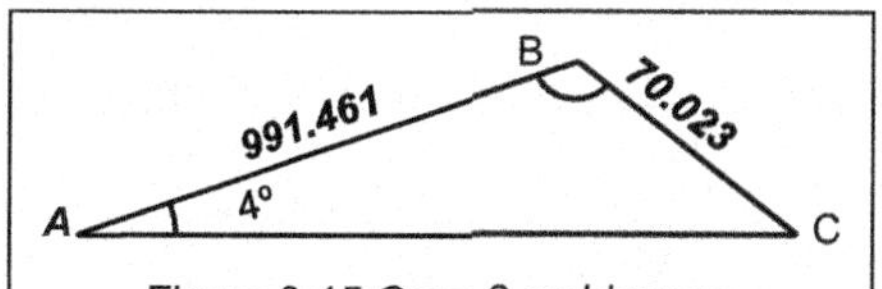

Figure 3-15 Case 2 ambiguous

thus two POSSIBLE answers: See Figure 3-15 | Figure 3-16

$\angle C = 80°59'58''$ or, (Subtracting from 180°) $\angle C' = 99°00'02''$

Step (b) Find $\angle B$ from $(180° - \angle B - \angle C)$

$\therefore$ $\angle B = 95°00'02''$ or | $\angle B' = 76°59'58''$

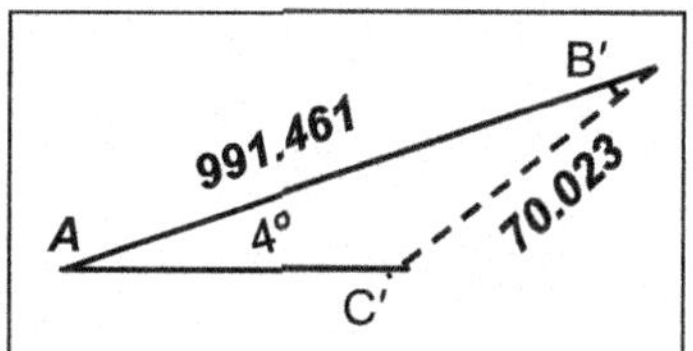

Figure 3-16 Case 2 ambiguous

With the ambiguous case the two sets of results are equally valid.

Step (c) Find sides b and b′ by SIN rule for SIDES (Eqn 3-1a)

$$\frac{b}{\sin B} = \frac{c}{\sin C} \quad \text{then} \quad b = \frac{c \sin B}{\sin C}$$

where $\sin C = 0.987687$ from Step (a)

Using $\angle B = 95°00'02''$

$$b = \left(\frac{991.461 \times (0.996194)}{0.987687}\right)$$

$b = 1000.000$

$\angle B' = 76°59'58''$ (both from Step (b))

$$b' = \left(\frac{991.461 \times (0.974368)}{0.987687}\right)$$

$b' = 978.091$

Step (d) Check $\angle B$ using Cos Rule $\quad \angle B = \arccos\left(\frac{a^2 + c^2 - b^2}{2ac}\right)$

$$= \arccos\left(\frac{991.461^2 + 70.023^2 - 1000^2}{2 \times 991.461 \times 70.023}\right)$$

$= \arccos(-0.087158)$

$\angle B = 95°00'00''$ (checks)

$$\text{or} = \arccos\left(\frac{991.461^2 + 70.023^2 - 978.091^2}{2 \times 991.461 \times 70.023}\right)$$

$= \arccos(0.224963)$

$\angle B' = 76°59'57''$ (checks)

The slight angular misclose in the checks comes from loss of precision in the calculation of side a. (1000.000 or 978.091, using only 3dp.). In practice a surveyor has some foreknowledge of the shape of the triangle and can therefore select the correct solution. When in doubt, good survey practice would be to take another measurement in the field to avoid the ambiguity.

3.6.3 Worked Example 3.4a. Alternate solution using a quadratic equation.

Case 2. Given 2 sides and opposite angle (AMBIGUOUS case)
find the missing angles and distances

This is an alternate solution to finding the sides ***b and b′*** using a quadratic equation from the cos rule. This allows the parallel solutions of side b and b'.
The two angles, C and C' for C can then be calculated.

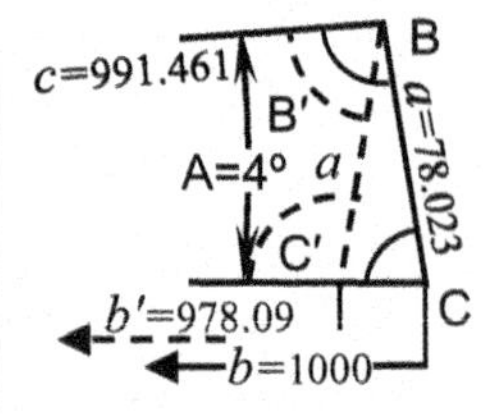

Figure 3-17 Case 2 quadratic

Given: Figure 3-17. Solution diagram

$a = 70.023$

$c = 991.461$

$\angle A = 4^\circ$

Solution:

Step (a) Solve for two values of side b in quadratic form by COS Rule (Eqn 3-2 b.2)

$$b = c\cos A \pm \sqrt{a^2 - c^2 \sin^2 A}$$

1) Evaluate $c\cos A = 991.461 \times \cos(4^\circ) = 989.0458$
2) Evaluate the determinant: $\pm\sqrt{a^2 - c^2 \sin^2 A}$

$$\pm\sqrt{70.023^2 - 991.461^2 \times (\sin(4^\circ))^2} \Rightarrow \pm\sqrt{4903.2205 - 982994.9145 \times 0.004866} \Rightarrow \pm\sqrt{120.0011}$$

± 10.9545

Find the two values of side b:

$b = 989.0458 + 10.9545$	$b' = 989.0458 - 10.9545$
$\boldsymbol{b = 1000.000}$	$\boldsymbol{b' = 978.091}$

Step (b) Solve for two values of angle **C** by cyclic COS Rule

No ambiguity as COS is resolved in first and second quadrants

$\angle C = \arccos\left(\dfrac{a^2 + b^2 - c^2}{2ab}\right)$	$\angle C' = \arccos\left(\dfrac{a^2 + b'^2 - c^2}{2ab'}\right)$
$\angle C = \arccos\left(\dfrac{70.023^2 + 1000^2 - 991.461^2}{2 \times 70.023 \times 1000}\right)$	$\angle C' = \arccos\left(\dfrac{70.023^2 + 978.091^2 - 991.461^2}{2 \times 70.023 \times 978.091}\right)$
$\boldsymbol{\angle C = \arccos(0.156436) = 80^\circ 59' 59''}$	$\boldsymbol{\angle C' = \arccos(-0.156874) = 99^\circ 00' 02''}$
$\angle B = \arccos\left(\dfrac{a^2 + c^2 - b^2}{2ac}\right)$	$\angle B' = \arccos\left(\dfrac{a^2 + c^2 - b'^2}{2ac}\right)$
$\angle B = \arccos\left(\dfrac{70.023^2 + 991.461^2 - 1000^2}{2 \times 70.023 \times 991.461}\right)$	$\angle B' = \arccos\left(\dfrac{70.023^2 + 991.461^2 - 978.091^2}{2 \times 70.023 \times 991.461}\right)$
$\boldsymbol{\angle B = \arccos(-0.087163) = 95^\circ 00' 01''}$	$\boldsymbol{\angle B' = \arccos(0.224958) = 76^\circ 59' 59''}$
Check A + B + C = 180	Check A′ + B′ + C = 180

3.6.4 Worked Example 3.5.

Case 3. Two sides and the included angle
find third side and other two angles.

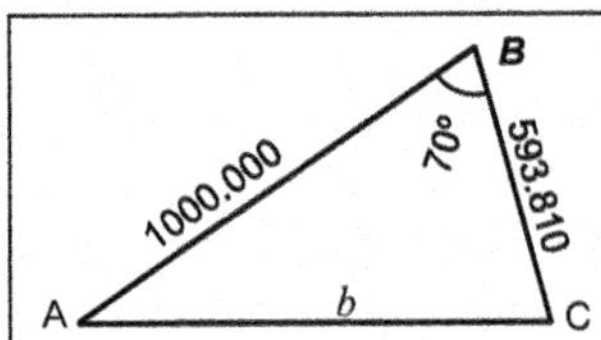

Figure 3-18 Case 3

Given: Figure 3-18

a = 593.810

c = 1000.000

$\angle B = 70^\circ$

Often the angle will have been derived from the known bearings AC and AB.
Angles B and C can be calculated directly by tan rule.

Solution:

Step (a) Solve side opposite included angle by cos rule (Eqn 3-2b)

$$b = \sqrt{a^2 + c^2 - 2ac\cos B}$$

$$= \sqrt{593.810^2 + 1000.000^2 - 2\times 5983810 \times 1000.000 \times (0.342020)} = \sqrt{946420.353}$$

$\boldsymbol{b = 972.841}$

Step (b) Solve other two angles by **tan** rule **OR** use this side ***b***, with given side ***a***, for **sin** rule solution.

$$\tan\left(\frac{A-C}{2}\right) = \left(\frac{a-c}{a+c}\right)\cot\left(\frac{B}{2}\right) \quad \text{(Eqn 3-3 a)}$$

$$\tan\left(\frac{A-C}{2}\right) = \left(\frac{\left(\frac{593.810-1000.000}{593.810+1000.000}\right)}{\tan\left(\frac{70}{2}\right)}\right) = -0.363970$$

$$\left(\frac{A-C}{2}\right) = \arctan(-0.363970) = -20^\circ 00'00''$$

Note that the TAN Rule (also known as the TAN HALF Rule) can solve for angles B and C directly.

Side ***b*** is **not** needed in solution.

The solution uses ONLY the original data, angle A and sides b and c.

(A-C)/2 = –20°

From the calculated $\tan\left(\frac{A-C}{2}\right)$ above

$$\tan\left(\frac{A+C}{2}\right) = \left(\frac{a+c}{a-c}\right)\tan\left(\frac{A-C}{2}\right) \quad \text{(Eqn 3-3b)}$$

(A+C)/2 = arctan(1.428148) (A+C)/2 = 55°

(A+C)/2 + (A–C)/2 = A = 55 + (–20) = 35°

A = 35°00′00″

(A+C)/2 – (A-C)/2 = C = 55 – (–20) = 75°

C = 75°00′00″

OR, using *b* = 972.841381, solve by sin Rule

$$\frac{\sin A}{a} = \frac{\sin B}{b} = \frac{\sin C}{c} \quad \text{(Eqn 3-1b)}$$

$$\sin A = \frac{a\sin B}{b} = \frac{593.810\times\sin(70)}{972.841381} \quad \text{(solve } A\text{)}$$

$\sin A = 0.573576$ $\quad A = 35^\circ$

A = 35°00′00″

$$\sin C = \frac{c\sin B}{c} = \frac{1000\times\sin(70)}{972.841381} \quad \text{(solve } C\text{)}$$

$\sin C = 0.965926$ $\quad C = 75^\circ$

C = 75°00′00″

For BOTH answers: Check ∠A + ∠B + ∠C = 70° + 75° + 35° = 180°

3.6.5 Worked Example 3.6.

Case 4. Two angles and an included side find third angle and two sides.

Given: Figure 3-19

∠A = 45°

∠C = 50°

AC = b =783.514

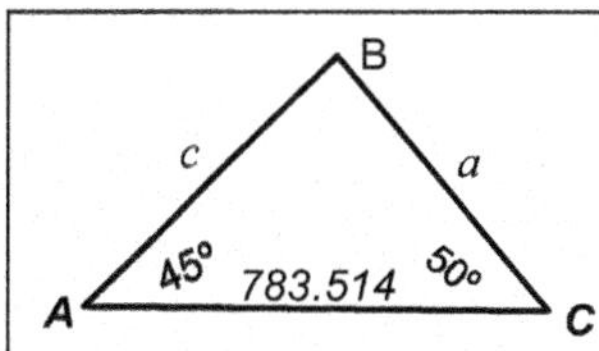

Figure 3-19 Case 4

Solution:

Step (a) Third Angle is the supplement of the sum of the other two:

∠B = 180° – (∠A + ∠C)

= 180° – (45 + 50)

∠B = 85°

Step (b) (1) Solve for each of the other sides by Sin Rule (Eqn 3-1a)

$$\frac{a}{\sin A} = \frac{b}{\sin B} \text{ then } a = \frac{b\sin A}{\sin B}$$

$$\text{and } \frac{b}{\sin B} \text{ is a constant } = \frac{783.514}{0.996195} = 786.506986$$

$$a = \frac{b\sin A}{\sin B} = \text{Sin}(45^\circ)\times 786.506896 = (0.707107)\times 786.506896$$

a = 566.144

If it's not checked it's wrong.

Step (b) (2) Solve for second side by Sin Rule

$$c = \frac{b\sin C}{\sin B} = \text{Sin}(50^\circ)\times 786.506896 = (0.766044)\times 786.506896$$

c = 602.499

Step (c) Check third angle by Cos rule.

$$\angle B = \arccos\left(\frac{a^2+b^2-c^2}{2ac}\right) = \arccos\left(\frac{556.144^2+602.499^2-783.514^2}{2\times 556.144\times 602.499}\right)$$

$$= \arccos(0.087156)$$

$$\angle \mathbf{B} = \mathbf{85^\circ 00'00''} \quad \text{checks}$$

Step (c) We can also check using the cos rule for sides. In this case we will VERIFY the value a that was given against calculated answers. (Eqn 3-2b)

$$b = \sqrt{a^2+c^2-2ac\cos B}$$

$$= \sqrt{556.144^2+602.499^2-2\times 556.144\times 602.499\times 0.087156}$$

$$= \sqrt{613893.563}$$

***b* = 783.514** (checks)

Checks on other sides may reveal small discrepancies due to rounding errors.

3.7 Examples of Triangle Solutions incorporating Area.

For the 5 area solutions we will use the same triangle (see Figure 3-20) which will be the model for the solutions. In these solutions there will be minor variations in answers as distances are quoted to the mm (0.001m)

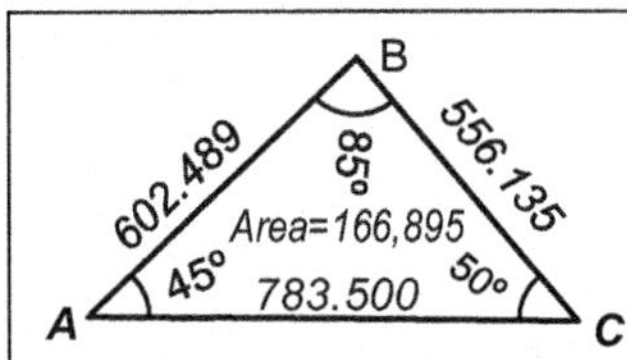

Figure 3-20 Case 5 data

Case 5. Solutions using area rules.

3.7.1 Worked Example 3.7.

Case 5.1 Area and 3 angles, find SIDES

Given: Figure 3-21

Area = 166895

Angles A = 45°, B = 85°, C = 50°.

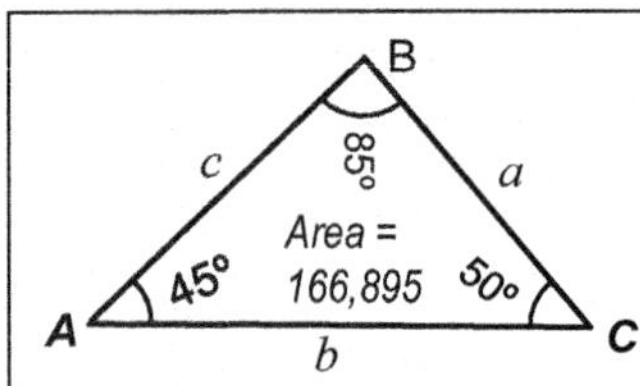

Figure 3-21 Area. 3 angles

Solution:

Use formula in Eqn 3-10 for sides:

$$a = \sin A\sqrt{\frac{2\times Area}{\sin A\sin B\sin C}} \quad \text{and solve cyclically.}$$

where $\sqrt{\frac{2\times Area}{\sin A\sin B\sin C}}$ is a constant (**k**) for the triangle. k= 786.493365

The sides are found cyclically as the product of the Sines of the three angles.

$a = k\sin A$ = 786.493365×Sin(45) = 786.493365×0.707107 = 556.1348

$b = k\sin B$ = 786.492843×Sin(85) = 783.5005

$c = k\sin C$ = 786.492843×Sin(50) = 602.4889. All check.

Check: Area by Perimeter Rule (Eqn 3-7) where semi-perimeter $S = \frac{a+b+c}{2}$ = 976.0621. (Eqn 3-7)

$$\text{Area} = \sqrt{S(S-a)(S-b)(S-c)} = 166895.009$$

Area by Sine Rule (Eqn 3-9)

$$\text{Area} = \frac{bc\sin A}{2} = 166895.003$$

3.7.2 Worked Example 3.8.

Case 5.2 Area and 2 angles,
find third angle and 3 SIDES

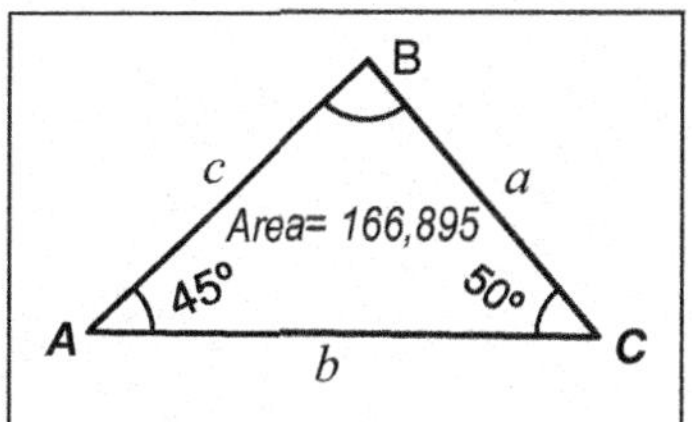

Figure 3-22 Area. 2 angles

Given: Figure 3-22
Area = 166895
Angles A = 45°, C = 50°.

Solution: find angle B = 180 – B – C = 180 – 50 – 45,
$B = 85°$

Use formula in Eqn 3-10 and solve as for Case 5.1 for sides:

$$a = \sin A \sqrt{\frac{2 \times Area}{\sin A \sin B \sin C}}$$ and solve cyclically.

As in § 3.7.1 (Worked Example. 3.7), the sides are found cyclically as products of the sines of the three angles.

$a = 566.1348$
$b = 783.5005$
$c = 602.4885$. All check.

Checks similar to Case 5.1 for area.
Check angle B by cos rule for angles (Eqn 3-2a).

3.7.3 Worked Example 3.9.

Case 5.3 Area and 2 sides,
find third side, and 3 ANGLES

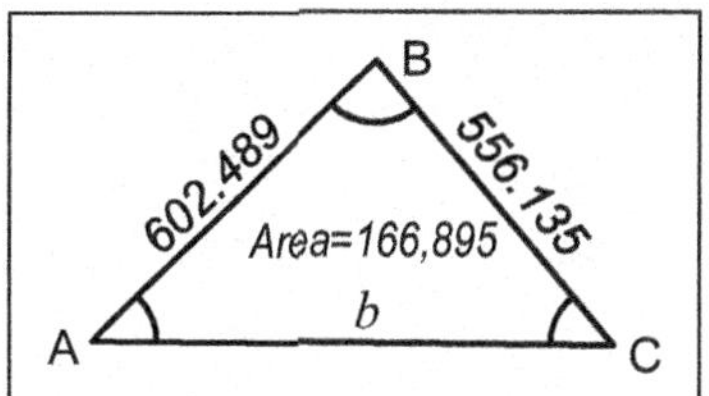

Figure 3-23 Area. 2 sides

Given: Figure 3-23.
Area = 166895
Sides $a = 556.135$, $c = 602.489$.

Solution:

Use Eqn 3-9 to find angle B the included angle between the two sides a and c.

$$Area = \frac{ac \sin B}{2} \qquad \text{(Eqn 3-9)}$$

$$B = \arcsin\left(\frac{2 \times Area}{a\,c}\right) = \arcsin(0.996141)$$

B = 84.99962° = 84°59′58″ (85°)

Having two sides and included angle, the problem is now solved as § 3.5.3, Case 3, using formula 4-2a, cos rule for sides.

$$b = \sqrt{a^2 + c^2 - 2ac\cos B} = 783.5004$$

The other two angles can be found cyclically either by:

3.7.3.1 Method 1)

Using Eqn 3-1b, sin rule for angles

$$\frac{\sin A}{a} = \frac{\sin B}{b} = \frac{\sin C}{c}$$ where we have values for angle B and side b.

now $\frac{\sin B}{b}$ is a constant, $k = \frac{0.9961941}{783.5004} = 0.001271$

$$A = \arcsin\left(\frac{a \sin B}{b}\right) = \arcsin(a \times k) = \arcsin(0.7071070)$$

$A = 45.0000°$

$$C = \arcsin\left(\frac{c \sin B}{b}\right) = \arcsin(c \times k) = \arcsin(0.766045)$$

$C = 50.0000°$
Check: A + B + C = 180°

Method 2)

Eqn 3-2a, cos rule for angles.

$$A = \arccos\left(\frac{b^2 + c^2 - a^2}{2bc}\right) = \arccos(0.707107)$$

$A = 45.0000°$

$$C = \arccos\left(\frac{a^2 + b^2 - c^2}{2ab}\right) = \arccos(0.7666044)$$

$C = 50.0000°$
Check $A + B + C = 180°$

3.7.4 Worked Example 3.10.

Case 5.4 Area, 1 side and an adjacent angle.
Find the other 2 sides and ANGLES.

Given: See Figure 3-24
Area = 166895
Side c = 602.489. ∠B = 85°.

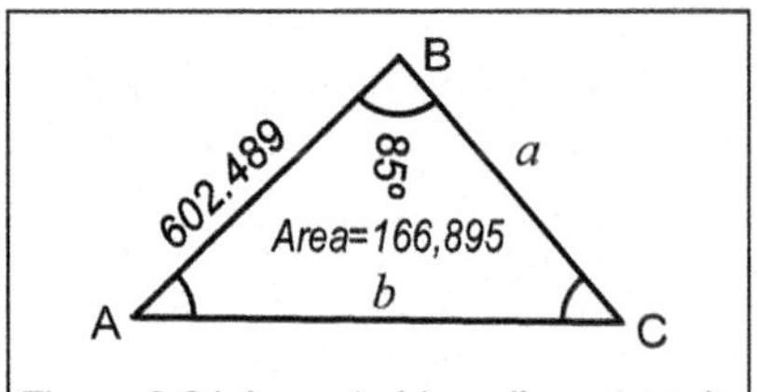

Figure 3-24 Area. 1 side, adjacent angle

Solution:

Use Eqn 3-9 to find side a, which forms the included angle B between the two sides a and c.

$$\text{Area} = \frac{ac\sin B}{2}$$

$$a = \left(\frac{2\times Area}{c\sin B}\right) = \frac{333790}{602.948\times 0.996195}$$

$a = 556.1348$

Having two sides and included angle, the problem is solved as § 3.5.3, Case 3 using Eqn 3-2b, cos rule for sides.

$$b = \sqrt{a^2 + c^2 - 2ac\cos B} = 783.5004$$

Now, as in § 3.7.3, Case 5.3, find the other two angles cyclically, using either method, or by using Eqn 3-3, the tan rule, with re-arranged values:

$$\tan\left(\frac{A-C}{2}\right) = \left(\frac{a-c}{a+c}\right)\cot\left(\frac{B}{2}\right) \qquad \text{(called TAN HALF rule)} \qquad \text{(Eqn 3-3)}$$

$$\tan\left(\frac{A-C}{2}\right) = \left(\frac{\dfrac{556.1348-602.489}{556.1348+602.489}}{\tan\left(\dfrac{85°}{2}\right)}\right) = \frac{-0.040008}{0.916331} = -0.043661$$

$$\left(\frac{A-C}{2}\right) = \arctan(-0.043661) = -2.5°$$

Now, angles $A + B + C = 180$, thus $A + C = 180 - B = 95°$

$$\left(\frac{A+C}{2}\right) = \left(\frac{95}{2}\right) = 47.5°$$

$$A = \left(\frac{A+C}{2}\right) + \left(\frac{A-C}{2}\right) = 47.5 + (-2.5) = 45°$$

$$C = \left(\frac{A+C}{2}\right) - \left(\frac{A-C}{2}\right) = 47.5 - (-2.5) = 50°$$

3.7.5 Worked Example 3.11.

Case 5.5 Area, 1 side and an **opposite** angle.
Find the other 2 sides and ANGLES.

As with the Case 2 (§ 3.6.2, Worked Example 3.4), the answer is ambiguous and, although the two angles and sides are resolved, which is which is the question.

Comment: This is a difficult calculation as we have not explored the trigonometric tricks needed for a solution. The problem is that we don't know the other two angles and sides, and thus have no idea of its shape. Or do we? I am indebted to Mr. Bob Atkinson (Curtin University) for showing me his method.

Given: Figure 3-25,
Area = 166895
Side b = 783.500. ∠B = 85°.

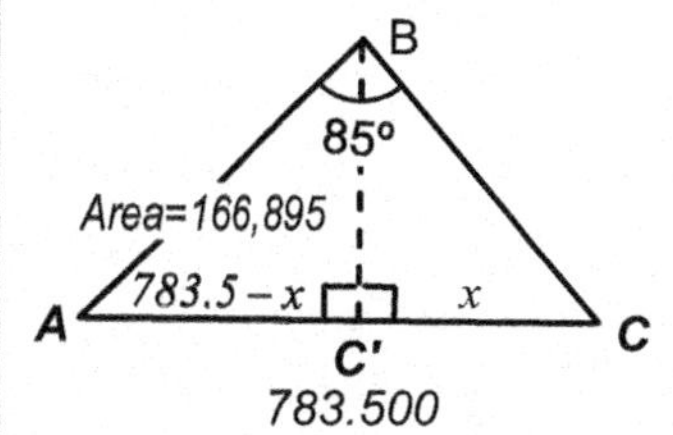

Figure 3-25 Area. 1 side, opposite angle

By finding the height of the triangle, we can split the area into two triangles, with two unknown bases. The product of the two unknown bases times the height must add to 2×area. BC′ is perpendicular to AC.

$$Area = \left(\frac{AC \times BC'}{2}\right) \Rightarrow BC' = \frac{2 \times Area}{AC}$$

Height $BC' = 333790 \div 783.5 = 426.02425$

In Triangle BCC′:

$$\tan C = \frac{BC'}{CC'} \text{ and let } CC' = x$$

$$\tan C = \frac{426.02425}{x}$$

Similarly, in triangle ABC′,

$$\tan A = \frac{BC'}{AC'} \text{ and } AC' = 783.5 - x$$

$$\tan A = \frac{426.02425}{783.5 - x}$$

Compound angles: $\tan(A+C) = \dfrac{\tan A + \tan C}{1 - \tan A \tan C}$, designated $\dfrac{a}{b}$. (See following § 3.8).

Now (A + C) = 95° (180 − B)
tan(95) = −11.430052

Let $a = \tan A \tan C$:

$$a = \frac{426.02425}{783.5 - x} \times \frac{426.02425}{x} = \frac{426.02425x \ + \ 426.02425 \times (783.5 - x)}{783.5x \ - \ x^2} = \frac{426.02425 \times 783.5}{783.5x \ - \ x^2}$$

$$a = \frac{333790.0}{783.5x \ - \ x^2}$$

Let $b = 1 - \tan A \tan C$:

$$b = 1 - \frac{426.02425}{783.5 - x} \times \frac{426.02425}{x} = \frac{1}{1} - \frac{181496.662}{783.5x - x^2}$$

$$b = \frac{783.5x - x^2 - 181496.662}{783.5x - x^2}$$

Now, $\tan(A{+}C) = -11.430052 = \dfrac{a}{b} = a \times \dfrac{1}{b}$, so, invert b and cross multiply

$$-11.430052 = \frac{333790.0}{783.5x - x^2} \times \frac{783.5x - x^2}{783.5x - x^2 - 181496.662}$$

$$783.5x - x^2 - 181496.662 = \frac{333790.0}{-11.430052}$$

$$783.5x - x^2 - 181496.662 = -29202.841$$

Form the quadratic equation: $ax^2 + bx + c = 0 \Rightarrow x^2 - 783.5x + 152293.821 = 0$

Solving the quadratic equation, where $a = 1, b = -783.5, c = 152293.821$

$$x = \frac{-b \pm \sqrt{b^2 - 4ac}}{2a} = \frac{-(-783.5) \pm \sqrt{(-783.5)^2 - 4 \times 1 \times 152293.821}}{2 \times 1}$$

Solving the determinant of the quadratic:

$$\det = \sqrt{(-783.5)^2 - 4 \times 1 \times 152293.821} = \sqrt{4696.9671}$$
$$= 68.5344$$
$$x_1 = \frac{-(-783.5) + 68.5344}{2} = 426.017 \textbf{ or}$$
$$x_2 = \frac{-(-783.5) - 68.5344}{2} = 357.483\,.$$

Note that $x_1 + x_2 = \text{AC} = 783.5$

Find angles A and C. We are unable to define which is which!

If we defined angle C by $x_1 = 426.017$ then:

$$C = \arctan\left(\frac{BC'}{AC}\right) = \arctan\left(\frac{426.0242}{426.017}\right) = \arctan(1.000016)$$

$C = 45^\circ \Rightarrow A = 50^\circ$ (within rounding errors). Figure 3-26

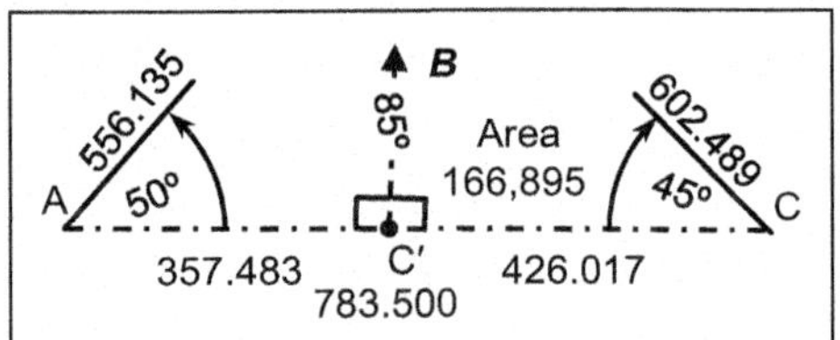

Figure 3-26 Angle C from x_1 =426.017

If we defined angle C by $x_2 = 357.483$ then:

$$C = \arctan\left(\frac{BC}{C''C'}\right) = \arctan\left(\frac{426.0242}{357.843}\right) = \arctan(1.191734)$$

$C = 50^\circ \Rightarrow A = 45^\circ$ (within rounding errors) Figure 3-27

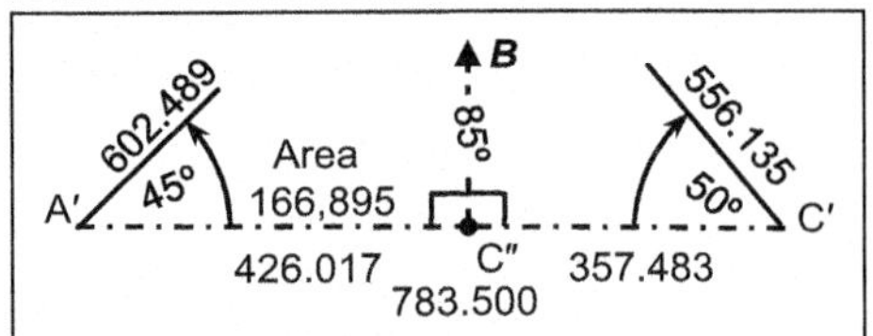

Figure 3-27 Angle C' from x_2 = 357.483

Note: We are unable to **define** which angle is which unless we had further data from field notes, say a sketch or description of the plot. Attention to computation sequence on a known figure **may** help.

The remaining sides are now found from Eqn 3-10: (§ 3.7.1, § 3.7.2 Worked Examples. 3.7, 3.8)

For sides $a = \sin A \times \sqrt{\dfrac{2 \times Area}{\sin A \sin B \sin C}} \Rightarrow k = 786.492843$ (Eqn 3-10)

$b = k\sin B = 786.493365 \times \sin(85) = .0.996195 = 783.500$ (check)

a or $c' = k\sin A = 602.489$, depending on chosen value for angles A and C

c or $a' = k\sin C = 556.135$. All check. See , Figure 3-27.

Check: Area by Heron's semi-perimeter Rule (Eqn 3-7) where: $S = ½(a+b+c) = 976.062$.

$$\text{Area} = \sqrt{S(S-a)(S-b)(S-c)} = 166895.$$

3.8 Summary of some trigonometric identities.

Remember that **csc** *is* cosecant (cosec) = 1/**sin**

3.8.1 Pythagorean identities:

$$\sin^2 A + \cos^2 A = 1$$
$$\tan^2 A + 1 = \sec^2 A$$
$$1 + \cot^2 A = \csc^2 A$$

3.8.2 Compound identities:

$$\sin(A \pm B) = \sin A \cos B \pm \cos A \sin B$$
$$\cos(A \pm B) = \cos A \cos B \mp \sin A \sin B$$
$$\tan(A \pm B) = \frac{\tan A \pm \tan B}{1 \mp \tan A \tan B}$$
$$\cot(A \pm B) = \frac{\cot B \cot A \mp 1}{\cot B \pm \cot A}$$

3.8.3 Double angle identities:

$$\sin 2A = 2\sin A \cos A = \frac{2\tan A}{1 + \tan^2 A}$$
$$\cos 2A = \cos^2 A - \sin^2 A = 2\cos^2 A - 1$$
$$= 1 - 2\sin^2 A = \frac{1 - \tan^2 A}{1 + \tan^2 A}$$
$$\tan 2A = \frac{2\tan A}{1 - \tan^2 A} = \frac{2\cot A}{\cot^2 A - 1}$$
$$= \frac{2}{\cot A - \tan A}$$

3.8.4 Half angles identities:

$$\sin\frac{A}{2} = \pm\sqrt{\frac{1 - \cos A}{2}}$$
$$\cos\frac{A}{2} = \pm\sqrt{\frac{1 + \cos A}{2}}$$
$$\tan\frac{A}{2} = \pm\sqrt{\frac{1 - \cos A}{1 + \cos A}} = \frac{\sin A}{1 + \cos A} = \frac{1 - \cos A}{\sin A}$$

Sum to product identities:

$$\sin A \pm \sin B = 2\sin\left(\frac{A \pm B}{2}\right)\cos\left(\frac{A \mp B}{2}\right)$$
$$\cos A + \cos B = 2\cos\left(\frac{A + B}{2}\right)\cos\left(\frac{A - B}{2}\right)$$
$$\cos A - \cos B = -2\sin\left(\frac{A + B}{2}\right)\sin\left(\frac{A - B}{2}\right)$$

3.8.5 Product to sum identities:

$$\sin A \sin B = -\frac{1}{2}\left[\cos(A + B) - \cos(A - B)\right]$$
$$\cos A \cos B = \frac{1}{2}\left[\cos(A + B) + \cos(A - B)\right]$$
$$\sin A \cos B = \frac{1}{2}\left[\sin(A + B) + \sin(A - B)\right]$$
$$\cos A \sin B = \frac{1}{2}\left[\sin(A + B) - \sin(A - B)\right]$$

3.8.6 Compound identity proof.

See Figure 3-28 below.

$$\angle RPQ = \angle RQO = A$$
$$\sin(A + B) = \frac{PS}{OP} = \frac{RS + PR}{OP} = \frac{RS}{OP} + \frac{PR}{OP}$$
$$= \frac{RS}{OQ} \times \frac{OQ}{OP} + \frac{PR}{PQ} \times \frac{PQ}{OP}$$
$$= \sin A \cos B + \cos A \sin B$$
$$\cos(A + B) = \frac{OS}{OP} = \frac{OT - ST}{OP} = \frac{OT}{OP} - \frac{ST}{OP}$$
$$= \frac{OT}{OQ} \times \frac{OQ}{OP} - \frac{ST}{PQ} \times \frac{PQ}{OP}$$
$$= \cos A \cos B - \sin A \sin B$$

Now $\tan(\alpha) = \dfrac{\sin(\alpha)}{\cos(\alpha)}$ thus:

$$\tan(A + B) = \frac{\sin(A + B)}{\cos(A + B)} = \frac{\sin A \cos B + \cos A \sin B}{\cos A \cos B - \sin A \sin B}$$

Divide through by $\cos A \cos B$

$$\tan(A + B) = \left(\frac{\frac{\sin A \cos B}{\cos A \cos B} + \frac{\cos A \sin B}{\cos A \cos B}}{\frac{\cos A \cos B}{\cos A \cos B} - \frac{\sin A \sin B}{\cos A \cos B}}\right) \text{then cancelling}$$
$$= \left(\frac{\frac{\sin A \cancel{\cos B}}{\cos A \cancel{\cos B}} + \frac{\cancel{\cos A} \sin B}{\cancel{\cos A} \cos B}}{\frac{\cancel{\cos A}\,\cancel{\cos B}}{\cancel{\cos A}\,\cancel{\cos B}} - \frac{\sin A \sin B}{\cos A \cos B}}\right)$$
$$= \left(\frac{\tan A + \tan B}{1 - \tan A \tan B}\right)$$

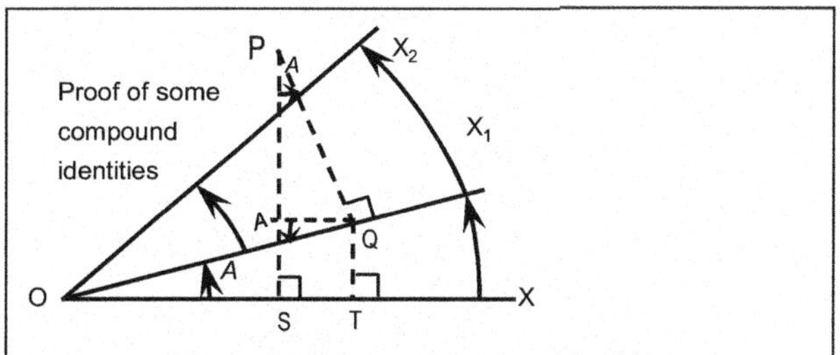

Figure 3-28 Prove compound identities

Similarly, the negative compounded angles of § 3.8.2 can be proved. Shepherd, F.A. (1968)

3.8.7 Tan half angle proof

Show in plane ΔABC that $\tan\left(\dfrac{A-C}{2}\right)=\left(\dfrac{a-c}{a+c}\right)\cot\left(\dfrac{B}{2}\right)$

In ΔABC, by sin rule $\dfrac{a}{\sin A}=\dfrac{b}{\sin B}=\dfrac{c}{\sin C}=k$ (constant)

$a=k\sin A; b=k\sin B; c=k\sin C$

$$\frac{a-c}{a+c}=\frac{k\sin A-k\sin C}{k\sin A+k\sin C}=\frac{\sin A-\sin C}{\sin A+\sin C}$$

$$\because \sin A-\sin C\Leftrightarrow 2\cos\left(\frac{A+C}{2}\right)\sin\left(\frac{A-C}{2}\right)$$

$$\because \sin A+\sin C\Leftrightarrow 2\sin\left(\frac{A+C}{2}\right)\cos\left(\frac{A-C}{2}\right)$$

$$\frac{\sin A-\sin C}{\sin A+\sin C}=\frac{\cos\left(\frac{A+C}{2}\right)\sin\left(\frac{A-C}{2}\right)}{\sin\left(\frac{A+C}{2}\right)\cos\left(\frac{A-C}{2}\right)}\text{ (2s cancel)}$$

because $\dfrac{\sin\left(\frac{A-C}{2}\right)}{\cos\left(\frac{A-C}{2}\right)}=\tan\left(\dfrac{A-C}{2}\right)$ and $\dfrac{\cos\left(\frac{A+C}{2}\right)}{\sin\left(\frac{A+C}{2}\right)}=\cot\left(\dfrac{A+C}{2}\right)$

$$\frac{a-c}{a+c}=\frac{\sin A-\sin C}{\sin A+\sin C}=\tan\left(\frac{A-C}{2}\right)\times\cot\left(\frac{A+C}{2}\right)\text{ and }\cot\left(\frac{A+C}{2}\right)=\tan\frac{B}{2}$$

$$\frac{a-c}{a+c}=\tan\frac{B}{2}\times\tan\left(\frac{A-C}{2}\right)$$

$$\therefore \tan\left(\frac{A-C}{2}\right)=\left(\frac{a-c}{a+c}\right)\cot\frac{B}{2}$$

3.9 Chapter Exercise 3.2

3.9.1 Solve the following General Triangles

(1)	A = 80°	b = 130.000	c = 110.000
(2)	B = 50°	C = 60°	a = 110.000
(3)	B = 70°	b = 110.000	c = 115.281
(4)	a = 10000.000	b = 10465.140	c = 9458.750
(5)	a = 82.990	b = 73.580	c = 101.920

(6) In Figure 3-29 Solve for the missing elements of the triangle.

∠A = 58°44′15″

∠B = 73°38′35″

Side AD = 400m.

∠BAD = ∠DAC

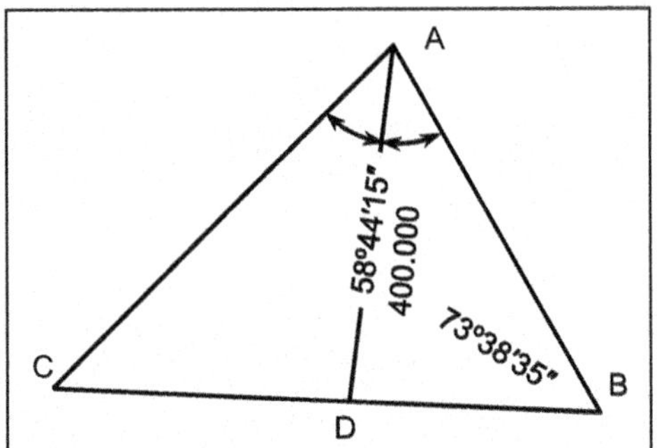

Figure 3-29 Problem (6)

3.9.2 Answers to Chapter Exercise 3.2

(1)	B = 55°40′18″	C = 44°19′42″	a = 155.028
(2)	A = 70°00′00″	b = 89.673	c = 101.377
(3)	A = 30°00′00″	C = 80°00′00″	a = 58.530
(4)	A = 60°00′00″	B = 65°00′00″	C = 55°00′00″
(5)	A = 53°32′02″	B = 45°28′53″	C = 80°59′04″
(6)	BD = 204.45	BA = 406.17	DC = 265.57
	BC = 470.02	AC = 527.60	

3.10 References for Chapter 3

Shepherd, F.A. Surveying Problems and Solutions. London. Edward Arnold (Publishers) Ltd, 1968

Abbott, P and H. Neill. Trigonometry. A complete introduction. John Murray Learning, London 2018. ISBN978 1 473 67849 1

Bradley, J.M. Mathematical formulae and statistical tables for tertiary institutions Murdoch University 2014. ISBN 0-86905-786-3

If it's not checked it's wrong.

Chapter 4 TRIANGLE SOLUTIONS, ADVANCED EXAMPLES

As briefly introduced in Chapter 1, engineering and surveying structures are set out (established) either on a plane earth or in space (where a sphere/ellipsoid defines the system). This chapter focuses on the plane system exemplified by the Australian Map Grid (MGA) system and shows you how one can convert measured angles to metric units in what is called join calculation (i.e., from rectangular/metrics to polar (bearings)). Given coordinate differences, one is expected to obtain bearings. This is often necessitated by the fact that during setting out in the field, one used already known coordinates (points) and as such needs to use them to establish bearings (directions from true north) to the points to be set out. The chapter shows how this can be achieved using a calculator.

In chapter 4 we introduced formulae that are used to solve general triangles. In this section we will attempt to draw together all the material so far covered to help solve more advanced surveying problems.

4.1 Use of Bearings.

When the bearings of some or all of the sides of a figure are known, the angles of the figure can be deduced.

In Figure 4-1 the angles at A, B, C, D and E can be calculated by the method shown in Chapter 2.

The angles are:

$\angle A$ 80°08′30″
$\angle B$ 159°12′50″
$\angle C$ 60°34′50″
$\angle D$ 140°18′40″
$\angle E$ 99°45′10″
Sum 540°00′00″

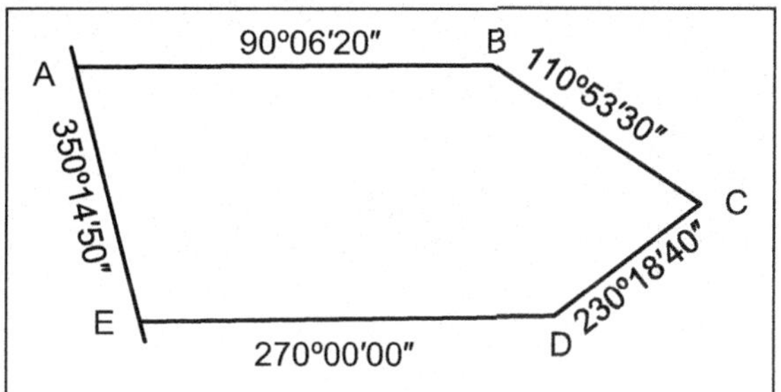

Figure 4-1 Bearings to angles

We will make use of this type of calculation to solve the following problem.

4.1.1 Worked Example 4.1. Area solution.

Find the Area of A B C D and dimensions AB and AD. See Figure 4-2.

Step (1) Find all angles in the quadrilateral

$\angle A =$ 59°40′55″
$\angle B =$ 110°08′15″
$\angle C =$ 100°10′40″
$\angle D =$ 90°00′10″
Sum 360°00′00″

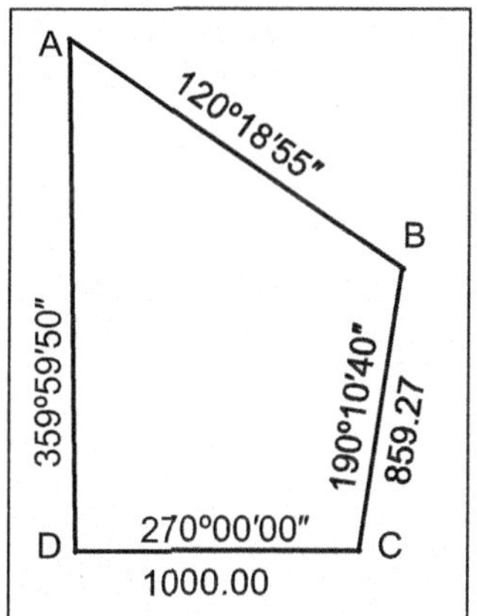

Figure 4-2 Worked Example 4.1

Step (2) The figure really consists of two triangles ABD and DBC shown in Figure 4-3.

In the Δ DBC, we know the angle at C and the sides CD and BC. Therefore, we can solve for the side DB and the angles $\angle BDC$ and $\angle DBC$

$\angle C = C =$ 100°10′40″
$CD = b =$ 1000.00
$CB = d =$ 859.27. See Figure 4-4

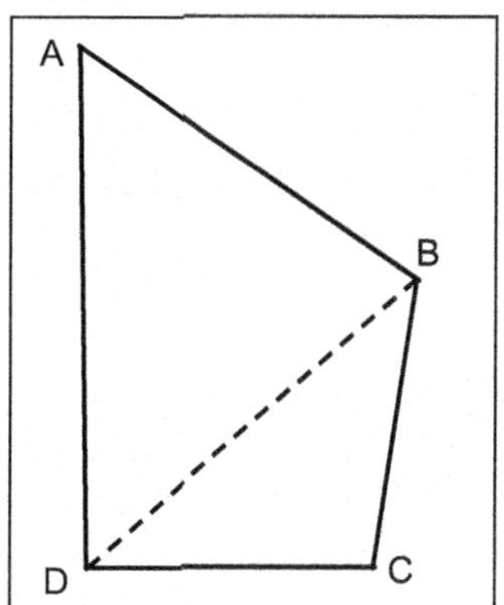

Figure 4-3 Two triangles

Solution: Given two sides and an included angle.

Step (a) by Cos Rule for sides: find $DB = c$

$$c = \sqrt{b^2 + d^2 - 2bd\cos C}$$

$$= \sqrt{1000.000^2 + 459.27^2 - 2 \times 1000.000 \times 859.27 \times \cos(100°10'40'')}$$

$$= \sqrt{2042016}$$

DB = c = 1428.99

Step (b) (1) by sin rule find $\angle BDC = \angle D$

$$\frac{d}{\sin D} = \frac{c}{\sin C} \Rightarrow \sin D = \frac{d \sin C}{c}$$

$$D = \arcsin\left(\frac{d \sin C}{c}\right) = \arcsin\left(\frac{859.27 \times \sin(100°10'40'')}{1428.99}\right) = \arcsin(0.5918)$$

∠BDC = ∠D = 36°17′18″

If it's not checked it's wrong.

J. Walker and J. Awange, *Surveying Computations*,
https://doi.org/10.1007/978-3-031-94382-9_4

(2) by sin Rule find ∠DBC = ∠B

$$B=\arcsin\left(\frac{b\sin C}{c}\right)=\arcsin\left(\frac{1000.000\times\sin(100^\circ10'40'')}{1428.99}\right)$$

$$=\arcsin(0.688783)$$

∠DBC = ∠B = 43°32′02″

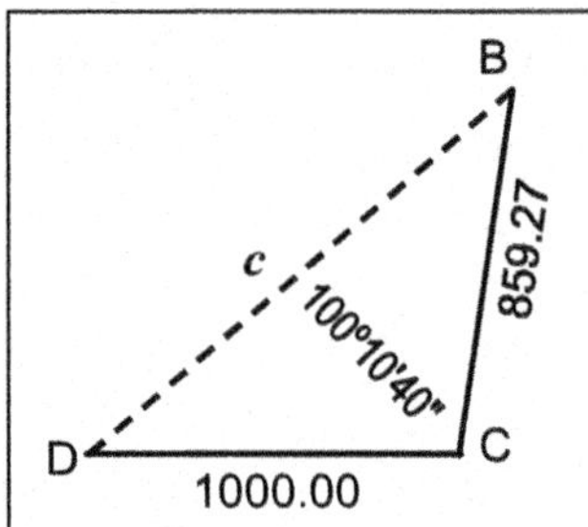

Figure 4-4 Triangle BCD

Check ∠BDC +∠DBC + BCD = 36°17′18″ + 43°32′02″ + 100°10′40″ = 180°

Step (c) Find Area of BDC by area Eqn 3-10 in Chapter 3.

$$\text{Area}=\frac{c^2\sin B\sin D}{2\sin C}=\left(\frac{1428.99^2\times\sin(43^\circ32'02'')\times\sin(36^\circ17'18'')}{2\times\sin(100^\circ10'40'')}\right)$$

$$\text{Area}=422873\text{m}^2$$

Area BDC = 42.2873 ha (1 hectare = 10,000m^2)

$$\text{Check area}=\frac{1}{2}bd\sin C=\frac{1}{2}(1000.000\times859.27\times\sin(100^\circ10'40''))$$

Area BDC = 42.2874 ha ✓

Solve the second ΔABD

Step (d) As we already know the angles at D (∠ADC) and B (∠ABC) we can calculate ∠ADB and ∠ABD

∠ADB = Angle at D (∠ADC) – ∠BDC)

= 90°00′10″ – 36°17′18″

∠ADB = 53°42′52″ (see Figure 4-5)

Similarly,

∠ABD = 66°36′13″ (see Figure 4-6).

Check ∠ABD + ∠ABD + ∠BAD = 180°

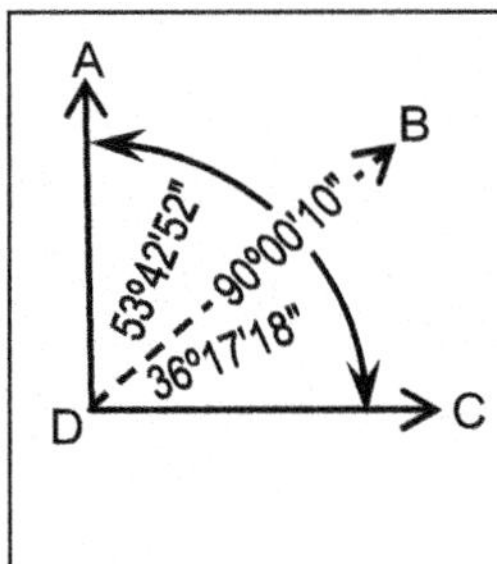

Figure 4-5 Angle ABD

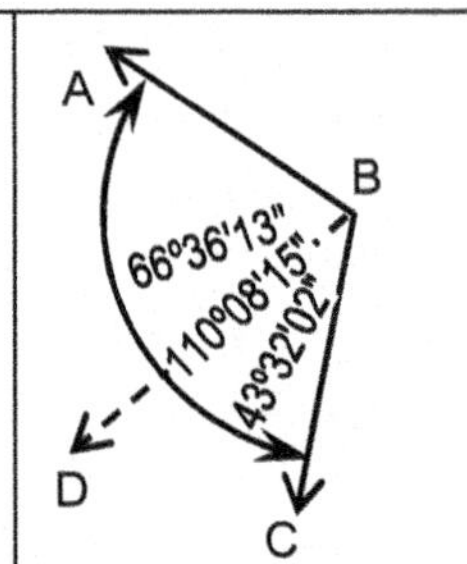

Figure 4-6 Angle ABD

In Figure 4-7 ∠BAD = A = 59°40′55″ from given bearings

∠ADB = D = 53°42′52″

∠ABD = B = 66°36′13″

BD = a = 1428.99

Solution: Given two angles and an included side. (see Figure 4-7).

Step (e) (1) by sin rule find AD = b

$$\frac{a}{\sin A}=\frac{b}{\sin B}\Rightarrow b=\frac{a\sin B}{\sin A}$$

$$b=\left(\frac{1428.99\times\sin(66^\circ36'13'')}{\sin(59^\circ40'55'')}\right)$$

AD = *b* = 1520.28

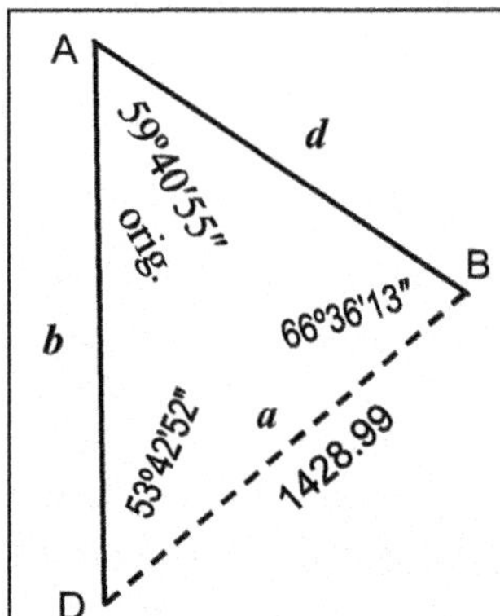

Figure 4-7 Triangle BAD

(2) by sin rule find AB = d

$$\frac{a}{\sin A}=\frac{d}{\sin D}\Rightarrow d=\frac{a\sin D}{\sin A}$$

$$d=\left(\frac{1428.99\times\sin(53^\circ42'52'')}{\sin(59^\circ40'55'')}\right)$$ Figure 4-7

AB = *d* = 1334.37

Check: ∠A by Cos rule:

$$\cos A=\left(\frac{b^2+d^2-a^2}{2bd}\right)=\frac{1519.28^2+1334.37^2-1428.99^2}{2\times1519.28\times1334.37}$$

$$A=\arccos(0.504800)$$

∠A = 59°40′55″ ✓ (original data from bearings)

If it's not checked it's wrong.

Step (f) Find Area of Δ ABD by area using Eqn 3-10

Area ABD = 87.5012 ha (1 hectare = 10,000m^2)

Check area: $\frac{1}{2} bd \sin A = \frac{1}{2}\left(1519.28 \times 1334.37 \times \sin\left(59°40'55''\right)\right)$

Area = 87.5012 ha ✓

Step (g) Area ABCD = Area ABD + Area BDC = 42.2873 + 87.5012

Area ABCD = 129.7885 Ha

4.1.2 Worked Example 4.2. Solve similar triangles

Some problems in surveying can be solved by using known areas and similar triangles. This is a special case where lots may have to be subdivided.

N.B. General case does not have parallel sides, therefore triangles are not similar.

In Figure 4-8 the area of ABCD is 100Ha.

Find the dimensions AD, AB and BC.

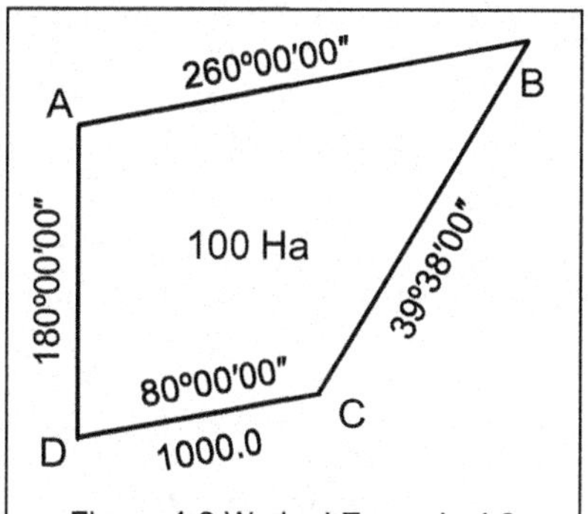

Figure 4-8 Worked Example 4.2

Solution: The above figure can be extended so that two similar triangles are formed as in Figure 4-9.

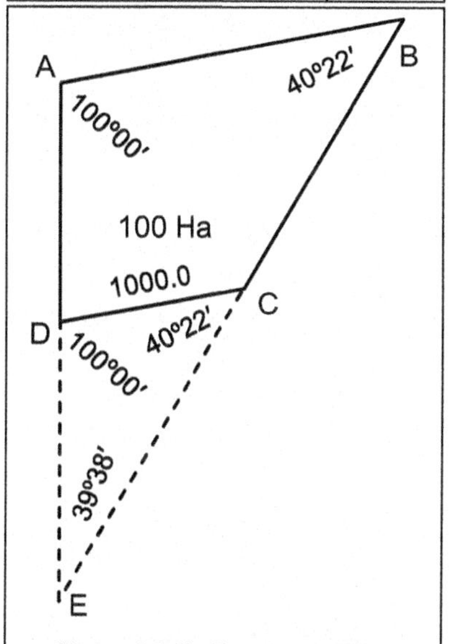

Figure 4-9 Similar triangle DCE

Triangle ABE is **similar** to triangle DCE

Solution: Area Δ DCE by Sin rule given three angles and a side.

∠EDC = D = 100°00′

∠DCE = C = 40°22′

∠CED = E = 39°38′

BC = e = 1000.0

Step (a) Area of ΔDCE by area formula

$$\text{Area} = \frac{e^2 \times \sin C \times \sin D}{2 \times \sin E} = \frac{1000.0^2 \times \sin\left(40°22'\right) \times \sin\left(100°\right)}{2 \times \sin\left(39°38'\right)}$$

Area = 49.9972Ha

Step (b) The given area of ABCD = 100 Ha

and area DCE = 49.9972 Ha

area ABE = 149.9972 ha. = 1 499 972m^2 (see Figure 4-10)

$$\text{Area} = \frac{AB^2 \times \sin\left(40°22'\right) \times \sin\left(100°\right)}{2 \times \sin\left(39°38'\right)}$$ and, by transposition

$$AB = \sqrt{\frac{2\text{Area} \times \sin\left(39°38'\right)}{\sin\left(40°22'\right) \times \sin\left(100°\right)}} = \sqrt{\frac{2 \times 1499972\text{Area} \times \sin\left(39°38'\right)}{\sin\left(40°22'\right) \times \sin\left(100°\right)}}$$

AB = 1732.08

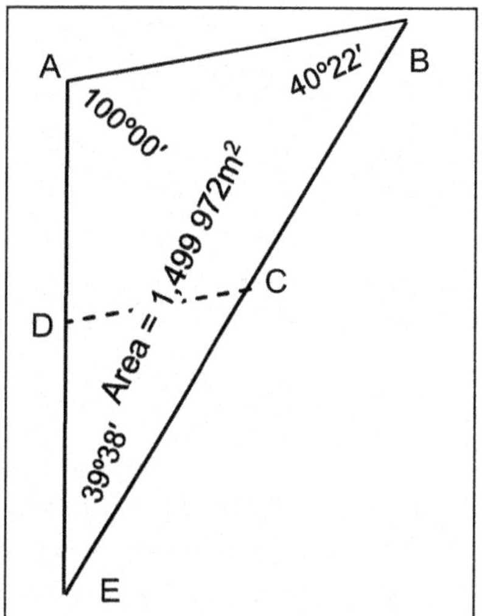

Figure 4-10 Area triangle ABE

Step (c) Solve remaining sides of ΔDCE by sin rule

(1) Find DE

$$DE = c = \frac{e \sin C}{\sin E} = \frac{1000.0 \times \sin\left(40°22'\right)}{\sin\left(39°38'\right)}$$

DE = 1015.37

(2) Find CE

$$CE = d = \frac{e \sin D}{\sin E} = \frac{1000.0 \times \sin\left(100°\right)}{\sin\left(39°38'\right)}$$

CE = 1543.90

Step (d) Solve remaining sides of ΔABE by sin rule

(1) Find $AE = \dfrac{1732.08 \times \sin\left(40°22'\right)}{\sin\left(39°38'\right)}$

AE = 1758.70

(2) Find $BE = \dfrac{1732.08 \times \sin(100°)}{\sin(39°38')}$

BE = 2674.15

Step (e) Calculate sides of ABCD by subtraction: (see Figure 4-11).

(1) AD = AE – DE	(2) BC = BE – CE
= 1758.70 – 1015.37	= 2674.15 – 1543.90
AD = 743.33	**BC = 1130.25**

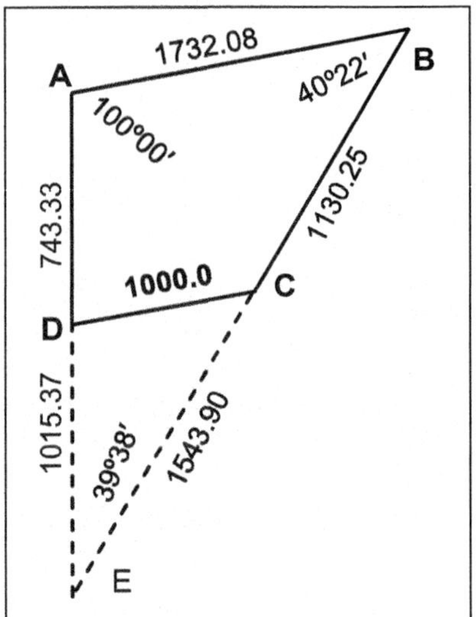

Figure 4-11 Individual sides

Step (f) Check

(1) $\dfrac{DE}{AE} = \dfrac{DC}{AB}$ similar triangles

$DE = \dfrac{AE \times DC}{AB} = \dfrac{1758.70 \times 1000.00}{1732.08}$

DE = 1015.37 Checks

(2) $\dfrac{CE}{BE} = \dfrac{DC}{AB}$ similar triangles

$CE = \dfrac{BE \times DC}{AB} = \dfrac{2674.15 \times 1000.00}{1732.08}$

CE = 1543.90 Checks ✔

4.1.3 Worked Example 4.3. Quadratic solution

It is also sometimes necessary to use quadratic equations to solve some triangle problems especially where town planning regulations have to be invoked.

Problem: Subdivide ABCD into 2 equal areas such that CF = EB

Step (a) Produce CD and AB to intersect at G. (see Figure 4-12).

Angle at G is 45°

∴ BG = BD = 275.00 and (AG = AC, BG = AG – AB)

(b) In Δ AGC AG = AC×tan(45°) = 525.00

Area Δ AGC = ½×525×525 = 13.78125 Ha.

(c) Area Δ DGB = ½×275×275 = 3.78125 Ha.

(d) Area ABDC = 13.78125 – 3.78125 = 10 Ha.

(e) Side $CG = \dfrac{AG}{\cos(45°)} = \dfrac{525}{0.707107}$

CG = 742.462

(f) Side $DG = \dfrac{BG}{\cos(45°)} = \dfrac{275}{0.707107}$

DG = 388.909

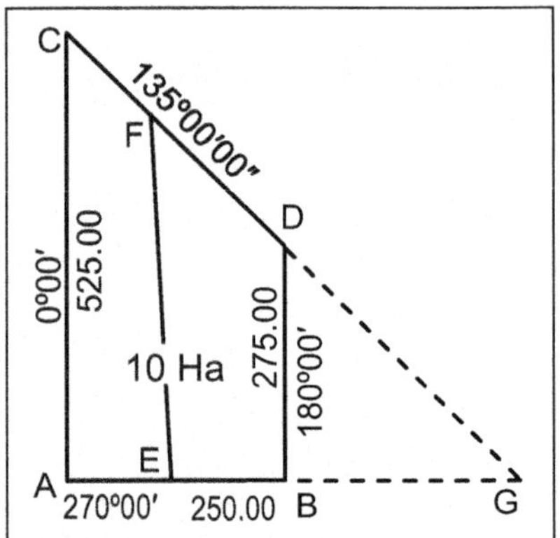

Figure 4-12 Worked Example 4.3

(g) Area Δ EFG = Area Δ DGB + ½ Area ABDC (= 3.78125 + 5)
= 8.78125 Ha = 87812.5m^2.

let CF = EB = x

now **Area** ΔEFG = ½×EG×FG×sin(45°) ∵(½absin$C$). Rearrange formula:

½×(EG×FG)·= **Area** ΔEFG ÷ sin(45°)

and EG = BG + x = 275 + x

FG = CG – x = 742.462 – x

∴ ½ (275 + x)(742.462 – x) = 87812.5÷sin(45)

½ (204177.083 – 275 x + 742.462 x – x^2) = 124185.628

which reduces to the quadratic equation form: $ax^2+bx+c = 0$

$x^2 - 467.462x + 44194.174 = 0$

Step (h) solve the quadratic equation for $x = \dfrac{-b \pm \sqrt{b^2 - 4ac}}{2a}$

Now: $a = 1$, $b = -467.462$, $c = 44194.174$

$$x = \frac{-(-467.462) \pm \sqrt{(-467.462)^2 - 4 \times 1 \times 44194.174}}{2 \times 1}$$

$$x = \frac{467.462 + 204.314}{2 \times 1} \qquad \text{first root}$$

or $$x = \frac{467.462 - 204.314}{2 \times 1} \qquad \text{second root}$$

$\therefore \quad x = 335.890$ or 131.574

Only the latter is a feasible result since we know that EB = x must be less than AB (250.00m)

x = CF = EB = 131.574

Substituting back into previous equations.

EG = BG + x = 275.000 + 131.574 = 406.574
FG = CG – x = 742.462 – 131.574 = 610.888
AE = AG – EG = 525.000 – 406.572 = 118.428
FD = FG – DG = 610.888 – 388.909 = 221.979

EG = 406.574
FG = 610.888
AE = 118.426
FD = 221.979

Step (i) Solve ΔEFG: 2 sides (EG, FG) and included angle G. See Figure 4-13

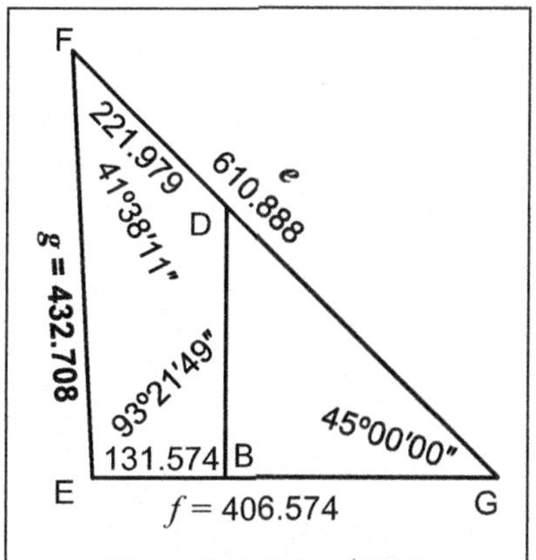

Figure 4-13 Solve ΔEFG

(a) Solve for EF by cos rule. Side $g = \sqrt{e^2 + f^2 - 2ef\cos G}$

$$EF = \sqrt{FG^2 + EG^2 - 2 \times FG \times EG \cos(45°)}$$

EF = 432.708 = g

(b) Solve for angles by Sin Rule $\dfrac{e}{\sin E} = \dfrac{f}{\sin F} = \dfrac{g}{\sin G}$

(Solve smaller **∠F** first to help resolve **∠E**)

(1) $$\sin F = \frac{f \sin G}{g} = \frac{EG \times \sin(45)}{EF} = \frac{406.574 \times \sin(45)}{432.708}$$

$F = \arcsin(0.664399) = 41°38'11''$

∠EFG = F = 41°38′11″ Brg FE = 176°38′11″

(2) $$\sin E = \frac{e \sin G}{g} = \frac{FG \times \sin(45)}{EF} = \frac{610.80 \times \sin(45)}{432.708}$$

$E = \arcsin(0.998277) = 86°38'11''$

(Check $E + F + G = 173°16'22'' \neq$ **180° invalid result)**

∠FEG = E = 93°21′49″

(E in second quadrant: $E = 180 - 86°38'11''$)

Note: sin value is **positive** in quadrant I and II

∠EFG and ∠FEG can be solved directly by cos or tan rule from distances.

Step (j) Check ∠EFG + ∠FEG + ∠FGE = 180°00′00″

The subdivision dimensions are summarised in Figure 4-14.

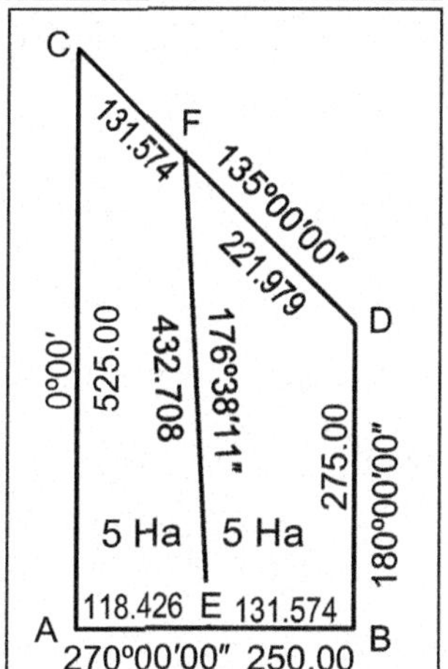

Figure 4-14 Subdivision dimensions

If it's not checked it's wrong.

4.2 Chapter Exercise 4.1

Q1 Compute the Area of the quadrilateral in Figure 4-15.
Find AB and BC and the area of the two triangles ADC and ABC

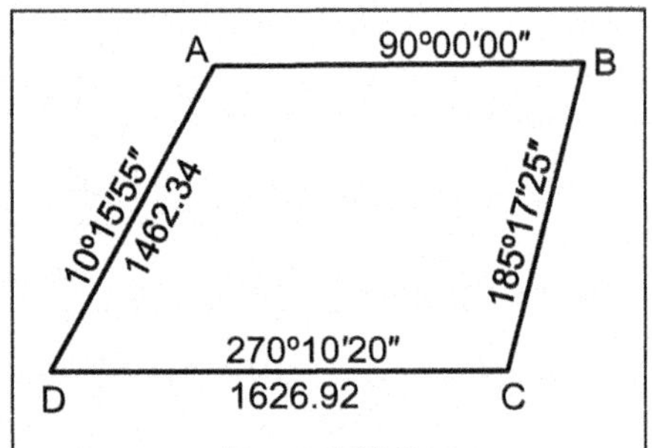

Figure 4-15 Q1

Q2 In Figure 4-16 subdivide ABCD into 2 equal areas such that AE = EB.
Find AE, EB, EF and Brg EF

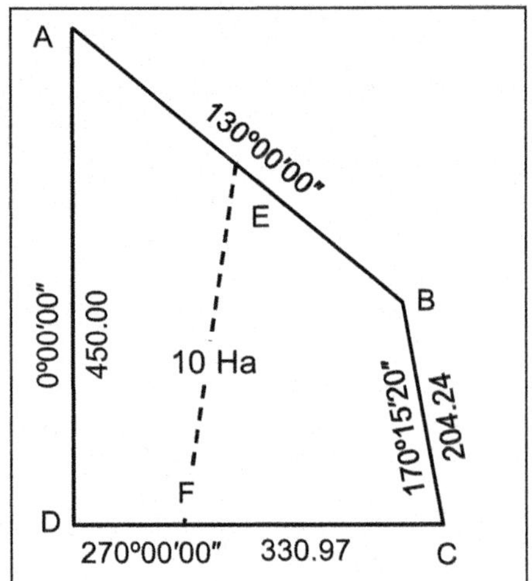

Figure 4-16 Q2

Q3 In the Figure 4-17 find the lengths AB, BC and CD.

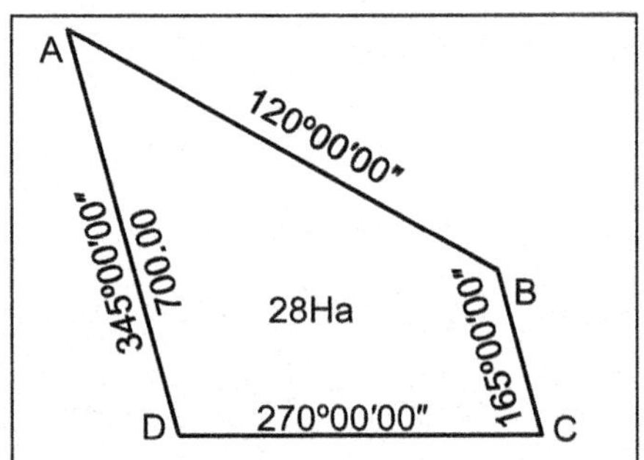

Figure 4-17 Q3

Q4 In the Figure 4-18 find the lengths AD and CB and Brg CB.

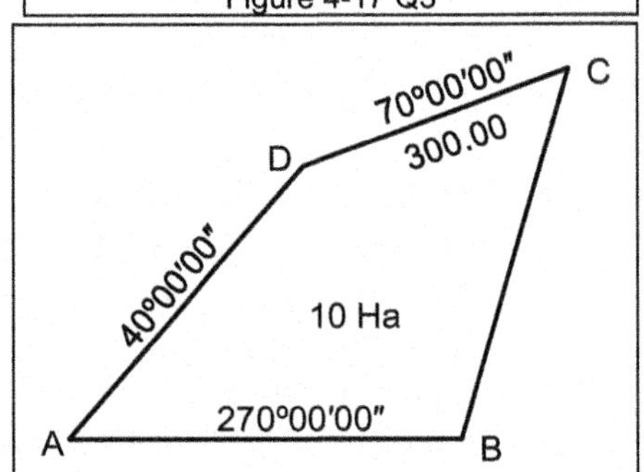

Figure 4-18 Q4

4.2.1 Answers to Chapter Exercise 4.1.

Q1 AB = 1500.01
BC = 1450.00
AREA ACD = 117.1146 Ha
AREA ABC = 108.2872 Ha
AREA ABCD = 225.4018 Ha

Q2 AE = EB = 193.46
DF = 102.30
FC = 228.67
EF = 328.87
Brg EF = 188°01′30″

Q3 AB = 805.71
BC = 282.93
CD = 589.82

Q4 AD = 384.272
CB = 384.243
Brg CB = 195°58′42″

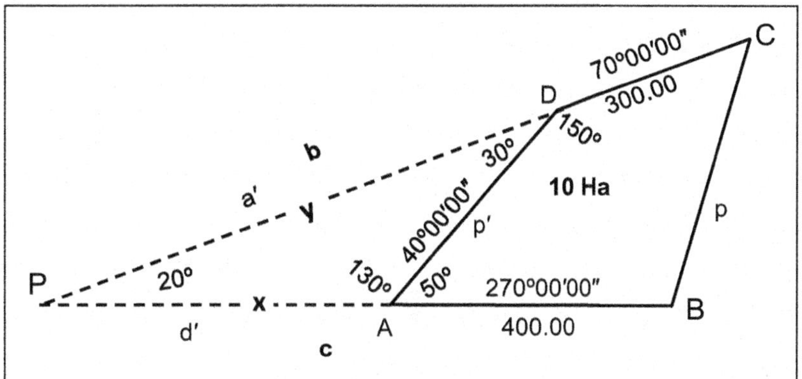

Figure 4-19

Produce BA and CD to intersect at P. ∠P = 20°. Let PA = x and PD = y
DRAW the diagram as in Figure 4-19.
Triangle PDA + ABCD (10Ha = 100 000m^2) = triangle PCB
PDA = ½xysin$P$, ABCD = 100 000, and thus PCB = ½(x+400)(y+300)sinP.
Multiply ×2 to get rid of the half area of triangle.
xysin$P$ + 200 000 = (x+400)(y+300)sin$P$.
Divide by sin$P$ and gather terms to form the equality
200 000/sin$P$ = (x+400)(y+300) – xy; ⇒200000/sin(20) = 12000 + 300x + 400y
Now express x in terms of y (or y in terms of x) from triangle PDA,
∠A = 130°, ∠D = 30°, by Sin rule: y/sin$A$ = x/sin; y = $x$sin$A$/sin$D$
y = (1.532x), or x = (0.653)y
Substitute your chosen x or y in the equality (approximate answers)
solve for x (=509), y (=780) and p (AD = 348, answer) via sin rule using ∠P
calculate PC(y + 300), PB(x + 400)
Solve triangle PCB (2 steps)
Sin rule for side p (CB = 348, answer)
∠C, ∠B from: 1) sin rule for angles (2 sides, included angle. Ambiguity?)
2) cos rule for angles (3 sides) or
3) tan half formula (2 sides, included angle)
Check (Σ (∠P + ∠C + ∠B) = 180?)
Brg CB from angle C, brg BC from angle B (reciprocal?)
Check
area PDA + 100 000 = area PCB (16.8Ha)

4.3 References for Chapter 4

Shepherd, F.A. Surveying Problems and Solutions. London. Edward Arnold (Publishers) Ltd, 1968

Clark, David, and James Clendinning. Plane and Geodetic Surveying for Engineers. Volume 1, Plane Surveying. 6th ed. Constable, 1969.

Allan, A.L., Hollwey, J.R., and Maynes, J.H.B. Practical Field Surveying and Computations. William Heinemann Ltd, London. 1971

Chapter 5 TRAVERSES COMPUTATIONS AND ADJUSTMENT

Traversing is the procedure and technique that surveyors use to obtain the coordinates (positions) of new points from the measured angles and distances. Whereas the field observation procedures are presented in our other book Walker and Awange (2020), in this chapter, we introduce to you the various types of traverses and show you the reader the computational aspects together with the relevant checks to ensure that the correct positions of the new points are obtained. Starting with the corrections to the bearings, various approaches are presented starting with use of formulae sheets to undertaking a BOWDITCH adjustment where the linear misclose (i.e., the difference between the position of the starting point and the resulting value from the measured field values) is distributed in a logical way to all the points to be positioned. More advanced techniques such as the least squares adjustment will be considered in our advanced book.

5.1 Traverses.

The reader should check back to Chapter 2 concerning the calculation of traverse bearings.

A traverse is a means of coordinating a number of widely spaced points by means of angle and distance measurements.

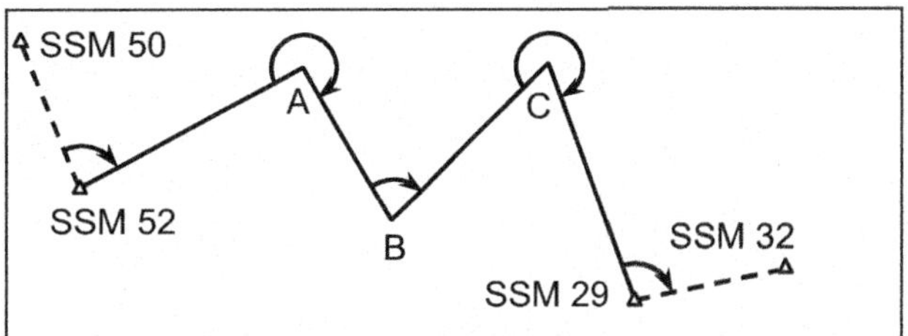

Figure 5-1 Closed connected traverse

5.2 Types of Traverses.

5.2.1 Closed connecting (or link) Traverse

In this case, (Figure 5-1), the traverse starts on the fixed coordinated point (SSM 52) and finishes on coordinated point (SSM 29). The bearings and distances from SSM 52 - SSM 50 and SSM 29 - SSM 32 can be calculated.

Angles at SSM 52, A, B, C and SSM 29 are measured and the angular misclose can be found and apportioned amongst them.

The distances of the traverse sides starting on SSM 52 and ending on SSM 29 are measured. While not required to establish the traverse, it is good practice to also including the distance measurements SSM 52 - SSM 50 and SSM 29 - SSM 32. This allows the scale factor of the measured distances to be checked against the known initial and final legs (e.g. between the coordinated SSMs). The main aim of the traverse is to calculate the coordinated positions of A, B and C.

Only this type of traverse can be fully checked. In particular a closed connecting traverse checks the:

(1) correctness of all angle and distance measurements.
(2) correctness of control point coordinates (e.g. transcription error).
(3) correctness of start and end bearings derived from coordinates.
(4) correctness of all calculations.

Whenever possible a closed connecting traverse should be selected. Other types of traverses should be treated with care as they may not be fully checked.

5.2.2 Closed Loop Traverse. – In Australia this is the most common traverse type.

In Figure 5-2 all angles and distances in the traverse are read. Beginning from an initial point (say A) of known co-ordinates, the traverse takes the form of a number of lines of known bearing and distance, closing back on to the initial point. The traverse is closed in order to check that the distance and angle readings have been made correctly. However systematic errors in distances cannot be detected.

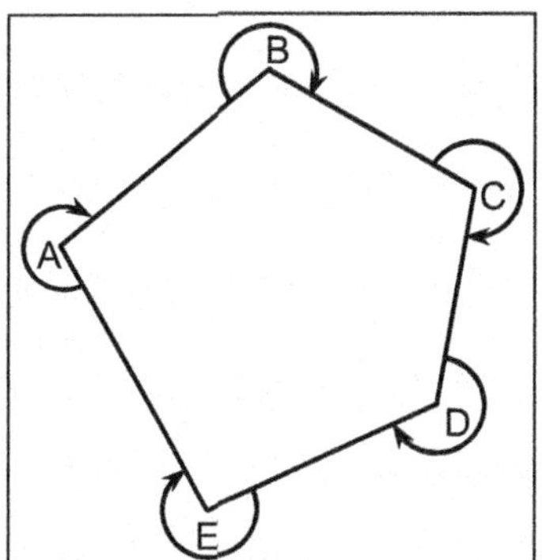

Figure 5-2 Closed loop traverse

In particular a closed loop traverse checks the:

(1) correctness of all angle measurements.
(2) correctness of all distance measurements.
(3) correctness of all calculations.

Distance measurements instead are only partially checked. Systematic errors such as a scale factors or prism constants are not fully checked. Therefore, it is important that the distance(s) between coordinated points (e.g. SSMs) are measured whenever possible and checked against the known distance(s).

If it's not checked it's wrong.

J. Walker and J. Awange, *Surveying Computations*,
https://doi.org/10.1007/978-3-031-94382-9_5

5.2.3 Open Connecting Traverse.

Figure 5-3 is a diagram of the open connecting traverse form but, in this case, the traverse begins only on a fixed point, A, and does not finish on a fixed station point. This would be a rare case and might happen where the traverse is later to be fixed to a known coordinated point. Importantly, if the traverse is not closed onto a known coordinated point it is not checked and should be avoided whenever possible. However, a common application of an open connecting traverse can be found in tunnel work where naturally one end of the traverse cannot be closed.

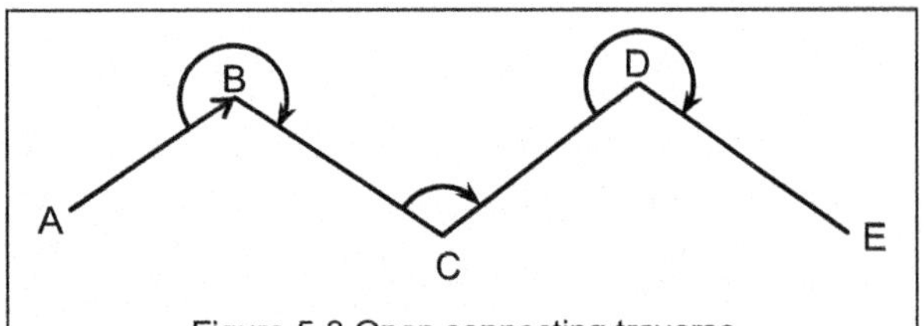

Figure 5-3 Open connecting traverse

5.3 Initial bearing.

As the object of traversing is to coordinate the stations of the traverse, then it is necessary to determine the bearing (e.g. orientation) of one of the traverse sides. Usually this is the starting side in a traverse (e.g. A - B in Figure 5-2 and Figure 5-3 or SSM 50 – SSM 52 in Figure 5-1).

In Figure 5-4, if the bearings and distances of the traverse ABCDEA are known, all stations can be coordinated. As only the horizontal angles are measured, the bearing of one traverse side must be found. Suppose this is side AB. Bearing AB (θ_{AB}) can be found in a number of ways

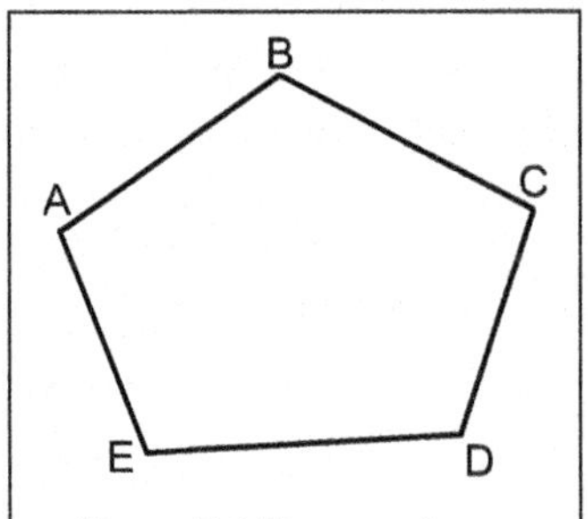

Figure 5-4 Traverse figure

(1) It can be assumed. (Rotated to suit calculations.)
(2) Can be obtained from a previous survey, e.g. from given coordinated points such as SSMs.
(3) Can be obtained from a plan of the area.
(4) Can be found from astronomical observations. Solar; sun observations, Stellar; star observations.
(5) Can be obtained from another fixed point and a clockwise angle (α). (known bearing between two reference marks, i.e. two SSM's)
(6) Can be obtained from a compass.

For example, in Figure 5-5 a bearing AB can be obtained from another fixed point and a clockwise angle (α).

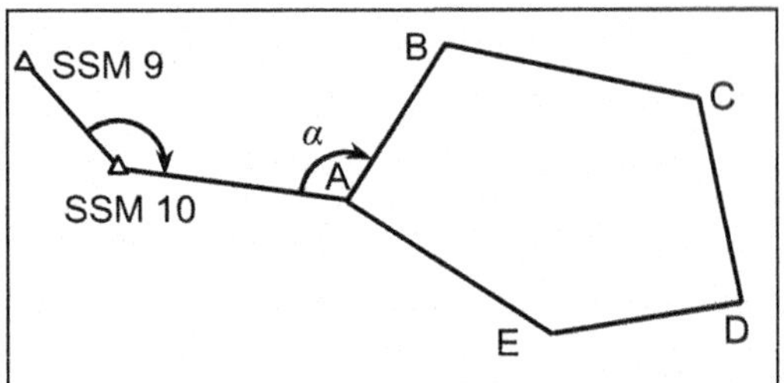

Figure 5-5 Traverse bearing sources

(known bearing between two reference marks, i.e. SSM 9 and SSM 10)

i.e. $\theta_{AB} = \theta_{(A - SSM\ 10)} + \alpha$

5.4 Distribution of Angular Misclose, closed loop traverse.

Angles are usually read in a traverse and these angles then adjusted prior to the calculation of the bearings of each traverse leg.

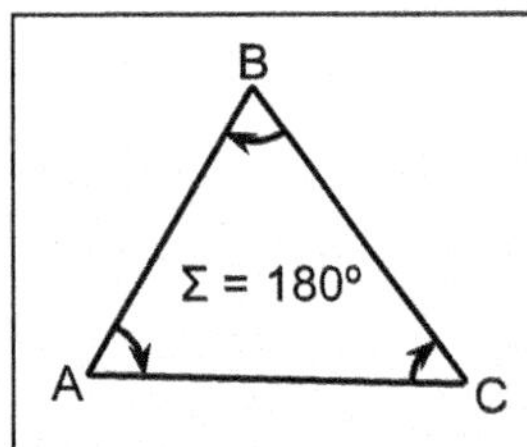

Figure 5-6 Sum triangle

5.4.1 Sum of the interior angles of any polygon.

$\Sigma^{o} = (n - 2)\times 180$ (n = number of sides)

e.g. In a triangle, n = 3 sides (See Figure 5-6).

Interior angles = $(n - 2)\times 180 = (3 - 2)\times 180 = 180^{o}$

e.g. In a quadrilateral, n = 4 sides. (See Figure 5-7).

Interior angles = $(4 - 2)\times 180 = 360^{o}$

In a closed loop traverse we generally read the internal angles.

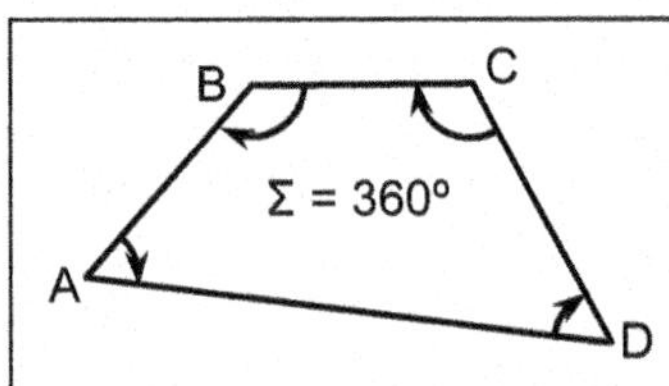

Figure 5-7 Sum quadrilateral

5.4.2 Sum of the exterior angles of any polygon.

$= n\times 360 - \text{Interior angles}$

$= n\times 360 - (n - 2)\times 180$

$= n\times 360 - n\times 180 + 360$

$= n\times 180 + 360$

e.g. Examining Figure 5-7 where n = 4, Sum of exterior angles = $4 \times 180 + 360 = 1080^{o}$

5.4.2.1 Worked Example 5.1 Closed Loop Traverse.

Given θ_{AB} = 254°09′15″ together with the measured angles of the traverse at A, B, C and D, find the adjusted bearings of the traverse. (Figure 5-8).

Measured angles:

∠B = 210°28′45″
∠C = 276°24′30″
∠D = 242°16′00″
∠A = 350°49′30″
Sum 1079°58′45″

(n = 4 angles)

Sum of exterior angles = (n×180) + 360 = (4×180) + 360 = 1080

There is a misclose of (1080° – 1079°58′45″) = 1′15″ = 75″.

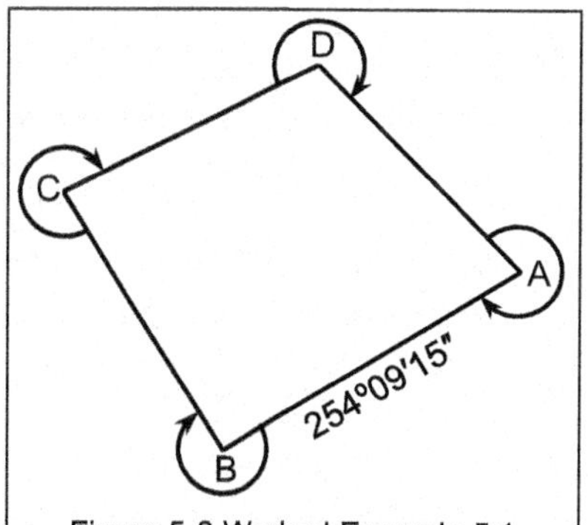

Figure 5-8 Worked Example 5.1

It can be assumed that errors are evenly distributed amongst the four angles so that a correction of 75/4″ must be added to each of the four field angles. i.e. field angles adjusted by 19″ each.

4 x 19″ = 76″, thus one angle adjustment will be reduced to +18″ to bring the total of the required adjustment to 75″. Distribute only whole values.

The bearings, angles, corrections and final adjusted bearings are summarized in Table 5-3.

Apply only whole second corrections. Your angular measurement, even with the 1" instruments, does not justify decimal seconds (18.75" above). When you **distribute** the corrections make sure that the sum of the angles adds up to the required figure. It will be checked. The errors are considered random and the whole second corrections are the best **estimate** of the errors encountered.

5.4.2.2 Adjustment of observed angles. Closed loop traverse

It is customary to calculate bearings clockwise around a traverse, or in such a direction that the clockwise angles are added. To achieve this in Figure 5-8 it is necessary to start at Station B and add the exterior, clockwise, angle B to the direction between B and A. Since the bearing given, θ_{AB}, is 254°09′15″ the bearing θ_{BA}, the **back** or reverse bearing is used as the initial bearing.

Table 5-1 Traverse adjustment

Station At:	Observed Angle	Adjustment	Bearing + Adjusted Angle	Adopted Bearing
		Adopted Bearing BA	74 09 15	74°09′15″ θ_{BA}
B	210 28 45	+ 19″	210 29 04	
		Sum	284 38 19	
		Mod 360	284 38 19	284°38′19″ θ_{BC}
		Back bearing	104 38 19 (C→B)	
C	276 24 30	+ 19″	276 24 49	
		Sum	381 03 08	
		Mod 360	21 03 08	21°03′08″ θ_{CD}
		Back bearing	201 03 08 (D→C)	
D	242 16 00	+ 19″	242 16 19	
		Sum	443 19 27	
		Mod 360	83 19 27	83°19′27″ θ_{DA}
		Back bearing	263 19 27 (A→D)	
A	350 49 30	+ 18″	350 49 48	
		Sum	614 09 15	
		Mod 360	254 09 15	254°09′15″ θ_{AB}
Sum observed angles=	1079 58 45	Sum Correction = + 75		Check
Sum exterior angles=	1080 00 00			
Misclose = (obs – true)	–1′15″= – 75”			
Adjustment =	75/4 = 19 (–1)			

In the adjustment of closed loop traverses you must use either exterior, Figure 5-8, or interior Figure 5-6, Figure 5-7 angles.

5.5 Distribution of Angular Misclose, Open connecting traverse.

The angular misclose between the initial fixed and final fixed bearing is calculated by:

1) summing the angles and adding the reciprocal of the first bearing
2) subtracting ((n+1)×180) where n = number of angles

Final bearing = reciprocal of first bearing + sum of angles – ((n+1)×180°)

where n = number of angles in the traverse.

The misclose is the difference between the observed final bearing and the fixed final bearing. The misclose is then spread over the *n* observed angles.

Using the first bearing instead of the reciprocal of the first bearing the term ((n+1)×180°) simplifies to (n×180°), thus the final bearing is derived as

Final bearing = first bearing + sum of angles – (n×180°).

5.5.1 Worked Example 5.2. is an open connecting traverse,

Given: Fixed station coordinate information in the open connecting traverse, Figure 5-9

	E	N
SSM 18	916.925	1471.142
SSM 20	1000.000	1000.000
SSM 35	2406.500	906.700
SSM 52	2495.632	313.357

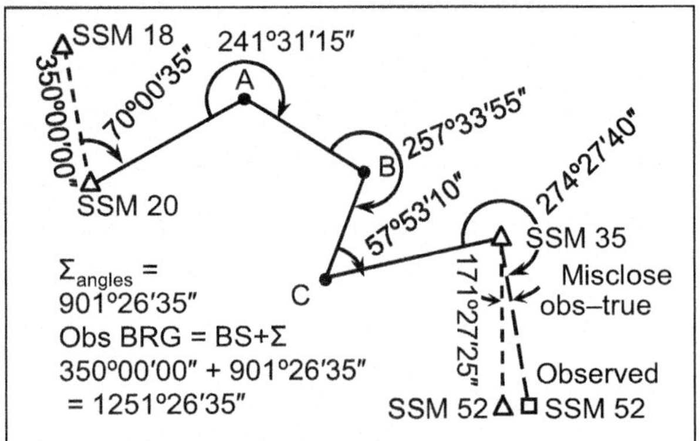

Figure 5-9 Open connecting traverse

Aim: find the adjusted bearings of the traverse sides.

Note: The bearings from SSM 20 - SSM 18 and SSM 35 - SSM 52 are fixed and we required the adjusted bearings of the remaining traverse sides.

Method:

1) Calculate the bearings:
   SSM 20 - SSM 18 and
   SSM 35 - SSM 52.

Use the JOIN method.

Bearing (SSM 20 - SSM 18) = 350°00′00″

Bearing (SSM 35 - SSM 52) = 171°27′25″

2) Angle Adjustment (here: n = 5, the angles observed being at SSM20, A, B, C and SSM35).

Check Observed Bearing (SSM 35 - SSM 52)

= Bearing (SSM 20 - SSM 18)

+ ∠SSM 20 + ∠A + ∠B + ∠C + ∠SM 35 (5 angles)

– ((n + 1)×180)

= 350°00′00″ + 901°26′35″ – 1080°

= 171°26′35″.

Compared with the true (calculated) bearing, $\theta_{(35\text{-}52)}$ = 171°27′25″ the observed bearing (35-52), 171°26′35″, is 50″ too small. A correction of 50″ must be added evenly over the 5 observed angles, giving 10″ correction per angle.

This adjustment procedure requires the initial back bearing and the clockwise forward angles (see Figure 5-9).

5.5.1.1 Adjustment of observed angles. Open connecting traverse

Table 5-2 Open connecting traverse

Station At:	Observed Angle	Adjustment	Bearing +Adj Angle	Adopted Bearing
BRG (20 – 18)	350 00 00	Adopted BRG (20 – 18)	350 00 00	350°00′00″ $\theta_{(20\text{-}18)}$
SSM 20	70 00 35	+ 10″	70 00 45	
		Sum	420 00 45	
		Mod 360	60 00 45	60°00′45″ $\theta_{(20\text{-}A)}$
		Back Bearing	240 00 45 (A→20)	
A	241 31 15	+ 10″	241 31 25	
		Sum	481 32 10	
BRG (A – B)		Mod 360	121 32 10	121°32′10″ θ_{AB}
		Back bearing	301 32 10 (B→A)	
B	257 33 55	+ 10″	257 34 05	
		Sum	559 06 15	
BRG (B – C)		Mod 360	199 06 15	199°06′15″ θ_{BC}
		Back bearing	19 06 15 (C→B)	
C	57 53 10	+ 10″	57 53 20	
		Sum	76 59 35	
BRG (C –35)		Mod 360	76 59 35	76°59′35″ $\theta_{(C\text{-}35)}$
		Back bearing	256 59 35 (35→C)	
SSM 35	274 27 40	+ 10″	274 27 50	
SUM obs. ∠	1251 26 35	Sum	531 27 25	
–((N+1)×180)	–1080	Mod 360	171 27 25	171°27′25″ $\theta_{(35\text{-}52)}$
Sum =	171 26 35	Corr'n = + 50		Checks with original
BRG (35 – 52)	171 27 25			
Misclose =(obs – true) =	–50″			
Adjustment =	50/5 = +10″			

5.6 Chapter Exercise 5.1

Q. 1

Given that the bearing A - SSM 82 = 36°16′20″, find the adjusted bearing of the traverse sides. (Hint: There are 7 angles but only 6 external angles). (Figure 5-10).

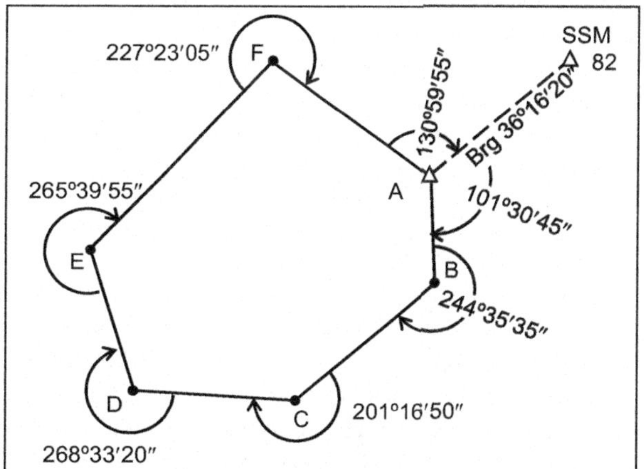

Figure 5-10 Closed traverse

Q. 2

Given that bearings (Figure 5-11).

A - SSM 10 = 335°15′55″, and

F - SSM 90 = 168°25′50″.

Find the adjusted bearings of the traverse sides.

5.6.1 Answers to Chapter Exercise 5.1.

Q1. Bearing AB = 137°47′10″
BC = 202°22′50″
CD = 223°39′45″
DE = 312°13′10″
EF = 37°53′10″
FA = 85°16′20″

Q. 2 Bearing AB = 63°51′20″
BC = 155°10′45″
CD = 78°47′05″
DE = 156°42′15″
EF = 74°21′00″

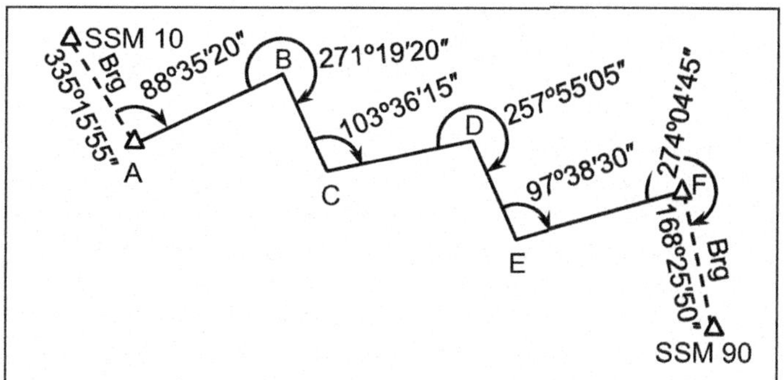

Figure 5-11 Open linked traverse

5.7 Traverse Closures in ΔE and ΔN.

When a traverse starts and finishes on the same point or starts and finishes on two points whose spatial relationship is defined, we can employ a mathematical check to see whether the work in the field is sufficiently accurate for the job in hand. This process is called taking out a close.

To understand the mechanics of a close we should refer to the rectangle ABCD in Figure 5-12. Here you will notice that all bearings follow one another around in a clockwise direction. Also, you will note that a larger rectangle WXYZ has been drawn to contain ABCD. The latter rectangle lies in a north-south-east-west orientation.

The following Table 5-3 deals with trigonometric function of bearings in the four quadrants. The mnemonic **A S T C** (ALL, SIN, TAN, COS) refers to the positive trigonometric function in each of the four quadrants.

Table 5-3 Sign of trigonometric functions by quadrant

ANGLE OR BEARING θ	0° - 90°	90° - 180°	180° - 270°	270° - 360°
QUADRANT	1 (I)	2 (II)	3 (III)	4 (IV)
SIGN OF TRIGONOMETRIC FUNCTIONS	SIN + COS + TAN +	SIN + COS – TAN –	SIN – COS – TAN +	SIN – COS + TAN –

Now if you refer to Table 5-3 you will notice that each traverse line (e.g. AB, BC etc.) can be resolved in a North-South and an East-West direction.

For example, traverse line AB has a North component AW and East component WB.

Let us now examine the east-west lines WX which is reckoned positive and YZ which is reckoned negative.

If this is so then WX + YZ =. 0

WX consists of $AB \times \sin(B_1) + BC \times \sin(B_2)$

YZ consists of $CD \times \sin(B_3) + DA \times \sin(B_4)$

$AB \times \text{Sin}(B_1) + BC \times \sin(B_2) + CD \times \sin(B_3) + DA \times \sin(B_4) = 0$ (1)

Now if we examine the north-south lines ZW and XY in like manner we will see that ZW + XY =. 0

ZW consists of $DA \times \cos(B_4) + AB \times \text{c os}(B_1)$

YZ consists of $BC \times \cos(B_2) + CD \times \cos(B_3)$

$AB \times \cos(B_1) + BC \times \cos(B_2) + CD \times \cos(B_3) + DA \times \cos(B_4) = 0$ (2)

We introduce the concept terms:

Sum of Differences in East co-ordinates, (Sigma Delta E), $\Sigma\Delta E = 0$

Sum of Differences in North co-ordinates, (Sigma Delta N), $\Sigma\Delta N = 0$

Equation 1 and 2 only hold if the sign conventions for the trignometric functions in Table 5-3 actually apply. With modern electronic calculators, these sign conventions are automatically applied so that, e.g.

$\sin(210°) = -0.500000$

$\cos(150°) = -0.866025$

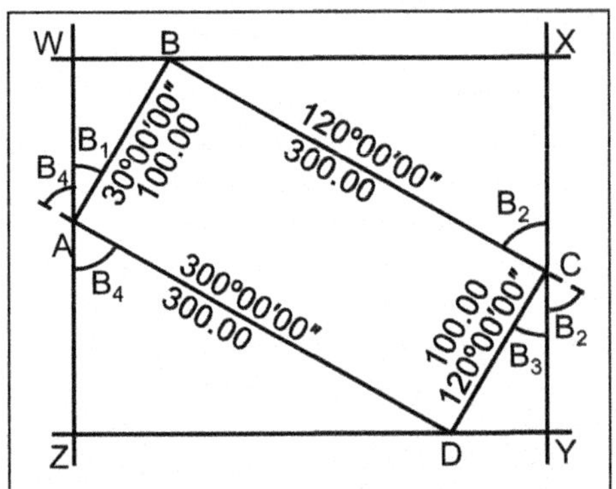

Figure 5-12 Ideal traverse

Three important facts have thus far emerged.

1) All the east-west components of lines are a function of the SINE and
2) All the north-south components of lines are a function of the COSINE.
3) If all lines are resolved into their north-south and east-west components, designated as ΔE and ΔN, then we know that: $\Sigma\Delta E = 0$
$\Sigma\Delta N = 0$

We can arrange a computation form and carry out the above operation in systematic fashion.

5.8 Example Close.

Referring to Figure 5-12 we can set out a close as follows:

Table 5-4 Ideal traverse

LINE	BEARING	DISTANCE	SIN	COS	ΔE	ΔN
AB	30 00 00	100.00	0.500000	0.866025	50.00	86.60
BC	120 00 00	300.00	0.866025	–0.500000	259.81	–150.00
CD	210 00 00	100.00	– 0.500000	–0.866025	–50.00	–86.60
DA	300 00 00	300.00	–0.866025	0.500000	–259.81	150.00
	Σ	800.00		Σ	0.00	0.00

If it's not checked it's wrong.

Scientific calculators carry the sine (SIN) and cosine (COS) functions automatically and hence the *sin* and *cos* columns are not really required but are added here for illustration purposes.

We have examined the "ideal" case where $\Sigma\Delta E$ and $\Sigma\Delta N = 0$ exactly. In practice this seldom occurs due to small errors which accumulate during the measuring process. The size of the $\Sigma\Delta E$ and $\Sigma\Delta N$ is the misclose and the precision of the measurements in the field can be estimated by the misclose ratio.

The misclose ratio is defined as $\dfrac{\text{Misclose distance}}{\text{Perimeter}} = \dfrac{\sqrt{\Sigma\Delta E^2 + \Sigma\Delta N^2}}{\text{Perimeter length}}$

Note, the formula above is defined for a closed traverse. For an open connecting traverse, the perimeter is replaced by the total length of the measured traverse (not including the opening and closing joins).

The misclose ratio can then be expressed as $1:\left(\dfrac{1}{\text{Misclose ratio}}\right)$

e.g. the perimeter (or total length) in Figure 5-12 = 800.00.

Suppose, instead of being zero, the misclose distance is: $\sqrt{\Sigma\Delta E^2 + \Sigma\Delta N^2} = 0.20$

misclose ratio $= \dfrac{0.2}{800} = 0.00025$ or

misclose ratio $= 1:\left(\dfrac{1}{0.00025}\right) = 1:4000 = 1$ IN 4000, or: $1:\left(\dfrac{\text{Perimeter}}{\text{Misclose}}\right)$, or $1:\left(\dfrac{800}{0.2}\right) = 1:4000$

Most survey authorities have minimum standard for traverse misclose and therefore it is important that the calculated misclose contains only the measurement errors and NOT calculation errors.

5.9 Methods of Checking a Close.

Using an electronic calculator, there are four sources of error when taking out a close.

(a) Using an incorrect bearing. A blunder, or not converting D M S to decimal degrees.
(b) Using an incorrect distance. A blunder.
(c) Interchanging ΔE and ΔN (e.g. putting them in the wrong column).
(d) Performing the $\Sigma\Delta E$ or $\Sigma\Delta N$ incorrectly.

(a) and (b) values used in a close must be checked from the original information. i.e. from the field book. It is a waste of time to take out a close using wrong information.

(c) The sequence should always be

DIST×sin (BRG) = ΔE
DIST×cos (BRG) = ΔN

Your calculator should have the rectangular conversion function [→y,x] built in where $y = \Delta E$ and $x = \Delta N$. Be sure this function is used correctly.

If this procedure is followed as a matter of habit, the error will rarely, if ever, occur.

(d) The addition of ΔE and ΔN is a common area of error, especially for students unfamiliar with an electronic calculator. You should learn to make use of the statistical summation key, [Σ+] to check a for misclose.

You **must** be thoroughly familiar with the polar/rectangular functions built in to your calculator when using it, or with the methodology you use for the HP 35s.

A combined Traverse Calculation/Check sheet can be used to show an independent check of your traverse calculations. As such, calculation/check sheets can come in many different formats so you should familiarize yourself with them first before using them. A modified form is used here to show how the check is derived and used.

Table 5-4 shows the traverse section of the traverse/check sheet. This is where the traverse is calculated in ΔE and ΔN and then checked by rotation and scaling in § 5.10.

5.9.1 Traverse **Calculation/~~Check~~** Sheet. Calculating the traverse ΔE and ΔN.

Table 5-5 Traverse closure check

Party No: Job No:
Traverse: **WORKED EXAMPLE 8.1**
Azimuth: F.B: Survey Date: Surveyor:
Co-ords: Computer:

STN	Angle	Bearing	Dist	ΔE	ΔN	E	N
~~STN~~	~~Angle~~	~~Brg+45~~	~~Dist/√2~~	~~S~~	~~C~~	~~□E=S-C~~	~~□N=S-C~~
A							
to		*30 00 00*	*100.00*	*50.00*	*86.60*		
B							
to		*120 00 00*	*300.00*	*259.81*	*-150.00*		
C							
to		*210 00 00*	*100.00*	*-50.00*	*-86.60*		
D							
to		*300 00 00*	*300.00*	*-259.81*	*150.00*		
A		***Σ Distance***	*800.00*				
			Σ ΔE =	*0.00*			
				Σ ΔN =	*0.00*		
	Misclose	$\sqrt{\Sigma\Delta E^2+\Sigma\Delta N^2}$ =		*0.000*	1 *IN* Perfect		

5.10 Checking a Close, Rotation & Scaling.

A mathematical method can be devised that obviates the previously mentioned errors. Importantly this is only a mathematical check for the correct calculation of ΔE and ΔN and does not check for measurement errors.

The check is performed by **rotating** the traverse by 45°,
i.e. add 45° to all bearings, scale the distances by 1/√2
and take the close out again to check ΔE and ΔN.

Let $\Delta\mathbf{E}' = \text{Dist}_{AB}\times\sin((\theta_{AB})+45)$.

Recall: $\sin(A+B)=\sin(A)\cos(B)+\cos(A)\sin(B)$

Expanding: $\Delta E' = \text{Dist}_{AB}\times\sin(\theta_{AB})\times\cos(45)+\text{Dist}_{AB}\times\cos(\theta_{AB})\times\sin(45)$

and: $\sin(45)=\cos(45)=1/\sqrt{2}=0.70710678$

$$\Delta E' = \frac{D_{AB}\times\sin\theta_{AB}}{\sqrt{2}} + \frac{D_{AB}\times\cos\theta_{AB}}{\sqrt{2}} = \frac{1}{\sqrt{2}}\left(D_{AB}\times\sin\theta_{AB} + D_{AB}\times\cos\theta_{AB}\right) = \frac{\Delta E+\Delta N}{\sqrt{2}}$$

$$i.e: \Delta E' = D_{AB}\times\sin(\theta_{AB}+45) = \frac{\Delta E+\Delta N}{\sqrt{2}}$$

Multiply both sides by $\frac{1}{\sqrt{2}}$

$$\frac{D_{AB}\times\sin(\theta_{AB}+45)}{\sqrt{2}} = \frac{D_{AB}}{\sqrt{2}}\times\sin(\theta_{AB}+45) = \frac{\Delta E+\Delta N}{2} = S$$

Let $\Delta\mathbf{N}' = \text{Dist}_{AB}\times\cos((\theta_{AB})+45)$.

Similarly, because cos(A + B) = cos(A)sin(B) – sin(A)cos(B), we have:

$$\frac{D_{AB}\times\cos(\theta_{AB}+45)}{\sqrt{2}} = \frac{D_{AB}}{\sqrt{2}}\times\cos(\theta_{AB}+45) = \frac{\Delta N-\Delta E}{2} = C$$

The **traverse check** then becomes:

S – C = ΔE
S + C = ΔN

The following traverse **check** sheet shows an independent check of each line by adding 45° to each bearing, and by dividing each distance by √2. (1.4142135624)

The resultant values of S and C are subtracted, and added, to check the **original** ΔE **and** ΔN. The mathematical check should be exact within the limits of rounding. See Table 5-6.

5.10.1 Traverse ~~Calculation~~/**CHECK** Sheet. **Checking** the traverse ΔE and ΔN.

Table 5-6 Rotation and scaling check

Party No: Job No:
Traverse: **WORKED EXAMPLE 8.1**
Azimuth: F.B: Survey Date: Surveyor:
Co-ords: Computer:

~~STN~~	~~Angle~~	~~Bearing~~	~~Dist~~	~~ΔE~~	~~ΔN~~	~~E~~	~~N~~
STN	Angle	Brg+45	Dist/√2	S	C	ΔE=S–C	ΔN=S+C
A			*100/√2=*				
to		*75 00 00*	*70.71*	*68.30*	*18.30*	*50.00*	*86.60*
B			*300/√2=*				
to		*165 00 00*	*212.13*	*54.90*	*–204.91*	*259.81*	*–150.00*
C							
to		*255 00 00*	*70.71*	*–68.30*	*–18.30*	*–50.00*	*–86.60*
D							
to		*345 00 00*	*212.13*	*–54.90*	*204.91*	*–259.81*	*150.00*
A	***Check: Σ Dist/√2 =***		*565.68*				
			ΣΔE =	*0.00*			
				ΣΔN =	*0.00*		

5.11 Chapter Exercise 5.2

State the misclose in ΔE and ΔN and the misclose ratio for each of the following traverses. Using a suitable traverse sheet, show the 45° rotation and $1/\sqrt{2}$ scaling check.

1)	BEARING	DISTANCE
AB	33 40 51	180.808
BC	108 15 57	185.395
CD	182 09 45	180.116
DE	257 27 58	164.650
EA	318 31 59	164.524

2)	BEARING	DISTANCE
AB	40 39 29	561.100
BC	88 39 14	498.480
CD	133 03 15	841.239
DE	219 46 55	665.701
EF	270 44 56	586.876
FG	327 37 55	596.816
GA	313 03 50	200.276

3)	BEARING	DISTANCE
AB	38 08 14	689.778
BC	76 51 49	743.774
CD	153 38 13	794.909
DE	216 18 48	722.762
EF	266 23 02	699.828
FA	328 59 52	731.828

4)	BEARING	DISTANCE
AB	25 33 24	493.374
BC	90 43 43	510.228
CD	153 44 33	489.290
DE	205 27 30	494.259
EF	273 31 09	496.720
FG	331 09 23	474.919
FA	269 02 30	2.100

5.11.1 Answers to Chapter Exercises 5.2

(1)	–0.145	–0.086	1 IN	5180
(2)	0.008	0.011	1 IN	281928
(3)	–0.132	0.085	1 IN	27910
(4)	0.042	0.001	1 IN	70500

5.12 The BOWDITCH adjustment.

Survey Regulations in Western Australia require that a traverse must be adjusted by BOWDITCH method prior to any area calculation. This method of adjustment has a long history and ensures that real calculations are carried out only on closed figures. No attempt is made to adjust the input bearings and distances - only the ΔE and ΔN values are adjusted. However, before applying the Bowditch adjustment the traverse angles must be adjusted. From these adjusted angles the adjusted bearings for each traverse leg are then derived. (Chapter 5 and § 5.4, page 48).

The theory of the adjustment is that each ΔE and ΔN receives an adjustment in proportion to the length of each traverse line divided by the total length of the traverse (e.g. perimeter for a closed loop traverse). This follows the error propagation principle that measurement errors cause larger errors in the ΔE and ΔN values the further advancing in the traverse (e.g. errors are smaller at the station B and larger at C and D; cf. Figure 5-13.

Here the adjustment principle is explained diagrammatically using a closed loop traverse. Similarly, the principle can be illustrated for an open connecting traverse. Figure 5-13 represents a traverse with an exaggerated misclose.

The distance A to A′ is the misclose of the closed traverse ABCDA

To make the A′ coincide with A the position of A′ must be moved by:

–ΣΔ**E** (the **minus** of the misclose in easting) and,

–Σ Δ**N** (the **minus** of the misclose in northing).

B′, C′ and D′ are moved in proportionate amounts against the misclose vectors to their adjusted positions at B, C and D.

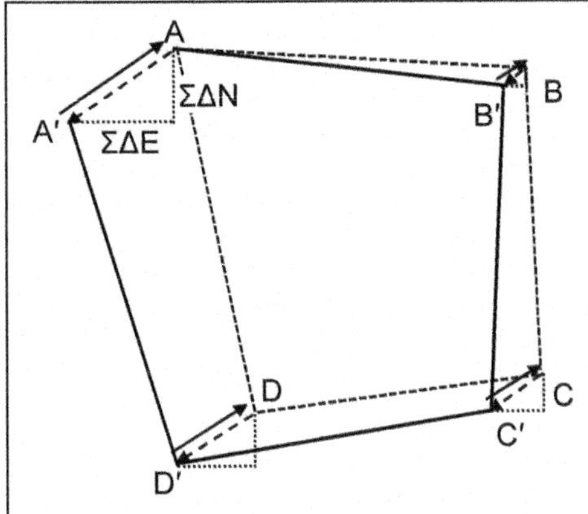

Figure 5-13 BOWDITCH adjustment

5.12.1 Graphical illustration of the Bowditch adjustment.

A, B, C, D, A represents the perimeter of the traverse points.

(a) Plot the length (e.g. perimeter) of the traverse as one straight line A B C D A, where AB, BC, CD and DA are the lengths of each side. (See Figure 5-14).
(b) To a suitable scale at the end of the line plot a point, at right angles, which represents the misclose in easting ΣΔE (A - A′)
(c) Join A to A′
(d) Plot perpendiculars from B, C and D
(e) The amounts BB′, CC′, DD′, AA′ are the amounts that points B, C, D and A must be shifted to close the figure
(f) The amounts BB′, C″C′, D″D′, A″A′ are the amounts of ΔE adjustment for **each** line segment
(g) Repeat steps (a) – (f) to adjust the misclose in northing ΣΔN (A - A′).

Mathematically we can express the adjustment this way:

Suppose ΣS = sum of the traverse lengths

The correction, vE_i (v, nu) to each ΔE_i, ΔN_i is given by; (for each line. i)

$$vE_i = \frac{D_i}{\Sigma S} \times vE \qquad vE_i \text{ (Correction to } \Delta E_i)$$

$$vN_i = \frac{D_i}{\Sigma S} \times vN \qquad vN_i \text{ (Correction to } \Delta N_i)$$

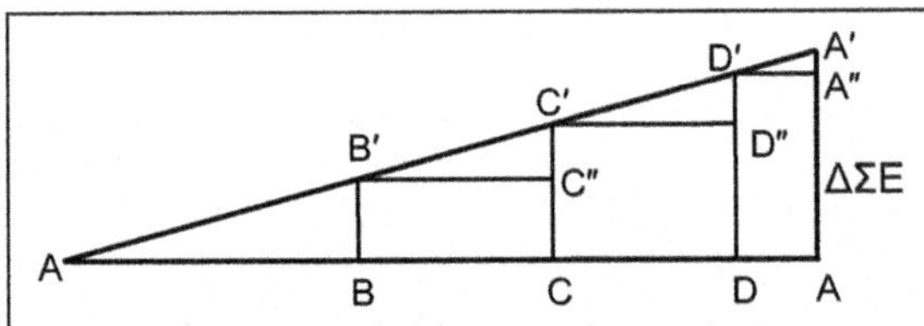

Figure 5-14 Graphic misclose in ΔE

Remember that:

vE = MINUS ΣΔE

vN = MINUS ΣΔN.

They are the CORRECTIONS which have the opposite sign from the misclose.

The adjusted co-ordinate differences can now be given as;

$$\Delta E_{AB} = \Delta E_{AB'} + v\Delta E_{AB}$$

$$\Delta N_{AB} = \Delta N_{AB'} + v\Delta N_{AB}$$

5.13 Adjustment procedure.

The procedure for finding final closure co-ordinates in any traverse is:

1) Determine the adjusted bearings (e.g. from adjusted angles)
2) From the adjusted bearings and measured distances find ΔE′, ΔN′ for each line
3) Adjust the initial co-ordinate differences by the Bowditch adjustment.
4) Find the adjusted co-ordinates.
5) Take out check calculations.

5.13.1 Worked Example 5.3

In Figure 5-15 the misclose is large, this demonstrates the corrections more clearly. The bearings have been adjusted.

$\Delta E = \text{Dist} \times \sin(\text{Brg})$ $\Delta N = \text{Dist} \times \cos(\text{Brg})$

Find the unadjusted co-ordinate differences of B, C, D, E and A'. (Table 5-7)

Table 5-7 Misclose

STN	Adj Brg	Dist	ΔE	ΔN	STN
A					**A**
to	*48 50 20*	*233.610*	*175.876*	*153.757*	
B					**B**
to	*121 00 30*	*250.060*	*214.325*	*–128.822*	
C					**C**
to	*183 55 50*	*254.310*	*–17.432*	*–253.712*	
D					**D**
to	*281 25 00*	*300.670*	*–294.721*	*59.515*	
E					**E**
to	*335 14 30*	*186.000*	*–77.895*	*168.903*	
A'					**A**
	Σ Distance (ΣS)	*1224.65*			
	E misclose	*ΣΔE =*	*0.153*		
Misclose	N misclose	*ΣΔN =*		*-0.359*	
Distance = $sqrt(\Sigma\Delta E^2 + \Sigma\Delta N^2)$			*0.390*	1 *IN 3140*	
BRG = arctan (ΣΔE / ΣΔN) =			*156°55'*		

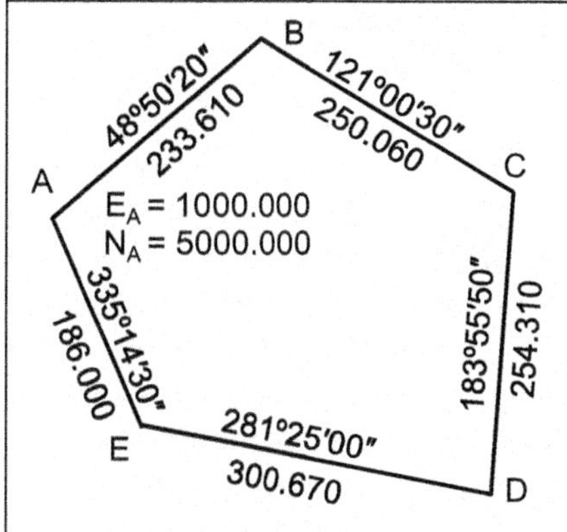

Figure 5-15 Coordinate differences

Adjust co-ordinates of B, C, D and E. Extra columns in the table for the **corrections** for corrected ΔE and ΔN. Table 5-8, coordinates adjusted by Bowditch method. (Correction method below).

Table 5-8 Coordinate adjustment

STN	Adj Brg	Dist	ΔE	ΔN	vE	vN	Adj ΔE	Adj ΔN	Adj. E	Adj. N	STN
A									1000.000	5000.000	**A**
to	*48 50 20*	*233.610*	*175.876*	*153.757*	*–0.029*	*0.069*	175.847	153.826			
B									1175.847	5153.826	**B**
to	*121 00 30*	*250.060*	*214.325*	*–128.822*	*–0.031*	*0.073*	214.294	–128.749			
C									1390.141	5025.077	**C**
to	*183 55 50*	*254.310*	*–17.432*	*–253.712*	*–0.032*	*0.074*	–17.464	–253.638			
D									1372.677	4771.439	**D**
to	*281 25 00*	*300.670*	*–294.721*	*59.515*	*–0.038*	*0.088*	–294.759	59.603			
E									1077.918	4831.042	**E**
to	*335 14 30*	*186.000*	*–77.895*	*168.903*	*–0.023*	*0.055*	–77.918	168.958			
A'									1000.000	5000.000	**A**
	Σ Dist (ΣS)	*1224.65*									
E misclose		*ΣΔE =*	*0.153*	*ΣvE =*	***–0.153***						
	N misclose		*ΣΔN =*	*–0.359*	*ΣvN =*	***0.359***					
Dist = $\sqrt{\Sigma\Delta E^2 + \Sigma\Delta N^2}$			*0.390*	Ratio: 1 *IN 3140*							
BRG = arctan(ΣΔE / ΣΔN)			*156°55'*								

Method: Calculate the **corrections** for line A to B′ to make it coincide with corrected B. *ΣS* = perimeter

$$vE_{AB} = \frac{D_{AB}}{1225} \times -0.153 \quad vN_{AB} = \frac{D_{AB}}{1225} \times +0.359$$

N.B. These are **corrections** (opposite misclose):

$$vE_{AB} = \frac{233.61}{1225} \times -0.153 \qquad vN_{AB} = \frac{233.61}{1225} \times +0.359$$

$vE_{AB} = -0.029$ $vE_{AB} = +0.069$

Three points should be noted with respect to the Bowditch adjustment.

1) if the misclose is positive the correction is applied to the ΔE or ΔN in the opposite sense. i.e. as negative.
2) The sum of the corrections = – misclose (equal but opposite sign).
3) The final check is that ΣΔE = 0.000 and ΣΔN = 0.000
   i.e. We start and finish on the same co-ordinate value.

If it's not checked it's wrong.

Note: $vE/\Sigma S$ and $vN/\Sigma S$ give constants in EASTING and NORTHING which are correction per metre of perimeter. They can be stored in the calculator as variables. Being constants, they only need the input of line distance to calculate corrections for each line.

Then, for each line, ***i*** $vE_i = (vE/\Sigma S) \times D_i$, correction in Easting and similarly: $vN_i = (vN/\Sigma S) \times D_i$, correction in Northing.

The misclose ratio of 1:3140 is too large but is used to provide a reasonable demonstration of the corrections to be applied.

5.13.2 Adjustment of a closed connected (linked) traverse.

A closed connecting, or linked, traverse, § 5.2.1, page 47, joins two known networks of control. It is a method used to provide extra control points, perhaps needed on a building site. Figure 5-9, page 50 also shows the linked traverse, including the angular misclose between the observed bearing, from observed angles to the final observed bearing to the known, calculated (true) bearing of the final control line.

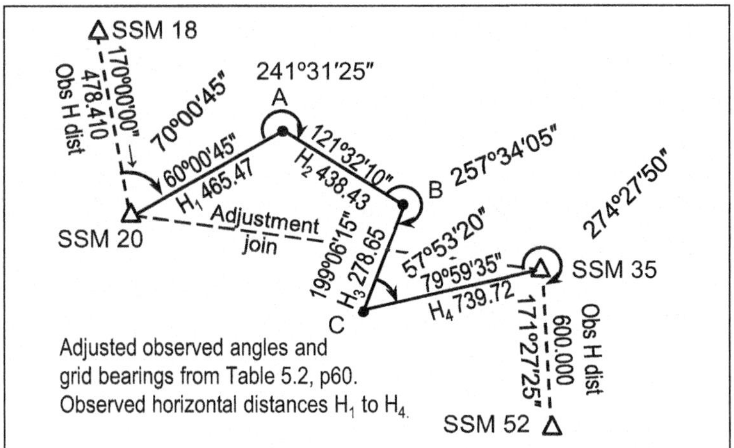

Figure 5-16 Unadjusted linked traverse with reduced adjusted observations

The scenario presented in Figure 5-16 shows densification of an existing control network between SSM18-SSM20 and SSM35-SSM52 where 3 additional control points are introduced between SSM20 and SSM35.

The reduced field book observations show:

- mean observed angles SSM20, A, B, C, SSM35;
- mean horizontal observed distances $H_{SSM20\text{-}SSM18}$, H_1 to H_4, $H_{SSM35\text{-}SSM52}$;
- observed BRG SSM35 – SSM52′.

Refer to Table 5-2, page 51 and , for the adjusted bearings.

5.13.2.1 Determine coordinate misclose and distribute corrections proportionally.

The adjusted bearings and observed horizontal distances, reduced by projection scale factor, are used to calculate the coordinate differences, ΔE and ΔN for each line to be adjusted.

Line SSM20-SSM18 and SSM35-SSM52 are **not** part of the misclose corrections as they are considered fixed.

Once the misclosures have been calculated, a vector hypotenuse can be found over the traverse. Divided into the sum of the horizontal observed distances, adjusted by S.F., the figure produced represents the "accuracy" of the survey. Quoted as 1:nnnnn, the result can give an indication of the precision of the survey before any more calculations are attempted.

A survey accuracy of, say, 1:5,000 would be met in this Worked Example as its "accuracy" approaches 1:9,000.

If the accuracy specification is not met, then it is time to check the survey observations and reductions.

5.13.2.2 Calculate final adjusted coordinates.

Table 5-9 Proportional linear misclose distribution. Adjusted coordinate differences.

Line	Adj BRG	H dist	ΔE	ΔN	vE	vN	Adj. ΔE	Adj. ΔN
20 – A	60 00 45	465.47	403.160	232.647	+0.029	+0.044	403.189	232.691
A – B	121 32 10	438.43	373.678	–229.315	+0.027	+0.042	373.705	–229.273
B – C	199 09 15	278.65	–91.198	–263.303	+0.017	+0.026	–91.181	–263.277
C – 35	76 59 35	739.72	720.741	166.488	+0.046	+0.071	720.787	166.559
Σ obs H dist.		1922.27			Adj. ΣΔE, ΣΔN		1406.500	–93.300
20 – 35′	Obs. ΣΔE, ΣΔN		1406.381	–93.483				
20 - 35	Calc. ΔE, ΔN		1406.500	–93.300	True ΔE, ΔN between 20 & 35			
	Misclose ΔE, ΔN		–0.119	–0.183	+0.119	+0.183	Check Σcorr's	
Corrections per metre $vE = -\Delta E/m$, $vN = -\Delta N/m$			$+6.21^{-5}$	$+9.51^{-5}$	Corr'n to ΔE, ΔN per m. H distance			
Misclose vector		213.15°	0.218	1:8,805	Accuracy = Σ obs dist/misclose vector			

The adjusted coordinates are presented in table form (Table 5-9) using the adjusted coordinate differences between the stations. Check that the coordinates of the end point agree with the original data.

Table 5-10 is a continuation of Table 5-9, providing final adjusted coordinates of the new control points A, B and C. Figure 5-17 is a diagram of the misclose between SSM35 and SSM35′.

Table 5-10 Adjusted coordinates A, B, C.

Point	Adj ΔE	Adj ΔN	E	N
SSM20			1000.000	1000.000
A	403.189	232.691	*1403.189*	*1232.691*
B	373.705	-229.273	*1776.894*	*1003.418*
C	-91.181	-263.277	*1685.713*	*740.141*
SSM35	720.787	166.559	2406.500	906.700

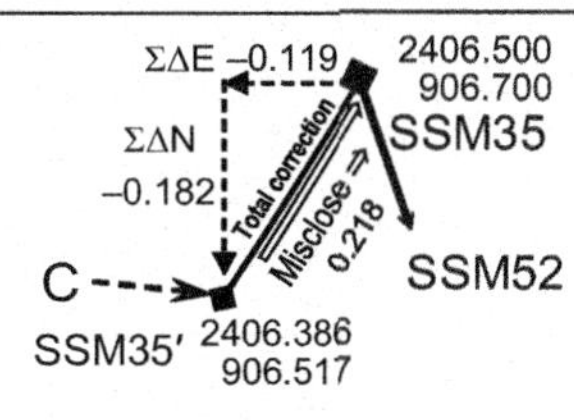

Figure 5-17 Misclose diagram.

5.13.2.3 Bowditch adjustment calculation form.

Table 5-12 is a Bowditch traverse calculation form, or a summary of results. The form is different from the previous table methods where the adjustment was broken down into three activities. In this method still follow the process, adjusting the angles, finding the misclose, adjusting coordinated differences, adding adjusted differences sequentially. A blank worksheet for reproduction is shown in Table 5-14.

1. Angular misclose adjustment to produce adjusted bearings. Figure 5-11 is a repeat of Table 5-2 to provide the adjusted bearings used in Table 5-9.

Table 5-11 Open connecting traverse

Station At:	Observed Angle	Adjustment	Bearing +Adj Angle	AdoptedBearing
BRG (20 – 18)	350 00 00	Adopted BRG (20 – 18)	350 00 00	350°00′00″ $\theta_{(20\text{-}18)}$
SSM 20	70 00 35	+ 10″	70 00 45	
		Sum	420 00 45	
		Mod 360	60 00 45	60°00′45″ $\theta_{(20\text{-}A)}$
		Back Bearing	240 00 45 (A→20)	
A	241 31 15	+ 10″	241 31 25	
		Sum	481 32 10	
BRG (A – B)		Mod 360	121 32 10	121°32′10″ θ_{AB}
		Back bearing	301 32 10 (B→A)	
B	257 33 55	+ 10″	257 34 05	
		Sum	559 06 15	
BRG (B – C)		Mod 360	199 06 15	199°06′15″ θ_{BC}
		Back bearing	19 06 15 (C→B)	
C	57 53 10	+ 10″	57 53 20	
		Sum	76 59 35	
BRG (C –35)		Mod 360	76 59 35	76°59′35″ $\theta_{(C\text{-}35)}$
		Back bearing	256 59 35 (35→C)	
SSM 35	274 27 40	+ 10″	274 27 50	
SUM obs. ∠	1251 26 35	Sum	531 27 25	
–((N+1)×180)	–1080	Mod 360	171 27 25	171°27′25″ $\theta_{(35\text{-}52)}$
Sum =	171 26 35	Corr'n = + 50		Checks with original
BRG (35 – 52)	171 27 25			
Misclose =(obs – true) =	–50″			
Adjustment =	50/5 = +10″			

The initial fixed bearing is the **inbound** bearing from the back sight to the start of the adjustment station. (From STN 18 **to** STN 20).

Does angular misclose meet acceptable limits?

Ensure final bearing agrees with original data.

2. Calculation of coordinate differences (ΔE, ΔN) to determine coordinate misclose over the adjustment figure.
   a. Calculate misclose ratio. Does it meet acceptable limits?
   b. Calculate misclose corrections (vE, vN) in proportion to traverse leg lengths. (Negative of misclose.)
   c. Distribute corrections to each traverse leg to provide adjusted coordinate differences vE+ΔE, vN+ΔN
3. Sum adjusted coordinate differences from start point to end point.
   a. Ensure end coordinates agrees with original data.
   b. Complete misclose diagram.

The sheet requires additional undocumented steps such as finding the back bearing AT the station to allow addition of the adjusted angle.

Inbound BRG + 180 + angle correction + unadjusted angle = forward adjusted bearing.

170 00 00	+ 180	+ 0 00 10	+ 70 00 35	= 60 00 45
60 00 45	+ 180	+ 0 00 10	+ 241 31 15	= 121 32 10
121 32 10	+ 180	+ 0 00 10	+257 33 55	= 199 09 15
199 09 15	+ 180	+ 0 00 10	+57 53 10	= 76 59 35
76 59 35	+ 180	+ 0 00 10	+274 27 40	= **171 27 25**

Table 5-12 Traverse calculation form. (NOTE: BRG is FROM 18 TO 20)

TRAVERSE CALCULATION FORM

Project: Survey comps example Topic 5/10 Survey date: 3/10/2019—Surveyor:—S.W—Calculation: date: 5/10/2019

Calculator: B.C Page 1 of 1

Station Line	Corr. ν Angle θ	Bearing ϕ	Horizontal distance D	Corr. νE $\Delta E = D \times \sin\theta$	Corr. νN $\Delta N = D \times \cos\theta$	Easting	Northing	Station Line
SSM 18								SSM 18
	+10	170 00 00	---					
SSM 20	70 00 35			+.029	+.044	1000.000	1000.000	SSM 20
	+10	60 00 45	465.47	403.160	232.647			
A	241 31 15			+.027	+.042	1403.189	1232.691	A
	+10	121 32 10	438.43	373.678	-229.315			
B	257 33 55			+.017	+0.026	1776.894	1003.418	B
	+10	199 09 15	278.56	-91.198	-263.303			
C	57 53 10			+.046	+0.071	1685.713	740.141	C
	+10	76 59 35	739.72	720.741	166.488			
SSM 35	274 27 40					2406.500	906.700	SSM 35
		171 27 25						
SSM 52								SSM 52

[1] $\Sigma\alpha$: 901 26 35
[2] n×180: 900
[3] $\theta_{end} - \theta_{start}$ 1 27 25

ΣD: 1922.27 | [4] $\Sigma\Delta E$: 1406.381 | [5] $\Sigma\Delta N$: -93.483 | [6] $E_{end} - E_{start}$ 1406.500 | [7] $N_{end} - N_{start}$ -93.300

Misclose (δE)=[4]-[6] -0.119
Misclose (δN)=[5]-[7] -0.183

Angular misclose ($d\alpha$)=[1]–[2]–[3]: +0 00 50"
Acceptable angular misclose: $30\sqrt{5}=67"$

Misclose distance (δD)=$\sqrt{(\delta E^2+\delta N^2)}$: 0.218
Misclose ratio 1:($\Sigma D/\delta D$) =1:8805
Misclose bearing (sf)=atan($\delta E/\delta N$): 213.15°
Acceptable misclose ratio: =1:5000 ✓

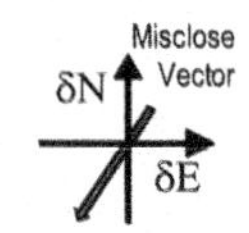

5.14 Check for scale factor in horizontal distances.

Total stations (TS) display slope, S, horizontal, H and vertical, V distances. H and V are generated from the slope distance and zenith angle. Horizontal distance (H dist.) may, or may not, incorporate scale factor. This depends on the TS software (Sokkia SET530). And whether the TS is in basic (display only) or program (data collection) mode (Sokkia FX101). Slope distance is not affected by scale factor.

If your reference coordinates are modelled on, say UTM (e.g. MGA2020) then there will be a point or line scale factor in the horizontal distances. Your observed horizontal distances must also incorporate the scale factor. The scale factor may be introduced via:

1. Adjusting the scale factor in the JOB parameters in the total station program. The displayed H distance then incorporates the scale factor,
2. Applying the scale factor to the horizontal distance calculated from ZA and slope dist., this is the H distance with a S.F. = 1. It is recommended to use S.F. = 1 in the field and reduce the H distance to the projection S.F. in the office calculations.

The opening and closing observations allow for checking the presence of horizontal scale factor (S.F.) in the coordinate frame. The calculated S.F. should agree with the area's scale factor. Disagreements should be resolved before proceeding with the adjustment.

Observations over the base line should incorporate zenith angle (ZA) and slope distance (S) measurements. This will allow verification of the horizontal distance with the (H) distance displayed on the instrument. The vertical distance (V) will allow rudimentary checking of station RLs.

From Figure 5-16

$H_{18\text{-}20(calc)} = 478.410$, $H_{18\text{-}20(obs)} = 478.601$. S.F. = 478.410/487.601 = 0.9996

$H_{35\text{-}52(calc)} = 600.000$, $H_{35\text{-}52(obs)} = 600.240$. S.F. = 600.000/600.240 = 0.9996

If it's not checked it's wrong.

Use the projection S.F for the area = 0.9996 to reduce observed H distances to projection. The observed horizontal distances (H) in Figure 5-16 have been reduced to projection distances using S.F. 0.9996 as shown in Table 5-9 and .Table 5-13 shows the effect of using observed horizontal distances of S.F. = 1 when, in fact, the coordinates of the control stations are UTM coordinates with a scale factor of 0.9996 (zone central meridian, k_0.) and the observed distances must be reduced by that scale factor.

Table 5-13 Observed and S.F. reduced H distances S.F. = 0.9996 (C.M.)

Line	Adj BRG	Observed H dist	Reduced H dist	Unadjusted ΔE	Unadjusted ΔN			
18 – 20	170 00 00	~~478.610~~	~~*478.41*~~	-----	-----			
20 – A	60 00 45	465.656	*465.47*	403.321	232.740			
A – B	121 32 10	438.605	*438.43*	373.828	–229.406			
B – C	199 09 15	278.762	*278.65*	–91.235	–263.409			
C – 35	76 59 35	740.016	*739.72*	721.029	166.555			
35 – 52	171 27 25	~~600.240~~	~~*600.00*~~	-----	-----			
Σ obs H dist.		1923.039	1922.27					
20 – 35′	Obs. ΣΔE, ΣΔN			1406.943	–93.520			
20 – 35	Calc. ΔE, ΔN		1406.500	–93.300	True ΔE, ΔN between 20 & 35			
	Misclose δE, νN		0.443	–0.220				
Corr'n *ν*E = –δE/m, *ν*N = –δN/m			–2.30^{-4}	+1.14^{-4}	Corr'n *ν*E, *ν*N per m. H distance			
Misclose vector		116.41°	0.495	1:3,890	Accuracy = Σ obs dist/misclose vector			

The accuracy of 1:3890 is unexpectedly large compared with both the acceptable value (1:5000) and the achieved value (1: 8800) in Table 5-9 (page 58). The difference between the observed and calculated distances on the two fixed legs, SSM18 – SSM20 and SSM35 – SSM52 shows a S.F.(k) = 0.9996. This is the UTM central meridian scale factor at 117°E, out near Beverley, Western Australia.

The example is extreme to illustrate the points being made.

A blank traverse calculation sheet is provided in Table 5-14 to reflect Table 5-12. Enlarged by about 10%, it should fit an A4 sheet.

Table 5-14 Blank Traverse calculation form.

TRAVERSE CALCULATION FORM

Project: Survey date: Surveyor: Calculation: date: Calculator: Page of__

Station		Corr. $v\theta''$	Bearing ϕ	Horizontal distance D		Corr. vE		Corr. vN	Easting	Northing	Station
Line	Angle θ				$\Delta E = D \times \sin\theta$		$\Delta N = D \times \cos\theta$				Line

[1] $\Sigma\alpha$:

[2] $n \times 180$:

[3] $\theta_{end} - \theta_{start}$

ΣD:

[4] $\Sigma\Delta E$:

[5] $\Sigma\Delta N$:

[6] $E_{end} - E_{start}$ [7] $N_{end} - N_{start}$

Misclose $(\delta E) = [4] - [6]$

Misclose $(\delta N) = [5] - [7]$

Angular misclose $(d\alpha) = [1] - [2] - [3]$:+

Acceptable angular misclose:

Misclose distance $(\delta D) = \sqrt{(\delta E^2 + \delta N^2)}$:

Misclose ratio 1:$(\Sigma D/\delta D)$–1:

Misclose bearing (sf)=atan$(\delta E/\delta N)$:

Acceptable misclose ratio:——1:

Misclose vector

If it's not checked it's wrong.

5.15 Chapter Exercise 5.3

Find the misclose in the following traverses and adjust by Bowditch method. The answer is to include:
Perimeter Distance
Misclose ΣΔE =
ΣΔN =
Misclose vector Brg Distance
Misclose ratio 1:Distance/vector

1)

Line	BEARING	DISTANCE
A-B	29°59′35″	396.658
B-C	78°01′36″	378.513
C-D	122°02′27″	400.336
D-E	184°57′20″	404.390
E-F	232°05′23″	384.138
F-G	269°29′08″	443.628
G-A	343°44′03″	451.499

2)

Line	BEARING	DISTANCE
A-B	41°12′59″	269.644
B-C	103°19′49″	267.524
C-D	172°40′04″	314.454
D-E	258°40′40″	316.646.
E-A	324°16′37″	286.987

3)

Line	BEARING	DISTANCE
A-B	23°17′01″	581.547
B-C	70°03′20″	635.521
C-D	129°19′24″	664.406
D-E	186°01′24″	631.908
E-F	229°09′36″	608.861
F-G	292°09′24″	582.348
G-A	329°59′42″	550.416

5.15.1 Answers to Chapter Exercise 5.3

1) Perimeter 2859.162
Misclose Σ ΔE = -0.161
Σ ΔN = 0.218
Misclose vector 323°280.271
Misclose ratio 1:10 556

Line	ΔE'	ΔN'
A-B	198.310	343.510
B-C	370.300	78.496
C-D	339.376	–212.418
D_E	–34.910	–402.909
E-F	–303.053	–236.054
F–G	–443.585	–4.017
G-A	–126.437	433.392

2) Perimeter 1455.255
Misclose Σ ΔE = 0.071
Σ ΔN = 0.093
Misclose vector 37°18′0.117
Misclose ratio 1:12 406

Line	ΔE'	ΔN'
A-B	177.657	202.816
B-C	260.303	–61.699
C-D	40.116	–311.903
D-E	–310.499	–62.186
E-A	–167.577	232.972

3) Perimeter 4255.007
Misclose $\Sigma \Delta E = -0.278$
$\Sigma \Delta N = -0.370$
Misclose vector 216°48′ 0.462
Misclose ratio 1:9203

Line	ΔE'	ΔN'
A-B	229.913	534.236
B-C	597.446	216.837
C-D	514.016	–420.974
D-E	–66.267	–628.364
E-F	–460.587	–398.111
F-G	–539.307	220.678
G-A	–275.214	476.698

5.16 Introduction – Scientific calculators.

The course requires the use of a scientific calculator with hardwired trigonometric function. Additional functionality, in terms of unit conversion, angular mode conversion, statistics summation, polar ↔ rectangular coordinate conversion and memory storage and manipulation will speed the calculation process. The ability to write small program routines is considered useful.

Regardless of the type of suitable calculator that you use, it is essential that you are totally familiar with its functions. Do you have the manual?

In this course the important functions needed are:

[+], [–], [x], [÷]	operators
[x^2], [√x], [1/x]	conversion for squared, square root and inversion
[π]	constant key (3.14159265359...)
[SIN], [COS], [TAN]	trigonometric functions
[ASIN], [ACOS], [ATAN]	arc trigonometric functions
[→ DEG], [→ RAD]	conversion, convert between degrees and radians
[→ HR], [→ HMS] [° ′ ″]	conversion, to change between decimal degrees and degrees, minutes and seconds
[→ θ,r], [→ y,x]	conversion, polar ↔ rectangular coordinates
[Σ+], [Σ–] σ_x σ_y s_x s_y, $\bar{x}$, $\bar{y}$	statistical summation, stat function
STO, RCL	operations from available memory, the ability to use memory arithmetic (STO +, RCL ÷)
FIX	number of decimal places displayed
[+/–] [(–)] [E] [EXP]	change sign of a number, exponentiation
[C], CLEAR, [←]	input cancellation, clear and backspace edit keys

5.16.1 Problems and Solutions.

Many of the problems to be solved require a number of incremental answers to be derived during the solution process. You will have to devise a method of recording these intermediate answers so that they may be applied later in the process; e.g. an angle, expressed in degrees, minutes and seconds must be converted to decimal degrees before it can be converted to, say, radians or one of the trigonometric function. It may be used repeatedly.

This means that intermediate results should be stored in memory locations. However this action implies some recordkeeping during the solution process so that a record of memory location contents is kept.

Similarly, the written proof of the solution requires that answers be written down. The problem then becomes one of how many decimal places should be used to record results.

If recorded results are to be re-introduced later in the solution, then you need to record 6 to 7 decimal places. If truncated results are re-inserted, then final accuracy of the result will suffer. By using storage registers, the full precision of the calculation is maintained.

However, at the end of the problem, the answer should be provided in real world numbers. There is no point in giving the final calculated length to more than 3 decimal places, which represents 1mm.

5.16.2 Algebraic calculators. HP10S+, Casio fx-82AU, Sharp EL-531

Normal algebraic entry calculators, some approved by the School of Civil and Mechanical Engineering (CME), will be able to handle the calculations in this course. However, some processes such as traverse summation using rectangular coordinates will be more cumbersome and time consuming through a lack of integration of processes.

The HP10s+ and similar Sharp and Casio calculators can do the work.

§ 18.16. Survey Calculations on HP 10s+, HP300s+, Casio, Sharp, page 298 enhances the owner's manual with help on using the calculator in the surveying environment. The other two calculators are similar, but slightly different, and some of the differences are noted in 18.16, § 18.20.3, page 301.

5.17 Chapter Exercise 5.4

Q. 1 Make certain you are familiar with your calculator and in particular with the following keys:

(a) arithmetic keys: +, -, x, ÷

(b) clear x key

(c) **x** x^2 $^1/_x$ keys

(d) To decimal degrees. To degrees, minutes and seconds.

(e) SIN, COS and TAN keys. ASIN, ACOS and ATAN keys.

(f) Storage and Recall keys.

Q. 2 Find the values of the following trig functions:

TAN	26°40′42″
COT	316°41′17″
SIN	17°51′35″
COS	152°14′36″
COSEC	277°35′25″
SEC	191°51′39″
SEC	1°00′14″
SIN	89°39′20″

Q. 3 Find the two values of the angle A in each case.

TAN A =	0.431 065
TAN A =	–1.261 071
COT A =	0.777 055
SEC A =	–1.737 114
SIN A =	0.557 379
SIN A =	–0.673 106
COS A =	0.833 198
COS A =	–0.261 235

Q. 4 Convert:.

0.002756 radians to DMS

ATAN($-1.2E^{-7}$) to radians

TAN((π/2)radians) to its value

5.17.1 Answers to Chapter Exercise 5.4

Q. 2.		Q. 3.		
	0.502 474	23°19′09″	or	203°19′09″
	–1 060 731	308°24′49″	or	128°24′49″
	0.306 688	52°09′03″	or	232°09′03″
	–0.884 933	125°08′47″	or	234°51′13″
	–1.008 839	33°52′29″	or	146°07′31″
	–1.021 816	317°41′34″	or	222°18′36″
	1.000 154	33°34′17″	or	326°25′43″
	0.999 982	105°08′36″	or	254°51′24″

Q. 4 0°09′28″

–0.00000012 or 3.14159253

–190 985 931 710 (note the 12 digit answer) (Tan 90° ((π/2)radians) →∞)

TAN(90) displays INVALID DATA, ((π/2) converted to degrees SHOWS 90.0000000003 and evaluates to –190 985 931 710. That's IEEE574 at work).

The closest we can get to 90° without error is:

90.0000000001 which evaluates to –572 957 795 131. ($-5.72957795131 \times 10^{11}$).

In MS Excel the closest we can get to TAN(90°) is $1.63245522776191 \times 10^{16}$.

5.18 Co-ordinate Systems

The concept of co-ordinate systems should not be entirely new since you will have covered this subject at school. However, as surveyors, we tend to treat co-ordinates a little differently from mathematicians and we systematize our usage so that ambiguities are avoided.

In Australia the system of co-ordinates is designated as a North-East system where the north and east directions are treated positively. In 3 dimension coordinates the system is known as **NEZ**. Be very careful of the coordinate order used in Total Stations. Most instruments, on a **reset** or initialisation, default to an NEZ. This follows the order often used in the United States to mimic their "latitude and departure" plane system. Generally, the only configuration to change in the default settings is in the observation settings for Coordinates. Refer to the Total Station manufacturers manual, e.g., (Sokkia SET FX1, section 20.1, page 104).

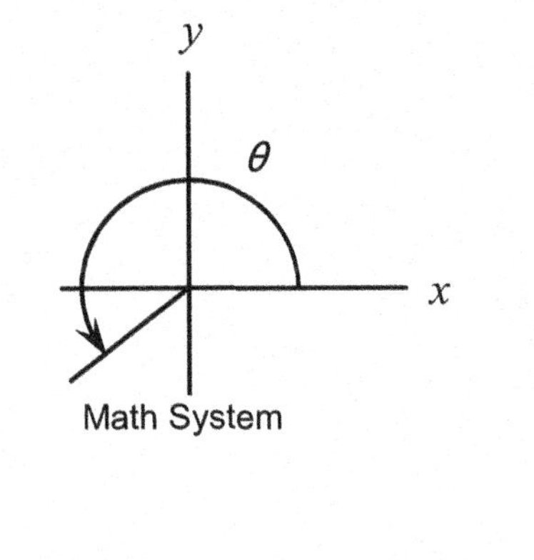

Figure 5-18 Math system. Anti-clockwise from horizontal x axis

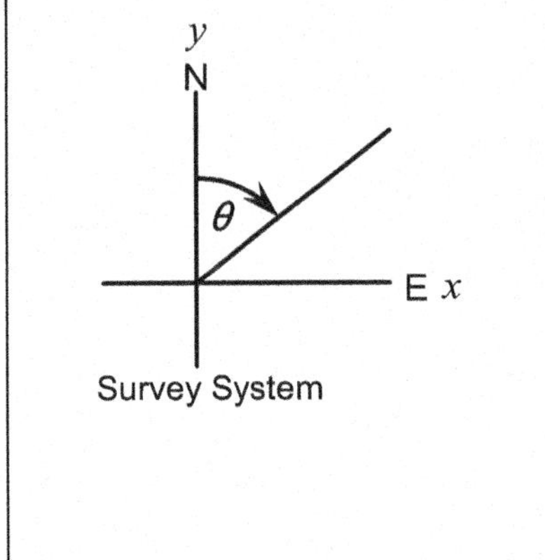

Figure 5-19 Survey and navigation system. Clockwise from y, North axis.

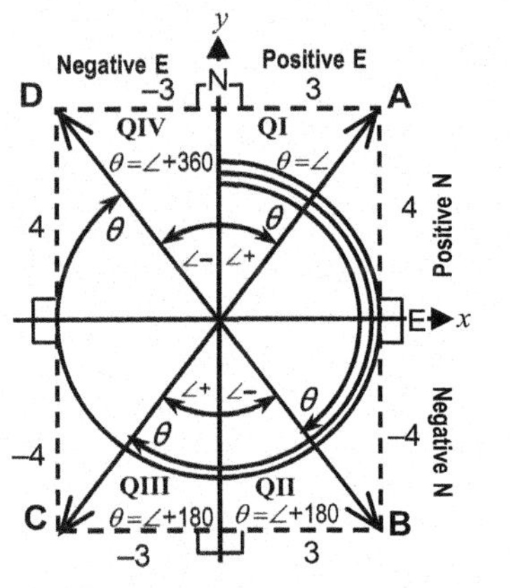

Figure 5-20 Quadrants

Bearings are treated as being clockwise from the north axis (Figure 5-19) (as distinct from anticlockwise from E axis as in maths (Figure 5-18)). This angular convention is adopted so that bearings are consistent with surveyors' theodolite circles and compass readings in navigation.

In a planar co-ordinate system co-ordinates of **A** can be given as the right-angled distances from the east and north axes respectively.

The co-ordinates of **A** in Figure 5-20 are E3 (e.g. E = 3), N4 (e.g. N = 4), adopting the co-ordinates convention commonly used in Australia. Similarly, **B** = E3, N–4, **C** = E–3, N–4 and **D** = E–3, N4.

5.19 The Map Grid of Australia (MGA)

A high percentage of co-ordinates quoted in Survey Practice these days refer to the Map Grid of Australia (MGA), or the former Australian Map Grid (AMG), both of which you will study in Coordinate and Mapping Systems. It suffices to say that these co-ordinates are part of a worldwide system which allows us to uniquely define the position of any surveyed point.

MGA co-ordinates are expressed in metres from an origin system designed to make all co-ordinate values positive. All Easting co-ordinates fall in the range 100,000 to 800,000 and Northing co-ordinates 5,000,00 to 8,000,000. MGA co-ordinates will often be used throughout this book, so that you become familiar with their usage. The false origin is 500,000m West of the Central Meridian and 10,000,000m South of the equator.

5.20 Coordinate system calculations. Joins and Polars

There are two basic survey calculations which confront us when using co-ordinates:

(1) The Join Calculation

Given the co-ordinates of two points A and B determine the bearing and distance between them. Convert RECTANGULAR coordinates to POLAR vector.

(2) The Polar Calculation

Given the co-ordinates of a point A (E_A, N_A) together with the bearing and distance from A to point B, determine the co-ordinates of point B(E_B, N_B). Convert POLAR vector to RECTANGULAR coordinates.

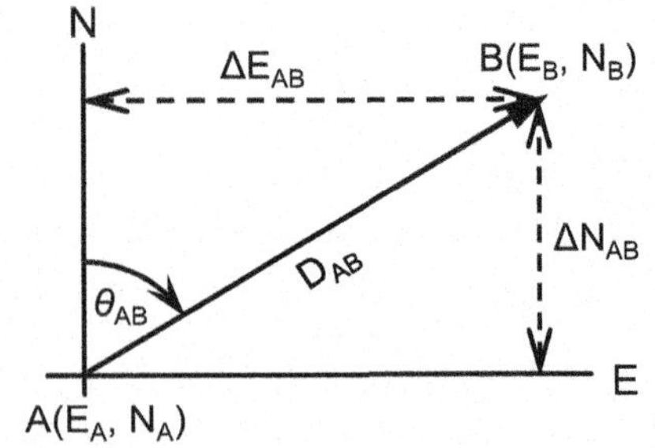

Figure 5-21 POLAR/RECTANGULAR coordinates

If it's not checked it's wrong.

In Figure 5-21, the co-ordinates of A and B are represented by (E_A, N_A) and (E_B, N_B) respectively. The bearing of AB is shown as θ_{AB} (i.e. an angle clockwise from the north). Distance AB = D_{AB}.

5.21 The Join Calculation (RECTANGULAR to POLAR).

In all vector calculations the rule for the difference (Δ) between two coordinates is:
"the coordinates you are going **to** minus the coordinates you are coming **from**".

Given: The co-ordinates of A (E_A, N_A) and the co-ordinates of B (E_B, N_B). (See Figure 5-21).
Find: The bearing and distance A to B
i.e. θ_{AB}, D_{AB}.
Solution: $\Delta E_{AB} = E_B - E_A$
$\Delta N_{AB} = N_B - N_A$

$$\tan\theta_{AB} = \frac{\Delta E_{AB}}{\Delta N_{AB}} \Rightarrow \theta_{AB} = \arctan\left(\frac{\Delta E_{AB}}{\Delta N_{AB}}\right) \text{ in the correct quadrant.}$$

$$D_{AB} = \frac{\Delta E_{AB}}{\sin(\theta_{AB})} = \frac{\Delta N_{AB}}{\cos(\theta_{AB})} = \sqrt{\Delta E^2 + \Delta N^2} \text{ hypotenuse from Pythagoras.}$$

5.21.1 From the arctangent, how do we obtain the bearing of AB in all 4 quadrants?

Mathematically, the arctangent function is only defined for $-90° < \theta < +90°$, thus results from scientific calculators need to be adjusted for the correct quadrant. This can be done by the following Quadrant Rule, see Figure 5-22. Examine the **sign** of the result. Tan is positive in the 1st and 3rd quadrants. Thus, the Quadrant Rule applied here uses the **sign** of the NORTH component (ΔN).

ΔN > 0 **(positive):** $\theta_{AB} = \theta_{AB} + 360$. If $\theta > 360$: $\theta = MOD(\theta, 360)$

ΔN < 0 **(negative):** $\theta_{AB} = \theta_{AB} + 180$ $(90 < \theta < 270)$

The MOD(θ,360), or MODULO 360 (θ), operator reduces the value of θ to a range 0 → 360°. $(0 < \theta < 360)$

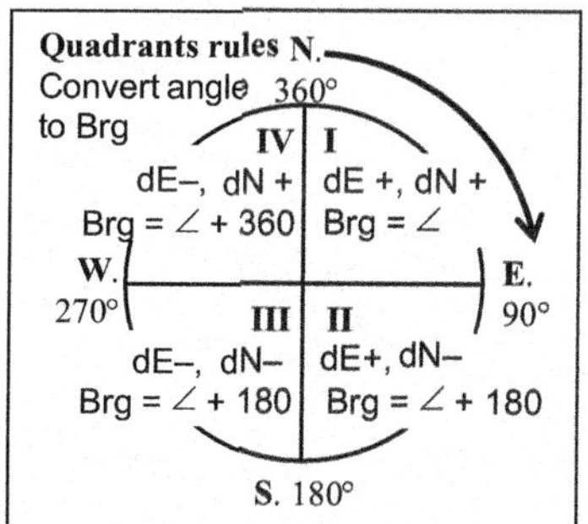

Figure 5-22 Quadrant rules

ΔN = 0 (**zero**): $\arctan\left(\frac{\Delta E}{0}\right)$ cannot be evaluated;

calculators give a "Divide by 0" error message.

Tan(90°) is undefined because $\lim_{\theta \to 90°}[\tan\theta] \to \infty$ (tanθ tends to infinity),

thus for ΔN = 0, θ = 90° or 270°.

(For convenience we say tan(90°) = ∞).

ΔE = 0: $\arctan\left(\frac{0}{\Delta N}\right)$ is evaluated correctly as 0°

Test statement **if ΔN = 0 AND if ΔE > 0 then** θ = 90°
else if ΔE < 0 then θ = 270°

Quadrant 1: (0 - 90°). tanθ is POSITIVE Figure 5-23

$$\tan\theta_{AB} = \left(\frac{+\Delta E}{+\Delta N}\right) \Rightarrow \Delta N > 0\ (+\text{ve})$$

$$\theta_{AB} = 360 + \arctan\left(\frac{\Delta E}{\Delta N}\right)$$

$$\theta_{AB} = 360 + (+\theta)$$

$$\theta_{AB} = MOD(\theta, 360) \text{ MS Excel function}$$

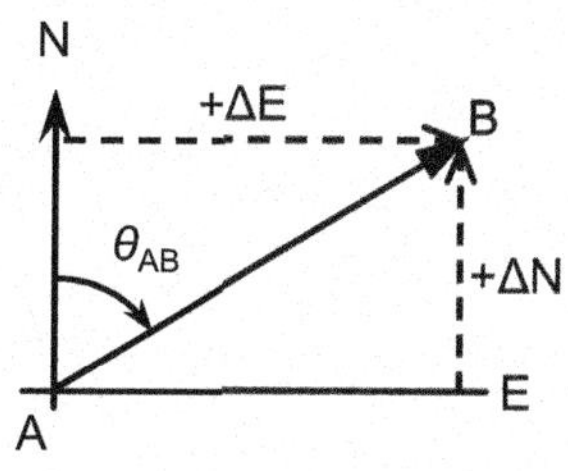

Figure 5-23 Quadrant 1

Quadrant 2: (90° - 180°). tanθ is NEGATIVE Figure 5-24

$$\tan\theta_{AB} = \left(\frac{+\Delta E}{-\Delta N}\right) \Rightarrow \Delta N < 0\ (-\text{ve})$$

$$\theta_{AB} = 180 + \arctan\left(\frac{\Delta E}{-\Delta N}\right)$$

$$\theta_{AB} = 180 + (-\theta)$$

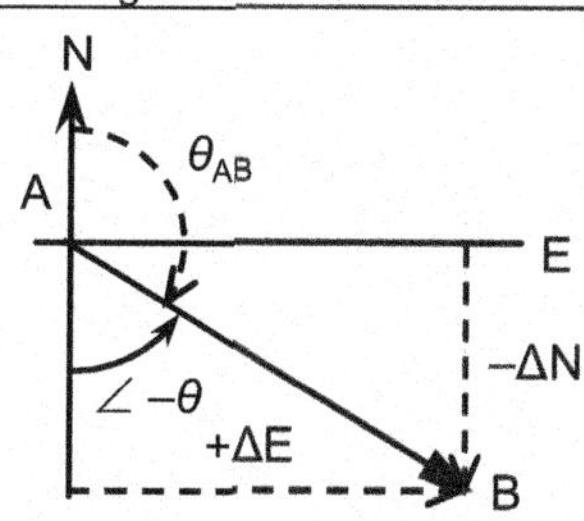

Figure 5-24 Quadrant 2

If it's not checked it's wrong.

Quadrant 3: (180° - 270°). $\tan\theta$ is POSITIVE Figure 5-25

$$\tan\theta_{AB} = \left(\frac{-\Delta E}{-\Delta N}\right) \Rightarrow \Delta N < 0 \ (\text{-ve})$$

$$\theta_{AB} = 180 + \arctan\left(\frac{-\Delta E}{-\Delta N}\right)$$

$$\theta_{AB} = 180 + (+\theta)$$

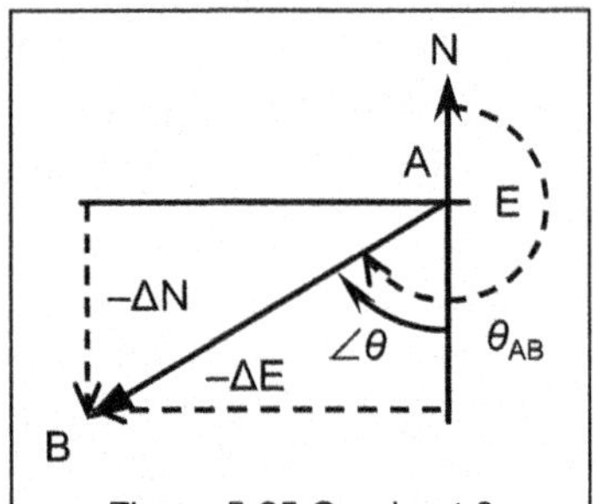

Figure 5-25 Quadrant 3

Quadrant 4: (270° - 360°). $\tan\theta$ is NEGATIVE Figure 5-26

$$\tan\theta_{AB} = \left(\frac{-\Delta E}{+\Delta N}\right) \Rightarrow \Delta N > 0 \ (\text{+ve})$$

$$\theta_{AB} = 360 + \arctan\left(\frac{-\Delta E}{+\Delta N}\right)$$

$$\theta_{AB} = 360 + (-\theta)$$

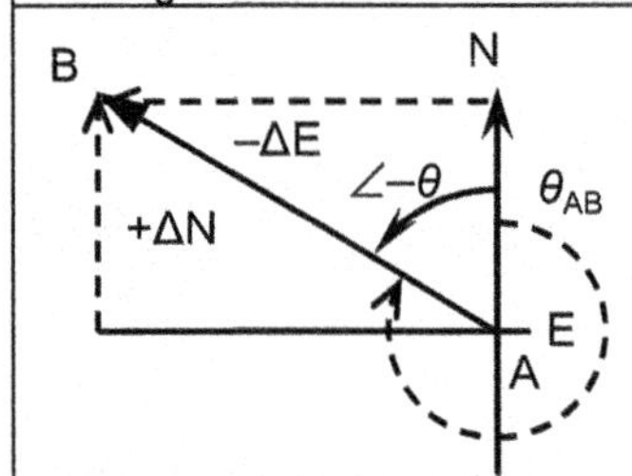

Figure 5-26 Quadrant 4

5.21.2 Join Calculation.

Given: The co-ordinates A (E_A, N_A) and B (E_B, N_B)

Find: The bearing and distance A to B

Format: (remember: coordinate **to** minus coordinate **from**).

To:	E_B.	N_B	, $\tan\theta_{AB} = \frac{\Delta E_{AB}}{\Delta N_{AB}} \Rightarrow \theta_{AB} = \arctan\left(\frac{\Delta E_{AB}}{\Delta N_{AB}}\right)$
− From:	E_A	N_A	$D_{AB} = \sqrt{\Delta E_{AB}^2 + \Delta N_{AB}^2}$
Difference	ΔE_{AB}	ΔN_{AB}	

CHECK:

$E_B = E_A + D_{AB} \times \sin(\theta_{AB})$

$N_B = N_A + D_{AB} \times \cos(\theta_{AB})$

Method: **to find θ_{AB}, D_{AB}.**

1) Find the coordinate difference $\Delta E_{AB} = E_B - E_A$ and $\Delta N_{AB} = N_B - N_A$

2) calculate the Tangent of the bearing: $\tan\theta_{AB} = \frac{\Delta E_{AB}}{\Delta N_{AB}}$,

3) find the arctangent of $\theta_{AB} = \arctan\left(\frac{\Delta E_{AB}}{\Delta N_{AB}}\right)$ in correct quadrant

4) and distance $D_{AB} = \sqrt{\Delta E_{AB}^2 + \Delta N_{AB}^2}$ using Pythagoras' theorem.

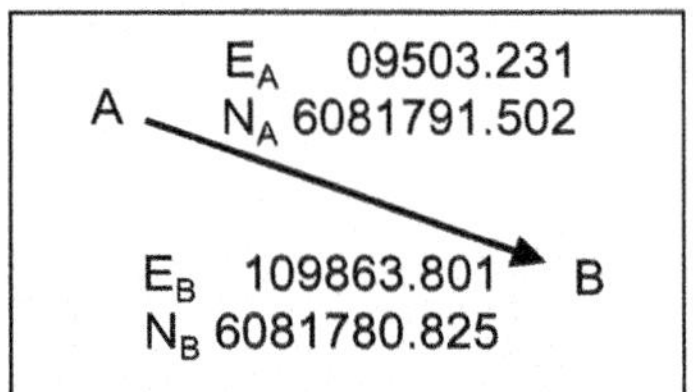

Figure 5-27 Worked Example 5.4

5.21.2.1 Worked Example 5.4

Compute the bearing and distance from A to B (see Figure 5-27)

	Easting	Northing
B	109863.801	6081780.825
− A	109503.231	6081791.502
	360.570	− 10.677 **ΔN < 0**

$$\theta_{AB} = \arctan\left(\frac{360.570}{-10.677}\right)$$

$\theta_{AB} = -88°18'14''$

$\theta_{AB} = 180 + (-88°18'14'')$

Determine quadrant QII (Figure 5-28) $\boldsymbol{\theta_{AB} =}$ **91°41′46″**

Calculate distance $D_{AB} = \sqrt{360.57^2 + (-10.677)^2}$

D_{AB} = 360.728

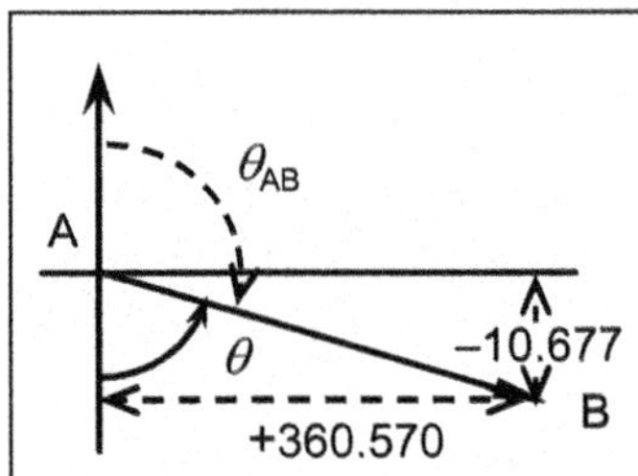

Figure 5-28 Determine quadrant

CHECK.

$E_B = E_A + D_{AB}\times\sin(\theta_{AB})$	$N_B = N_A + D_{AB}\cdot\cos(\theta_{AB})$
$= 109503.231 + 360.728\times\sin(91°41'46'')$	$= 6081791.502 + 360.728\times\cos(91°41'46'')$
$= 109503.231 + 360.570$	$= 6081791.502 + (-10.677)$
$E_B = 109\ 863.801$ ✓	$N_B = 6\ 081\ 780.825$ ✓

5.21.3 Worked Example 5.4 Polar Trigonometric Conversion using the [Pol(] Function Key

Using the [Pol(] function key: – converting rectangular(dN, dE) to polar (**r**,**θ**) coordinates

Note: we have to input dN, then dE,

`Pol(6081780.825-6081791.502` [,] `109863.801-109503.231)` [=] `360.728`

distance, **r**, is on the screen and stored automatically in memory, [$\cos^E$]

then, to retrieve bearing, **θ**,

RCL [$\tan^F$] F= 91.6961, bearing, **θ**, is positive,

this means **θ** is in the eastern hemisphere, 0° - 180°;

91.6961 [°'"] 91°41°46.02

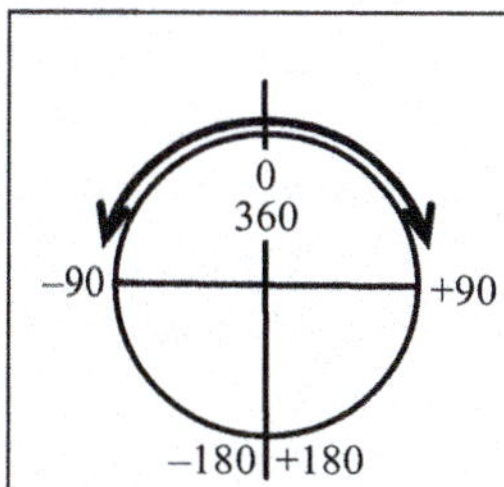

Figure 5-29 ATAN2() hemispheres

The [Pol(] function uses the ATN2(dN,dE) function, giving a bearing between:

0 and +180 clockwise for quadrants 1 and 2 and

0 and −180 anti-clockwise for quadrants 4 and 3. (See Figure 5-29)

Thus a negative result requires the addition of 360°.

r and **θ** will be retained in [$\cos^E$] and [$\tan^F$] until overwritten.

5.21.4 Coordinate conversion in different calculators.

Apart from the HP 10S+ features in this chapter, there are two other popular calculators, the Casio fx-82AU and the Sharp D.A.L. EL-531 used in the course. They have similar functions to the HP 10S+, but their keyboards and memory locations are slightly different. The comma separator between the two-input field is hard to find, especially when non-contrasting fore/background colours are used (Casio).

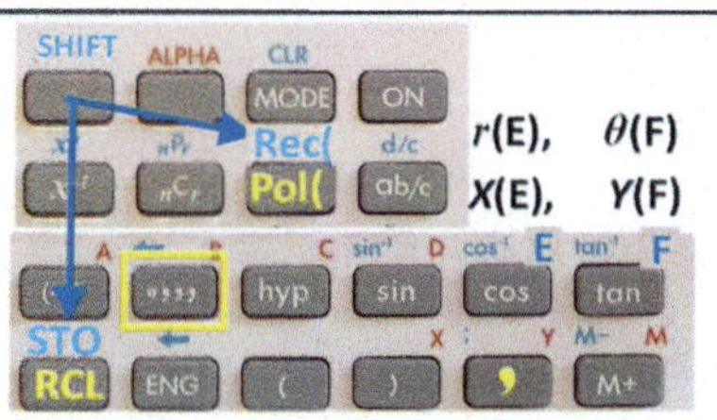

Figure 5-30 HP 10S+ memory locations

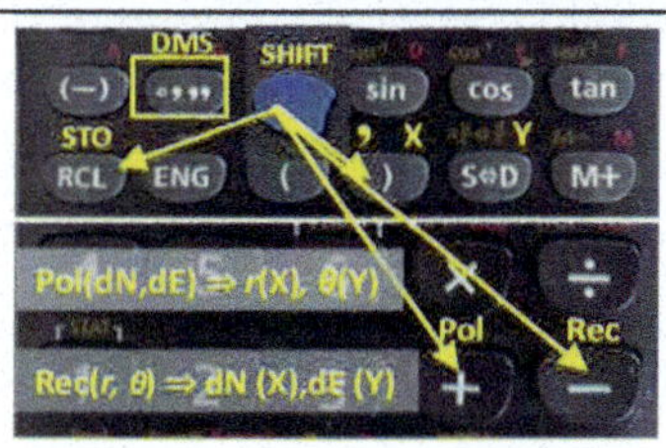

Figure 5-31 Casio fx-82AU memory locations

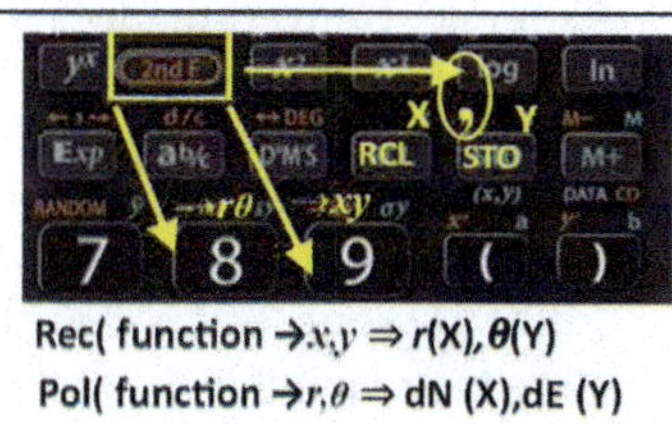

Figure 5-32 Sharp EL-531 memory locations

Coordinate conversion between rectangular and polar coordinates is a basic skill in surveying that is your constant companion in the field. You MUST be comfortable with your calculator.

Following conversion, the values are STORED in the appropriate registers. You have to use the RCL F on the HP10S+ to see the second result. The Casio displays the X, Y or r, θ registers on the screen. It's a bit easier.

5.21.4.1 Result: θ_{AB} 91°41′46″ , D_{AB} 360.728 ✓ with Worked Example 5.4.

5.22 The Polar Calculation (POLAR to RECTANGULAR).

5.22.1 Polar Calculation.

Given: A (E_A, N_A) together with bearing/distance A to B. (Figure 5-33).

Find: The coordinates of B (E_B, N_B)

Now,

$$\Delta E_{AB} = D_{AB}\times\sin(\theta_{AB})$$
$$\Delta N_{AB} = D_{AB}\times\cos(\theta_{AB})$$
$$E_B = E_A + \Delta E_{AB}$$
$$N_B = N_A + \Delta N_{AB}$$

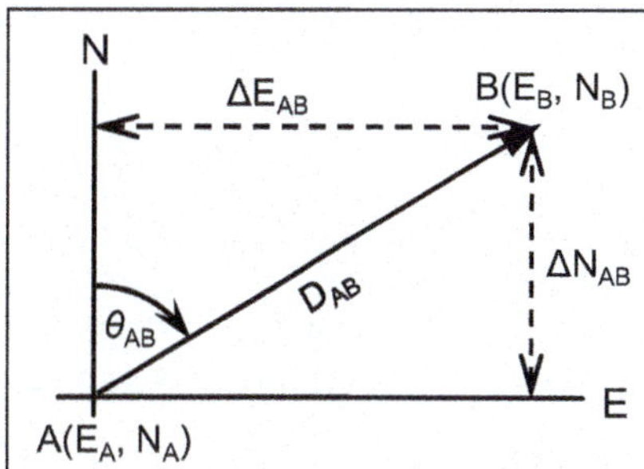

Figure 5-33 Polar calculation

If it's not checked it's wrong.

Format:

	Brg/Dist	E	N
A	θ_{AB}	E_A	N_A
	D_{AB}	$\pm\Delta E_{AB}$	$\pm\Delta N_{AB}$
B	SUM	E_B	N_B

The calculation of the coordinate differences using the sine or cosine functions is free of ambiguities and is calculated correctly for each quadrant.

5.22.1.1 Worked Example 5.6 (see Figure 5-34).

Given: $E_A = 548979.504$, $N_A = 6444421.068$

$\theta_{AB} = 221°03'00''$

$D_{AB} = 522.211$

Find: Co-ordinates of B (E_B, N_B)

	Brg/Dist	E	N
A	221°03′00″	548979.504	6444421.068
	522.211	ΔE −342.945	ΔN −393.818
B		**548636.559**	**6444027.250**

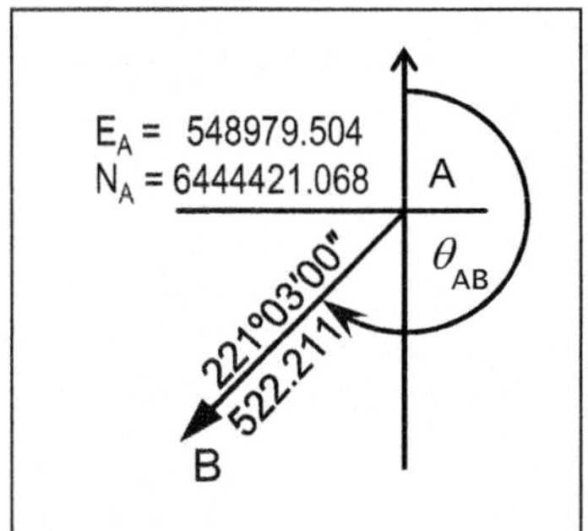

Figure 5-34 Worked Example 5.6

5.22.2 Worked Example 5.7. Rectangular Trigonometric Conversion using the Rec(Function Key

To convert from the polar BRG/DIST coordinates to dE, dN you should use the Rec(function; (SHIFT Pol( ). However, there are a couple of gotchas with these two functions:

1. the input is in the sequence **r,θ**, i.e. distance, bearing. Note the comma separator.
2. the answers, dE and dN, are held in two memory locations, E and F.

   cos^E holds the dN and tan^F holds dE. The display show the dN value (memory E)

   dE is accessed by recalling memory F (RCL tan^F)

   dN is accessed by recalling memory E (RCL cos^E)

Example: Convert the following BRG/DIST to dE, dN:

BRG 125°15′ 20″, distance 102.536m.

5.22.2.1 Conventional method:

522.211xsin 221°03°00° = -342.945044, dE, then re-keying for dN

522.211xcos 221°03°00° = -393.818519, dN.

Note the requirement to repeatedly key the same data.

5.22.2.2 Using Rec(:

Input the distance, the comma, then the BRG in either DMS, using the °’” key for each value of D, M and S; or use decimal degrees.

There is no need to close the parenthesis) key before enter, = :

Rec(522.211,221°03°00° = -393.818519, dN, (stored in cos^E) then, to retrieve dE

RCL tan^F F= -342.945044, dE.

dN and dE will be retained in cos^E and tan^F until overwritten.

Check: 548979.504+ RCL tan^F = 548636.559

6444421.068+ RCL cos^E = 6444027.249

Result: E_B, 548636.559 N_B 6444027.249 ✓ with Worked Example 5.6.

5.23 Chapter Exercise 5.5

Q. 1

Determine the bearing and distance between the following pairs of co-ordinates:

	E	N		E	N
(a) From	682 403.256	5 420 225.061	To	678 811.406	5 422 851.459
(b) From	353 076-721	6 852 373.486	To	355 668.143	6 851 660.119
(c) From	318 889.711	5 186 963.009	To	326 825.597	5 185 472.189
(d) From	370 531.855	5 367 547.937	To	368 153.029	5 369 316.498

Q. 2
Given the co-ordinates of A (326 886.021, 6953 910.858) together with the bearing and distance of AB in each case, determine the co-ordinates of B.

	θ_{AB}	D_{AB}
(a)	4°16′20″	3416.840
(b)	115°35′35″	1305.501
(c)	246°51′40″	4632.328
(d)	319°09′10″	116.333

Q. 3
Given the co-ordinates of A (1000.000, 1000.000), bearing AB (319°53'55" ±10") and distance AB (4301.665 ±0.115), find the co-ordinates of B and the standard deviation of those co-ordinates.
Review § 11.12.1, Worked Example 11.6, page 178.

5.23.1 Answers to Chapter Exercise 5.5

Q 1.	θ_{AB}	D_{AB}
(a)	306°10′29″ for 4449.646	
(b)	105°23′28″ for 2687.817	
(c)	100° 38′22″ for 8074.703	
(d)	306° 37′46″ for 2964.223	

Q 2.	E	N
(a)	327 140.559	6 957 318.204
(b)	328 063.433	6 953 346.912
(c)	322 626.341	6 952 090.532
(d)	326 809.934	6 953 998.859

Q 3.	E	N
A	−1770.884	4290.368

$\sigma_{\Delta E} = \pm 0.176$
$\sigma_{\Delta N} = \pm 0.160$

5.24 References for Chapter 5

Elfick, M.H, J.G Fryer, R.C. Brinker and P.R Wolf. Elementary Surveying, Eighth edition. Longman, 1994. ISBN 0-00-499001-3

Bannister, A and S. Raymond. Surveying. Third edition, Pitman Paperbacks, 1972. ISBN 0-273-36148-1

Chapter 6 AREAS OF REGULAR AND RECTILINEAR FIGURES

Before an engineering tender for construction of e.g., a road is given, it is a must that the engineering company works out not only the surface area of the work but also the volume of materials for the earthworks (i.e., to be removed or borrowed). Mine engineers will want to know the volumes of their stockpiles whereas the surveyors will want to know the areas of land parcels to which titles are to be granted. In all these examples, sufficient skills in areas and volume computations is a must. This chapter divides the working area into two; regular (i.e., those figures of regular shapes such as triangles, squares, rectangles etc) and irregular (i.e., those areas that do not take on the commonly known shapes) and shows you various means of computing the areas.

6.1 Circle

$$Circumference = 2\pi \times R \quad or \quad \pi \times D$$

$$Area = A = \pi \times R^2 \quad or \quad \pi\left(\frac{D^2}{4}\right)$$ See Figure 6-1

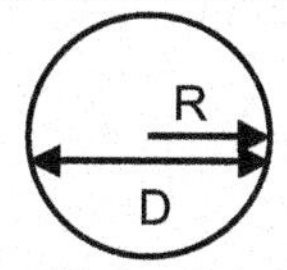

Figure 6-1

6.2 Square

$$Area = A = a \times a = a^2$$ See Figure 6-2

6.3 Rectangle

$$Area = A = a \times b$$ See Figure 6-3

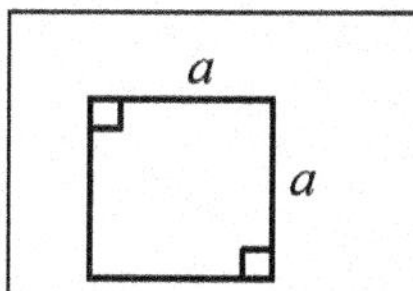

Figure 6-2 Square

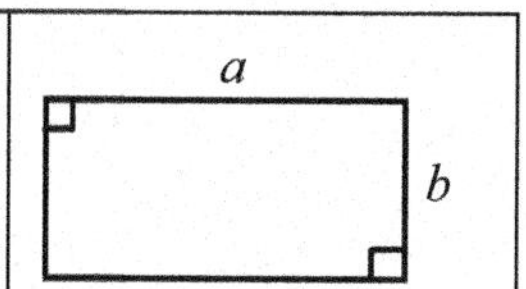

Figure 6-3 Rectangle

6.4 Regular polygon, equal length sides

See Figure 6-4

R = circum radius, passing through the vertex if the polygon

r_a = inradius, perpendicular to each side (the chord of each vertex pair).

$N = number\ sides$
$s = length\ side$
$\alpha = centre\ deflection\ angle$
$\alpha = 360/N$
$R = radius$ (circum radius)
$s = side\ length$
$s = 2R \times \sin(\alpha/2)$
$r_a = inradius$ (apothem)
$r_a = R \times \cos(\alpha/2)$
$S = perimeter = s \times N$

$$Area = A = \frac{1}{2}\left(R^2\, N \times \sin\left(\frac{360}{N}\right)\right)$$

$$Area = A = r_a^{\ 2}\, N \times \tan\left(\frac{180}{N}\right)$$

$$Area = A = \frac{1}{4}\left(\frac{s^2 N}{\tan\left(\frac{180}{N}\right)}\right)$$

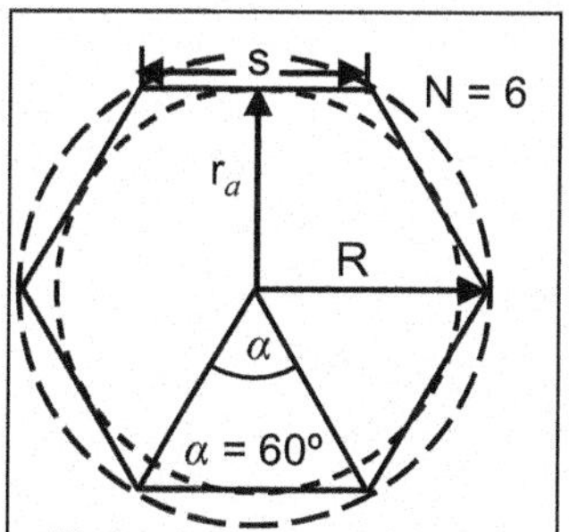

Figure 6-4 Regular polygon

Triangle (N = 3): $\alpha = 120°$ s = 1.732·R, S = 3s×2R·Sin(120/2) = 3×2×0.866 = 5.196
Hexagon (N = 6): $\alpha = 60°$ s = 1·R, S = 6s×(2R·Sin(60/2) =3×2×0.5 = 6
Octagon (N = 8): $\alpha = 45°$ s = 0.7654·R, S = 6.1229×R = 6.1229
for N = 1000: $\alpha = 0.36°$ s = 0.006283, S = 6.28317500 (2π = 6.28318531)
for N = 10,000: $\alpha = 0.036°$ s = 0.000628, S = 6.28318520 (2π = 6.28318531)
This technique finds 2π by successive reduction.
The perimeter of a regular polygon can never exceed 2πR

6.5 Irregular triangle

See Figure 6-5

$$Area = A = \tfrac{1}{2} \times h \times DC + \tfrac{1}{2} \times h \times BD$$

$$= \tfrac{1}{2} h \times (DC + BD) = \tfrac{1}{2} h \times (BC) = \tfrac{1}{2} h \times a$$

$But\ \ h = b \sin C \quad and\ \ thus:$

(1) $$A = \tfrac{1}{2} ab \sin C.$$

$$Now \frac{b}{\sin B} = \frac{a}{\sin A}\ thus\ b = \frac{a \sin B}{\sin A}\ \ and,\ substituting\ in\ (1),$$

$$A = \tfrac{1}{2} a \times \frac{a \sin B}{\sin A} \times \sin C$$

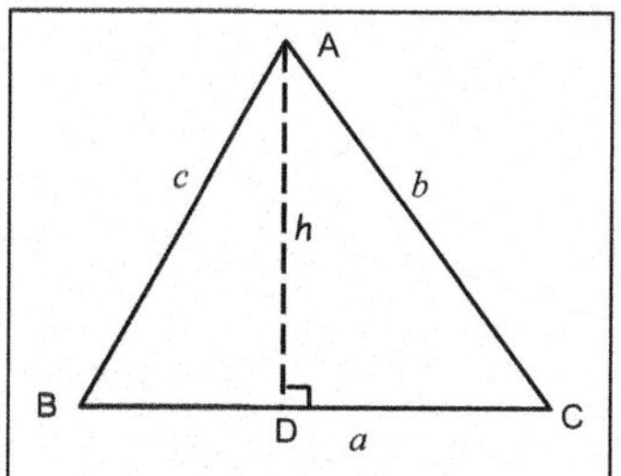

Figure 6-5 Triangle

J. Walker and J. Awange, *Surveying Computations*,
https://doi.org/10.1007/978-3-031-94382-9_6

If it's not checked it's wrong.

(2) $A = \frac{1}{2} a^2 \frac{\sin B \sin C}{\sin A}$

(3) $A = \sqrt{S(S-a)(S-b)(S-c)}$

where $S = \frac{1}{2}(a+b+c)$ [Semi-perimeter]

6.6 Parallelogram

See Figure 6-6

This is a figure in which the opposite sides are equal, and the opposite angles are equal.

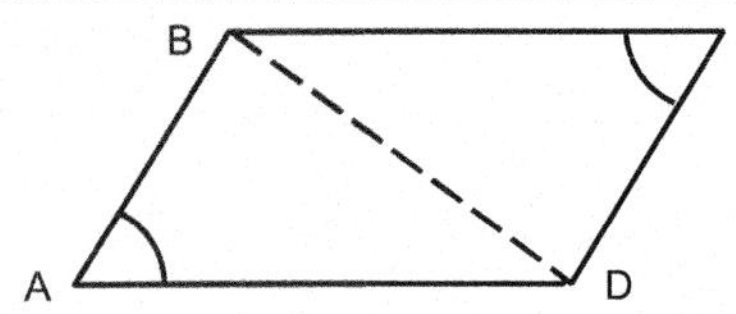

Figure 6-6 Parallelogram

$$Area = A = \frac{1}{2} \times AB \times AD \times \sin A + \frac{1}{2} CB \times CD \times \sin C$$

but $\angle A = \angle C$ and opposite sides are equal

$Area = A = AB \times AD \times \sin A.$

In general;

$Area = \text{Product}(adj.\ sides) \times \sin(incl.\ angle).$

6.7 Trapezium

See Figure 6-7

This is a figure where two sides are parallel and all angles and sides may be unequal, but:

A + B = 180°

C + D = 180°

$$Area = A = \left(\frac{BC + AD}{2}\right) \times h = \left(\frac{a+b}{2}\right) \times h$$

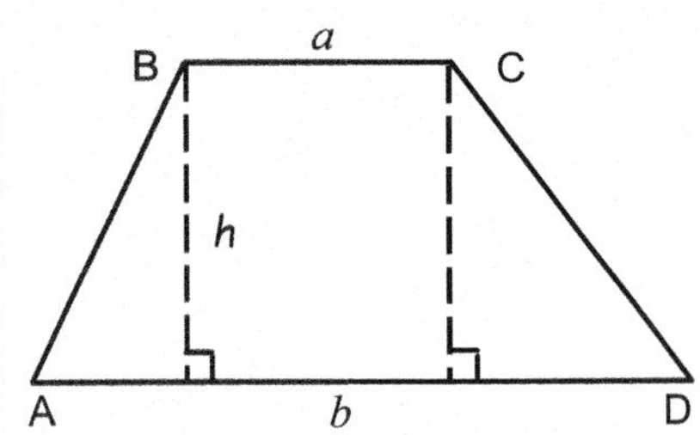

Figure 6-7 Trapezium

6.8 Arc of a Circle

All angles in circular measure are expressed in **radians**.

In plane measure, Figure 6-8 and Figure 6-9, the angle θ is in degrees. If the angle has a function (sin, etc) it is a plane angle in degrees, otherwise it is in radians.

6.8.1 Sector:

See Figure 6-8

Area between the arc and the radii of circle.

Area ABDCA = $\frac{1}{2} \theta^{Radians} \times R^2$

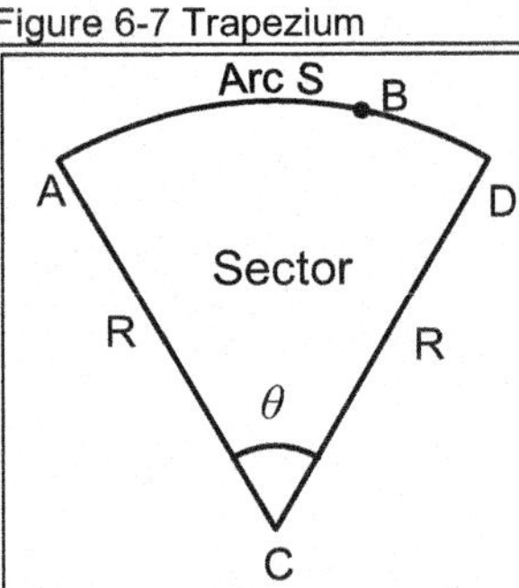

Figure 6-8 Sector

6.8.2 Segment:

See Figure 6-9

Area between chord and arc of a circle

i.e. To find area of ABDA

Given AD = c (the chord length)

Radius = R

Area ΔADC = $\frac{1}{2} R^2 \times \sin\theta^\circ$

Area ABDA = Area Sector ABDCA – Area ΔADC

Area ABDA = $\frac{1}{2}\theta^R \times R^2 - \frac{1}{2} \times R^2 \times \sin\theta^\circ$

$= \frac{R^2}{2}(\theta - \sin\theta)$. Note: $\sin\left(\frac{\theta}{2}\right) = \frac{c}{2R}$ and $\theta = 2 \times \arcsin\left(\frac{c}{2R}\right)$

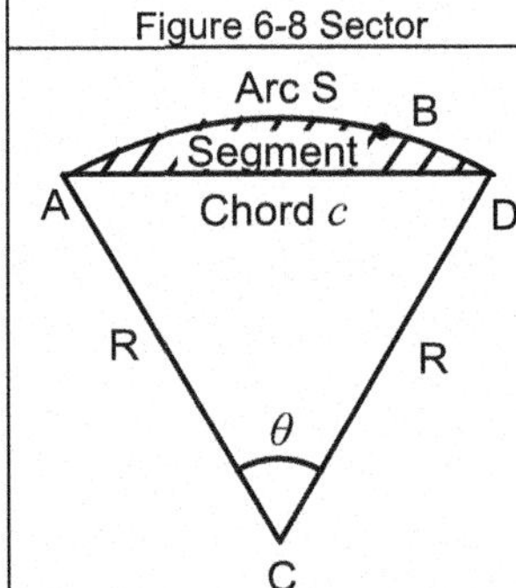

Figure 6-9 Segment

6.9 Areas Calculation Requirements.

In Western Australia surveyors are required to calculate areas enclosed by boundaries by the method of Double Longitudes. Areas can be calculated by using Double Longitude Coordinates or Double Longitude Coordinate Differences. Whichever method is used, the adjusted values are used in the calculation. i.e. the adjusted coordinates by Bowditch or the adjusted coordinate differences by Bowditch.

6.10 Area of Irregular Figures by Double Longitude Coordinates.

In Figure 6-10, 8 trapeziums are formed.

In ANY area calculations the figure MUST CLOSE back to the start point.

Aim: Given the coordinates of points 1, 2, 3 & 4
find the area bounded by 1, 2, 3, 4. 1 (close back to start).

From points 1, 2, 3, 4 drop perpendiculars to the respective North and East axes.

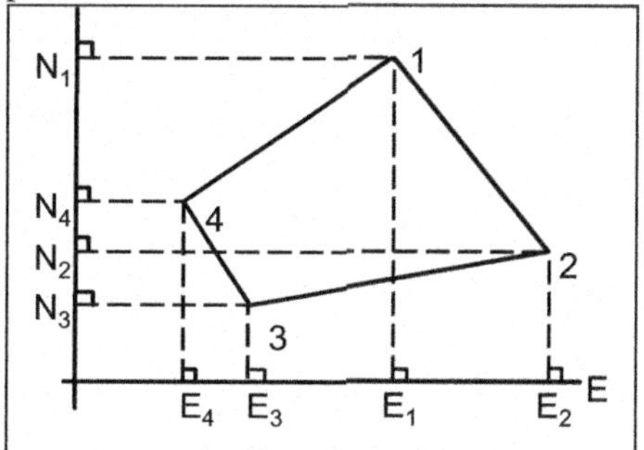

Figure 6-10 Coordinated trapezium

Area bounded by (1, 2, 3, 4, 1)

= A = Trapezium (1, E1, E4, 4) + Trapezium (1, 2, E2, E1)
– Trapezium (4, 3, E3, E4) – Trapezium (3, 2, E2, E3)

$$= \left(\frac{N_4+N_1}{2}\right)\times(E_1-E_4)+\left(\frac{N_1+N_2}{2}\right)\times(E_2-E_3) -\left(\frac{N_4+N_3}{2}\right)\times(E_3-E_4)-\left(\frac{N_3+N_2}{2}\right)\times(E_2-E_3)$$

multiply through by 2 to remove the half values, swap negatives, re-order.

i.e. $2A = (N_4 + N_1)(E_1 - E_4) + (N_1 + N_2)(E_2 - E_1) - (N_4 + N_3)(E_3 - E_4) - (N_3 + N_2)(E_2 - E_3)$

$2A = (N_4 + N_1)(E_1 - E_4) + (N_1 + N_2)(E_2 - E_1) + (N_4 + N_3)(E_4 - E_3) + (N_2 + N_3)(E_3 - E_2)$ (swap)

$2A = (N_4 + N_1)(E_1 - E_4) + (N_1 + N_2)(E_2 - E_1) + (N_2 + N_3)(E_3 - E_2) + (N_3 + N_4)(E_4 - E_3)$ (re-order)

And, in general,

$$2A=\sum_{R=1}^{N}(N_R+N_{R+1})(E_{R+1}-E_R) \text{ where } (N_R+N_{R+1})=\Sigma N \; and (E_{R+1}-E_R)=\Delta E$$

OR
$$2A=\sum_{R=1}^{N}(E_R+E_{R+1})(N_{R+1}-N_R) \text{ where } (E_R+E_{R+1})=\Sigma E \; and (N_{R+1}-N_R)=\Delta N$$

This is the method of area determination by Double Longitude Coordinates.

6.10.1 Worked Example 6.1. Double Longitude Coordinate Method.

Given the coordinates of the figure ABC find the area enclosed by ABCA (Figure 6-11) using the Double Longitude co-ordinate method.

$$2A=\sum_{R=1}^{N}(E_R+E_{R+1})(N_{R+1}-N_R)=\Sigma E^S\times\Delta N^S$$

In order to determine the area it is best to set out the workings in a neat systematic manner. The coordinates are written down, then the differences and sums are calculated on the next line.

	E Diff E^S	N Diff N^S	Sum E^S	Sum N^S
A	3000.00	1500.00		
	–1000.00	500.00	5000.00	3500.00
B	2000.00	2000.00		
	707.06	593.29	4707.06	4593.29
C	2707.06	2593.29		
	292.94	-1093.29	5707.06	4093.29
A (Start)	3000.00	1500.00		
	$\sum\Delta E = 0.0$	$\sum\Delta N = 0.0$		

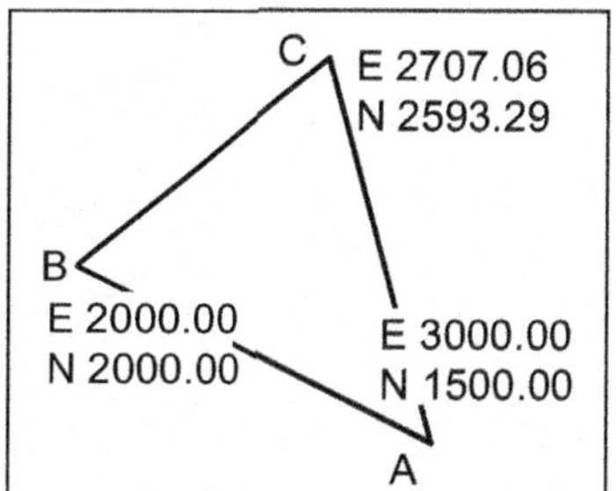

Figure 6-11 Coordinates

The coordinates are first set out as shown above. The coordinate differences are now obtained (note: $\Sigma\Delta E$ = 0.0, $\Sigma\Delta N$ = 0.0). Next adjacent Eastings are summed and this figure placed under the sum Easting column. The same procedure is adopted under the sum Northing column. The area can now be obtained as:

$2A = (500.00 \times 5000.00) + (593.29 \times 4707.06) + (-1093.29 \times 5707.06)$

$A = -473\,410\text{m}^2$ (Note the negative area.)

The calculation of the area must be checked using the alternate $\sum N^S$ and ΔE^S.

CHECK $$2A = \sum_{R=1}^{N} (N_R + N_{R+1})(E_{R+1} - E_R) = \Sigma N^S \times \Delta E^S$$

$2A = (-1000.00 \times 3500.00) + (707.06 \times 4593.29) + (292.94 \times 4093.29)$

$A = 473\,410\text{m}^2$ (Note the positive area in this check.)

6.11 Area of Irregular Figures by Double Longitude Coordinate Differences.

Quite frequently, when obtaining the area of an irregular figure, only the coordinate differences are available. (i.e. ΔE, ΔN only are available).
In this case it is far more convenient to determine the area from these ΔES, ΔNS.

6.11.1 Worked Example 6.2. Double Longitude Coordinate Difference Method.

Consider Figure 6-12.
You may be given a set of coordinate differences between points in a traverse from which you are required to find the area. A method can be devised where it is unnecessary to know the coordinates of the points to find the area.

Let us shift the origin of the traverse to A(0.00, 0.00), so that the coordinates differences between A, B and C are as shown.
The area of the traverse is unchanged.

The area can be determined as before with exactly the same format used. However, we use the coordinate differences to generate the coordinate sums. It involves manipulating and adding coordinate differences.

The value of the origin is not needed, and it is best to set out the workings in a neat systematic manner showing just the coordinate differences.

	ΔE Diff E^S	ΔN Diff N^S	Sum E^S	Sum N^S
A				
Line AB	-1000.00	500.00		
B				
Line BC	707.06	593.29		
C				
Line CA	292.94	-1093.29		
A				

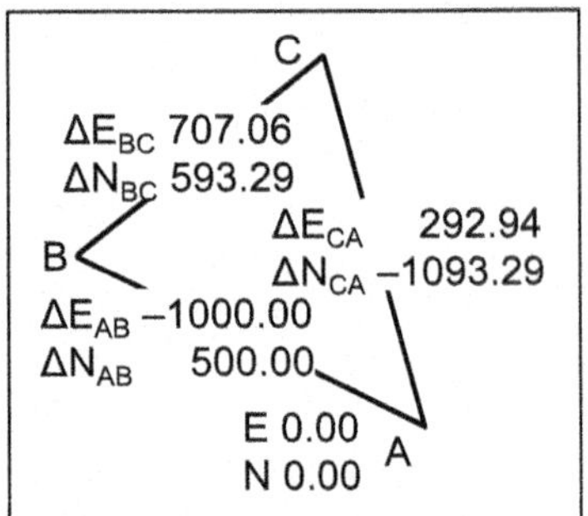

Figure 6-12 Coordinate differences

Note that we are working purely with ΔE and ΔN, which are the same as the Diff E^S and Diff N^S calculated from coordinates in Worked Example 6.1 (§ 6.10.1).

6.11.1.1 Calculating the SUMS of the E^S and the N^S.

$$2A = \sum_{R=1}^{N} (E_R + E_{R+1})(N_{R+1} - N_R) = \Sigma E^S \times \Delta N^S$$

		ΔE Diff E^S		ΔN Diff N^S		Sum E^S		Sum N^S
A								
AB	[1]	−1000.00	[2]	500.00	[9]	−1000.00	[10]	500.00
B								
BC	[3]	707.06	[4]	593.29	[11]	*−1292.94*	[12]	*1593.29*
C								
CA	[5]	292.94	[6]	−1093.29	[13]	*−292.94*	[14]	*1093.93*
A								

[7] ΣΔE = 0.0 [8] ΣΔN = 0.0

Question; given the co-ordinate differences [1], [2], [3], [4], [5] and [6] how do we go about getting [9] → [14] ?

Now [9] = [1]
[11] = [9] + [2] + [1]
[13] = [11] + [5] + [3]

Similarly:

[10] = [2]

[12] = [10] + [4] + [2]

[14] = [12] + [6] + [4]

The check that the sum ES and Sum NS figures are correct is:

[13] = – [5] and [14] = – [6]

The area is now obtained in exactly the same manner as for double longitude coordinates:

$$2A = \sum_{R=1}^{N} (E_R + E_{R+1})(N_{R+1} - N_R) \quad = \Sigma E^S \times \Delta N^S$$

i.e. 2A = (500.00 × –1000.00) + (593.29 × –1292.94) + (-1093-29 × –292.94)

A = –473 410m^2

$$\text{CHECK: } 2A = \sum_{R=1}^{N} (N_R + N_{R+1})(E_{R+1} - E_R) \quad = \Sigma N^S \times \Delta E^S$$

2A = (–1000.00 × 500.00) + (707.06 × 1593.29) + (292.94 × 1093.93)

A = 473 410m^2 (Note the positive area in this check.)

6.11.2 Worked Example 6.3

Refer to Figure 6-13.

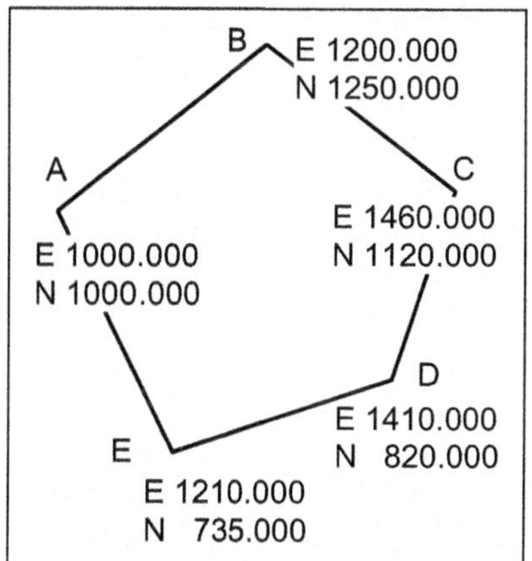

Figure 6-13 Traverse coordinates

6.11.2.1 Area by Double Longitude Coordinates.

	E Diff E^S	N Diff N^S	Sum E^S	Sum N^S
A	1000.000	1000.000		
	200.000	*250.000*	2200.000	2250.000
B	1200.000	1250.000		
	260.000	*−130.000*	2660.000	2370.000
C	1460.000	1120.000		
	−50.000	*−300.000*	2870.000	1940.000
D	1410.000	820.000		
	−200.000	*−85.000*	2620.000	1555.000
E	1210.000	735.000		
	−210.000	*265.000*	2210.000	1735.000
A	1000.000	1000.00		

Inner pair Outer pair

ΣΔE = 0.000 ΣΔN = 0.000 Check

CALCULATE the twice area by "Sums of the inner columns (solid outline)"

2A = (250.0 × 2200.0) + (–130.0 × 2660.0) + (–300.0 × 2870.0) + (–85.0 × 2620.0) + (265.0 × 2210.0)

= –293 850

A = –146 925.0m^2

CHECK the twice area by "Sums of the outer columns (dashed outline)"

2A = (200.0 × 2250.0) + (260.0 × 2370.0) + (–50.0 × 1940.0) + (–200.0 × 1555.0) + (–210.0 × 1735.0)

= 293 850

A = 146 925.0m^2

NOTE: Same area, different sign.

6.11.2.2 Area by Double Longitude Coordinates Differences.

Review the methodology of finding Sum E^S and Sum N^S and checking methods in § 6.11.1, Worked Example 6.2 Consider the case where this same area as Worked Example 6.3 is to be found but we are given the co-ordinate differences only.

ΔE	ΔN	ΣE^S	ΣN^S
200.000	250.000	200.000	250.000
260.000	–130.000	660.000	370.000
–50.000	–300.000	870.000	–60.000
–200.000	–85.000	620.000	–445.000
–210.000	+265.000	+210.000	–265.000

Opposite signs

$\sum \Delta E = 0.0 \quad \sum \Delta N = 0.0$

$2A = (250.0 \times 200.0) + (-130.0 \times 660.0) + (-300.0 \times 870.0) + (-85.00 \times 620.0) + (265.0 \times 210.0)$

$A = -146\,925.0\text{m}^2$

CHECK

$2A = (200.0 \times 250.0) + (260.0 \times 370.0) + (-50.0 \times -60.0) + (-200.0 \times -445.0) + (-210.0 \times -265.0)$

$A = 146\,925.0\text{m}^2$

6.12 General Remarks.

(a) The area figure MUST close back to the START point.

(b) The calculation of areas as performed by electronic calculators will be made easier if the workings are set out as in the given formats.

If the coordinates are very large (MGA94) look at shifting the origin close to the work area, so that you have fewer numbers to crunch. e.g. By striking out 345000 and 6458000 we can have:

	E	N				
A =	345200.025	6458463.212	~~345~~200.025	~~6458~~463.212	200.025	463.212
B =	345215.147	6458468.987	~~345~~215.147	~~6458~~468.987	215.147	468.987
C =	345226.854	6458470.200	~~345~~226.854	~~6458~~470.200	226.854	470.200
A =	345200.025	6458463.212	~~345~~200.025	~~6458~~463.212	200.025	463.212

the numbers are now much easier to work with.

(c) When taking out areas, keep a watch on the significant figures in the area calculations.

(d) A check area must always be calculated; it will be EXACTLY the same, with the opposite sign.

(e) The answer will always be a DOUBLE area, don't forget to halve it.

(f) Metric Areas:

Areas are computed in metric units (in m^2 or hectares (ha))

1 ha = 100 m x 100 m = 10 000 m^2

Areas written in hectares are given to 4 decimal digits (to m^2).

6.13 Area by Coordinates.

The area by double longitudes is the only method whereby the calculations are checked within the solution procedure. Any transcription or math errors will result in a misclose somewhere.

Referring to § 6.10, there is another way to calculate the area by cross multiplication of the coordinate pairs. The structure has been created, now put it into practice.

In Figure 6-14 the trapezoidal areas are added and subtracted to calculate the area of the bounded figure.

This is also known as Gauss's shoelace formula because the cross multiplication diagram represents shoelaces. (Figure 6-15).

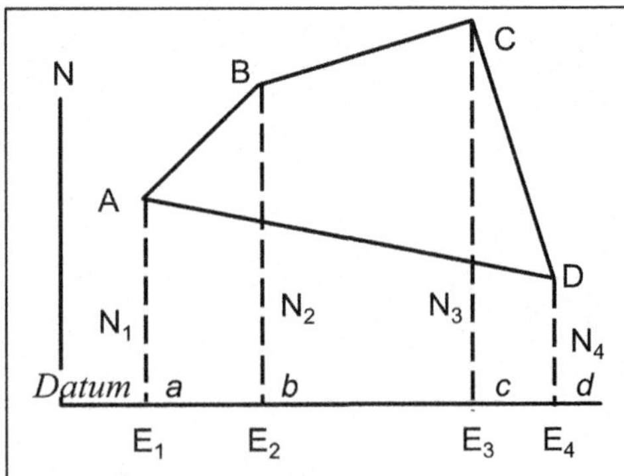

Figure 6-14 Area by coordinate cross multiplication

AREA = A = ABCDA

= ABbaA + BCcbB + CDdcC – ADdaA

$= \frac{1}{2}(N_1 + N_2)(E_2 - E_1) + \frac{1}{2}(N_2 + N_3)(E_3 - E_2) + \frac{1}{2}(N_3 + N_4)(E_4 - E_3) - \frac{1}{2}(N_1 + N_4)(E_4 - E_1)$

Multiplying out

$= \frac{1}{2}(N_1E_2 - N_1E_1 + N_2E_2 - N_2E_1 + N_2E_3 - N_2E_2 + N_3E_3 - N_3E_2 + N_3E_4$
$- N_3E_3 + N_4E_4 - N_4E_3 - N_1E_4 + N_4E_1 + N_1E_1 - N_4E_4)$

Gathering Like (similar) Terms:

AREA $= \frac{1}{2}(N_1E_2 - N_2E_1 + N_2E_3 - N_3E_2 + N_3E_4 - N_4E_3 - N_1E_4 + N_4E_1)$

Writing in matrix cross multiplication form.

Cross multiply [n×2] matrix = $Area = \left|\frac{1}{2}\left(\sum_{i=1}^{n}\left(E_iN_{i+1} - N_iE_{i+1}\right)\right)\right|$

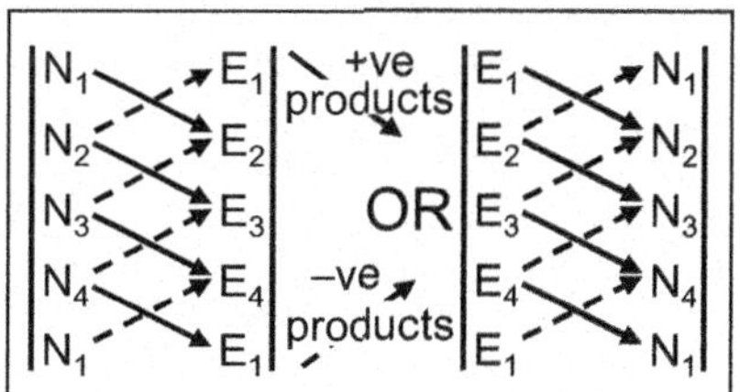

Figure 6-15 Shoelace formula

The difference in the order of N, E and E, N sequence calculation is the SIGN of the result. The answer also depends on whether the sequence is clockwise or anticlockwise around the figure.

In cross sectional roadworks area calculations, the result, where the cross section is in both cut and fill, will show the difference between the cut and the fill areas (the coordinate vectors cross). Knowing the sign of the result will determine cut or fill. As long as you work your coordinate order in the same direction (clockwise or anti-clockwise) you will arrive at cut or fill difference area.

Figure 6-16 represents a vertically exaggerated cross section coordinated about the formation centreline level at STN 4 (0,0) where the dashed line 1 – 2 represents the surface and 3, 4, 5 is the formation.

By matrix cross multiplication the area = 9m^2.

However, by finding the intercept between 1 – 2 and 4 – 5 (*Pxy*) the 1, *P*, 5, 1 cut area is – 9.2m^2 and the fill area (*P*, 2, 3, 4, *P*) is 18.2m^2. The sum (9m^2) agrees with the cross multiplication method.

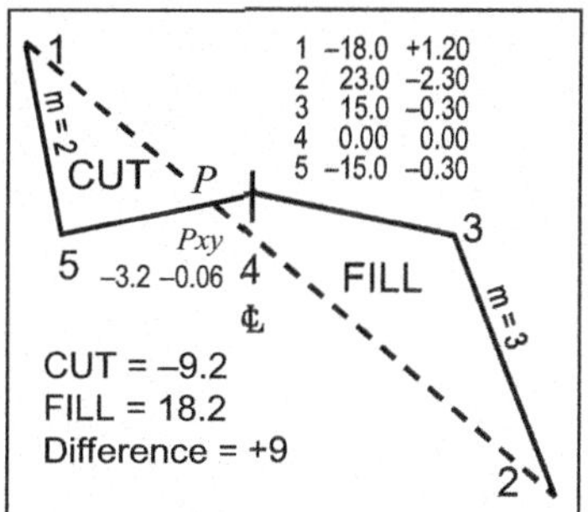

Figure 6-16 Cut and fill

6.13.1 Worked Example 6.4. Area by Coordinates.

Using coordinates in Worked Example 6.3 (6.11.2.1), use Figure 6-15 method.

	East	North	+ve	–ve
A	1000.000	1000.000		1200000
B	1200.000	1250.000	1250000	1825000
C	1460.000	1120.000	1344000	1579200
D	1410.000	820.000	1197200	9922000
E	1210.000	735.000	1036350	735000
A	1000.000	1000.000	1210000	
			Σ+ve **6037550**	Σ–ve **–6331400**

2A = 6037550 – 6331400 = –293850,

A = **146 925**m^2 Check with Worked Example 6.3

The method is particularly suited to RPN calculators where the +ve products are summed during +ve calculation and the –ve products are subtracted during the –ve calculation. Then the answer is divided by 2 for area.

The sign is relatively unimportant; we can't have a negative area. However, there are ways to use the sign to determine cut or fill in cross sectional areas calculations. ALWAYS go around the figure in one direction. Figure 6-16 in § 6.13 Illustrates the point.

The use of UTM (MGA94, MGA2020) coordinates REQUIRES that you shift the origin (§ 6.11 above) to stop your calculator overflowing. Excel can only calculate to the millimetre before it runs out of internal precision (Blame IEEE 754 and 64bit memory).

6.14 Cross section area for earthworks

This chapter has focussed on the traditional surveying area calculation methods. However, there is a need to calculate the area of cross section profiles, the "end area", as a precursor to volume calculations, especially in earthworks. The earthworks of interest will generally be longitudinal structures: roads, drainage lines, pipeline excavations etc.

6.14.1 Longitudinal and cross sections (Irvine and Maclennan 2006, Chapter 6):

- Longitudinal section – Vertical section along the centre line of the complete length of the work.
- Cross sections – Vertical sections drawn at right angles to the centre line of the works.

Information given by the sections provides data for:

(a) Determining suitable gradients for construction work.
(b) Calculating volumes of earthworks.
(c) Supplying details of depth of cutting or heights of filling required.

Example: Levelling is undertaken at the centre line and cross sections. See Table 6-1.

Table 6-1 Reduced levels for a road design.

BATTER INTERCEPTS FROM CROSS SECTION LEVELLING											
RL	RMK CH/offset	Above DATUM RL 12.000	m = 4 b = 5	$h_L = (W_L - b/2)/m$	W_L	Slope k @ 5m or offset interval	CL final grade level	h = CUT/FILL	Slope k @ 5m or offset interval	W_R	$h_R = (W_R - b/2)/m$
13.191	00 15L	1.191	15L			15.9		1.19			
12.876	00 10L	0.876	10L			8.4		0.88			
12.281	00 5L	0.281	5L	0.18	3.2	17.8	0.00	0.28			
12.00	CH00 CL	0.000	0					0.00			
12.00	00 5R	0.000	5R					0.00	level	2.5	0
12.80	CH15 CL	0.800	CH15				0.36	0.44			
14.585	30 15L	2.585	15L			11.1		1.86			
14.132	30 10L	2.132	10L			13.0		1.41			
13.747	30 5L	1.747	5L	1.20	7.3	14.4		1.02			
13.40	CH30 CL	1.400	0				0.72	0.68			
12.96	30 5R	0.960	5R					0.24	-11.4	3.9	0.35
12.725	30 7R	0.725	7R					0.00	-8.5		
13.90	CH45 CL	1.900	CH45				1.09	0.81			
14.614	60 6L	2.614	6L			33.3		1.17			
14.584	60 5L	2.584	5L	1.20	7.3	17.6		1.14			
14.30	CH60 CL	2.300	0				1.45	0.85			
14.129	60 5R	2.129	5R					0.68	-29.2	5.2	0.68
13.884	60 10R	1.884	10R					0.50	-20.4		
13.711	60 15R	1.711	15R					0.32	-28.9		
14.50	CH75 CL	2.500	CH75				1.81	0.69			
15.177	90 9.5L	3.117	9.5L			14.1		1.05			
14.858	90 5L	2.585	5L	0.73	5.4	9.0		0.73			
14.30	CH90 CL	2.300	0				2.17	0.13			
14.153	90 1.8R	2.153	1.8R					-0.02	-12.2	2.5	0
14.25	CH93.3	2.250	CH93.3				2.25	0.00			
Grade slope = 1:41.5	Grade slope = run/rise = 93.3/2.25 = 41.5										

6.14.2 Formation and Inverted Levels (Irvine and Maclennan 2006, Chapter 6):

- Formation level of any construction works is the level to which the earth is excavated or deposited to accommodate works (e.g., the straight line rising from CH00 with RL of 12m to CH93.3 with RL of 14.25m in Figure 6-17). Formation width is the width of the design feature, e.g., a road.
- Inverted level is the inside of the bottom of a pipe but for practical purposes maybe considered to be the bottom of excavation.

The plotted longitudinal profile of reduced data in Table 6-1 is shown in Figure 6-17.

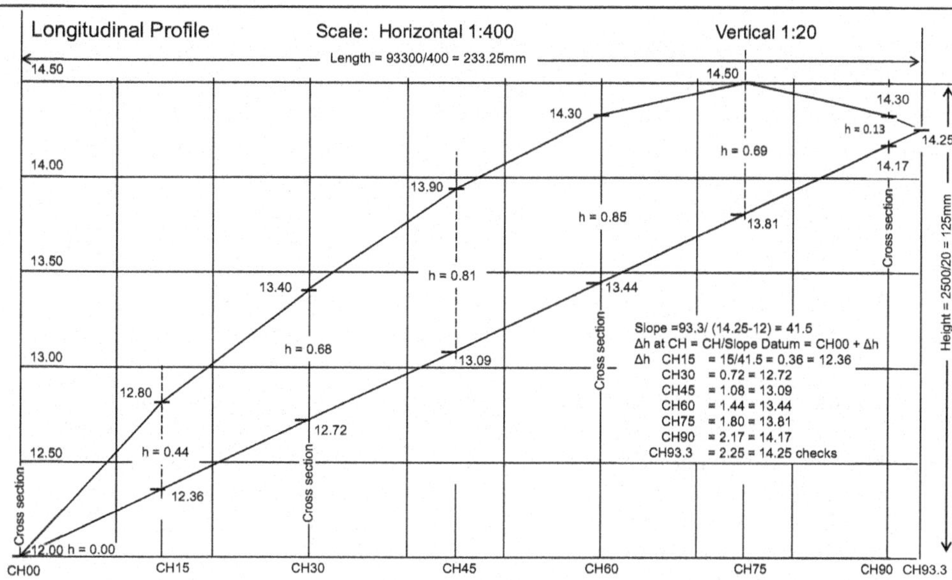

Figure 6-17 Longitudinal profile (not to scale) from the centreline measurements in Table 6-1

6.14.3 Embankment and Cutting

- Embankment – The formation level of the proposed works is higher than the actual ground level such that the side slope (batters) slope downwards from the formation level to the ground (see e.g., Figure 6-18).
- Cutting is the opposite, i.e., the formation level is lower than the actual ground (see e.g., Figure 6-19).

6.14.4 Design gradient, batter gradient, and surface gradient

Design gradient is the slope of the design feature, e.g., a road (rising or falling). Batter gradients are the side slopes in Table 6-1. Reduced levels for a road design. above, while surface slope (*k*) is the slope of the actual ground surface (see §6.14.7).

6.14.5 Chainage

- This is the linear measurement from the start of the structure, e.g., road along the centreline at given intervals. They are normally specified at the design stage and useful in specifying points at which reduced levels should be taken for the generation of longitudinal and cross-sectional profiles.

6.14.6 Digital Elevation and Digital Terrain Models

- DEM (Digital Elevation Model) is the representation of terrain surface numerically by very dense network of points of known X, Y, Z coordinates. It includes natural terrain features, vegetation, roads, buildings etc. Triangulation Irregular Network (TIN) are favoured for DEM modelling. Delaunay triangulations are used to create the best conditioned triangular network. DEM input required to generate surfaces for engineering design can be ordered (grids), semi-ordered (profiles or contours) or random pointing.
- DTM (Digital Terrain Model) is a filtered version of DEM and contains the "bare Earth".

Uses of DEM and DTM

- DEM can be used for:
  - ✓ Landscaping
  - ✓ City modelling
  - ✓ Visualization applications.
- DTM can be used for
  - ✓ Route design (from contours and cross sections)
  - ✓ Flood modelling (from contour lines)
  - ✓ Drainage modelling (from contours and cross sections)
  - ✓ Land use studies
  - ✓ Slope models or volume calculations

6.14.7 Application: Route Design

Design of new roads and drainage routes will generally fall into either (i) roads following existing routes, e.g., those involved with either upgrading of an existing road or a short deviation or (ii) roads following new routes, e.g., those roads linking two points for the first time. For road design along existing routes, a cross section survey of the existing road is generally carried out as a first step in the design. This survey includes, e.g., road cross sections every 25m along the formation where the measured locations will include:

- Centre line, edge of bitumen, edge of formation, bottom of drain, top of drain, any other changes of grade, fence, etc.
- Road furniture (to all fixtures in the road and road reserve).
- Improvements such as buildings and bridges.

In the example of Australia, a road design for an existing route will generally be undertaken in 3 stages as follows:

1. Stage 1 (Horizontal Control) - Control points are placed approximately every 250 m in the existing bitumen road and connected to Australian Map Grid (AMG) via adjacent Standard Surveying Marks (SSM's). The requirements are generally traversing with an accuracy of 1:10,000 to 1:20,000 or better.
2. Stage 2 (Vertical Control) - Control points are connected to the Australian Height Datum (AHD) via adjacent Benchmarks. The requirements are generally a level traverse with an accuracy of 12mm$\sqrt{k}$, where k is the traverse distance in kilometres.
3. Stage 3 (Detail Survey) - A cross sectional survey is carried out every 20 - 50m along the alignment of the road. From the control points established in stages 1 and 2 above, an EDM radiation survey of not more than 125 m is carried out. Distance and angular measurements are undertaken together with the heights of the instrument and target. From these measurements, a topographical map showing features and contours/cross sections can be generated. These types of surveys could be carried out using GNSS surveying. For Australia, the main roads department (MRD; www.mainroads.wa.gov.au) has formulated a set of procedures covering a cross section survey of this nature.

6.15 Calculation of Road Cross Sections in Cut and Fill Operations

The calculation of road cross section area was presented as a series of materials developed for the Curtin University Civil Engineering Surveying workshops to try to distil and simplify material presented in standard textbooks. A simplified profile of an embankment, in either cut or fill, is used as an illustration.

The basic components of embankment design are:

1. Pavement, design width. This is given the label ***b***. It is symmetrical about the centre line.
2. Design finish level of the formation centre line, the datum.
3. Surface level at formation centre line.
4. The difference between centreline design finish level and surface level, either the depth of cut or fill, ***h***.
5. The formation batter slope, the constructed gradient of the cut or fill operation, labelled ***m***. The slope is expressed as a rise and run ratio of 1:*m*., e.g. 1:3 means that for every vertical change in height there will be a 3-fold increase in horizontal deviation. Normally the batter slope, *m*, is designed to mimic the natural repose of the formation construction material. In road construction using natural materials the slope may be different for cut operations as opposed to fill.
6. The cross fall of the natural surface from the design centre line is designated ***k***. It goes outwards beyond the batter slope intercepts. For calculations, *k* has a sign such that $+k$ represents a rise from the centreline outwards while $-k$ represents a fall from the centreline outwards. If there is no cross fall, then the slope is flat, and k = 0:1 or $k \to \infty$ (k = 10,000 will suffice in general calculations).
7. Formation width, the distance from the centreline at which the batter slope intercepts the surface is designated ***w*** and expressed as a distance left, W_L, and right, W_R, of the centreline. The formation width depends on *h*, *b*/2, *m* and *k*.

The calculations are expressed as distances either side, left or right of the centre line. The designation of left and right is taken in the sense of observing in the direction of the centreline progress, i.e., of increasing chainage. Figure 6-18 shows the definitions given above for a section in fill. Note that the cross-section distances are measured horizontally away from the centreline. The vertical distances are labelled, in this case, from some arbitrary vertical datum starting at 10.0. The finish level is at 10.3 and the centre line surface level is about 10.08. Thus h = 0.22.

Cross section diagrams are often shown with a vertical exaggeration which helps to illustrate the formation section. In Figure 6-18 a vertical exaggeration of 10:1 is evident.

Vertical exaggeration at a nominated scale also allows one to plot the road section, showing natural surface and the pavement width at design level. By plotting the formation batters (at the exaggerated scale) it is possible to scale off the formation width intercept with the natural surface.

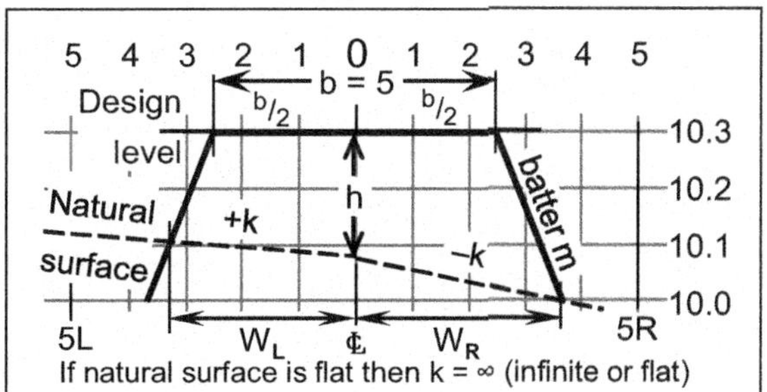

Figure 6-18 Definitions of a sectional fill.

6.15.1 Calculation of Cut Formation Width W_L, and W_R, on flat ground

The development of the calculation of formation width will start with a cut embankment referenced to a flat natural surface (see Figure 6-19).

6.15.1.1 Worked Example 6.5.

The design components are:

- pavement width, b = 5
- batter slope, m = 3,
- depth (cut/ fill), h = 0.25,
- cross slope is flat,
- k = ∞ = 10,000. It is ignored here.

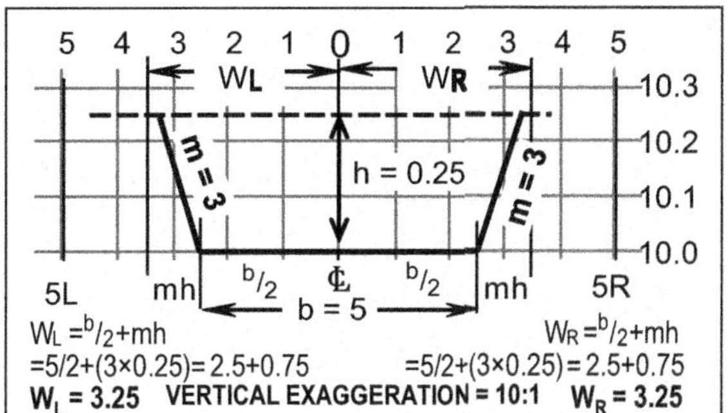

Figure 6-19 Flat formation surface

Flat formation surface:	$W_L / W_R = \frac{b}{2} + mh$	Eqn 6-1

Using Eqn 6-1, $W_L = 5/2 + 3 \times .25 = 3.25$ and $W_R = 3.25$

6.15.2 Calculation of Formation Width, W_L, and W_R, Incorporating Cross Fall on Sloping Ground

Figure 6-20 introduces a constant surface slope, *k*, that accounts for determining formation width W_L and W_R. As seen from the exaggerated diagram, compared with a level surface, the batter width increases as we "chase" an upward *k*+ (positive) slope to intercept the ground. Likewise, the batter width will decrease against falling *k*– (negative) slope.

6.15.2.1 Formation width with constant cross fall.

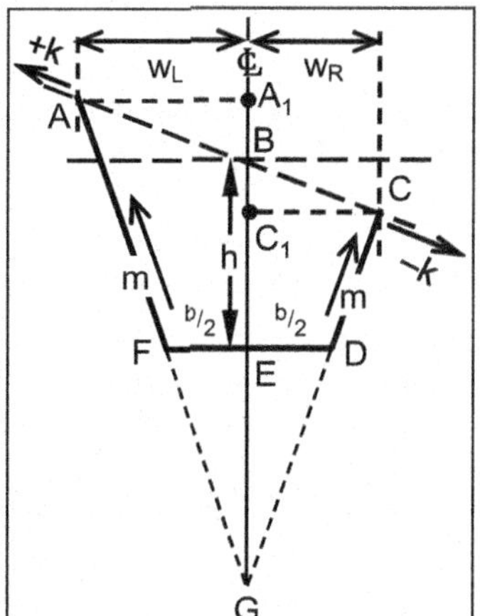

Figure 6-20 Formation width for a constant sloping surface

In the diagram, Figure 6-20:	$A_1B = \frac{W_L}{k}$	Eqn 6-2

The difference in height between the ground at the centreline and the formation batter intercept is due to the slope of 1:*k* over the distance $\mathbf{w_1}$.

Similarly:	$C_1B = \frac{W_R}{k}$	Eqn 6-3

The extended batter, of slope ***m***, over a horizontal base ${}^b/_2$ intercepts the centreline at **G** with a height: $GE = \frac{b}{2m}$

Triangles AA_1G and FEG are similar. Using similar triangles:

$\frac{AA_1}{FE} = \frac{GA_1}{GE}$ and substituting:	$\frac{W_L}{\frac{b}{2}} = \left\{ \frac{\frac{b}{2m} + h + \frac{W_L}{k}}{\frac{b}{2m}} \right\} \Leftrightarrow w_1 = m\left[\frac{b}{2m} + h + \frac{W_L}{k}\right]$ $W_L\left(1 - \frac{m}{k}\right) = \frac{b}{2} + mh \Rightarrow W_L = \left(\frac{b}{2} + mh\right)\left(\frac{k}{k-m}\right)$	Eqn 6-4

As the denominator is less than the numerator, then $\left(\dfrac{k}{k-m}\right) > \mathbf{1}$. The batter "chases" the slope.

Note that the slope of **k** has been taken as positive. If the slope is falling from the centre line (and is kept positive) then, by inspection, $w < \left(\dfrac{b}{2} + mh\right)$.

Thus, $\mathbf{w_2}$ must be expressed in terms of triangles CC_1G and DEG.

$\dfrac{CC_1}{DE} = \dfrac{GC_1}{GE}$ and substituting:	$\dfrac{W_R}{\frac{b}{2}} = \left\{\dfrac{\frac{b}{2m} + h - \frac{W_R}{k}}{\frac{b}{2m}}\right\} \Leftrightarrow w_2 = m\left[\dfrac{b}{2m} + h - \dfrac{W_R}{k}\right]$ $W_R\left(1 + \dfrac{m}{k}\right) = \dfrac{b}{2} + mh \Rightarrow W_R = \left(\dfrac{b}{2} + mh\right)\left(\dfrac{k}{k+m}\right)$	Eqn 6-5

As the denominator is greater than the numerator, then $\left(\dfrac{k}{k+m}\right) < \mathbf{1}$. The batter "retreats". This then means applying one of the two rules, determined by inspection of $\boldsymbol{k}$, for a rising or a falling slope from the centre line.

6.15.2.2 Unified formula using sign of k

However, if we apply the **sign** of the slope to $\boldsymbol{k}$ then we can express the left and right formation widths with one formula (keeping $\boldsymbol{m}$ positive):

From Eqn 6-4	$W_{L/R} = \left(\dfrac{b}{2} + mh\right)\left(\dfrac{k}{k-m}\right)$ and, if k is negative then $W_{L/R} = \left(\dfrac{b}{2} + mh\right)\left(\dfrac{-k}{-k-m}\right) \Leftrightarrow \left(\dfrac{b}{2} + mh\right)\left(\dfrac{k}{k+m}\right)$ $W_{L/R} = \left(\dfrac{b}{2} + mh\right)\left(\dfrac{k}{k-m}\right)$	Eqn 6-6

This returns to the original statement for $\boldsymbol{W_R}$, as stated above, where the slope is falling from the centreline. It is best to have only **one** formula to work with and use the **sign** of $\boldsymbol{k}$ to control the denominator.

$$W_{L/R} = \left(\frac{b}{2} + mh\right)\left(\frac{k}{k-m}\right)$$

6.15.3 Calculation of Formation Width, W_L, and W_R, Constant Cross Fall Ground

The development of the calculation of formation width now moves on to calculating formation width against a constantly sloping natural surface (Figure 6-21).

In Figure 6-21 the cross slope has been calculated from a level profile measured in the field:

10.5 RL at 5m Left
10.25 RL centreline, ℄
10.0 RL at 5m Right

The slope k is calculated using run/rise from the centreline:

$\boldsymbol{k}_L$ = 5/(10.5 – 10.25) ℄ – 5L
= 5/0.25 = 20,

- a **rise** away from the centreline:

$\boldsymbol{k}_R$ = 5/(10.0 – 10.25) ℄ – 5R
= 5/–0.25 = –20,

- a **fall** away from the centreline.

The components of the design are:

pavement width, $\boldsymbol{b} = 5$
batter slope, $\boldsymbol{m} = 3$
depth (cut or fill), $\boldsymbol{h} = 0.25$
cross slope is **1:20**, $\boldsymbol{k}_L = +20$, $\boldsymbol{k}_R = -20$.

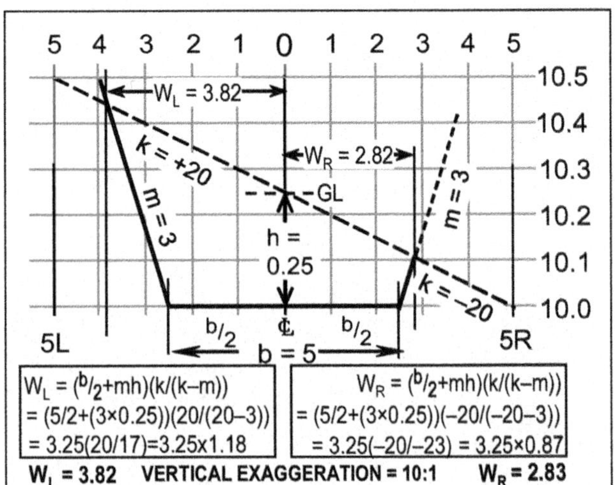

Figure 6-21 Formation width for a constant crossfall

If it's not checked it's wrong.

Applying Eqn 6-6: $W_{L/R} = \left(\dfrac{b}{2} + mh\right)\left(\dfrac{k}{k-m}\right)$ results in $\boldsymbol{W}_L = 3.82$ and $\boldsymbol{W}_R = 2.83$.

The calculation of $\boldsymbol{W}_L$ and $\boldsymbol{W}_R$ are also shown in the Figure 6-22 and Figure 6-23, demonstrating the operation of the **sign** of $\boldsymbol{k}$ in the results.

6.15.4 Calculation of Formation Width, W_L, and W_R, Varying Cross Fall Ground

The development of the calculation of formation width then moves on to calculating formation width against a varying sloping natural surface (see Figure 6-22).

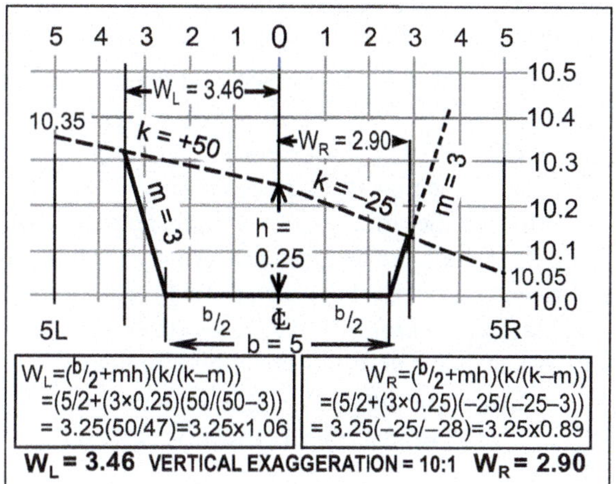

Figure 6-22 Formation width for a varying crossfall.

6.15.4.1 Worked Example 6.6.

In Figure 6-22 the cross slope has been calculated from a level profile measured in the field:

10.35 RL at 5m Left
10.25 RL centreline
10.05 RL at 5m Right

Slope $\boldsymbol{k}$ is calculated using run/rise **from** the centreline:

$\boldsymbol{k}_L = 5/(10.35 - 10.25) \Rightarrow 5/0.1 = 50$,

- a **rise** away from the centreline:

$\boldsymbol{k}_R = 5/(10.05 - 10.25) \Rightarrow 5/{-0.2} = -25$,

- a **fall** away from centreline.

The components of the design are:

pavement width, **b** = 5
batter slope, **m** = 3
depth (cut or fill), **h** = 0.25
cross slope varies, $\mathbf{k}_L = +50$, $\mathbf{k}_R = -25$.

Applying Eqn 6-6 gives $\boldsymbol{W}_L = 3.46$ and $\boldsymbol{W}_R = 2.90$.
(See Figure 6-22).

*Another case to consider is where the depth, **h**, is such that the batter intercept falls outside the first cross section.*

6.15.5 Batter Width Falls Outside the First Cross Section

Calculations with varying different cross falls. A big grade change at 5L.

Using k=50 in Figure 6-23, the batter intercept, $\boldsymbol{W}_L$ = 5.76m, which is outside the 5L section. The intercept is between 5L and 10L.

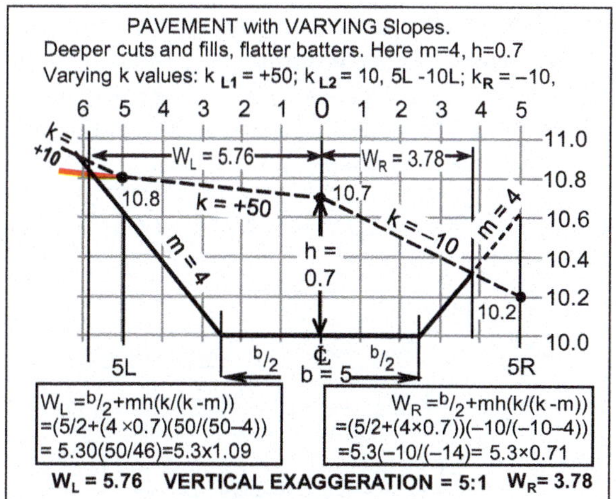

Figure 6-23 Formation width when batter intercept falls outside the cross section.

6.15.5.1 Worked Example 6.7.

The field notes reveal, RLs at:

10L = 11.3.
5L = 10.8
CL = 10.7
Datum = 10.0

Resultant slopes, from CL to 5L (K_{L1}) and 5L to 10L (K_{L2}) are:

$\boldsymbol{k}_{L1(CL-5L)} = 5/(10.8 - 10.7) = +50$
$\boldsymbol{k}_{L2(5L-10L)} = 5/(11.3 - 10.8) = +10$.

What is the height of the batter slope above the datum at profile offset 5L?

Find the height at which a batter of $\boldsymbol{m} = 4$ intercepts the 5L cross section. Here $\boldsymbol{b} = 5$, so $^b/_2 = 2.5$.

From the edge of the formation the batter slope rises from 2.5L at slope of $\boldsymbol{m} = 4$ until it intercepts 5L, at distance of 2.5m. (See .Figure 6-24).

As $w = mh$ then

$h = w/m$

The height of batter intercept at 5L:

$h_{wL=5} = (5 - 2.5)/4 = 0.625$ at 5L.

Turning to the design:

h_{5L} is the surface height at 5L design datum.

From field book and design level:

$h_{5L} = (10.8 - 10.0) = 0.8$

The depth of the batter intercept at 5L, called dh_{5L}, between the 5L surface and the 5L batter intercept is:

$dh_{5L} = 0.8 - 0.625 = 0.175$ below the surface at 5L.

What is the formation width increment, dw_L, from 5L, where the batter is still below the slope, to the rising $k_{L2} = +10$ slope surface intercept?

Reiterating: $dh_{5L} = 0.175$, $k_{L2} = +10$, $m = 4$.

dw_L is then calculated using the formula:

$dw_L = (m \times dh_{5L})(k_{L2}/(k_{L2} - m))$

$= (4\times0.175)(10/(10-4)) = 0.7\times1.67$

$dw_L = 1.17$.

Total $W_L = W_{L=5} + dw_L$

$= 5.0 + 1.17 = 6.17$ as shown in Figure 6-23 and Figure 6-24.

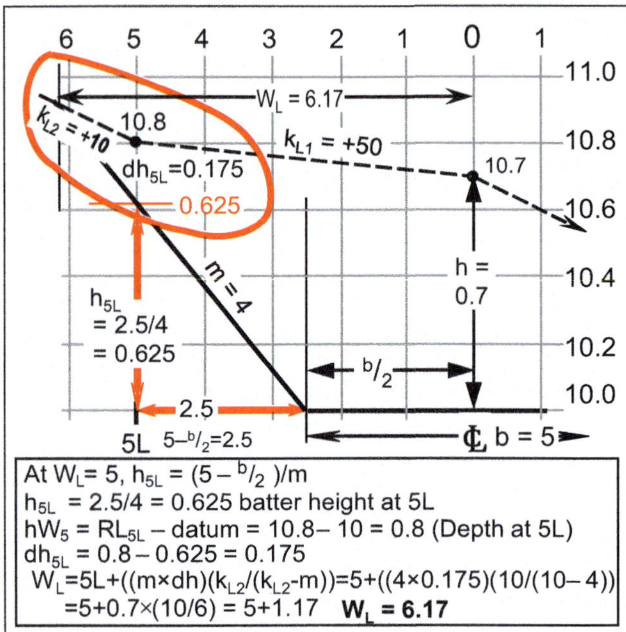

Figure 6-24 Extended batter intercept.

6.15.6 Derivation of AREA Formula using Formation Widths

Having calculated the formation widths, W_L and W_R, it is a simple matter to calculate the AREA of the section.

A **CONSTANT** slope is assumed from the batter intercept to the centreline. Individual slopes, **I**, can vary as they only affect the formation widths W_L and W_R. This means that the slope would be *slightly* incorrect in § 6.15.1.1.

In practical terms **assuming** a constant slope makes little difference to the calculated area compared with methods that more closely follow the surface profile, which rely on using a matrix cross multiplication method. In Figure 6-25:

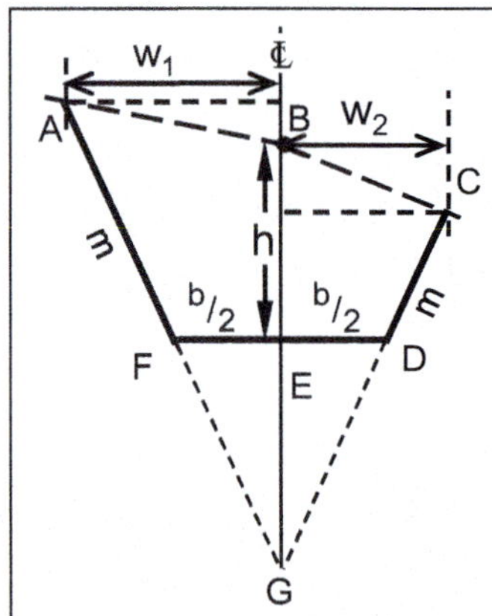

Figure 6-25 Area computations.

Area ABCDFA = Area ABG + Area CBG – Area DFG

$$A = \frac{1}{2}W_1(h+EG) + \frac{1}{2}W_2(h+EG) - \frac{1}{2}FD\times EG$$

$$EG = \frac{b}{2m},\ FD = b$$

Producing the general expression for area

$$A = \left\{\frac{1}{2}W_1\left(\frac{b}{2m}+h\right) + \frac{1}{2}W2\left(\frac{b}{2m}+h\right)\right\} - \frac{1}{2}b\left(\frac{b}{2m}\right).\ \text{Factor out } \frac{1}{2}$$

$$= \frac{1}{2}\left\{\left(\frac{b}{2m}+h\right)(W_1+W_2) - \left(\frac{b^2}{2m}\right)\right\}.\ \text{Factor out } \frac{1}{m}$$

$$Area = \frac{1}{2m}\left(\left(\frac{b}{2}+mh\right)(W_1+W_2) - \left(\frac{b^2}{2}\right)\right) \qquad \text{Eqn 6-7}$$

Note the process in Eqn 6-7:

You have to calculate $\left(\frac{b}{2}+mh\right)(W_1+W_2)$, subtract $\frac{b^2}{2}$ and then divide the result by $\frac{1}{2m}$.

6.15.6.1 Worked Example 6.8.

Using the values in Worked Example 6.6, Figure 6-22:

$$Area = \frac{1}{2\times 3}\left\{\left(\frac{5}{2}+3\times 0.25\right)(3.46+2.9)-\frac{5^2}{2}\right\} = \frac{1}{6}\{(3.25)(6.36)-12.5\} = 1.36$$

6.15.7 Area by Coordinates

Referring to Figure 6-26 we can see that the section provides an XY coordinate set if we can determine the $\mathbf{h_L}$ and $\mathbf{h_R}$ above the datum. This drawing replicates Figure 6-22.

Having determined W_L and W_R, we can then calculate the height of the batter intercept with the ground slope above the datum level.

Basically $w = mh$

($^b/_2$ merely allows for the formation width; **k** introduces the effect of slope.)

Now $w = {}^b/_2 + mh$,

thus $\mathbf{w - {}^b/_2 = mh}$, and

$\mathbf{h = (w - {}^b/_2)/m}$.

By following around the XY coordinates of the cross section can be tabulated:

Remember, centreline is at **W = 0.**

Coordinates are **negative** to the **left** of centreline and positive above the zero-height datum.

Recapitulating § 6.13, page 78 the area of a closed, non-intersecting polygon can be found from:

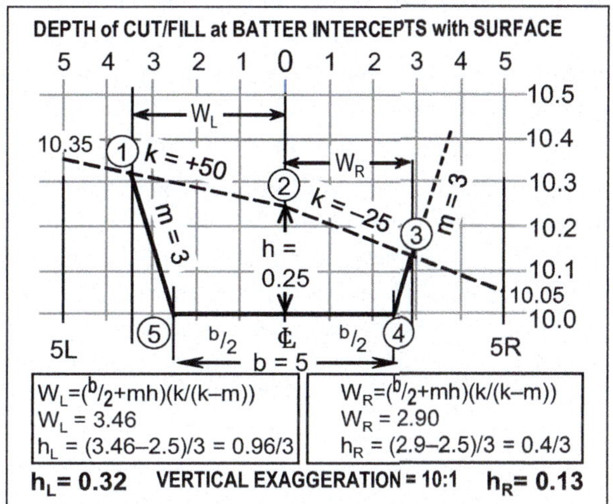

Figure 6-26 Area calculation, finding surface intercepts.

$Area = \left\lvert\frac{(x_1y_2 - x_2y_1)+(x_2y_3 - x_3y_2)+(x_3y_4 - x_4y_3)\ldots(x_ny_{n+1} - x_{n+1}y_n)}{2}\right\rvert$	Eqn 6-8

{w, h} represents the {X, Y} coordinates above the datum in .

The coordinates of an n vertex polygon are best written as a 2 × (n+1) matrix where (n+1)=1, the start vertex coordinate. The cross multiplication of the matrix represents **twice** the area of the polygon.

And remember to **close** back to the start point.

6.15.7.1 Worked Example 6.9.

Using calculations in .

n	Point	w	h	above datum
1	1	–3.46	0.32	w_L, h_L
2	2	0.0	0.25	CL, h
3	3	2.9	0.13	w_R, h_R
4	4	2.5	0.0	$^b/_{2R}$, datum
5	5	–2.5	0.0	$^b/_{2L}$, datum
6	1	**–3.46**	**0.32**	w_L, h_L (return to 1)

Matrix cross (shoelace) multiplication method

All the zero elements drop out.

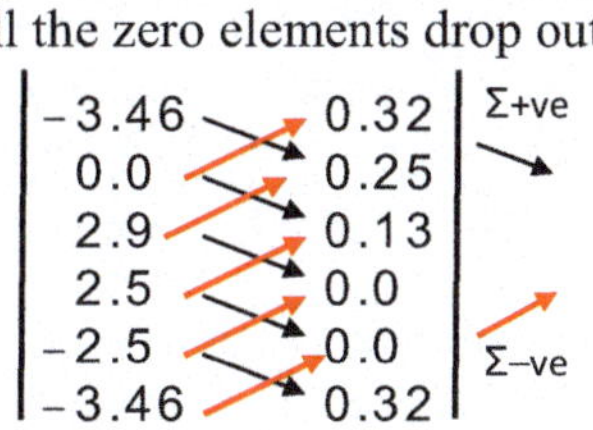

2 Area =

(–3.46·0.25 + –2.5·0.32)

– (2.9·0.25 + 2.5·0.13).

= –1.66 – 1.05 = **–2.71**

taking absolute value:

Area = ½×2.71m². = **1.36.**

6.15.8 Cut and fill batter widths.

So far we have made calculations on batter width, W_L and W_R as being all cut (excavation) or fill (embankment). Eqn 6-6, page 84 uses a **positive *m*.** But what happens if the formation level (the design level) is the same as the centreline surface level?

Whilst we have defined the cross slope, k, with a sign ($\pm k$) for calculation, we haven't done it for the formation batter slope, m. The easiest way to do this is to examine the **sign** of the batter from the ground surface level to the formation (the same method we use for determining k).

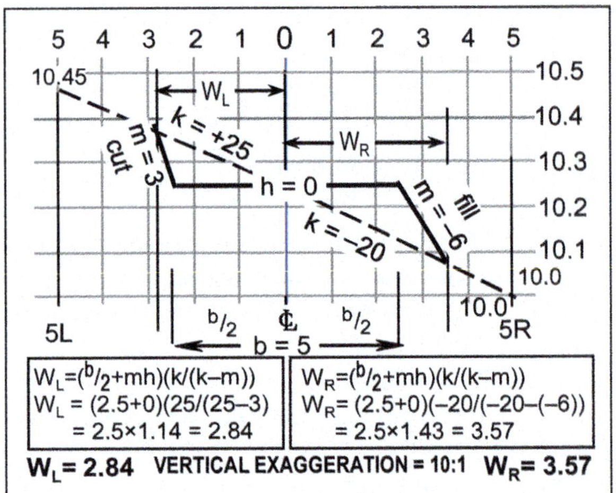

Figure 6-27 Cut/fill at h=0.

6.15.8.1 Worked Example 6.10.

Figure 6-27 shows a cross section where the formation level coincides with the centreline level, $h = 0$. The left side is in cut; the right side is fill. In cut the embankment batter slope is generally steeper than the fill (repose) batter. It may depend on traffic safety considerations.

Formation width $b = 5$. In cut, $k_L = +25$, $m_L = +3$

In fill, $k_R = -20$, $m_R = -6$.

As the calculations in Figure 6-27 show:

$W_L = 2.84$ with a shallow cut of 0.11m and $W_R = 3.57$, a fill of 0.18m.

Refer to Walker & Awange Chapter 9 (2020) for a fuller treatment of this section.

6.16 Calculation of Embankment Volume from Route Cross Sections in Cut and Fill Operations

After calculating the various cross section areas there are a number of ways to find the volume of the earthwork.

a) Calculation of volume from mean areas.

$$\text{Volume} = V = \left(\frac{A_1 + A_2 + A_3 \ldots A_{n-1} + A_n}{n}\right) \times L$$

. The method is **not** recommended.

b) Calculation of volume from end areas. This is the recommended method for most practical cases

$$Volume = \Sigma V = D_{12}\left(\frac{A_1 + A_2}{2}\right) + D_{23}\left(\frac{A2 + A3}{2}\right) + D_{34}\left(\frac{A3 + A_4}{2}\right) + \ldots D_{n-1,n}\left(\frac{A_{n-1} + A_n}{2}\right) \qquad \text{Eqn 6-9}$$

And, where D is the **common** distance **between** sections:

$$Volume = \frac{D}{2}\left(A_1 + A_n + 2\left(A_2 + A_3 + \ldots A_{n-1}\right)\right) \qquad \text{Eqn 6-10}$$

c) Calculation of volume from Simpsons Rule. You need an ODD number of sections.

$$Volume = \frac{D}{3}\left(A_1 + A_n + 4A_2 + 2A_3 + 4A_4 + 2A_5 \ldots 4A_{n-1}\right) \qquad \text{Eqn 6-11}$$

It is a variation of the prismoidal formula.

n must be an **odd** number of sections. This means an **even** number of bays.

D is the common distance **between** sections.

Main Roads WA require volumes by the prismoidal formula to be used but accept end area volumes for Mass Haul Diagrams. https://www.mainroads.wa.gov.au/Documents/67-08-90_Earthworks%20Volume%20Calculations%20Standard.RCN-D12%5E23434852.DOCX

6.17 Chapter Exercise 6.1

Question 1 (Answers provided as a check)
The traverse bearings have been adjusted

(a) Find adjusted coordinates of all points in Figure 6-28
(b) Using Double Longitude Coordinates, compute the area enclosed by the traverse in hectares.

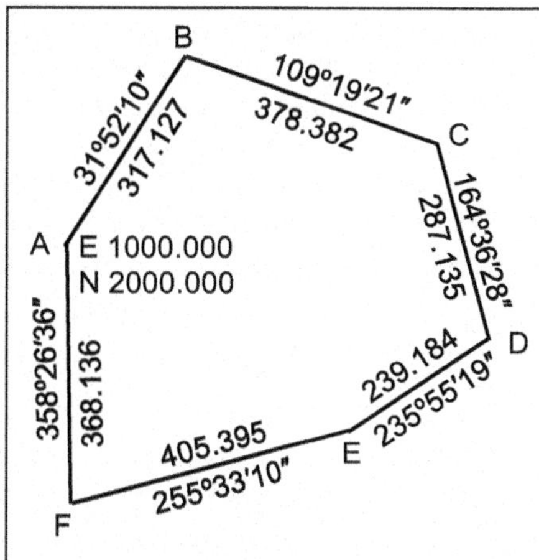

Figure 6-28 Chapter exercise 6.1

Answer: adjusted coordinates

A	1000.000	2000.000
B	1167.433	2269.302
C	1524.495	2144.078
D	1600.703	1867.224
E	1402.589	1733.189
F	1010.007	1632.023

Area = 26.2803 ha

6.18 Chapter Exercise 6.2

Q 1. From the following lengths, using previously adjusted bearings, the ΔEs and ΔNs were obtained from the observed legs of a closed traverse ABCDEFA.

LINE	LENGTH	ΔE	ΔN	Answer: adjusted coordinates		
		Unadjusted differences		A	1000.000	2000.000
AB	183.790	+183.790	0.000	B	1183.859	1999.944
BC	160.020	+98.051	+128.722	C	1281.970	2128.617
CD	226.775	−140.850	+177.762	D	1141.205	2306.310
DE	172.523	−154.440	−76.666	E	986.831	2229.592
EF	177.091	0.000	−177.091	F	986.898	2052.447
FA	53.957	+13.082	−52.431	A	1000.000	2000.000

Assume coordinates of A are: E 1000.000, N 2000.000

(a) Adjust the traverse by the Bowditch method.
(b) Use the adjusted coordinates to determine the area by Double Longitude Coordinates in square metres.
(c) Use the adjusted co-ordinate differences to check the area by using Double Longitude Co-ordinate Differences in hectares.

6.19 Answers to Chapter exercises

6.19.1 Exercise 6.1 Area = 26.2803 ha

6.19.2 Exercise 6.2 Area = 65321.4m^2 Area = 6.5321 ha

6.20 Introduction to missing data in a traverse.

In Survey practise there are many occasions when information in a close is missing. An example would be when a surveyor cannot measure a boundary directly but instead traverses between the terminals of the boundary. The surveyor can compute the bearing and distance of the boundary from this traverse.

Similarly in Subdivision design, a surveyor has to design the dimensions of the lots on his plan to meet Town Planning Regulations.

A maximum of two elements can be missing from a traverse i.e. a bearing and a distance, two bearings or two distances.

Note: The student should be familiar with his calculator keys and in particular with the To Polar and To Rectangular keys.

6.21 Missing Data in a Traverse

In a traverse there can be no more than two unknown elements of the traverse. The following four cases can arise:

(a) Bearing and distance of 1 line missing.
(b) Two distances in a traverse are missing.
(c) Two bearings in a traverse are missing.
(d) The bearing of one line and the distance of another line are missing.

The solution for each case follows the same format.

(1) Assign coordinates of (0, 0) to the end of one line which contains the missing data.
(2) Layout the problem using the usual traverse format.
(3) Determine the co-ordinate differences and find $\Sigma\Delta E$, $\Sigma\Delta N$ from the known traverse information. Then $\Sigma\Delta E = E_u$, $\Sigma\Delta N = N_u$.
(4) Solve for the missing elements in a triangle.
(5) Place the missing data back into the traverse and close the traverse.

6.22 Missing Bearing and Distance.

This is a common problem encountered in survey practise. The boundary of a lot is usually fenced, and the surveyor traverses the boundary on a small offset (about 2m minimum due to TS vertical angle restrictions).

6.22.1 Worked Example 6.11 Missing Bearing and Distance.

Given: Bearing and distance of sides DC, CB, and BA (See Figure 6-29)

Aim: To find bearing and distance of line AD.

If we allocate arbitrary coordinates (say 0,0) to station D then, knowing the bearing and distances of DC, CB and BA, the coordinates of station A can be found. The bearing and distance from A to D is found by a join between these points. The missing elements are computed in the following manner. Set out the close as shown.

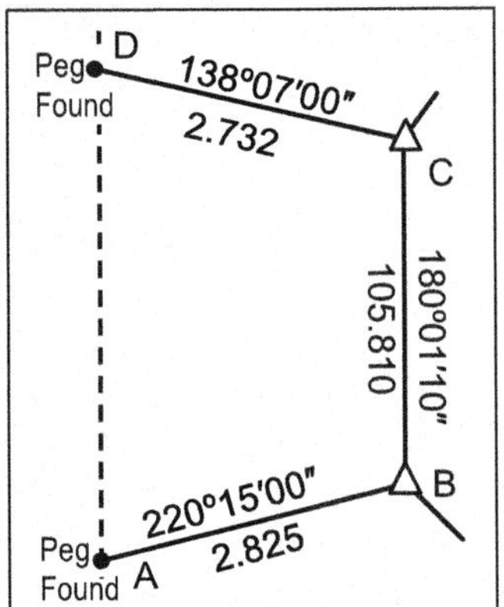

Figure 6-29 Offset traverse. Worked Example 6.11

6.22.1.1 Traverse Calculation Sheet.

Table 6-2 Offset traverse calculation

STN	Bearing	Dist	ΔE	ΔN	E	N	STN
D							D
to	138 07 00	2.732	1.824	−2.034			
C							C
to	180 01 10	105.810	−0.036	−105.810			
B							B
to	220 15 00	2.825	--1.825	−2.156			
A							A
to							
D							D
	E misclose	ΣΔE =	−0.037				
		N misclose	ΣΔN =	−110.000			

Note: If the coordinates of D are taken as (0, 0) then the coordinates of A are (− 0.037, −110.000) i.e. coordinates of A are ($\Sigma\Delta E$, $\Sigma\Delta N$) (See Figure 6-30)

The missing bearing and distance of AD is obtained by the standard join.

Line	ΔE	ΔN	
AD	0.037	110.000	(note the change of sign of ΔE & ΔN as we now refer to the line **AD**, NOT DA)

$\theta_{AD} =$ 0°01′09″ $D_{AD} =$ 110.000

Table 6-3 shows these values placed back into the original close, Table 6-2.

Table 6-3 Closed traverse

STN	Bearing	Dist	ΔE	ΔN	STN
D					D
to	138 07 00	2.732	1.824	−2.034	
C					C
to	180 01 10	105.810	−0.036	−105.810	
B					B
to	220 15 00	2.825	--1.825	−2.156	
A					A
to	**0 01 09**	**110.000**	**0.037**	**110.000**	
D					D
	E misclose	ΣΔE =	**0.000**		
		N misclose	ΣΔN =	**0.000**	

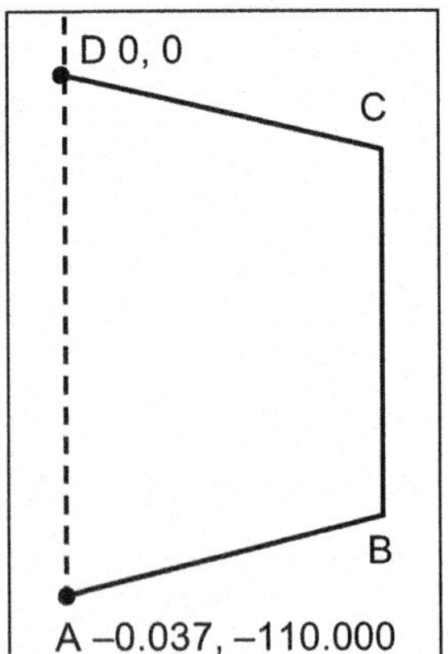

Figure 6-30 Misclose at A

Note: (1) Any error in the field measurements will result in an error to the bearing and distance of the missing line.

(2) If any error occurs in the determination of $\Sigma\Delta E$, $\Sigma\Delta N$ then the bearing and distance of AD will be incorrect. When substituting the (potentially incorrect) bearing and distance of AD back into the original close, a perfect closure may still occur.

6.23 Two Missing Distances.

Consider the general case where two distances are missing from a traverse. Consider traverse ABCDEA. (See Figure 6-31)

In this traverse the missing information is distance CD and distance EA. Assign coordinates of (0, 0) to point A. Starting from A the coordinates of C can be found as we know the bearing and distance of AB, BC. We cannot determine the coordinates of D because distance CD is missing. Construct CU parallel and equal to DE. Then CDEU is a parallelogram with CD parallel and equal to EU. (See Figure 6-32).

The coordinates of U can be calculated by the traverse A, B, C then to U using Brg/Dist DE (moved up to C). See Figure 6-32.

Consider now the triangle AUE (See Figure 6-33).

Data: coordinates A and coordinates U

Now, θ_{UA}, D_{UA} can be calculated knowing the coordinates of U and A.

$\theta_{EU} = \theta_{DC}$ is given. θ_{EA} is also given.

Thus, triangle AUE (3 angles, A, E, U and one distance, AU) can be solved for the missing two distances EA and EU

(EU = DC). (Figure 6-33).

Angle $A = \theta_{AE} - \theta_{AU}$,

angle $E = \theta_{EU} - \theta_{EA}$,

angle $U = (180 - A - E)$

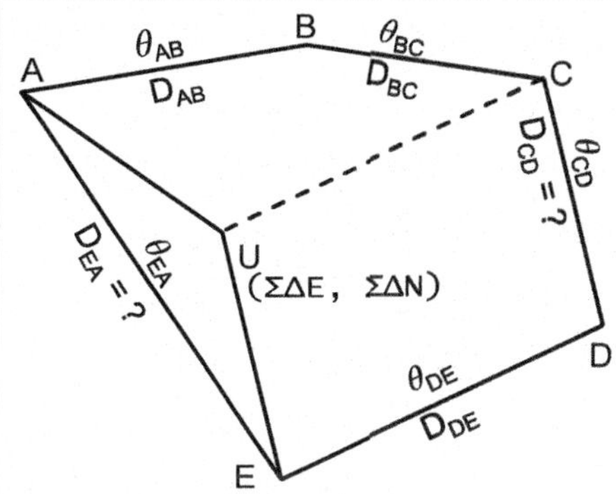

Figure 6-31 Two missing distances

6.23.1 General solution to missing data problems

The easiest way to visualize the problem is to draw up the traverse calculation sheet and fill in all the data. Then calculate the coordinate differences of each line.

This will show the missing data and will allow you to complete and check the traverse at the end of your calculations. E.g. § 6.23.2 Worked Example 6.6. Table 6-4 and Table 6-5 is an illustration in point.

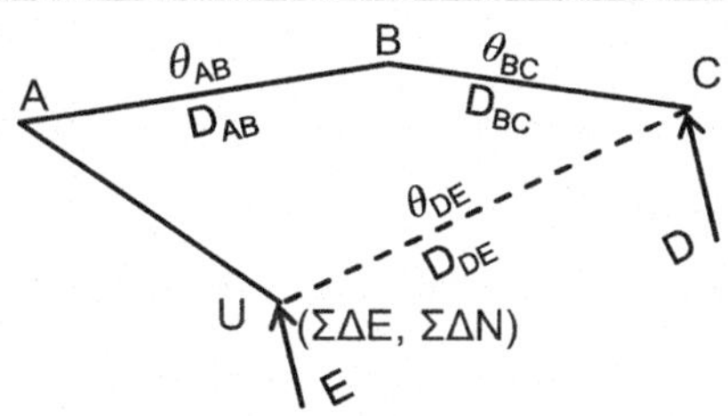

Moving D-E parallel up to C, closing the traverse from known data.
Figure 6-32

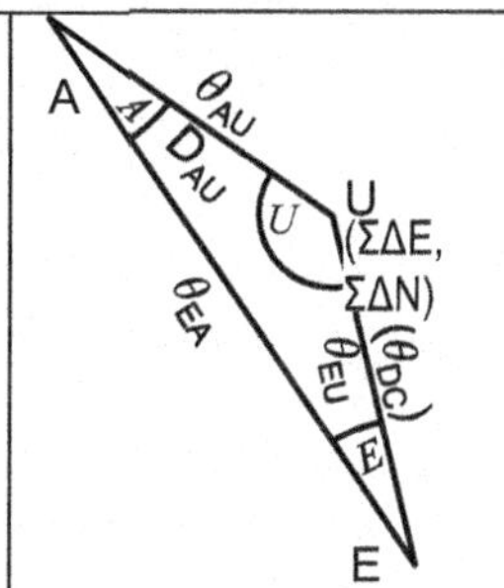

Figure 6-33

Where two distances are missing in a traverse, the solution is as follows.

(1) Assign coordinates (0, 0) to the end of one line which contains the missing data (Figure 6-31). In this case either point A or D could have been chosen.

(2) Layout the problem in the usual traverse format and solve for the ΣΔE, ΣΔN using all the known traverse information.

(3) Now construct the triangle to be solved. This is done by taking the beginning of the line from which coordinates (0, 0) were assigned (point E) and constructing a line through that point which is parallel and equal to the other line with missing data.
This is the triangle AEU, the sides of which contain the two missing distances and the third side has a bearing and distance from U(ΣΔE, ΣΔN) to A(0, 0)

(4) Solve for the missing elements in triangle AEU.

(5) The missing elements can now be placed back into the original close and this closure checked,

6.23.1.1 Examples of constructing triangle of unknown elements.

Traverse ABCDEFA has all bearings observed and all distances observed except CD and EF. (Figure 6-34)

Assign coordinates (0, 0) to point D say, now through C construct a line equal and parallel to

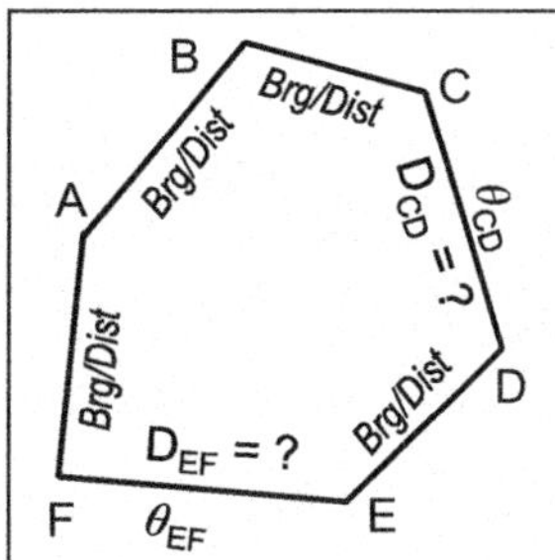

Figure 6-34 Unknown sides

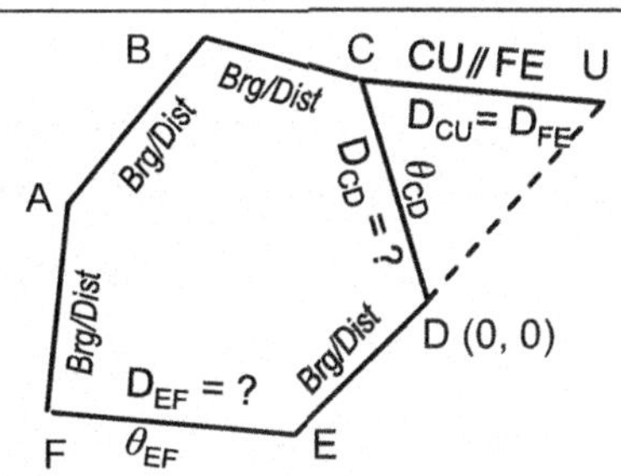

Figure 6-35 Traverse construction

FE. (i.e. line CU). U will contain the coordinates ΣΔE, ΣΔN. Now solve triangle CDU. (See Figure 6-35).
Note: The line through C to U is parallel to FE, not EF.

CU // FE (Bearing parallel)
CU = FE (Distance equal)

If however, coordinates (0, 0) are assigned to F, the triangle to be solved is FEU. (See Figure 6-36).

EU // DC (Bearing parallel)
EU = DC (Distance equal)

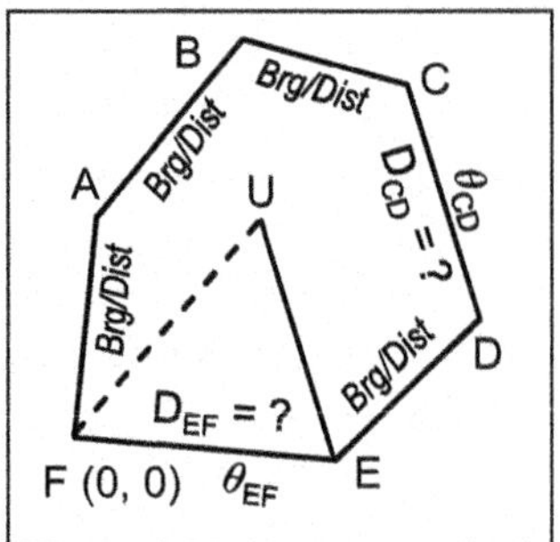

Figure 6-36 Alternate method

6.23.2 Worked Example 6.12. Two missing distances.

This problem frequently occurs in practice during sub-division design, where the lot boundaries are defined as being parallel to some previous boundary (therefore known), but the distances are unknown. In a common case the southern and western sides of the lots must be found. (See Figure 6-37).

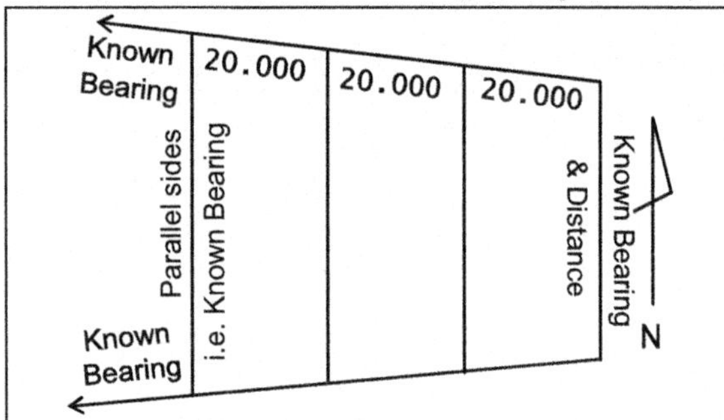

Figure 6-37 Subdivision problem Worked Example 6.12

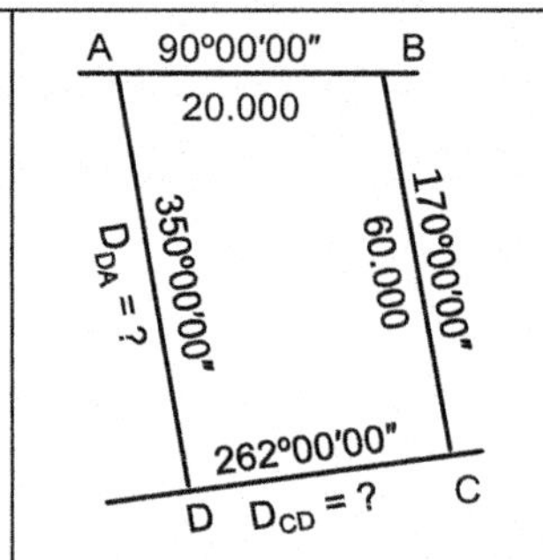

Figure 6-38 Traverse

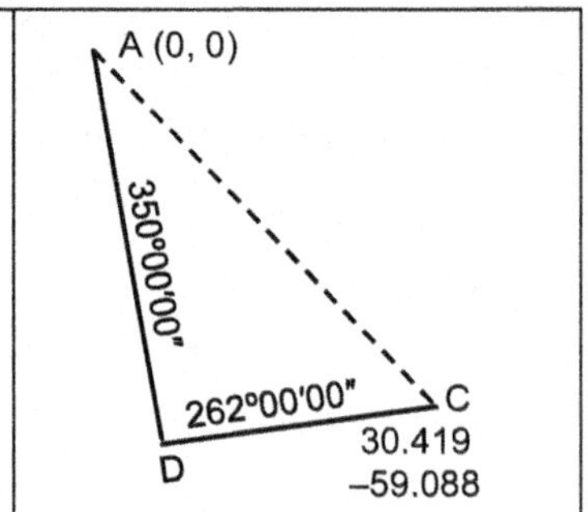

Figure 6-39 Misclose

Consider the following example. (Figure 6-38).
Find: distances CD, DA
Use data from Figure 6-38 calculate a traverse of know components (Table 6-4). Draw the resulting misclose as shown in Figure 6-39.

If A has coordinates (0, 0), point C has coordinates (30.419, –59.088) (Figure 6-39)

The missing bearing and distance of AC is now obtained by the standard join calculation.

Table 6-4 Traverse calculation

STN	Bearing	Dist	ΔE	ΔN	STN
A					D
to	*90 00 00*	*20.000*	*20.000*	*0.000*	
B					C
to	*170 00 10*	*60.000*	*10.419*	*–59.088*	
C					B
to	*262 00 00*				
D					A
to	*350 00 00*				
A					D
	E misclose	*ΣΔE =*	*30.419*		
		N misclose	*ΣΔN =*	*–59.088*	

(a) Line ΔE ΔN
CA –30.419 59.088
(note the change of sign of ΔE & ΔN as we now refer to the line CA, NOT AC)
θ_{CA} = 332°45′37″
D_{CA} = 66.458

(b) By differences in bearings, find the angles of the triangle
∠C = 70°45′37″
∠D = 92°00′00″
∠A = 17°14′23″ Σ = 180°00′00″ (Check)

(c) Using Sine Rule

$$AD = \left(\frac{66.458 \times \sin(70°45'37''')}{\sin(92°00'00'')}\right) = \mathbf{62.784} \qquad DC = \left(\frac{66.458 \times \sin(17°14'23'')}{\sin(92°00'00'')}\right) = \mathbf{20.70}$$

(d) Check close. Substitute values back into the original close and complete the traverse. (Table 6-5).

Table 6-5 Traverse closure

STN	Bearing	Dist	ΔE	ΔN	E	N	STN
A							D
to	*90 20 00*	*20.000*	*20.000*	*0.000*			
B							C
to	*170 00 10*	*60.000*	*10.419*	*−59.088*			
C							B
to	*262 00 00*	*20.708*	*-20.516*	*-2.742*			
D							A
to	*350 00 00*	*62.784*	*-10.902*	*61.830*			
A							D
	E misclose	*ΣΔE =*	*0.001*				
		N misclose	*ΣΔN =*	*0.000*			

The 0.001m misclose is caused by rounding off errors.

6.23.3 Checking the traverse close

Note: Should errors occur in the original closure it is still possible to solve the triangle AEU for two missing distances. This missing information can be placed back into the original traverse and a perfect close obtained. In this case the two missing distances would be incorrect. For example, ΔE, or ΔN, could be incorrectly calculated or ΣΔE, ΣΔN could be incorrect etc. The triangle so formed could be solved for the missing distances, these distances placed back into the original close and a closure affected. For this reason, it is important to carry out careful checks of the input data and mathematics.

6.24 Two Missing Bearings

The solution of this problem is similar to the case of two missing distances. The difference is that different elements of the standard triangle are solved. The triangle formed in this case is, as shown in Figure 6-40.

D_{EA} is known

D_{EU} is known

D_{UA} and θ_{UA} can be calculated.

Hence knowing the three sides of the triangle, the angles and thus the bearings of the triangle sides can be determined.

This problem sometimes occurs in subdivision design and also in engineering problems which involve the design of horizontal curves.

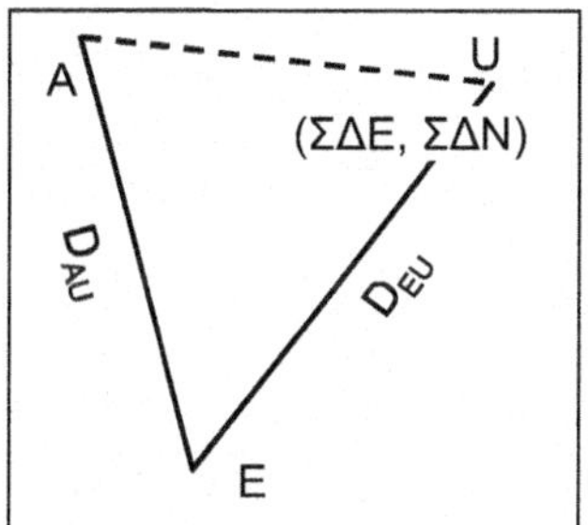

Figure 6-40 Missing bearings

6.24.1 Worked Example 6.13. Two Missing Bearings

Design: Lot 1 is to have a depth of 60.000 m on its eastern boundary. The dimension of the southern boundary is 26.000 m. (See Figure 6-41).

Aim: To determine bearings BC and CD of Lot 1.

First take out a close with the known traverse data. (Table 6-6).

Table 6-6 Closed traverse

STN		Bearing	Dist	ΔE	ΔN	STN
D						D
to		*350 00 00*	*60.000*	*−10.419*	*59.088*	
C						C
to		*90 00 00*	*25.000*	*25.000*	*0.000*	
B						B
to			*60.000*			
A						A
to			*26.000*			
D						D
		E misclose	*ΣΔE =*	*14.581*		
			N misclose	*ΣΔN =*	*59.088*	

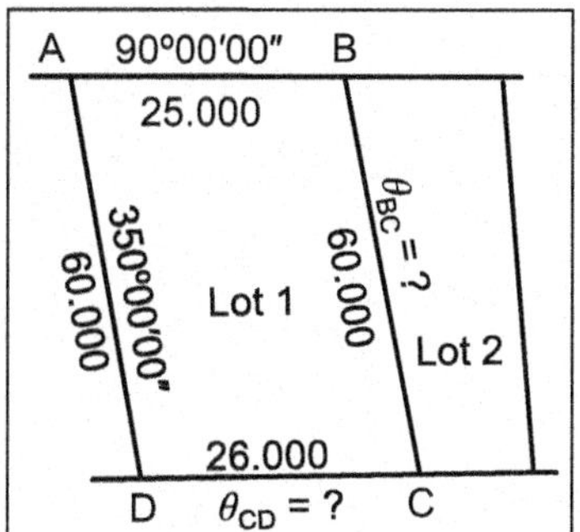

Figure 6-41 Missing bearings Worked Example 6.13

Consider triangle BCD. (See Figure 6-42).

If D has coordinates (0, 0), point B has coordinates (14.581, 59.088)

The missing bearing and distance of BD is now obtained by the standard join calculation.

(a) Line ΔE ΔN (note change of sign of ΔE & ΔN, we
BD −14.581 −59.088 now refer to the line BD, NOT DB)

θ_{BD} = 193°51′42″

D_{BD} = 60.861 (side *c*)

(b) Solve for missing elements in triangle BDC (see Figure 6-43).

(1) Solve for ∠C by Cos Rule.

$$\cos C = \frac{BC^2 + DC^2 - BD^2}{2 \times BD \times DC} = \frac{60^2 + 26^2 - 60.861^2}{2 \times 60 \times 26} = 0.183314$$

∠C = 79°26′14″

(2) Solve for ∠B, ∠D by Sin Rule using ∠C and side *c*.

$$\frac{60}{\sin B} - \frac{26}{\sin D} = \frac{60.861}{\sin(79°26'14'')}$$. And check sum = 180°

∠D = 75°43′50″
∠B = 24°49′56″
∠C = 79°26′14″ Σ = 180°00′00″

(c) find bearings BC, CD

θ_{DB} = 13°51′42″. (reciprocal of line BD)
+ ∠D = 75°43′50″
θ_{DC} = 89°35′32″
θ_{CD} = **269°35′32″** (reciprocal)
+ ∠C = 79°26′14″
θ_{CB} = 349°01′46″
θ_{BC} = **169°01′46″** (reciprocal)
+ ∠B = 24°49′56″
θ_{BD} = **193°51′42″** CHECK with join in (a)

(d) Check traverse closure (see Table 6-7).

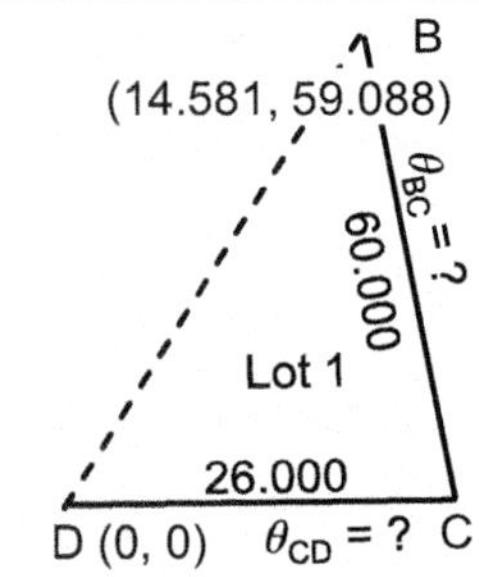

Figure 6-42 Misclose triangle

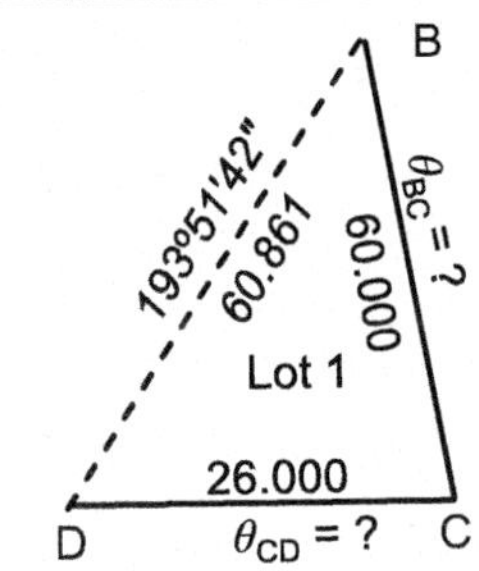

Figure 6-43 Solve join DB

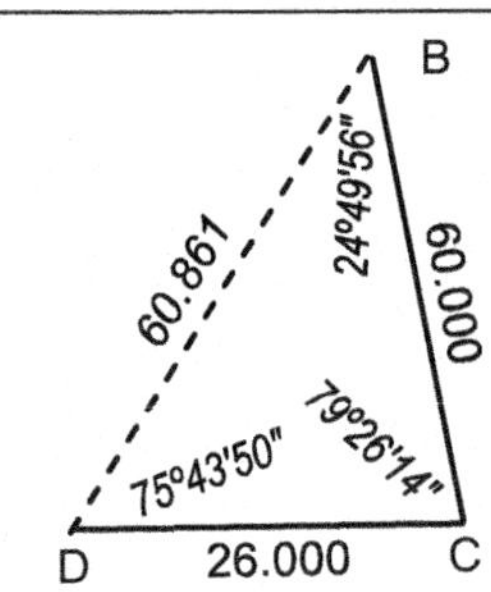

Figure 6-44 Solve angles B, C, D

Figure 6-44 shows the angle solution and Figure 6-45 shows the traverse solution.

Table 6-7 Check traverse closure.

STN	Bearing	Dist	ΔE	ΔN	STN
D					D
to	*350 00 00*	*60.000*	*−10.419*	*59.088*	
C					C
to	*90 00 00*	*25.000*	*25.000*	*0.000*	
B					B
to	***169 01 46***	*60.000*	***11.418***	***-58.904***	
A					A
to	***269 35 32***	*26.000*	***-25.999***	***-0.185***	
D					D
	E misclose	ΣΔE =	*0.000*		
		N misclose	ΣΔN =	*0.001*	

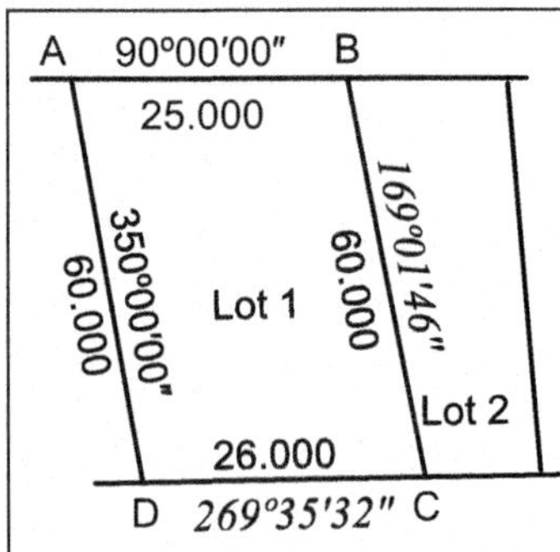

Figure 6-45 Traverse solution

Note: Neither of the last two problems is affected by having the missing elements on non-adjacent lines.

Note: The remarks in § 6.23.3 still apply here.

6.25 Bearing of one side and distance of another side missing.

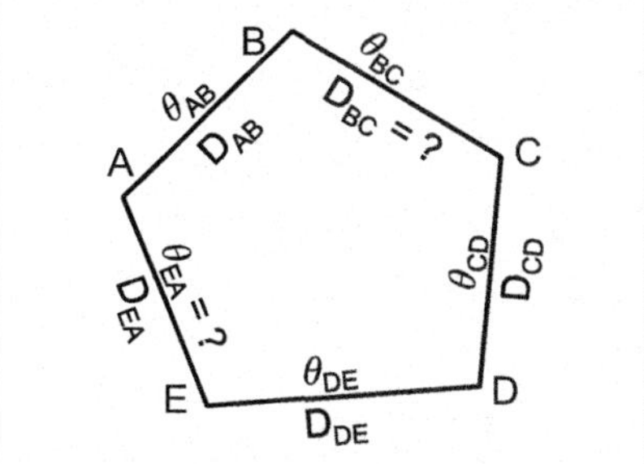

Figure 6-46 Missing Brg/Dist

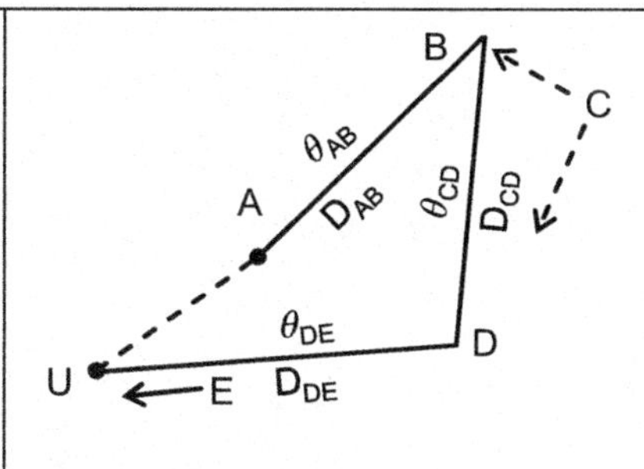

Figure 6-47 Traverse

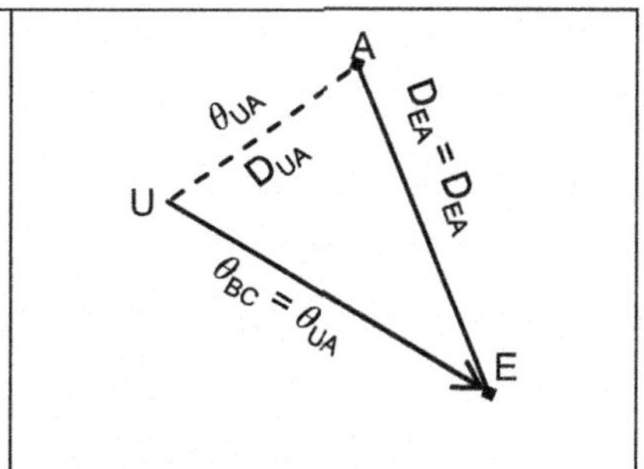

Figure 6-48 Traverse vector

The solution is similar to previous cases. The triangle is the **ambiguous** case. (Refer Chapter 4, Solution of triangles)

Figure 6-46 deconstructs to Figure 6-47 by removing the two unknown vector BC and EA. Figure 6-48 is the triangle to be solved.

The solution of this triangle is solved by the Sin rule (two sides and the angle opposite one side). This is the **ambiguous** case. (Refer to Chapter 3 and § 3.6.2, page 27). More information is needed to resolve the correct answer. This data would normally come from the field book.

Note: Checking procedures (§ 6.23.3) apply in the above case.

Traverse calculations of Figure 6-46 in Table 6-8 shows the determination of ΔE_{AE} and ΔN_{AE}. Figure 6-47 shows it graphically with the vector diagram in Figure 6-48.

6.26 Worked Example 6.14. Bearing of one side and distance of another side missing.

In Figure 6-49 find θ_{EA} and Distance BC

First take out a close with the known traverse data. See Table 6-8.

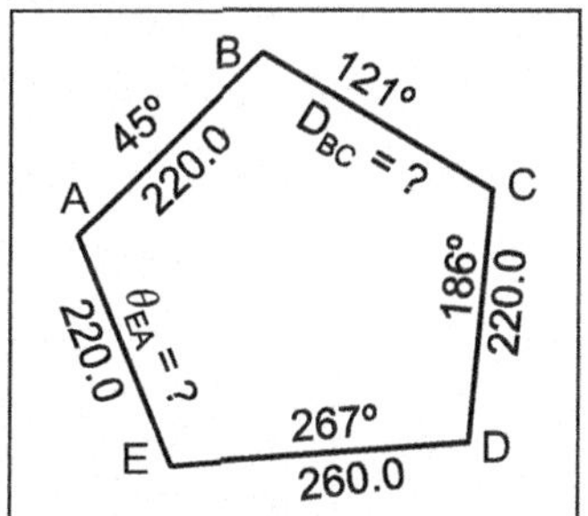

Figure 6-49
Worked Example 6.14

Table 6-8 Traverse misclosure.

STN	Bearing	Dist	ΔE	ΔN	STN
A					D
to	*45 00 00*	*220.000*	*155.563*	*155.563*	
B					C
to	*121 00 00*				
C					B
to	*186 00 00*	*220.000*	*−22.996*	*−218.795*	
D					A
to	*267 00 00*	*260.000*	*−259.644*	*−13.607*	
E					
to		*220.000*			
A					D
	E misclose	*ΣΔE =*	*−122.077*		
		N misclose	*ΣΔN =*	*−76.839*	

If A has coordinates (0, 0), point E has coordinates (–127.077, –76.839) (See Figure 6-50)

The missing bearing and distance of UA is found by the standard polar.

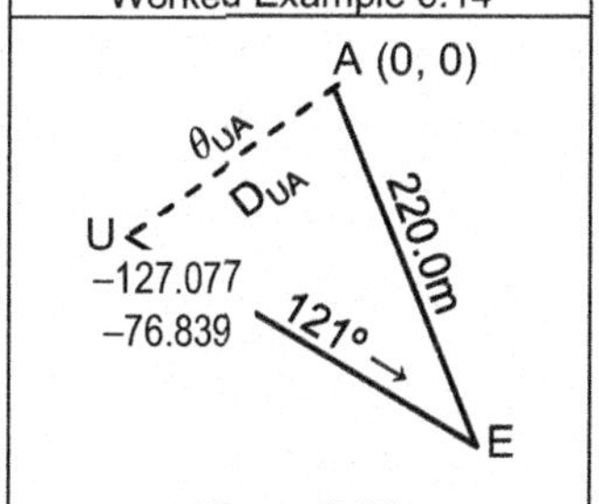

Figure 6-50

(a) Line ΔE ΔN

UA –127 077 –76.839

θ_{UA} = 58°50′24″

D_{UA} = 148.502

Consider triangle AEU, Figure 6-51

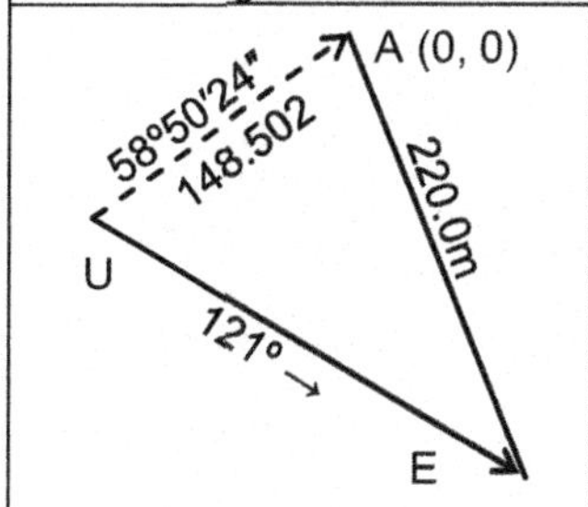

Figure 6-51

(d) Solve for missing elements in triangle EUA by Sine rule

θ_{UE} = 121°00′00″

θ_{UA} = <u>– 58°50′24″</u>

∠U = 62°09′36″

By Sine rule, find ∠E:

$$\frac{148.502}{\sin E} = \frac{220.0}{\sin(62°\,09'36'')}$$

∠E = 36°38′48″

∠A = 180° – (∠U + ∠E) = 81°11′36″

If it's not checked it's wrong.

By Sine rule, find side UE (side a)

$$\frac{a}{\sin(81°11'36")} = \frac{220.0}{\sin(62°09'36")}$$

a = UE = 245.864

BC = 245.864

(c) find bearings EA

θ_{EU} (reciprocal θ_{BC})

θ_{EU} = 301°00′00″

+ ∠E = 36°38′48″

θ_{EA} = 337°38′48″

Table 6-9 Check traverse closure.

STN	Bearing	Dist	ΔE	ΔN	STN
A					D
to	*45 00 00*	*220.000*	*155.563*	*155.563*	
B					C
to	*121 00 00*	***245.864***	***210.747***	***-126.629***	
C					B
to	*186 00 00*	*220.000*	*−22.996*	*−218.795*	
D					A
to	*267 00 00*	*260.000*	*−259.644*	*−13.607*	
E					
to	***337 38 48***	220.000	***-83.669***	***203.468***	
A					D
	E misclose	*ΣΔE =*	*0.000*		
		N misclose	*ΣΔN =*	*0.000*	

(d) Check closure by reworking the traverse. (See Table 6-9).

All the rules of checking apply. The "correct" answer may indeed be wrong, and only transcription errors will be found in the traverse. The completed misclose triangle is shown in Figure 6-52.

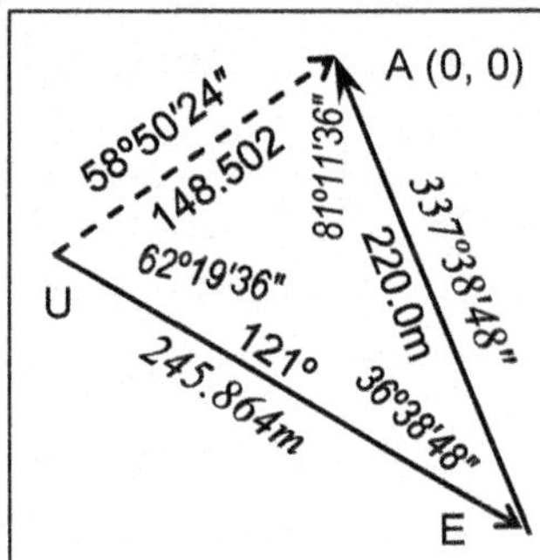

Figure 6-52 Misclose triangle

6.27 Chapter Exercise 6.3

Find the missing elements in the following traverse closes. Draw diagrams at a reasonable representation of the figure using a protractor and scale rule. Label carefully.

1)

BEARING	DISTANCE
41°21′37″	693.435
136°48′03″	714.558
( )	( )
319°28′52″	703.855

2)

BEARING	DISTANCE
( )	( )
104°07′28″	265.747
174°27′44″	300.864
252°16′10″	320.662
326°11′07″	210.287

(3)

BEARING	DISTANCE
20°08′11″	825.308
87°07′50″	804.708
117°32′30″	816.205
177°47′20″	893.623
232°59′52″	898.376
293°53′07″	( )
330°32′30″	( )

(4)

BEARING	DISTANCE
38°06′10″	( )
111°27′38″	411.882
178°33′28″	400.906
243°29′45″	( )
328°36′04″	481.647

(5)

BEARING	DISTANCE	
42°47′32″	434.962	
103°21′03″	420.787	
186°00′51″	415.087	
(249°)	395.892	require solution in this direction
(318°)	440.949	

(6)	BEARING	DISTANCE	
	35°14′32″	608.103	
	(78°)	644.353	require solution in this direction
	112°04′16″	653.711	
	176°51′16″	736.813	
	234°37′49″	736.776	
	(288°)	686.042	require solution in this direction
	326°11′55″	680.017	

6.27.1 Answers to Chapter Exercise 6.3

1)	BEARING	DISTANCE	(4)	BEARING	DISTANCE
	222°30′33″	725.259		As given	441.176
				As given	463.423
2)	BEARING	DISTANCE	(5)	BEARING	DISTANCE
	43°03′28″	279.018		249°14′55″	As given
				318 40′23″	As given
(3)	BEARING	DISTANCE	(6)	BEARING	DISTANCE
	As given	825.550		78°41′39″	As given
	As given	759.876		288°41′03″	As given.

6.28 References for Chapter 6

Irvine, W and F. Maclennan, Surveying for construction. Fifth edition, McGraw-Hill, London. ISBN 978-0-07-711114-4h

Walker, J and J. Awange, Surveying for civil engineers.2nd edition, Springer Nature Switzerland, 2020. ISBN 978-3-030-45802-7

Bannister, A and S. Raymond. Surveying. Third edition, Pitman Paperbacks, 1972. ISBN 0-273-36148-1

Chapter 7 INTERSECTIONS

Consider a case where you are unable to occupy a point (e.g., a cross on top of a roof of a church building) but whose position is needed, e.g., to monitor whether the church building is deforming. The usual practice is to occupy two places whose positions are known and then measure angles and/or distances to the cross. This chapter provides you with the skills to compute the position of the cross from either the measured angles or distances. This procedure is called intersection and can be done using either measured angles or bearings (angles measured from the true north) or measured distances.

7.1 Introduction

In survey practice, particularly in engineering works and photogrammetric control work, not all stations are occupied in a survey. Instead, the position of some of the stations is intersected by two rays of known bearings from known points, two angles off a known base, or by two horizontal distances from two known points. Distances are denoted as **H** to emphasize that they are **horizontal** distances.

The computation procedures are different in each case.

7.2 Intersections by Bearings.

7.2.1 Derivation

H_{21} and H_{31}, the horizontal distances from 2 - 1 and 3 - 1, are not observed. They are introduced to the derivation and then eliminated.

7.2.2 Two bearings

In the Figure 7-1 two bearings B_2 and B_3 have been observed intersecting at point **1**. Note that Points 2 and 3 do not need to be intervisible.

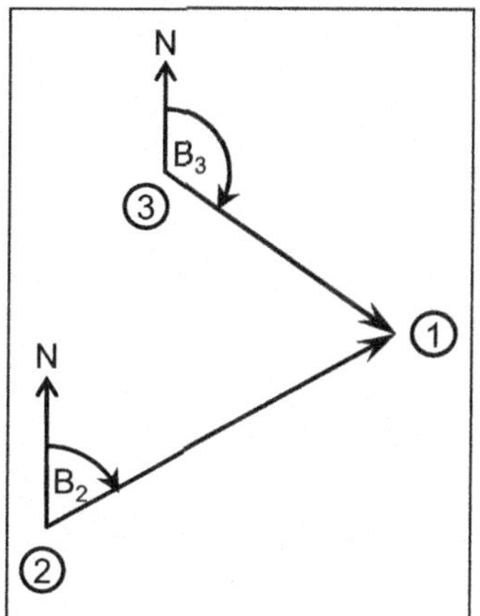

Figure 7-1 Bearing interception

From Point 2

$$E_1 = E_2 + H_{21}\times\sin B_2$$
$$N_1 = N_2 + H_{21}\times\cos B_2$$
$$E_1 - E_2 = H_{21}\times\sin B_2 \qquad (1)$$
$$N_1 - N_2 = H_{21}\times\cos B_2 \qquad (2)$$

Dividing (1) by (2) (Remember $\sin\theta/\cos\theta = \tan\theta$)

$$E_1 - E_2 = (N_1 - N_2)\times\tan B_2 \qquad (3)$$

From Point 3

$$E_1 = E_3 + H_{31}\times\sin B_3$$
$$N_1 = N_3 + H_{31}\times\cos B_3$$
$$E_1 - E_3 = H_{31}\times\sin B_3 \qquad (1a)$$
$$N_1 - N_3 = H_{31}\times\cos B_3 \qquad (2a)$$
$$E_1 = E_3 + H_{31}\times\sin B_3$$
$$N_1 = N_3 + H_{31}\times\cos B_3$$
$$E_1 - E_3 = H_{31}\times\sin B_3 \qquad (1a)$$
$$N_1 - N_3 = H_{31}\times\cos B_3 \qquad (2a)$$

Dividing (1a) by (2a)

$$E_1 - E_3 = (N_1 - N_3)\times\tan B_3 \qquad (4)$$

Subtracting (3) from (4) to eliminate E_1

$$E_3 - E_2 = (N_1 - N_2)\times\tan B_2 - (N_1 - N_3)\times\tan B_3$$
$$E_3 - E_2 = N_1\times\tan B_2 - N_2\times\tan B_2 - N_1\times\tan B_3 + N_3\times\tan B_3$$ and, collecting terms
$$E_3 - E_2 + N_2\times\tan B_2 - N_3\times\tan B_3 = N_1(\tan B_2 - \tan B_3)$$

N_1	$N_1 = \dfrac{E_3 - E_2 + N_2\tan B_2 - N_3\tan B_3}{\tan B_2 - \tan B_3}$	Eqn 7-1

E_1 by substitution in equation (3)

E_1	$E_1 = (N_1 - N_2)\tan B_2 + E_2$ *a* or $E_1 = (N_1 - N_3)\tan B_3 + E3$ *b*	Eqn 7-2

Note: **If either B_2 or B_3 is 90° or 270° the above formulæ are invalid since $\tan(90°) \to \infty$ (infinity, see § 5.21.1, page 67) and N_1 cannot be evaluated from Eqn 7-1.**

J. Walker and J. Awange, *Surveying Computations*,
https://doi.org/10.1007/978-3-031-94382-9_7

N_1 is then logically evaluated from:

If $B_2 = 90°$ or $270°$ $N_1 = N_2$

If $B_3 = 90°$ or $270°$ $N_1 = N_3$

Having found the value of N_1 a value for E_1 can then be calculated by the following method.

Note: Since tan 90° → ∞ it follows that cot(90°) → 0 (zero).

Remember this fact because tan(90) or cot(90) cannot be evaluated from your calculator. (Try it.)

If we divide equation (2) by equation (1) we obtain

$$N_1 - N_2 = (E_1 - E_2)\cot B_2 \qquad (5)$$

similarly

$$N_1 - N_3 = (E_1 - E_3)\cot B_3 \qquad (6)$$

subtracting (6) from (5) to eliminate N_1:

$$N_3 - N_2 = (E_1 - E_2)\cot B_2 - (E_1 - E_3)\cot B_3$$

$$N_3 - N_2 = E_1\cot B_2 - E_2\cot B_2 - E_1\cot B_3 + E_3\cot B_3$$

collecting terms

$$E_1(\cot B_2 - \cot B_3) = N_3 - N_2 + E_2\cot B_2 - E_3\cot B_3$$

E_1	$E_1 = \dfrac{N_3 - N_2 + E_2\cot B_2 - E_3\cot B_3}{\cot B_2 - \cot B_3}$	Eqn 7-3

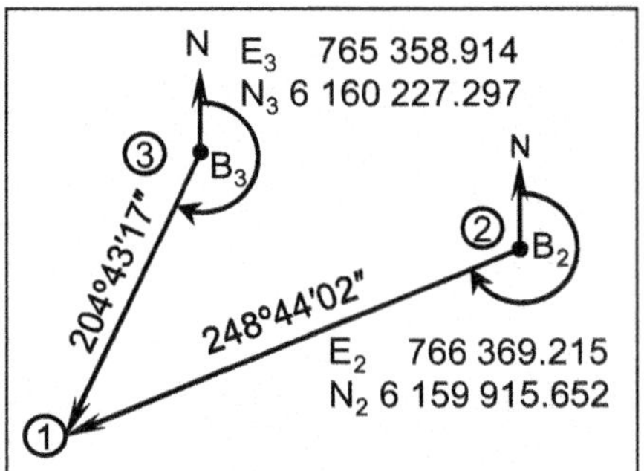

Figure 7-2 Intersection by bearings. Worked Example 7.1

7.2.3 Worked Example 7.1 Intersection by Bearings.

Given data: (See Figure 7-2)

Stn 2	E_2 766 369.215	N_2 6 159 915.652
Stn 3	E_3 765 358.914	N_3 6 160 227.297
Bearing from 2 to Stn 1		$B_2 = 248°44'02''$
Bearing from 3 to Stn 1		$B_3 = 204°43'17''$

(1) Find N_1

Eqn 7-1 $$N_1 = \frac{E_3 - E_2 + N_2\tan B_2 - N_3\tan B_3}{\operatorname{Tan} B_2 - \tan B_3}$$

$$N_1 = \frac{765358.194 - 766369.215 + 6159915.652\times\tan(248°44'02'') - 6160227.297\times\tan(204°43'17'')}{\tan(248°44'02'') - \tan(204°43'17'')}$$

N_1 = 6 159 368.565 Using full precision, the calculator gives the result: (MS Excel™ spreadsheet).

(2) Find E_1

Eqn 7-2a $E_1 = (N_1 - N_2)\tan B_2 + E_2$ and check using:

Eqn 7-2b $E_1 = (N_1 - N_3)\tan B_3 + E3$

E_1 = 764 963.553 Again, using full precision, the **result** from Eqn 7-3 is: (MS Excel™ spreadsheet).

The examples above show the absolute necessity of using full precision calculations, especially using trigonometric functions. Note that Eqn 7-2 for E_1 uses **calculated** N_1. You should get the same results using either B_2, E_2 or B_3, E_3.

(3) Check Calculations

Check 1 Join from Pt 2 to Pt 1

	Pt 1	764 963.553	6 159 368.569
Subtract	Pt 2	766 369.215	6 159 915.652
	ΔE	−1 405.662	ΔN −547.083

which yields **BEARING 21 = 248°44'02"** Check✓

DISTANCE 21 = 1 508.372

Check 2 Join from Pt 3 to Pt 1

	Pt 1	764 963.553	6 159 368.569
Subtract	Pt 3	765 358.914	6 160 227.297
	ΔE	−395.361	ΔN −858.728

BEARING 31 = 204°43'17" Check✓

DISTANCE 31 = 945.370

If it's not checked it's wrong.

7.3 Intersection by Angles

Sometimes the two ends of the baseline are intervisible in which case it is useful to be able to use the ends as back-sights and read a clockwise angle to the intersected point. **Points 2 and 3 are numbered CLOCKWISE from intersection point 1.**

7.3.1 Calculation of N_1

Construct a chord on the E/W axis through point 1. See Figure 7-3.
Since ∠3R1 and ∠321 stand on same chord they are equal.

∠3R1 = ∠321 = ∠α.

So too are ∠2R1 and ∠231 equal;

∠2R1 = ∠231 = ∠β

In Figure 7-3

$N_1 - N_2 = \text{S-2} = RS\times\tan\beta$

$N_3 - N_1 = \text{3-Q} = RQ\times\tan\alpha$

$E_3 - E_2 = \text{S-Q} = RQ - RS$

$E_3 - E_2 = (N_3 - N_1)\cot\alpha - (N_1 - N_2)\cot\beta$

$E_3 - E_2 = N_3\times\cot\alpha - N_1\times\cot\alpha - N_1\times\cot\beta + N_2\times\cot\beta$

Collecting terms:

$N_1(\cot\alpha + \cot\beta) = E_2 - E_3 + N_3\times\cot\alpha + N_2\times\cot\beta$

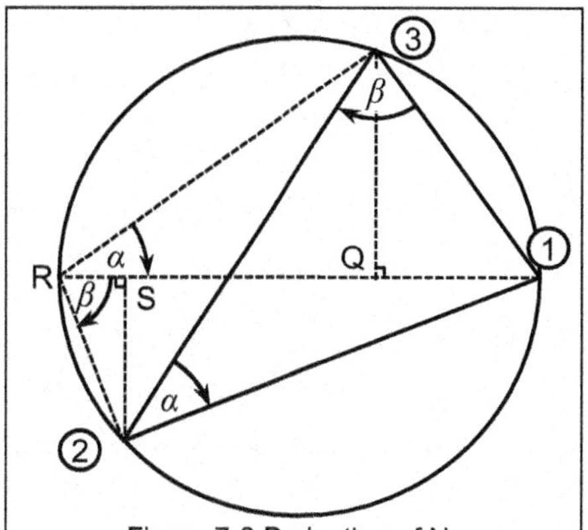

Figure 7-3 Derivation of N_1

N_1	$N_1 = \dfrac{E_2 - E_3 + N_2\cot\beta + N_3\cot\alpha}{\cot\alpha + \cot\beta}$	Eqn 7-4

7.3.2 Calculation of E_1

We use a similar approach for E_1. See Figure 7-4.
Construct a chord on the N/S axis through point 1.
Since ∠2R1 and ∠231 stand on same chord they are equal.

∠2R1 = ∠231 = ∠β.

Since ∠3RQ and ∠321 stand on same chord they are equal.

∠3RQ = ∠321 = ∠α.

(∠3R2 and ∠312 = 180 – ∠α – ∠β. Thus ∠3RQ = ∠α.)

In Figure 7-4

$E_1 - E_2 = \text{2-S} = RS\times\tan\beta$

$E_1 - E_3 = \text{3-Q} = RQ\times\tan\alpha$

$N_3 - N_2 = \text{S-Q} = RS + RQ$

$N_3 - N_2 = (E_1 - E_2)\cot\beta + (E_1 - E_3)\cot\alpha$

$E_3 - E_2 = E_1\times\cot\beta - E_2\text{ Cot }\beta + E_1\text{ Cot }\alpha - E_3\text{ Cot }\alpha$

Collecting terms:

$E_1(\cot\alpha + \cot\beta) = N_3 - N_2 + E_2\times\cot\beta + E_3\times\cot\alpha$

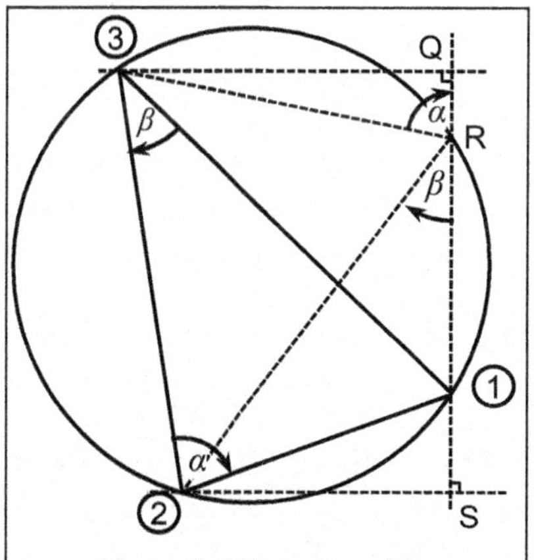

Figure 7-4 Derivation of E_1

E_1	$E_1 = \dfrac{N_3 - N_2 + E_2\cot\beta + E_3\cot\alpha}{\cot\alpha + \cot\beta}$	Eqn 7-5

Note: These formulæ will only work if the diagram is drawn such that Pt 1 is the point to be fixed and point 2 and 3 are numbered clockwise around the figure from 1.

7.3.3 Worked Example 7.2 Intersection by angles.

In Figure 7-5

Pt 2: E_2 586 775.319 N_2 6 958 950.320

Pt 3: E_3 585 720.107 N_3 6 958 693.887

Angle at 2 from 3 to 1: $\alpha = 36°58'36''$

Angle at 3 from 1 to 2: $\beta = 97°17'54''$

The examples must be evaluated using full precision calculations, especially when using trigonometric functions.

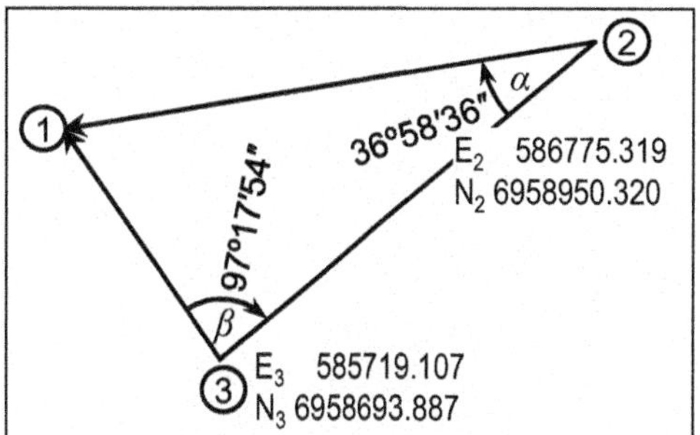

Figure 7-5 Intersection by angles. Worked Example 7.2

(1) Find E_1 using Eqn 7-3 $$E_1 = \frac{N_3 - N_2 + E_2 \cot\beta + E_3 \cot\alpha}{\cot\alpha + \cot\beta}$$ **E1 = 585 392.712**

(2) Find N_1 using Eqn 7-4 $$N_1 = \frac{E_2 - E_3 + N_2 \cot\beta + N_3 \cot\alpha}{\cot\alpha + \cot\beta}$$ **N1 = 6 959 546.627**

(3) Find Angle 1. Find Bearings from Joins

Join from Pt 2 to Pt 1

	Pt 1	585 392.712		6 959 546.627
Subtract	Pt 2	586 775.319		6 958 950.320
	ΔE_1	−1 382.607	ΔN_1	596.307

which yields **BEARING 21 = 293°19′48″ DISTANCE 21 = 1 505.712**

Join from Pt 3 to Pt 1

	Pt 1	585 392.712		6 959 546.627
Subtract	Pt 3	585 720.107		6 958 693.887
	ΔE_1	−326.395	ΔN_1	852.740

BEARING 31 = 339°03′18″ DISTANCE 31 = 913.071

Angle at 1 = 339°03′18″ – 293°19′48″ = **45°43′30″**

(4) Check: Sum angles = 36°58′36″ + 97°17′54″ + 45°43′30″ = 180°00′00″ ✓

7.4 Intersection by bearings or angles - Triangle solution.

This section shows the amount of calculation needed to solve the intersection by using the general triangle methods studied in Chapter 5.

Given the co-ordinates of points ② and ③ and either angles α, β at ② and ③ or the bearings from ② - ① and ③ - ①, the co-ordinates of ① can be found by triangle solutions. (See Figure 7-6)

Method:

(a) Join ② - ③ find bearing and distance of line ② - ③

(b) In either case (bearings or angles) we now have a triangle in which we can determine all 3 angles, and side ② - ③ is known. Solve the triangle to obtain distances ③ - ① and ② - ①

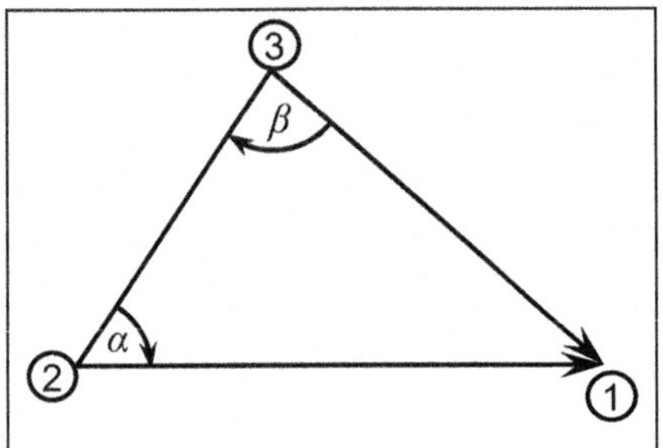

Figure 7-6 Intersection. Triangle solution

7.4.1 Worked Example 7.3

7.4.1.1 Triangle solution – angles or bearings.

This is the same example as § 7.2.3 Worked Example 7.1 (Figure 7-2).

	Stn 2	E_2 766 369.215	N_2 6 159 915.652
Subtract	Stn 3	E_3 765 358.914	N_3 6 160 227.297

Bearing from 2 to Stn 1 $B_2 = \theta_{21}$ = 248°44′02″

Bearing from 3 to Stn 1 $B_3 = \theta_{31}$ = 204°43′17″ (See Figure 7-7)

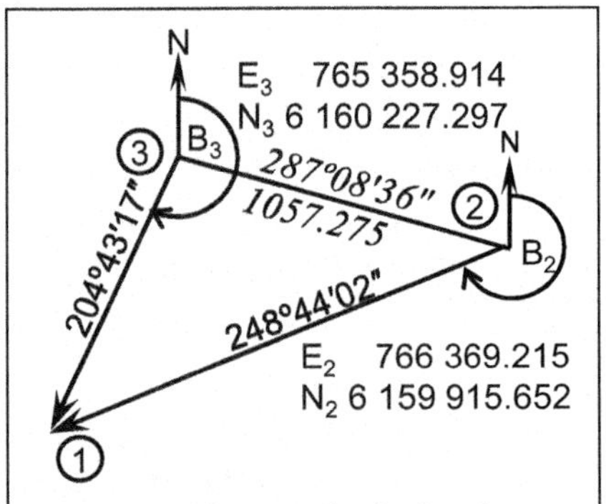

Figure 7-7 Intersection by bearings. Worked Example 7.3

(1) Find Bearings from Joins. JOIN from Stn 2 to Stn 3

	Stn 2	766 369.215		6 159 915.652
Subtract	Stn 3	765 358.914		6 160 227.297
	ΔE	−1 010.301	ΔN	+311.645

BEARING 2 - 3 = 287°08′36″ DISTANCE 2 - 3 = 1 057.275

(2) Determine angles of triangle 123

	287°08′36″	(θ_{23})
	−248°44′02″	(θ_{21})
∠2 =	38°24′34″	∠2
	204°43′17″	(θ_{31})
	– 107°08′36″	(θ_{32})
∠3 =	97°34′41″	∠3
	248°44′02″	(θ_{21})
	204°43′17″	(θ_{31})
∠1 =	44°00′45″	∠1

If it's not checked it's wrong.

CHECK SUM

∠1	44°00′45″		
+ ∠2	38°24′34″		
+ ∠3	97°34′41″	CHECK	180°00′00″

(3) Solve distances 3 - 1, 2 - 1 using Sine rule:

$$\frac{H_{31}}{\sin(38°24'34")} = \frac{H_{21}}{\sin(97°34'41")} = \frac{1057.275}{\sin(44°00'45")}$$

$H_{31} =$ 945.374

$H_{21} =$ 1508.374

(4) Calculate coordinates Stn 1 by POLAR from Stn 2:

Polar from Stn 2 - Stn 1

$E_1 = E_2 + \Delta E_{21}$

$= E_2 + H_{21} \times \sin(248°44'02")$

= 766 369.215 + 1058.373×(−0.931906)

= 766 369.215 + (−1405.662)

E_1 = 764 963.552

$N_1 = N_2 + \Delta N_{21}$

$= N_2 + H_{21} \times \cos(248°44'02")$

= 6 159 915.652 + 1058.373×(−0.362700)

= 6 159 915.652 + (−547.087)

N_1 = 6 159 368.565

(5) CHECK Polar from Stn 3 – Stn 1

E_1 = 764 963.553 ✓

N_1 = 6 159 368.565 ✓

The examples must be evaluated using full precision calculations, especially when using trigonometric functions. (MS Excel™ spreadsheet).

Note: The Intersection by Triangle solution is much less sensitive to rounding errors compared with Intersection by Bearings/Angles because of the smaller products involved.

$0 > \cot(\beta) > \infty$ compared with $0 >$ (sin or cos) > 1.

Below 45° $\text{Cot}(\beta) > 1$ and increases quickly (try cot 40°, 20°, 10° and 5°).

e.g. Eqn 7-4; $N_2 \times \cot(\beta)$ versus the POLAR form $H_{21} \times$SIN(Brg),

In § 7.3.3 Worked Example 7.2, N_2 is 6000 times larger than D_{21} even before $\cot(\beta)$ is evaluated. The use of coordinate differences (relocated origin) still does not overcome the magnitude of cot problem at angles <45°.

7.5 Resection/intersection by distances. Direct solution of horizontal distances.

This method presumes the distances from an unknown station are measured to two (or more) known points. Point 1 is one of the intersections of two circles, radius H_2 (from point 3) and H_3 (from point 2). In Figure 7-8 there are **two** intersections of the circles, reflected either side of the line 2 to 3.

The surveyor must note the position of 1 relative to the line 2 - 3.

Below, see Figure 7-9, where distances H_1, H_2 and H_3 are the sides opposite the points 1, 2 and 3.

Points 2 and 3 have known coordinates with calculated distance H_1 between them. Horizontal distances H_2 and H_3 have been measured to 2 and 3 from 1. You will need to know the location of point 1 relative to the base line through points 2 & 3. In Figure 7-9 point 1 is to the **right** of the line 2 - 3.

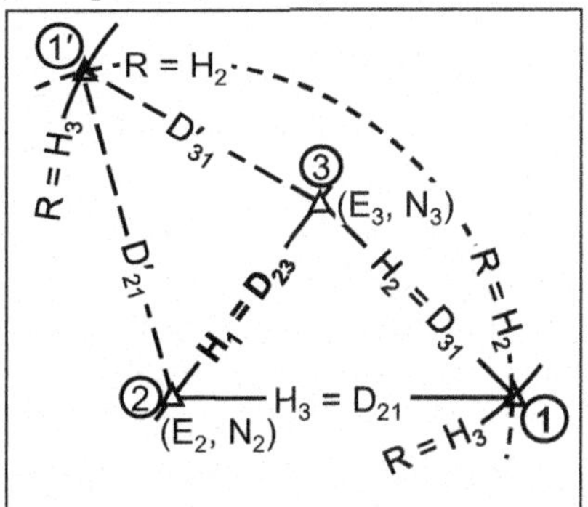

Figure 7-8 Intersection by distance

7.5.1 Derivation:

Draw a line perpendicular to line 2 - 3 (Figure 7-9) through point 1 and label p, p1 and h.

p is the horizontal distance **from** point 2 to the base of the perpendicular through vertex point 1. **h** is the perpendicular distance through point 1.

Now, $H_2^2 = h^2 + p1^2$

But $p1 = H_1 - p$

$\therefore$ $H_2^2 = h^2 + (H_1 - p)^2$ (expand equation)

$= h^2 + H_1^2 + p^2 - 2pH_1$

$H_2^2 = h^2 + p^2 + H_1^2 - 2pH_1$

Now, $H_2^2 = h^2 + p1^2$

But $p1 = H_1 - p$

$\therefore$ $H_2^2 = h^2 + (H_1 - p)^2$ (expand equation)

$= h^2 + H_1^2 + p^2 - 2pH_1$

$H_2^2 = h^2 + p^2 + H_1^2 - 2pH_1$ (1)

And, $H_3^2 = p^2 + h^2$. Thus $h^2 = H_3^2 - p^2$

Substituting in H_2^2 ((1) above), and, cancelling out:

$H_2^2 = (H_3^2 - \cancel{p^2}) + \cancel{p^2}\ H_1^2 - 2pH_1$.

$H_2^2 = H_3^2 + H_1^2 - 2pH_1$. (2)

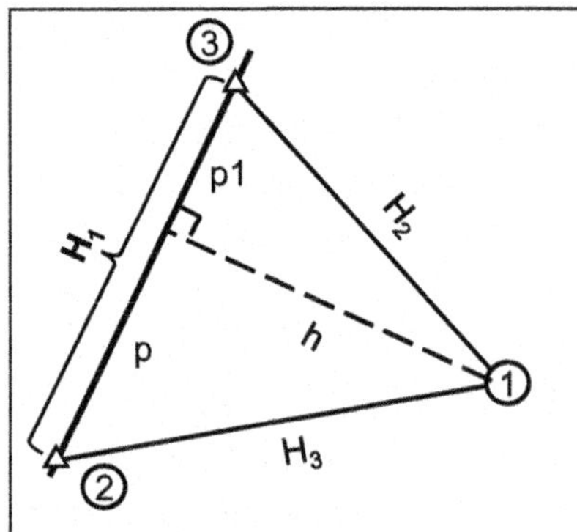

Figure 7-9 Derivation diagram

Solve for **p** in terms of the 3 distances. **p** may be negative (Figure 7-10) or positive (Figure 7-11, Figure 7-9).

Distance p	$p = \dfrac{H_3^2 + H_1^2 - H_2^2}{2H_1}$ where $H_1 = \sqrt{(E_3 - E_2)^2 + (N_3 - N_2)^2}$ and $h = \sqrt{H_3^2 - p^2}$	Eqn 7-6

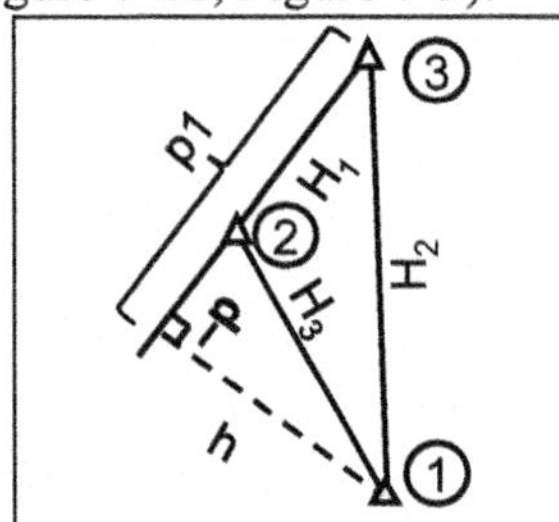

Figure 7-10 p negative

7.5.2 Bearing method.

How, in polar co-ordinate terms, do we traverse from Point 2 **to** Point 1 via the base distance, **p,** and the vertical height, **h**?

$E_1 = E_2 + p \times \sin\theta_{23} + h \times \sin(\theta_{23} + 90)$

$N_1 = N_2 + p \times \cos\theta_{23} + h \times \cos(\theta_{23} + 90)$

Note that the positive (+) sign between the last two terms only applies if Point 1 lies to the right of Point 2 on the line 2 to 3. If not, i.e. Point 1 is to the left, then a negative (–) sign must be used here. Figure 7-8 shows that point **1** is to the **right** of 2 – 3 whereas point **1′** is on the **left**.

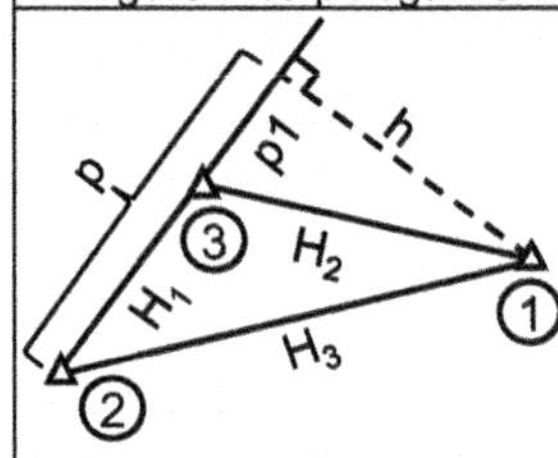

Figure 7-11 p positive

also remember $\sin(\theta + 90) = \cos\theta$

and $\cos(\theta + 90) = -\sin\theta$ N.B. the change of SIGN.

Generalizing the formulae to account for Point 1 being to **left** or **right** of Point 2 on the line through 2 to 3, removing the addition of 90° to the θ_{23} and including change of sign for $\cos(\theta + 90) = -\sin\theta$, produces the two formula:

$\mathbf{E_1 = E_2 + p\ Sin\ \theta_{23} \pm h\ Cos\ \theta_{23}}$	Point 1 **right/left** line 2-3: use $\pm$ **(+ or -)**
$\mathbf{N_1 = N_2 + p\ Cos\ \theta_{23} \mp h\ Sin\ \theta_{23}}$	Point 1 **right/left** line 2-3: use $\mp$ **(- or +).**

7.5.3 Coordinate method.

It may be more convenient to use a coordinate form rather than the bearing from 2 to 3. The polar function on modern scientific calculator makes this step redundant, however the problem can be solved with a four function plus [√] calculator.

The $\sin\theta_{23}$ and $\cos\theta_{23}$ can found directly from coordinate differences:

$E_3 - E_2$ = opposite side;

$N_3 - N_2$ = adjacent side;

H_1 is the hypotenuse found by Pythagoras.

$$H_1 = \sqrt{(E_3 - E_2)^2 + (N_3 - N_2)^2}$$

$$E_1 = E_2 + p\left(\frac{E_3 - E_2}{H_1}\right) \pm h\left(\frac{N_3 - N_2}{H_1}\right)$$

$$N_1 = N_2 + p\left(\frac{N_3 - N_2}{H_1}\right) \mp h\left(\frac{E_3 - E_2}{H_1}\right)$$

and if we put $a = \frac{E_3 - E_2}{H_1}$ we can calculate **sin θ_{23}** directly as ($E_3 - E_2$) = opposite side, H_1 = hypotenuse

and $b = \frac{N_3 - N_2}{H_1}$. Here we are calculating **cos θ_{23}** directly as ($N_3 - N_2$) = adjacent side, H_1 = hypotenuse

restating $p = \frac{H_3^2 + H_1^2 - H_2^2}{2H_1}$ see Eqn 7-6

where $H_1 = \sqrt{(E_3 - E_2)^2 + (N_3 - N_2)^2}$ (distances H_2, H_3 already measured from point 1.)

and $h = \sqrt{H_3^2 - p^2}$

$E_1 = E_2 + ap \pm bh$	Point 1 to **right/left** line 2-3: use $\pm$ (**+ or –**)
$N_1 = N_2 + bp \mp ah$	Point 1 to **right/left** line 2-3: use $\mp$ (**– or +**).

In practice, the surveyor in the field can always choose which of the two solutions to use because of the knowledge of point 1 relative to the line through points 2 and 3. In Figure 7-8 point 1 is to the right, and point 1′ is to the left, of line through 2 to 3.

Always draw a diagram. Most Total Station on-board programs require a clockwise sequence of pickup to allow calculation of correctly placed resection coordinates. You will then always be on the RIGHT of 2 to 3. See point 1 in Figure 7-8 and Figure 7-9. Note that the given examples defy this rule.

Notes on the use of $\pm$ (plus/minus) and $\mp$ signs (minus/plus):

For the bearing format the + and – signs depend on whether point (1) lies:

(i) to the **right** of the line through points 2 to 3 E1 = ... + $h\cos\theta_{23}$, N1 = ... – $h\sin\theta_{23}$ or:

(ii) to the **left** of the line through points 2 to 3 E1 = ... – $h\cos\theta_{23}$, N1 = ... + $h\sin\theta_{23}$

For the coordinate format the + and – signs depend on whether point (1) lies:

(i) to the **right** of the line through points 2 to 3 E1 = ... + bh, N1 = ... – ah or:

(ii) to the **left** of the line through points 2 to 3 E1 = ... – bh, N1 = ...+ ah

7.5.4 Worked Example 7.4 Intersection by distances, Direct solution.

Figure 7-12 illustrates the reduced field book notes.

This type of calculation allows us to use the storage registers on the calculator. Using 5 (I've numbered them [1] to [5] or [A] to [F]) registers ensures that you maintain precision throughout the computational process.

Given: E_2 = 479334.510, N_2 = 6637464.274

E_3 = 478503.099. N_3 = 6637818.588

H2 = 976.518

H3 = 1250.124

Solve: $H_1 = \sqrt{(E_3 - E_2)^2 + (N_3 - N_2)^2}$ **H_1 = 903.760** STO [1]

$p = \frac{H_3^2 + H_1^2 - H_2^2}{2H_1}$ **p = 788.929** STO [2]

$h = \sqrt{H_3^2 - p^2}$ **h = 969.743** STO [3]

$a = \frac{E_3 - E_2}{H_1}$ **a = –0.919947** STO [4]

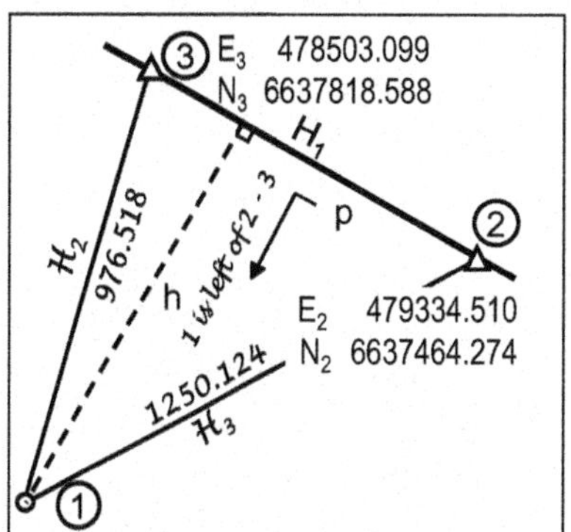

Figure 7-12 Worked Example 7.4

$$b = \frac{N_3 - N_2}{H_1}$$ **b = 0.392044** STO [5]

TEST: Is point 1 to **left** or **right** of line 2 - 3?

In Figure 7-12, point 1 to the **left** of the line through 2 - 3

Use − and +.

$E_1 = E_2 + ap - bh$ (pt 1 to the left of line 2 - 3, use –)

$N_1 = N_2 + bp + ah$ (pt 1 to the left of line 2 - 3, use +)

E_1 = 478 228.555

N_1 = 6 636 881.457

Checks:

Join From 2 to 1

Bearing$_{21}$ 242°12′42″ Distance$_{21}$ **1250.124** = H_3 ✓

Join From 3 to 1

Bearing$_{31}$ 196°19′43″ Distance$_{31}$ **976.518** = H_2 ✓

7.6 Intersection by distances - Triangle solution.

With the ability of electronic calculators to solve joins and polar calculations rapidly, intersection by distances is probably best carried out in the following manner.

Given the co-ordinates of Stn 2 and Stn 3 together with the horizontal distances D_{21}, D_{31}, find the co-ordinates of Stn 1. (See Figure 7-13).

(a) Join Stn 2 - Stn 3 (Brg23/Dist23)

(b) Solve triangle 123 (Given 3 sides)
Cos Rule and then Sin Rule ($\angle 1 + \angle 2 + \angle 3 = 180°$)

(c) Polar Stn 2 - Stn 1 (Brg23 + ∠2 = Brg21/D21)

(d) Check Polar Stn 3 - Stn 1 (Brg32 - ∠3 = Brg31/D31)

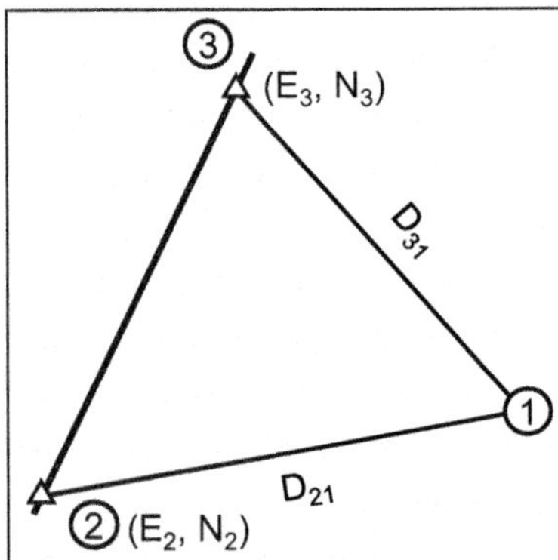

Figure 7-13

7.6.1 Worked Example 7.5 Intersection by Distances, triangle solution

Given: E_2 479 334.510 N_2 6 637 464.274 (See Figure 7-14)

E_3 478 503.099 N_3 6 637 818.588

D_{21} = 1250.124 = H_3 The side opposite station 3

D_{31} = 976.518 = H_2 The side opposite station 2

Find: Coordinates Point 1

(1) Find Bearing/Distance Pt 2 to Pt 3.
JOIN from Pt 3 to Pt 2

Stn 2	479334.510	6637464.274
Stn 3	478503.099	6637818.588
	ΔE −831.411	ΔN +354.314

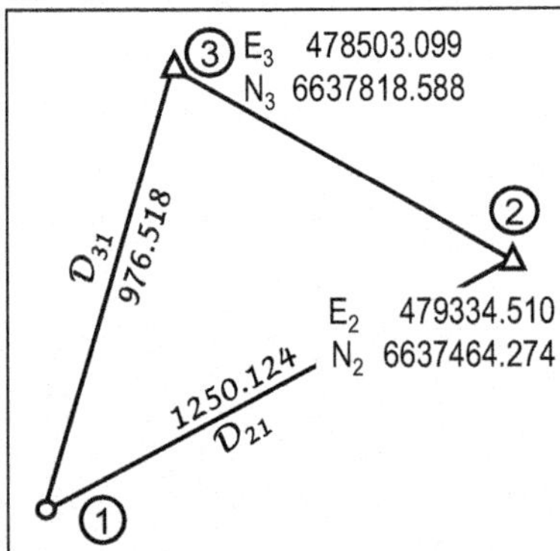

Figure 7-14 Worked Example 7.5

BEARING 2 - 3 = 293°04′54″ DISTANCE 2 - 3 = 903.760

(2) Using Cos rule solve for angle opposite longest side, ∠3.
(See § 3.6.1, page 26 Worked Example 3.3)

$$\cos(\angle 3) = \frac{H_1^2 + H_2^2 - H_3^2}{2H_1H_2}$$. Solving angle opposite longest side.

∠3 = 83°14′49″

(3) Solve angles 1 and 2 using Sine rule:

$$\frac{903.760}{\sin(\angle 1)} = \frac{976.518}{\sin(\angle 2)} = \frac{1250.124}{\sin(83°14'49'')}$$

∠1 = 45°52′59″

∠2 = 50°52′12″

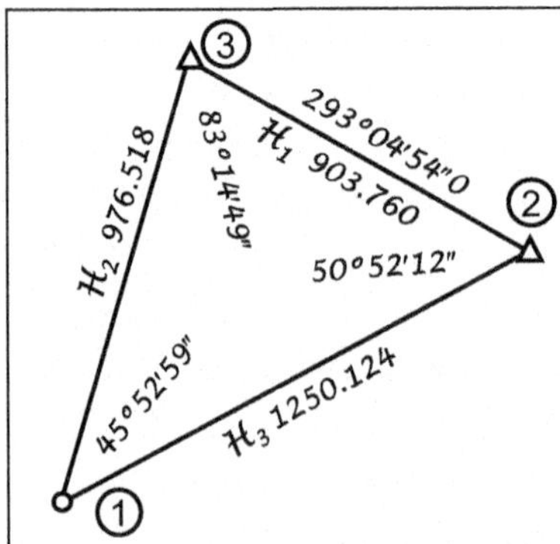

Figure 7-15 Worked Example 7.5

If it's not checked it's wrong.

CHECK SUM

$\angle 1 + \angle 2 + \angle 3$ 45°52′59″ + 50°52′12″ + 83°14′49″ = <u>180°00′00″</u> CHECK

(4) Calculate Bearings of triangle 123. Start with back bearing 3 to 2.

	113°04′54″	θ_{32}
+	<u>83°14′49″</u>	$\angle 3$
	196°19′43″	θ_{31}
	16°19′43″	θ_{13}
+	<u>45°52′59″</u>	$\angle 1$
	62°12′42″θ_{12}	
	<u>242°12′42″</u>	θ_{21}
+	<u>50°52′12″</u>	$\angle 2$
	<u>293°04′54″</u>θ_{23}	CHECK ✓

CALCULATE Stn 1 by POLAR from Stn 2 - Stn 1

Bearing/Distance 2 - 1 = 242°12′42″, 1250.124m

$E_1 = E_2 + \Delta E_{21}$ = 478 228.555

$N_1 = N_1 + \Delta N_{21}$= 6 636 881.458

The solution is shown in Figure 7-15.

CHECK Stn 1 by POLAR from Stn 3 - Stn 1

Bearing/Distance 3 - 1 = 196°19′43″, 976.518m

$E_1 = E_3 + \Delta E_{31}$ = 478 228.555✓

$N_1 = N_3 + \Delta N_{31}$ = 6 636 881.458✓

7.7 Chapter Exercise 7.1

Q. 1

Compute Coordinates of P. (See Figure 7-16).

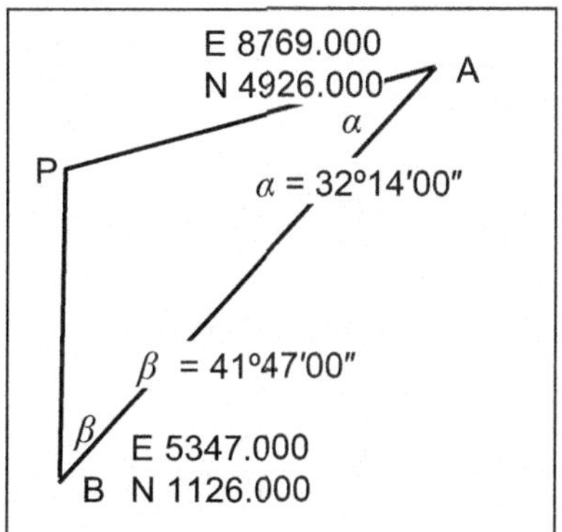

Figure 7-16 Compute P by angles

Q. 2

Compute Coordinates of P. (See Figure 7-17).

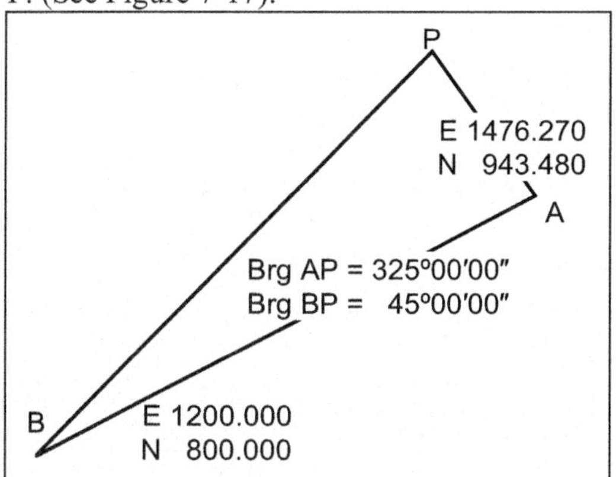

Figure 7-17 Compute P by bearings

Q. 3 Given: Stn 1 E 57434.160 N 8354.590

Stn 2 E 2787.160 N 5416.540

θ_{13} = 273°04′31″

θ_{23} = 46°42′29″

Compute the coordinates of Stn 3. (See Figure 7-18).

This problem has some bigger numbers.

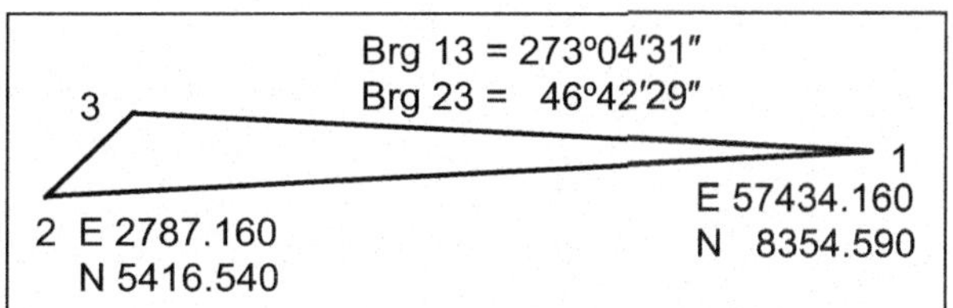

Figure 7-18 Compute Stn 3 by bearings, big numbers

Q. 4

Compute the coordinates of Stn P. (See Figure 7-19).

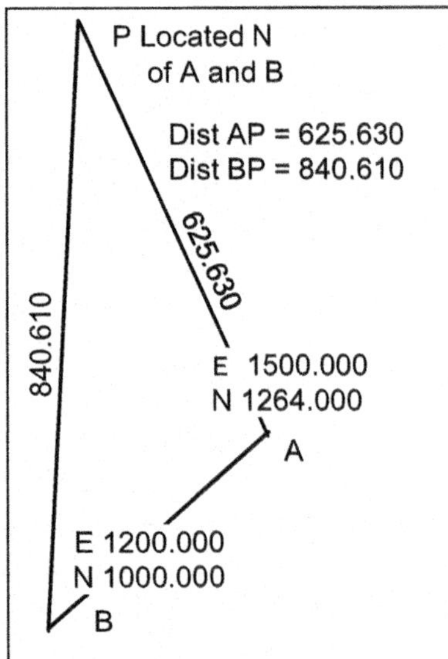

Figure 7-19 Compute P by distances

Q. 5

Given the information in Figure 7-20, compute E_B, N_B.

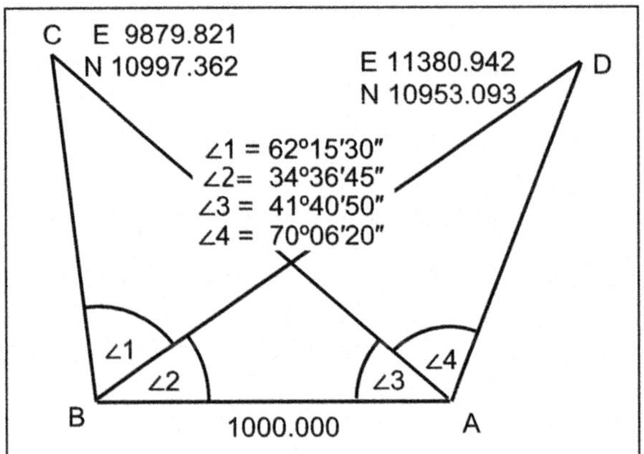

Figure 7-20 Compute Stn B, complex problem

7.7.1 Answers to Chapter Exercise 6.1

Q. 1 $E_P = 5357.920$

$N_P = 3963.154$

Q. 2 $E_P = 1421.582$

$N_P = 1021.582$

Q. 3 $E_3 = 8685.839$

$N_3 = 10973.609$

Q. 4 $E_P = 1253.263$

$N_P = 1838.921$

Q. 5 $E_B = 10000.000$

$N_B = 10000.000$

From Figure 7-21, in the two triangles find missing angles, ∠BCA and ∠BDA.
Use SIN rule to calculate missing sides.
Join C to D to find BRG/dist CD.
Use COS rule for angles to find ∠ACD and ∠BDC.
Check all angles add to 360°.
Find BRGS CA, CB. Polar from C to A and C to B.
Check distance A to B.
Find BRGS DA, DB. Polar from D to A and D to B.
Check distance A to B.
Confirm coordinates of B.
Note the answer only requires coordinates of B. However, you also have to calculate the coordinates of A to perform the check.
E_A = 11 000.000, N_A = 10 000.000.

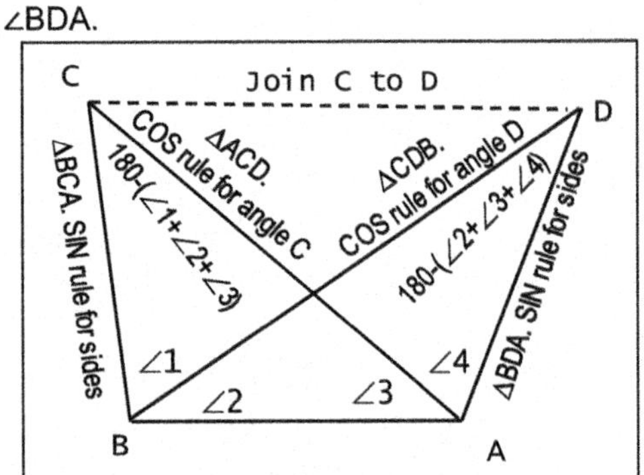

Figure 7-21 Q 5 solution

7.8 Resection to a Point.

The RESECTION, or Three Point Problem, is a method of solving the position of a point by making BEARING observations to at least three other known points, A, B and C. The bearings allow intersection angles to be derived.

In Figure 7-22 to Figure 7-24 the point P can lie inside or outside the figure

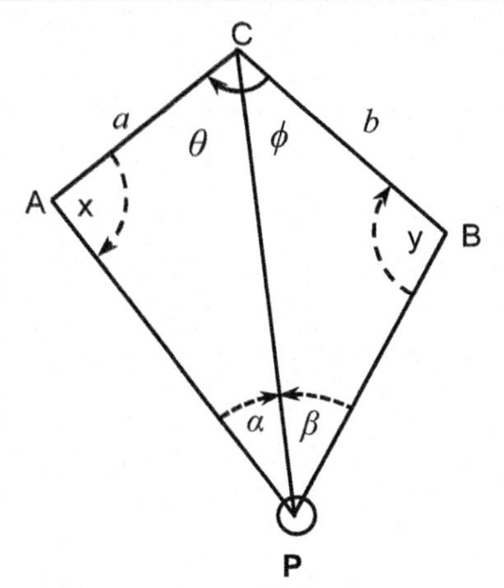

Figure 7-22

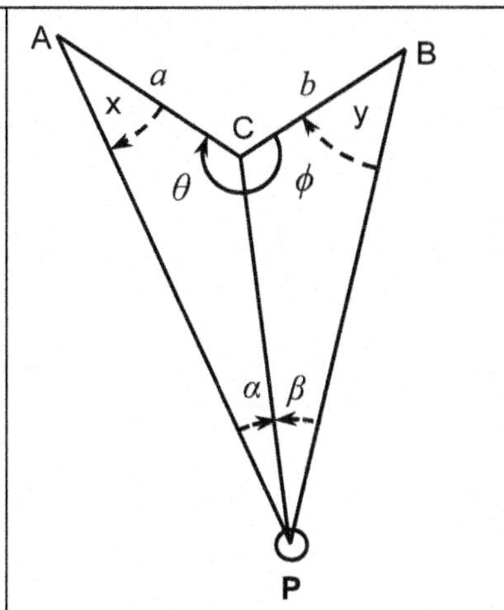

Figure 7-23

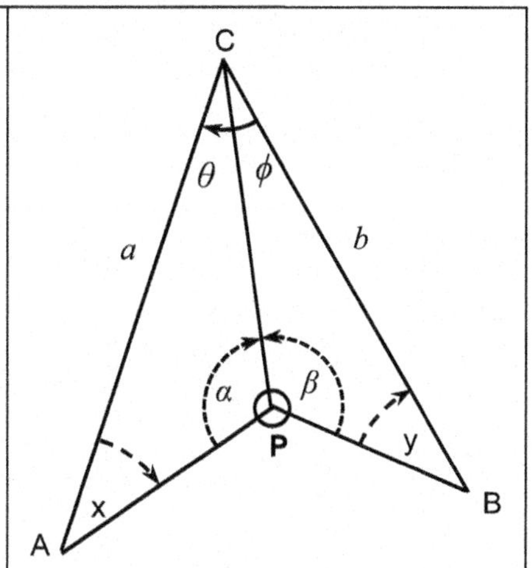

Figure 7-24

The solution of the position of P is possible because P is at the intersection of circles circumscribing the triangles ACP, CBP and ABP (not illustrated in this solution). Any two of these circles, in this case ACP and CBP, will provide an intersection.

In Figure 7-25 the circle passing through ACP is defined by the chord AC (side a) and ∠APC (∠α). Similarly, the circle through CBP is defined by the chord CB (side b) and ∠BPC (∠β).

The chords a and b can be calculated from the coordinates of A, B and C. The observation of the angles α and β will allow the calculation of intersection of the circles at P.

Angles x and y are calculated from a tan (½) formula, which will be derived.

There is no solution if figure ACBP forms a cyclic quadrilateral, i.e., the two circles coincide.

The method requires caution if point P lies close to a circle circumscribing the three points A, B and C.

Caution: When Σ(x+y)→180° the solution becomes unstable.

Observation of three points will give an exact solution. However, it is more common to observe extra stations to provide checking and elimination of unsatisfactory solutions. A least squares solution of coordinates produced from an over-determined observation is possible.

The solution offered here is one of many analytical solutions. It is based on the fact that, if the **sum** of x and y is known, (Σ(x+y)), and the **ratio** of the Sines, i.e sin(x)/sin(y) is also known, then ∠x **and** ∠y can be found. Knowing ∠x and ∠y at A and B respectively allows ΔACP and ΔBCP to be solved. Hence the coordinates of P can be calculated.

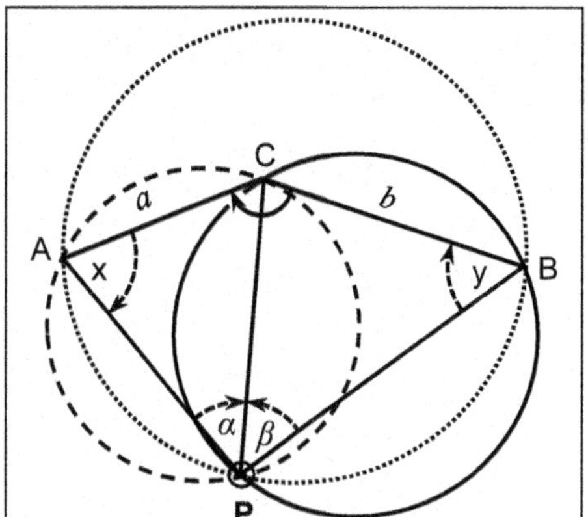

Figure 7-25 Three circumscribing circles through A, B, C and P.

Show $\tan\left(\frac{x-y}{2}\right)=\left(\frac{k-1}{k+1}\right)\times\tan\left(\frac{x+y}{2}\right)$

In Figure 7-25 the sum of the angles $x+y+\alpha+\beta+\angle C=360^\circ$. Thus $(x+y)=360^\circ-(\alpha+\beta+\angle C)$.

Let $\frac{\sin(x)}{\sin(y)}=k$.

Derivation: $\frac{k-1}{k+1}=\left(\frac{\frac{\sin(x)}{\sin(y)}-1}{\frac{\sin(x)}{\sin(y)}+1}\right)=\left(\frac{\sin(x)-\sin(y)}{\sin(x)+\sin(y)}\right)$

And, from trig identities (Bradley, 2014, page5.)

$$\frac{\sin(x)-\sin(y)}{\sin(x)+\sin(y)}=\frac{2\sin\frac{1}{2}(x-y)\times\cos\frac{1}{2}(x+y)}{2\cos\frac{1}{2}(x-y)\times\sin\frac{1}{2}(x+y)}=\frac{\tan\frac{1}{2}(x-y)}{\tan\frac{1}{2}(x+y)}$$

$\frac{k-1}{k+1}=\frac{\tan\frac{1}{2}(x-y)}{\tan\frac{1}{2}(x+y)}$. Thus:

Tan half formula, for angles	$\tan\left(\frac{x-y}{2}\right)=\left(\frac{k-1}{k+1}\right)\times\tan\left(\frac{x+y}{2}\right)$	Eqn 7-7

Solution:

1. From the coordinates of A, B and C the bearing/distances CA and CB can be found.
   $\angle$C =Brg CA – Brg CB

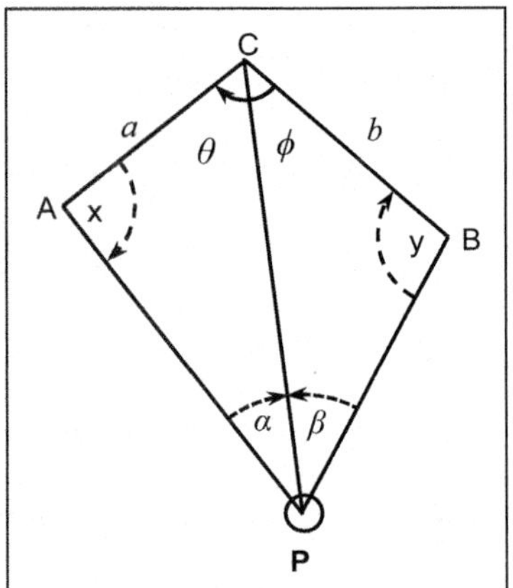

Figure 7-26 Solution diagram

2. Find $\frac{\sin(x)}{\sin(y)}=k$; using Figure 7-26 and sin rule. CP is the common side.

   In ΔACP: $\frac{CP}{\sin(x)}=\frac{CA}{\sin(\alpha)}\Rightarrow \quad CP=\frac{a\times\sin(x)}{\sin(\alpha)}$

   and ΔBCP: $\frac{CP}{\sin(y)}=\frac{CA}{\sin(\beta)}\Rightarrow \quad CP=\frac{b\times\sin(y)}{\sin(\beta)}$

   So: $\frac{\sin(x)}{\sin(y)}=\frac{b\times\sin(\alpha)}{a\times\sin(\beta)}=k$

3. Substitute k to solve tan half formula (Eqn 3-3, page 24 and § 3.6.4, page 28)
   Solve tan half formula (Eqn 7-7):

$$\tan\left(\frac{x-y}{2}\right)=\left(\frac{k-1}{k+1}\right)\times\tan\left(\frac{x+y}{2}\right)$$

$$\left(\frac{x-y}{2}\right)=\arctan\left(\left(\frac{k-1}{k+1}\right)\times\tan\left(\frac{x+y}{2}\right)\right),\text{ and}$$

$$\left(\frac{x+y}{2}\right)=\frac{1}{2}\left(360^\circ-(\alpha+\beta+\angle C)\right)$$

4. Solve for x and y by addition and subtraction.

$$x=\left(\frac{x+y}{2}\right)+\left(\frac{x-y}{2}\right)$$

$$y=\left(\frac{x+y}{2}\right)-\left(\frac{x-y}{2}\right)$$

5. Solve angles in ΔACP and ΔBCP
   θ = $\angle$ACP = 180 – (x + α)
   ϕ = $\angle$BCP = 180 – (y + β)

6. Solve sides in ΔACP and ΔBCP using sin rule.

   ΔACP $\quad AP=\frac{a\times\sin(\theta)}{\sin(\alpha)}$ $\qquad$ ΔBCP $\quad BP=\frac{a\times\sin(\phi)}{\sin(\alpha)}$

   Check that $CP=\frac{a\times\sin(x)}{\sin(\alpha)}=\frac{b\times\sin(y)}{\sin(\beta)}$

7. Find bearings to P from A and B
   Brg AP = Brg AC + x.
   Brg BP = Brg BC – y

8. Calculate coordinates of P by a join from A, check from B

If it's not checked it's wrong.

7.8.1 Worked Example 7.6 Resection by Angles

A set of observations were made at point P to Stations A, B and C.

The observed angles are: (Figure 7-27).

$\angle APC = \alpha = 44°57'27''$

$\angle BPC = \beta = 52°17'29''$

Calculate the coordinates of P.

Solution:

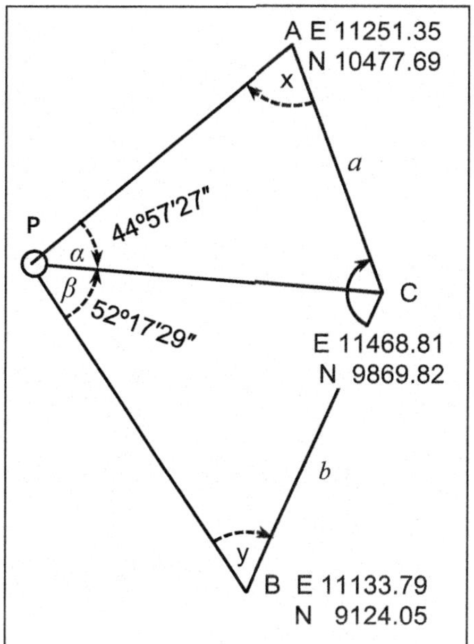

Figure 7-27 Worked Example 7.6

1. From coordinates A, B and C
   Brg CA= 340°18′57″
   Dist CA = a = 645.596
   Brg CB = 204°11′27″
   Dist CB = b = 817.564
   $\angle C = 136°07'30''$

$(x + y) = 360° - (44°57'27''+52°17'29''+136°07'30'') = 126°37'34''$

$\left(\frac{x+y}{2}\right) = 63°18'47''$

2. Find $\frac{\sin(x)}{\sin(y)} = k$; using Figure 7-27 and sin rule.

CP is the common side.

$$k = \frac{\sin(x)}{\sin(y)} = \frac{b\times\sin(\alpha)}{a\times\sin(\beta)}$$

$b\times\sin(\alpha) = 577.6762$

$a\times\sin(\beta) = 510.7518$

$k = 577.6762/510.7518 = 1.1310$

3. Substitute k to solve tan formula

$$\tan\left(\frac{x-y}{2}\right) = \left(\frac{k-1}{k+1}\right)\times\tan\left(\frac{x+y}{2}\right)$$ see Eqn 7-7.

$\left(\frac{k-1}{k+1}\right) = 0.131031/2.131031 = 0.061487$

$\left(\frac{x+y}{2}\right) = 63°18'47''$. $\tan\left(\frac{x+y}{2}\right) = 1.989413$

$\tan\left(\frac{x-y}{2}\right) = 0.061487\times1.989413 = 0.122324$ $\left(\frac{x-y}{2}\right) \Rightarrow \arctan(0.122324) = 6°58'26''$

4. Solve for x and y by addition and subtraction.

$$x = \left(\frac{x+y}{2}\right)+\left(\frac{x-y}{2}\right) = 63°18'47'' + 6°58'26'' = 70°17'13''$$

$$y = \left(\frac{x+y}{2}\right)-\left(\frac{x-y}{2}\right) = 63°18'47'' - 6°58'26'' = 56°20'21''$$

5. Solve angles in ΔACP and ΔBCP

$\theta = \angle ACP = 180 - (x + \alpha) = 180° - (70°17'13'' + 44°57'27'') = 64°45'20''$

$\phi = \angle BCP = 180 - (y + \beta) = 180° - (56°20'21'' + 52°17'29'') = 71°22'10''$

6. Solve sides in ΔACP and ΔBCP using Sin rule.

ΔACP $AP = \frac{a\times\sin(\theta)}{\sin(\alpha)} = 583.939040/0.706582 = 826.4277$

ΔBCP $BP = \frac{b\times\sin(\phi)}{\sin(\beta)} = 774.722535/0.791132 = 979.2587$

If it's not checked it's wrong.

By SIN rule, check that $CP = \frac{a \times \sin(x)}{\sin(\alpha)} = \frac{b \times \sin(y)}{\sin(\beta)} = 860.1421$

7. Find **bearings** to P from A and B
   Brg AP = Brg AC + x = 160°18′57″ + 70°17′13″ = 230°36′10″
   Brg BP = Brg BC – y = 24°11′27″ – 56°20′21″ = 327°51′06″

Calculate coordinates of P by a join from A,

	E		N
A	11251.35		10477.69
dE_{AP}	–638.835	dN_{AP}	–524.527
P	**10612.715**		**9953.163**

Check from B.

	E		N	
B	11133.79		9124.05	
dE_{BP}	–521.075	dN_{BP}	829.113	
P	**10612.715**		**9953.163**	Check ✓

7.9 Chapter Exercise 7.2

Q. 1 Further angles have been read at Station P to Station D as shown in Figure 7-28.
Given the coordinates of Stations A, B and D, and the two angles DPA and DPA.
Calculate the coordinates of P.
Refer to Figure 7-27.

Hint: review the diagrams in § 7.8, on page 109.

7.9.1 Answer to Chapter Exercises 7.2

Q. 1 E_P = 10612.712 N_P = 9953.150

Compare the coordinates with § 7.8.1 to see the effect of observational errors that can be incorporated in the least squares adjustment of an overdetermined figure.

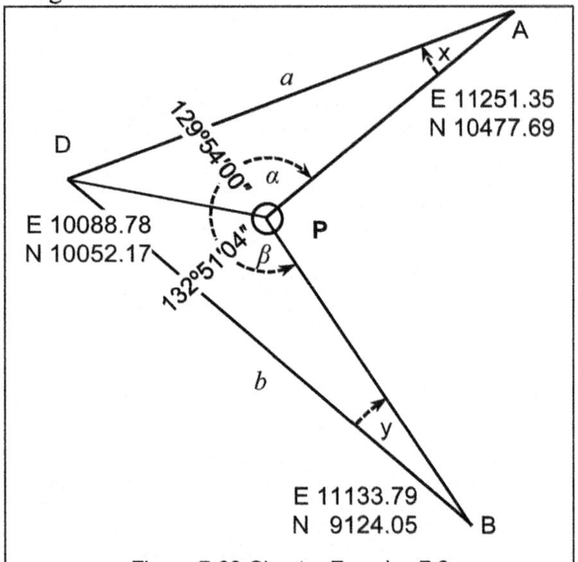

Figure 7-28 Chapter Exercise 7.2

7.10 References for Chapter 7

Uren, J. and B. Price. Surveying for engineers. Fifth edition. Palgrave Macmillan, London 2010. ISBN978-0-230-22157-4

Bradley, Jennifer M, Tables and Formulae, Murdoch University, 2014, ISBN 0-86905-786-3.

Chapter 8 Simple Horizontal Curves

Consider a portion of a road that is straight connecting to another portion of a straight road. If a vehicle is to connect these straight roads, a vehicle would have to make a sharp corner which in practice is impossible. Normally, these straight portions of a road are joined by a smooth curve to make the vehicle motion smooth. This is normally known as horizontal alignment, which is normally simplified by a circle known as simple curve in whose radius is uniform. In a more complicated case, the radius varies in what is known as a complex or transitional curve. This chapter provides you with the skills of computing elements of the simple circular curve. More complicated computations of the complex curve are treated in Chapter 12.

8.1 Introduction.

A surveyor's concern with curves is twofold. The first is with the design of curves on roads and the second is with the responsible for setting out the road curves. This chapter is limited to calculations on the simple horizontal curve.

8.2 Types of curves.

8.2.1 Simple Curve

Figure 8-1 is a simple curve which is formed by part of an arc of a circle.

AB is a simple curve tangent to two straights at A and B. The tangent point at A is referred to as the P.C. (Point of curvature) and B as the P.T. (Point of Tangency). The intersection of the tangents at C is called the P.I. (Point of Intersection) or the I.P.

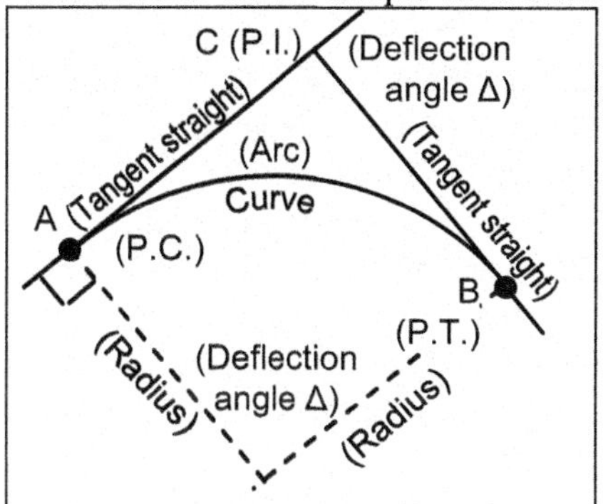

Figure 8-1 Simple curve

8.2.2 Compound Curve

Figure 8-2 shown here is where two or more simple curves are combined to form a compound curve between three straights. The tangents between the two curves on the common straight form the "point of common curvature" or P.C.C.

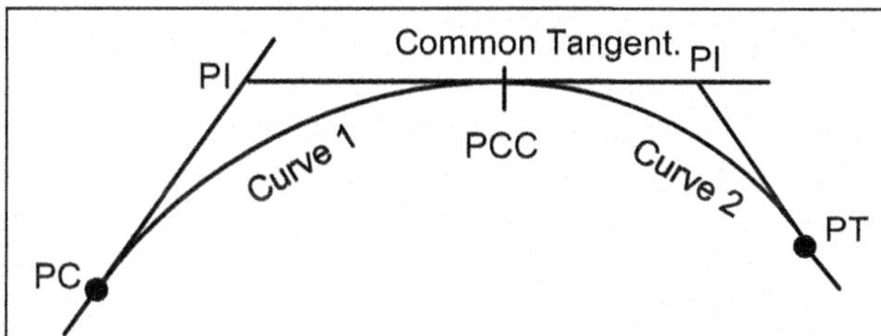

Figure 8-2 Compound curve

8.2.3 Reverse Curve

A reverse curve (see Figure 8-3) consists of two circular simple curves tangent to each other at a point but the curves are on opposite sides of the tangent.

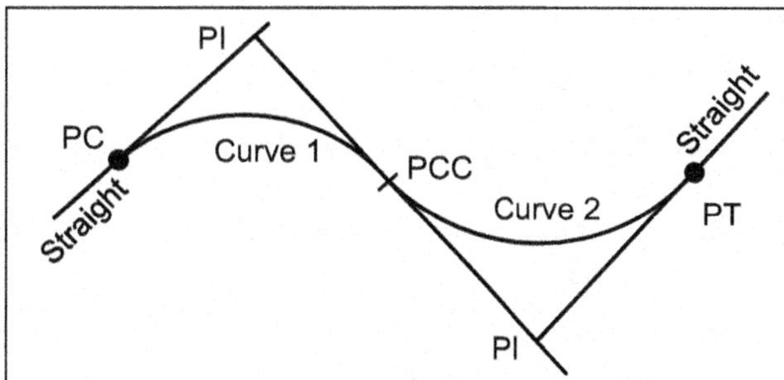

Figure 8-3 Reverse curve

8.2.4 Transition or Spiral Curve

Because the change from a straight section of road to a curve may be too abrupt, a spiral curve (see Figure 8-4) is introduced between the straight and curve and again between the curve and straight. The spiral gradually introduces super-elevation from zero at the TS to its maximum at SC. The spiral from the CS to the ST reverses this procedure.

The transition curve is presented in Chapter 12, page 181.

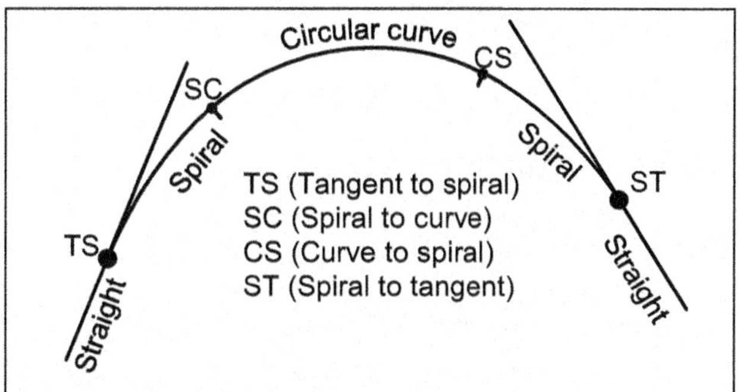

Figure 8-4 Transition curve

J. Walker and J. Awange, *Surveying Computations*,
https://doi.org/10.1007/978-3-031-94382-9_8

8.2.5 Vertical Curve

The vertical curve (see Figure 8-5) connects two straights in profile (elevation). It is required at the intersection of gradients over rises at a summit, or at sags. Vertical curves are not treated in this chapter.

For completeness, the parabolic vertical curve is presented in Chapter 14, page 217.

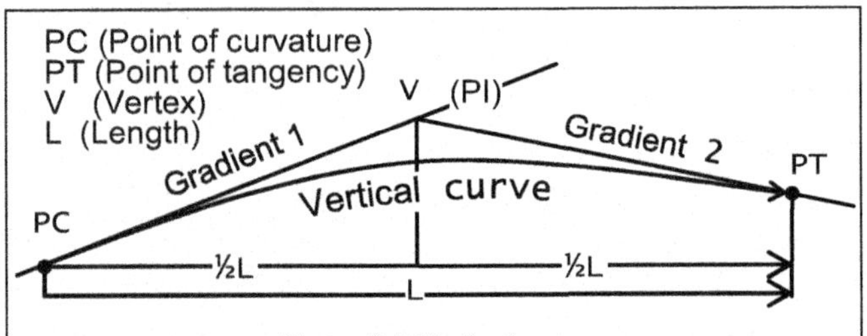

Figure 8-5 Vertical curve

8.3 Defining simple curves.

How are curves defined? – There are two methods of calculating a simple curve (i.e., a curve of fixed radius). This chapter introduces the two methods and will concentrate on the curve defined by radius.

The second method, the degree method, is a specialised railway lay-out process whereby the curve is set out using simple tools of tape and transit (theodolite) to achieve satisfactory accuracy commensurate with the task. Modern survey calculations and techniques have superseded the method.

Simple curves are defined:

a) In terms of its radius

The usual method of defining a simple curve is by reference to its radius. i.e. a 100m curve where the radius of the curve is 100m. The longer the radius, the flatter the curve and vice versa. From the deflection angle between two tangent (straight) lines, the curve is calculated in terms of design radius and resultant arc length. From the arc length the intermediate arcs between lay out points (chainages) define the layout deflection angles and **chord** distances.

b) Degree of curve (see Figure 8-6)

The "degree of a curve" is a method used for setting out large radius curves, e.g., in railways, roads and the like. It is based on the length of arc between chainage stations being the (almost) same length as the chord. Standard US practice was to use a 100-foot chord (the length of the tape measure) around a curve of defined central deflection. Some British practices use a 66ft chord length

The "degree of curve" deflection angle was a whole number of degrees, thus allowing the surveyor to lay out the curve by simple tangent deflection angles, they being half the degree of curvature.

As the radius is large, the arc and the chord are almost the same. Thus, the layout is simplified in the field. The radius is relatively inconsequential except for computation purposes.

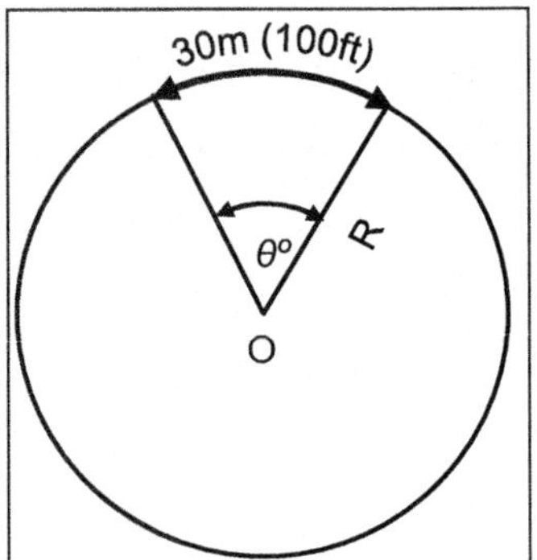

Figure 8-6 Degree curve

A curve may be defined by the central angle subtended by an arc of, say, 30m. The angle at the centre of the curve subtended by an arc of 30m is called the degree of the curve. Thus, a 4° curve is a curve in which an arc of 30m is subtended by a central angle of 4°. The degree of a curve simplifies set out by deflection angle, but results in a difficult radius. Given the degree of the curve, how do we find its radius?

8.3.1 Worked Example 8.1 Degree of a curve.

Suppose the degree of 30m (100ft) curve = 4° (See Figure 8-7)

Arc = $R \cdot \theta^R$ (θ^R in Radians)

i.e. $30 = R \cdot \theta^R$

$$\theta = 4^\circ = \frac{4 \times \pi}{180} = 0.070^R$$

$$R = \frac{30}{0.070}$$

R = 429.7m

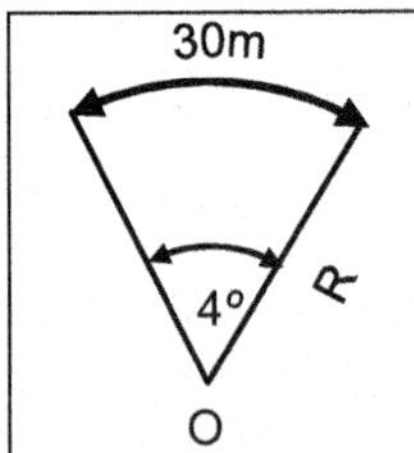

Figure 8-7 4° curve

For the 4° curve the surveyor uses a deflection angle of 2° around the curve.

The chord length between 30m arc stations is:

Chord = 2RSin(2°) = 29.993, very close to the arc length.

Minimum radius of high-speed rail systems, with speeds of up to 350km.hr^{-1}, require radii to 7,500m.

Part of Joondalup Drive, in a northern Perth suburb, has a radius of over 6,000m.

8.4 Tangent and chords to circle.

A tangent is a straight line that touches a circle. A chord is a straight line that cuts the circle.

a) The radius and the chord are perpendicular. (Figure 8-8).
b) Radius is perpendicular to the chord at its midpoint. (Figure 8-9).

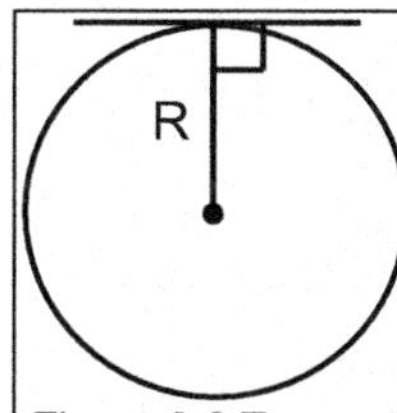

Figure 8-8 Tangent perpendicular

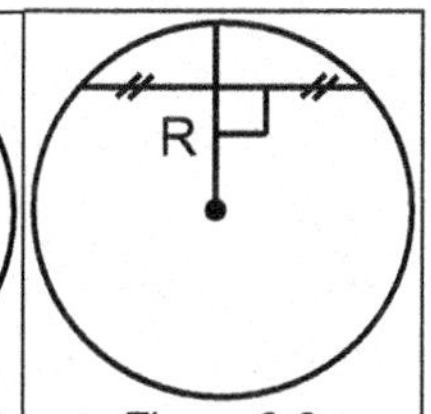

Figure 8-9 Chord bisects

8.5 Angular Relationship between the Centre and the Circumference.

One of the most important properties of a circle is that an angle at the **circumference** of a circle, subtended to a chord, is **half** the angle subtended at the **centre** of the circle. (See Figure 8-10).

$\angle O = 2\angle C$

The MINOR $\angle AOB$ (Δ) = $2\angle ACB$

$\angle ACB$, $\angle ADB$ = ½Δ

The deflection angle on the circumference of a circle is **half** the deflection angle at the centre of that circle.

$\angle O' = 2\angle C'$

The MAJOR $\angle AOB$ (Δ') = $2\angle AC'B$

If the centre of the circle and the point on the circumference are adjacent (same side) the chord then $\angle ACB$, $\angle ADB$ = ½minor $\angle AOB$ (Δ).

If the centre of the circle and the point on the circumference are opposite (other side) the chord then $\angle AC'B$ = 180 – ½minor $\angle AOB$ (Δ)

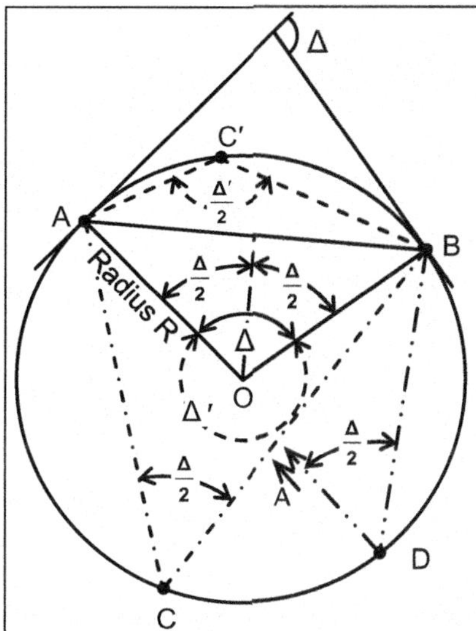

Figure 8-10 Deflection angles on a circle

8.6 Visualising a Circular Curve

The circular curve is the simple figure that joins two straight lines together. It is a thing of beauty once it has been laid out in the field. But it is also a bit of a mystery in its generation, until the basics are mastered. The development of the circular curve is best illustrated by a build-up of the elements of the curve.

8.6.1 Basic Formulae. Arc length

All CIRCULAR calculations are carried out in RADIANS. 360° = 2π (6.2832) radians.

Arc length	$S = R\theta^{radians}$	Eqn 8-1

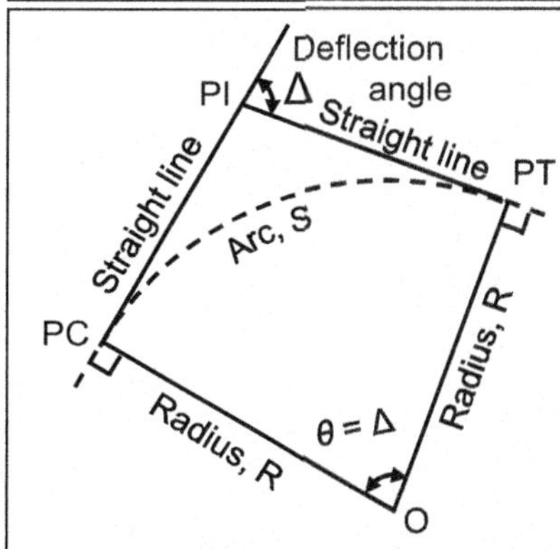

Figure 8-11 Simple curve definitions

Figure 8-11 shows a simple curve of given RADIUS, R, joining two intersecting straight lines.

The straight lines differ in direction by a DEFLECTION angle, Δ (delta) degrees, at the INTERSECTION POINT (PI or IP).

The curve has a CENTRE at the intersection of two RADII that are PERPENDICULAR to two straight lines that are the TANGENT POINTS to the curve. The radii are PERPENDICULARS to the straight lines AT the TANGENT POINTS.

The DEFLECTION angle, θ (theta), at the centre of the arc is equal to the deflection angle, Δ, at the intersection point (PI or IP).

8.6.2 Initial geometry of the curve.

The LENGTH, S, of the ARC between the two tangent points, TPs, is given by the expression: Eqn 8-1 and shown in Figure 8-12.

$S = R\theta$ radians

The ingoing curve starts at the tangent point,
the POINT OF CURVATURE, PC,
and its outgoing ends at the other tangent point,
the POINT OF TANGENCY, PT.

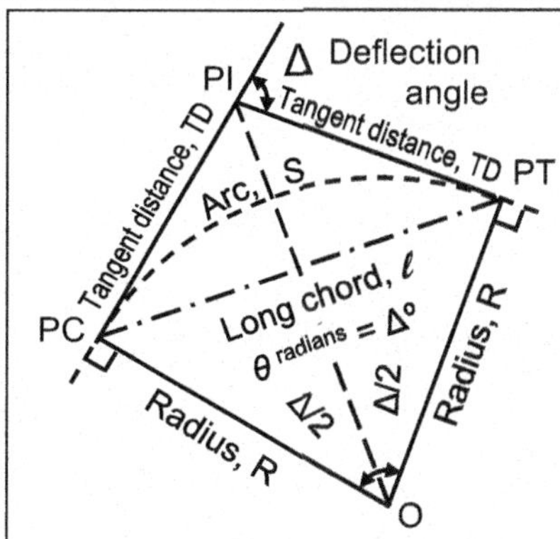

Figure 8-12 Initial curve geometry

Example:
Radius, R = 200
Deflection angle at IP, Δ = 80° then $\theta = 80\times\pi/180 = 1.3963$ radians
Arc S = 200×1.3963= 279.253

8.6.3 The Tangent Distance

The line O-IP, joining the IP and the centre, O, bisects the deflection angle, Δ.
The triangles O-PC-IP and O-PT-IP are equal.
The length of the straight line from the PC to the IP and IP to PT is the TANGENT DISTANCE, TD.

From plane trigonometry, the TD is the opposite side, and the radius, R, is the adjacent side of the triangle PC- O-IP from the half deflection angle **Δ**/2.

In Figure 8-12 and Figure 8-13, in triangle O-PC-IP,

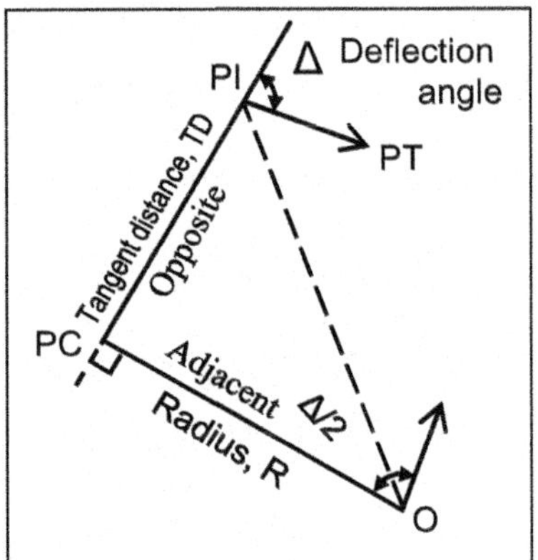

Figure 8-13 Tangent distance

Derive TD, Tangent distance.	$\tan\left(\frac{\Delta}{2}\right)=\left(\frac{opposite}{adjacent}\right)=\frac{TD}{R}$ $TD = R\tan\left(\frac{\Delta}{2}\right)$	Eqn 8-2

From Eqn 8-2:
TD = R tan(Δ/2) = 200×tan(40°)
= 167.820
The two tangent distances are equal: PC to IP = IP to PT.

8.6.4 The Long Chord

Figure 8-12, from PC to PT is made up of two half chords, PC - P, and P - PT, shown in Figure 8-14, in triangle O-PC-PI.
The long chord, length ℓ, joining PC to PT:
ℓ = 2R sin(Δ/2) = 400×sin(40°)
= 257.115

The ANGLE between the TANGENT STRAIGHT and the CHORD at the tangent point is the CHORD DEFLECTION ANGLE, it is half the TANGENT deflection angle at the tangent intersection, PI. = Δ°/2.

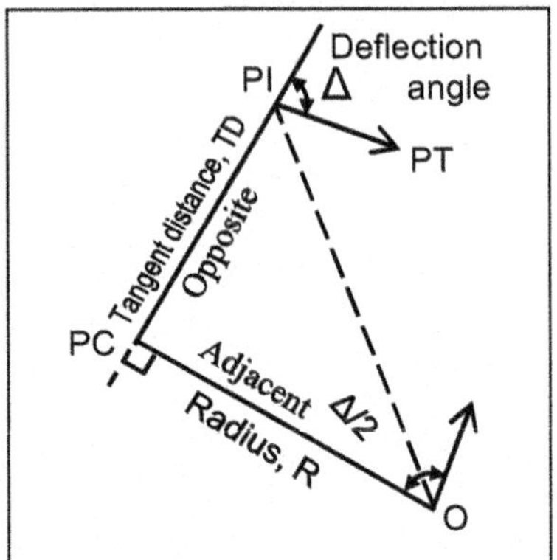

Figure 8-14 Long chord derivation

Notice in Figure 8-12, the difference between the arc length, S, and the chord, ℓ.
From § 8.6.2 Arc length, S = 279.253
From § 8.6.4 Long chord, ℓ = 257.115

8.6.5 Other Chords

The other chords we are concerned with join any two points on the arc of circle radius R. Consider the chord starting at the tangent point, the PC.

Derive ℓ, Long chord.	$\sin\left(\frac{\Delta}{2}\right)=\frac{opp}{hypot}=\frac{\ell/2}{R}$ Long chord $\ell = 2R\sin\left(\frac{\Delta}{2}\right)$	Eqn 8-3

Thus, in Figure 8-15 any point, **p**, on the arc at an arc distance, *s*, from the tangent point, is tangential to the arc at **p** towards the centre, **O.** It intercepts the tangent at **I**.

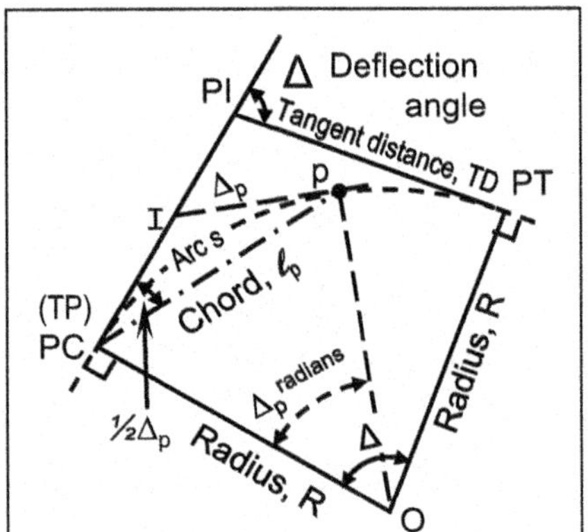

Figure 8-15 Deflection angle and chord length for a given arc distance

8.6.6 The Chord Deflection Angle

It follows that, from Eqn 8-1, then the TANGENT deflection:

Tangent deflection angle:	$\theta_{p(radians)}=\frac{s_p}{R}$	Eqn 8-4

and the angle, in degrees, is:

Radians to degrees:	$\Delta^{\circ}{}_{p} = \theta_{p}\dfrac{180}{\pi}$	Eqn 8-5

The CHORD deflection from the tangent point to **p** is *half* the tangent deflection angle:

Chord deflection angle:	$\delta^{\circ}{}_{p} = \dfrac{\Delta_{p}}{2}$	Eqn 8-6

and the chord DISTANCE is (from):

$\ell_p = 2R\sin(\tfrac{1}{2}\Delta_p{}^{\circ})$

From these equations we can now start to find some solutions to common circular curve problems.

8.6.6.1 Worked Example 8.2

Suppose point **p** is 150m around the curve from the PC, see Figure 8-16. Find coordinates of p.

arc length, s = 150m

s_p = 150m, R = 200m:

remember, as $S_p = R\theta_p$, then $\theta_p = s_p/R$

θ_p = 150/200 = 0.75radians

Δ_p = 42.972° = 42°58′19″

$\tfrac{1}{2}\Delta_p$ = 21.486° = 21°29′09″ (half Δ)

ℓ_p = 2R sin($\tfrac{1}{2}\Delta_p$) = 400×sin(21°29′09″)

= 146.509m

The BEARING of the CHORD at the tangent point is the BEARING from the PC to the IP PLUS the TANGENT deflection angle, $\tfrac{1}{2}\delta_p$.

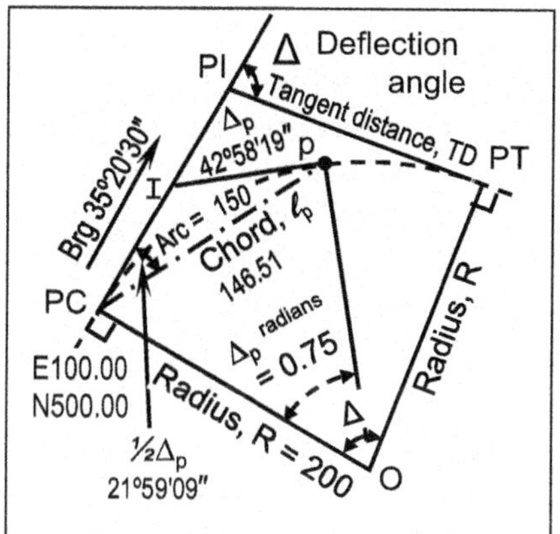

Figure 8-16 Worked Example 8.2

BRG PC – IP = 35°20′30″

+ $\tfrac{1}{2}\Delta_p$ = <u>21°29′09″</u>

BRG PC - p = 56°49′39″

Chord ℓ_p = 146.509

The vector between the PC and p can be converted to rectangular coordinates:

δEp = ℓ_p sin(BRG),

δNp = ℓ_p cos(BRG)

If the coordinates of the PC are known, then: Ep = EPC + δEp and Np = NPC + δNp.

Note that the points along the curve RADIATE from the PC.

All the calculated δEs and δNs are added to the PC coordinates.

Find coordinates of p from PC E100.00, N500.00

BRG_p = 56°49′39″, ℓ_p = 146.509 converts to δE_p = 122.632, δN_p = 80.16

E_p = 222.632

N_p = 580.164

8.7 Circular or Radian Measure.

When working with curves it is often necessary to work with the radian measure of an angle. A definition of radian measure follows:

1 radian is the angle subtended at the centre of a circle by an arc equal in length to the radius. (See Figure 8-17)

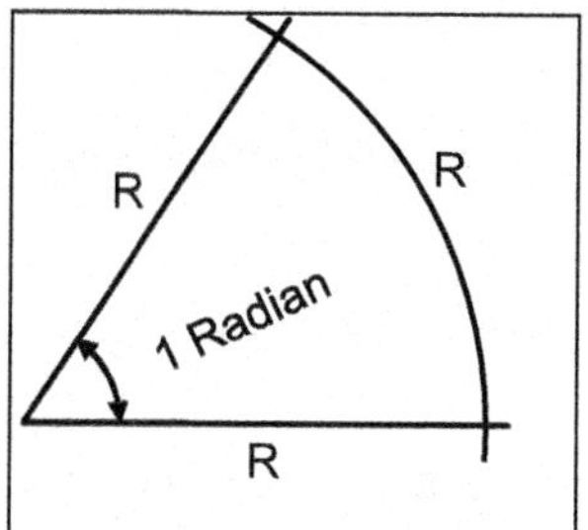

Figure 8-17 1 radian

The radian measure of any angle θ is defined as:

$$\theta = \frac{arc}{radius} = \frac{S}{R} \qquad \text{(See Figure 8-18)}$$

Consider the case of semi-circle (See Figure 8-19)

$$\theta = \frac{arc}{R} = \frac{\pi \times R}{R} \text{ Radians}$$

i.e. 180° = π Radians or 180° ≅ 3.1415927 radians

Thus 1° = 0.0174533 radians

1′ = 0.0002909 radians

1″ = 0.0000048 radians and,

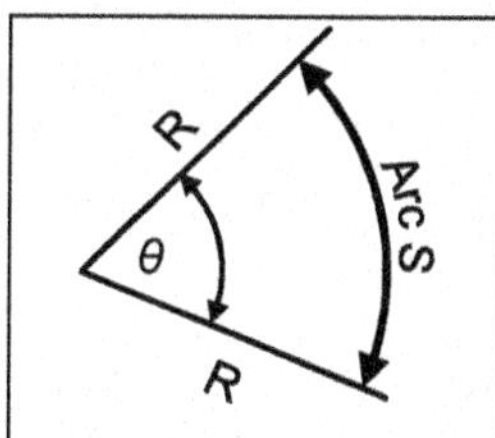

Figure 8-18 Deflection

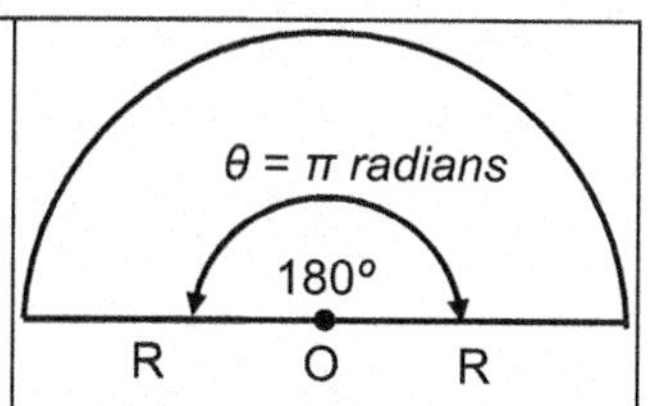

Figure 8-19 Semicircle

If it's not checked it's wrong.

1 radian $= 57.2957795^\circ$
$= 3437.7467'$ (3438")
$= 206264.8062''$ (206265")

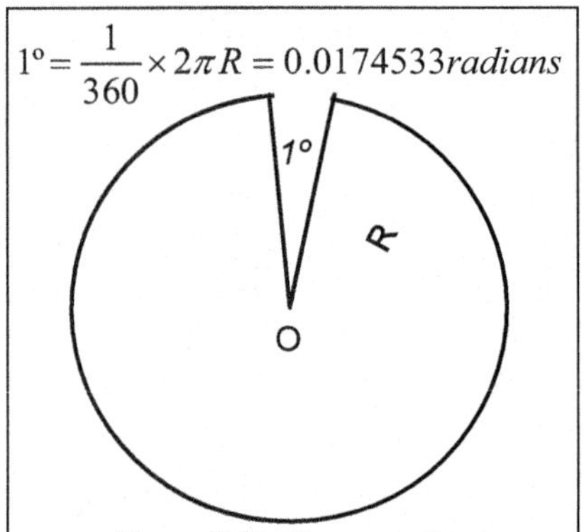

Figure 8-20 1° as a radian

8.8 Degree Measure.

A degree is defined as the angle subtended at the centre of a circle by $^1/_{360}$ of the circumference. (See Figure 8-20).

$360^\circ = 1$ Circle
$60' = 1^\circ$
$60'' = 1'$
$3600'' = 1^\circ$

8.9 The simple curve.

AB is a simple curve, radius **R** connecting two straights EA and BF (See Figure 8-21)

A and B are the tangent points, and the straights are produced to meet in D (the intersection point or point of intersection, PI).

AD, DB are tangents to the simple curve and AD = BD.

Δ is the deflection angle, the exterior angle between the two straights.

Deflection angle (Δ) equals angle at the centre, $\angle ACB$

Triangles ADC and BDC are congruent. Thus,

$\angle ACD = \angle BCD = ^{\Delta}/_2$

Point S, on arc ASB, has deflection angle $\angle DAS = ^{\delta}/_2$.

$\angle ACS = \delta$ (The angle between the straight AB and the tangent to the arc at S).

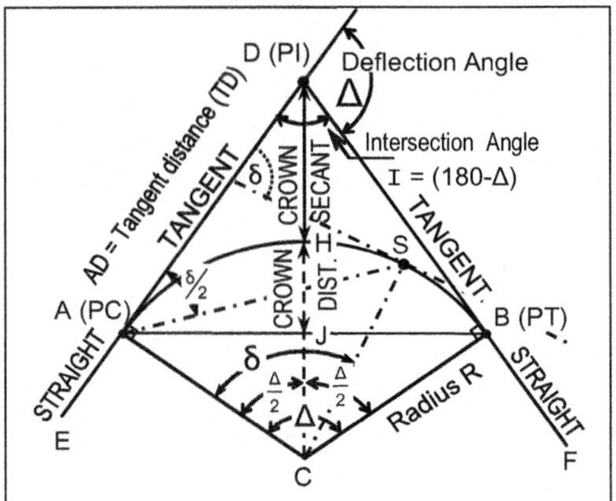

Figure 8-21 Trigonometry of a simple curve

By definition: The DIRECTION to the RADIUS is PERPENDICULAR to the TANGENT.

Bearing Radius = Bearing Tangent $\pm 90^\circ$

8.9.1 Formulae.

Tangent Distance AD	$TD = R\tan\left(\frac{\Delta}{2}\right)$	Eqn 8-7
Radius (R)	$R = TD\cot\left(\frac{\Delta}{2}\right) \because \cot = \frac{1}{\tan}$	Eqn 8-8
Arc length AB (S)	$S = R\Delta^{radians}$	Eqn 8-9
Secant Distance DC	$DC = R\sec\left(\frac{\Delta}{2}\right) \because \sec = \frac{1}{\cos}$	Eqn 8-10
Crown Secant DH	$DH = R\sec\left(\frac{\Delta}{2}\right) - R = R\left(\sec\left(\frac{\Delta}{2}\right) - 1\right)$	Eqn 8-11
Crown Distance JH	$DH = R\sec\left(\frac{\Delta}{2}\right) - R = R\left(\sec\left(\frac{\Delta}{2}\right) - 1\right)$	Eqn 8-12
Long Chord AB	$AB = 2R\sin\left(\frac{\Delta}{2}\right)$	Eqn 8-13
Chord AS (chord to point S)	$AS = 2R\sin\left(\frac{\delta}{2}\right)$	Eqn 8-14

8.9.2 Angular Relationship between the Centre and the Circumference.

One of the most important properties of a circle is that an angle at the **circumference** of a circle, subtended to a chord, is **half** the angle subtended at the **centre** of the circle. (See Figure 8-22)

$\angle O = 2\angle C$

The MINOR $\angle AOB$ (Δ) = $2\angle ACB$

$\angle ACB$, $\angle ADB = \frac{1}{2}\Delta$

The deflection angle on the circumference of a circle is **half** the deflection angle at the centre of that circle.

$\angle O' = 2\angle C'$

The MAJOR $\angle AOB$ (Δ') = $2\angle AC'B$

If the centre of the circle and the point on the circumference are adjacent (same side) the chord then $\angle ACB$, $\angle ADB$ = ½minor $\angle AOB$ (Δ).

If the centre of the circle and the point on the circumference are opposite (other side) the chord then $\angle AC'B$ = 180 – ½minor $\angle AOB$ (Δ)

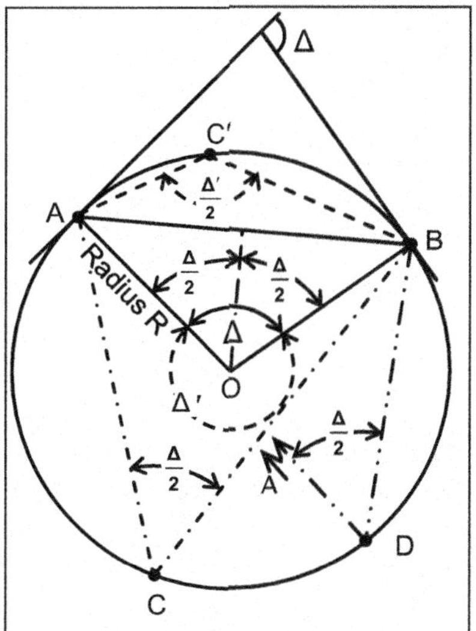

Figure 8-22 Deflection angles on a circle

8.9.3 Area Formulæ.

Additional formulæ dealing with areas generated by components of the simple curve are presented to provide a convenient reference. (Figure 8-23).

Area	Circle	$= \pi R^2$.
	Sector ASBOA	$= \pi R^2 \times (\Delta/2\pi)$
		$= R^2 \times \Delta/2$ (Δ in radians)
	Triangle AOB	$= \frac{1}{2} R^2 \times \sin\Delta$ ($\frac{1}{2} ab \times \sin(C)$)
	Crown Sector ASBA	$= R^2\, \Delta/2 - \frac{1}{2} R^2 \cdot \sin\Delta$
		$= R^2\, (\Delta/2 - \frac{1}{2} \cdot \text{Sin}\,\Delta)$
	Tangent figure AOBDA	$= R \times \tan\Delta/2 \times R$
		$= R^2 \times \tan\Delta/2$ (Tangent Distance × R)
	Truncation ADBSA	$= R^2 \times \tan\Delta/2 - R^2 \times \Delta/2$
		$= R^2(\tan\Delta/2 - \Delta/2)$
	Road reserve	$= \Delta/2(R^2 - R_2^2)$ (Δ in radians)

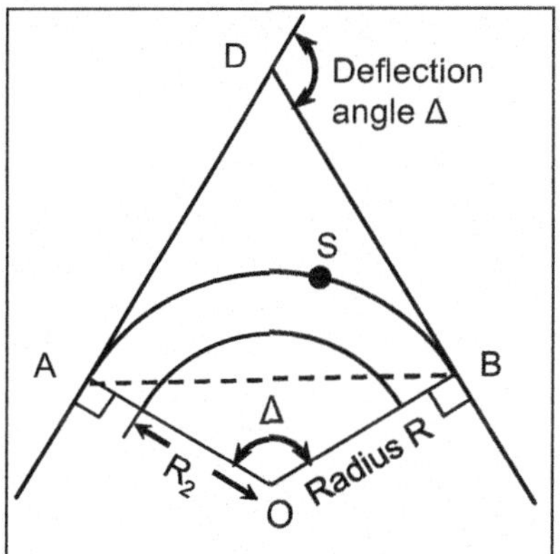

Figure 8-23 Areas

8.10 Worked Examples demonstrate fundamental quantities of a simple curve.

8.10.1 Worked Example 8.3

Two straights intersect at D as shown.

Bearing AD = 60°20′

Bearing DB = 124°40′

The straights are to be joined by a 600m curve. (Figure 8-24)

Calculate:

Tangent length

Length of curve

Crown secant DH

Long chord AB

Crown Distance HJ

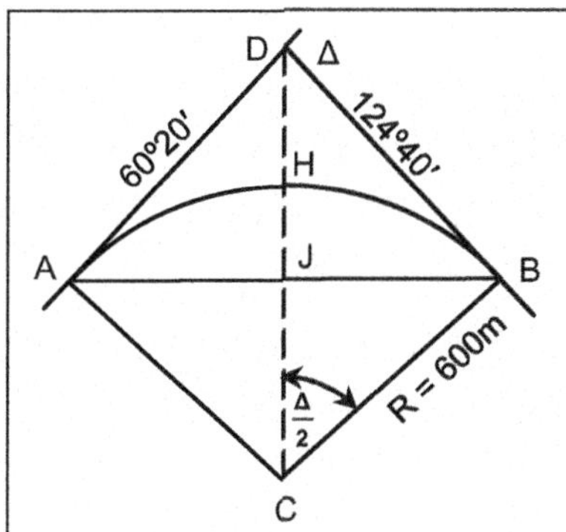

Figure 8-24 Worked Example 8.3

Given:

Deflection angle = Δ = (124°40′ – 60°20′) Δ = 64°20′

$\Delta/2$ = 32°10′

R = 600

(i) Tangent (AD) = $R \cdot \tan\Delta/2$ = 600·tan(32°10′) = 377.35m

(ii) Length of curve = $R \cdot \Delta$ = 600·(64°20′)^r (Radian measure) = 673.70m.

(iii) Crown Secant (DH) = $R \cdot \sec\Delta/2 - R$ = 600/cos(32°10′) – 600 = 108.80m

(iv) Long chord (AB) = $2R\sin\Delta/2$ = 2×600·Sin(32°10′) = 1200×(sin(32°10′)) = 638.86m

(v) Crown Distance (HJ) = $R - R\cos\Delta/2$ = 600 – 600×cos(32°10′) = 92.10m

If it's not checked it's wrong.

8.10.2 Worked Example 8.4

Two straights intersect as shown above and are joined by a simple curve AB. (See Figure 8-25)

Given: Curve (arc) length, S = 1076.30m

Radius (R) = 2000.00m

Calculate:

Deflection angle (Δ)

Tangent distance

Arc = R·Δ = 1076.30

Δ (radians) = (1076.30/2000.00) = 0.538150R

(i) Δ = 30°50′01″ (deflection angle)

(ii) TD = R×tan Δ/2 = 2000.0·Tan(15°25′)

= 551.52m

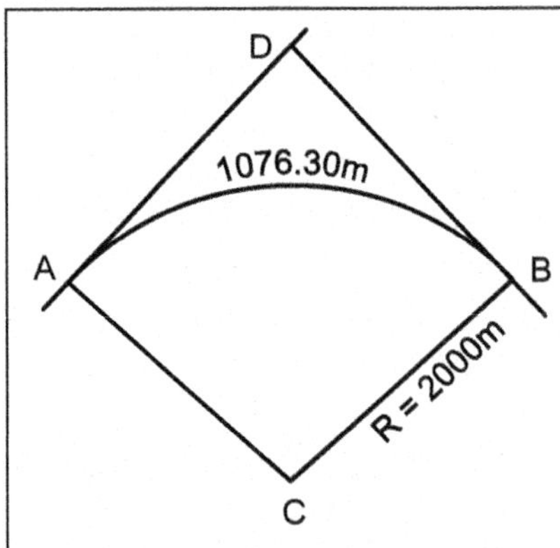

Figure 8-25 Worked Example. 8.4. Δ and TD

Worked Example 8.5

Lot 12 is a corner block which is to have its corner truncated, and the area (ADBHA) given over to the road. Given the above information, find this truncated area. (See Figure 8-26)

Find tangent distance (TD) AD = R×tan Δ/2

Δ = Brg BD – Brg AD = 85°20′ – 0°00′

Δ = 85°20′, Δ/2 = 42°40′

AD = 50.0×tan(42°40′) = 50×(0.921697)

AD = 46.085m

Area ADBCA = AD×R = 46.085×50.0

(2×area ΔADC = 2× ½(AD×R))

Area = 2304.25m²

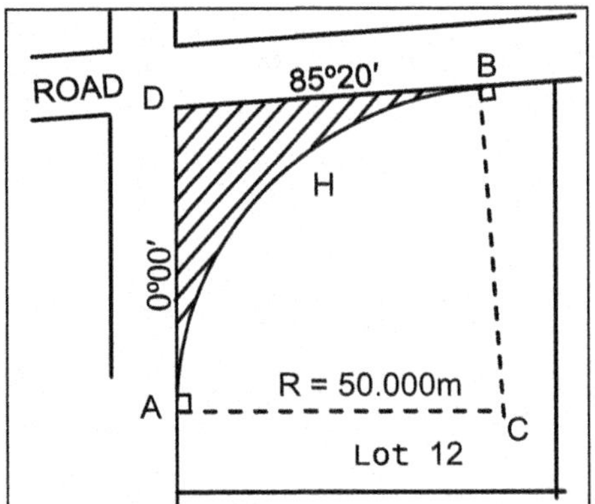

Figure 8-26 Worked Example 8.5 Area

Area sector CAHB = $\dfrac{85^\circ 20'}{360^\circ}\times(2\pi R^2) = (0.237037\times\pi\times 50.0^2)$

Area = 1861.68m²

Truncated area = 2304.25 – 1861.68 = 442.56m2

Area = 0.0442 ha

The general formula for a truncated area is:

$$\text{Area} = \text{Area} = R^2\left(\tan\left(\frac{\Delta}{2}\right)-\frac{\Delta}{2}\right) \quad (\text{ADBCA} - \text{sector AHBCA})$$

$$(AD\times R - R^2\times\Delta/2\ (\Delta \text{ in radians}) \Rightarrow (R\times\tan(\Delta/2)\times R - R^2\times\Delta/2) \Rightarrow (R^2\tan(\Delta/2) - R^2\times\Delta/2)$$

8.11 Setting out a simple curve.

There are a number of methods using angles or distances by which a curve may be set out in the field. The popular method is to set out a curve by deflection angles from a point or points on the curve.

However, since the introduction of electronic computers, it is a simple matter to coordinate points along the curve, thus allowing it to be set out by radiation from any coordinated adjacent point. It is evident that this latter method has certain advantages. You will learn the methods and techniques for setting out curves in your surveying classes.

8.11.1 Setting out by deflection angles.

Setout of the centreline of a road around a curve generally involves setting out points on the curve at intervals (chainages) corresponding to the road longitudinal chainage. This may depend on the complexity of the road. The distance between the chainage points is the arc distance. Over short distances the difference between the arc and chord distance is small.

The method outlined will calculate the deflection angle and the chord length directly. The X (Run) and Y (Offset) from the straight tangent will also be calculated.

In this case the deflection will start at the PC.

8.11.2 Worked Example 8.6

In Figure 8-27 a curve is to be calculated for set out using both deflection and chord method and the run/offset method.

Chainage at PC = 1000.00

Arc length per interval, s = 30

Radius, R = 500

Deflection angle, $\angle D = \Delta = 42°20'$

Arc length (S) of curve

$S = R\times\Delta^R$

$= 500\times(42°20')^R = 500\times0.738856$

$S = 369.428\text{m}$

Deflection angle, δ, from centre, O

for arc segment PC to S_{1030}

$s = 30$

$\delta = s/R = 30/500 = 0.06R$

$= 3°26'16''$

But deflection angle, DAS_{1030}, $\delta/2$, from the PC is the one used for the set out.

$\delta/2 = 0.03R = 1°43'08''$

The chord, ℓ, PC to S_{1030}

$\ell = 2R\times\sin(\delta/2)$

$= 2\times500\times\sin(1°43'08'')$

$= 1000\times(0.029996)$

$\ell = 29.996\text{m}$

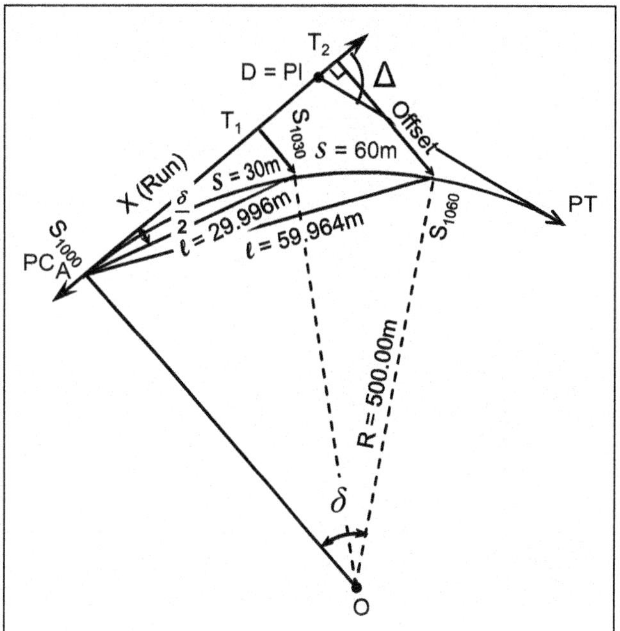

Figure 8-27 Worked Example 8.6. Road set out

The two other values that may be of interest in setting out are the "Run", the X distance along the tangent from the PC and the "Offset" distance, Y, perpendicular from the tangent to the curve.

Run $= X = \ell\cos(\delta/2)$

$= 29.996\times\cos(1°43'08'') = 29.996\times(0.99955)$

$X = 29.982\text{m}$

Offset $= Y = \ell\times\sin(\delta/2)$

$= 29.996\times\sin(1°43'08'') = 29.996\times(0.02996)$

$Y = 0.900\text{m}$

Similarly, for the second chainage for arc segment TC to S_{1060}

$s = 60$

$\delta = s/R = 60/500 = 0.12R$

$= 6°52'32''$

$\delta/2 = 3°26'16''$

The chord, PC to S_{1060}

$\ell = 2R\times\sin(\delta/2)$

$= 2\times500\times\sin(3°26'16'') = 1000\times(0.059964)$

$= 59.964\text{m}$

Run $= X = \ell\cos(\delta/2)$

$= 59.964\cdot\text{Cos}(3°26'16'') = 59.962\cdot(0.998201)$

$X = 59.856\text{m}$

Offset $= Y = \ell\sin(\delta/2)$

$= 59.962\times\sin(3°26'16'') = 59.962\times(0.059964)$

$Y = 3.596$

A table of set out, Table 8-1, can be prepared from the PC. This is suited to EDM using a total station. It is evident that, from a practical point of view, only three or four points could be pegged before moving to a new set out point.

It should be noted that the offset distances increase rapidly and rely on an accurate determination of the 90° offset at the run distance. Better suited to the band and tape using a perpendicular offset.

Table 8-1 Worked Example 8.6

Station Chainage	Arc Dist. S	Deflection (δ)	Chord ℓ	Run = X = ℓcos(δ/2)	Offset = Y = ℓsin(δ/2)
PC = 1000	0	0° 00'	0.000	0.00	0.00
1030	30	1° 43'	30.00	29.98	0.90
1060	60	3° 26'	59.96	59.86	3.60
1090	90	5° 09'	89.88	89.52	8.08
1120	120	6° 53'	120.71	118.85	14.33
1150	150	8° 36'	149.44	147.76	22.33
1180	180	10° 19'	179.03	176.14	32.05
1210	210	12° 02'	208.46	203.88	43.46
1240	240	13° 45'	237.70	230.89	56.50
1270	270	15° 28'	266.73	257.07	71.14
1300	300	17° 11'	295.52	282.32	87.33
1330	330	18° 54'	324.04	306.56	105.00
1360	360	20° 38'	352.27	329.69	124.10
PT = 1369.428	369.428	21° 10'	361.08	336.72	130.38

8.11.3 Worked Example 8.7 Curve set out

A new circular 20m road formation is designed to replace an existing road bounded by PC, PI and PT. (See Figure 8-28).

The chainage at the PC is 635.00m.

The bearing PC – PI is 30°, with coordinates of the curve centre, O, 1000.00, 1000.00,

coordinates of PI are 794.32, 1293.75

The task is to peg the centre line of the new road to correspond with a running chainage interval of 50m on the road centre line. (e.g. C9 is point for chainage 900.00 on ℄).

Prepare:

1) a list of components of the curve
2) a table of centreline coordinates
3) a radiation setout table from the PI to the centreline coordinates and the crown secant.

The calculated deflection angle, Δ, and the radius are both integer values and will be used throughout the design calculations.

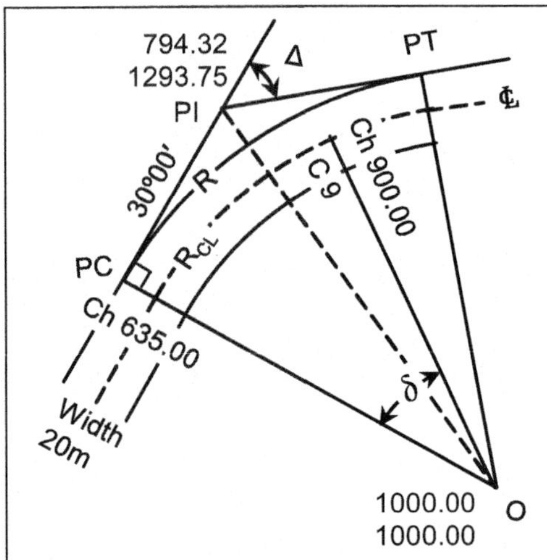

Figure 8-28 Worked Example 8.7. New road

The chainage of the PC is 635m from some start point. This means that the chainage at any point on the radius PC – O (the centre) is also 635m.

The task is to peg the centre line of the 20m road. This means that the chainage of the centreline PT will be ChPC + centreline arc length.

Solution:

1) find deflection angle, Δ

a) join PI to O to find the bearing and the secant distance from PI.

	E		N
O	1000.00		1000.00
– PI	794.32		1293.75
ΔE	205.68	ΔN	–293.75

Brg PI-O 145° (integer value)

Secant dist PI-O 358.598 m (See Figure 8-29)

b) by definition, Δ/2 = Brg PI-O – Brg PC-O

Brg	PI-O	145
– Brg	PC-O	120 (being perpendicular to initial Brg.
Δ/2 =		25°

Deflection angle, Δ = 50°

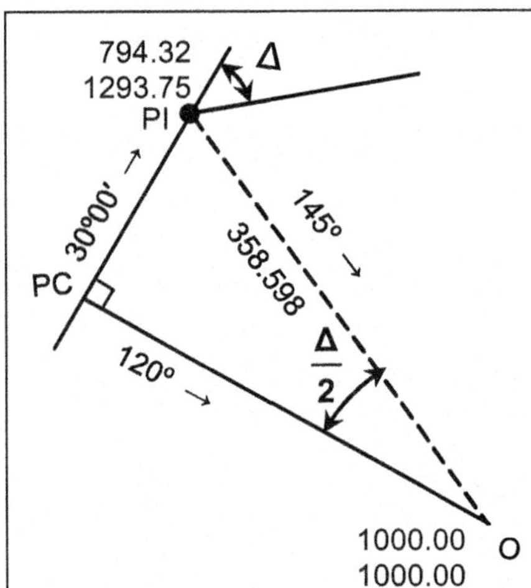

Figure 8-29 Find Δ, deflection angle

2) find radius to PC, R_{PC}
   a) secant distance = $R\times\sec(\Delta/2) = R/(\cos(\Delta/2)$; thus
      R = Secant distance$\times\cos(\Delta/2) = 358.598\times\cos(25) = 358.598\times(0.906308) = 325.000$
   Radius to PC = 325m (integer value)
   b) tangent distance = $R\times\tan(\Delta/2) = 325\times\tan(25) = 325\times(0.466308)$
      TD = 151.550 (tangent distance along boundary from PC to PI)
   c) Chainage PI = $Ch_{PI} = Ch_{PC} + TD_{Boundary} = 635 + 151.550$
      $Ch_{PI} = 786.550$m
3) find radius to centreline PC, R_{CL}
   a) $R_{CL} = R_{PC} - \text{Width}/2 = 325 - (20/2) = 315$
   Radius to CL = 315m
4) find other components of centreline
   a) $Arc_{CL} = R_{CL}\times\Delta(\text{radians}) = 315\times(50^{\circ})^{R}$– $= 315\times(0.872665)$
      $Arc_{CL} = 274.890$
   b) Chainage PT_{CL} = Ch PC + $Arc_{CL} = 635 + 274.890$
      Chainage PT_{CL}= 909.890m
   c) Tangent Distance (TD_{CL}) = $R_{CL}\times\tan(\Delta/2) = 315\times\tan(25) = 315\times(0.466308)$
      $TD_{CL} = 146.887$m
   d) Long chord PC_{CL} - PT_{CL}= $2R\times\sin(\Delta/2) = 2\times315\times\sin(25) = 630\times(0.422618)$
      $TD_{CL} = 266.250$
   e) Chainage crown $\text{secant}_{CL} = PT_{CL} + Arc_{CL}/2 = 635 + 274.890/2$
      Chainage crown $\text{secant}_{CL} = 772.445$
5) find coordinates of components of centreline:
   a) Coordinates PC = Coords O at Brg 300°, Distance 315m from O

	E	N
O	1000.00	1000.00
$\Delta E_{PC}, \Delta N_{PC}$	–272.80	157.50
PC_{CL}	727.20	1157.50

   Similarly, find centreline coordinates for:
      Crown secant_{CL} at Brg/Dist 325°, 315 = 820.32, 1258.03
      PT_{CL} at Brg/Dist 350°, 315 = 945.30, 1310.21
6) table of coordinates by deflections to chainages on centreline. The next 50m chainage from the PC is at Ch_{650}
   Deflection angle(δ) = $((arc_{CL}/radius_{CL}))^{degrees}$
   Ch_{650} arc(s) = 650 – 635 = 15
      deflection(δ) = $(15/315) = (0.047619)^{R} = 2°43'52''$
      chord(ℓ) = $2R\times\sin(\delta/2) = 630\times\sin(1°21'51'')$
      chord(ℓ) = 14.999
   Continue for all chainages and tabulate, as shown in Table 8-2,
      calculate coordinates from $Brg_{PC\text{-}PI} + \delta/2$, dist ℓ and check

Table 8-2 Curve set out from PC and chainage coordinates

Chainage		**Radius = 315m**				**COORDS PC**		**727.20**	**1157.50**	
Centre line		**arc S**	**defl'n δ/2**	**Bearing**	**chord ℓ**	ΔE	ΔN	E	N	
PC	635	0	0°00′00″	30°00′00″	0.000	0.000	0.000	727.20	1157.50	✓
	650	15	1°21′51″	31°21′51″	14.999	7.806	12.807	735.01	1170.31	
	700	65	5°54′41″	35°54′41″	64.885	38.057	52.552	765.26	1210.05	
	750	115	10°27′32″	40°27′32″	114.362	74.210	87.015	801.41	1244.52	
CS	772.44	137.55	12°30′00″	42°30′00″	136.352	92.118	100.530	820.32	1258.03	✓
	800	165	15°00′22″	45°00′22″	163.120	115.355	115.331	842.56	1272.83	
	850	215	19°33′12″	49°33′12″	210.851	160.460	136.787	887.66	1294.29	
	900	265	24°06′02″	54°06′02″	257.254	208.388	150.845	935.59	1308.34	
PT	909.89	274.89	25°00′00″	55°00′00″	266.250	218.099	152.714	**945.31**	**1310.21**	✓

If it's not checked it's wrong.

7) Create table of Brg/Dist from PI (794.32, 1293.75) to ℄ chainages. (Table 8-3)

8) Checks from PI include:
   a) equal distance PI to PC and PT ✓
   b) Crown Secant Brg 145°✓ (reciprocal Brg O to PI)
   c) Crown Secant dist
      = Sec dist$_{PI}$ – R_{CL}
      = R_{PC}×sec(δ/2) – R_{CL}
      = 325/cos(25°) – 315
      = 358.60 – 315
      = 43.60. ✓

Table 8-3 Check from PI

COORDS PI		794.32	1293.75	Setout		
Ch. ℄		E	N	Bearing	Distance	
PC	635	727.20	1157.50	206°13′29″	151.88	✓
	650	735.01	1170.31	205°39′46″	136.95	
	700	765.26	1210.05	199°08′47″	88.60	
	750	801.41	1244.52	171°47′57″	49.74	
CS	772.44	820.32	1258.03	145°00′00″	43.60	✓
	800	842.56	1272.83	113°26′22″	52.58	
	850	887.66	1294.29	89°40′04″	93.35	
	900	935.59	1308.34	84°06′01″	142.03	
PT	909.89	**945.31**	**1310.21**	83°46′31″	151.88	✓

8.12 Problems involving Simple Curves.

8.12.1 The case where the PI is inaccessible

If the curve is to be laid out using deflection angles, the PC and PT are first laid out by measuring from the PI back the tangent distance to get the PC; and measuring from the PI forward the tangent distance to get the PT.

8.12.1.1 Worked Example 8.8. PI inaccessible.

It is desired to lay out a 1000.00m radius curve joining two straights as shown in Figure 8-30. Point D, the PI, is inaccessible. The problem reduces to laying out the PC and PT from known points A and B on the straights.

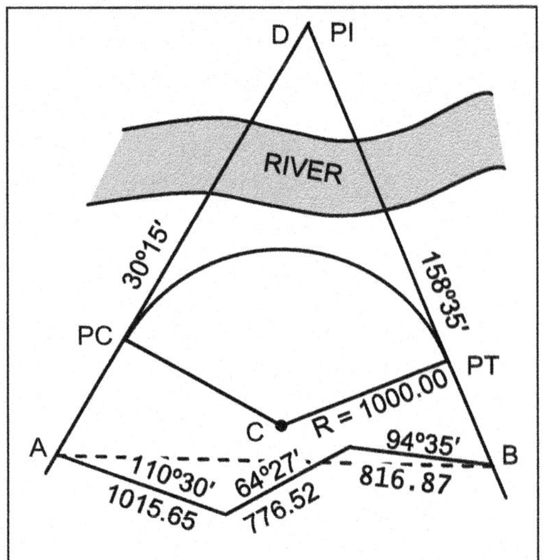

Figure 8-30 Worked Example 8.8 Inaccessible PI

Method:
(1) Determine the bearing and distance of AB. (This can be done by traversing between these points)
(2) Solve the triangle ADB for AD, BD.
(3) Determine tangent distances.
(4) Solve for A-PC, B-PT

Solution:
(1) Find missing brg/dist. AB by traverse.
    Bearing AB = 91°59′55″
    Distance AB = 2467.67m.
(2) Solve triangle ADB. Figure 8-31.

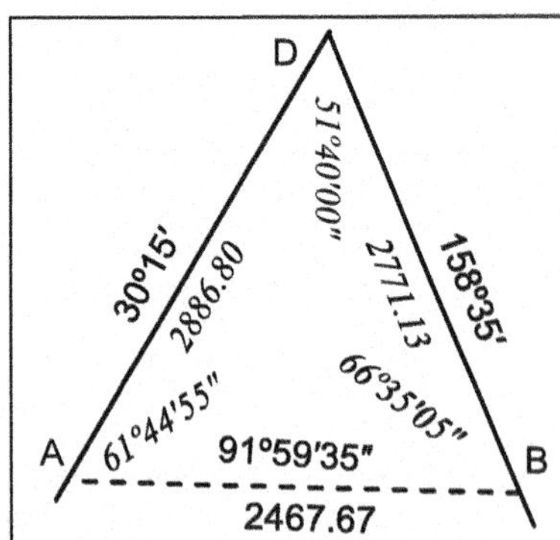

Figure 8-31 Traverse result

   (a) Find angles A, B, D
   ∠A = 91 59 55 – 30 15 00   ∠A = 61 44 55
   ∠B = 158 35 00 – 91 59 55   ∠B = 66 35 05
   ∠D = 210 15 00 – 158 35 00 ∠D = 51 40 00
   ∠A + ∠B + ∠D = 180 00 00 CHECK ✓
   (b) Find missing distances using Sin Rule. d = 2467.67m

$$\frac{b}{\sin(66°35'05'')} = \frac{a}{\sin(61°44'55'')} = \frac{2467.67}{\sin(51°40'00'')}$$

b = AD = 2886.80m
a = BD = 2771.13m

   (c) Check by solving for ∠A using Cos Rule for angles

$$\cos A = \frac{b^2 + d^2 - a^2}{2bd} \qquad ∠A = 61°44'55'' \quad ✓$$

(3) Find tangent distance AD = DB
    Tangent (AD) = R×tan (Δ/2)

Deflection Δ = (180 – Intersection angle (D))
Δ = (180 – 51 40) = 128°20′
$\Delta/2$ = 64°10′
Tangent (AD/DB) = 1000.00×tan(64 10)
= 2065.53

(4) A-PC = AD – TD = 2886.80 – 2065.53 = 821.27
B-PT = BD – TD = 2771.13 – 2065.53 = 705.60

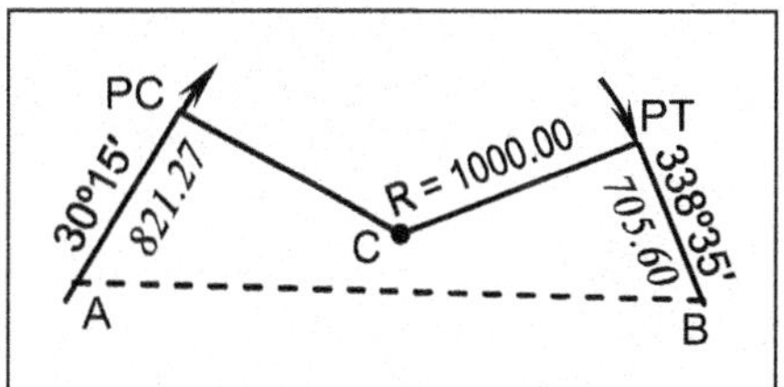

Figure 8-32 Set PC and PT

The PC and PT can be laid out from the traverse stations A and B. (Figure 8-32).

Note: There will be no problem in this case if the curve is coordinated at points along its length. It will be a simple matter to lay it out by radiations (polars) from any convenient coordinated point.

8.12.2 Case where the PC, PT or both are inaccessible.

PI is accessible. Points on the curve have to be set out from the PI. (See Figure 8-33)

(a) This problem will not arise if the curve is coordinated at points along its length.
(b) If the curve is to be set out by deflection angles.

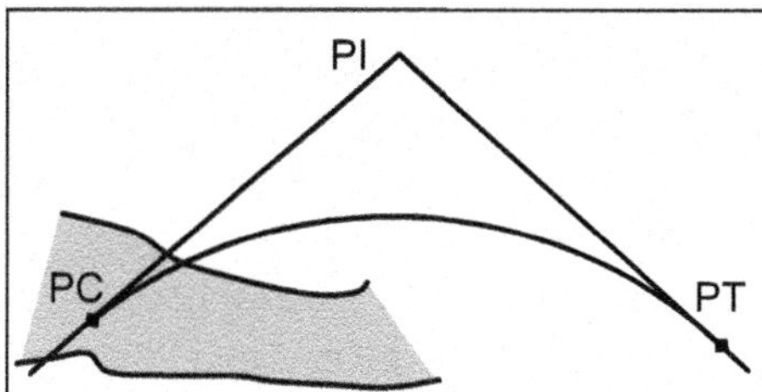

Figure 8-33 Inaccessible PC

8.12.3 Worked Example 8.9. PC/PT inaccessible.

Set out by intermediate deflection angle.

Figure 8-34 shows the Δ = 42°20′ deflection angle.

Set out point S from an inaccessible PC.

Many curves are quite "flat", i.e. a large radius or Δ is a small angle, so an exaggerated drawing such as Figure 8-35 gives a clearer illustration.

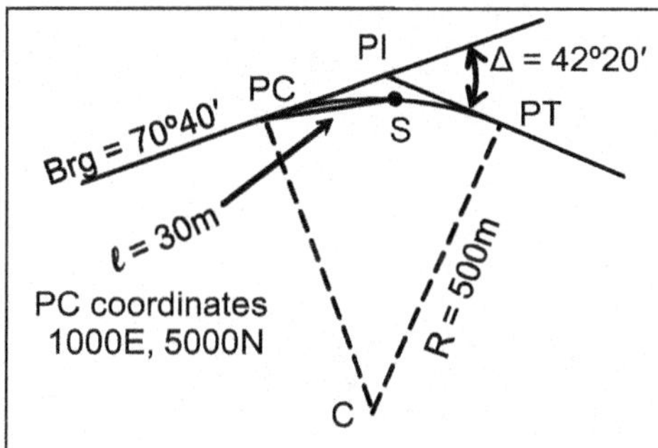

Figure 8-34 Δ = 42°40′

Given: PC coordinates 1000E, 5000N
Δ = 42°20′
R = 500m

Aim: Lay out a point S such that:
Chord distance, PC to S, ℓ = 30.0m
Calculate coordinates of S.

(1) Find tangent distance from PC to PI

TD_{AD} = R×tan ($\Delta/2$)
Δ = 42°20′
$\Delta/2$ = 21°10′
TD_{AD} = 500.00×tan(21°10′) = 500.0×(0.387205)
= 193.60

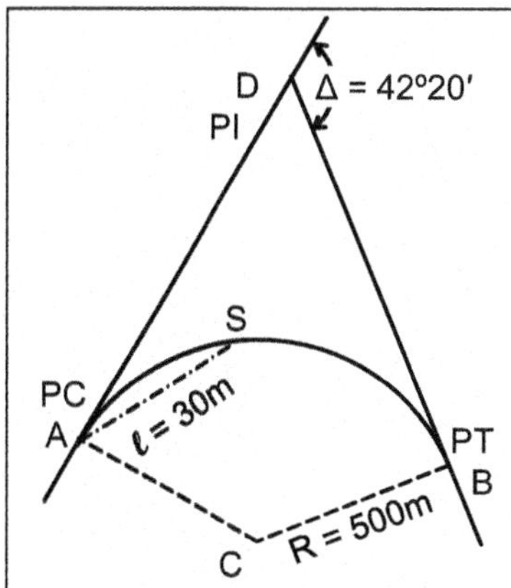

Figure 8-35 Exaggerated figure

(2) Find deflection angle α from the straight at PC to S where we can define the deflection as $\alpha = \frac{1}{2}\delta$ and δ is the deflection angle at the centre between the end of the chord PC-S = 30m.

Chord ℓ = 2R×sin ($\frac{1}{2}\delta$) or 2R×sin(α)

$$\sin(\alpha) = \frac{1}{2}\left(\frac{\ell}{R}\right)$$

$$\alpha = \arcsin\left(\frac{1}{2}\left(\frac{30.0}{500.0}\right) = 0.03\right)$$

α = 1°43′09″

(3) TS = square offset from T (on tangent AD) to S on arc ASB
TS = ℓ×sin(α) = 30×sin(1°43′09″) = 30×(0.03)
= 0.900m

Distance from PI to T (DT)
= TD_{AD} – AT

AT $= \ell\times\cos(\alpha) = 30\times\cos(1°43'09'') = 30\times(0.999550)$
$= 29.986$m
DT $= 193.60 - 29.986$
$= 163.614$

Thus point S can be established from the PI

Note: Arc dist. AS = $R\cdot 2\alpha^r = 500\times 2\times(1°43'09'')^r = 30.004 = 30.0 = \ell$
in practical terms arc AS = ℓ = AT = 30.0m

Thus DT = 193.6 – 30 = 163.6, TS = 0.9m at the decimetre level. (Figure 8-36).

Addendum: If PC coordinates are known then the following coordinates are generated:
(PC = 1000E, 5000N)
Brg from PC to PI = 70°40′
PI = 1000 + dE, 1000 + 182.68 = 1182.68E
= 5000 + dN, 5000 + 64.09 = 5064.09N
Brg from PC to S = 70°40′ + α = 70°40′ + 1°43′09″ = 72°23′09″
S = 1000 + dE = 30×sin(72°23′09″) = 1000 + 28.59 =1028.59
= 5000 + dN = 30×cos(72°23′09″) = 5000 + 9.08 = 5009.08
Consequently Brg/Dist from PI to S. dE = –154.09, dN = –55.01
Brg = 250°21′10″, Dist = 163.61m

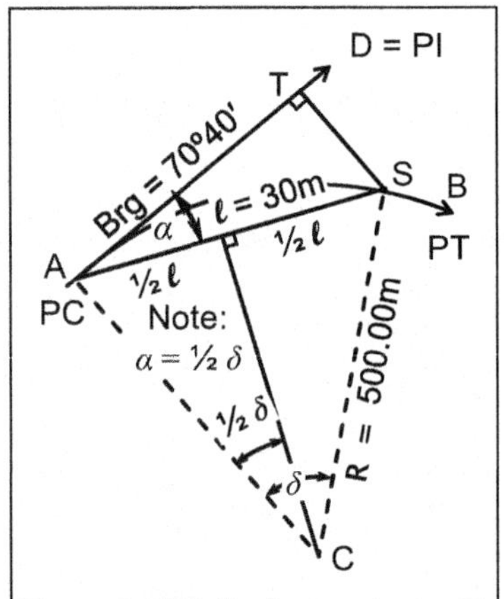

Figure 8-36 Deflection angle, ℓ = 30

8.12.4 Radius of curve though a point

Aim: To find the radius of a curve to pass through P (Brg/Dist to D) and be tangential to AD and DB. (See Figure 8-37)

Data: 1) Bearing AD, DB
2) Bearing and distance DP

Method:
1) Join D and P to centre C.

In ΔDCP; ∠PDC = α, ∠DCP = β
DC = R×sec Δ/2

by Sin Rule: $\dfrac{DC}{\sin P} = \dfrac{PC}{\sin\alpha}$

$$\sin P = \frac{DC\sin\alpha}{PC} \Rightarrow \frac{R\sec\left(\frac{\Delta}{2}\right)\sin\alpha}{R}$$

$$= \sec\left(\frac{\Delta}{2}\right)\sin\alpha$$

$$P = \arcsin\left(\frac{\sin\alpha}{\cos\left(\frac{\Delta}{2}\right)}\right)$$ (Note: there are two angles for P)

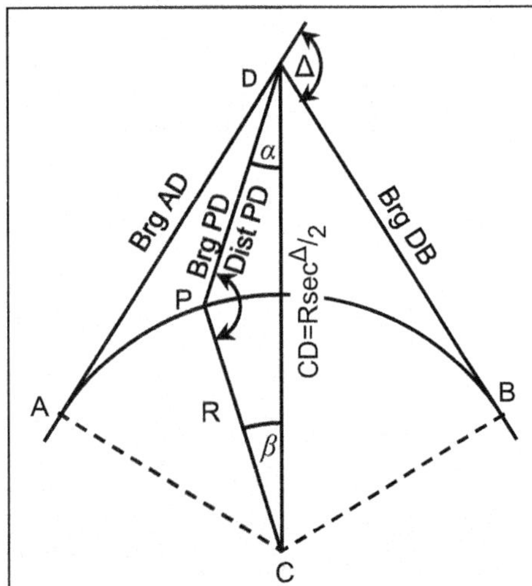

Figure 8-37 Radius of a curve through a point

Hence $\beta = 180 - (P + \alpha)$

and by Sin Rule $\dfrac{R}{\sin\alpha} = \dfrac{DP}{\sin\beta}$ But $\sin(\beta) = \sin(P + \alpha)$

$$R = \frac{d\sin\alpha}{\sin(P+\alpha)}$$

Note: P is the obtuse angle, not the acute angle.
Use $(P - \alpha)$ for acute angle P.

8.12.5 Worked Example 8.10 Radius through a point.

Aim: Given the following information, find the radius of the curve which will pass through P. (See Figure 8-38).

Data: Brg AD = 50°30′, Brg DB = 131°10′,
Brg/Dist PD = 38°20′, 50.35m

1. $\Delta = 131°10' - 50°30' = 80°40'$

$\angle ADC$ = Interception angle $I = 180 - \Delta = 99°20'$

$\Delta/2$ = **40°20′** $I/2$ = **49°40′**

2. Brg DP = Brg PD + 180 = 218°20′ (reciprocal)

$\alpha = \theta_{DP} - \theta_{DC} = (218°20' - (131°10' + 49°40'))$

α = 37°30′

3. In ΔDCP, By Sin Rule: $\dfrac{DC}{\sin P} = \dfrac{PC}{\sin\alpha}$

$$\sin P = \frac{DC\sin\alpha}{PC} \Rightarrow \frac{R\sec\left(\frac{\Delta}{2}\right)\sin\alpha}{R}$$

$$\Rightarrow \sec\left(\frac{\Delta}{2}\right)\sin\alpha \Rightarrow \frac{\sin\alpha}{\cos\left(\frac{\Delta}{2}\right)}$$

$$\sin P = \left(\frac{\sin(37°30')}{\cos(40°20')}\right) = \frac{0.608761}{762292} = 0.798594$$

P = arcsin(0.798594) = 52°59′46″, (2 values of P as sin*P* is positive in QI and QII)
but P is obtuse so P = (180 – P):

P = **127°00′14″**

β = 180 – (P + α)

β = **15°29′46″**

4. $$R = \frac{d\sin\alpha}{\sin(P+\alpha)} = \frac{50.35\times\sin(37°30')}{\sin(164°30'14'')} = \frac{50.35\times 0.608761}{0.267173}$$

R = 114.72m

5. Check: $DC = R\cdot\sec\Delta/2 = 150.494$

$$\cos A = \frac{DP^2 + DC^2 - PC^2}{2\times DP\times DC} \qquad \text{(Cos rule for angles)}$$

α = 37°30′ ✓

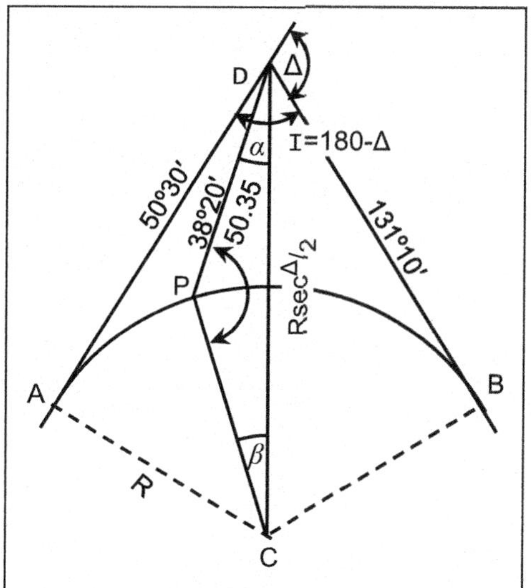

Figure 8-38 Worked Example 8.10

8.12.6 Radius of curve offset from a point.

Figure 8-39 shows a situation where a circular curve is to clear a point by a particular amount.

The direction of the two tangents is known. The bearing and distance from the PI to the point, P, is known.

And the offset distance is known.

Calculate the radius of the curve through tangent points A and B.

There are two methods available to solve the problem.

1. Construction of an auxiliary circle, concentric with the required circle, to pass **through** point P. Solve radius through P and add offset.
2. Solved by a quadratic equation.

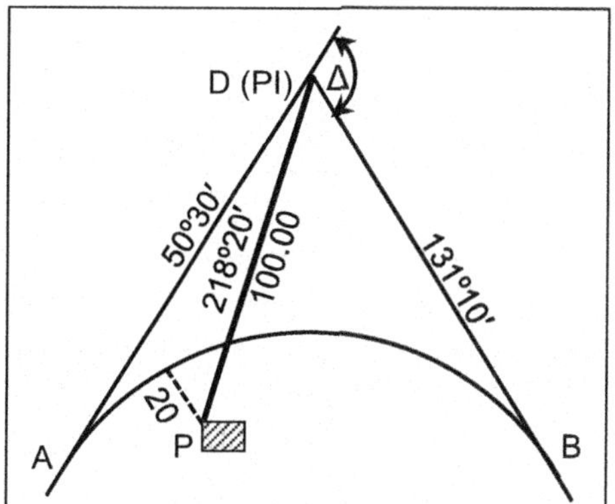

Figure 8-39 Radius offset from a point

8.12.6.1 Worked Example 8.11 Radius through an offset point.

<u>Solution 1 Solved by using an auxiliary circle.</u>

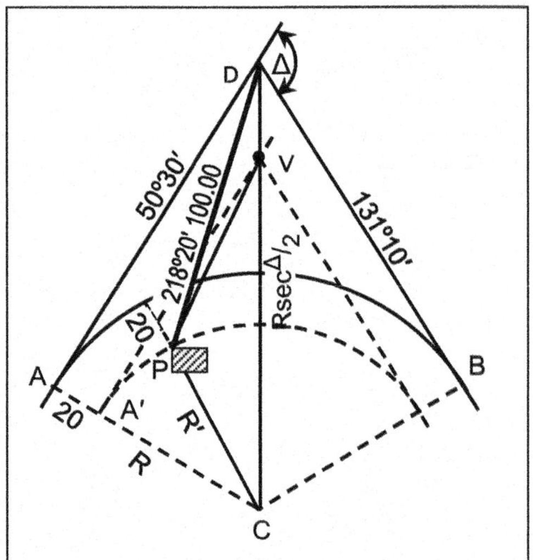

Figure 8-40 Worked Example 8.11

Figure 8-40 shows the pick-up of point P from the PI of two tangent straights. Find the radius of a curve offset from P by 20m.

1. Construction of an auxiliary circle through P.
   a) Construct a line parallel and 20m offset from tangent AD. It intersects the secant distance at V.
2. Solve ΔDPV for DV (See Figure 8-41).
   a) $\Delta = 131°10' - 50°30' = 80°40'$
      $\angle ADC = I = 180 - \Delta = 99°20'$
      $^{\Delta}/_2 = \mathbf{40°20'}$ $\quad ^{I}/_2 = \mathbf{49°40'}$
   b) $\angle ADP = 230°30' - 218°20'$
      $= 12°10'$
      $\angle PDV = {^{I}/_2} - 12°10'$
      $= 37°30'$
   c) $DV = 20 \times \sec(^{\Delta}/_2) = 26.237$
   or $DV = 20 \times \operatorname{cosec}(^{I}/_2) = 26.237$

Solve ΔDPV for VP and ∠DVP

Two sides, included angle:

use Cos rule for sides (See Figure 8-42)

$\angle PDV = 37°30'$ (from 2).)
$DV = p = 26.2367$
$PD = v = 100.00$

a) By Cos rule for side

$$d = (v^2 + p^2 - 2vp\cos(D))^{1/2}$$
$$= (100^2 + 26.2367^2 - (100 \times 26.2367 \times \cos(37°30'))^{1/2}$$

$VP = 80.7798$

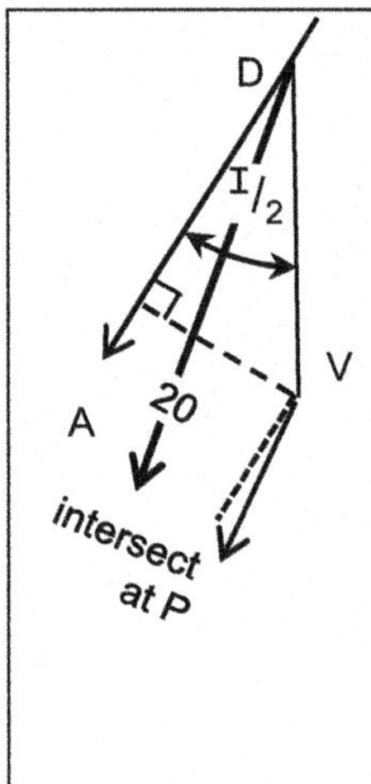

Figure 8-41

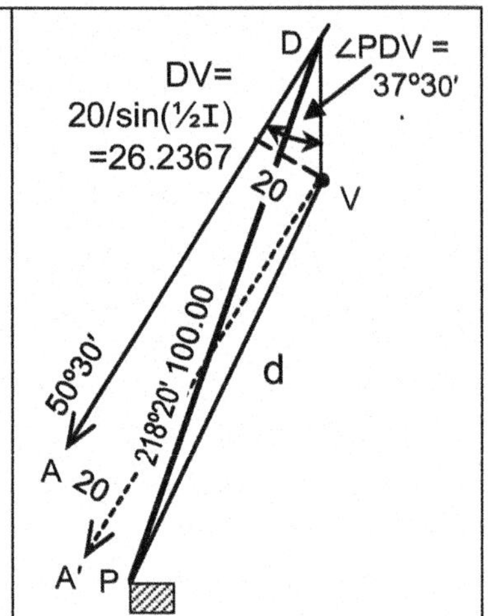

Figure 8-42

Two sides, opposite angle:

b) By Sin rule (ambiguous case)

$\sin(V) = v \times \sin(D)/d$
$= 100 \cdot \sin(37°30')/80.7798$
$= 0.753606$

$V = 48°54'13''$, ambiguous, and, by inspection, $PVD = 180 - V$

$\angle PVD = 131°05'47''$

Similarly, to find ∠DPV (angle at P)

$\sin(P) = p \times \sin(D)/d$
$= 0.197721$
$P = 11°24'13''$

Check $P + V + D = 11°24'13'' + 131°05'47'' + 37°30' = \underline{180°}$ ✓

Note: 3a) solves for side **d** using Cos rule for sides, then uses the Sin rule.

However, with 3 sides (d = VP) solved, find remaining 2 angles using Cos rule The Cos rule for angles will solve for V and P unambiguously.

$\cos(V) = (p^2 + d^2 - v^2)/(2pd)$
$\cos(P) = (d^2 + v^2 - p^2)/(2dv)$

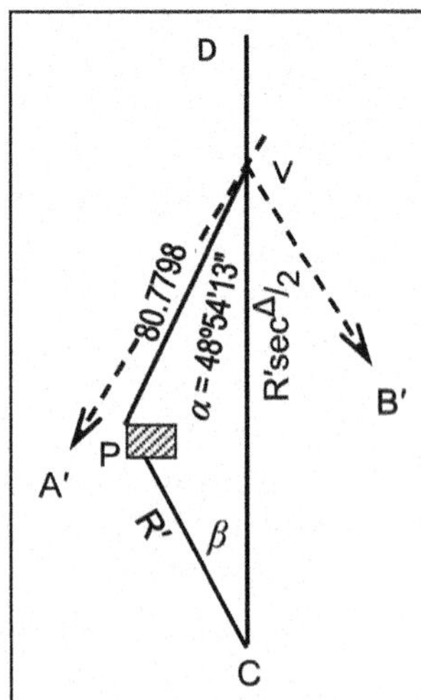

Figure 8-43

3. Now solve ΔPVC for ∠CPV and ∠PVC

   Refer § 8.12.4, page 126 (∠P, ∠α, ∠β)

   $\angle\alpha = 48°54'13''$ see Figure 8-43
   $^{\Delta}/_2 = 40°20'$ (VB′ parallel to DB)

a) $\sin P = \left(\dfrac{\sin\alpha}{\cos\left(\frac{\Delta}{2}\right)} \right)$ from $\dfrac{\sin P}{R'\sec\left(\frac{\Delta}{2}\right)} = \dfrac{\sin\alpha}{R'}$ (cancel R′ each side)

sin(P) = 0.988606

P = 81°20′33″

but P is obtuse so P = (180 – P): ambiguity

P = 98°39′27″

β = 180 – (P + α)

β = 32°26′20″

b) By sine rule

R′ = PV×sin(α)/sin(β)

= 80.7798×sin(48°54′13″)/sin(32°26′20″)

= 113.4907

4. R = R′ + offset

R = 113.491 + 20

R = 133.491m

Solution 2, Solved by a quadratic equation.

1. The cosine rule can be applied to the diagram to solve for R. (See Figure 8-44).

$PC^2 = PD^2 + CD^2 - 2\times PD\times CD\times\cos(\angle PCD)$	Eqn 8-15

PC = R – 20

PD = 100

CD = Rsec(Δ/2)

∠PDC = 37°30′ (Solution 1, 2(b) where ∠PDC = ½I–∠ADP)

Δ/2= 40°20′

a) Form quadratic equation from Eqn 8-15:

$(R-20)^2 = (100^2+(R\sec(\Delta/2))^2 - 2\times100\times R\sec(\Delta/2)\times\cos(37°30'))$

Rsec(Δ/2) = R/cos(40°20′)

= 1.311833R

Cos(37°30′) = 0.793353

$(R-20)^2 = 100^2+(1.311833R)^2 - 200\times1.311833R\times0.793353$

$R^2 - 40R + 400 = 10000 + 1.720907R^2 .- 208.149483R$

b) Gather terms

$0.720907R^2 - 168.149483R + 9600 = 0$

c) Solve quadratic equation, two roots of R

$$R = \frac{-b\pm\sqrt{b^2-4ac}}{2a}$$

$$= \frac{-(-168.149483)\pm\sqrt{(-168.149483)^2 - 4\times0.720907\times9600}}{2\times0.720907}$$

$$R = \frac{168.149483\pm\sqrt{24.319211}}{1.4418144}$$

R = 133.491 **OR** R = 99.756

d) Inspect two roots of R for correct solution

Tangent distance: TD = RTan(Δ/2)

1) R = 133.491, TD = 113.341

2) R = 99.756, TD = 84.7

Since distance DP = 100m,

the TD has to be greater than DP

Thus R = 133.491

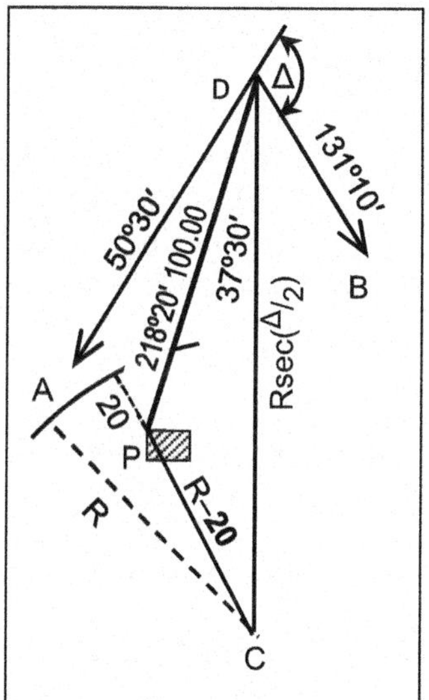

Figure 8-44

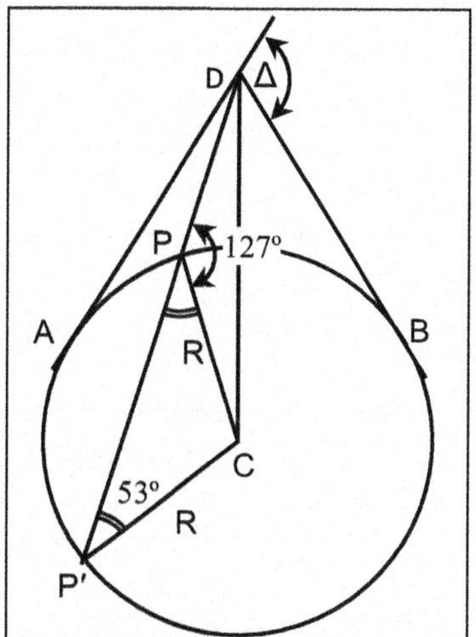

Figure 8-45 Two position of P

This confirms the two positions of P.

From both Worked Examples, note that there can be two positions of point P. This is shown in Figure 8-45.

These points are P or P′

The angle at P in ΔDCP, from Sin Rule, has two values (§ 8.12.5 Worked Example 8.10).

i.e. ∠P = 127°00′14″ or 52°59′46″

Both solutions are valid, but the usual solution would be for P.

8.13 Compound Curves

8.13.1 Definition.

If two circular curves, having different radii, are tangential to each other at a point, with both curves/centres being on the same side of the tangent, then the combination is called a compound curve (see Figure 8-46).

8.13.2 Uses.

i) Occur in sub-divisional situations i.e. changes of boundaries, changes in roads.
ii) Where a road must fit the contour of the ground.
iii) Where the road is designed to avoid obstacles.

8.13.3 Properties.

When two circles intersect, the line joining the centres passes through the common tangent point.

(See Figure 8-47, Figure 8-48)

C.T.P. = Common tangent point

P.C.C. = Point of common curvature.

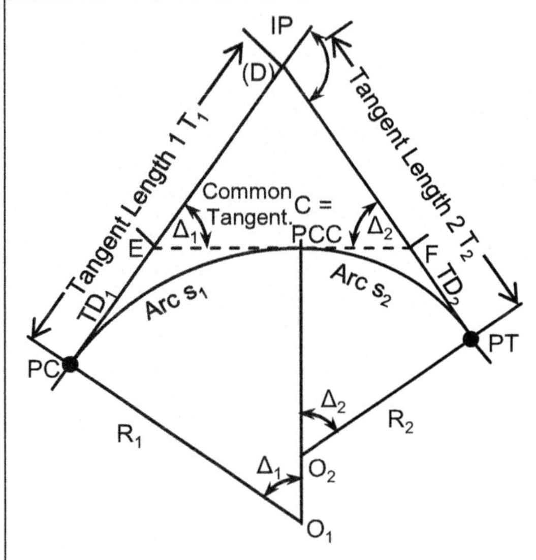

Figure 8-46 Definition of compound curve components

Refer also to Figure 8-46 and note in § 8.15.2 page 134 relating to the intercept angle ***I***. (I = 180° – Δ)

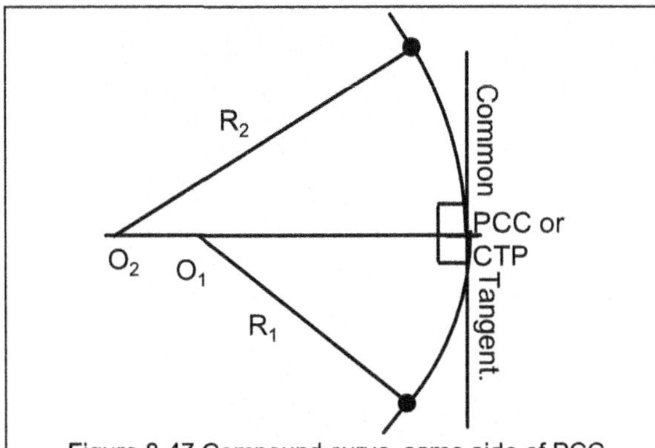

Figure 8-47 Compound curve, same side of PCC

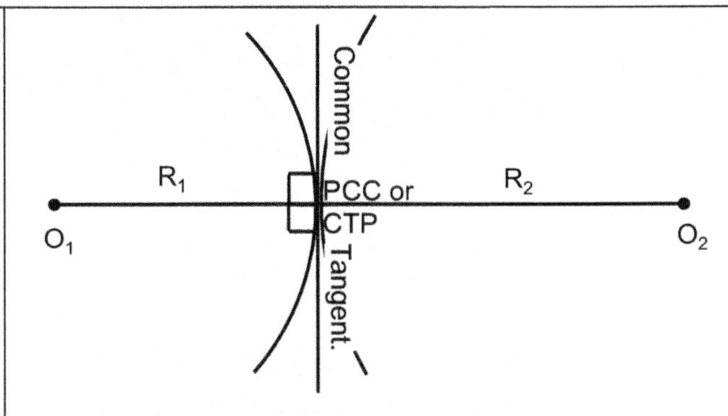

Figure 8-48 Compound curve, opposite sides of PCC

Two curves of radii R1, R2 (centres O_1, O_2) are joined at the P.C.C. to form a compound curve.

The tangents to the compound curve, at the PC and the PT, intersect at the IP (D). They are called the tangent lengths T1 and T2 respectively. (See Figure 8-46).

The two curves are joined at a common tangent point (P.C.C.)

The tangents lengths, through the P.C. or P.T., may intersect or be parallel.

Centres and P.C.C. are co-linear, i.e. O_1, O_2 and P.C.C. are co-linear

$\Delta = \Delta_1 + \Delta_2$

Length of Common Tangent (EF)

$= TD_1 + TD_2 = R_1\tan(\Delta_1/2) + R_2\tan(\Delta_2/2)$.

There are several independent elements;

Tangent lengths T_1 andT_2,

Radii R_1 and R_2,

Arc lengths arc s_1 and arc s_2 and

Total deflection angle Δ.

Four of these elements are necessary to define both curves.

If it's not checked it's wrong.

8.13.4 To set out a Compound Curve.

If a computer is available, stations around the compound curve can be coordinated and laid out by polars from any other coordinated position.

Deflection angles.

If possible, peg P.C., P.T. and P.C.C. Then peg by deflection angles from P.C., P.T., P.C.C. or any other points on the curve.

Note: Problems on compound curves are overcome as with simple curves.

8.14 General Methods for the Solution of Compound Curves.

Solution methods are shown for each case. Worked examples are not used in some instances.

8.14.1 Case 1 Δ, R_1, R_2, and T_1 or T_2

Given: Δ, R_1, R_2, and T_1 **or** T_2

Solve by simple close for bearing of one side and distance of another.

Review Chapter 12 – Missing Data in Traverses

Traverse sheet, with missing data

Assume Brg IA, find Brg IB from Δ

Point	Brg	Dist	ΔE	ΔN
I				
A	I – A	T_1	dE	dN
O_1	A – O_1	R_1	dE	dN
O_2	(------)	R_1–R_2		
B	O_2-B	R_2	dE	dN
I	B-I	(-----)		
		Misclose	ΣdE	ΣdN

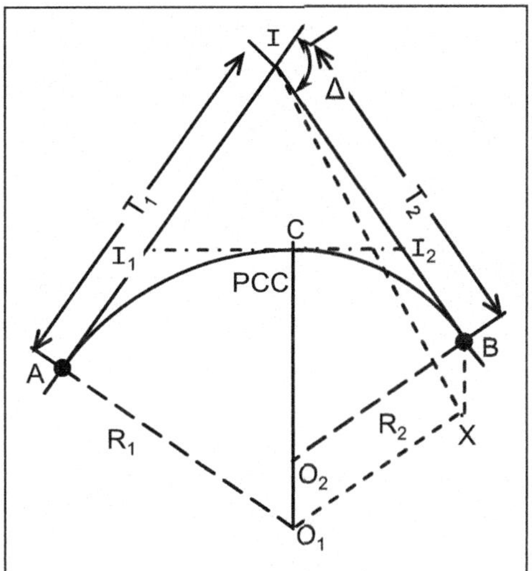

Figure 8-49 Case 1

Solve for Brg IX, Dist IX using $\boxed{\rightarrow\theta,r}$ function.

In ΔIBX, (Figure 8-50)

i = $R_1 - R_2$, **b** = misclose dist IX,

$\angle I$ = Brg IB – Brg IX.

Two sides and angle opposite one side:

solve for $\angle B$ by Sin rule (ambiguous, >90°?)

Deduce $\angle X$, solve for **x** = T_2 by Sin or Cos rule.

In Figure 8-49, traverse IAO_1O_2BI and solve for missing non adjacent bearing and distance i.e., bearing O_1O_2 and distance IB.

Then, in ΔIBX, there are two sides and one angle. Use Sin rule to solve angles. (Watch for ambiguous case.). [Use a traverse table.]

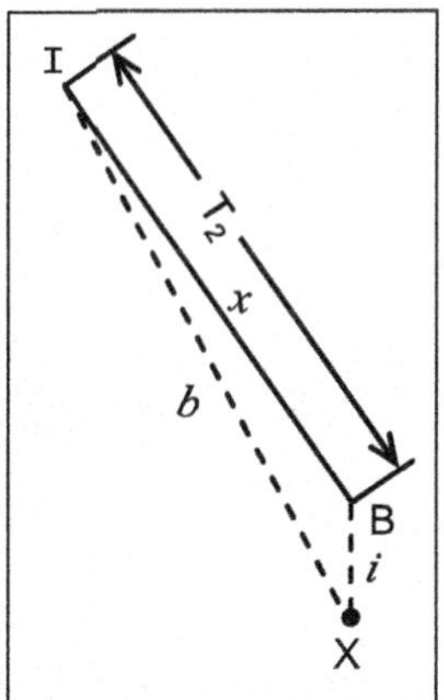

Figure 8-50 ΔIXB

From bearing O_1O_2 and bearing O_1A

Calculate Δ_1

Deduce Δ_2

Solve for arc lengths.

Check by closing figure $AIBO_2O_1A$.

8.14.2 Case 2a Δ, R_1, Arc_1, and T_1

Given: Δ, R_1, Arc_1, and T_1 (See Figure 8-51)

Solve: R_2, Arc_2, and T_2

Solution: Using Arc_1 and R_1 calculate Δ_1

$\therefore$ deduce $\Delta_2 = \Delta - \Delta_1$

Draw in common tangent.

Calculate tangent distance for $curve_1$ i.e. AI_1 and I_1C

$$td_1 = R_1 \times \tan\left(\frac{\Delta_1}{2}\right)$$

$II_1 = IA\ (T_1) - AI_1\ (td_1)$

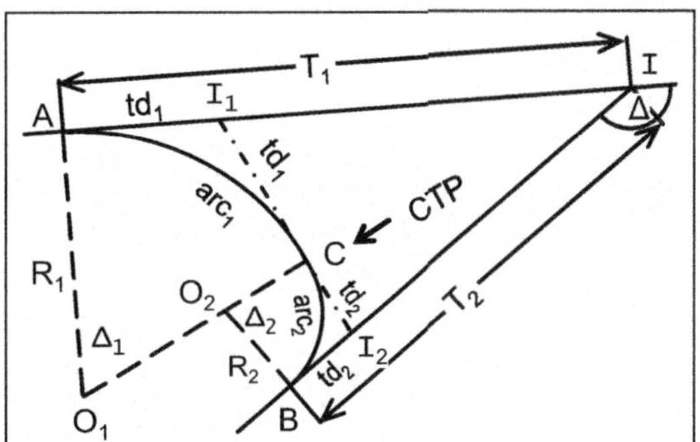

Figure 8-51 Case 2a

If it's not checked it's wrong.

In ΔII_1I_2 has three angles and one side (II_1).
Solve for two the sides I_1I_2 and II_2

Then $td_2 = CI_2 = I1I_2 - td_1$

$T_2 = IB = II_2 + td_2$

$$R_2 = \frac{t_2}{\tan(\Delta_2)}$$

$Arc_2 = R_2 \times \Delta_2^{radians}$

Check by closing figure $AIBO_2O_1A$.

8.14.3 Case 2b Δ, T_1, R_1 and arc_2

Given Δ, T_1, R_1 and arc_2

This case is rare and must be solved iteratively.

i.e. Nominate R_2 and calculate Arc_2. Compare this calculated value with the nominated value for arc_2; adjust the value of R_2 as nominated.

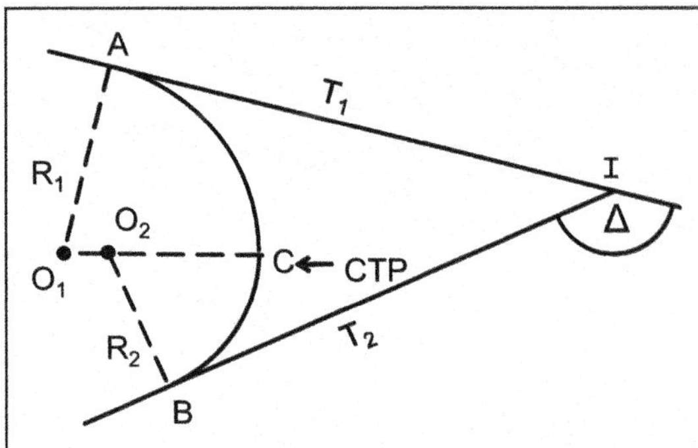

Figure 8-52 Case 3

8.14.4 Case 3 Δ, T_1, R_1, T_2 where $R_1 \neq R_2$

(See Figure 8-52)

Given Δ, T_1, R_1, T_2 where $R_1 > R_2$ or $R_2 > R_1$

Solve: R_2

Close IAO_1BI by traverse and solve for missing bearing and distance O_1B .

Use Cos formulae to solve for R_2 (See Figure 8-53).

$$(R_1 - R_2)^2 = R_2^{\ 2} + S^2 - 2R_2S\cos\alpha$$

In this equation R_2 is the only unknown,

expand $(R_1 - R_2)^2$

$$R_1^{\ 2} - 2R_1R_2 + R_2^{\ 2} = R_2^{\ 2} + S^2 - 2R_2S\cos\alpha$$

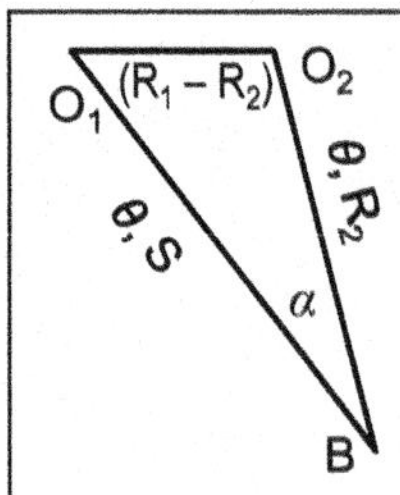

Figure 8-53

multiply through, cross multiply and eliminate common terms.

$$R_1^{\ 2} - 2R_1R_2 + \cancel{R_2^{\ 2}} = \cancel{R_2^{\ 2}} + S^2 - 2R_2S\cos\alpha$$

$$\Rightarrow 2R_2S\cos\alpha - 2R_1R_2 = S^2 - R_1^{\ 2}$$

gather LHS

$$2R_2(S\cos\alpha - R_1) = S^2 - R_1^{\ 2}$$

$$\therefore R_2 = \frac{S^2 - R_1^{\ 2}}{2(S\cos\alpha - R_1)}$$

Check by closing figure $AIBO_2O_1A$.

8.14.5 Worked Example 8.12 (Case 2)

*Note: Interception angle at P.I,

$I = 180 - \Delta$

Now: Sum internal angle = (n – 2) sides×180.

The sides are AB, BC, R_2, O_2O_1, R_1 = 5 = n.

$180^\circ - \Delta + 90^\circ + \Delta_1 + 180^\circ + \Delta_2 + 90^\circ = (5-2)\times 180 = 540^\circ$

gather terms: $(\cancel{180 + 90 + 180 + 90}) - \Delta + \Delta_1 + \Delta_2 = \cancel{540^\circ} = 0$

$\therefore \quad \Delta = \Delta_1 + \Delta_2$

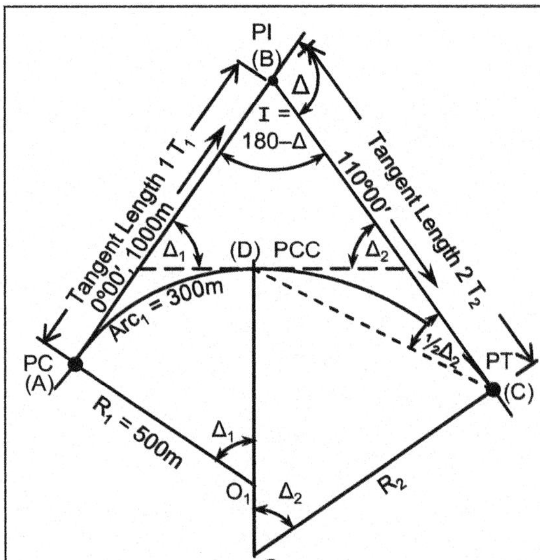

Figure 8-54 Worked Example 8.12

Aim: Given an existing curve (R_1) to find another curve (R_2) which will be tangential to a given line and to an existing curve at a given point. (See Figure 8-54)

Problem: The arc AD, 300m long, is part of an existing curve of radius 500m and is tangential to AB at A(P.C.).

At a point B(P.I.), 1000m from A, a new tangent BC bears 110°.

Find the radius of a curve which is tangential to the existing curve at D(P.C.C) and tangential to BC. Compute also the tangent length BC.

If it's not checked it's wrong.

Construction: Join points A, D, C to their respective centres. Note that O_1, O_2 and D are co-linear.
Solution:

1. Calculate Δ_1 ($Arc_1 = R_1 \times \Delta_1$)

$$\Delta_1 = \frac{Arc_1}{R_1} = \frac{300.00}{500.00} = 0.60000^{radians} \qquad \Delta_1 = \mathbf{34°22'40''}$$

2. Deduce $\Delta_2 = \Delta - \Delta_1$
   $= 110°00' - 34°22'40''$ **= 75°37'20''**
3. Angle BCD = $\frac{1}{2}\Delta_2$ = **37°48'40''**

Hence Brg DC = 290°00'00'' – 37°48'40'' **= 252°11'20''**

4. Brg **DO_1** = Brg $AO_1 + \Delta_1$
   = 90 + 34°22'40'' **= 124°22'40''**
5. Starting at D, take a close on figure DO_1ABCD to solve for missing distances BC, CD.

Table 8-4 Traverse for misclose of unknown components

STN	Bearing	Dist	ΔE	ΔN	E	N	STN
D							D
to	*124 22 40*	*500.00*	*412.666*	*-282.323*			
O_1							O_1
to	*270 00 00*	*500.00*	*-500.00*	*0.00*			
A							A
to	*0 00 00*	*1000.00*	*0.00*	*1000.00*			
B							B
to	*110 00 00*	*( )*					
C							C
to	*252 11 20*	*( )*					
D							D
	E misclose	*ΣΔE =*	*-87.334*				
		N misclose	*ΣΔN =*	*717.677*			

6. Solve ΔBCD
   a) Solve Join B to D (see Figure 8-55) from traverse Table 8-4.

	ΔE	ΔN
Line DB	-87.334	717.67

$\boldsymbol{\theta_{BD}}$ = **173°03'42''**

$\mathbf{D_{BD}}$ = **722.971**

   b) Find angles B, C, D.

Brg $_{BD}$ 173 03 42 (θ_{BD})
– 110 00 00 (θ_{BC})
∠B 63°03'42''

252 11 20 (θ_{CD})
– 173 03 42 (θ_{BD})
∠D 79°07'38''

290 00 00 (θ_{CB})
– 252 11 20 (θ_{CD})
∠C 37°48'40''

∠B + ∠D + ∠C = 180° CHECK

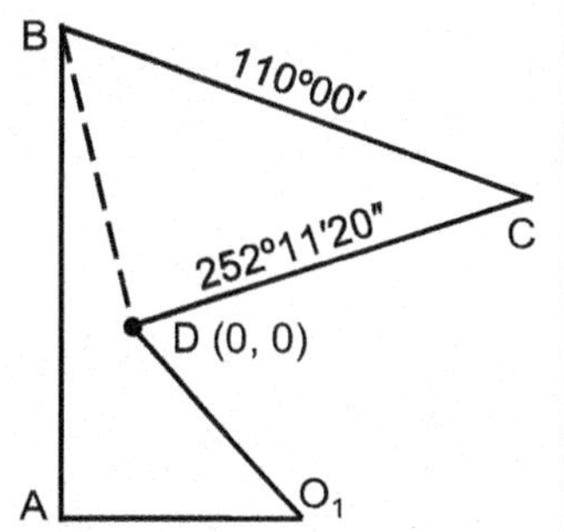

Figure 8-55 Missing line BD

   c) Solve for BC, CD (Sine Rule)

$$\frac{BC}{\sin D} = \frac{DC}{\sin B} = \frac{BD}{\sin C}$$

$$\frac{BC}{\sin(79\ 07\ 38)} = \frac{DC}{\sin(63\ 03\ 42)} = \frac{722.971}{\sin(37\ 48\ 40)}$$

BC = 1158.112
DC = 1051.328

7. Check close.

Table 8-5 Check close of figure for missing values

STN	Bearing	Dist	ΔE	ΔN	E	N	STN
D							D
to	*124 22 40*	*500.00*	*412.666*	*-282.323*			
O_1							O_1
to	*270 00 00*	*500.00*	*-500.00*	*0.00*			
A							A
to	*0 00 00*	*1000.00*	*0.00*	*1000.00*			
B							B
to	*110 00 00*	**1158.112**	**1088.269**	**-396.098**			
C							C
to	*252 11 20*	**1051.328**	**-1000.938**	**-321.580**			
D							D
	E misclose	*ΣΔE =*	*-0.003*				
		N misclose	*ΣΔN =*	*-0.001*			

8. Calculate R_2 in triangle DCO_2.DC is the chord of Arc_2. Thus Δ_2 must be split in two to use the chord formula $\ell = 2R\sin(\Delta/2)$ From $\sin(\Delta/2)$ and opposite side DC/2 (see Figure 8-56).

$$R_2 = \frac{DC/2}{\sin(\Delta_2/2)} = \frac{1051.328/2}{\sin(75°37'20''/2)} = \frac{525.664}{0.613030}$$

R_2 = 857.443m

BC = 1158.112m, tangent length (from 6 c))

Other derived values are:

TD_1 = 92.8m, TD_2 = 665.37m, Common tangent = 758.17m, Arc_2 = 1131.7m

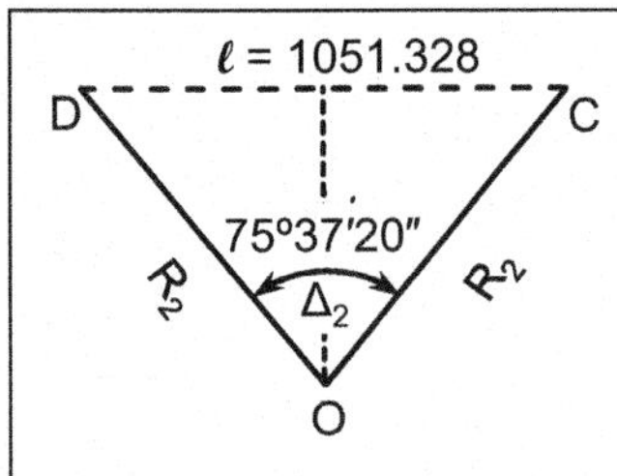

Figure 8-56 Find radius 2

8.15 Reverse Curves.

Curves comprising two discrete simple circular curves meeting at (PCC) of common tangency on opposite sides of the tangent.

In reverse curves the centre of the circular arcs lies on opposite sides of the common tangent.

Usually a reverse curve is employed to join tangents as in Figure 8-57.

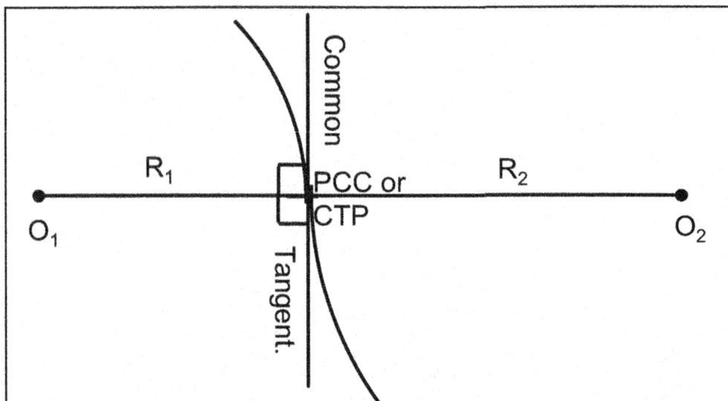

Figure 8-57 Reverse curve

8.15.1 Elements of a Reverse Curve.

As with compound curves, there are seven (7) elements in a reverse curve:

Δ R_1 R_2 Arc_1 Arc_2

Tangent length T_1

Tangent length T_2

Any four of these define the curve.

8.15.2 Reverse Curves - non Parallel Tangents.

*Note: Interception angle at P.I,

I = 180 – Δ (See Figure 8-58)

$180° - \Delta + 90° + (360 - \Delta_1) + 180° + \Delta_2 + 90° = 540°$

$\therefore$ **$180 - \Delta = \Delta_1 - \Delta_2$**

i.e. Δ_1 and Δ_2 are dependant.

The solution for the remaining 3 elements is treated in the same manner as the compound curve.

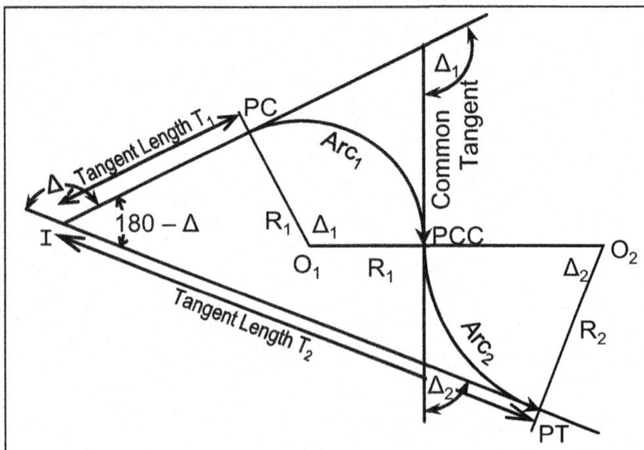

Figure 8-58 Reverse curve, non-parallel tangents

8.15.3 Reverse Curves: Parallel Tangents

*Note: in Figure 8-59, $\Delta_1 = \Delta_2 = \Delta$

$C_1 = R_1 \text{ Sin } \Delta_1 = R_1 \text{ Sin } \Delta$

$C_2 = R2 \text{ Sin } \Delta_2 = R_2 \text{ Sin } \Delta$

$M_1 = R_1 \text{ Vers } \Delta_1 = R_1 \text{ Vers } \Delta$
$= R_1 (1 - \text{Cos } \Delta)$

$M_2 = R_2 \text{ Vers } \Delta_2 = R_2 \text{ Vers } \Delta$
$= R_2 (1 - \text{Cos } \Delta)$

$M_1 + M_2$ = Perpendicular distance between parallel tangents
$= (R_1 + R_2) \text{ Vers } \Delta$
$= (R_1 + R_2)(1 - \text{Cos } \Delta)$

$C_1 + C_2$ = Perpendicular distance between tangent points
$= (R_1 + R_2) \text{ Sin } \Delta$

Note: The problems of Reverse Curves are similar to those of Compound Curves.

Versine = (1 – Cosine)

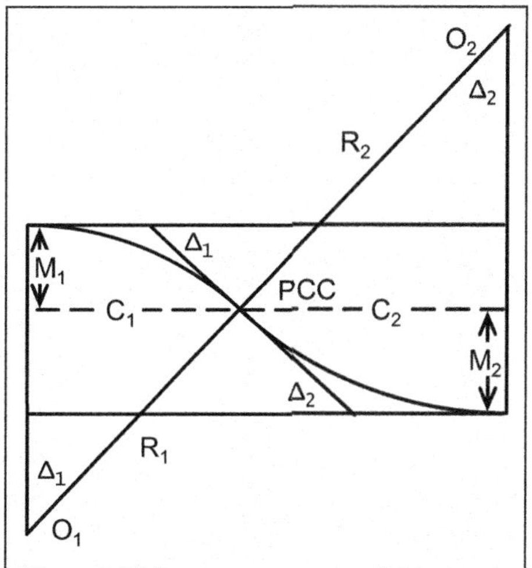

Figure 8-59 Reverse curve, parallel tangents

8.15.4 General Methods for Solution of Reverse Curves.

Case 1 Given Δ, R_1, R_2 and either tangent length T_1 or T_2, see Figure 8-60.

Solution:

Figure IAO_1O_2BI has missing bearing O_2O_1 and missing distance T1 i.e., non-adjacent information.

Close IXO_2BI for IX.

In ΔIAX we have angle AIX and distances IX and XA Solve for angles and T_1

Solve for missing elements.

Beware!! This case may yield an ambiguous solution and field data is required to resolve the ambiguity.

Check by closing figure IAO_1O_2BI.

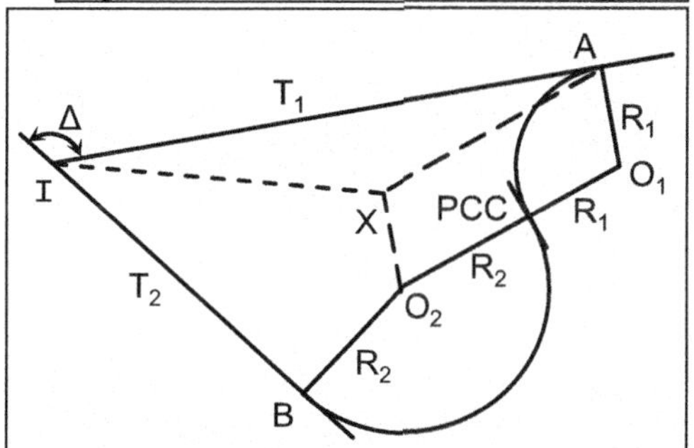

Figure 8-60 Reverse curve solutions

Note:

In Figure 8-61 there are 2 tangent lengths, T_2 and $T_{2'}$, and 2 bearings for the common radii line O_1O_2 that will satisfy requirements.

$\Delta_2 = 360 - \Delta_2$ for the respective cases.

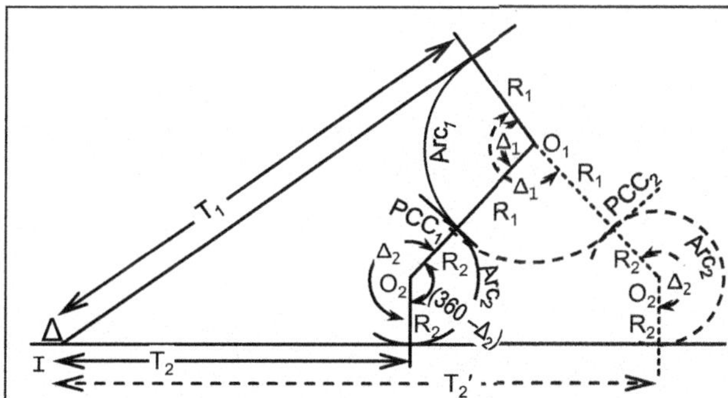

Figure 8-61 Reverse curve solutions, 2 values T_2

8.15.4.1 Worked Example 8.13

Case 1

Given: $\Delta_1 = 105°$, $T_1 = 1170$m,
$R_1 = 220$m, $R_2 = 430$m (see Figure 8-62)

Calculate T_2
(T_2 is to be a maximum value)

Adopt an orientation

$\theta_{IA} = 0°$, $\theta_{IB} = 0.0 + (180 - \Delta) = 75°$

traverse known sides, solve for missing Brg/Dist:

IA	= 0°, 1170.0	
AO_2	= 90°, 220.0	
O_1O_2	= (…)°,	650.0 (// line XB)
O_2B	= 165°,	430.0 (// line O_1X)
BI	= 255°,	(…)
ΔE_{IX}	= 331.292	ΔN_{IX} = 754.652

IX = 23°42′05″, 824.169

$\angle X = \theta_{IB} - \theta_{IX} = 51°17'55''$

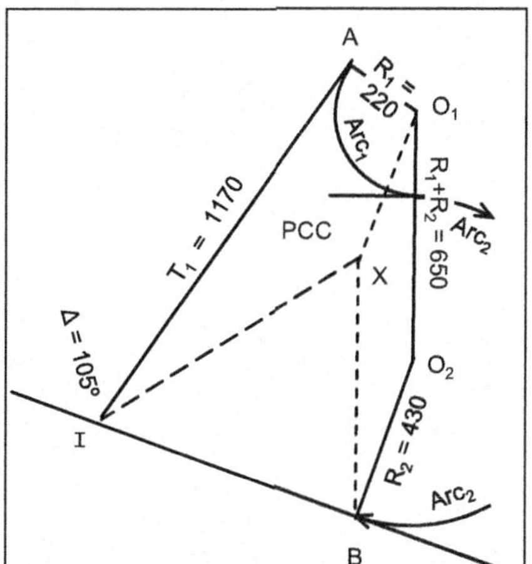

Figure 8-62 Reverse curve Worked Example 8.13

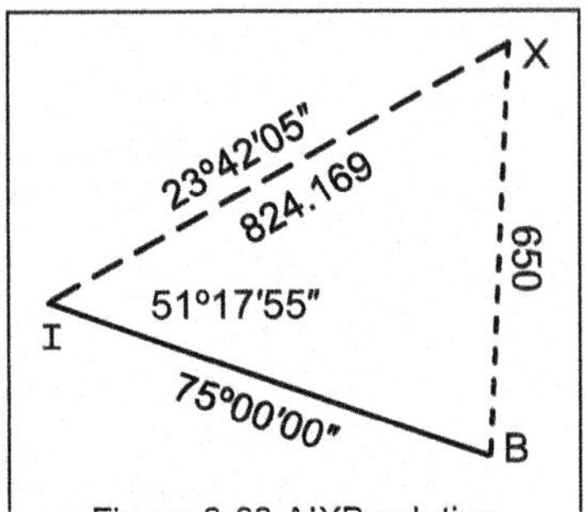

Figure 8-63 ΔIXB solution

In ΔIXB, solving ANGLE B by Sin Rule. (See Figure 8-63)

$\angle B = 81°42'05''$

$\angle X = 180 - \angle B - \angle X = 47°00'00''$

$\theta_{XB} = \theta_{XI} - \angle X = 203°42'05'' - 47°$

Brg O_1O_2 =156°42'05",

In ΔIXB, solving for SIDE T_2 by Sin Rule

IB = T_2 = 609.137m

Check: solving SIDE T_2 by Cos Rule

IB = 609.137m ✓

Case 2A

Given Δ, R_1, arc_1 and T_1. (Figure 8-64)

Using arc_1 and radius R_1 calculate Δ_1 and hence deduce Δ_2.

$180 - \Delta = \Delta_1 - \Delta_2$

Draw in the common tangent and calculate TD_1 AX; hence IX.

In the ΔIXY.

Solve line CY = Tangent distance in circular curve 2, hence solve for R_2 and arc_1.

Check by closing figure IAO_1O_2BI.

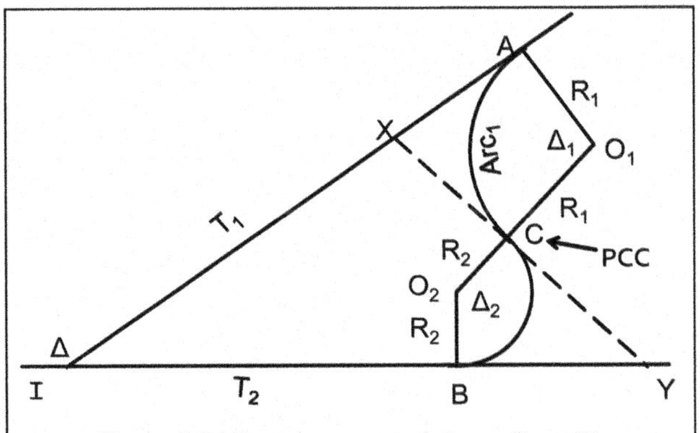

Figure 8-64 Reverse curve solutions. Case 2A

Case 2B

Given Δ, one tangent distance, the radius of one curve and the arc length of the other.

The solution is iterative: Nominate a second radius and calculate its arc length. Compare this value with the required value and adjust the nominated radius accordingly.

Case 3 Given Δ, R_1 (or R_2), T_1 and T_2.

Traverse known sides, see Figure 8-65 and Figure 8-66.

solve for missing Brg/Dist:

Brg O_1B, Dist O_1B

$\angle B = \alpha = |\text{Brg } BO_2 - \text{Brg } BO_1|$

(See derivation in § 8.14.4, page 132)

$$(R_1 + R_2) = R_2^2 + S^2 - 2R_2 S\cos\alpha$$

$$\therefore R_2 = \frac{S^2 - R_1^2}{2(S\cos\alpha - R_1)}$$

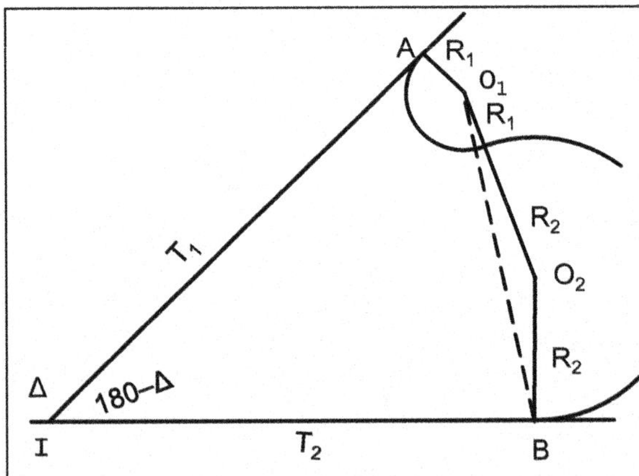

Figure 8-65 Reverse curve solutions. Case 3

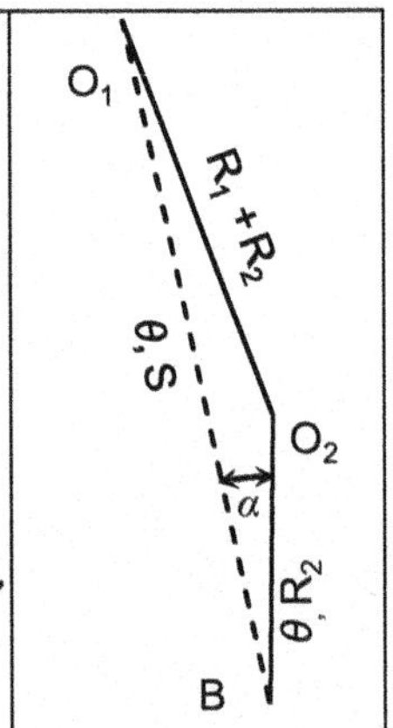

Figure 8-66. Case 3

8.16 Chapter Exercise 8.1

Q. 1

Given: Boundaries as shown with curved street

R = 200.000m, Δ =17°39′00″.

Distances shown are in metres. (See Figure 8-67)

Calculate:

(1) Distances BC (two solutions)

(2) Arc and chord length AB. Use lesser length BC

(3) Area ABCEFA. Use lesser length BC

Note: There are two solutions.

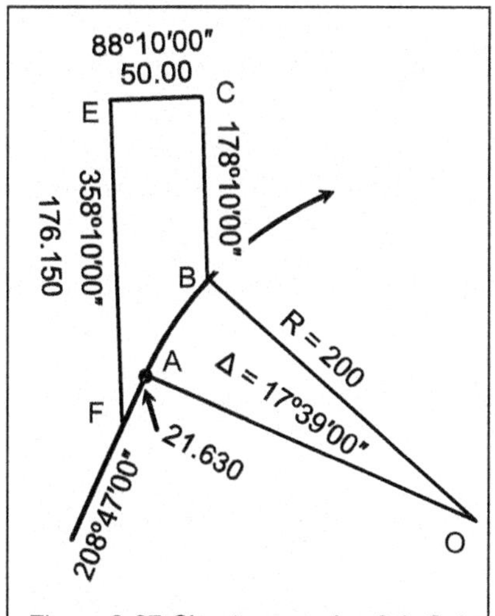

Figure 8-67 Chapter exercise 8.1, Q 1

Q. 2

Given: The information in the Figure 8-68.

A and B are the tangent points to R_1 an R_2

Calculate: Radius length R_2 ($R_2 = BO_2$)

Hint: Draw O_1X where BX = By Cos Rule using R_1 as shown.

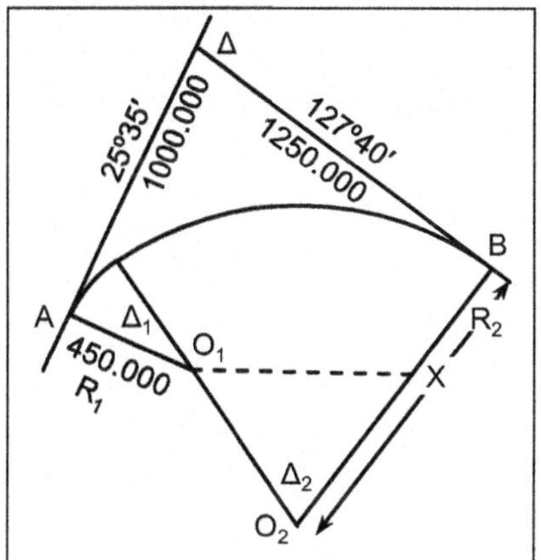

Figure 8-68 Chapter exercise 8.1, Q 2

Q. 3

Using data as shown in Figure 8-69.

Distance BC is measured as 225.000m and distance CD measured as 715.180m.

Brgs BA = 6°26′, AE = 90°00′

Aim: Compute the two tangent distances AB, AE.

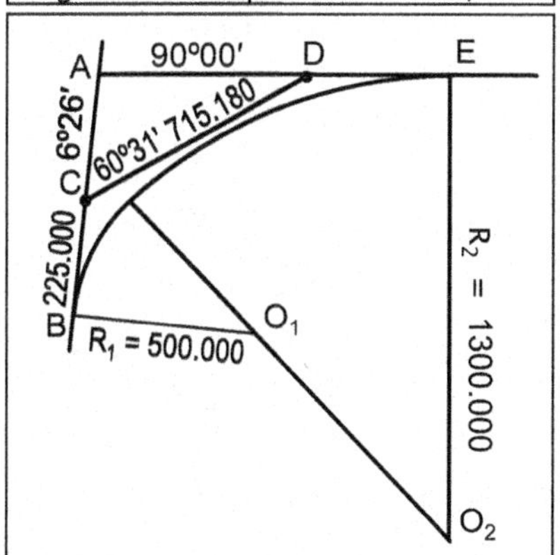

Figure 8-69 Chapter exercise 8.1, Q 3

Q. 4

Given the following data in a Reverse Curve; Figure 8-70

$R_1 = 500.000$m

$R_2 = 200.000$m 0

$\Delta_1 = 110°$

$\Delta_2 = 80°$

Compute the tangent distances T1, T2.

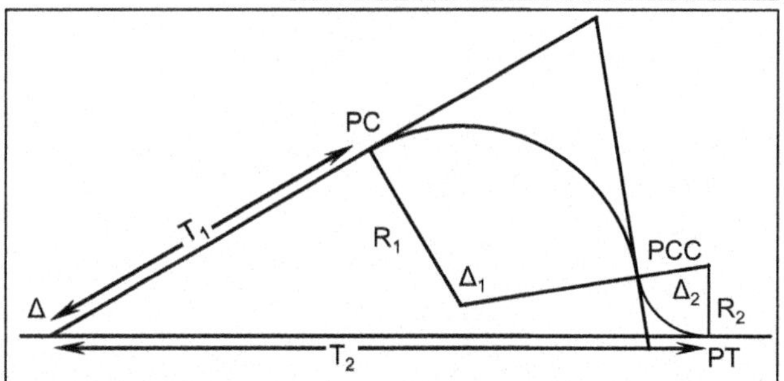

Figure 8-70 Chapter exercise 8.1, Q 4

8.16.1 Answers to Chapter Exercise 8.1

Q. 1 (1) BD can have either of two lengths

BD = 110.146m (case 1)

OR BD = 408.644m

(2) Arc_{AB} = 61.610m AND $Chord_{AB}$ = 61.367m (Case 1 only)

(3) Area = 0.6958 ha (Case 1 only)

Q. 2 $R_2 = BO_2$ = 1082.789m

Q. 3 AB (T_1) = 579.230m,

AE (T_2) = 871.560m

Q. 4 T_1 = 1022.922m,

T_2 = 1825.240m.

8.17 References for Chapter 8

Abbott, P and H. Neill. Trigonometry. A complete introduction. John Murray Learning, London 2018. ISBN 978-1-473-67849-1

Hickerson, T.F. Route Survey and Design. Fourth edition. McGraw-Hill Book Company, Inc. New York. 1959

Bannister, A and S. Raymond. Surveying. Third edition, Pitman Paperbacks, 1972. ISBN 0-273-36148-1

If it's not checked it's wrong.

Chapter 9 ROAD RESERVE INTERSECTIONS

When setting out of the road alignments discussed in Chapter 8, normally portion of lands belonging to private individuals must be taken for public infrastructure. This portion taken is normally called a reserve and the property owners are normally compensated. In this chapter, we provide the computational skills employed to compute these reserves.

9.1 Introduction.

This topic is examined in some detail as road intersections are not usually covered in textbooks. It is of particular interest to licensed survey work related to property boundaries.

The boundaries of reserves for roads, railways, pipelines, drainage channels, of easements (R.O.W's), etc. are usually formed by a series of pairs of parallel lines at some fixed distance apart.

E.g.: a 60 m railway reserve, a 5 m drainage reserve, a 20 m road reserve.

Road reserves will be referred to simply as roads.

When the reserves are less than 100 metres wide, only one side of the reserve may have been completely set out and pegged. The other boundaries of the reserve are set out by offsets from this surveyed line.

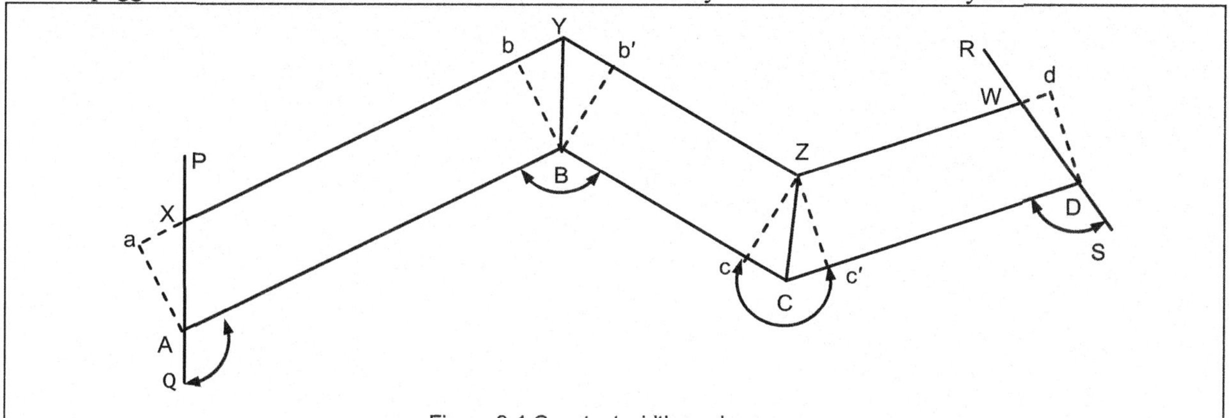

Figure 9-1 Constant width road reserve

In Figure 9-1 a line ABCD (the southern boundary of the road) has been surveyed between two boundaries, PQ and RS. Measurements could consist of the angles PAB, CBA and BCD, and of the distances AB, BC and CD.

It is usually required to calculate:

i) the lengths of the lines forming the northern boundary, XY, YZ, and ZW.
ii) The area of the road, the area of the figure AXYZWDCBA.
iii) The data necessary for setting out the northern boundary, X, Y, Z and W. The bearing and distance of each intersection join.

To calculate these the road width must be known, (dist. aA, Bb, Zc, Dd).

i) and ii) require knowledge of the distances Xa, bY, cC and Wd, the cotangent distances. (The boundary extension distances)

iii) requires calculation of the distances AX, BY, ZC and WD, the cosecant distances, (the intersection diagonal distances) and angles YXA = QAB, XYB = ½ABC, YZC = ½BCD and ZWD = CDS.

The COSECANT method always produces a positive bisector (diagonal) across the road. SIN is positive 0 – 180° (quadrant I and II).

9.2 Calculation of boundary extension and cosecant (diagonal) distances.

9.2.1 Roads of constant width

In Figure 9-2: Bb = Bb′ = the road width.

∠YbB = ∠Bb′Y = 90° (perpendicular distances).

The triangles YbB and Yb′B are congruent, (2 sides, 1 angle) and therefore Yb = Yb′, and ∠BYb = ∠b′YB.

When ∠ABC is known, ∠BYb = ½∠ABC = ∠b′YB

∠ABY = ∠YBC = ½(360° - ∠ABC)

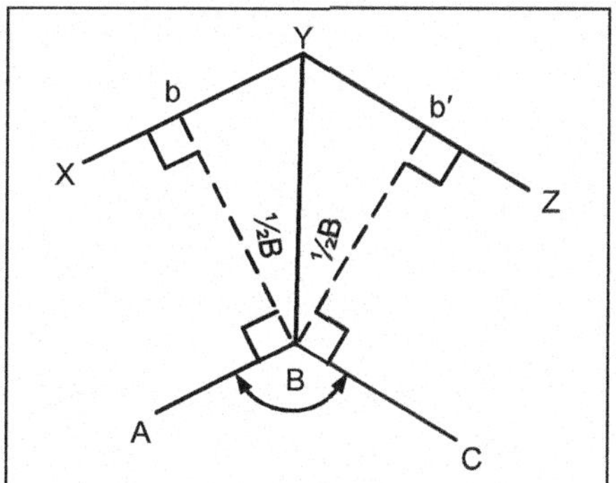

Figure 9-2 Intersection geometry

If it's not checked it's wrong.

J. Walker and J. Awange, *Surveying Computations*,
https://doi.org/10.1007/978-3-031-94382-9_9

Intersection distance

$YB = Bb \times Cosec(\angle BYb) = Bb \times Cosec(½\angle ABC)$

Boundary offset distance

$bY = b'Y = Bb \times Cot(\angle BYb) = Bb \times Cot(½\angle ABC)$

9.2.2 Worked Example 9.1 Constant width road.

Let the data of boundaries and measurements of the traverse in Figure 9-3 be as follows:

$\theta QP = 360°$	$\theta RS =$	170°	
$\theta AB = 79°02'$	AB =	810.0 m	
BC = 504.7 m	$\angle ABC =$	130°30′	
CD = 601.3 m	$\angle BCD =$	220°34′	

ABCD is the traversed southern road boundary, and the road width is 20 metres. Calculate the lengths of the lines forming the northern boundary and the intersection diagonal distances.

Solution:

First calculate all required angles, see Figure 9-3.

It is advantageous to calculate the angles on the outside of the road reserve; this will automatically give the boundary offset distances, the cotangent (1/tan) distances, their proper sign for addition or subtraction, and obviates the necessity for a scaled drawing.

Similarly, the distance between the apices of the corner will be best calculated from the cosecant of the external angle. The cosecant (1/sin) will be positive between 0° and 180° (first and second quadrant).

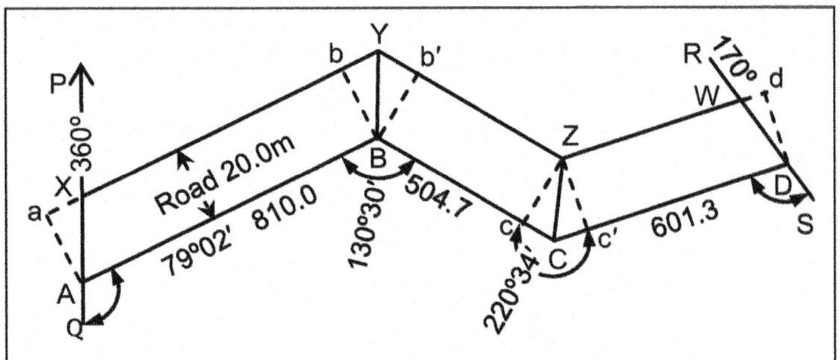

Figure 9-3 Worked Example 9.1

$\angle QAB = 180°00' - 79°02' = 100°58'$

$\angle CBA = 130°30'$ (Given)

$\angle DCB = 220°34'$ (Given)

$\angle SDC = \theta DC - \theta DS$

θDC must first be calculated through the traverse ABCD.

θAB	= 79°02′	
θBA	= 259°02′	(reciprocal)
$- \angle CBA$	= 130°30′	
θBC	= 128°32′	
θCB	= 308°32′	(reciprocal)
$- \angle DCB$	= 220°34′	
θCD	= 87°58′	
θDC	= 267°58′	(reciprocal)

$\angle SDC = \theta DC - \theta DS$

θDC	= 267°58′
$- \theta DS$	= 170°00′
$\angle SDC$	= 97°58′

NOTE: The EXTERNAL angles $\angle B$ and $\angle C$ must be halved to produce the required deflection angles.

Cosecant (intersection) Distances	Cotangent (offset) Distances
AX = 20/sin(100°58′)	aX = 20/Tan(100° 58′)
AX = 20.372	aX = –3.876
BY = 20/sin(½(130°30′))	bY = 20/Tan(½(130°30′))
BY = 22.023	bY = Yb′ = +9.220
CZ = 20/sin(½(220° 34′))	cC = 20/Tan(½(220°34′))
CZ = 21.322	cC = Cc′= –7.392
DW = 20/sin(97° 58′)	Wd = 20/Tan(97°58′)
DW = 20.195	Wd = –2.799

The northern boundary can be calculated:

XY = AB + aX + bY

= 810.00 + (–3.876) + 9.220

XY = 815.344

YZ = BC + Yb′ + cC

= 504.70 + 9.220 + (– 7.392)

YZ = 506.528

ZW = CD + Cc′ + Wd

= 601.30 + (– 7.392) + (– 2.799)

ZW = 591.109

Note that the original boundary distances were defined to 4 significant digits (one decimal place in this case). The final pegging may only be valid at the decimetre (0.1m) level.

Check the results by putting the boundaries of the road into a traverse close.

Table 9-1 Traverse check

TRAVERSE LEG	Angle	Bearing	Distance	ΔE	ΔN
AX		0°00′	20.372	0.000	20.372
XY		79°02′	815.344	800.454	155.109
YZ		128°32′	506.528	396.229	–315.552
ZW		87°58′	591.109	590.737	20.973
WD		170°00′	20.195	3.507	–20.888
DC		267°58′	601.300	–600.921	–21.335
CB		308°32′	504.700	–394.799	314.413
BA		259°02′	810.000	–795.208	–154.093
			Misclose	–0.001	0.000

45° Check. Rotation and scaling of results gives a check of original close.

$S = (D/\sqrt{2})\times\sin(\theta + 45)$ $C = (D/\sqrt{2})\times\cos(\theta + 45)$ $S - C = \Delta E$ $S + C = \Delta N$

Table 9-2 45° rotation and √2 scaling traverse check

LEG	Bearing +45°	Dist./√2	S	C	ΔE = S–C	ΔN=S+C
AX	45°00′	14.405	10.186	10.186	0.000	20.372
XY	124°02′	576.535	477.782	–322.672	800.454	155.109
YZ	173°32′	358.169	40.339	–355.891	396.229	–315.552
ZW	132°58′	417.977	305.855	–284.882	590.737	20.973
WD	215°00′	14.280	-8.191	–11.698	3.507	–20.888
DC	312°58′	425.183	–311.128	289.793	–600.921	–21.335
CB	353°32′	356.877	–40.193	354.606	–394.799	314.413
BA	304°02′	572.756	–474.650	320.558	–795.208	–154.093

9.3 Calculation of Areas.

9.3.1 Roads of constant width

The area of the road consists of the sum of the areas of the trapezia formed between the pairs of parallel road boundaries.

e.g., in Figure 9-4 road area = area ABYX + area BCZY + area CDWZ.

The area of each of these trapezia is equal to half the sum of the parallel sides × the road width.

$$\text{Area } ABYX = \frac{AB + YX}{2} \times \text{ road width}$$

$$+\text{Area } BCZY = \frac{BC + ZY}{2} \times \text{ road width, etc}$$

When the road width is constant, then:

Road area = road width × half the sum of the parallel boundaries.

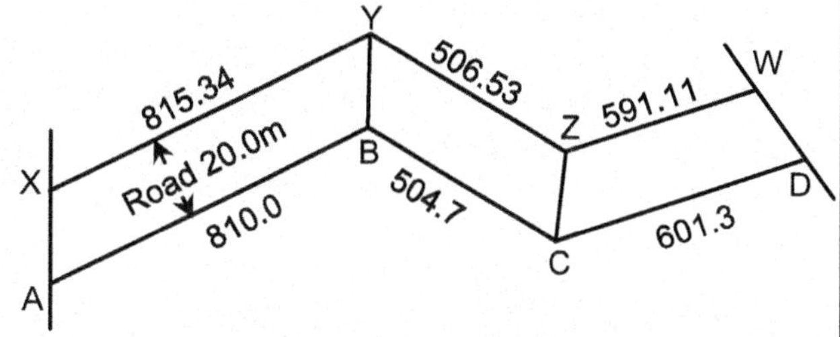

Figure 9-4 Area by trapezium. Worked Example 9.2

9.3.2 Worked Example 9.2 Road area.

Calculate the area of the road in § 9.2.1, page 139, Worked Example 9.1

Sum of all parallel boundary sides

AB	810.00
BC	504.70
CD	601.30
WZ	591.11
ZY	506.53
YX	815.34
	3828.98

∴ Area $= \frac{1}{2}\cdot(3828.98) \times 20 = 38289.8\text{m}^2$

Area = 3.8290 ha

9.4 Calculation of boundaries and cosecant (intersection) distances, different widths.

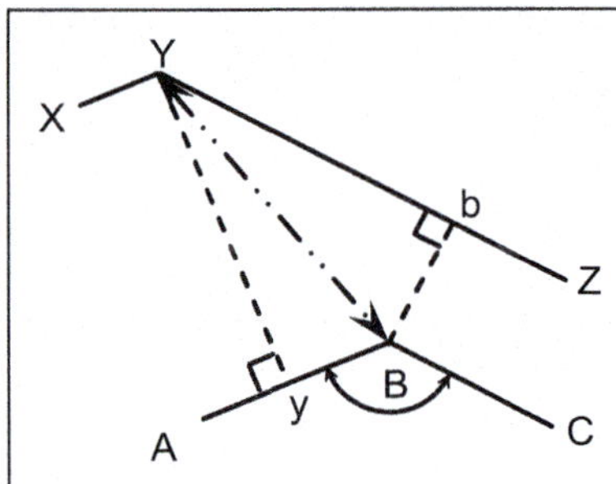

Figure 9-5 Different width roads

9.4.1 Intersection of roads of different widths.

In Figure 9-5 let ABC and XYZ be the boundaries of 2 roads intersecting in points B and Y. Since the two intersecting roads are not of equal width, the problem is slightly different from the previous case.

Note: Cosecant (1/sin) (intersection) distance is BY (∠ABY).
Cotangent (1/tan) (offset) distances are By (∠ABY), Yb (∠BYZ).

9.4.2 Intersection Construction.

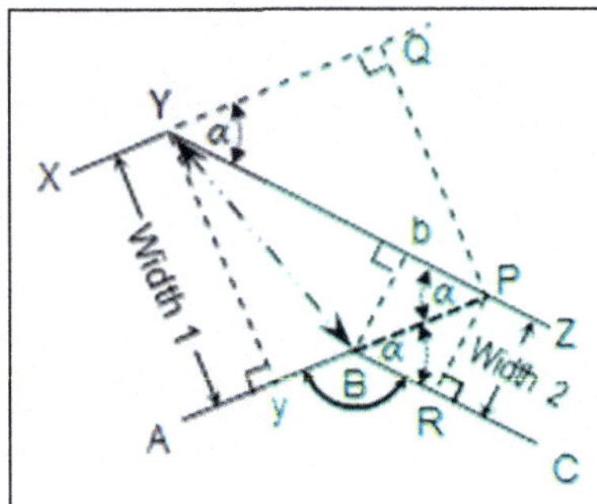

Figure 9-6 Construction case (ii)

Produce one boundary until it meets the other boundary. The two possible cases in this case are: (Figure 9-6).

(i) Produce BC to meet XY
(ii) Produce AB to meet YZ

The latter case is chosen, let AB meet YZ at P.

From P drop perpendiculars on the two opposite (extended) road boundaries meeting them in Q and R respectively.

Now, ∠QYP = ∠BPY = ∠PBR
= ∠α = 180 – External ∠ABC (180 – B)

Note: PQ (W_1), PR (W_2) are the respective road widths.

In triangle PYQ, YP = PQ×cosec(∠QYP) = W_1 ×cose (α)

In triangle PBR, PB = PR×cosec(∠PBR) = W_2 × cosec(α)

9.4.2.1 Solution.

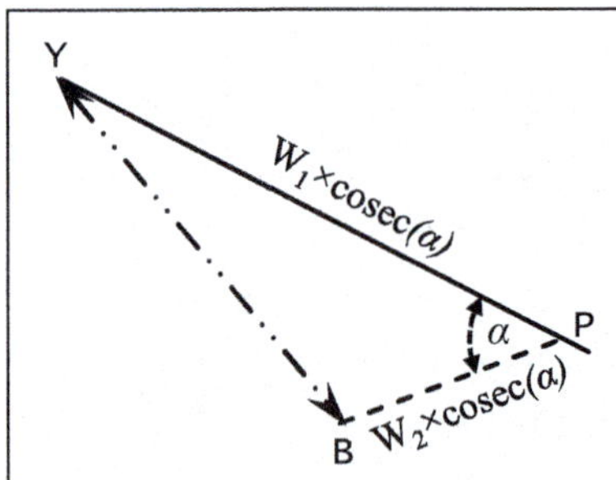

Figure 9-7 Solve ΔYPB

In triangle YPB, (See Figure 9-7)

From § 9.4.2 above, YP, PB and ∠BPY (α) are known thus the remaining elements of this triangle can be solved (given 2 sides plus included angle)

Cosecant (intersection) distance YB. (See Figure 9-8)

(i) solve cosecant distance YB (using Cos Rule for sides).
(ii) solve ∠YBP [δ] and ∠BYP [∠γ] (using Sin Rule where ∠YBP [δ] is in quadrant II. Or use Cos rule for angles).
(iii) check Σ interior angles = 180°

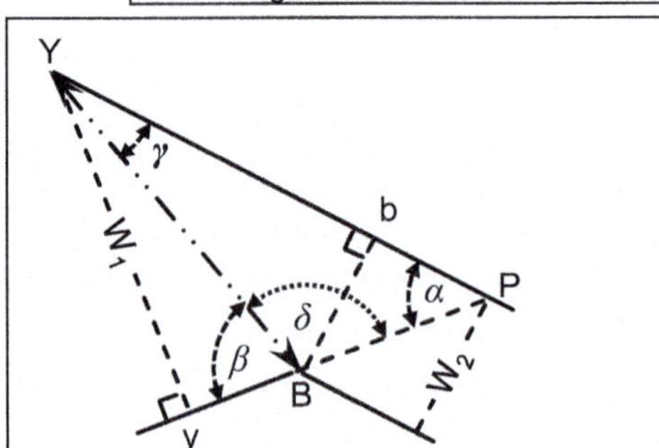

Figure 9-8 Solve ΔYBy

Cotangent (offset) distances (See Figure 9-8)

∠YBy = β = 180 – δ (or δ from the QI solution of ΔYPB)

By = Yy×cot(∠YBy) [∠β] (Yy = Width Road 1 (W_1))
(∠YBy = ∠YPB + ∠PYB) [∠β= ∠α + ∠γ]

Yb = Bb×cot(∠PYB) [∠γ] (Bb = Width Road 2 (W_2))

Checks:

Yb = YP – Pb (Pb = BP×cos(∠YPB)) [∠α]

By = Py – PB (Py = Yy·×cot(∠PYy)) [∠α]
(Py = Yy×tan(∠YPB)) [90 – ∠α]

9.4.2.2 Worked Example 9.3 Different widths.

In the two roads of widths 60m and 20m intersect at Y and B.

Aim:

To calculate the cosecant (intersection) distance YB, bearing YB and cotangent (offset) distances Yb, By from the information in Figure 9-9.

Construction:

The construction is similar to that as in Figure 9-5.

Solution:

1. Cosecant (1/sin) distance:
   YB, the intersection distance.

Note: ∠QYP = ∠YPB = ∠PBR = 110°00′ – 55°00′ = 55°00′ = ∠α

In ΔQYP, YP = QP×cosec(∠QYP) = W_1×cosec(55) = 60/Sin(55)
= 73.246

In ΔPBR, PB = PR×cosec(∠PBR) = W_2×cosec(55) = 20/Sin(55)
= 24.415 (this is, of course, 1/3 YP)

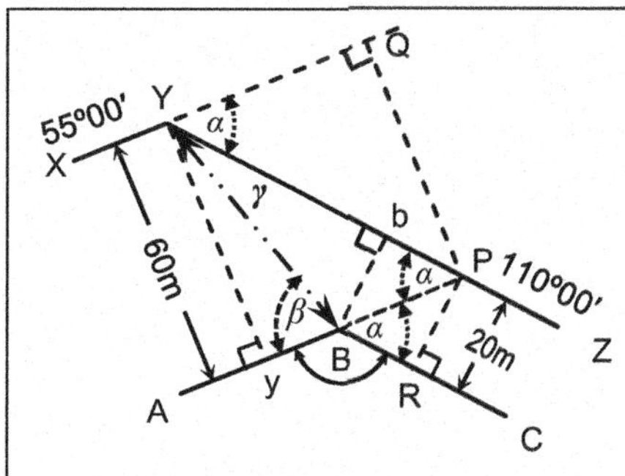

Figure 9-9 Worked Example 9.3

(i) Solve for YB by Cos Rule for sides

$$\text{YB} = p = \sqrt{y^2 + b^2 - 2yb\cos(\alpha)}$$

YB = **62.527m**

(ii) Solve for angles using Sin Rule,

$$\frac{\sin(Y)}{y} = \frac{\sin(B)}{b} = \frac{\sin(P)}{p}$$

$$\sin(Y) = \frac{y \times \sin(\alpha)}{p} = \frac{24.415 \times \sin(55)}{62.527} = 0.319855$$

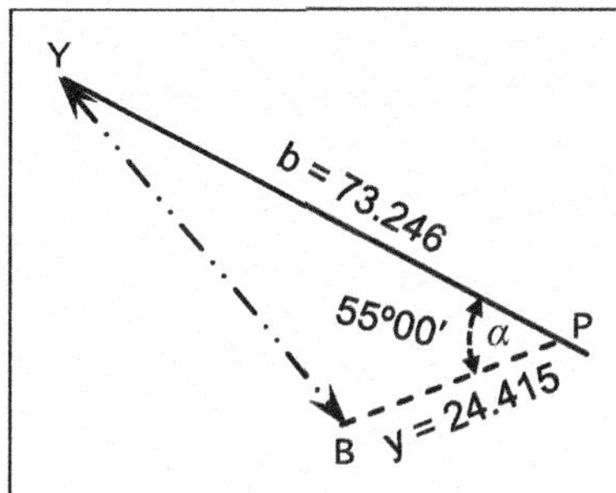

Figure 9-10 Solve ΔYPB

Y = γ (gamma) = 18°39′16″

Bearing YB = θ_{YC} + γ = 110°00′00″ + 18°39′16″

$\boldsymbol{\theta_{YB}}$ **= 128°39′16″** and $\boldsymbol{\theta_{BY}}$ **= 308°39′16″** (See Figure 9-10)

Similarly:

$$\sin(B) = \frac{b \times \sin(\alpha)}{p} = \frac{73.246 \times \sin(55)}{62.527} = 0.959579 \quad \text{AMBIGUOUS CASE}$$

B = 73°39′16″ OR 106°20′44″

By inspection B = 106°20′44″ (∠Y + ∠P < 90°).

Or Cos rule. $\cos(Y) = \dfrac{b^2 + p^2 - y^2}{2bp}$ & cyclically for cos(B)

(iii) Check:

	∠Y	18°39′16″
	∠B	106°20′44″
	∠P	55°00′00″
SUM:		180°00′00″

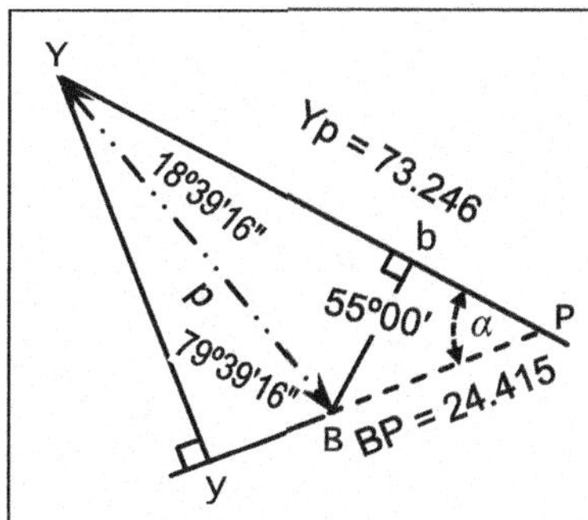

Figure 9-11 Solve ΔYBy

2. Cotangent (offset) distances:
   Find By, Yb (See Figure 9-11)

(i) By = Yy×cot(∠YBy) (∠YBy = β) = W_1×cot(β)
= 60/tan(73°39′16″)
= 17.597

Check: By = Py – PB	PB = 20×cosec(55) = 24.415
= 17.597	Py = Yy·Cot(P) = W_1/tan(α) = 42.012

(ii) Yb = Bb×cot(∠Y) ⇒ W_2×cot(γ)
= 20/tan(18°39′16″)
= 59.242

Check: Ab = YP – bP	YP = 60·Cosec(55) = 73.246
= 59.242	bP = Bb·Cot(55) = 20/Tan(55) = 14.004

YB = 62.527 $\boldsymbol{\theta_{YB}}$ **= 128°39′16″**
By = 17.597
Yb = 59.242

If it's not checked it's wrong.

> Note: In deciding whether the cotangent distances are positive or negative, it's advisable to draw the road intersection to scale.

9.4.2.3 Worked Example 9.4 Different widths, area calculation

Two intersecting roads of width 20m and 60m intersect in AB.

From the information given in Figure 9-12 construct the intersection and calculate:

(i) Intersection distance AB, Bearing AB
(ii) Offset distances Ba, Bb
(iii) Distances of the southern side of the road reserve (DB, BF)
(iv) Area of the road reserve CAEFBDC

Note: represents the road intersection of Figure 9-12 drawn to a larger scale. ΔAHB

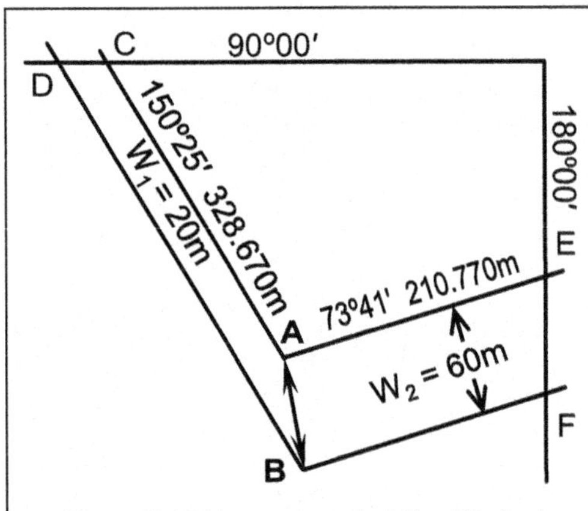

Figure 9-12 Unequal road widths. Worked Example 9.4

<u>Construction</u>:

Produce CA to meet BF in H. (See Figure 9-13)

From H drop perpendiculars to the opposite sides of the road to meet them in G and J respectively.

AB is the cosecant distance, the intersection distance

Ba, Bb are the cotangent (1/tan), or offset, distances.

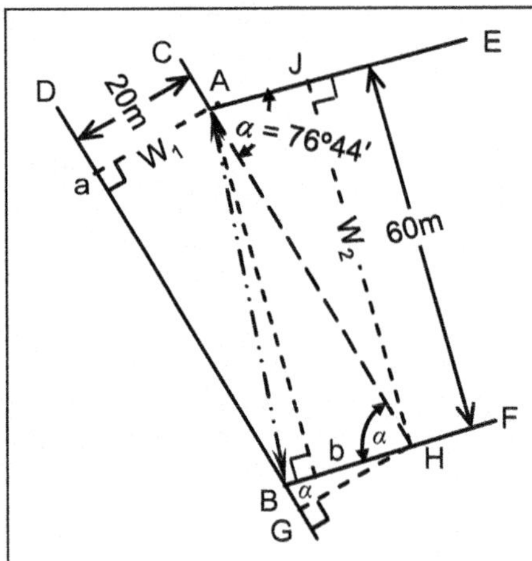

Figure 9-13 Enlarged intersection construct

<u>Solution</u>:

1. <u>Intersection or cosecant (1/sin) distance</u>: AB

∠JAH = ∠AHB = ∠HBG = 150°25′ – 73°41′ = 76°44′ = ∠α

In ΔJAH, AH = JH×cosec(∠JAH) = W_2×cosec(α)
= 60/sin(76°44′)
= 61.645

In ΔHBG, BH = HG·Cosec(∠HBG) = W_1×cosec(α)
=20/sin(76°44′)
= 20.548

In triangle ABH, (see Figure 9-14)

(i) solve for AB by Cos Rule

$$AB = h = \sqrt{a^2 + b^2 - 2ab\cos(H)}$$

AB = 60.340m

(ii) solve for angles using Sin Rule,

$$\frac{a}{\sin(A)} = \frac{b}{\sin(B)} = \frac{h}{\sin(H)}$$

$$\sin(A) = \frac{a \times \sin(H)}{h} = \frac{20.548 \times \sin(76°44')}{60.340} = 0.331452$$

A = 19°21′25″

Bearing AB = θ_{AH} + ∠A =150°25′00″ + 19°21′25″

θ_{AB} = 169°46′25″

Similarly

$$\sin(B) = \frac{b \times \sin(H)}{h} = \frac{61.465 \times \sin(76°44')}{60.340} = 0.994356$$

AMBIGUOUS CASE

B = 83°54′35″ OR 96°05′25″

By inspection ∠B = 83°54′35″ (∠A + ∠H > 90°)

(iii) Check: ∠A 19°21′25″
∠B 83°54′35″
<u>∠H 76°44′00″</u>

SUM: 180°00′00″

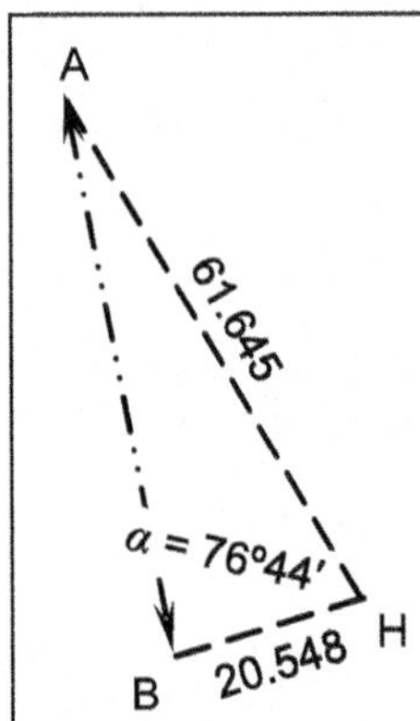

Figure 9-14 ΔAHB

If it's not checked it's wrong.

2. Offset or cotangent (1/Tan) distances: Ba, Bb

i) In ΔBAa, ∠ABa = ∠A = ∠HAB = γ = 19°21′25″ (See Figure 9-15)

Ba = Aa×cot(∠HAB) = W_1×cot(γ)
= 20/tan(19°21′25″)
= 56.930

Check: Ba = Ga - GB
= 56.930

Ga = AH = 61.645
GB = W_1×cot(α)
= 20/Tan(76°44′) = 4.715

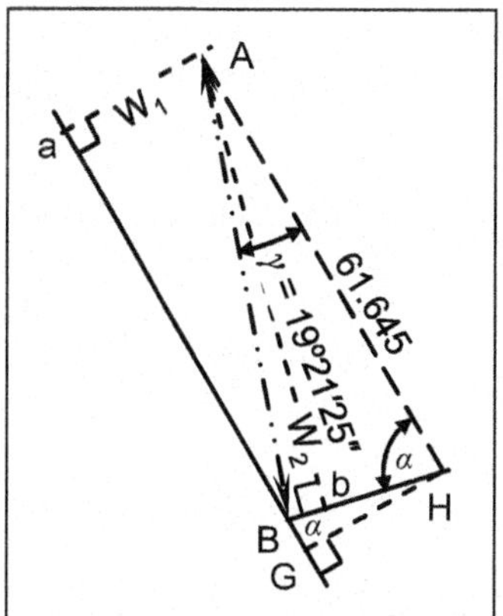

Figure 9-15 Offset distances

ii) In ΔABb, ∠ABb = ∠B = 83°54′35″

Bb = Ab×cot(∠B) = W_2×cot(B)
= 60/tan(83°54′35″)
= 6.402

Check: Bb = BH – bH
= 6.402

BH = 20.548
bH = W_2×cot(α)
= 60/Tan(76°44′) = 14.146

AB = 60.340 $\quad$ θ_{AB} = **169°46′25″**
Ba = 56.930
Bb = 6.402

3. Distances of southern side of road: DB, BF.

Find remaining cotangent and cosecant distances (See Figure 9-16)

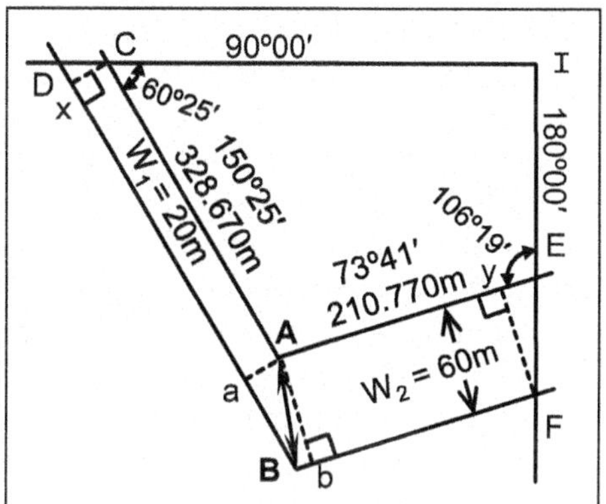

Figure 9-16 South side of road

i) DB = CA + Dx + Ba

Dx = Cx×cot(∠ICA) (External angle with boundary)
∠ICA = 150°25′ – 90°00′ = 60°25′
Dx = 20/tan(60°25′) = 11.354
DB = 328.670 + 11.354 + 56.930
DB = 396.945
CD = W_1×cosec(60°25′) = 20/sin(60°25′) = **22.998**

ii) BF = AE + Bb + Ey

Ey = Fy·Cot(∠IEA) (External angle) Figure 15.7c
∠IEA = 360°00′ – 253°41′ = 106°19′
Ey = W_2×cotan(106°19′) = 60/Tan(106°19′)
= –17.564
BF = 210.770 + 6.402 + (– 17.564)
BF = 199.608
EF = W_2×cosec(106°19′) = 60/Sin(106°19′) = **62.518**

Note: The external angles to provide correct sign for cotangent distances.

4. Check: Traverse using the directions and dimensions in a close. (See Table 9-3).

Table 9-3 Check: Traverse using the directions and dimensions in a close

TRAVERSE LEG	Angle	Bearing	Distance	ΔE	ΔN
CD		270°00′	22.998	–22.998	0.000
DB		150°25′	396.954	195.972	–345.207
BF		73°41′	199.608	191.568	56.079
FE		360°00′	62.518	0.000	62.518
EA		253°41′	210.770	–202.281	–59.215
AC		330°25′	328.870	–162.261	285.824
			Misclose	0.000	–0.001

5. 45° Check: Rotation and scaling of results gives a check of original close. (See Table 9-4).

Table 9-4 45° Check: Rotation and scaling of close

LEG	Brg +45°	Dist./√2	S	C	ΔE	ΔN
CD	315°00′	16.262	–11.499	11.499	–22.998	0.000
DB	195°25′	261.597	–74.617	–270.589	195.972	–345.207
BF	118°41′	141.144	123.824	–67.745	191.568	56.079
FE	45°00′	44.207	31.259	31.259	0.000	62.518
EA	298°41′	149.037	–130.748	71.533	202.281	–59.215
AC	15°25′	232.405	61.782	224.042	–162.261	285.824
S = (D/√2)×sin(θ + 45)		C = (D/√2)×cos(θ + 45)		S – C = ΔE	S + C = ΔN	

6. Area of Road reserve.

Area CDBAC $= ½(CA + DB)×W_1$ (Aa_
$= ½(328.67 + 396.954)×20$
$= 7256.24m^2$

Area ABFEA $= ½(AE + BF)×W_2$ (Ab)
$= ½(210.77 + 199.608)·60$
$= 12311.34m^2$

Total Area $= 19567.58m^2$
= 1.9568 ha

Answers:

(1) AB = 60.340m
$\theta_{AB} = 169°46'25''$

(2) Ba = 56.930m
Bb = 6.402m

(3) DB = 396.954m
BF = 199.608m

(4) Area = 1.9568 ha

9.5 Differing road widths, differing offsets.

When two roads of differing widths intersect the problem becomes more complex. The deflection angle does not create a simple half angle diagonal, and the two Cotangent offsets are different.

A more general approach, using calculation methods similar to constant width intersections, is developed from the **cosecant** intersection and **cotangent** offset calculation based on the intersection **external** (deflection) angle.

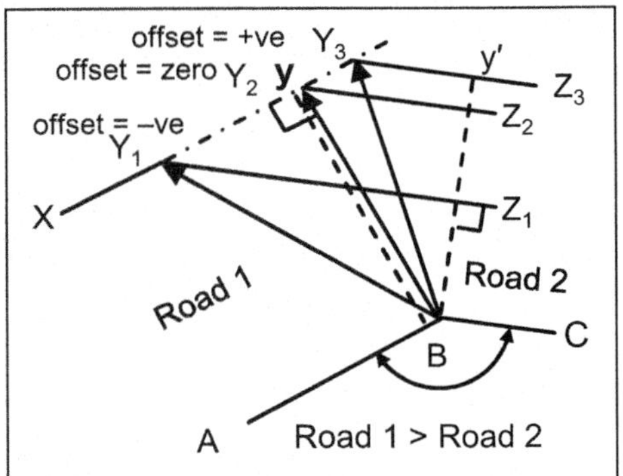

Figure 9-17 Differing widths and offsets

Referring to Figure 9-17, it can be seen that the boundary offsets, in keeping with previous notation, are

$Y_n y$ for Road 1 and

$Y_n y'$ for Road 2.

If the external deflection angle B is **less than 180°** then the **narrower** road, Road 2, will have an offset that is **always** positive.

However, for the **wider** road, Road 1, the offset, $\mathbf{Y_n y}$, may be: negative, **zero**, or positive.

The **sign** of the offset depends on the **size** of the external angle. The **critical** angle, where the offset is zero, is the arc cosine of the ratio of the narrow to the wider roads.

$\angle \mathbf{B_{critical}} = \mathbf{180 \pm arccos(Rd1/Rd2)}$

Similarly, at a given $\angle B$, the width, $W_{critical}$ at which the offset = **zero** is:

$W_{critical}$ Road 2 = Road 1×|cos ($\angle$B)| (absolute value of cos$B$)

In Figure 9-18, the setout boundary in ABC:

Bearing AB = 60°
$\angle$ABC = 140°
Road 1 width = 30m
Road 2 width = 15m

Examine the critical width and critical angle where **offset = 0**:

Width: $W_{critical}$ Road 2
Angle: $\angle B_{critical}$

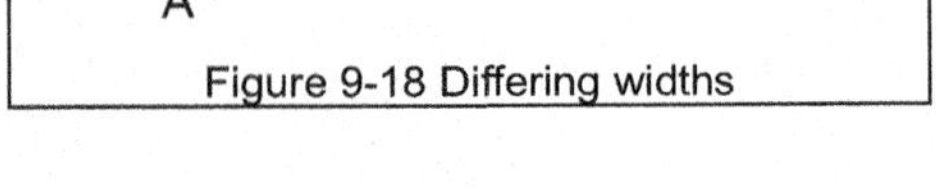

Figure 9-18 Differing widths

Road 2 $W_{critical}$ = Road 1×|cos(140°)|
= 30×0.7660 (ABS(cos(140))
= 22.98

$\angle B_{critical}$ = 180 ± arccos(Rd1/Rd2)
= 180 – arccos(15/30)
= 180 – 60
= 120°

9.5.1 Worked Example 9.5.

Differing road widths, external angle method.

We require to find: (See Figure 9-19)

1 Intersection diagonal BY (cosecant distance)

2. Boundary offsets Yy (YQ and Yy′ (Yb)

Project boundary AB to meet the boundary YZ at P, (See Figure 9-20)

y is the perpendicular offset from Y on to AB

b is the perpendicular offset from B on to YP

∠ABC = 140°

Solve for cosecant (hypotenuse) and cotangent (adjacent or offsets)

1 ΔYyP (Road 1, 30m)

YP = 30×cosec(140) = 46.672 Py = 30×cot(140) = –35.753

2 ΔBy′P (Road 2,150m)

BP = 15×cosec(140) = 23.336 Py = 15×cot(140) = –17.876

Solve ΔBYy (Figure 9-20)

The offset distance for Road 1,

BY = Py + PB = –35.753 + 23.336 = –12.417 and

Yy = –30 (Road 1)

In Figure 9-21 (Note, this is the intersect point B is on the axis AP)

Let yY = ΔE = –30

By = ΔN = –12.417

Then ∠YBA = arctan(ΔE/ΔN) and YB = hypotenuse

∠B = **67°30′57″** and YB = **32.468**

Brg BY = BRG BA + ∠B =240° + 67°30′57″ = **307°30′57″**

Or, using the polar conversion function (→***θ, r***)

∠PBY = **–112° 29′ 03″** YB = **32.468.**

Brg BY = Brg AB + ∠PBY = 60° + (–112° 29′ 03″)

Brg BY = 420° + (–112° 29′ 03″) = **307° 30′ 57″**

The offset road distance for Road 2 (See Figure 9-22)

Yb = YP + Py = 46.672 + (–17.876) = **28.796**

In summary

Brg BY	=	307° 30′ 57″
Brg BY	=	307° 31′ (3″ = 0.5mm)
BY	=	32.468m
By	=	−12.417m
Yb	=	+28.796m

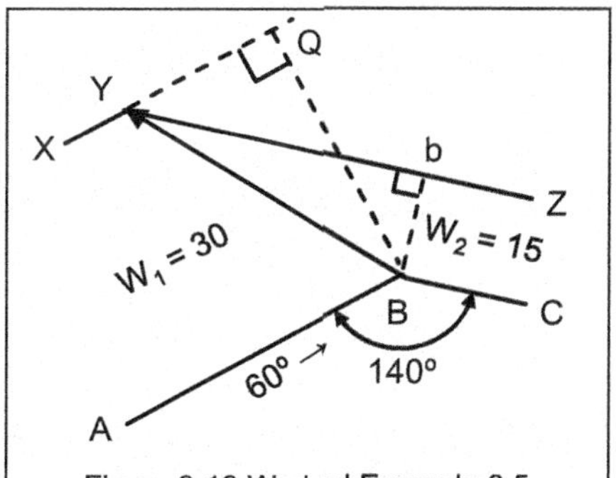

Figure 9-19 Worked Example 9.5

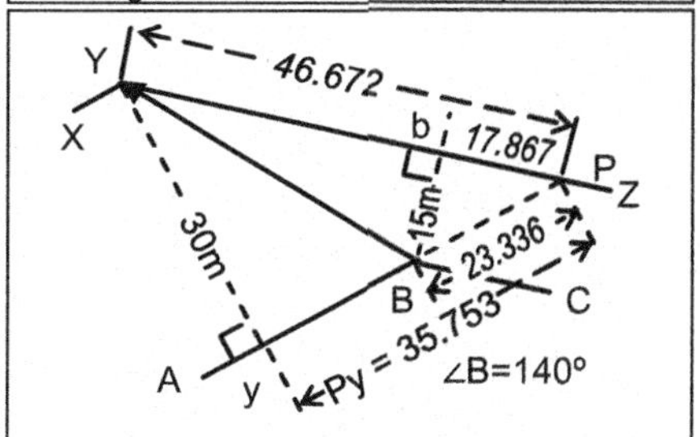

Figure 9-20 Projecting line AB

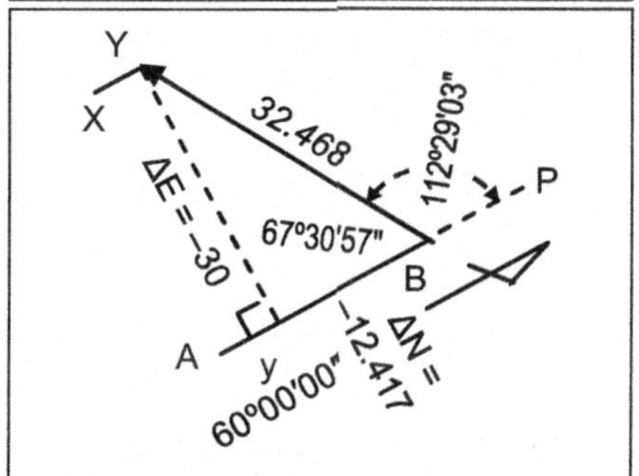

Figure 9-21 Rectangular framework.

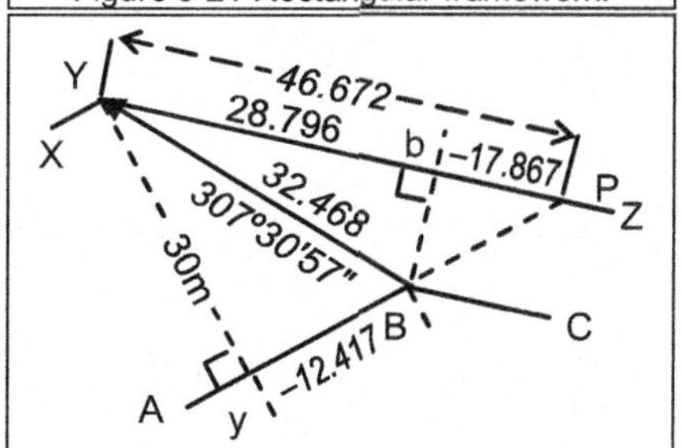

Figure 9-22 Offset for Road 2

9.5.2 Worked Example 9.6.

Three different external angles (150°, $\angle_{critical}$, 120°) are summarised.

The Worked Example shows the **critical angle** (§ 9.5) for a 20m and 30m road intersection, equals 180 – arccos(20/30) = **131° 48′ 37″**

∠ABC = ∠B = 150°. Solve for cosec (intersection) and cot (offsets) in Figure 9-23.

1) ΔYyP (Road 1, 30m)

YP = 30×cosec(150) = 60.000 .

Py = 30×cot(150) = –51.961

2) ΔBbP (Road 2, 20m)

BP = 20×cosec(150) = 40.000 .

Pb = 20×cot(150) = –34.641

ΔE = 30.000

ΔN = By = Py + PB = −11.961

∠PBY = 111° 44′ 17″ YB = 32.297.

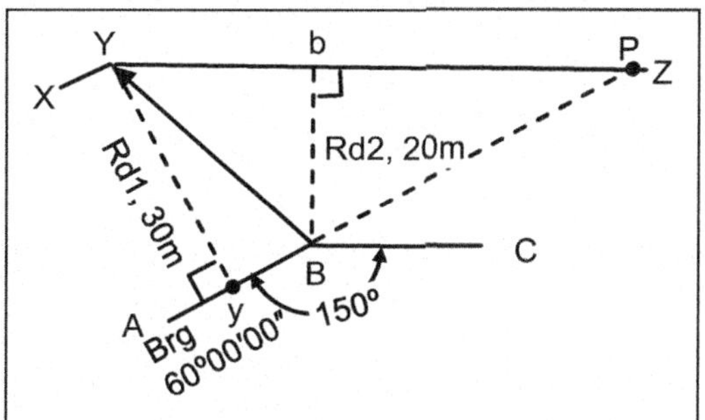

Figure 9-23 Offsets for ∠B > ∠Critical

Brg BY	=	308° 15′43″ BY = 32.297m
By	=	−11.961m (the negative offset on the wide road [∠B > $\angle B_{critical}$])
Yb	=	+25.359m (the offset on the narrower road is positive)

∠ABC = ∠B = 141°48′37″. ([∠$B_{critical}$]) Solve for cosec (intersection) and cot (offsets) in Figure 9-24.

1) ΔYyP (Road 1, 30m)
   YP = 30×cosec(B) = 40.349 .
   Py = 30×cot(B) = −26.833
2) ΔBbP (Road 2, 20m)
   BP = 20×cosec(B) = 26.833 .
   Pb = 20×cot(B) = −17.888
   ΔE = 30.000
   ΔN = By = Py + PB = 0.000

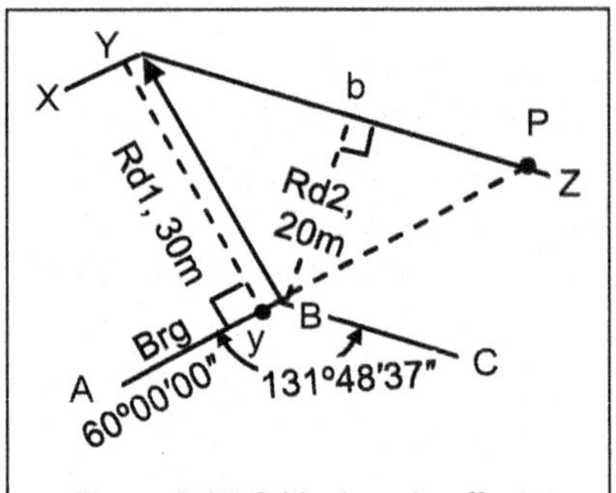

Figure 9-24 Critical angle offsets

∠PBY = 90°00′00″ YB = 30.000.

Brg BY	=	**330° 00′ 00″**
BY	=	**30.000m**
By	=	0.000m (∠$B_{critical}$)
Yb	=	+22.361m (the offset on the narrower road is positive).

∠ABC = ∠B = 120° Solve for cosec (intersection) and cot (offsets) in Figure 9-25.

1) ΔYyP (Road 1, 30m)
   YP = 30×cosec(120) = 34.641 .
   Py = 30×cot(120) = −17.320
2) ΔBbP (Road 2, 20m)
   BP = 20×cosec(120) = 23.094 .
   Py′ = 20×cot(120) = −11.547
   ΔE = 30.000
   ΔN = By = Py + PB = 5.773

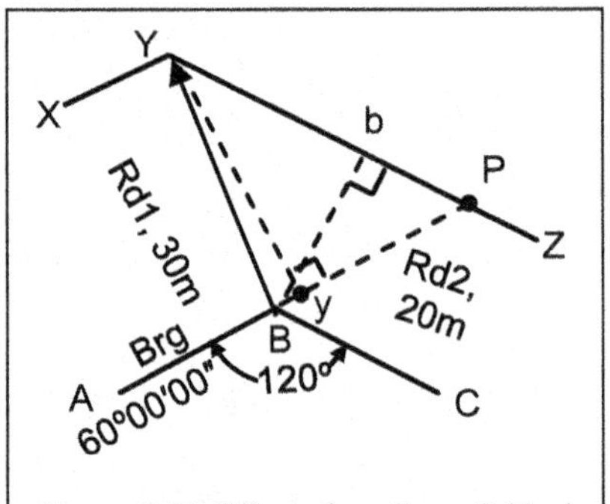

Figure 9-25 Offsets for ∠B < ∠Critical

∠PBY = 79°06′24″ YB = 30.551.

Brg BY	=	340° 53′ 36″
BY	=	30.551m
By	=	+5.773m (the **positive** offset on the wide road [∠B < ∠$B_{critical}$])
Yb	=	+23.094m (the offset on the narrower road is positive).

9.6 Chapter Exercises 9.1

Q1. In Figure 9-26,
Calculate AB, BC and cosecants AF, EB and CD.

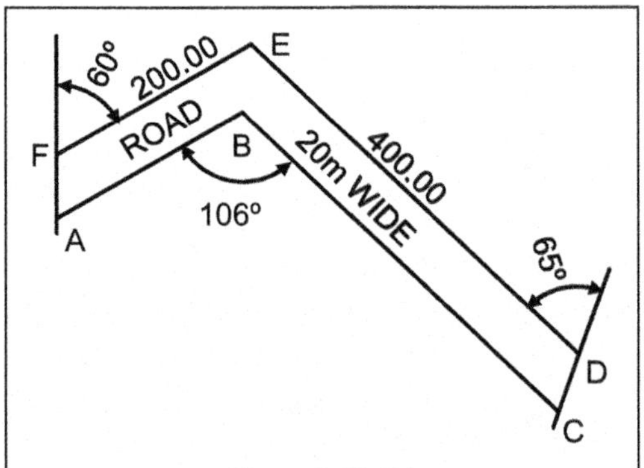

Figure 9-26 Q1

Q2. In Figure 9-27
Find EF, FG, GH,
Cosecant distances AE, FB, CG and DH.
Also find the area of the road.

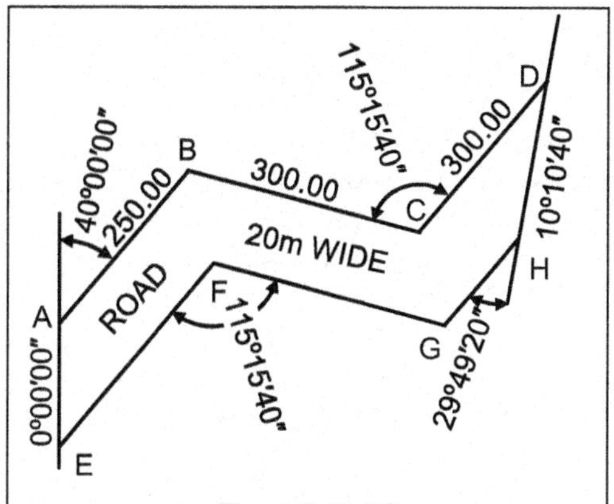

Figure 9-27 Q2

Q3. In Figure 9-28
A 20 m road joins a 60 m road, and the dimensions are as shown. Calculate the length of the southern boundary EFGH and the area of the road.

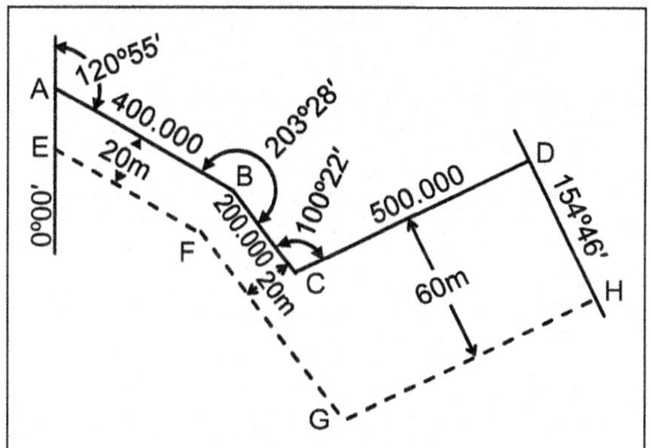

Figure 9-28 Q3

Q4. Three roads intersect as shown in Figure 9 29.
Given the data in this figure,
calculate the bearing and distance from A to B.

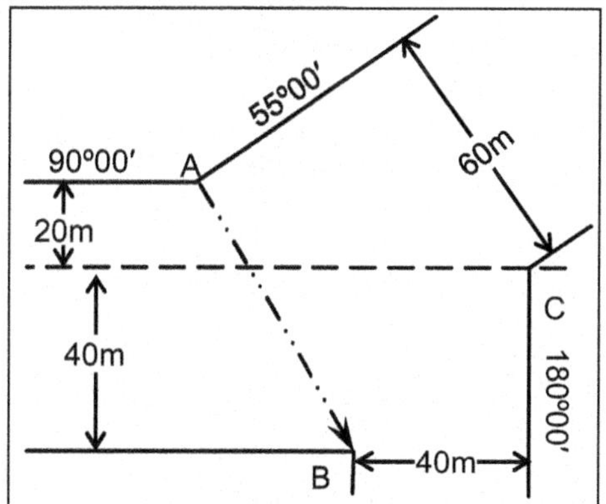

Figure 9-29 Q4

Hint: Calculate the bearing and distance AC, CB and thus the bearing and distance AB.

9.6.1 Answers to Chapter Exercises 9.1

Q1. AF = 23.094
AB = 196.476
BC = 394.255
CD = 22.068
EB = 25.043

Q2 FG = 300.000
GH = 277.787
AE = 31.114
FB = 23.679
CG = 23.679
DH = 40.216
Area = 1.6889 ha

Q3 EF = 383.868
FG = 253.183
GH = 509.338
Area = 4.2651 ha

Q 4 θAB = 149°00′20″
DAB = 69.994

A solution method for Q4.
Split the problem into two parts:

1). Differing width roads intersecting at A with (See Figure 9-30)

W1 = 20 and W2 = 60

∠A = 235 – 90 = 145

Project both boundaries from A across the two roads to a′ and c′

<u>Road1</u> Cosec dist Aa′ = W_1×cosec A

= 20/sin(145) = 34.869 (not used)

Offset aa′ = W_1×cot A

= 20/tan(145) = –28.563

<u>Road2</u> Cosec dist Ac′ = W_2×cosec A

= 60/sin(145) = 104.607

Offset cc′ = W_2×cot A

= 60/tan(145) = –85.689 (not used)

aC = Ac′ + aa′ (ΔE_{AC} = aC, Ac′ = a′C [parallel 20m width])

= 104.607 – 28.563 = 76.044

ΔE_{AC} = 76.044

ΔN_{AC} = –20.000 (W_1)

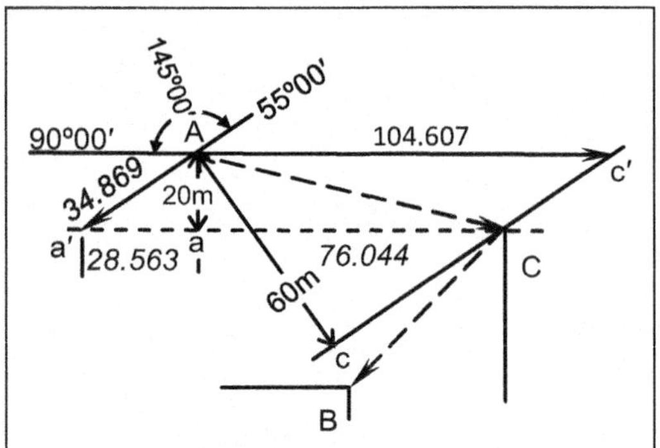

Figure 9-30 Part 1 of solution

2). Constant road width intersecting at B with (See Figure 9-31)

W1 = W2 = 40 = W

∠B = 270 – 180 = 90

no calculations needed, by inspection (B = 90°)

ΔE_{CB} = –40.000 (W)

ΔN_{CB} = –40.000 (W)

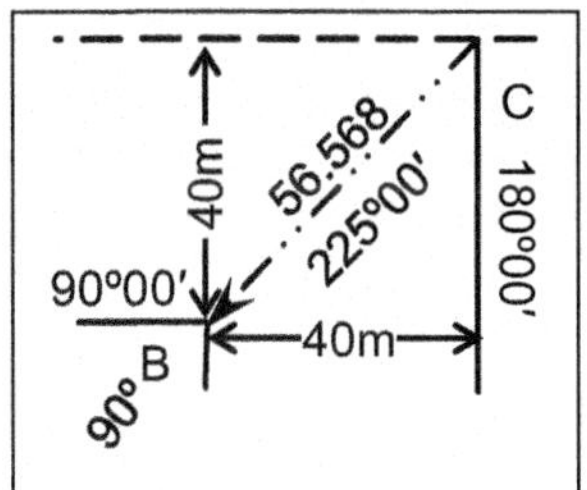

Figure 9-31 Part 2 solution

3) Closed traverse around intersection (See Figure 9-32)

Starting at A = 0, 0 and summing ΔEs, ΔNs

ΔE_{AC}	= 76.044	ΔN_{AC}	= –20.000
ΔE_{CB}	= –40.000	ΔN_{CB}	= –40.000
E_B	= 36.044	N_B	= –60.000

Brg/dist AB by polar formula

θ_{AB} = 149°00′20″

D_{AB} = 69.994

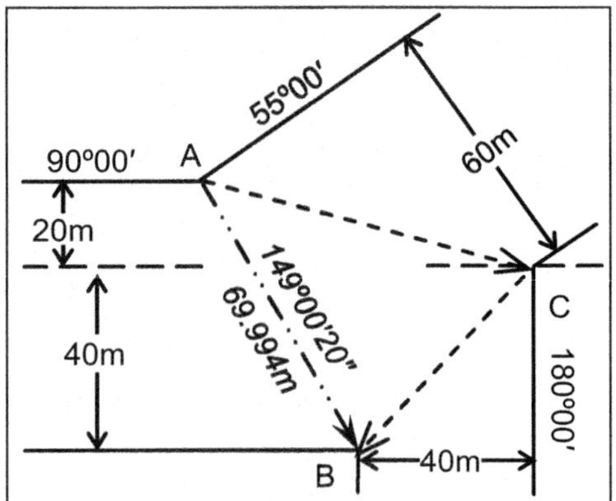

Figure 9-32 Intersection traverse

9.7 References for Chapter 9

School of Surveying & Land Management. Surveying Computations 185. Curtin University (1998)

Chapter 10 Property Subdivisions

In life, people aspire to buy land to develop, i.e., build a structure of use it for various purposes. This is normally done once the surveyor has excised this land from the larger parcel to give it its own title deed (a document detailing the size, number and owner of the land). The survey procedure used to achieve this is called subdivision. This chapter is essential for surveyors who aim at becoming licensed in order to undertake such parcel subdivisions. It provides essential computational skills that underpin the tasks involved in the subdivision survey.

10.1 Introduction.

Computational techniques of the subdivision of land are covered extensively in this topic as the methods applied are not generally available in survey texts.

A parcel of land can be subdivided into smaller areas called **lots**. The number of lots able to be created generally depends on planning regulations; it may cover minimum lot size.

From this the **ratio** of one lot size to another determines the area of each lot. If a parcel is to be split in half the Lot 1 = Lot 2 then the ratio of Lot 1 to Lot 2 = 1:1. The Lots are equal area.

If one lot is to be twice as big as the other we say Lot 1:Lot 2 = 1:2; $\frac{Lot1}{Lot2}=\frac{1}{2}$ and,

by cross multiplying 2×Lot 1 = 1×Lot 2.

By summing the portions (1 + 2 = 3), Lot 1 = 1/3 and Lot 2 = 2/3 of total area (the ratio of 1:2 is maintained).

The problems arising here are the subdivision of areas of land into portions which have a given ratio to each other. Usually roads are involved, so that the calculations involving the areas of portions are only possible after the road dimensions have been calculated. Initially the sub-division of areas without roads will be used to demonstrate various techniques used in problem solving.

The information which must be calculated in any subdivision:

(i) Bearing and distance of the dividing lines

(ii) The data needed so that a surveyor may set out the dividing lines in the field.

10.2 Subdivision of Triangles.

It is not usual to sub-divide triangles. However, many lots can be transformed into triangles by the addition or subtraction of simple figures.

Review Chapter 5, Triangle Solutions, advanced examples, W.E. 5.2 on page 47 which introduced the subdivision of triangles.

Consider for example the area ABCD in Figure 10-1, which is to be sub-divided into two equal portions by a line through P. Addition of triangle CDE produces triangle ABE which can then be treated in a particular manner.

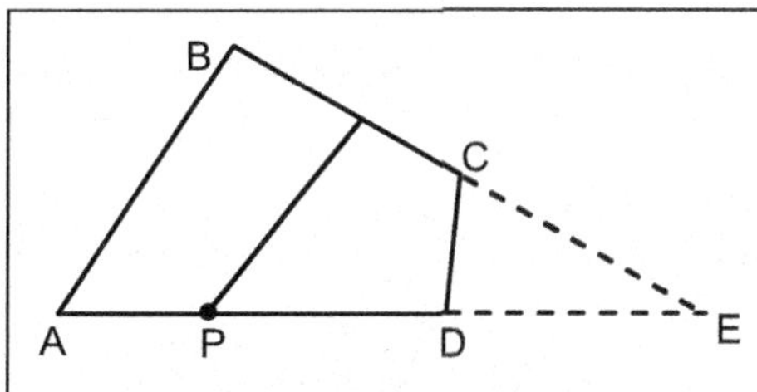

Figure 10-1 Triangle subdivision

The subdivision of triangles in this course will be carried out by using lines which fulfill certain conditions:

10.2.1 Pass through a given point,

(a) The point is one of the apex points

(b) The point lies on one of the sides OR

10.2.2 The line has a given bearing.

10.3 Subdivision by a line through a given point.

10.3.1 The given point is one of the apex points. (§ 10.2.1 (a))

Consider triangle ABC in Figure 10-2 which is to be subdivided by a line BX such that each triangle is equal.

Area ABX = ½AX×BX×sinα

Area CBX = ½ CX×BX×sinβ

and $\frac{\text{Area } ABX}{\text{Area } CBX}=\frac{AX}{CX}$; (sin$\alpha$ = sinβ, Sin being +ve in Quadrants I and II).

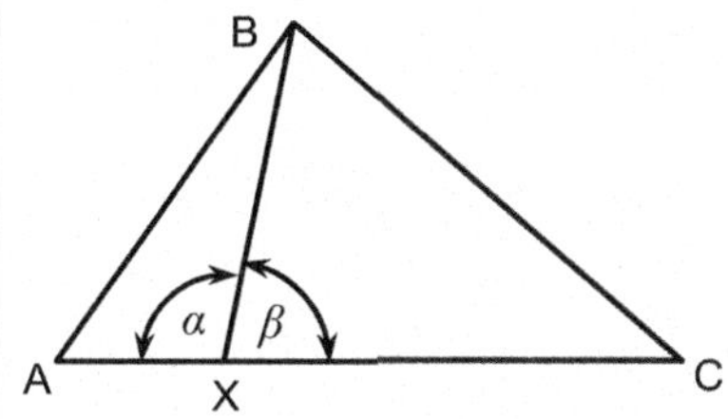

Figure 10-2 Subdivide through a point

If, in Figure 10-2, the area ABC is to be subdivided in the ratio 1:2 then AC must be divided in the ratio 1:2. Then (ABX:CBX = 1:2).

J. Walker and J. Awange, *Surveying Computations*,
https://doi.org/10.1007/978-3-031-94382-9_10

$$\frac{\text{Area } ABX}{\text{Area } CBX} = \frac{AX}{CX} = \frac{1}{2}, \text{ and } \frac{AX}{AX + CX} = \frac{1}{(1+2)} \text{ i.e. } AX = \frac{AC}{3}$$

10.3.1.1 Worked Example 10.1 Line through apex.

Given: In Figure 10-3, AB = 98.820, θ_{AB} = 35°20′00″
AC = 155.770, θ_{AC} = 88°44′00″

Aim: To subdivide the triangle ABC by a line through B such that Area ABX = 3/7 Area ABC

Solution:

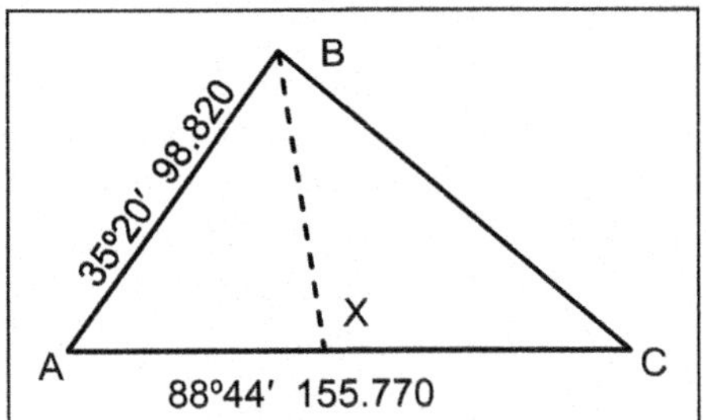

Figure 10-3 Line through the apex. Worked Example 10.1

(1) $\frac{AX}{AC} = \frac{3}{7} \Rightarrow AX = \frac{3}{7}AC = \frac{3 \times 155.770}{7} = 66.759\text{m}$

(2) ∠BAC = 88°44′00″ – 35°20′00″ = 53°24′00″

(3) Solve triangle ABX (2 sides + included angle)

(4) Hence calculate bearing and distance BX.

Thus, we have the bearing and distance of the dividing line BX and the data necessary to set out this dividing line.

10.3.2 The given point lies on one of the sides. (§ 10.2.1 (b))

10.3.2.1 Worked Example 10.2. Point on a side.

See Figure 10-4

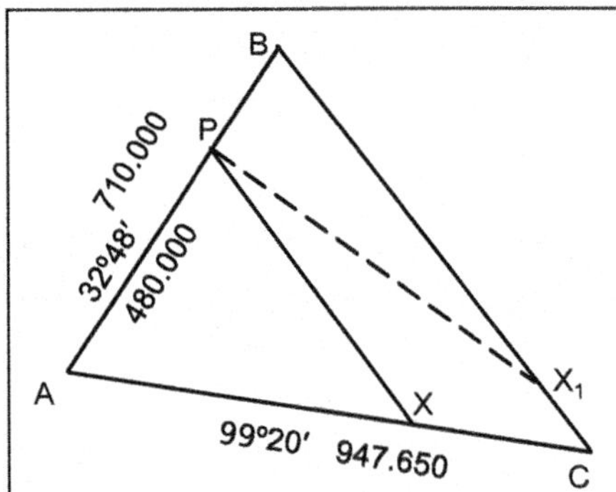

Figure 10-4 Through point on one side Worked Example 10.2

Given: AB = 710.000m θ_{AB} = 32°48′00″
AP = 480.800m
AC = 947.650m θ_{AC} = 99°20′00″

Aim: Subdivide the triangle in Figure 10-4 into two portions in the ratio 2:5, the larger portion near A, by a line through P.

Area APX = Lot 1: Area Lot 2. Lot 1:Lot 2 = 5:2

Area APX = $^5/_{(5+2)}$×Area ABC = $^5/_7$ Area ABC

Note: With this type of problem, the position of P must be known. The triangle APX can be solved given the area, one side (AP) and an angle (BAC). Thus PX, AX and ∠APX can be calculated.

There is every chance that AX will be greater than AC which means the triangle BPX_1 should be solved and not triangle APX.

∠BAC = 99°20′00″– 32°48′00″ = 66°32′00″

Area ABC = ½·AB×AC×sin(∠BAC) = 308,591m^2

Area APX = $^5/_7$ Area ABC = 220,422m^2

But Area APX = ½AP×AX×sin(∠BAC) cross multiply to find AX

$$AX = \frac{2 \times AreaAPX}{ap \times \sin(\angle BAC)} = 999.567\text{m}$$

Now, AX > AC (999.567 > 947.650) and thus a wrong assumption has been made. It is the triangle BPX_1 to be solved and not triangle APX.

Using triangle BPX^1

1) Solve for BC (Cos Formula for side)
BC = 930.740m

2) Solve for ∠B, ∠C (Sin Rule)
∠B = 69°03′37″
∠C = 44°24′23″
∠A = 66°32′00″

CHECK: = 180°00′00″

3) BP = AB – AP = 710.000 – 480.800 = 229.200m

4) Area BPX_1 = $^2/_7$ Area ABC = 88,169m^2

5) $$BX_1 = \frac{2 \times Area\, BPX_1}{BP \times \sin(\angle CBA)} = 823.766\text{m}$$

If it's not checked it's wrong.

6) $CX_1 = BC - BX_1 = 930.740 - 823.770 = 106.974$m
7) In triangle BPX_1 solve for the missing elements. (Figure 10-5)

(Given 2 sides plus an included angle)

Using Cos Rule, $PX_1 = 772.123$m

Solve for $\angle P$, $\angle X_1$ in same triangle using Sin Rule.

$$\frac{\sin P}{BX_1} = \frac{\sin X_1}{BP} = \frac{\sin B}{PX_1}$$

$\angle P = 94°50'38''$

$\angle X_1 = 16°05'44''$

$\angle B = 69°03'38''$

CHECK: $= 180°00'00''$

$\theta_{PX1} = \theta_{AB} + \angle P = 32°48'00'' + 94°50'37''$
$= 127°38'37''$

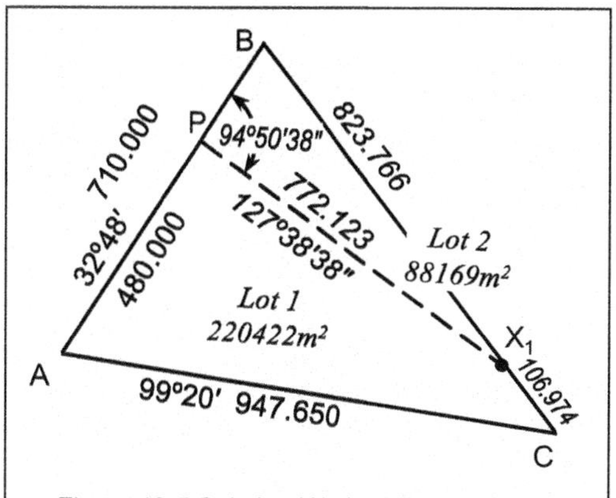

Figure 10-5 Solution Worked Example 15.2

8) The necessary data to set out line PX_1 is:
$\theta_{PX1} = 127°38'38''$
$PX_1 = 772.121$
$BX_1 = 823.766$
$CX_1 = 106.974$
9) Check area: Lot 1 + Lot 2 $= 220422\text{m}^2 + 88169\text{m}^2 = 308591\text{m}^2$. ✓

10.4 Subdivision by a line in a given direction.

(Line in a given bearing, situation § 10.2.2)

The direction of the dividing line must be known with respect to one of the sides of the given triangle.

In Figure 10-6 the area of the triangle PQC can be calculated given the known area of ABC and the ratio of subdivision plus the angles of triangle PQC.

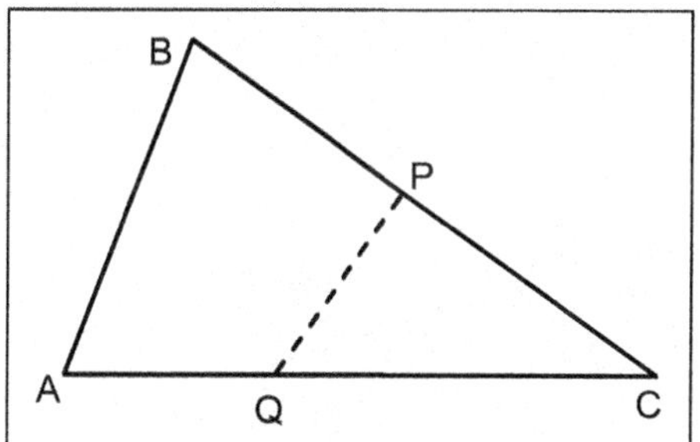

Figure 10-6 Line in a given direction

In any triangle ABC, any one of 3 sides can be calculated given the triangle **area** plus the 3 angles.

i.e. $$a = \sin A \sqrt{\frac{2 \times Area}{\sin A \sin B \sin C}}$$

thus, in triangle PQC

$$PQ = \sin C \sqrt{\frac{2 \times Area}{\sin P \sin Q \sin C}}$$

$$PC = \sin Q \sqrt{\frac{2 \times Area}{\sin P \sin Q \sin C}}$$

$$CQ = \sin P \sqrt{\frac{2 \times Area}{\sin P \sin Q \sin C}}$$

There is one problem that can arise in this case; the position of the dividing line with respect to one of the apex points is not clear, and the wrong triangle may be solved. Consider the triangle in Figure 10-7.

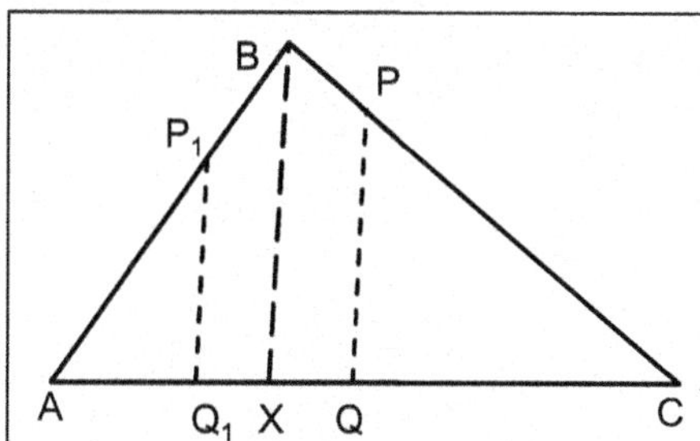

Figure 10-7 Which line?

For a given ratio, the dividing line is either PQ or P_1Q_1.

Note that P_1Q_1 // BX // PQ. (both parallel to BX)

Construct the triangle with the dividing line passing through one apex in the given direction.

In order to determine on which side of BX is the dividing line, the area of triangle ABX (3 angles and 1 side) is solved and this area compared with the proposed subdivided area.

The dividing line will be PQ or P_1Q_1.

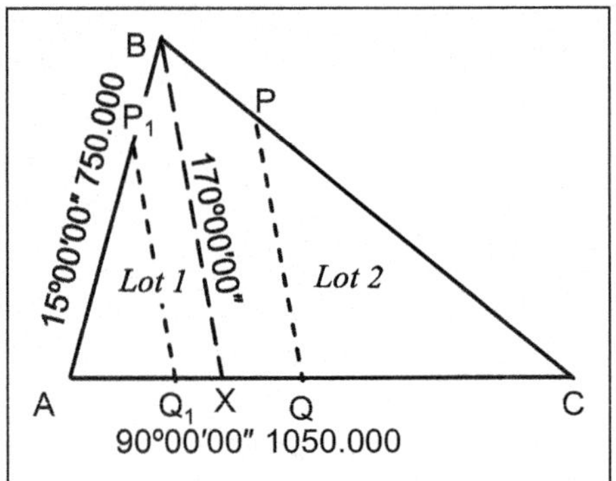

Figure 10-8 Worked Example 10.3

10.4.1 Worked Example 10.3

AB= θ_{AB} = 15°00′00″ 750.000m

AC= θ_{AC} = 90°00′00″ 1050.000m

ABC (Figure 10-8) is to be subdivided into 2 Lots in the ratio Lot 1: Lot 2 = 1:2, the smaller portion near A, by a line PQ of bearing 170°00′00″.

Area ABX = $^{1}/_{(2+1)}$ Area ABC = ⅓ Area ABC

Solution:

1. Construct line BX at θ_{BX} = 170°00′00″. (See Figure 10-9)
   Ascertain on which side of BX is the dividing line.
   $\angle A = \angle BAC = \theta_{AC} - \theta_{AB}$ = 75°00′00″
   $\angle B = \angle ABX = \theta_{BA} - \theta_{BX}$ = 25°00′00″
   $\angle X = \angle AXB = \theta_{BX} - \theta_{AC}$ = 80°00′00″ Check Σ = 180°
   Area ABC = ½·AB×AC×sin(75) = 380333m²

$$\text{Area } ABX = \frac{x^2 \sin A \sin B}{2\sin X} = \frac{1}{2}\left(\frac{AB^2 \times \sin A \times \sin B}{\sin X}\right) = 116582\text{m}^2 \text{ (See Eqn 3-9 b, page 24)}$$

Triangle ABC is to be subdivided so that ⅓ area ABC lies near to A.

⅓ Area ABC = (380333)/3 = 126778m²

Area of proposed subdivision (⅓ ABC) > Area ABX because 126778 > 116582.

Thus, PQ must lie to the right of BX, further away from A. (See Figure 12-9).

2. Solve triangle ABC for bearing and distance BC.
   Compute BC using Cos Rule:
   BC = 1121.321m
   Solve for ∠B, ∠C using Sin Rule.

$$\frac{\sin B}{AC} = \frac{\sin C}{AB} = \frac{\sin A}{BC}$$

   ∠B = 64°45′16″
   ∠C = 40°14′44″
   ∠A = 75°00′00″
   CHECK: = 180°00′00″
   Solve for bearing BC.
   θ_{CA} = 270°00′00″
   + ∠C = 40°14′44″
   $\theta_{CB} = \theta_{CA} + \angle C$ = 310°14′44″
   θ_{BC} = 130°14′44″ reciprocal θ_{CB}
   + ∠B = 64°45′16″
   CHECK: θ_{BA} = 195°00′00″ reciprocal θ_{AB}

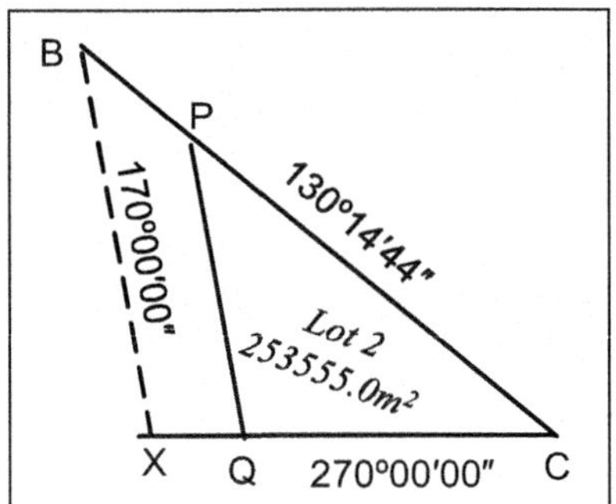

Figure 10-9

3. Area PQC = ⅔×380333 = 253555.0m²
   θ_{PQ} = 170°00′00″
   $\theta_{PC} = \theta_{BC}$ = 130°14′44″
   $\theta_{CQ} = \theta_{CA}$ = 270°00′00″
   ∠C = 40°14′44″
   ∠P = 39°45′16″
   ∠Q = 100°00′00″
   CHECK: = 180°00′00″
4. Solving for sides of triangle PQC
   Area PQC = 253555.0

$$c = PQ = \sin C\sqrt{\frac{2\text{Area}}{\sin P \sin Q \sin C}}$$

$$= \sin C\sqrt{\frac{2\times 253550}{0.639499\times 0.984808\times 0.646056}} \Rightarrow \sin C\sqrt{\frac{507110}{0.406881}} = (0.646065)\times 1116.394$$

   PQ = 721.263m

If it's not checked it's wrong.

Note: $\sqrt{\dfrac{2Area}{\sin P \sin Q \sin C}} = 1116.394$ is a constant for each of the calculations of PQ, PC and CQ

5. Results:

PQ = 721.263, θ_{PQ} = 170°00′00″
PC = 1099.433, θ_{PC} = 130°14′44″
CQ = 713.932, θ_{CQ} = 270°00′00″

Check: $\cos Q = \dfrac{PQ^2 + CQ^2 - PC^2}{2 \times PQ \times CQ}$

(cos rule for angles, Eqn 3-2 a)

∠Q = 100°00′00″.

Also, BP = BC – PC = 21.888
AQ = AC – QC = 336.068

10.5 Chapter Exercise 10.1

Q. 1 See Figure 10-10.

Given that θ_{AB} = 35°50′00″
θ_{BC} = 105°35′20″

Subdivide the triangle ABC in ratio 1:3 by a line through B so that the smaller portion is nearest to A. Calculate the bearing and distance of the dividing line BX.

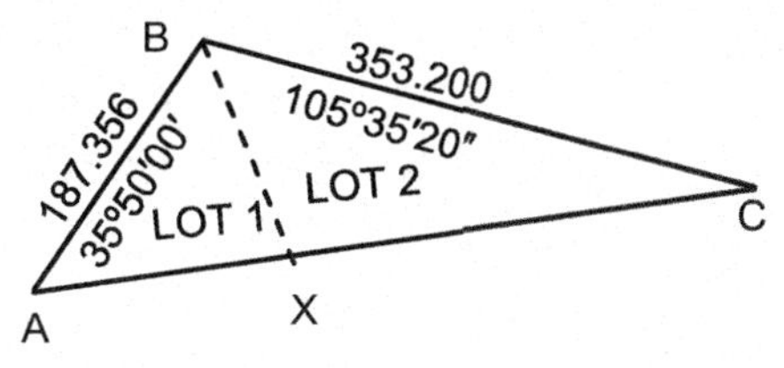

Figure 10-10 Chapter exercise 10.1, Q1

Q. 2 See Figure 10-11.

A triangular lot ABC has the following boundaries:
θ_{AB} = 122°30′, AB = 400.350
θ_{BC} = 20°16′, BC = 530.365

Divide the lot into 2 equal portions by a line bearing 270°00′ from the boundary BC.

Calculate the bearing and distance of this boundary and the data for setting out.

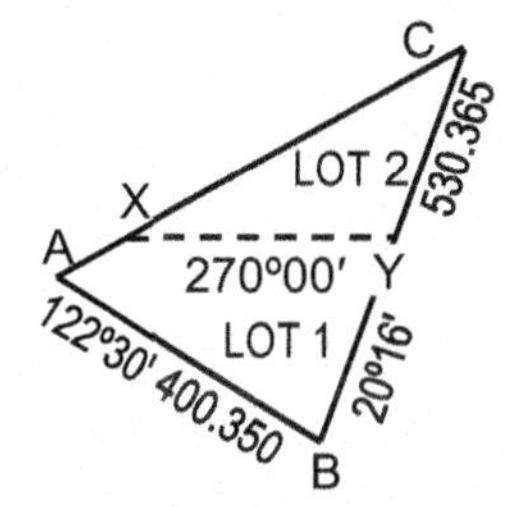

Figure 10-11 Q2

10.6 Subdivision of rectangles and parallelograms by lines parallel to one boundary.

10.6.1 Minimum lot area subdivision

A typical example might comprise the following subdivision.

In Figure 10-12 ABCD is a parallelogram in which the bearings and distances of the sides are known.

Aim: To subdivide the Lot 1 area into the largest number of lots of equal size so that none will be less than, say, 1000m². The dividing lines are to be parallel to AB and DC.

Solution: Determine the area of ABCD, then find the number of 1000m² lots which will fit into ABCD (say ***n***).
Divide AD, BC into ***n*** equal parts. Solve for the boundary of each lot.

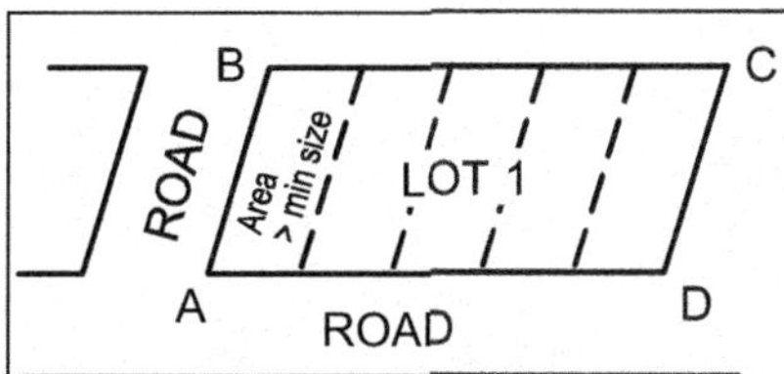

Figure 10-12

10.6.2 Multi-lot subdivision distributed at various ratios.

As a precursor to subdivision with access roads, it is best to re-iterate that the subdivision is made in the ratio of the lots. In many cases there may be adjacent lots as part of the subdivision which must be compared with a third lot to make up the whole subdivision area.

In Figure 10-13 a plot of land of 9000m2 is to be subdivided in the ratio of Lot 1:Lot 2:Lot 3 = 2:3:4. The variable in this case is the position of the boundary between the common Lots 1 and 2 and the adjacent Lot 3.

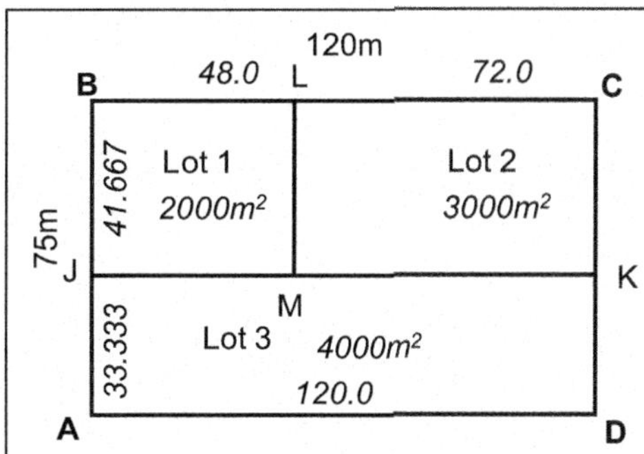

Figure 10-13 Multi-lot subdivision

If it's not checked it's wrong.

For a proper distribution of the areas, we must combine the proportions of Lots 1 and 2 and compare it with Lot 3.

Lot 1 + Lot 2 = 2 + 3 =5. Lot 3 = 4

Thus (Lot 1 + Lot 2): Lot 3 = 5:4 and by cross multiplication:

4×(Lot 1 + Lot 2) = 5×Lot 3.

The total area = 9000m^2 so each portion = 9000/(4+5) = 1000m^2.

Lot 1 = 2×1000 = 2000m^2.

Lot 2 = 3×1000 = 3000m^2.

Lot 3 = 4×1000 = 4000m^2

Test the distribution against the equality 4×(Lot 1 + Lot 2) = 5×Lot 3

4×(Lot 1 + Lot 2) = 4×(2000 + 3000) = 20000

5×Lot 3 = 5×4000 = 20000

Side AJ = KD = 4000/120 = 33.333m.

JB = LM = KC = 75 – 33.333 = 41.667m.

BL = 48m and LC = 72m. Note that these distances are in proportion 2:3 of AD = 120m.

10.7 Subdivision of Rectangles with Access Roads.

10.7.1 Worked Example 10.4 Access road method.

In Figure 10-14, the length of the access road (x) depends on the position of XY. The position of XY will also determine the area of the access road. It is not possible to calculate the area of each portion from their ratios, though it is possible to express a relationship between the areas.

e.g. If the area of ABCD in Figure 10-14 was to be subdivided into two portions so that Area 1:Area 2 = 1:3.

i.e. $\frac{\text{Area 1}}{\text{Area 2}} = \frac{1}{3}$ OR 3×Area 1 = Area 2

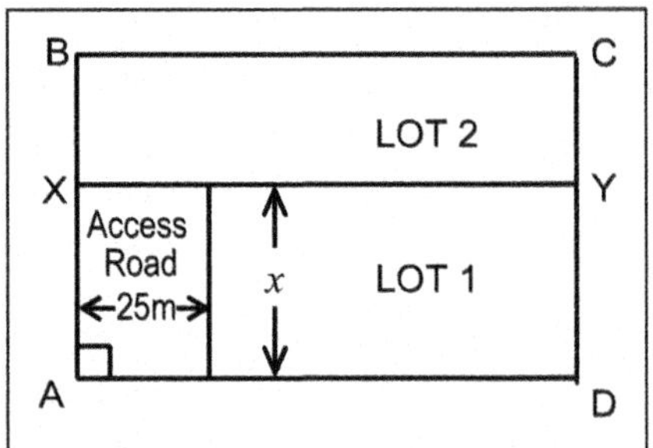

Figure 10-14 Worked Example 10.4

Now Area 1 and Area 2 can be expressed in terms of the unknown x. This enables equation (1) to be formed with x as the only unknown.

i.e. Area 1 = x×(AD – 25)

Area 2 = BC×(AB – x)

Taking the ratio of 3×Area 1 = Area 2

3x×(AD – 25) = BC×(AB – x) so that x can be found.

10.7.2 Worked Example 10.5 Access road.

A rectangular area ABCD in Figure 10-15 is to be subdivided in the ratio 2:3, the smaller portion to be adjacent to the road. A 25 metre access road to the rear portion must be provided.

AD = 300.000m, AB = 150.000m. Calculate dimensions and areas of the access road of the two portions.

Solution:

$\frac{\text{Area 1}}{\text{Area 2}} = \frac{2}{3}$ ∴ 3×Area 1 = 2×Area 2

now, Area 1 = x(300 – 25) = 275x

Area 2 = 300×(150 – x) = 45000 – 300x

Thus 3×(275x) = 2×(45000 – 300x)

i.e. 825x = 90000 – 600x

1425x = 90000

x = 63.158

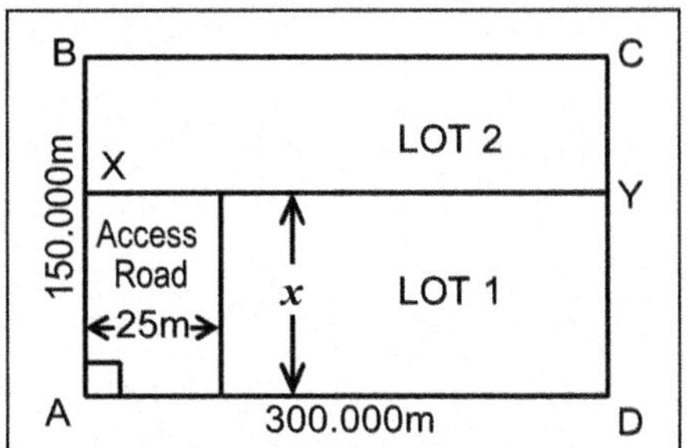

Figure 10-15 Worked Example 10.5

AX = DY = 63.158

BX = CY = 86.842

Area 1 = 275.000×63.158 = 17368.45m^2

Area 2 = 300.000×86.842 = 26052.60m^2

Road Area = 25.000×63.158 = <u>1578.95m^2</u>

Sum = 45000.00m^2

CHECK: Area ABCD = 300.000×150.000 = 45000.00m^2

If it's not checked it's wrong.

10.7.3 Worked Example 10.6. Access road, adjacent bounding lots method.

Given: Distances AB, BC and GH of rectangular block ABCD (See Figure 10-16)

Aim: Area ABCD is to be subdivided in the following ratios

Area 1:Area 2:Area 3 = 1:3:2

With this type of problem, first divide ABCD into 2 portions i.e. Compare Area AXYD against Area XBCY.

Thus Area (1 + 2):Area 3 = 4:2

OR $\frac{\text{Area }(1+2)}{\text{Area }3} = \frac{4}{2}$

and 2×Area (1 + 2) = 4×Area 3

Area (1 + 2) = x×(AG + HD)

= x×(AD – 10)

Area 3 = BC×(AB – x)

The required equation is 2×(AD – 10) = 4×BC×(AB – x) from which x may be solved.

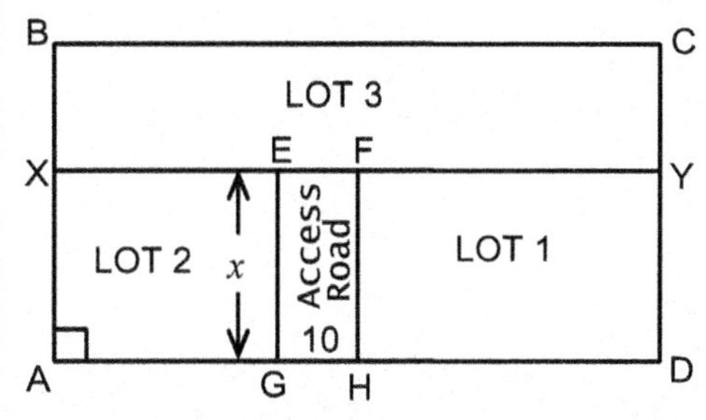

Figure 10-16 Worked Example 10.6

Now,

$$\frac{\text{Area }1}{\text{Area }2} = \frac{HD\cdot x}{AG\cdot x} = \frac{HD}{AG} = \frac{1}{3}$$

$$\frac{HD}{HD+AG} = \frac{1}{1+3}$$

HD = ¼ (HD + AG) = ¼ (AD – 10)

Also AG = 3HD

All elements of Lots 1, 2, and 3 can now be determined.

10.8 Subdivision of Parallelograms with Access Roads

Given:

AB, AD, ∠BAD

Area ABCD is a parallelogram. (See Figure 10-17)

AH = 20×cosec∠BAD = 20/sin(∠BAD)

(The access road is 20m wide)

Note that ∠BAD = ∠BHL = ∠KLF = ∠GED

Suppose that Area 1:Area 2:Area 3 = 1:1:1

i.e. Area 1 = Area 2 = Area 3

Area (1 + 2):Area 3 = 2:1

ORArea (1 + 2) = 2 Area 3

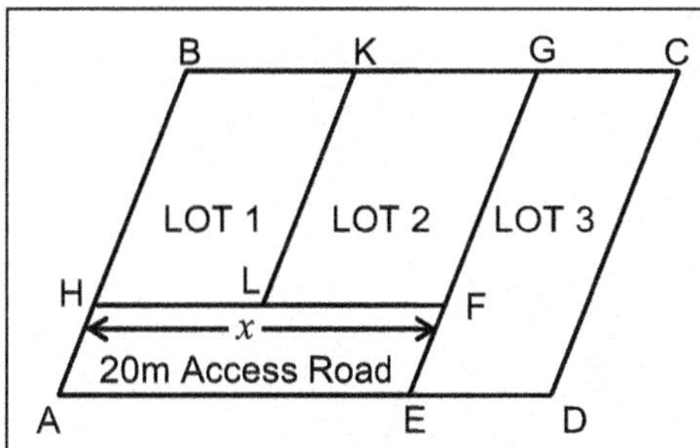

Figure 10-17

Now express all areas in terms of the road length x.

Area (1 + 2) = HB×HF×sin(∠BHL)

= x×(AB – 20/sin(∠BAD))×Sin(∠BAD)

Area 3 = EG×ED×sin(∠GED)

= AB×(AD – x)×sin(∠BAD)

And, as Area (1 + 2) = 2 Area 3, the equation is

x×(AB – 20/sin(∠BAD))×Sin(∠BAD) = 2 ×AB×(AD – x)×sin(∠BAD)

Sin(∠BAD) is common to each side of the equation and cancels out.

OR x×(AB – 20/sin(∠BAD)) = 2×AB×(AD – x)

from which x can be calculated.

Once x is known all other dimensions and areas can be calculated.

10.9 Chapter Exercises 10.2.

Q. 1

Divide the rectangle shown in Figure 10-18 into two areas - with the access road as shown - so that area LOT 1:area LOT 2 = 3:5

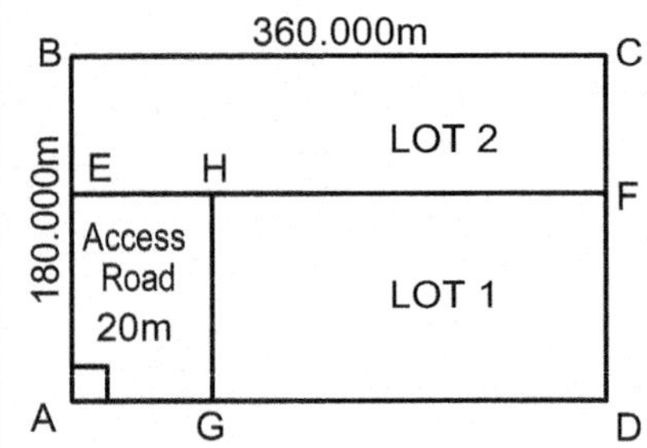

Figure 10-18 Chapter exercise 10.2, Q 1

Q. 2

Given: AD = BC = 265.42m

AB = CD = 610.74m

Subdivide the area in Figure 10-19 into the largest number of lots of equal size, so that none is less than 10 000m^2. The dividing lines are to be parallel to AD and BC.

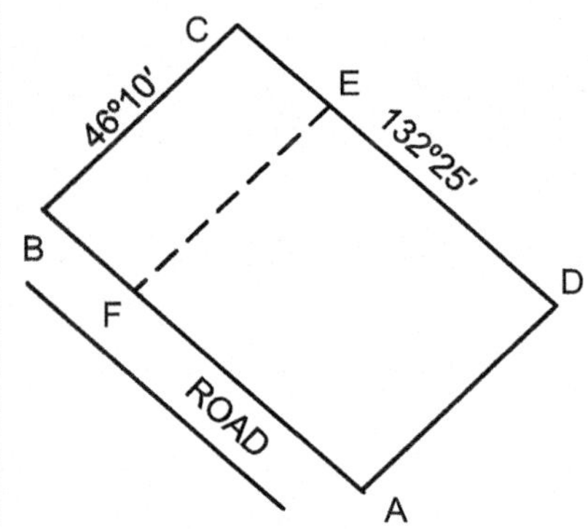

Figure 10-19 Chapter exercise 10.2, Q 2

Q. 3

Divide the rectangle in Figure 10-20 into three lots with access road as shown so that

Area 1:Area 2:Area 3 = 3:2:5

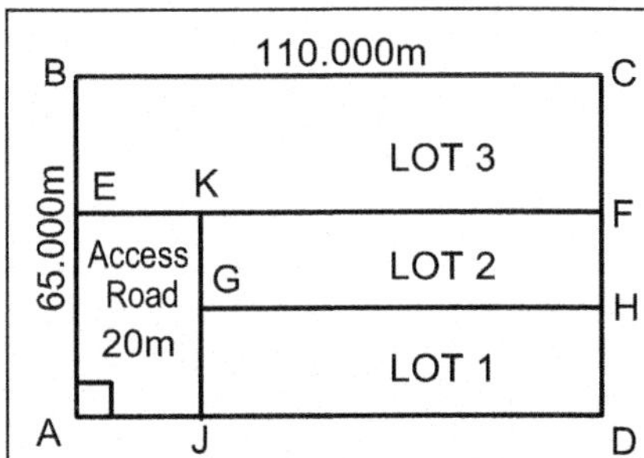

Figure 10-20 Chapter exercise 10.2, Q 3

Q. 4

Subdivide the parallelogram in Figure 10-21 into three equal portions with the access road as shown.

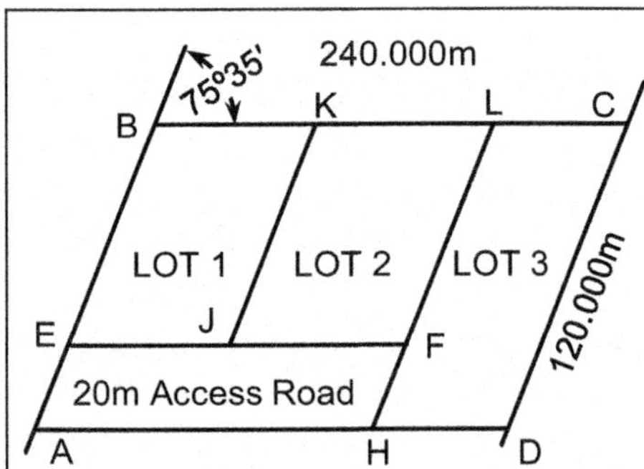

Figure 10-21 Chapter exercise 10.2, Q 4

10.10 Subdivisions using Trapeziums

The subdivisions of trapezium ABCD will be considered, see Figure 10-22, using four different strategies.

a) By lines perpendicular to the parallel sides
b) By lines joining the parallel sides and passing through a given point on one of those sides. e.g. line XY
c) By lines joining the parallel sides and having a given direction. e.g. line XY
d) By lines parallel to the parallel sides. e.g. line EF

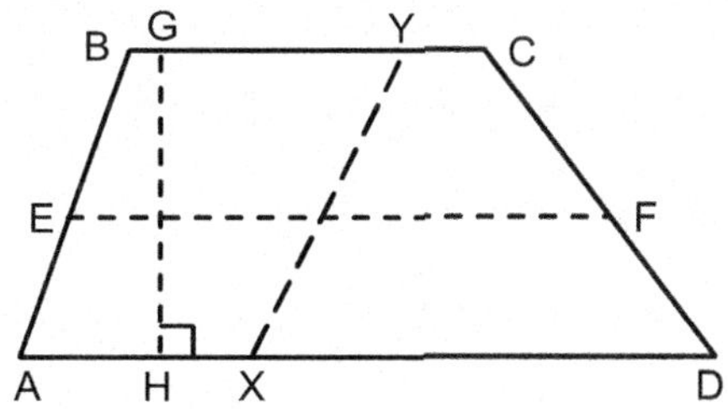

Figure 10-22 Subdivision of trapezium

10.10.1 Subdivision by a line perpendicular to the Trapezium Sides.

Trapezium ABCD has known dimensions on the four sides and area.

XY is perpendicular to BC, AD (See Figure 10-23)

Aim: To subdivide the trapezium so that Lot 1:Lot 2 = 1:2 by a line XY perpendicular to the parallel sides.

Solution: Calculate areas 1 and 2 using the given ratio.

Lot 1 = ⅓ Total area

Lot 2 = ⅔ Total area

Calculate distance XY from:

$$Area\ ABCD = \left(\frac{BC + AD}{2}\right) \times XY$$

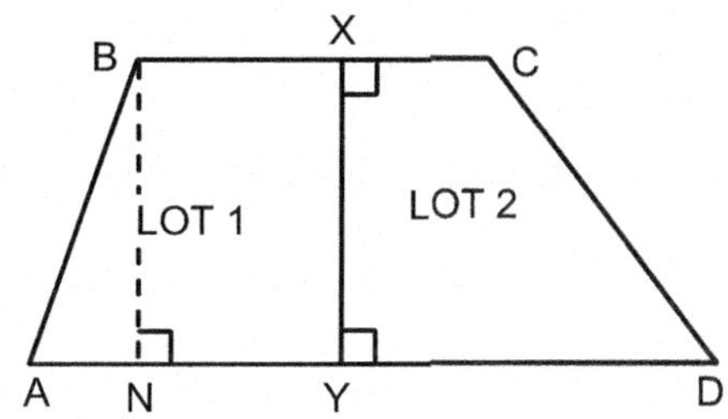

Figure 10-23 Perpendicular to parallel sides

In trapezium BXYA (Lot 1) calculate BX, AY (Drop a perpendicular BN on to AD and solve triangle ABN); calculate AN and, in rectangle BXYN, calculate BX = NY, enabling BX, AY to be found.

Then XC = BC – BX and YD = AD – AY

CHECK: Area Lot 2 = $\left(\frac{XC + YD}{2}\right) \times XY$; compare with the earlier area.

10.10.2 Subdivision by a Line through a Given Point.

Given: All boundary dimensions and area of trapezium ABCD. Also, distance AY is known. (See Figure 10-24)

Aim: To subdivide the trapezium into a given ratio (Lot 1:Lot 2 = p:q) with a line through Y (XY)

Solution: Calculate areas Lot 1 and Lot 2.

In triangle ABY calculate BE, BY, ∠ABY, Area ΔABY

Now in triangle BXY,

Area BXY = (Area Lot 1 – Area ABY)

BY is known from calculations of ΔABY.

∠XBY can be determined and BX can be calculated.

Hence solve for remaining elements in triangle BXY.

Thus, bearing and distance of YX is known

XC = BC – BX

YD = AD – AY

Check by calculating Area Lot 2 from:

$$\text{Area Lot 2} = \left(\frac{XC + YD}{2}\right) \times BE$$

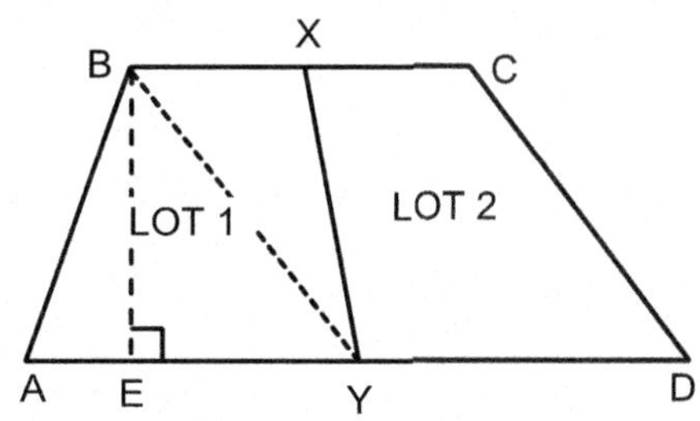

Figure 10-24 Line through a point

Note: Area BXY = (Area Lot 1 – Area ABY) may be negative and this is shown in Figure 10-25.

i.e. Area BXY = (Area ABY – Area Lot 1)

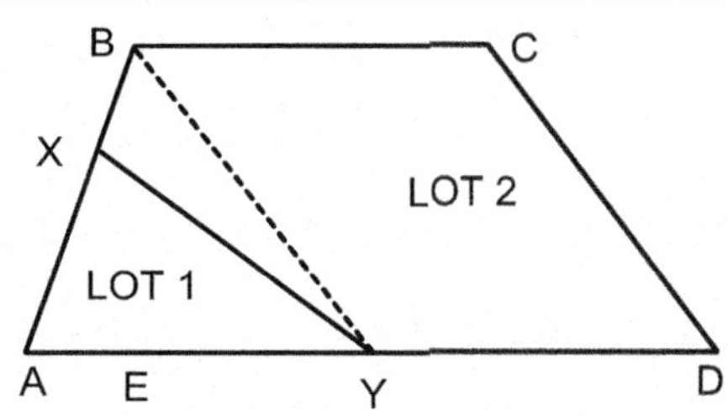

Figure 10-25 Area may be negative

10.10.3 Subdivision by a Line in a Given Direction.

10.10.3.1 Subdivision of entire lot

Given: Dimensions and area of trapezium ABCD.
Bearing YX. (See Figure 10-26).

Aim: Subdivide ABCD by a line YX of known bearing so that
Area 1:Area 2 = p:q

Solution: Calculate areas 1 and 2 from the given ratio.
Draw BE//XY
Solve for missing elements in triangle BAE
(3 angles and 1 side (AB))
solve also for area BAE.

Area BXYE = (Area 1 – Area BAE)
BXYE is a parallelogram of known area.
∴ BE×EY×sin(∠BEY) = Area BXYE
Solve for EY = BX
The remaining unknowns can now be calculated.
XC = BC – BX
YD = AD – AY

The check is obtained by calculating Area Lot 2.
Note: As in the previous example,
Area BXYE = (Area 1 – Area ABE) may be negative.
See Figure 10-27.

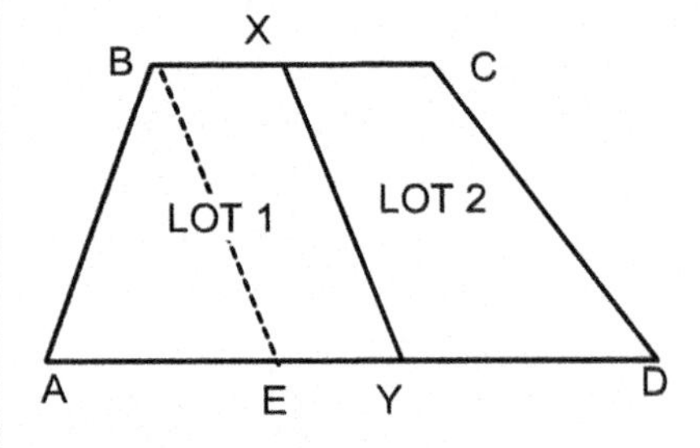

Figure 10-26 (c) Line in a given direction

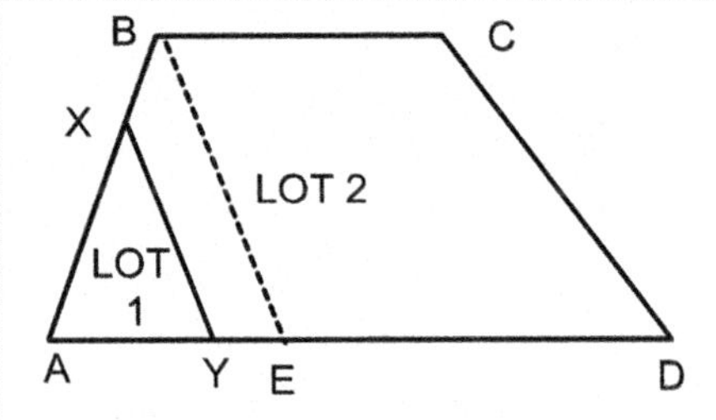

Figure 10-27 Area may be negative

10.10.3.2 Where the subdivision includes a road.

Trapezium ABCD in Figure 10-28 is to be subdivided so that Area Lot 1:Area Lot 2 = $p{:}q$ after provision for an access road of given width has been made. The trapezium is subdivide by a line XY in a given bearing.

Road Width = 10m.

Since the length of the road, and hence its area, is unknown, the areas of the portions 1 and 2 cannot be calculated directly.

Solution: Let the unknown road length XC = x.

Now EF $= x + EY\times\cot(\angle XEF) - 10\times\cot(\angle FCX)$
$= x + EY/\tan(\angle XEF) - 10/\tan(\angle FCX)$

DK $= CD\times\sin(\angle DCK)$

EL $= DK-10$

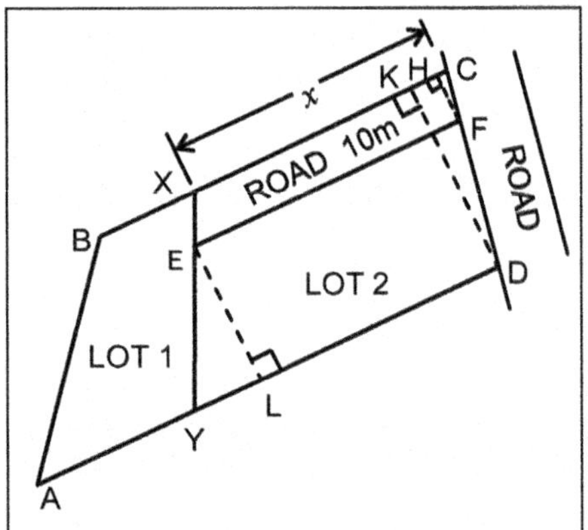

Figure 10-28 Trapezium with road reserve

$$\text{Area Lot 1}=\left(\frac{BX+AY}{2}\right)\times(BX+AY)\times\frac{DK}{2}$$

$$\text{Area Lot 1}=(BC-x+AD-AY)\times\frac{DK}{2}$$

$$=\left(BC-x+AD-\left(x+DK\times\cot(\angle XYD)+DK\times\cot(\angle DCB)\right)\right)\times\frac{DK}{2}$$

$$\text{Area Lot 2} = \text{Area Lot 2}=\left(\frac{(EF+YD)\times(DK-10)}{2}\right)$$

And as $$\frac{\text{Area Lot 1}}{\text{Area Lot 2}}=\frac{p}{q}$$

i.e. $q\times\text{Area Lot 1} = p\times\text{Area Lot 2}$

After substitution this leads to a linear equation with x as the only unknown.
When x has been solved, all other required information can be calculated.

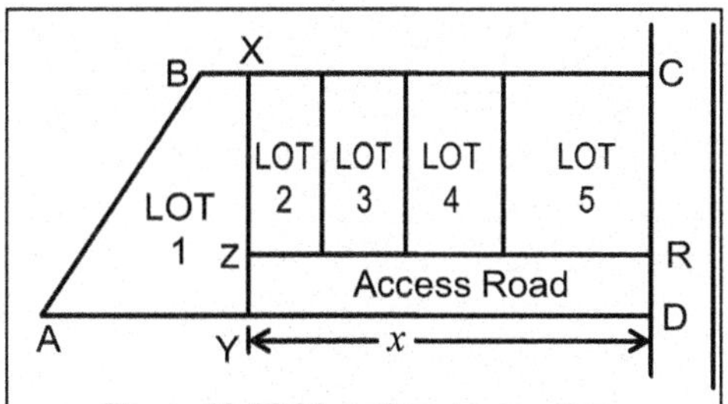

Figure 10-29 More than two portions

10.10.3.3 Subdivision into more than 2 portions.

In Figure 10-29 areas of lots 2, 3, 4 and 5 are taken together and considered as one portion. They are then compared to portion 1 and the dimensions of portion 1 first calculated.

The area of XCRZ is now calculated and then subdivided in the required ratios.

10.10.3.4 Worked Example 10.7 Multiple lots

Given: The dimensions of ABCD are shown in Figure 10-30.

The eastern boundary of Lot 1, XY, is parallel to CD.

The access road of width 20m and is parallel to AD.

Aim: To sub-divide ABCD such that Area Lot 2 = Area lot 3 and Area lot 1 = 2 × Area 2.

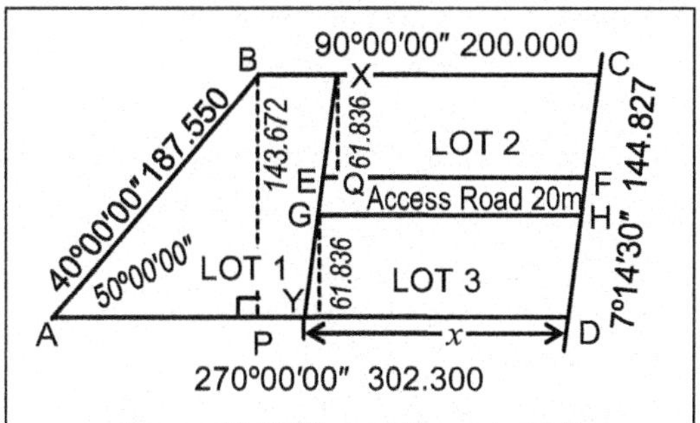

Figure 10-30 Worked Example 10.7

Solution: Area trapezium ABCD $=(BC+AD)\times\frac{BP}{2}$.

BP = AB×sin(∠BAP) = AB×sin(50°) = 143.672m

Area Trapezium ABCD = 36083.131m²

Area (2 + 3) = (CF + HD) × $\boldsymbol{x}$ × sin(∠XYD)

∠XYD = 90 – 7°14′30″ =82°45′30″

where: (CF + HD) = CD – FH

FH = 20×cosec(∠YDC) = 20÷sin(∠YDC)

∠YDC = 180 – ∠XYD = 97°14′30″

FH = 20.161m

Now, the unknown road length is YD (with its parallel sides), area CDYX = BP – Road (20m) × YD (or XC), x.

= 143.672 – 20.0 = 123.672 [1]

Area (Lot 2 + Lot 3) = 123.672 × x

$$\text{Area } Lot\,1 = (BX+AY)\times\frac{BP}{2} = (BC-x+AD-x)\times\frac{BP}{2}$$

$$= 71.836\times(502.300-2x) = 36083.2-143.672x$$

now Area 1 = Area (2 + 3)

71.836×(502.300 – 2x) = 123.672x ⇒ 36083.222 = 267.343x

x = 134.969m [2]

Dimensions:

BX = BC – x = **65.031**

AY = AD – x = **167.331**

XC = EF = GH = YD = x = **134.969**

now $$\text{Area Lot } 1 = (BX+AY)\times\frac{BP}{2} = (BC-x+AD-x)\times\frac{BP}{2}$$

Area 2= Area 3 = ½ Area 1 = 8345.936

Area 2= 8345.936 = x × XQ

XQ = 8345.936/134.696 = 61.836 (this is the perpendicular height).

EX=PQ×sec(XEF) = 61.836/cos(7°14′30″) = 62.333

EX = FC = GY = HD = **62.333m**

EG = FH = 20.161

Area ROAD = 20×x = 2699.386m²

Check CF+Fh+HD = 62.333+20.161+62.333 = 144.827 = DC

CHECK: Area 1	=	16691.87
Area 2	=	8345.94
Area 3	=	8345.93
Area Road = 20x	=	2699.39
Total	=	36083.13

Check Area ABCD = 36083.13 ✓

If it's not checked it's wrong.

10.10.4 Subdivision by a line parallel to the parallel sides of a trapezium.

10.10.4.1 Case (a) Two portions

The trapezium in Figure 10-31 is to be subdivided in a given ratio by a line XY parallel to BC and AD. Using the given ratio, Areas 1 and 2 can be found.

In trapezium AXYD the area, side AD and angles at A and D are known.

Thus sides XY, AX and YD can be calculated.

BX and CY can then be calculated.

As a check the individual areas can be found, summed and checked against area ABCD.

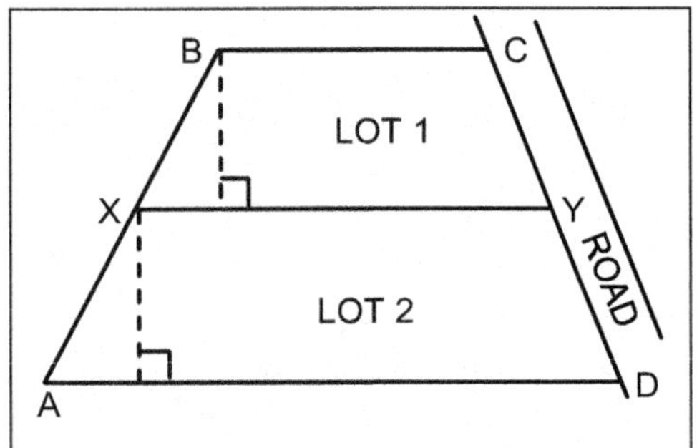

Figure 10-31 Subdivision by ratio

10.10.4.2 Case (b) Sub-division into more than two portions.

10.10.4.3 Worked Example 10.8 More than 2 portions.

In this method each dividing line is calculated separately from the original data.

Consider the figure ABCD (Figure 10-32). The trapezium ABCD is to be subdivided into 4 equal portions by lines parallel to BC and AD.

$\theta_{BC} = 90°00'00''$ $D_{BC} = 300.000$m

$\theta_{CD} = 167°50'00''$ $D_{CD} = 165.450$m

$\theta_{DA} = 270°00'00''$ $D_{DA} = 520.000$m

$\theta_{AB} = 80°32'05''$ $D_{AB} = 608.635$m

1. Find area ABCD

   $\angle BAD = 81°27'55''$

   Perpendicular distance BN = AB×sin(BAD)

   BN = 601.628

   Area ABCD = (BC + AD)×BN/2

   = (300.000 + 520.000)×601.628/2

   = 246667.5m^2.

2. Extend AB, DC to meet in E

   Now solve triangle BEC for area, BE and CE.

   $\angle B$ = 81°27′55″

   $\angle C$ = 77°50′00″

   $\angle E$ = .20°42′05″

   Sum = 180°00′00″

3. Solve BE and CE from Sin Rule: $\dfrac{BE}{\sin C} = \dfrac{CE}{\sin B} = \dfrac{BC}{\sin E}$

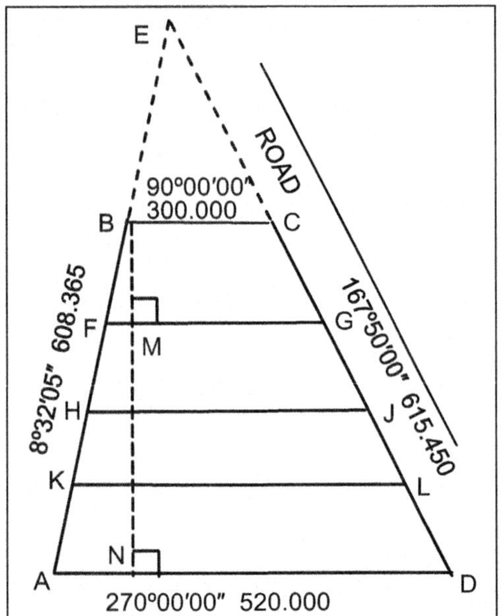

Figure 10-32 Multiple portions. Worked Example 10.8

BE = 829.600

CE = 839.264

Area BEC = ½×BE×BC×sin($\angle$B) = 123062.0m^2

4. Solve triangle FEG for FE, EG, FG

   Area FEG = Area BEC + ¼ ABCD = 184728.9m^2

$$\text{Area } FEG = \tfrac{1}{2} FG^2 \left(\frac{\sin F \sin G}{\sin E} \right)$$

Now,

$$FG^2 = \left(\frac{2 \times \text{Area}FEG \times \sin E}{\sin F \times \sin G} \right) \qquad FG = 397.559$$

EF, EG can be found from the Sin Rule: $\dfrac{FE}{\sin G} = \dfrac{GE}{\sin F} = \dfrac{FG}{\sin E}$

FE = 1016.422

GE = 1028.264

Now, BF = FE – BE = 186.822

CG = GE – CE = 189.000

5. Check: BM = BF×sin ∠F = 184.753

$$\text{Area } FBGC = (BC + FG)\times\frac{BM}{2} \qquad = 61666.8\text{m}^2$$

But, Area FBCG = ¼ Area ABCD = 61666.9m^2. ✓

6. Dimensions of Lot 1

BC = 300.000m

CG = 189.000m

FG = 367.559m

BF = 186.822m

In a like manner the dimensions of lots 2, 3 and 4 can be obtained.

10.11 Chapter Exercises 10.3

Q. 1

Given: (See Figure 10-33)

AB = 320.350m

AD = 710.000m

∠ADC = 61°35′00″

Subdivide the area into two portions:

so that Area LOT 1:Area LOT 2 = 2:3

Calculate all dimensions and areas.

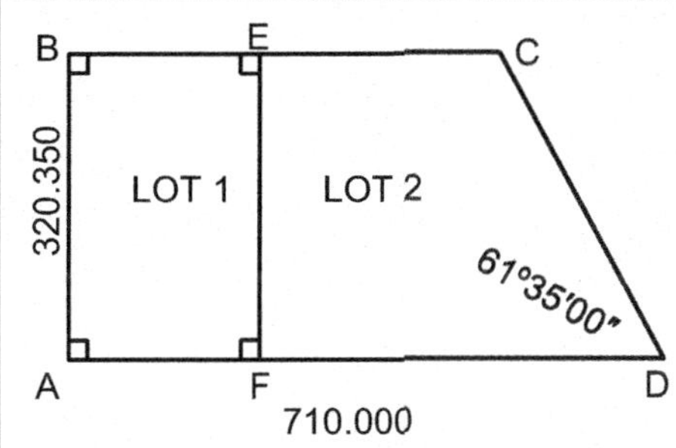

Figure 10-33 Chapter exercise 10.3. Q 1

Q.2

Given: (See Figure 10-34)

AB = 210.650m ∠ABC= 129°20′00″

BC = 1650.000 m ∠ADC = 137°22′40″

CD = 240.606 m

DA = 1960.560 m

Subdivide the area into 3 portions by lines perpendicular to AD and BC so that:

Area LOT 1:Area LOT 2:Area LOT 3 = 1:2:3

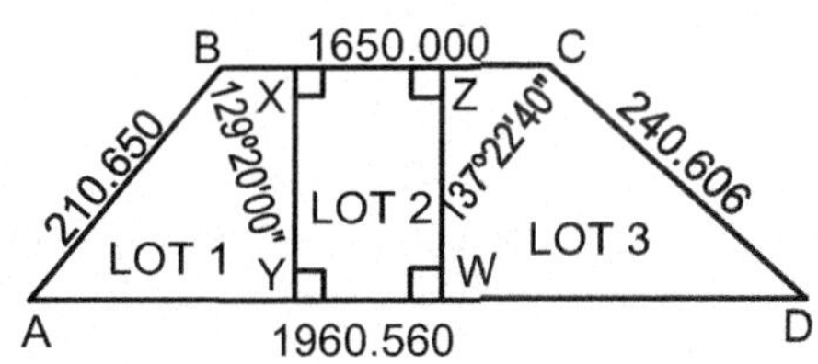

Figure 10-34 Chapter exercise 10.3. Q 2

Q. 3

Given: (See Figure 10-35)

AB = 400.240m

AP = 403.000m

PD = 188.350m

∠BAD = 50°45′, ∠ADC = 90°00′

Subdivide the trapezium into two portions so that:

Area LOT 1:Area LOT 2 = 2:3 by a line through P.

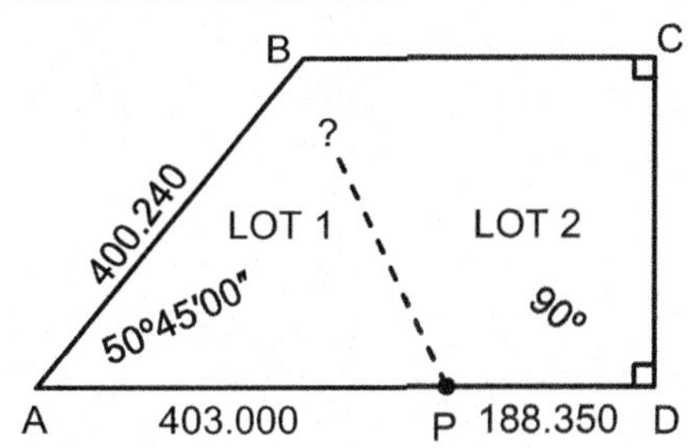

Figure 10-35 Chapter exercise 10.3. Q 3

Q. 4

Given: (See Figure 10-36)

AB = 250.000 m

BC = 300.000 m

AD = 610.000 m

∠BAD = 67°30′

Subdivided the trapezium into 4 equal lots by lines parallel to AB.

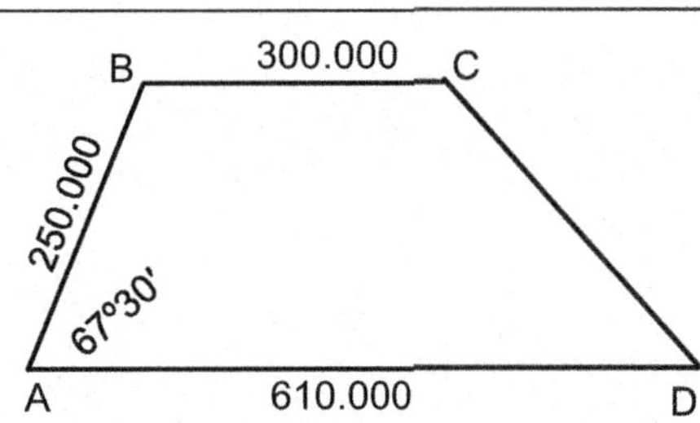

Figure 10-36 Chapter exercise 10.3. Q 4

Q. 5
Given: (See Figure 10-37)
AB = 620.000m
BC = 1235.500m
∠ADC = 70°15′
Lots 1, 2, 3 and 4 are to be of equal size. Lot 5 is six times area lot 1.
All angles except those at C and D are right angles.
Calculate all missing information, including areas.

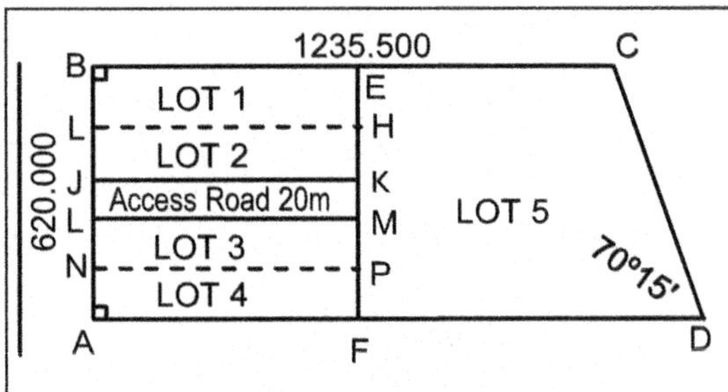

Figure 10-37 Chapter exercise 10.3. Q 5

10.11.1 Answers to Chapter Exercises

10.11.1.1 Chapter Exercise 10.1.

Q.1 See Figure 10-38
Area = 31043m²
LOT 1 = 7760.8m²; LOT 2 = 23282.2m²
AX = 113.371 θ_{AX} = 82°46′56″
BX = 137.678 θ_{BX} = 178°50′21″
CX = 340.114

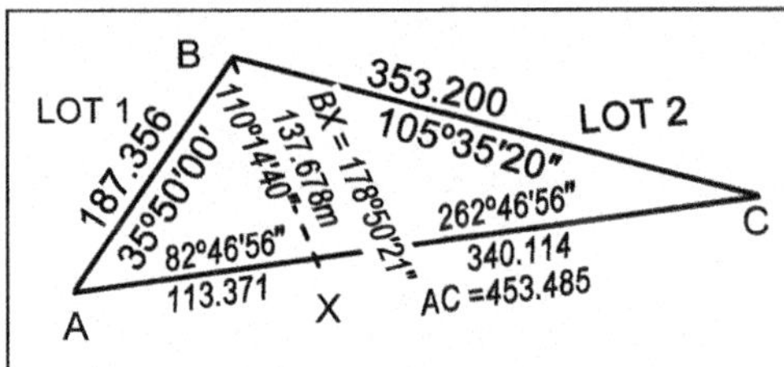

Figure 10-38 Chapter Exercise 10.1 Q 1

Q.2 See Figure 10-39.
Area = 103755m²
LOT 1 = 51877.5 = LOT 2 = 51877.5
AX = 36.452, θ_{AX} = 61°33′20″
CX = 556.493, θ_{CX} = 241°33′20″
CY = 282.553, θ_{CY} = 200°16′00″
BY = 247.812, θ_{BY} = 20°16′00″
YX = 391.440, θ_{XY} = 270°00′00″

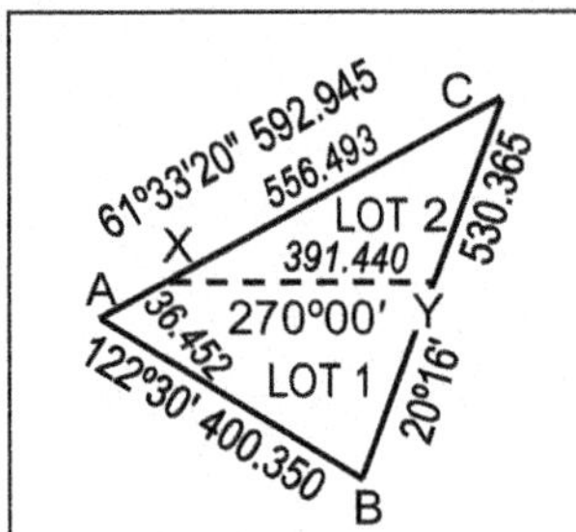

Figure 10-39 Chapter Exercise 10.1 Q 2

10.11.1.2 Chapter Exercises 10.2..

Q. 1 See Figure 10-40
Area = 64800m²
LOT 1 = 23775.5m²; LOT 2 = 39625.9m²;
Road = 1398.6m
GH = 69.928
HD = 340.000
BE = 110.072
BC = 360.000

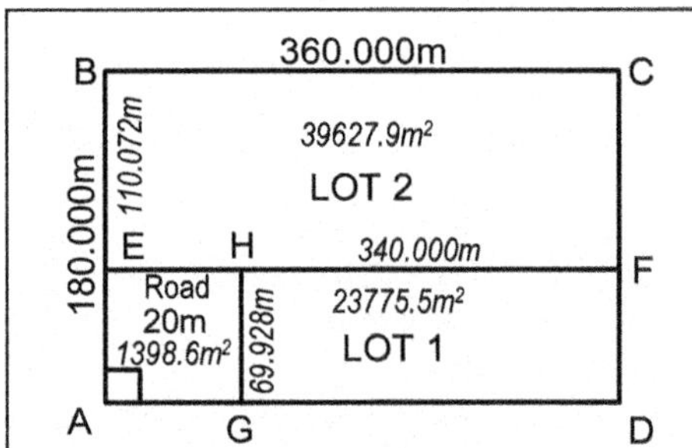

Figure 10-40 Chapter Exercise 10.2 Q 1

Q. 2
Area = 162106.6m²
16 Blocks
BC = FE = 265.420, θ_{BC} = 46°10′
CE = BF = 38.171, θ_{CE} = 132°25′

Q. 3
Area = 7150m²
LOT 1 = 1287.0m²; LOT 2 = 1930.5m²; LOT 3 = 3217.5m²;
Road = 715.0m².
BE = 29.250, BC = 110.000
GK = 14.300, KF = 90.000
JG = 21.450, GH = 90.000

Q. 4 See Figure 10-41

Area = 27893.1m^2

LOT 1 = 8166.13m^2; LOT 2 = 8166.13m^2;

LOT 3 = 8166.13m^2; Road = 3394.7m^2.

LD = AD×sin(B) =116.221m;

AE = Width (20)/sin(B) = 20.650

BE = 99.350, EJ = 84.868

KJ = 99.350, JF = 84.868

HL = 120.000, HD = 70.264

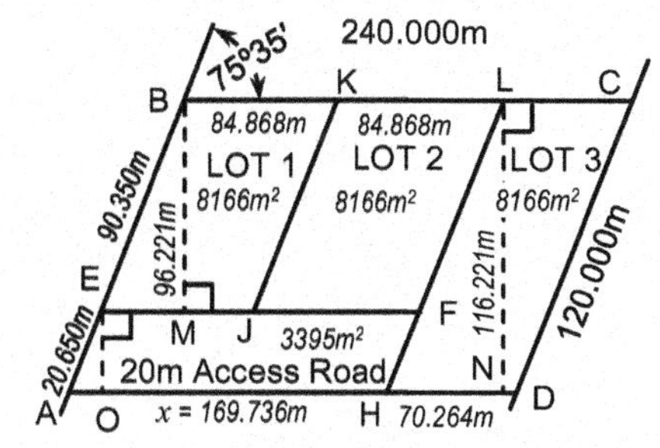

Figure 10-41 Chapter Exercise 10.2 Q 4

10.11.1.3 . Chapter Exercises 10.1

Q. 1 See Figure 10-42

Area = 27893.1m^2

Area 1	=	79873.83
Area 2	=	119811.00
AB=	320.350	
BE=	249.333	
BC=	536.667	
CD=	364.237	
DF=	460.667	

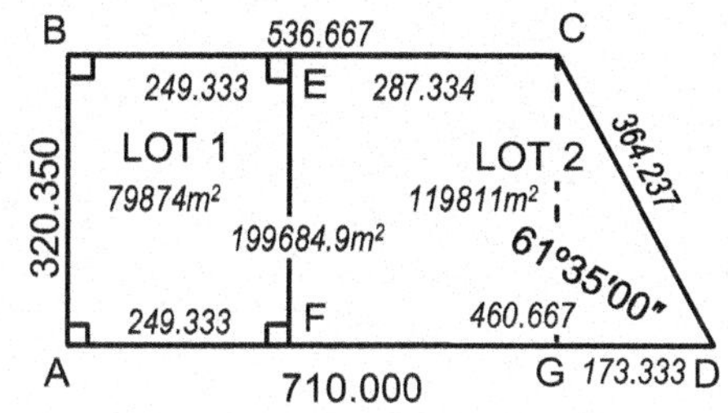

Figure 10-42 Chapter Exercise 10.3 Q 1

Q. 2 See Figure 10-43

Total Area = 294137.5m^2

Area 1 = 49022.9 m^2

Area 2 = 198045.8 m^2

Area 3 = 147068.8m^2

AB = 210.650

BX = 234.122

XY = 162.932

YA = 367.638

XZ = 601.760

ZC = 814.117

CD = 240.606

DW = 991.163

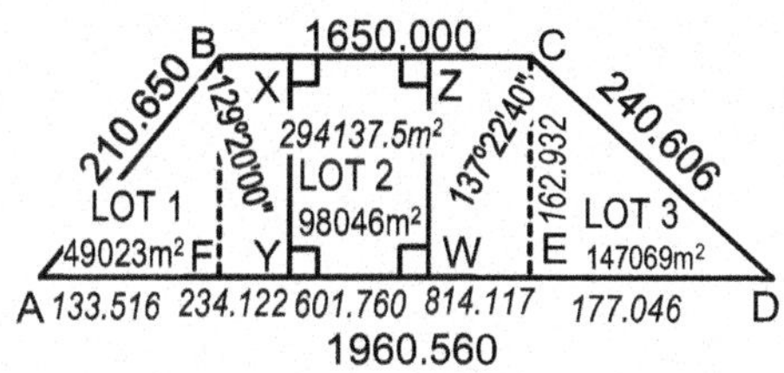

Figure 10-43 Chapter Exercise 10.3 Q 2

Q. 3 See Figure 10-44

Total Area = 144040.7m^2

Area 1 = 48013.6 m^2

Area 2 = 96027.1 m^2

AN = 307.700

NB = 92.540

BC = 338.116

CD = 309.943

DP = 188.350

AP = 403.000

∠ANP = 80°24′41″

∠APN = 48°50′19″

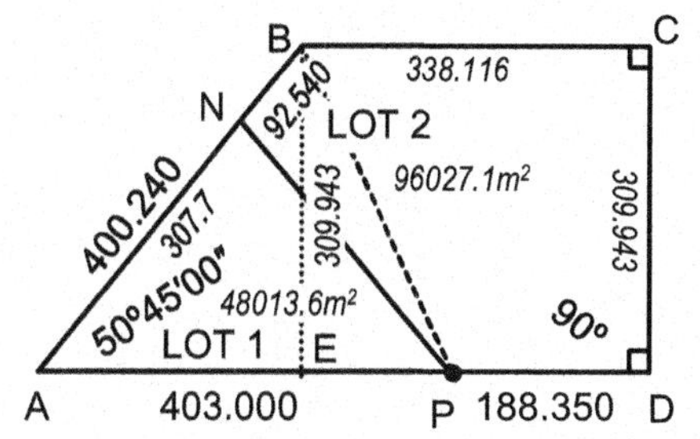

Figure 10-44 Chapter Exercise 10.3 Q 3

Q. 4 See Figure 10-45

Total area = 105091m^2.

Area 1 = Area 2 = Area 3 = Area 4 = 26272.8m^2

AB = 250.000 KJ = 116.934

BE = 113.750 KG = 214.166

EF = 113.750 ∠KGD = 65°21′36″

FC = 72.500 ∠GDK = 47°08′24″

CG = 45.165 ∠DKG = 67°30′

GD = 269.929 ∠A = 67°30′

DK = 265.565

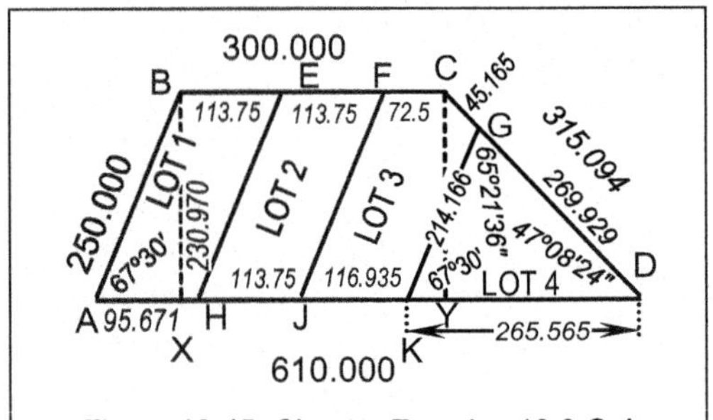

Figure 10-45 Chapter Exercise 10.3 Q 4

Q. 5 See Figure 10-46

Total area = 853106.8m^2.

Area 1 = Area 2 = Area 3 = Area 4 = 82403m^2 each.

Area 5 = 494417.8m^2.

Road = 10987m^2

BL = 150.000 CD = 658.750

BE = 549.353 DA = 1458.103

EC = 686.147 ∠BCD = 109°45′

FD = 908.750

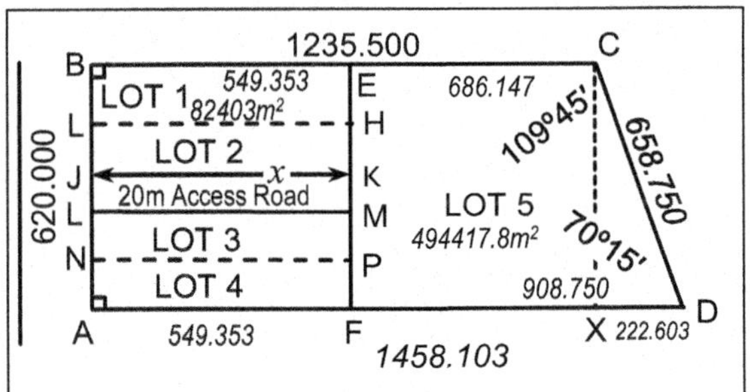

Figure 10-46 Chapter Exercise 10.3 Q 5

10.12 References for Chapter 10

Elfick, M.H, J.G Fryer, R.C. Brinker and P.R Wolf. Elementary Surveying, Eighth edition. Longman, 1994. ISBN 0-00-499001-3

Allan, A.L., Hollwey, J.R., and Maynes, J.H.B. Practical Field Surveying and Computations. William Heinemann Ltd, London. 1971

Clark, David, and James Clendinning. Plane and Geodetic Surveying for Engineers. 5th ed. Constable, 1958.

Shepherd, F.A. Surveying Problems and Solutions. London. Edward Arnold (Publishers) Ltd, 1968

PART 2: Intermediate undergraduate concepts

If it's not checked it's wrong.

Chapter 11 SIGNIFICANT FIGURES AND ERROR PROPAGATION

With the advent of computers, computations can now be achieved at very large digits of numbers. Calculators can also achieve large digits of numbers. When does one roundoff? What are significant figures? This chapter introduces the concepts of significant figures then proceeds to error propagation. Error propagation is a concept where some errors in measurable elements such as distances and angles eventually seep to the computed values such as coordinates/positions. We show in this chapter how these errors are propagated and eventually the analysis of their impact. An understanding of error propagation is must to be able to comprehend the final outcome of the processed data.

11.1 Introduction.

All survey computations require the understanding of the "significance" of the result. Surveyors can face ambiguous and embarrassing situations caused by other professions ignoring the significance of results.

Let us suppose a surveyor picks up a plan which shows distance of 77.33 metres. He would normally assume that the measurement has been taken to 2 decimal places of a metre (i.e. the nearest cm) and proceed accordingly. He may however find the 77.33 is the result of dividing a measured distance of 232 metres by three and that the original measurement was only taken to the nearest metre. A knowledge of the significance of results is therefore vital to good survey practice.

11.2 Significant Figures in Calculations.

When a distance is measured with a tape, divided in metres, then only the number of whole metres in that distance can be stated with certainty. A guess could be made at the number of tenths of a metre, and the result could for instance be given as 77.3m. The 0.3 metre however would be doubtful, and only the 77 metres could be stated with certainty. The 77 are therefore said to be the two "significant figures" of the measurement.

If, for more accurate measurement, a scale with subdivisions to hundredths of a metre (0.01m) is available, a closer approximation to the true length of the distance can be made, say 77.37m. There are now 4 significant figures in the measurement. More accurate means of measurement may give more significant figures in the result, but the true length cannot be measured with absolute certainty. Therefore, the numbers that are the results of measurements have a limited accuracy, depending on the type of measurement.

In contrast to these numbers there are numbers that are absolutely accurate, such as those denoting money, e.g. in transport costs of 9 cents per km. Hence the nine is not an approximation. Another example is the 2 used in squaring. Both kinds of numbers are used in survey computations.

The term "significant figures" refers to those digits in any number which have meaning, that is, whose values are definitely known to be correct. The following table will help to illustrate this.

Number	Significant Figures
693	3
693.7	4
692.68	5
0.000047	2
0.0000470	3
0.0100470	6
2.0000470	8

The number of significant figures in a whole number ending in zero, such as 73200, is doubtful unless more information is known. This number may contain 3, 4 or 5 significant figures. If the number is written 73200.0, then six significant figures are indicated. In any calculation, proper attention must be given to the number of significant figures in the result. A few examples may illustrate the possible pitfalls.

11.2.1 Worked Example 11.1

Add 1.23m and 690.49m

```
  1.23
690.49 +
------
691.72     The result can be stated to 5 significant figures.
```


J. Walker and J. Awange, *Surveying Computations*,
https://doi.org/10.1007/978-3-031-94382-9_11

11.2.2 Worked Example 11.2

Add 1.23m and 5m

1.23
5.?? +
6.?? The result can be stated to 1 significant figure only.

Worked Example 11.3

Add 1.23 m and 5.00 m

1.23
5.00 +
6.23 Three significant figures.

11.2.3 Worked Example 11.4

Multiply 1.23 m and 3.141 m (3 and 4 significant digits)

	1.23?			
	3.141? ×			
	3.69?	(3 × 1.23)	3.69	
	0.123???.	(0.1 × 1.23)	0.12	(rounded to 3 significant digits)
	0.0492???	(0.04 × 1.23)	0.05	(rounded)
Σ (SUM)	0.00123???	(0.001 × 1.23)	0.00	(rounded)
	3.86~~143~~???		3.86	(3 significant digits)

The addition in the third place of decimals is 1 + 9 + 1 + ?, which is 11 + ? All that can be said is that this is more than 10, so that there will be at least 1 to carry. We can state that if two numbers with significance, a and b are added or multiplied together, the result will contain a or b significant digits whichever is the smallest.

11.3 Precision versus Significant Figures in Calculations.

Many survey computations require a sequential chain of calculations to be carried out before arriving at the concluding answer. Most **results** will be reported at the measuring precision, generally the millimetre (0.001m or 3DP) level unless the number of significant digits is the limiting measure of the data.

Most students create errors by noting down the results of intermediate steps with too little precision, taking a calculator result (calculated to 9 decimal places (9DP), reasonably precise) and noting only two or three decimal places.

This truncated (as opposed to rounded) result is fed back into the calculator for the next step. It now becomes part of the 9DP answer, and again the result may be copied down to 2 or 3DP. and fed back in again. Precision is being lost all the time. Loss of precision.

If you MUST record intermediate steps, write down the answer to 6 - 8DP. This should maintain precision. Best answer is to use the memory locations on your calculator, recalling them for subsequent steps. The HP 10s+ calculator using proper calculation sequences and up to 9 memory locations (A – F, X, Y, M), will maintain full 9DP precision. If you store a value, note it down (1 or 2 DP) and note its memory location (say [A] for memory register A [[SHIFT] [STO] A]).

Once you have the answer you can ROUND the results to the appropriate significant figures.

11.3.1 Worked Example 11.5

Use the Cosine rule to calculate a missing side of a triangle:

side a = 105.234 m, side b = 125.476, angle C = 107° 25′ 23″

side $c^2 = (a^2 + b^2 - 2ab\cos(C))$

$a^2 = 11{,}074.194756$, $b^2 = 15{,}744.226576$, $a^2 + b^2 = 26{,}818.421332$

but only noting only 3DP, and probably TRUNCATING instead of ROUNDING

$a^2 = 11{,}074.194$, $b^2 = 15{,}744.226$, $a^2 + b^2 = 26{,}818.420$.

$2 \times a = 210.468$, $\times\, b = 26{,}408.682768$, truncated down to:

$2ab = 26{,}408.682$.

Now convert $\angle C$ to decimal degrees = 107.423056, cos(107.423056) = –0.299425

but cos(107.423) = –0.299424, truncated to – 0.299.

Gathering truncated intermediate results:

$c^2 = (26818.42 - 26408.682 \times -0.299) = 26818.42 + 7896.195$

$c = \sqrt{34{,}714.615} = 186.318$

Working to full precision:

$c = \sqrt{34725.83456655} = 186.348$.

Note that there is a 0.030m (30mm) discrepancy when your results had a significance at the 0.001m (1mm) level. All because of truncating to 3DP.

11.4 Rounding Off Numbers.

Rounding off a number is the process of dropping one or more digits so that the remaining value will contain only those digits that are significant, or that are necessary for continuing computations.

It is well to keep to a definite system so that the laws of chance operate and rounding-off should be done according to the following rules:

(a) When the digit to be dropped is less than 5, the number is written without this digit and any that follow it, (0.054 becomes 0.05).

(b) When the digit to be dropped is equal to 5, the nearest even number is used for the preceding digit, (0.055 becomes 0.06; 0.045 becomes 0.04; 0.035 becomes 0.04; 0.025 becomes 0.02).

(c) When the digit to be dropped is greater than 5. the number is written with the preceding digit increasing by one, (0.047 becomes 0.05).

(d) Dropped digits to the left of the decimal point are replaced with zeros.

(e) Dropped digits to the right of the decimal point are never replaced. (Why?)

11.4.1 Worked Example 11.6

	Number	Required number of significant figures	Rounded off number
(a)	2378.649	5	2378.6
(b)	2378649	5	2378600
(c)	792.850	4	792.8
(d)	792750	4	792800
(e)	4896.3	4	4896
(f)	3759.7	4	3760
(g)	3759.7	3	3760

(Note the ambiguity in the last two rounded off numbers.)

11.5 Accuracies involving the propagation of errors in the main arithmetic processes.

11.5.1 Error analysis of survey observations.

Error analysis aims to record and quantify the results of a set of measurements taken in the survey observation process. There will be an inevitable spread of results, and we will never know the exact value. Through repeated observations we attempt to find a value that is as close to the exact value as possible. By analysing these observed values we attempt to quantify the range of values within which the correct value lies, and to judge the results against acceptance/rejection criteria. (Standard for the Australian Survey Control Network (SP1), Version 2.2, December 2020. https://www.icsm.gov.au/publications/standard-australian-survey-control-network-v22)

Accepted surveying wisdom is that the more observations the better as it will lead to, hopefully, a more precise answer. But this is constrained by cost and time.

Errors in measurements are of three types. viz.;

(a) Gross, or blunders. These errors should have been eliminated from the observed data set prior to analysis.

(b) Systematic

(c) Random

11.5.2 Gross errors, or blunders

If equipment is operating, and operated, correctly then one of the main sources of gross errors is incorrect recording of observations. Typically, these are transcription errors in a field book. Modern survey instruments record the observations for later software analysis, so recording blunders should not arise. However, field observations may have to be recorded in a field book and then reduced in the field to ensure job specification compliance before leaving the field.

11.5.3 Systematic errors.

These errors are cumulative and will have the same sign and magnitude under the same given conditions. e.g. if sag corrections to a measuring band supported in catenary were omitted when determining the measurements of various distances, the measured distances would be longer than the actual distances.

Systematic errors arise because measurements are made in atmosphere (incorrect recording), or because of an error in the instrument used, or because of personal eccentricities. These errors can be isolated and have a correction applied to the measurements. They are minimized by using correct instrument adjustment, correct observation techniques etc.

11.5.4 Random errors.

These are errors of observation and as such they cannot be isolated. They are random in nature, usually small and have a tendency to compensate one another.

When we are discussing errors in Surveying, we are, in the main, talking about **random** errors, after ensuring that systematic errors have been minimized.

11.5.4.1 Introducing standard deviation.

The standard deviation of a measurement can be thought of as the likely deviation of that measurement from the mean of several such measurements. How far is an individual measurement from the central tendency of all the observations?

11.5.4.2 Worked Example 11.7. Statistics of random errors.

Nine independent measurements, x, made with a steel band between 2 points A & B, are shown Table 11-1 in with 4 columns, i, x, v and v^2:

Statistical calculations work with two population types: the **entire** population (observation ξ (xi) or a **sample** (x) of the entire population.

This Worked Example is a **sample** of all the measurements that **could** have been made (keep measuring). Surveying measurements normally use a **sample** population. Statistically we accept that **30** observations will be a minimum sample size. With more than 30 observations the difference between the **standard deviation** of the sample population and the entire population is insignificant.

Population mean is denoted $\boldsymbol{\mu}$, mu. Sample mean is denoted by $\bar{x}$, x_bar. The difference between each observation and the mean is called the *residual*, $\boldsymbol{v}$, nu. $v_i = (x_i - \bar{x})$ or $v_i = (\xi_i - \mu)$.

Table 11-1 Observations

1	2	3	4
i	x	$v=x-\bar{x}$	$v^2=(x-\bar{x})^2$
1	35.168	0.0008	6.05^{-7}
2	35.169	0.0018	3.16^{-6}
3	35.169	0.0018	3.16^{-6}
4	35.167	−0.0002	4.94^{-8}
5	35.163	−0.0042	1.78^{-5}
6	35.165	−0.0022	4.94^{-6}
7	35.170	0.0028	7.72^{-6}
8	35.166	−0.0012	1.49^{-6}
9	35.168	0.0008	6.05^{-7}
$\bar{x}$	35.1672	Σv 0.0000	Σv^2 3.96^{-5}

Thus, the check is that the sum of the residuals = 0. i.e., ($\sum_{i=1}^{n} v_i = 0$)

Use your calculator STATS functions to verify: ($\Sigma+$, $\bar{x}$, $\bar{y}$, S_x, σ_x, S_y, σ_y, n, Σ_x, Σ_y functions)

11.5.4.3 The mean (average)

The **mean** of a set of observations is the average of all the observations and is considered as probably the best value closest to the true value of the observation.

Population mean (average), $\boldsymbol{\mu} = 35.167$ Mean or $\mu = \frac{1}{n}\sum_{i=1}^{n} \xi_i$ = (Sum all obs)÷(number observations)

Sample mean (average), $\bar{x} = 35.167$ Mean or $\bar{x} = \frac{1}{n}\sum_{i=1}^{n} x_i$ = (Sum sample obs)÷(number observations)

11.5.4.4 Residuals, their sum and sum squared

Population mean is denoted $\boldsymbol{\mu}$, mu. Sample mean is denoted by $\bar{x}$, x_bar. The difference between each observation and the mean is called the *residual*, $\boldsymbol{v}$, nu. $v_i = (x_i - \bar{x})$ or $v_i = (\xi_i - \mu)$.

The sum of the residuals squared (Σv^2) from the **sample** mean, $\sum_{i=1}^{n}(x_i - \bar{x})^2 = 3.96\times10^{-5}$. (. Do it longhand).

If all these measurements are each made with the same care, there is no reason to prefer any one measurement over another. The laws of probability state that the **mean** measurement will be the best in these circumstances. And, since the nine measurements (n) represent the sample, and also the entire, population, the standard deviation of **any** (x_i) observation, S_x, is determined by n–1. (Since the mean $\bar{x}$ is determined from the sample set, it is no longer independent. So, the number of degrees of freedom is reduced by one.)

11.5.4.5 Standard deviation

Standard deviation of the observations, S_x, in the sample is: $S_x = \pm\sqrt{\frac{\sum_{i=1}^{n}(x_i - \bar{x})^2}{n-1}} = \pm\sqrt{\frac{\Sigma v^2}{n-1}} = \pm 0.0022.$

This means that most of the observations should be between 35.165 >$\bar{x}$ >35.169. Table 11-1 shows two observations falling outside this range so we have 7/9 or 78% of observations within $S_x = \pm 0.002$. This is a little better than the probability of 68% associated with the one standard deviation (1σ). (Gaussian normal distribution theory.)

Standard deviation of the observations, σ_x, in the population is: $\sigma_x = \pm\sqrt{\frac{\sum_{i=1}^{n}(\xi_i - \mu)^2}{n}} = \pm\sqrt{\frac{\Sigma v^2}{n}} = \pm 0.0021$

11.5.4.6 Standard deviation of mean (SDOM)

The standard deviation of the **mean** observation (35.1672), $\sigma_{\bar{x}}$, in the population is (sometimes labelled α):

$$\sigma_{\bar{x}} = \pm\sqrt{\frac{\sum_{i=1}^{n}(x_i - \mu)^2}{n(n-1)}} \quad \text{or}$$

$$S_{\bar{x}} = \pm\sqrt{\frac{3.9556\times 10^{-5}}{9\times 8}} = \pm\frac{\sigma_x}{n} = \pm\frac{0.0022}{\sqrt{9}} = \pm 0.0007$$

There is no **S$\bar{x}$** (standard deviation of the mean) on the calculator. In this example μ and $\bar{x}$ are the same. A good approximation for σ_x is to use S_x (From the stats registers use $S_x/\sqrt{n}$ to calculate $S_{\bar{x}}$).

11.5.4.7 Variance and weight

Whilst standard deviation is the spread the sample observations from the mean, the **variance** measures the average degree of variation from the sum of the squared residuals (Σv^2). Variance and weight are used in least squares adjustments of observations.

It is the square of the standard deviation. $S_x^2 = \frac{\Sigma v^2}{n-1}$, $\sigma^2 = \frac{\Sigma v^2}{n}$,

from the standard deviation of the sample, S_x the variance, $S_x^2 = 4.94^{-6}$

The **weight, P,** of the observation is the inverse of the variance. $P=1/S_x^2 = 1 \div 4.94^{-6} = 2.02^{+5}$.

11.5.4.8 Entire or sample population?

It depends on the circumstances.

You manufacture 100 prototype "widgets" and test them to destruction. You have now tested the **entire** population. Analyse the results to provide a statistical base for the desired test characteristics.

You manufacture batches of 100 and randomly select 6 for a destruction test. The six are a **sample** of the entire population (100). Analyse and compare the results against the baseline. (It's counter-productive to destroy all 100).

Microsoft™ Excel uses the functions **AVERAGE(x_1:x_n)** and **STDEV(x_1:x_n)** to calculate the Mean, $\bar{x}$, (average) and the standard deviation, S_x, of one value in the **sample** population (n-1).

Standard deviation of the **mean** value needs the standard deviation of the **whole** population (n).

Thus σ_x = ±**STDEVP(x_1:x_n)**. As the size of the sample approaches the entire population, so S_x approaches σ_x.

And a large standard deviation of the mean *could* signify an error in the data recording or reduction of one of the observations. (A **check** of the data caused by a recording blunder; transcription error, say.)

The use of symbols, especially for standard deviation of a sample, **S$_x$**, as opposed to whole population, σ_x, and consequent subscripts, depends on the text author. Be sure you know the meaning of the symbol used.

11.5.4.9 Independent observations

In our discussions we have spoken of independent observations. That is, one measurement is not a repeat of the previous, it is at least a re-read of that measurement.

The distances measured between A and B in § 11.5.4.2, Worked Example 11.7, can only be considered independent if the steel band (the tape) is lifted and re-laid over the points A and B and read, ideally, by different observers.

The collimation (2 peg) test of a level only becomes independent when the level is moved and relevelled for a second set of readings. Out and return levelling runs provide independent height differences between fixed points.

The distance between two points on the ground, measured by a total station (TS), can only be independent if the TS and the target reflector (prism) are upset and repositioned and levelled for another reading. The use of the forced centring of the TS and prism and face left/right readings could be considered independent for that one "round" of observations. And they are only between the instrument and the prism, not the points on the ground. The ground distance measurement is subject to the random errors of centring over the ground point, the measured height of the TS and the prism above the points and the errors associated with target aiming (vertical angle).

Independent GNSS measurements of a point require the receiver to be upset, repositioned, relevelled and its height remeasured. A two-hour AUSPOS observation session will have about 3000 and 8000 independent range observations. But that is only between the receiver and the satellites. It does not translate to the ground position which is subject to height measuring to the antenna phase centre and point centring errors. On a single observation session it is good practice to check the centring and remeasure the antenna height (it may have been disturbed).

11.6 Propagation of errors

The propagation of errors in functions (whether systematic or random) will depend upon whether the function is linear or non-linear.

11.6.1 Propagation of Systematic Errors.

Systematic Errors

Linear

$u = ax + by$

$\Delta u = a\Delta x + b\Delta y$

Non-linear

$u = F(x,y)$

$$\Delta u = \frac{\partial F}{\partial x}\Delta x + \frac{\partial F}{\partial y}\Delta y$$

Δu (delta u) is the error in the linear or non-linear function u arising from errors of Δx in x and Δy in y.

11.6.2 Propagation of Random errors.

The propagation of random errors in this chapter holds only if the variable, x, y, … are independent (uncorrelated).

However, in surveying, with correlated nonlinear functions the independent variables x and y means that there is no interdependence between the variables x and y. They are uncorrelated. Thus, σ_{xy} is usually treated as zero.

Random errors
Propagation of variances

Linear

$u = ax + by$

Uncorrelated

$$\sigma_u^2 = a^2\sigma_x^2 + b^2\sigma_y^2$$

$$\sigma_u = \sqrt{a^2\sigma_x^2 + b^2\sigma_y^2}$$

Correlated

$$\sigma_u^2 = a^2\sigma_x^2 + b^2\sigma_y^2 + 2ab\sigma_{xy}$$

Non Linear

$u = F(x,y)$

Uncorrelated

$$\sigma_u^2 = \left(\frac{\partial F}{\partial x}\right)^2 \sigma_x^2 + \left(\frac{\partial F}{\partial y}\right)^2 \sigma_y^2$$

Correlated

$$\sigma_u^2 = \left(\frac{\partial F}{\partial x}\right)^2 \sigma_x^2 + \left(\frac{\partial F}{\partial y}\right)^2 \sigma_y^2 + 2\left(\frac{\partial F}{\partial x}\right)\left(\frac{\partial F}{\partial y}\right)\sigma_{xy}$$

σ_u (sigma u) is the standard deviation of the function u, it comes from the standard deviations σ_x and σ_y.

This chapter will concentrate on how random errors are propagated in linear and nonlinear uncorrelated functions.

The standard deviation is derived from a set of measurements that represent a sample of the whole population of measurements (>30), not a sample of part of the set. The previous § 11.5.4.2, Worked Example 3.1, illustrated a measurement made from a sample set of nine (9) observations. It is a sample because we could have taken more measurements, say more than 30, and we could assume we had sampled to the entire population. What statistical difference would it make if we could have spent all day taking more, say hundreds, of measurements?

Thus, the mean, $\bar{x}$, the standard deviation a single observation (S_x) and the mean of the set of observations, $S_{\bar{x}}$ can be found. But is it a sample set or the whole set?

What do we mean by the propagation of errors?

Suppose two measurements, A and B, each contained random errors. What is the size of the random error which arises when these quantities are added together? If we took a guess at the size of this error our answer would be that it is greater than one of the random errors but less than the sum of both. Systematic errors accumulate. Random errors accumulate as the of the sum of the independent standard deviations squared (σ^2) of each variable.

In the following examples we will consider the effect of the propagation of errors in the main arithmetic processes. The random error in any function will be designated by the standard deviation symbol σ (sigma).

11.7 Addition of Values containing Random Errors.

Here we are dealing with linear functions.

$$U = x + y \qquad \text{thus } \sigma_u = \pm\sqrt{\sigma_x^2 + \sigma_y^2}$$

11.7.1 Worked Example 11.8. Random errors in addition.

If a line ABCD is comprised of 3 measured segments, each value being the mean of many measurements and measured independently from the other segments.

AB = 407.123 ±0.010 ($\sigma_{AB} = \pm 0.010$)
BC = 416.043 ±0.005 ($\sigma_{BC} = \pm 0.005$)
CD = 206.175 ±0.008 ($\sigma_{CD} = \pm 0.008$)

What is the length of the line AD and its standard deviation?

$$AD = 1029.341 \pm \sqrt{0.010^2 + 0.005^2 + 0.008^2}$$
$$= 1029.341 \pm 0.014$$

note if $A = x + y + z + b + c \ldots$

then $\sigma A = \pm\sqrt{\sigma_x^2 + \sigma_y^2 + \sigma_z^2 + \sigma_b^2 + \sigma_c^2 + \ldots}$

11.8 Subtraction of Values containing Random Errors.

This is identical to the addition case.

$$U = x - y \qquad \text{thus } \sigma_u = \pm\sqrt{\sigma_x^2 + \sigma_y^2}$$

11.8.1 Worked Example 11.9. Subtraction of two directions. (Figure 11-1)

$$\angle A = 97°34'15'' \pm 10''$$
$$-\angle B = \underline{42°16'10'' \pm 5''}$$
$$= \underline{55°18'05''} \pm\sqrt{5^2 + 10^2}$$
$$= \underline{55°18'05'' \pm 11''}$$

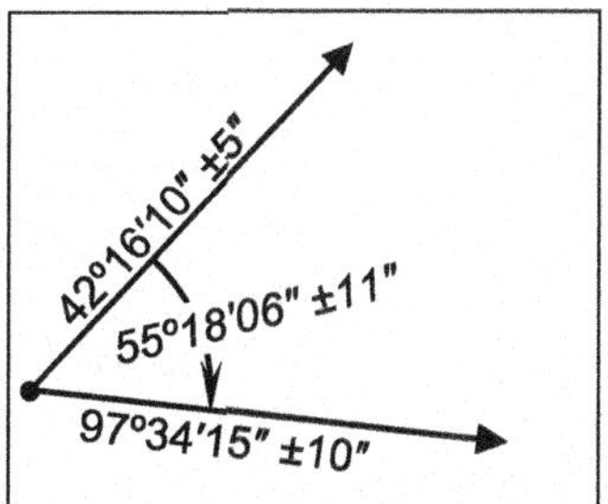

Figure 11-1 Angular difference

11.9 Multiplication of Values containing Random errors.

Suppose we multiply 2 quantities x and y each with a standard deviation, σ_x, σ_y respectively. We wish to determine the random error in x and y which we assume have been measured independently.

i.e. $u = (x.y)$ (u is a non linear function and, from § 11.6.2)

$$\sigma_u^2 = \left(\frac{\partial F}{\partial x}\right)^2 \sigma_x^2 + \left(\frac{\partial F}{\partial y}\right)^2 \sigma_y^2$$

and $\frac{\partial_u}{\partial_x} = y, \quad \frac{\partial_u}{\partial_y} = x$

Hence $\sigma_u = \pm\sqrt{(y\sigma_x)^2 + (x\sigma_y)^2}$

Generate relative errors in x and y

and $\sigma_u = \pm xy\sqrt{\dfrac{(y\sigma_x)^2 + (x\sigma_y)^2}{(xy)^2}}$

by multiplying σ_u by $\dfrac{xy}{\sqrt{(xy)^2}}$

$$\sigma_u = \pm xy\sqrt{\left(\frac{\sigma_x}{x}\right)^2 + \left(\frac{\sigma_y}{y}\right)^2}$$

This allows x and y to be factored out.

where $\dfrac{\sigma_x}{x}$, $\dfrac{\sigma_y}{y}$ are the relative errors in x and y respectively.

Similarly, the product of several factors has the error:

$$\sigma_u = \pm ABCD\cdots\sqrt{\left(\frac{\sigma_A}{A}\right)^2 + \left(\frac{\sigma_B}{B}\right)^2 + \left(\frac{\sigma_C}{C}\right)^2 + \left(\frac{\sigma_D}{D}\right)^2 + \cdots}$$

11.9.1 Worked Example 11.10. Area of a Rectangle.

In Figure 11-2 the rectangle has the following dimensions:

CD = 75.000 ±0.003m ($\sigma_{CD} = \pm 0.003$)

BC = 100.000 ±0.008m ($\sigma_{BC} = \pm 0.008$)

Find the area of the rectangle and the standard deviation of the area. (i.e. Random error in the area).

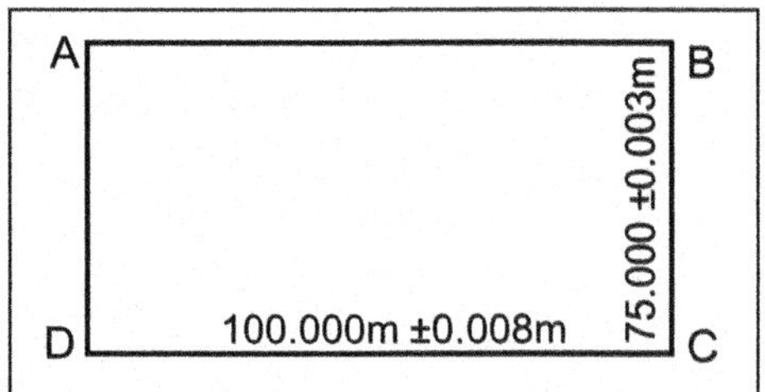

Figure 11-2 Rectangular area

Solved in two parts:

1) Area = A = CD × BC = 75.000 × 100.000 = 7500.0 m^2.

2) σ_A = Random error $= \pm 7500.0 \times \sqrt{\left(\dfrac{\sigma_{CD}}{CD}\right)^2 + \left(\dfrac{\sigma_{BC}}{BC}\right)^2}$

$$= \pm 7500.0 \times \sqrt{\left(\frac{0.003}{75.000}\right)^2 + \left(\frac{0.008}{100.000}\right)^2} \quad = \pm 0.67 \quad = \pm 0.7m^2$$

Area = A = 7500.0 ±0.7m^2

11.9.2 Worked Example 11.11. Area of a triangle

In Figure 11-3 the following dimensions have been observed:

a = 100.000 ±0.005m

b = 600.000 ±0.010m

∠C = 60° ±10″(the **random** error in angle C is ±10″).

Find the area of the triangle ABC and the random error in this area due to the random errors in the measured sides and the included angle.

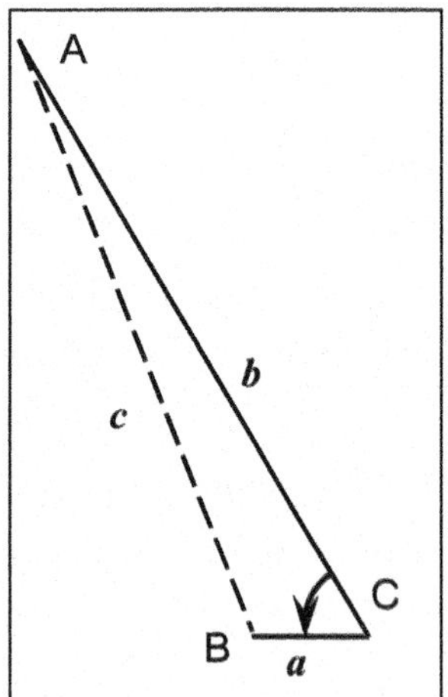

Figure 11-3 Area triangle

1) Area = AR = ½absin(C). A is non-linear, the angular error must be converted to RADIANS.

AR = 25980.8m^2

2) Random error $\sigma_{AR} = \pm AR \times \sqrt{\left(\dfrac{\sigma_a}{a}\right)^2 + \left(\dfrac{\sigma_b}{b}\right)^2 + \left(\dfrac{\sigma_{\sin C}}{\sin C}\right)^2}$

11.9.2.1 Evaluate the angular error

To evaluate $\left(\dfrac{\sigma_{\sin C}}{\sin C}\right)^2$ we must first evaluate $\sigma_{\sin C}$

(i.e. the error in sin(C) due to the random error of 10″ in angle C)

Suppose $y = \sin C$,

the **derivative** of y with respect to C is:

$$\frac{dy}{dC} = \cos C \text{, so, by cross-multiplication:}$$

$$dy = \cos C \times dC \text{ where}$$

$$dy = \sigma_{\sin C} = \cos C \times dC$$

and relative error $= \frac{dy}{y} = \frac{\cos C \cdot dC}{\sin C} = \frac{dC}{\tan C} \left(\tan\theta = \frac{\sin\theta}{\cos\theta} \right)$ (new derivation)

always remembering angular errors, **here** $\boldsymbol{dC}$, must be expressed in **radians**.

$dC = 10'' = 10/3600 \times \pi/180$ or $10''/206265 = 0.0000485$ (or, for small angles, sin(10″) or tan(10″)

$dy = \cos C \cdot dC = \cos(60°) \cdot 0.0000485$

$= 0.5 \times 4.85E^{-5}$

Thus $\sigma_{SinC} = 0.000024$ and

$\sin(60) = 0.866025$

Hence $\sigma_{AR} = \pm AR \times \sqrt{\left(\frac{0.005}{100.000}\right)^2 + \left(\frac{0.010}{600.000}\right)^2 + \left(\frac{0.000024}{0.866025}\right)^2}$

$= \pm 25980.8 \times \sqrt{2.5E^{-9} + 2.78E^{-10} + 7.83E^{-10}} = \pm 1.55$

Area $= 25{,}980.8 \pm 1.55\text{m}^2$.

Note: $\sigma_{\sin C}$ (the error in sin$C$ due to the random error of 10″ in C) can be derived by finding the difference between sin(C) and sin(C + dC).
In this case the difference between sin(60°) and sin(60°00'10") is (| 0.8660254 – 0.8660496 |) = 0.0000242, the same as $\sigma_{\sin C}$ above.

An **alternate** derivation simplifies $\sigma_{\sin C}$:

$\sigma_{\sin C} = \cos C \times dC$, thus $\left(\frac{\cos C \times dC}{\sin C}\right) = \left(\frac{dC}{\tan C}\right)$ so:

$$\sigma_{AR} = \pm AR \times \sqrt{\left(\frac{\sigma_a}{a}\right)^2 + \left(\frac{\sigma_b}{b}\right)^2 + \left(\frac{dC}{\tan C}\right)^2}$$

$dC = \sigma_{\angle C} = \pm 10''$ converted to radians $= 0.0000485$

$$\sigma_{AR} = \pm AR \times \sqrt{\left(\frac{0.005}{100.00}\right)^2 + \left(\frac{0.010}{600.00}\right)^2 + \left(\frac{0.0000485}{1.732051}\right)^2}$$

$$\Rightarrow \pm 25980.8 \times \sqrt{2.5E^{-9} + 2.78E^{-10} + 7.83E^{-10}} = 1.55$$

Area $= 25{,}980.8 \pm 1.55\text{m}^2$, the results as previously calculated.

11.10 Division of Values Containing Random Errors.

Suppose we divide a quantity x by a quantity y each containing random errors of σ_x and σ_y respectively.

i.e. $u = \frac{x}{y}$ (u is a non-linear function)

and $\sigma_u = \sqrt{\left(\frac{\partial u}{\partial x}\right)^2 \sigma_x{}^2 + \left(\frac{\partial u}{\partial y}\right)^2 \sigma_y{}^2}$ where $\frac{\partial u}{\partial x} = \frac{1}{y}$ and $\frac{\partial u}{\partial y} = -\frac{y}{x^2}$

now $\sigma_u = \pm \frac{x}{y} \sqrt{\frac{\frac{1}{y^2}\sigma_x{}^2 + \frac{x^2}{y^4}\sigma_y{}^2}{\left(\frac{x}{y}\right)^2}}$ reducing to $\sigma_u = \pm \frac{x}{y}\sqrt{\left(\frac{\sigma_x}{x}\right)^2 + \left(\frac{\sigma_y}{y}\right)^2}$

where $\frac{\sigma_x}{x}$ and $\frac{\sigma_y}{y}$ are the relative errors in x and y respectively. Similar to the multiplication formula

11.10.1 Worked Example 11.12. Division

In Figure 11-4 the following dimensions have been observed:

Given: ∠B = 60° ±10″.(see § 11.9.2, Worked Example 11.11)

AC = b = 1126.123 ±0.016m

Calculate side c (side AB) using $c = b/\sin(B)$ and the random error σ_c in that side

AB=1300.335

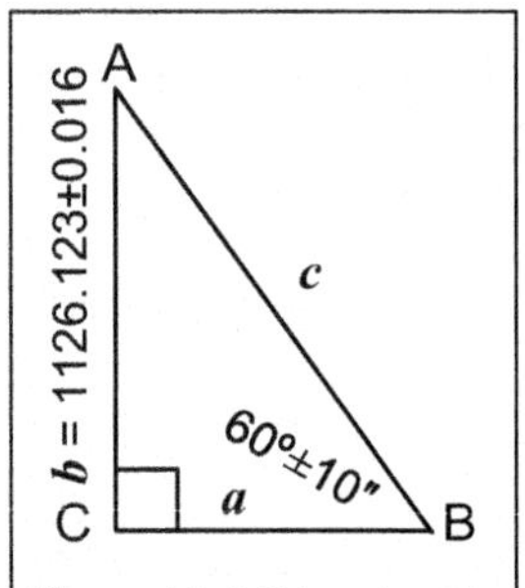

Figure 11-4 Triangle side

$$\sigma_c = \pm 1300.335 \times \sqrt{\left(\frac{\sigma_b}{b}\right)^2 + \left(\frac{\sigma_{\sin B}}{\sin B}\right)^2} \quad \text{where, from §3.8.2.1}$$

$$\sigma_{\sin B} = \cos B \times dB = \cos(60^\circ) \times 10'' / 206265 = 0.5 \times 4.85^{-5} = 2.42^{-5}$$

$$= \pm 1300.335 \times \sqrt{\left(\frac{0.016}{1126.123}\right)^2 + \left(\frac{0.0000242}{0.866025}\right)^2}$$

$$\sigma_c = \pm 1300.335 \times \sqrt{2.019E^{-10} + 7.83E^{-10}} = \pm 0.041$$

Alternately, using $\sigma_{\angle B}$ instead of $\sigma_{\sin B}$:

$$\sigma_c = \pm 1300.335 \times \sqrt{\left(\frac{\sigma_b}{b}\right)^2 + \left(\frac{\sigma_{\angle B}}{\tan B}\right)^2} \quad \text{where}$$

$$\sigma_{\angle B} = 10'' / 206265 = 4.85^{-5}$$

$$= \pm 1300.335 \times \sqrt{\left(\frac{0.016}{1126.123}\right)^2 + \left(\frac{0.0000485}{1.732051}\right)^2}$$

$$\sigma_c = \pm 1300.335 \times \sqrt{2.019E^{-10} + 7.83E^{-10}} = \pm 0.041$$

AB = 1300.335 ±0.041m.

11.10.2 Worked Example 11.13 Division – *a priori* (pre-determined) precision of measurement.

This case is different from the previous examples. We must measure a quantity to a specified precision.

e.g. Suppose a surveyor is required to lay out a rectangle which has an area of 200.000 ±0.165m². One of the sides of the rectangle has been measured and has a precision of 10.000 ±0.008m. Find the length of the other side and the allowable random error in that side.

In Figure 11-5 the rectangle has the following dimensions and specification:

Area = A = 200.000 ±0.165m².

AB= 10.000 ±0.008m.

Find AD and the random error in this side.

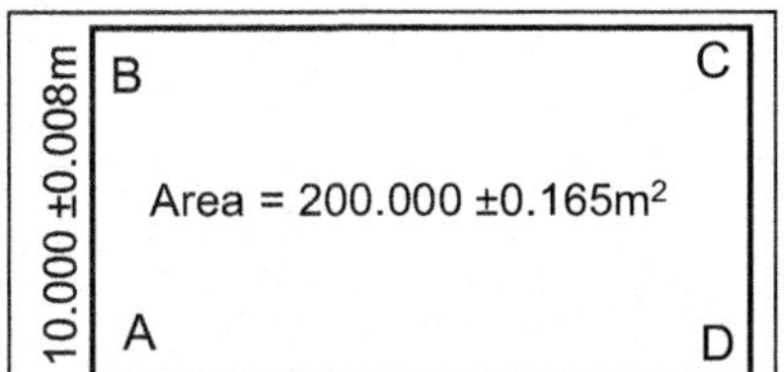

Figure 11-5 *a priori* determination of AD measurement precision

$$AD = \frac{200.000}{10.000} \pm \sigma_{AD}$$

$$\sigma_{AD} = \pm 20.000 \times \sqrt{\left(\frac{0.165}{200.000}\right)^2 + \left(\frac{0.008}{10.000}\right)^2}$$

$$= \pm 0.023$$

AD = 20.000 ±0.023m

Thus, a distance of 20.000m must be laid out in the field with a precision of ±0.023 m.

11.11 Multiplication of Values containing random Errors by an Exact Number.

11.11.1 Worked Example 11.14. Converting a measured value into other units.

Suppose the value of the product is $C\,(x \pm \sigma_x)$ where C is a constant and σ_x is the random error in x.

Then $C \cdot (x \pm \sigma_x) = Cx \pm C\sigma_x$, and the error in this product is seen to be $C\sigma_x$. As C is a constant then $\sigma_C = 0$.

Given ∠B = 60°

∠C = 45°

AC = b = 1126.123 ±0.011m

i.e. No error in the angles B & C. Find c and σ_c in Figure 11-6.

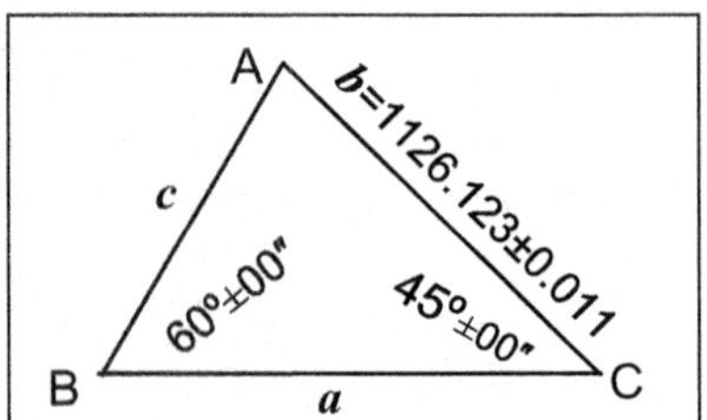

Figure 11-6 Find side AB and σ_{AB}

If it's not checked it's wrong.

Find the length AB and the random error in AB.

From the sin rule find length c:

$$\frac{c}{\sin C}=\frac{b}{\sin B}=\frac{a}{\sin A}$$

$$AB=\frac{b\sin C}{\sin B}=\frac{(1126.123\pm 0.011)\sin C}{\sin B}$$

$$=\frac{1126.123\times\sin C}{\sin B}\pm\frac{0.011\times\sin C}{\sin B}\quad\text{where }\frac{\sin(45)}{\sin(60)}=0.816497$$

$$=1126.123\times 0.816497\pm 0.011\times 0.816497$$

$$AB=919.476\pm 0.009\text{m}$$

11.11.2 Worked Example 11.15. Converting metres to feet.

A distance is measured as 802.316 ±0.027m

Convert this measurement to feet: (3.280839895 ft = 1m or 1ft = 0.3048m)

Distance = (1/0.3048)×(802.316 ± 0.027)
= 802.316/(0.3048) ± 0.027/(0.3048)
= 2632.270 ±0.089 ft.

Note that the answer is given to 7 significant figures, since the 8 in 802.316 could be regarded as 2 significant figures, being nearly equal to 10. (802 is close to 1000)

11.12 Multiplication of Values containing random errors in a traverse.

11.12.1 Worked Example 11.16. Converting a measured vector into plane units.

Vector AB (See Figure 11-7) has direction 335° 30′36″ ±10″ and magnitude 3500.255 ±0.115m.

From the origin X find the plane co-ordinates of Y and the standard deviation of those co-ordinates. (**Conventional** reference frame.)

$r = 3500.255 \qquad \sigma_r = \pm 0.115.$

$\theta = 335^\circ 30'36'' \qquad \sigma_\theta = \pm 10''$

$X = r\cos\theta = 3185.350$

$Y = r\sin\theta = -1450.976$

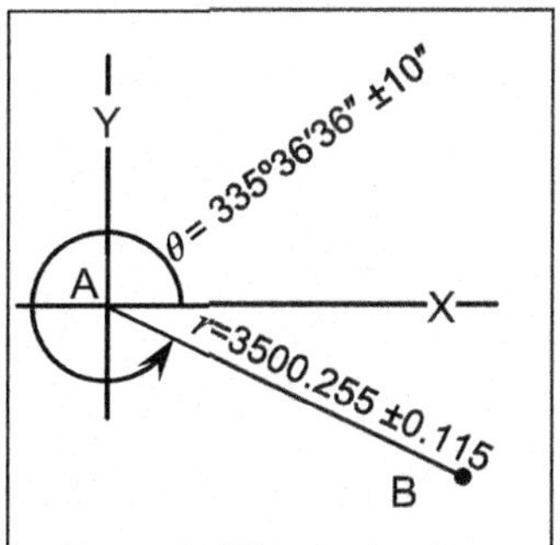

Figure 11-7 Vector to plane

[1] $$\sigma_x=\pm|X|\times\sqrt{\left(\frac{\sigma_r}{r}\right)^2+\left(\frac{\sigma_{\cos\theta}}{\cos\theta}\right)^2}$$

$$=\pm|X|\times\sqrt{dr^2+dx^2}$$

[2] $$\sigma_y=\pm|Y|\times\sqrt{\left(\frac{\sigma_r}{r}\right)^2+\left(\frac{\sigma_{\sin\theta}}{\sin\theta}\right)^2}$$

$$=\pm|Y|\times\sqrt{dr^2+dy^2}$$

Evaluate magnitude relative error in ***r***:

$$dr=\frac{\sigma_r}{r}=\frac{0.115}{3500.255}=3.28E^{-5}$$

$$dr^2=1.079E^{-9}$$

Evaluate direction relative errors: in $x \quad (dx=\frac{\sigma_{\cos\theta}}{\cos\theta})$

and $y \quad (dy=\frac{\sigma_{\sin\theta}}{\sin\theta})$

reviewing § 11.9.2, Worked Example 11.11, page 176,

$$x=r\cos\theta,\ \frac{dx}{d\theta}=-\sin\theta$$

$$y=r\sin\theta,\ \frac{dy}{d\theta}=\cos\theta$$

where $d\theta = 10'' = 10/206265$ radians

$d\theta = 0.00004848 = 4.85E^{-5}$ (shows as 48.48^{-06} on HP 10+)

$$dx = \frac{\sigma_{\cos\theta}}{\cos\theta} = \frac{-\sin\theta \cdot d\theta}{\cos\theta} = \frac{\sin(335°30'36''') \times 4.85E^{-5}}{0.91003} = 0.00002208$$

$$dx^2 = 4.877E^{-10}$$

$$dy = \frac{\sigma_{\sin\theta}}{\sin\theta} = \frac{\cos\theta \cdot d\theta}{\sin\theta} = \frac{\cos(335°30'36''') \times 4.85E^{-5}}{-0.41453} = 0.0001064$$

$$dy^2 = 1.133E^{-8}$$

Using formula [1]

$$\sigma_x = \pm|3185.35| \times \sqrt{1.079E^{-9} + 4.877E^{-10}} = \pm 3185.35 \times 3.9587E^{-5} = \pm 0.126$$

Xσx = 3185.350 ±0.126m

Using formula [2]

$$\sigma_y = \pm|-1450.98| \times \sqrt{1.079E^{-9} + 1.133E^{-8}} = \pm 1450.976 \times 1.1139E^{-4} = \pm 0.162$$

Yσy = -1450.976 ±0.162m

11.13 Laboratory Exercises 11.1

Q 1. Given the angles of a triangle ABC as

∠A = 60°00'30" ±10"

∠B = 82°25'45" ±15"

∠C = 37°33'15" ±15"

Determine the misclose in the triangle and the random error (standard deviation of the sum.)

Q 2. Four bays AB, BC, CD & DE were measured several times and gave the following results:

AB = 211.013 ±0.010m

BC = 196.331 ±0.008m

CD = 217.935 ±0.012m

DE = 253.516 ±0.009m

Compute the length AE and the random error (σ_{AE}) in that side.

Q 3. In a triangle ABC, angles A, B were measured and the following results recorded:

∠A = 60°00'35" ±10"

∠B = 72°41'40" ±08"

Compute the angle C and the random error (σ_C) in the angle C.

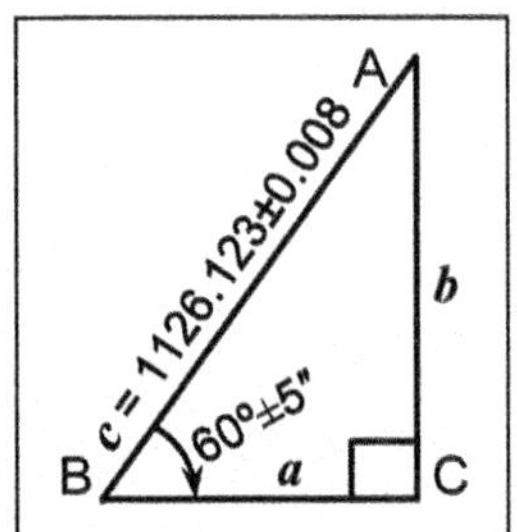

Figure 11-8 Q4

Q 4. In the right angle triangle ABC (see Figure 11-8).

∠B = 60° ±5"

AB = 1126.123 ±0.008m

Compute *AC* and the random error (σ_{AC}) in the side AC.

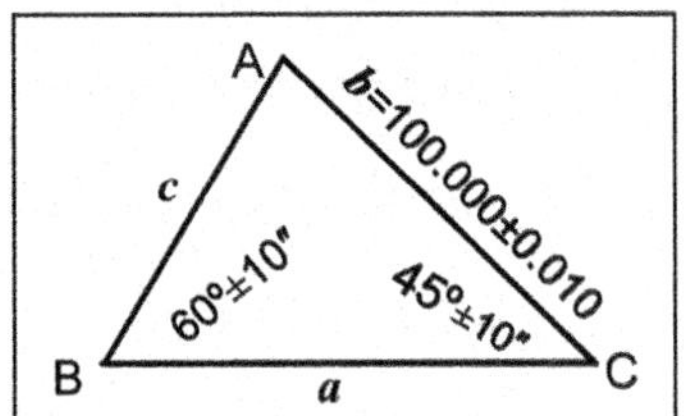

Figure 11-9 Q5

Q5. In triangle ABC (see Figure 11-9).

∠B = 60° ±10"

∠C = 45° ±10"

AC = 100.000 ±0.010m

Compute AB and σ_{AB}.

Q 6. For the vector AB (Figure 11-10).

The coordinate differences delta E, delta N of line AB are given by

ΔE = D·sin(θ)

ΔN = D·cos(θ)

where D is the distance of AB and
θ is the bearing of AB

Given that θ = 75° ±10" and D = 500.612 ±0.010m

Compute the coordinates differences of AB and the random error in those coordinates.

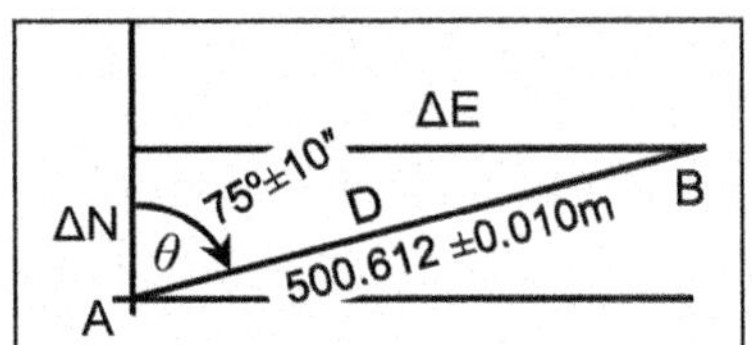

Figure 11-10 Q6

If it's not checked it's wrong.

Q 7. A distance AB was measured independently 6 times with the following results:

36.123 ± 0.010m
.127 ± 0.008
.125 ± 0.008
.126 ± 0.009
.126 ± 0.010
36.124 ± 0.009

Compute the mean distance AB and σ_{AB}.

11.13.1 Answers to Laboratory Questions 11.1. Some hints and checks are provided:

1. Misclose = 30″ σ_{SUM} = ±23″
(Σ∠ = 179°59′30″, σ_{SUM} = ±√($10^2 + 15^2 + 15^2$)″ = ±√(550)″ = ±23″)

2. AE = 878.795 ±0.020m
(σ_{SUM} = ±√($0.01^2 + 0.008^2 + 0.012^2 + 0.009^2$) = ±√(0.000389)m = ±0.020m)

3. ∠C = 47°17′45″ ±13″
(∠C = 180° - ∠A - ∠B, σ_{SUM} = ±√($10^2 + 8^2$)″ = ±√(164)″ = ±13″)
Addition of errors in subtraction.)

4. AC = 975.251 ±0.016m

(§ 11.10.1. Worked Example 11.6. You can use $\sigma_c = \pm B \times \sqrt{\left(\frac{\sigma_b}{b}\right)^2 + \left(\frac{\sigma_{\angle B}}{\tan B}\right)^2}$)

5. AB = 81.650 ±0.009m
(AB = side c, AC = b, Sin rule: c = $b$sinC/sinB = 100×sin(45)/sin(60)
from Worked Example 11.8 (§ 11.11.1), $\sigma_c = \pm\sigma_b \cdot$sinC/sinB = ±0.01×sin(45)/sin(60) = ±0.008m)
Now σ_B, σ_C = ±10″ = ±0.0000485
and $\sigma_{\sin c}$ = cos$C \cdot dC$, $\sigma_{\sin b}$= cos$B \cdot dB$

$$\sigma_c = \pm c \times \sqrt{\left(\frac{\sigma_c}{c}\right)^2 + \left(\frac{\sigma_{\sin C}}{\sin C}\right)^2 + \left(\frac{\sigma_{\sin B}}{\sin B}\right)^2}$$

$$= \pm c \times \sqrt{\left(\frac{\sigma_c}{c}\right)^2 + \left(\frac{\cos C \cdot dC}{\sin C}\right)^2 + \left(\frac{\sin B \cdot dB}{\sin B}\right)^2}$$

$$= \pm 81.65 \times \sqrt{\left(\frac{0.008}{81.65}\right)^2 + \left(\frac{\cos(45) \times 4.85E^{-5}}{\sin(45)}\right)^2 + \left(\frac{\cos(60) \cdot 4.85E^{-5}}{\sin(60)}\right)^2}$$

or, using Worked Example 11.5 (§ 11.9.2), where $\frac{\cos C \times dC}{\sin C} = \frac{dC}{\tan C}$

$$\sigma_c = \pm 81.65 \times \sqrt{\left(\frac{0.008}{81.65}\right)^2 + \left(\frac{4.85E^{-5}}{\tan(45)}\right)^2 + \left(\frac{4.85E^{-5}}{\tan(60)}\right)^2}$$

$$= \pm 81.65 \times \sqrt{\left(9.8E^{-5}\right)^2 + \left(4.85E^{-5}\right)^2 + \left(2.8E^{-5}\right)^2}$$

$$= \pm 81.65 \times \sqrt{1.27^{-8}}$$

σ_c = ±0.009m·

6. ΔE = 483.554 ±0.011m
ΔN = 129.568 ±0.024m
From Worked Example 11.10, (§ 11.12.1).
r = 500.612, σ_r = ±0.01, dr = 0.01/500.612, dr^2 = 4.0^{-10}

If it's not checked it's wrong.

$d\theta = \pm 10'' = 10/206265$ radians, $d\theta = 4.85E^{-5}$

$$dE = \frac{-\sin\theta \cdot d\theta}{\sin\theta} = \frac{0.259 \times 4.85E^{-5}}{0.696} = 1.7E^{-10}$$

$$\sigma_E = \pm|\Delta E| \cdot \sqrt{dr^2 + dE^2} = \pm|483.554| \times \sqrt{\left(4.0E^{-10}\right) + \left(1.7E^{-10}\right)} = \pm 0.011\text{m}$$

$$dN = \frac{-\sin\theta \cdot d\theta}{\cos\theta} = \frac{0.696 \times 4.85E^{-5}}{0.259} = -1.81E^{-4} \rightarrow dN^2 = 3.3E^{-8}$$

$$\sigma_N = \pm|\Delta N| \cdot \sqrt{dr^2 + dN^2} = \pm|129.568| \times \sqrt{\left(4.0E^{-10}\right) + \left(3.3E^{-8}\right)} = \pm 0.024\text{m}$$

7. AB = 36.125 ±0.010m

Mean: $\bar{x} = \frac{\Sigma x_1^n}{n} = \frac{216.751}{6} = 36.125$

but: $$\sigma_x = \pm\sqrt{\frac{\left(\Sigma \sigma_{x_1^n}\right)^2}{n-1}}$$

$$= \pm\sqrt{\frac{(0.01^2 + 0.008^2 + 0.008^2 + 0.009^2 + 0.01^2 + 0.009^2}{6-1}}$$

$$= \pm\sqrt{\frac{0.00045}{5}} = \frac{0.0221}{2.24}$$

$$\sigma_x = \pm 0.010\text{m}$$

11.14 References for Chapter 11

Bannister, A and S. Raymond. Surveying. Third edition, Pitman Paperbacks, 1972. ISBN 0-273-36148-1

Elfick, M.H, J.G Fryer, R.C. Brinker and P.R Wolf. Elementary Surveying, Eighth edition. Longman, 1994. ISBN 0-00-499001-3

Ogundare, J.O., Precision surveying: the principles and geomatics practice, John Wiley & Sons, Inc. 2016. ISBN 978-1-119-10251-9

Ghilani, C.D., Adjustment computations. Spatial data analysis. Fifth edition, John Wiley & Sons, Inc. 2010. ISBN 978-0-4701-46491-5

Chapter 12 TRANSITION CURVES AND SUPERELEVATION

In Chapter 8, page 113, the notion of simple curve was introduced. This chapter delves more into the compilated scenario where the radius of the curve varies, i.e., the transitional curves. Transitional curves are normally introduced to ease vehicles entry from one straight portion of a road to another while at the same time overcoming the effect of centrifugal force that tries to push the vehicle of the road. Another way of overcoming this effect is by raising one side of the road, i.e., super elevation. This chapter shows how both transition curve is calculated. The next chapter will show how superelevation is computed.

12.1 Road Design incorporating Transition Curves and Superelevation

12.1.1 Introduction

The circular curve is a fundamental element in road design. Its radius is a function of many factors which include operating (design) speed, curve super elevation, radial forces and tyre friction and occupant riding comfort.

These design criteria are published by: Austroads Ltd, Sydney, 2021. Edition 3.4 AGRD03-16: Guide to Road Design Part 3: Geometric Design. ISBN 978-1-922382-15-3.

The purpose of the guide is to collate modern road design practices of member organisations in Australia and New Zealand to provide safe and efficient road designs in urban and rural environments. It provides information on road alignments to cater for traffic requirements at chosen speeds in a variety of terrain types. The entirety of the guide is aimed at the traffic design engineering team. Different authorities may have local procedures that supplement the core document. The Introduction is commended to the reader.

The next three chapters are designed to give surveyors and civil engineers an appreciation of the basic principals on which the various design components of complex curves, superelevation and vertical curves.

Austroads is the association of Australian and New Zealand road transport and traffic authorities. Austroads members are the six Australian state and two territory road transport and traffic authorities, the Australian Government Department of Infrastructure and Regional, the Australian Local Government Association, and New Zealand Transport Agency.

12.1.2 The Fundamentals of Uniform Circular Motion

Uniform circular motion is the path taken by a body moving at constant linear speed around a centre point of constant radius. To maintain the circular path the body is subject to an **acceleration** towards the centre of the circle.

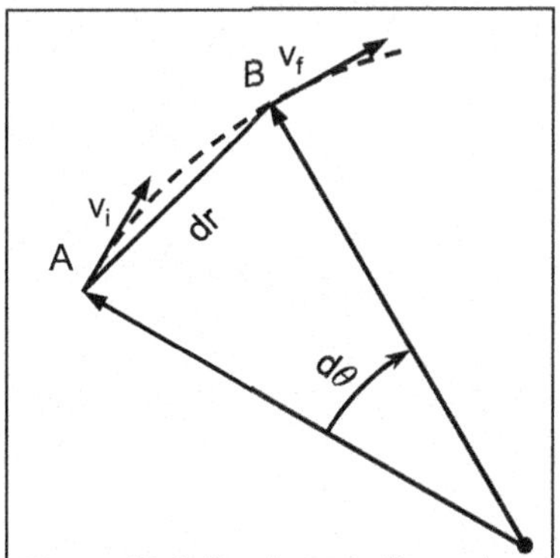

Figure 12-1 Circular velocity vectors

A particle moves, as in Figure 12-1, from
point A, at time t_i with velocity $\mathbf{v}_i$
to point B, at t_f, velocity $\mathbf{v}_f$.
$\mathbf{v}_i$ and $\mathbf{v}_f$ are the two velocity vectors. If the magnitudes are the same ($v_i = v_f = v$), then the only change is **direction**.
and Figure 12-2 shows the vector subtraction.

The average acceleration,	$\bar{a} = \frac{v - v}{t - t} = \frac{dv}{dt}$	Eqn 12-1

Consider Figure 12-1 with sides *dr* and *r* similar to Figure 12-2 so that:

$$\frac{dr}{r} = \frac{dv}{v}$$

And, as $\bar{a} = \frac{dv}{dt}$ by substitution we have: $\bar{a} = \frac{v \times dv}{t \times dt}$ (Figure 12-2)

In the limit, as $\lim_{t \to 0}\left(\frac{dr}{dt}\right) = \mathbf{v}$ and $\boldsymbol{dv}$ points to the centre of the circle.

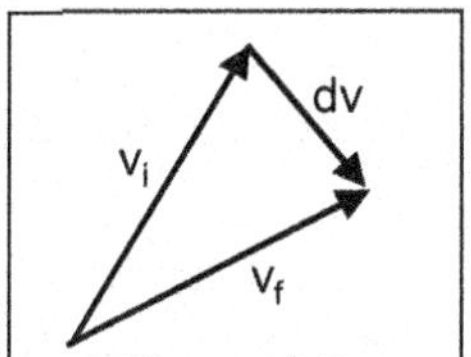

Figure 12-2
Vector subtraction.

The magnitude of the acceleration is:	$a_r = \frac{v^2}{r}$	Eqn 12-2

It is called centripetal (centre-seeking) acceleration.

J. Walker and J. Awange, *Surveying Computations*,
https://doi.org/10.1007/978-3-031-94382-9_12

12.1.3 Forces Involved in Uniform Circular Motion

From Newton's Second Law of Motion, we know that, to change the path of a body of mass, m, in uniform linear motion, we must apply an acceleration, a, to the body. The amount of **force, F,** needed to apply that acceleration is: $F = ma$

Newton's second law of motion:	$F = ma$	Eqn 12-3

A body in uniform **circular** motion is subject to a centripetal acceleration, a_r, towards the centre of the circle. If the body has mass, m, then the force needed to maintain that path is:

From Eqn 12-3 and Eqn 12-2:	$F = m\dfrac{v^2}{r}$	Eqn 12-4

The **force, F,** needed to maintain this circular path can be called:

a) tension, if the body is attached to a string,
b) gravity, if it is the moon circling the earth or
a) friction, if it is a vehicle driving around a circular curve.

12.1.4 The Forces on a Body in Uniform Circular Motion

Consider a body of mass, m, stationary on the surface of the earth. The force applied by the surface of the earth to maintain the body in its static position is $F = ma$. We are holding the body against the acceleration of gravity, $g \approx 9.8\text{m.sec}^{-2}$, towards the centre of the earth, opposing the force applied **by** the body **to** the surface.

We can call this force the **weight**, $W=mg$, in Newtons (kg.m.sec^{-2}). Please note that we normally call the **mass** of a body its **weight**.

Now let us call the body a vehicle and the surface of the earth a road. The forces on a vehicle (see Figure 12-3) following a circular (flat) curve are:

friction, F, keeping the vehicle on its circular path:
weight, W, of the vehicle on the road,
reaction, n, of road to weight of the vehicle

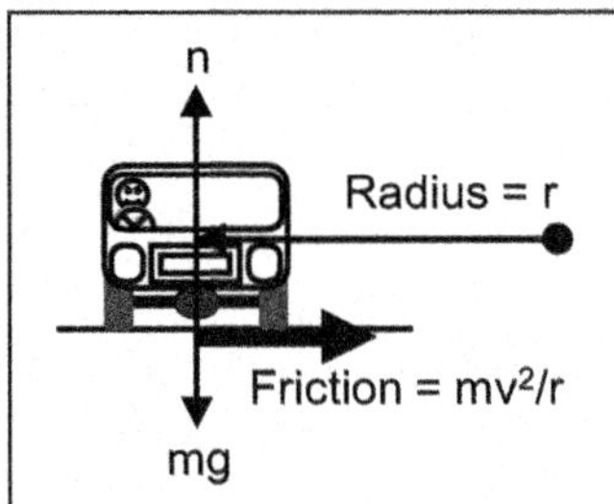

Figure 12-3 Flat curve

12.1.4.1 Worked Example 12.1.

Assume a vehicle of

Gross vehicle mass (GVM), m = 1500Kg
speed, $v = 70\text{Kmh}^{-1} = 70000/3600\text{ms}^{-1} = 70/3.6 = 19.4\ \text{ms}^{-1}$
radius, r = 100m

What is the frictional force needed to keep the vehicle on the road?

From Eqn 12-4:

$$F = m\frac{v^2}{r} = 1500\frac{19.44^2}{100} = 5671\,Newtons\,(kg.m.s^{-2})$$ See Figure 12-3.

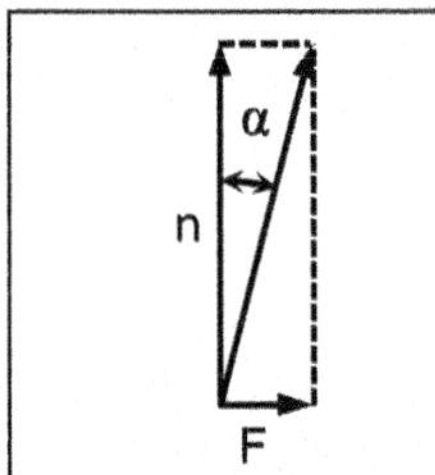

Figure 12-4
Static friction μ_S.

Now examine the resultant vector diagram of forces, Figure 12-4, n = W=mg:

From Eqn 12-4:	$\tan(\alpha) = \dfrac{F}{n} = \dfrac{\frac{mv^2}{r}}{mg} = \dfrac{\cancel{m}v^2}{r} \times \dfrac{1}{\cancel{m}g} = \dfrac{v^2}{rg} = \mu_S$	Eqn 12-5

In this case $\tan(\alpha)$ represents the **coefficient of static friction, μ s.** It is independent of mass.

In the previous example, and using Eqn 12-5 and where $n = mg = 14700N$

$$\tan(\alpha) = \frac{5671}{14700} = 0.385 \text{ or } = \frac{19.44^2}{100 \times 9.8} = 0.385$$
$$\mu_S = \tan(\alpha) = 0.385\ (\alpha = 21°)$$

The **maximum** coefficient of static friction between tyre and pavement depends very much on both surfaces but can be estimated at $\mu_{S(max)} = 0.7$ for a normal car tyre on an asphalt pavement. Specialist tyres can have $\mu_S > 1.0$. (Current F1 tyres have $\mu_S \cong 1.7$). In the above example (§ 12.1.4.1) the vehicle will maintain its circular path because $\mu_S = 0.2 < \mu_{S(max)} = 0.7$. (Below its static friction limit, after which it will slide off the curve.)

Static friction, as stylised in Figure 12-7, varies greatly, depending on the surface and the conditions. Dry asphalt, wet asphalt, water cover, ice, etc. Once the static friction, μ_S, is broken the moving friction, μ_K, is less, meaning a wet road may be OK, until aquaplaning takes over. See recommendations for μ_S in § 12.1.8.

Conversely, deciding on a design coefficient of static friction specifies a **maximum speed** for a given radius.

$$\mu_S = \frac{v^2}{rg} \Rightarrow v = \left(\mu_S \cdot rg\right)^{\frac{1}{2}} \left(g = 9.8ms^{-2}\right) \Rightarrow v^2{}_{\max} = \left(\mu_{S(\max)} \cdot rg\right) m \cdot \sec^{-1}$$

$$v_{(\max)} = \sqrt{127\left(\mu_{S(\max)} \cdot r\right)} Kmh^{-1}$$ where, factoring $m.\sec^{-1}$ to Kmh^{-1} and $g = 9.8ms^{-2}$ then $3.6^2 \times 9.8 = 127$

For: $\mu_S = 0.7$ and $r = 100$m, **Maximum** speed, $v_{(\max)} = 94\text{Kmh}^{-1}$.
Within limits of $\mu_S = 0.2$, $v_{(\max)} = 50Kmh^{-1}$, $\mu_S = 0.3$, $v_{(\max)} = 62Kmh^{-1}$

We are also able to specify **minimum radius** for a give speed for the design coefficient of static friction.

$$\mu_S = \frac{v^2}{rg} \quad \Rightarrow r_{(\min)} = \frac{v^2 m \cdot \sec^{-1}}{\mu_{S_{(\max)}} g} m = \frac{v^2 Kmh^{-1}}{127 \times \mu_{S_{(\max)}}} m$$

For: $\mu_S = 0.7$ and $v = 70\text{Kmh}^{-1}$, **Minimum** radius = 55m. Limits $\mu_S = 0.2$, $r_{(\min)} = 193$m, $\mu_S = 0.3$, $r_{(\min)} = 129$m.

It should be noted that operating a vehicle at maximum limits will result in very uncomfortable driving conditions, with no margin for safety. Beyond these figures the vehicle would skid tangentially from the curve.

12.1.5 Forces on a Canted Surface

If, however, the vehicle is placed on a cross fall (see Figure 12-5), or slope, that equals tan(α) as it traverses the curve then the frictional force, F, will be balanced by the resultant force through the vehicle in a direction perpendicular to the slope. The vehicle occupants will then be subject only to a force acting vertically through the seat. There will be no apparent side force.

This phenomenon is clearly demonstrated by a bicycles and motorbikes "leaning" into a turn, where the frictional force is reduced substantially.

The cross-fall slope needed to reduce F to 0 is:

From Eqn 12-5 and Figure 12-4: Converting to Kph	$\tan(\alpha) = \frac{v^2 m \cdot s^{-1}}{rg} = \frac{v^2 Kmh^{-1}}{\left(3.6^2 \times 9.8\right) r} = \frac{v^2 Kmh^{-1}}{127r} radians$

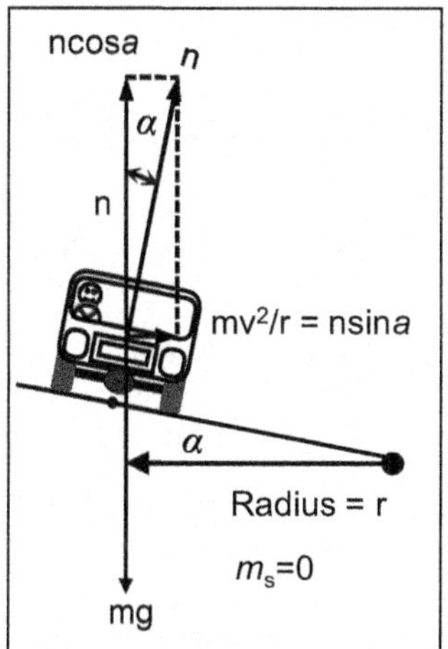

Figure 12-5
Vehicle on cross-fall.

Adverse cross fall (see Figure 12-6) on a circular curve refers to the normal cross fall, used on a straight road, being incorporated into a curve. Adverse Crossfall has been described as: "A slope on a curved pavement which generates forces detracting from the ability of a vehicle to maintain a circular path". Adverse cross fall, especially at speed, results in a very uncomfortable experience. It also adversely affects the stability of high load vehicles.

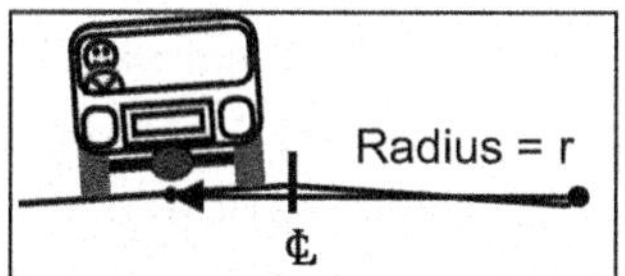

Figure 12-6 Adverse cross-fall.

12.1.6 Coefficients of Friction

Pavement design confines itself to the coefficient of **static** friction. This is the coefficient of friction that will prevent an object moving on a surface, up to the point where the applied force, F, overcomes the weight of the object, $n = mg$.

$$F_{\max} = \mu_S \cdot n$$

This situation occurs when the point of contact of the tyre on the pavement is not moving.

Once the body starts moving the coefficient of friction is called the coefficient of **kinetic** (moving) friction, μ_K. It is less than μ_S.

When a tyre starts to skid on a pavement (the point of contact is moving) it is under the influence of kinetic friction.

Figure 12-7 shows that an object will not move until the applied force matches the frictional resistance of the body at the **maximum** coefficient of static friction, $\mu_{S(\max)}$.

Once the body starts moving, a lesser applied force is needed to keep moving because the coefficient of kinetic friction, $\boldsymbol{\mu_K < \mu_{S(\max)}}$. drops to μ_K. Note the characteristically sharp break at $\boldsymbol{\mu_{S(max)}}$. A damp to wet surface may reduce to μ_K =0.4 and aquaplaning on a standing water surface may occur where μ_K =< 0.1. It is a dramatic example of friction threshold.

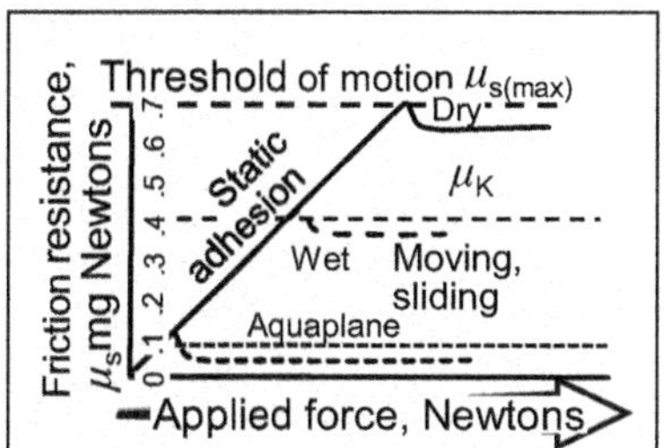

Figure 12-7 Static/kinematic friction

The third type of friction coefficient is **rolling** friction, $\boldsymbol{\mu_R}$. This is the friction needed to maintain the static contact between tyre and pavement as the tyre rolls along. Values of $\boldsymbol{\mu_R}$. generally vary between 0.02 and 0.06.

12.1.7 Practical Design Considerations for Circular Road Curves

The derivation of the formulae involved in uniform circular motion is developed and expanded into a practical set of road design standards.

The recommendations and considerations are contained in the document Austroads (2021. However, the previous versions can provide enough data for illustrative purposes.

Note that there are slightly different terminologies used in "Rural Road Design" AGRD03-16 2021.

Coefficient of static friction, μ_s, is called variously:

side friction factor, *f*, (Commentary 16, page 363 and Table 7.5, page 174) and

longitudinal friction factor (longitudinal friction factor), *d*, (Commentary 12, page 360). It is mainly concerned with stopping sight distance (SSD) deceleration.

Cross fall ($\tan(\alpha)$) is denoted as e (m.m^{-1}).

The sign indicates slope from the centreline to the edge of the pavement.

Minimum radius of a turn takes into account the cross fall, or elevation, *e*. (Austroads 2021, 7.4.1 (5))

This equation can then address the problem of adverse cross fall where e is negative.

12.1.8 Recommended Values for Coefficient of Static Friction

Whilst a value of $\mu_{s(max)} = 0.7$ may be achieved, much lower values are used in the design of curves. This is to take into account such factors as:

adequate safety margin against skidding,
road surface finish and texture (concrete, hotmix asphalt, two coat seal aggregate),
surface contamination from oil spill, loose aggregate etc,
wet weather,
tyre wear
driver behaviour
ride comfort (centripetal forces around curve)
stopping sight distance (SSD) deceleration.

12.1.8.1 Side friction and deceleration values.

Recommended static side friction coefficient values for cars and trucks are listed between 50 and 130Kmh^{-1} (Austroads 2021, Table 7.5, page 174).

Deceleration values, *d*, associated with stopping sight distances (SSD) are described in Austroads (2021) Table 5.3. Values for cars vary between $f = 0.61$ on dry bitumen to 0.27 on unsealed roads. Truck SSD is calculated on $f = 0.29$ and for a bus, $f = 0.15$ for comfort.

12.1.9 Recommended Minimum Radii Values of Horizontal Curves

Chapter 13 Calculations of Superelevation on a Composite Curve explores the application of superelevation on a transition curve developed in this chapter. The following calculated results are theoretical, but within the recommendations of Austroads (2021) published data.

The many variables in designing and deciding a curve radius are detailed in (Austroads, 2021) Table 7.6 which recommends minimum radii of horizontal curves based on differing values of superelevation, *e*, which, in turn, are based on terrain and vehicle type. Care must be exercised in choosing higher values of superelevation ($e = 0.1$, or 10%), recognising:

a higher level of driver discomfort,
tendency of slow vehicles to track towards the centre,
stability of high load commercial vehicles,
difference between inner and outer formation levels,
length available to develop superelevation.

Previous tables presented minimum values for a +3% for rural roads that mimic the current values for +5% on urban roads.

12.1.9.1 Minimum radii values of horizontal curves

Varying types of terrain undulation, classified between urban and mountainous have recommended super elevations between 5% and 10% have with associated desirable and absolute minimum values, $r_{(des)}$ and $r_{(abs)}$.

12.2 The Transition or Spiral Curve

Transitional curves are normally introduced to ease vehicles entry from one straight portion of a road to another while at the same time overcoming the effect of centrifugal force that tries to push the vehicle of the road.

Points around the transition curve (See Figure 12-8):

Straight to TS (Straight leading to start of the spiral curve)

TS (Tangent to Spiral) Start spiral

TS – SC. Spiral curve. ∞ to R (L)

SC (Spiral to Curve) Start circular

SC – CS. Circular curve, R constant (Shifted arc length S)

CS (Curve to Spiral) End circular

CS – ST. Spiral curve. R to ∞

ST (Spiral to Tangent) End spiral (L)

ST to straight (Straight continuing after the end of the spiral curve).

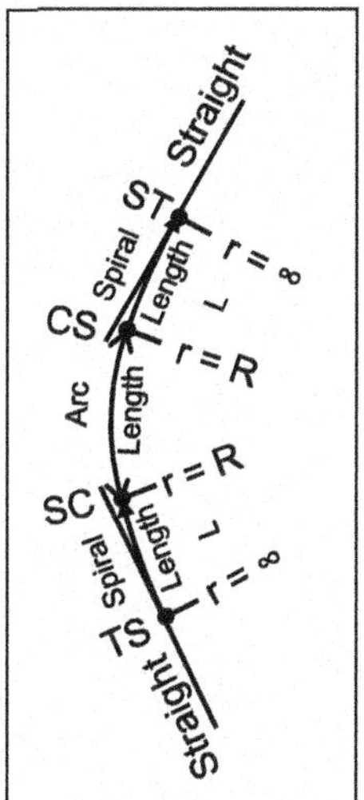

Figure 12-8
Transition curve

12.2.1 Introduction to Transition Curves

Much of the derivation of the transition curve presented here is based on the permitted use of the late J.A.L. Cavill's publication, "Survey Engineering, A Guide to First Principles."

The transition curve is a curve that joins a straight path to a circular curve in a gradual manner. The transition curve is part of a spiral. It is designed to provide uniform change of radial acceleration from zero, on the straight, to the design radial acceleration of the circular curve.

Because the change from a straight section of road directly to a curve may be too abrupt, a spiral curve is introduced between the straight and curve and again between the curve and straight. The spiral gradually introduces super-elevation from zero at the TS to its maximum at SC. The spiral from the CS to the ST reverses this procedure.

The distance between the TS and SC, and CS and TS, is called the **length,** L or L_p, of the transition curve.

The curve used to achieve this transition is the **Clothoid**, or Euler's Spiral, curve. This, , has the property that its curvature varies linearly with its arc length.

Clothoid Spiral Angle	$\Phi = \dfrac{L}{2R}\ (\text{radians})$	Eqn 12-6

12.2.2 Introducing the Transition Spiral

The transition curve is inserted into a space between the tangents of the circular curve and its arc by **shifting** the circular curve "inwards" away from the tangents. **The radius, *R*, of the curve does not change.** The centre of the curve moves away from the PI as shown in Figure 12-10.

In making such a shift the deflection angle, Δ, of the circular curve decreases by an amount **2Φ**, dependent on the length, ***L***, of the transition spiral.

The **arc length** of the shifted curve is then $L_{shifted\ arc} = S_\Delta R$.

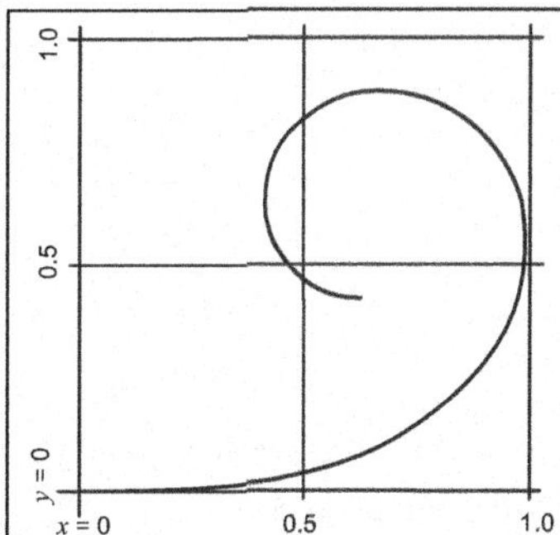

Figure 12-9 The Clothoid.

12.2.3 Equations for the Clothoid

The Clothoid, or Euler's spiral, is also called the Cornu curve. (See Figure 12-9).

Shifted circular deflection angle	$\Delta_S = (\Delta - 2\Phi)^r$	Eqn 12-7

The parametric form of the curve's equations is:

$x = \int_0^t \sin u^2 du \qquad y = \int_0^t \cos u^2 du$ These forms are derived from the Fresnel Integrals.

The clothoid spiral property is: *Curvature varies linearly with arc length.*

Thus, we will develop formulae that use the arc length (ℓ) along the transition curve between the start of the transition, TS and the start of the circular curve, SC. (See Figure 12-10).

At SC the arc length, ℓ, equals the **length** of the transition curve, L.

The equation for the Clothoid is:

$d\phi/d\ell = s$. The curvature (s) is linearly related to arc length ℓ.

ℓ any length around the spiral from TS

p any point on spiral at distance ℓ from TS

r radius of curvature at p

ϕ (phi), tangential angle (radians) between tangents through TS and p

L total **length** of spiral from TP to SC

R radius at L, radius of circular curve

Φ (phi), the **spiral angle** (radians) between tangents through TS and SC.

NOTE: both ϕ and Φ are measured in **RADIANS.**

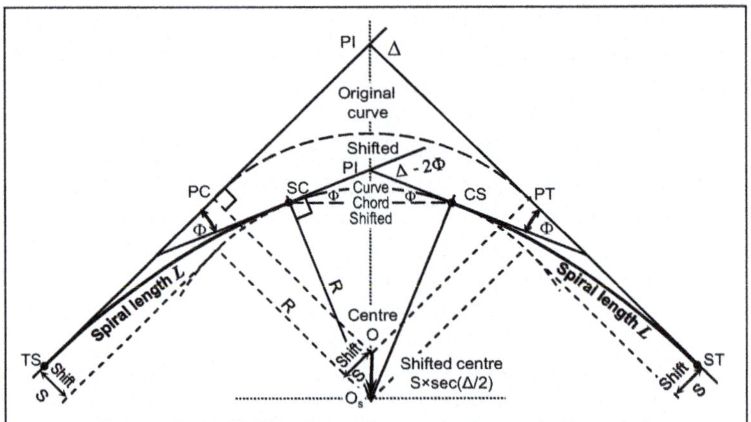

Figure 12-10 Shifting the radius centre to create the spiral.

12.2.4 Developing the Spiral Angle

Curvature increases with length (see Figure 12-11) so that:

$r\ell = RL$ = a constant, thus: $\quad \dfrac{1}{r} = \dfrac{\ell}{LR}$

At p, $\quad d\phi = \dfrac{d\ell}{r}$ so that $\quad \dfrac{d\phi}{d\ell} = \dfrac{\ell}{LR}$

and integrating: $\quad \phi = \dfrac{\ell^2}{2LR} + C$

When $\ell = 0$ then: $\phi = 0$ and $C = 0$ and

when $\ell = L$: then $\mathbf{\Phi = L/2R}$, (Eqn 12-6) the Spiral Angle in radians.

It follows that, for any point p on the spiral:

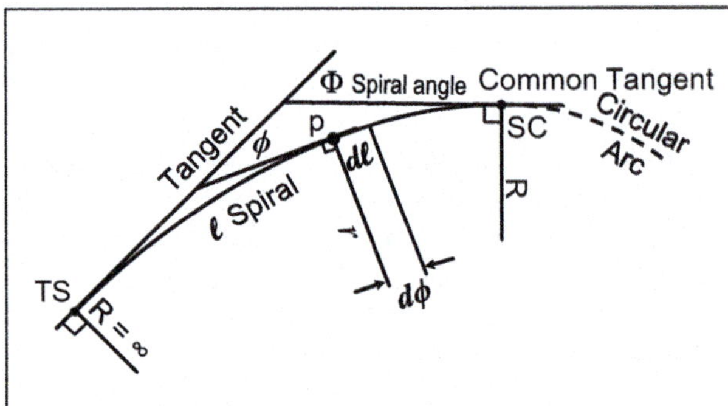

Figure 12-11 Developing the spiral angle.

	$\phi = \dfrac{\ell^2}{2LR}$ and, substituting $R = \dfrac{L}{2\phi} = \dfrac{\ell^2}{2L(L/2\phi)} = \dfrac{\ell^2}{2L^2/2\phi} \Rightarrow \phi = \left(\dfrac{\ell}{L}\right)^2 \Phi$	Eqn 12-8

12.2.5 Cartesian Coordinates of a Point on the Spiral

Cartesian coordinates of points along the spiral (see Figure 12-12), referenced to the tangent, can be developed from a power series.

In Figure 12-12 let $p(x, y)$ be any point on the spiral.
It is offset from the TS by:

x along the tangent

y perpendicular to the tangent.

Given: from § 12.2.4 and Eqn 12-8

$\mathbf{\Phi = L/2R}$, $\phi = \ell^2/2LR$, $\phi = (\ell/L)^2\Phi$

Near p, using *cos* and *sin* power series:

$$dx = \cos\phi\, d\ell = \left(1 - \frac{\phi^2}{2!} + \frac{\phi^4}{4!} - \frac{\phi^6}{6!} + \frac{\phi^8}{8!} - \cdots\right) d\ell$$

$$dy = \cos\phi\, d\ell = \left(1 - \frac{\phi^3}{3!} + \frac{\phi^5}{5!} - \frac{\phi^7}{7!} + \cdots\right) d\ell$$

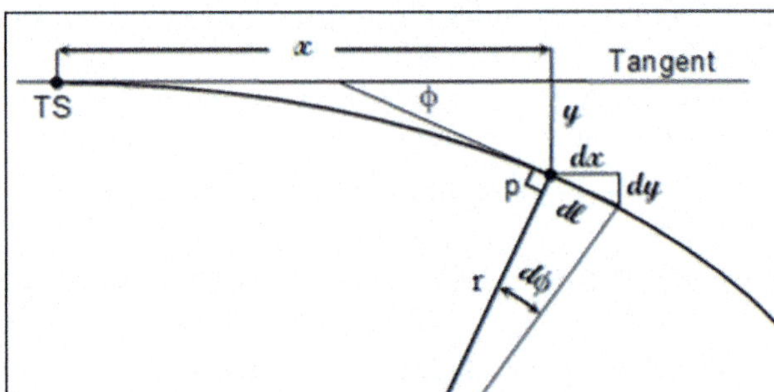

Figure 12-12 Coordinate points on the spiral.

By substituting $\phi = (\ell/L)^2\Phi$ in the above equations, integrating and then restoring ϕ the following expressions are derived:

$$x = \ell\left(1 - \frac{\phi^2}{10} + \frac{\phi^4}{216} - \frac{\phi^6}{9360} + \frac{\phi^8}{865540} - \cdots\right)$$

$$y = \ell\left(\frac{\phi}{3} - \frac{\phi^3}{42} + \frac{\phi^5}{1320} - \frac{\phi^7}{75600} + \cdots\right)$$

To find x and y at SC it is necessary to substitute Φ and L for ϕ and ℓ

	$x = \ell\left(1 - \frac{\Phi^2}{10} + \frac{\Phi^4}{216} - \cdots\right)$	Eqn 12-9
	$y = \ell\left(\frac{\Phi}{3} - \frac{\Phi^3}{42} + \cdots\right)$	Eqn 12-10

When the spiral length, L, and the radius, R, of the spiral curve are known the calculations of ϕ for various intermediate values of ℓ can be carried out.

12.2.5.1 Worked Example 12.2.

Given $L = 60$, $R = 150$ and $\ell = 25$: (25m from TS)

$$\phi = \frac{\ell^2}{2LR} = \frac{25^2}{(2\times60\times150)} = 0.03472^{radians}$$

The corresponding x and y are found:

$$x = 25\left(1 - \frac{0.03472^2}{10} + \frac{0.03472^4}{216} - \cdots\right)$$

$$= 25(1 - 0.000121 + 6.73^{-9} - \cdots) = 25\times0.99988$$

$x = 24.997$

$$y = 25\left(\frac{0.03472}{3} - \frac{0.03472^3}{42} + \cdots\right)$$

$$= 25(0.011574 - 9.97^{-7} + \cdots) = 25\times0.01573$$

$y = 0.289$

Generally, sufficient terms should be used in the determination of x and y so that the change in value is less than 1mm between successive terms.

The general formula for n terms is thus:

	$x = \ell\left(\sum_{n=0}^{\infty}(-1)^n \frac{\phi^{2n}}{(4n+1)(2n)!}\right)$	Eqn 12-11
	$y = \ell\left(\sum_{n=1}^{\infty}(-1)^{n-1} \frac{\phi^{2n-1}}{(4n-1)(2n-1)!}\right)$	Eqn 12-12

12.3 Worked Example 12.3. Design and set out of full transition curve.

The introduction of the transition spiral into a previously designed circular curve makes use of the original design parameters. This example provides a set of numbers that may exceed the minimum recommendations of Austroads (2021).

In fact, there are certain parameters where the combination of tangent deflection angle, Δ, speed, v, radius, R and radial acceleration, c that produce nonsensical results. There is no point in having a curve where the total spiral deflection, 2Φ, exceeds Δ. At best, if $2\Phi = \Delta$ then there is no circular curve. Just a transition from one straight to another.

Figure 12-13 is an existing circular curve that, for changed land planning purposes, has to incorporate a transition curve. The chainage of the IP and the tangent bearings of the two straights are known.

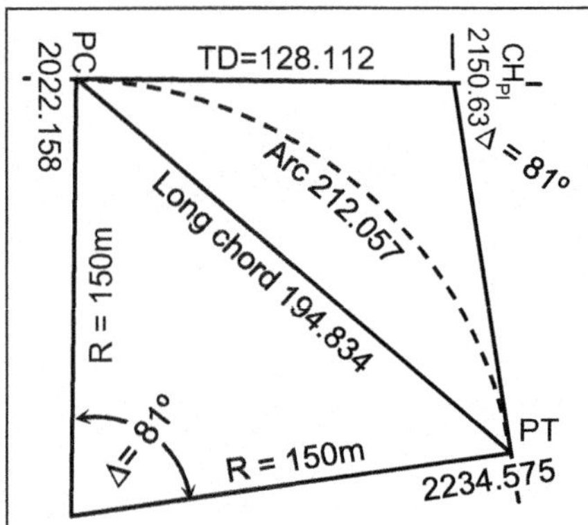

Figure 12-13 Existing circular curve

The basis for the design of spiral length is to provide drivers with a transition path accomplished between 2 and 3 seconds between the start of the spiral and the circular curve. A minimum transition time of 2 seconds allows drivers to anticipate a longer transition length. (Perco 2006). The design speed for the curve is known; thus the radius of the circular curve is determined from road design recommendations.

The design speed of 50Kmh^{-1} has a recommended minimum radius less than 60m (Austroads 2021 Table 7.6). Note that the design radius (150m) is about the recommended minimum at 70Kmh^{-1} maximum speed for this radius in undulating terrain, except for urban roads. The larger value for radius makes a better illustration of the computations.

Chainage IP	= 2150.63m	
Deflection angle, D	= 81°	
Design speed	= 50Kmh^{-1}	(minimum radius < 60m)
Radius R	= 150m.	

12.3.1 Length of Transition Spiral, L_p

The length of the transition spiral is calculated from:

Transition spiral length	$L = \dfrac{V^3}{cR}$	Eqn 12-13

where:

V in $m.s^{-1}$, speed
R in m, radius and
c in $m.s^{-3}$, radial acceleration.

c is an empirical figure for radial acceleration around the curve which does not result in discomfort for the human body.

c is generally 0.3 to a maximum of 0.6 except in mountainous areas where c_{max} is 0.9 for V_{max} 14m.sec^{-1}. (50Km/hr).

Transformed into Km/h the empirical formula becomes:

	$L = \dfrac{V^3}{3.6^3 cR} = \dfrac{V^3}{46.66cR} = \dfrac{0.0214V^3}{cR} \cong \dfrac{V^3}{47cR}$	Eqn 12-14

12.3.1.1 Calculate transition spiral length

At a design speed of 50Km/h, (14m.sec^{-1}) $c = 0.3$,
a circular curve is designed with radius $R = 150$m

The length of the transition spiral is thus:

$$L = \frac{V^3}{47cR} = \frac{50^3}{47 \times 0.3 \times 150} = \frac{125000}{2115} = 59.1\text{m}$$ using the approximation in Eqn 12-14.

The designer would then have the choice of using:

a) the calculated value
b) changing the value to, say, 60m for the design length
c) using a shorter value of 50m, with a consequent value of $c = 0.36$.

Choose:	$L = 60$m,
Design Radius:	$R = 150$m

This spiral length provides over 4 seconds for the drivers transition path at 50Kmh^{-1} and 3 seconds at 70Kmh^{-1}.

12.3.1.2 The spiral angle

The spiral angle Φ is calculated from Eqn 12-8:

Continuing the calculations for a spiral length:

$L = 60$
$R = 150$

$$\Phi = \frac{L}{2R} = \frac{60}{2\times150} = 0.2^R = 11°27'33''$$

12.3.1.3 Cartesian coordinates of spiral to curve point, SC, on the spiral

Cartesian coordinates of Spiral to Curve point, SC, along the spiral, referenced to the tangent, have been developed from a power series.

Referring to Figure 12-14,

To find X and Y at SC it is necessary to use Φ and L:

$L = 60$
$R = 150$
$\Phi = 0.2^R$
$\Phi = 11°27'33''$

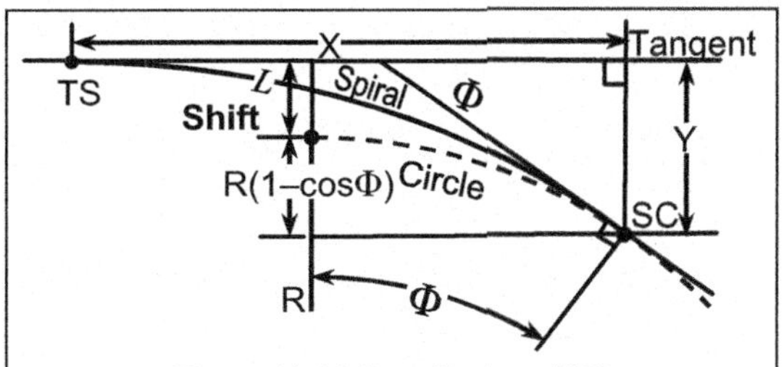

Figure 12-14 Coordinates of SC.

Φ radians in Eqn 12-9 and Eqn 12-10:

$$X = L\left(1-\frac{\Phi^2}{10}+\frac{\Phi^4}{216}-\ldots\right) = 60\left(1-\frac{0.2^2}{10}+\frac{0.2^4}{216}-\ldots\right) = 60(1-0.004+7.4^{-6})$$

$= 60\times(0.996)$ Note: 2nd term unnecessary.

X	$= 59.760$

$$Y = L\left(\frac{\Phi}{3}-\frac{\Phi^3}{42}-\ldots\right) = 60\left(\frac{0.2}{3}-\frac{0.2^3}{42}-\ldots\right) = 60(0.06667-0.00019)$$

$= 3.9886$ Note the second term is necessary (by a millimeter!).

Y	$= 3.989$

12.3.1.4 The shift distance

shows that:

Shift	$S = Y - R(1-\cos\Phi)$	Eqn 12-15

Where, from § 12.3.1.3:

Y= 3.989.

The mid ordinate, or crown distance (Chapter 8, § 8.9.1, Eqn 8-12)

$R(1-\cos\Phi) = 150\times(1-0.980067)$

$= 150\times0.019933 = 2.990$

Shift $S = 3.989 - 2.990 = 0.999 \quad = 0.999.$

The value for the shift can be found directly from the approximation:

Shift	$S = \dfrac{L^2}{24R}$	Eqn 12-16

Using the approximation (Eqn 12-16) and values from § 12.3.1.1:

$$\text{Shift} = \frac{L^2}{24R} = \frac{60^2}{24\times150} = \frac{3600}{3600}$$

Shift = 1.00

12.3.1.5 Derivation of the shift distance approximation

Shift $S = Y - R(1-\cos\Phi)$

By expanding cosΦ,

substituting $L/2R$ for Φ,
or $L/2\Phi$ for R where appropriate and
subtracting $R(1-\cos\Phi)$ from the series for Y it can be shown that:

$$\text{Shift} = \frac{L^2}{24R}\left(1-\frac{\Phi^2}{28}+\frac{\Phi^4}{1320}-\cdots\right) = 1\left(1-0.001429+1.26^{-6}\right)$$

$= 1\times0.99857 = 0.99986$ or 1.000

The approximation **L²/24R** (Eqn 12-16) finds the Shift sufficiently well from the Transition Length, L, and the Circular Radius, R, without using further terms.

12.3.1.6 Transition tangent distance; TS to PI. Tangent length.

Recall Tangent Distance (TD) for a circular curve (PC to PI): $TD = R\tan(\Delta/2)$ where Δ is the deflection angle between the two tangent straights.

By introducing a transition from the straight to the circular curve, the Tangent Length from the Transition Start (TS) to the Point of Intersection (PI) of the two curves will be significantly longer than the circular curve Tangent Distance from Tangent Point (TP) to the PI.

Continue with the problem already started:

Assume: $\Delta = 81^\circ$. (Existing circular curve)

$L = 60$m, given

$R = 150$m, given

$S = 1.000$m (§ 12.3.1.4)

$X = 59.760$m (§ 12.3.1.3)

$Y = 3.989$m (§ 12.3.1.3)

$\Phi = 0.2^R$ (§ 12.3.1.2)

$\Phi = 11^\circ 27' 33''$

From Figure 12-15 (not drawn to scale). It can be seen that:

Transition tangent Length (TL) from TS to PI;

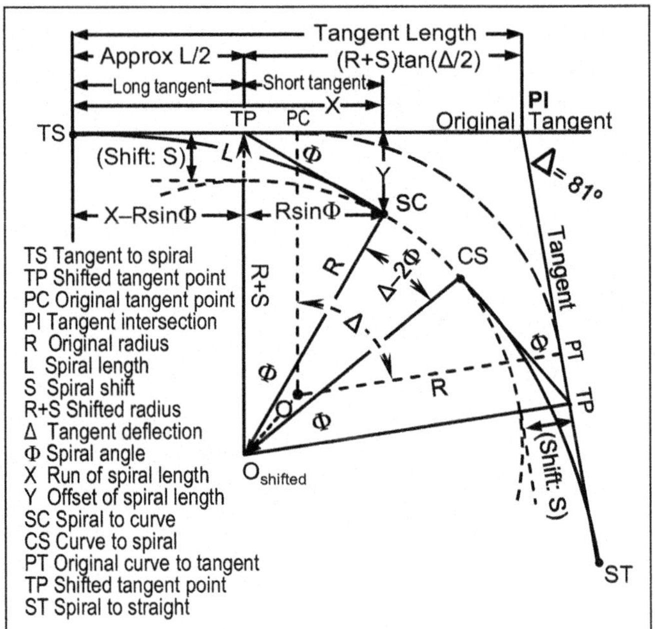

Figure 12-15 Tangent distances to PI.

$TL = (R+S)\tan\left(\dfrac{\Delta}{2}\right) + (X - R\sin\Phi)$	Eqn 12-17

$$(R+S)\tan\left(\frac{\Delta}{2}\right) = (150+1)\tan\left(\frac{81^\circ}{2}\right)$$

$$= 151 \times 0.8451 = 128.966\text{m}$$

$$(X - R\sin\Phi) \quad = 59.760 - 150 \times \sin(11^\circ 27' 33'')$$

$$= 59.970 - 150 \times 0.1987 = 59.970 - 29.800 = 29.960$$

$$TL \quad = 128.966 + 29.960$$

TL = 158.926m

Note: a **check** of the Transition Tangent Distance can be made using approximations:

TL = TD (circular curve) + shift + half Transition Length (L/2)

= 150×Tan(40.5°) + 1.0 + 60/2 = 128.11 + 1 + 30

= 159.11 approximate

12.3.1.7 Circular distance: SC to CS

From Figure 12-15 (not drawn to scale) it can be seen that:

Circular Distance, Arc:

$$\text{Arc} = R(\Delta - 2R) \text{ in radians}$$

$$= 150 \times ((81 - 2 \times (11^\circ 27' 33''))) = 150 \times (81 - 22^\circ 55' 06'')$$

$$= 150 \times (58^\circ 04' 54'') = 150 \times 1.10372$$

$$= 152.058$$

Arc = 152.058

12.3.1.8 Tangential components of transition curve: TS to SC

Setting out the basic points of the transition curve requires:

- Deflection of chord from TS to SC: the Tangential angle δ
- Distance from TS to SC: the Long Chord
- Distance from TS to the intercept of the common tangent at SC with the straight from TS to PI: the Long Tangent
- Distance from the straight/common tangent intercept to SC: the Short Tangent.

From Figure 12-16:

Tangential angle $\delta = \arctan((Y/X)$

Long Chord = $\sqrt{(X^2 + Y^2)}$

Long Tangent= X – Y×cotΦ

Short Tangent= Y×cosecΦ

Again, taking calculated values (see Figure 12-17):

X = 59.760m (§ 12.3.1.3)

Y = 3.989m (§ 12.3.1.3)

$\Phi = 0.2^R$ (§ 12.3.1.2) $\Phi = 11°27'33''$

$$\text{Tangential angle}\ \delta = \arctan(Y/X)$$

$$\delta = \arctan(3.989/59.760) = \arctan(0.06675) = 3.8188°$$

$$\boxed{\delta = 3°49'08''}$$

$$\text{Long chord} = \sqrt{X^2 + Y^2}$$

$$= \sqrt{59.76^2 + 3.989^2} = \sqrt{3587.170}$$

$$\boxed{\text{Long Chord} = 59.893}$$

$$\text{Long tangent} = X - Y\cot\Phi = 59.76 - \frac{3.989}{\tan(11°27'33'')}$$

$$= 59.76 - \frac{3.989}{0.20271} = 59.76 - 19.768$$

$$\boxed{\text{Long Tangent} = 39.992}$$

$$\text{Short tangent} = Y\csc\Phi \quad \left(\csc = \frac{1}{\sin}\right)$$

$$= \frac{3.989}{\sin(11°27'33'')} = \frac{3.989}{0.19867}$$

$$\boxed{\text{Short Tangent} = 20.078}$$

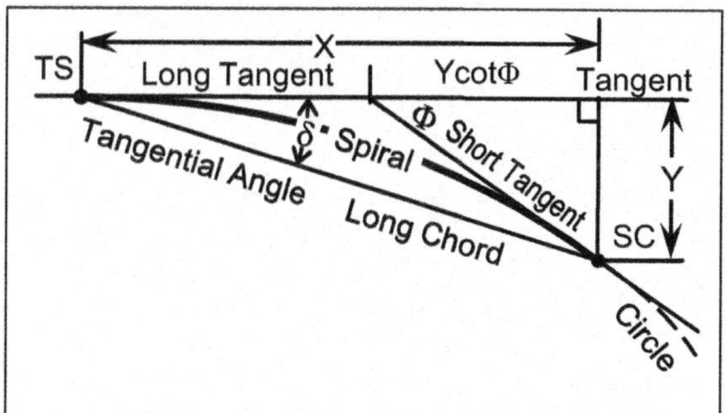

Figure 12-16 Tangential components.

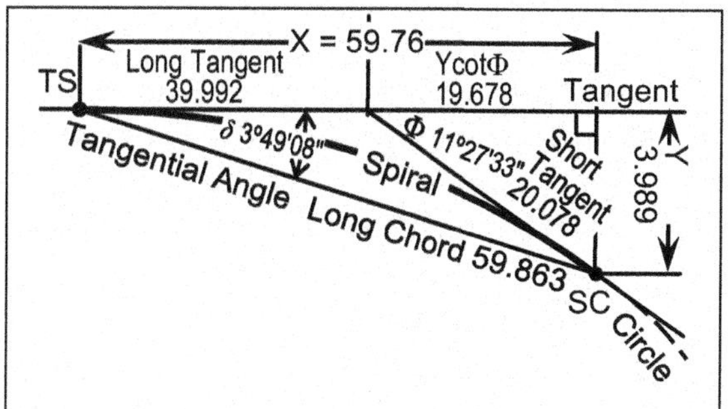

Figure 12-17 Tangential solution.

12.3.1.9 Chainages from TS to ST

The chainage around the curve (sometimes called the "through chainage") generally starts with the definition of the chainage of the PI. From this point the chainage of the TS is determined by subtracting the Tangent Distance.

Note that the two spirals are of equal length. The arc is not to scale.

The through chainages of the other points are then found by adding the lengths of the respective curves.

Continuing, assign a value for the through chainage of the PI. (see Figure 12-18)

Chainage PI = 2150.63.

PI	= 2150.63	2150.63	PI
– TL	= –158.926		
TS	= 1991.704	1991.704	TS
+ L	= +60.000		
SC	= 2051.704	2051.704	SC
+ Arc	= +152.058		
CS	= 2203.762	2203.762	SC
+ L	= +60.000		
ST	= 2263.762	2263.762	ST

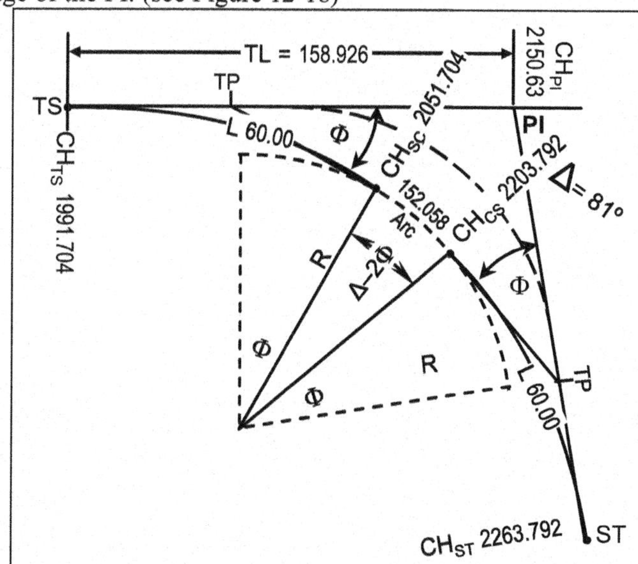

Figure 12-18 Chainages TS to ST.

12.3.2 Set out Calculations for the Spiral and the Arc

Practical set out methods for pegging the two transition spirals and the circular arc are covered in appropriate survey practical sessions.

Modern mapping systems, such as Civilcad, produce coordinates in accordance with the road design. Combined with control points, the centre line coordinates can be set out using by radiation using a Total Station.

However, it is possible to provide some initial set-out data which can be used in conjunction with setting out the TS, SC, CS and ST points.

In general, by setting up on the PI the TS can be set by measuring the Tangent Distance (TD) back along the straight. The ST is then set by measuring the TD forward, through the deflection angle, down the connecting straight.

12.3.2.1 Deflection calculations for the spiral and the arc

The Length of the spiral, $L = 60$m. Since the circular Arc Length (PC to TP) ≈ 152m (§12.3.1.7) it would make sense to use arc lengths of $\ell = 150$m. The extra 2m to finish the circular curve are immaterial.

12.3.2.1.1 Spiral arc

For the spirals, prepare a table of deflection angles and cumulative chords using the power series in § 12.2.5. See § 12.3.1.1.

A table of y_n and x_n allows the use of the rectangular to polar function to find deflection and cumulative chord distance (***d***). (See Table 12-1).

Modifying Eqn 12-8 and Eqn 12-9 for the power series to generate x and y is illustrated in § 12.2.5.1.

$$x = \ell\left(1 - \frac{\phi^2}{10} + \frac{\phi^4}{216} - \dots\right) \quad y = \ell\left(\frac{\phi}{3} - \frac{\phi^3}{42} + \dots\right)$$

Table 12-1 Spiral arc.

ℓ	$\phi = \ell^2/2LR$ radians	x	y	δ ° ′ ″	Cumulative chord d
0	0.0	0	0	0 00 00	0.000
15	0.0125	15.000	0.062	0 14 19	15.000
30	0.05	29.992	0.500	0 57 18	29.997
45	0.1125	44.943	1.686	2 08 54	44.975
60	0.2	59.760	3.989	3 49 06	59.893

12.3.2.1.2 Circular arc

For the arc, prepare a table of deflection angles and cumulative chords (ℓ) and deflection angles (δ) using the formulae in §12.2.4 and § 12.2.5 (PC to TP). (See Table 12-2).

Note: A difference in terminology between the symbols used for a clothoid and for a circle.

The clothoid parameters used are spiral length (ℓ) and cumulative chords (***d***).

The circle has used arc distance (S) and cumulative chords (ℓ). $\delta = \Delta/2$, both in radians.

Table 12-2 Circular arc.

Arc Dist. S	Deflection (δ) ° ′ ″	Cumulative Chord ℓ	Run = x (dN) $\ell\cos(\delta)$	Offset = y (dE) $\ell\sin(\delta)$
0	0 00 00	0.000	0.00	0.00
15	2 51 53	14.994	14.975	0.749
30	5 43 46	29.950	29.800	2.990
45	8 35 40	44.831	44.328	6.700
60	11 27 33	59.601	58.413	11.841
75	14 19 26	74.221	71.914	18.363
90	17 11 19	88.656	84.696	26.200
105	20 03 13	102.869	96.633	35.274
120	22 55 06	116.826	107.603	45.494
135	25 46 59	130.490	117.499	56.759
150	28 38 52	143.828	126.211	68.955
152.058	29 02 27	145.630	127.321	70.694

12.3.2.2 Setout Calculations for the shifted PI and shifted crown

The PI for the shifted arc, in the direction of the curve centre point, can also be set out. Referring to § 12.3.1.4 where S = shift = 1.000.

1. Shifted secant = S(sec(Δ/2)) = 1.000/Cos(81°/2) (see Figure 12-19)
   = 1.000/(0.760406)

[1]	Shifted Secant = 1.315

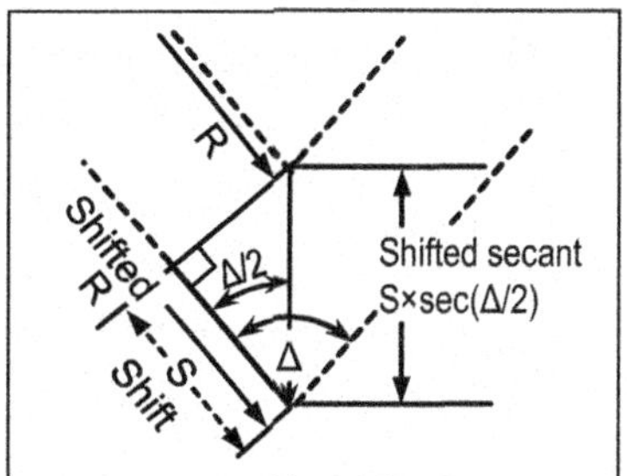

Figure 12-19 The shifted secant.

2. The Crown Secant for the original curve, (see Figure 12-20)
   = R×(sec(Δ/2) – 1) = 150×((1/(cos(40.5°) – 1)
   = 150×0.315087

[2]	Crown secant (orig) = 47.263

3. The Crown Secant for the shifted curve
   = R×(sec((Δ–2Φ)/2)–1) = 150×((1/(cos(29°02′27″))–1))
   = 150×0.143806

[3]	Crown secant (shifted) = 21.571

4. PI to $PI_{shifted}$
   = Crown Secant [2] + Shifted Secant [1] – Shifted Crown Secant [3]
   = 47.263 + 1.315 – 21.571

[4]	PI to $PI_{shifted}$ = 27.007

5. PI to $Crown_{shifted}$ This is the centre of the alignment of the shifted circular curve, E.
   = Crown Secant [2] + Shifted Secant [1]. The same as PI to $PI_{shifted}$ [4] + shifted crown secant [3]
   = 47.263 + 1.315 or 27.007 + 21.571

[5]	PI to $Crown_{shifted}$ = 48.578

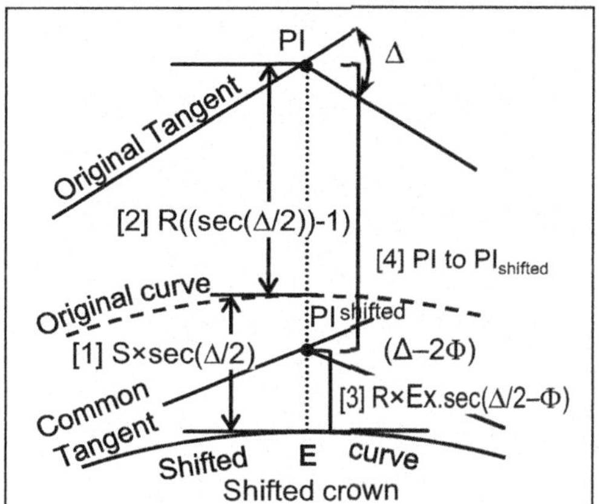

Figure 12-20 Shifted PI and shifted crown.

12.3.3 Diagram of a Circular Arc with Two Spirals

This general diagram, , was drawn with the shift, S, greatly exaggerated. Refer to Figure 12-20 for the relationship between the PI and shifted PI and Figure 12-19 for the Secant shift between the original and shifted centres.

Revise Chapter 8, page 113, § 8.9.1, Figure 8-21 and formula Eqn 8-10 and Eqn 8-11 to compare terminology.
Secant distance is from the PI to the centre of the curve, = R×sec(Δ/2). (Eqn 8-10)
Crown (external) secant distance is the distance from the PI to the crown, = R×ex.sec(Δ/2)
= R×(sec(Δ/2)–1) (Eqn 8-11)

Some texts refer to the ***exsec*** function, an unfamiliar term. Exsec(α) = (sec(α) – 1) = ((1/cos(α)) – 1).

12.4 Worked Example 12.4. Circular Curve with Transition Spirals

This Worked Example illustrates a problem set for a University Engineering Surveying unit. The nomenclature will follow the format use in the unit text, "Surveying for Engineers", Uren, J and W.F. Price, The McMillan Press Ltd, Fifth Edition. Chapter 13, Transition Curves.

Two straights, intersecting at IP, having a deflection angle of 14°26′28″, are to be joined by a composite curve consisting of a central circular arc and two transition curves of equal length. Figure 12-21 shows the deflection angle, Δ, greatly exaggerated.

The exercise will then continue with the application of super elevation in to, around and out of the compound curve.

Chainage at IP	= 461.34m
E_I	789.14
N_I	863.72
Bearing T to I, θ_{TI}	48°16′25″
Deflection, Δ	= 14°28′26″
Design speed	= 85Km/h

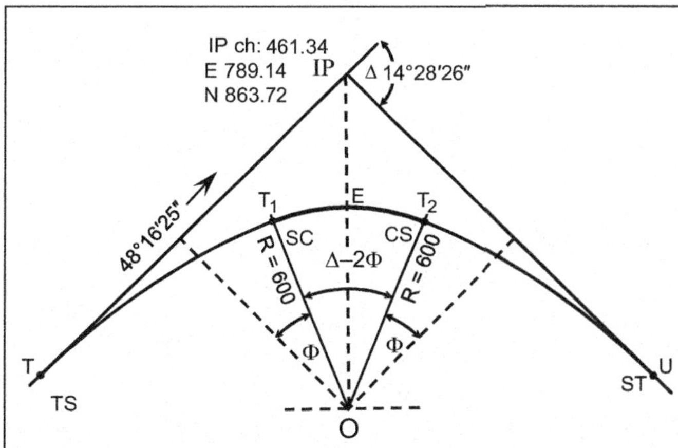

Figure 12-21 Worked Example 12.4

Circular radius = 600m
Radial acceleration (c) =0.3msec^{-3}
Crossfall = (–0.03) –3%
Superelevation = (+0.06) +6%
The curve is to be pegged at every even 20m chainage.
Curve superelevation is examined in Chapter 13.

12.5 Step by Step Solution:

12.5.1 Transition Curve Length

1) Find length of transition curves T – T$_1$, T$_2$ – U. Eqn 12-14, § 12.3.1.

$$L = \frac{V^3}{3.6^3 cR} = \frac{85^3}{46.66 \times 0.03 \times 600} = \frac{614125}{8398.08} = 73.127$$

L = 75m (5m rounding AGRD03-16 recommendation, §7.5.4)

12.5.2 Spiral angle

2) Find the Spiral **angle**, $\Phi = \dfrac{L}{2R}$ Eqn 12-6, § 12.3.1.2

= 75/(2·600) = 0.0625^r = 3°34′52″

Φ = 3°34′52″

12.5.3 Length of Circular Curve

3) Find length of the circular curve T$_1$ - T$_2$.

a) Shifted curve deflection angle = Δ – 2Φ § 12.3.1.7

Δ$_S$ = 14°28′26″ – 2×(3°34′52″) = 7°18′43″ = 0.127617^R

Δ$_S$ = 7°18′43″

b) Arc length, S, = RΔ$_S$^{r}

= 600·(0.127617) = 76.570m

Arc, S = 76.57m

12.5.4 Tangent Length

4) Find tangent length (distance) T – IP. § 12.3.1.6

TL = (R+S)tan(Δ/2) + (X–RsinΦ)

Three **components** are needed:

a) Spiral angle, Φ = 3°34′52″ (0.0625^R) § 12.5.2

b) X and Y Cartesian coordinate of T$_1$. § 12.3.1.3

Using radians for Φ $$X = L\left(1 - \frac{\Phi^2}{10} + \frac{\Phi^4}{216} - \ldots\right)$$

= 75×(1–(0.0625)^2/10+(0.0625)^4/216 – …) = 75×(1 – 0.000391 + 7.06^{-8} –…)

= 75 × 0.99961 = 74.971 Note: 2nd term unnecessary.

X = 74.971

$$Y = L\left(\frac{\Phi}{3} - \frac{\Phi^3}{42} + \ldots\right)$$ Y coordinate for Long Chord

= 75×(0.0625/3 – (0.0625)^3/42 +…) = 75×(0.020833 – 0.000006 + …)

= 75 × 0.020833 = 1.561 Note: 2nd term unnecessary.

Y = 1.561

c) Shift, S = L^2/24R § 12.3.1.5

= 75^2/(24×600) = 0.391m

Shift, S = 0.391

Combine the 3 components, and Δ = 14°26′28″, to find Tangent Length: § 12.3.1.6

TL = (R+S)tan(Δ/2) + (X–RsinΦ)

(R+S)tan(Δ/2) = (600 + 0.391)tan(14°28′26″/2) = 600.391×(0.126985) = 76.240m

(X–RsinΦ) = 74.971 – 600×sin(3°34′52″) = 74.971 – 37.476 = 37.495m

TL = 76.240 + 37.495 = 113.735

TL = 113.735

12.5.5 Find Through Chainages:

See Figure 12-22

1) Start of tangent spiral, T, from IP.
   T = IP – TL
   461.34 – 113.735
   T = 347.605
2) End of first spiral, T_1, from T
   $T_1 = T + L$
   347.605 + 75
   T_1 = 422.605
3) Chainage of shifted crown secant at E.
   $E = T_1 + S/2$
   422.605 + (76.57/2)
   E = 460.390

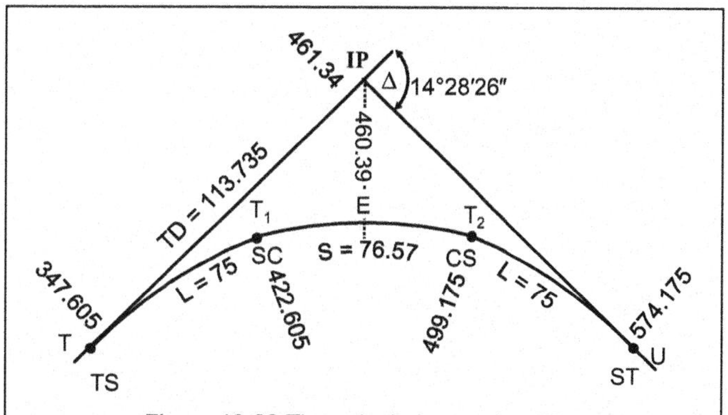

Figure 12-22 Through chainages. L = 75m

4) Find through chainage of T_2 around the circular curve
   $T_2 = T_1 + S$
   422.605 + 76.570
   T_2 = 499.175
5) Find through chainage of U along the second symmetrical spiral, length L
   $U = T_2 + L$
   499.175 + 75
   U = 574.175

12.5.6 Determine 20m Chainage Points on First Spiral:

Prepare a table of chords and deflection angles for set out of the spiral from T to T_1. (See Figure 12-23)
Using the power series in § 12.2.5
$\phi_{CH} = \ell_{CH}^2/2LR$ and ϕ is in radians

$$x = \ell\left(1 - \frac{\varphi^2}{10} + \frac{\varphi^4}{216} - \ldots\right) \quad y = \ell\left(\frac{\varphi}{3} - \frac{\varphi^3}{42} + \cdots\right)$$

$\phi_{CH} = \arctan(y_{CH}/x_{CH})$.
$d_{CH} = \sqrt{(y^2 + x^2)}$

A table of x_{CH} and y_{CH} (see Table 12-3) will allow the use of the rectangular to polar function to find deflection angle or tangential angle (δ) and cumulative chord distance (d).

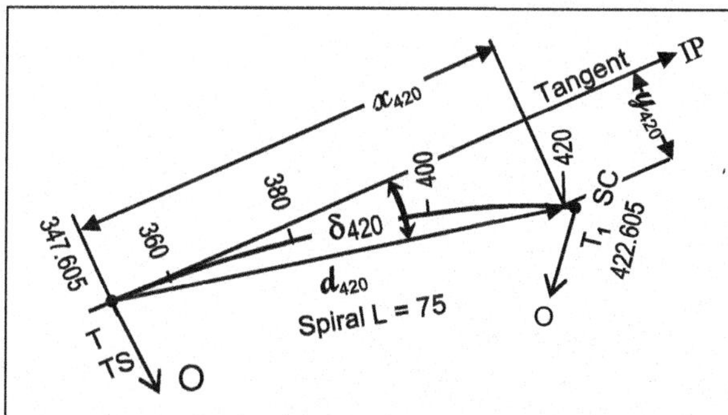

Figure 12-23 Components of CH420.

12.5.6.1 Chainage points on spiral curve, T to T_1

Table 12-3 Deflection angle and chord. T to T_1

Stn	Chain	Arc length	Dist. ℓ	$\phi = \ell^2/2LR$ radians	x run	y rise	δ ° ′ ″	Cumulative Chord d
T	347.605		0.00	0.0	0.00	0.00	0 00 00	0.00
360	360	12.396	12.395	0.001707	12.395	0.007	0 01 57	12.395
380	380	20	32.395	0.011661	32.395	0.126	0 13 22	32.395
400	400	20	52.395	0.030503	52.391	0.533	0 34 57	52.393
420	420	20	72.395	0.058234	72.371	1.405	1 06 44	72.385
T_1	422.605	2.604	75	0.0625	74.971	1.562	1 11 37	74.987

12.5.7 Determine 20m Chainage Points on Circular Curve, $T_1 - T_2$

Prepare a table of chords and deflection angles for set out of the circular curve from T_1 in direction of the common tangent. See § 12.3.2.1.2. (See Table 12-4)

Review formula: s_{CH} is the arc distance from SC.

$$\Delta_{CH} = \frac{s_{CH}}{R}. \text{ Thus } \delta_{CH} = \frac{\Delta_{CH}}{2}$$

$$\ell_{CH} = 2R\sin(\delta_{CH})$$

$$x_{CH} = \ell\sin\delta, y_{CH} = \ell\cos\delta$$

Values x and y are shown in , but do not figure in coordinate creation for set-out. The x, y components are aligned with the common tangent.

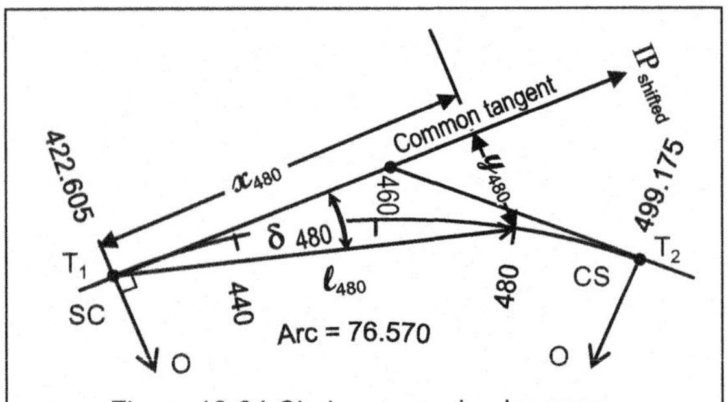

Figure 12-24 Chainage on circular curve.

They would have to be rotated to the bearing of the common tangent to produce coordinates.

Table 12-4 Points on circular curve.

Stn	Chain	Arc Dist. s	Deflection (δ=Δ/2) ° ′ ″	Cumulative Chord ℓ	Run = x $\ell\cos(\delta)$	Offset = y $\ell\sin(\delta)$
T_1	422.605	0.00	0 00 00	0.000	0.00	0.00
440	440	17.396	0 52 22	17.395	17.393	0.252
460	460	37.396	1 47 10	37.340	37.372	1.165
480	480	57.396	2 44 26	57.374	57.308	2.743
T_2	499.175	76.570	3 39 21	76.518	76.362	4.879

12.5.7.1 Chainage points on circular curve T_1 to T_2

Note that the deflection angle to T_2, $\delta_{T2} = ½(\Delta - 2\Phi)$, i.e. half the shifted deflection angle.
Thus, calculation check correct. $(\Delta - 2\Phi) = 7°18'43''$ § 12.5.3

12.5.8 Determine 20m Chainage Points on Second Spiral, T_2 to U:

See Figure 12-25.

Prepare a table of chords and deflection angles for set-out of the spiral from U in direction T2.

Using the power series in § 12.2.5

$\phi_{CH}=\ell_{CH}^2/2LR$ and ϕ is in radians

$$x = \ell\left(1 - \frac{\varphi^2}{10} + \frac{\varphi^4}{216} - \ldots\right) \quad y = \ell\left(\frac{\varphi}{3} - \frac{\varphi^3}{42} + \cdots\right)$$

$\delta_{CH} = \arctan(y_{CH}/x_{CH})$.

$d_{CH} = \sqrt{(y^2 + x^2)}$

It is necessary to set the second spiral out from the end of the spiral, U, back towards the circular curve, T_2. You cannot calculate **from** T_2 to U.

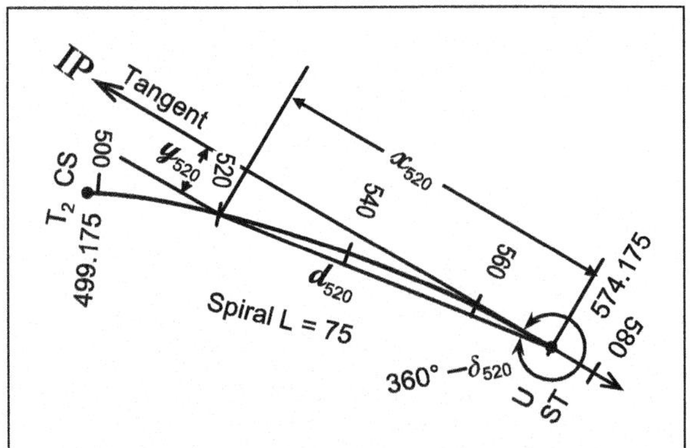

Figure 12-25 Components of CH520.

The sign of y, in this case, is negative as it is in the 4th quadrant with respect to the direction from U to IP. The tangential angles are thus negative and must be **subtracted** from the bearing U to I. See Table 12-5.

12.5.8.1 Chainage points on spiral curve T_2 to U

Table 12-5 Deflection angle and chord. U to T2.

Stn	Chain	Arc length	Dist. ℓ	$\phi = \ell^2/2LR$ radians	Run = x $\ell\cos(\delta)$	Offset = $-y$ $\ell\sin(\delta)$	δ ° ′ ″	Cumulative Chord ℓ
U	574.175		0.00	0.0	0.00	0.00	0 00 00	0.00
560	560	14.174	14.174	0.001928	13.174	−0.008	−0 02 13	13.174
540	540	20	34.174	0.012228	33.174	−0.135	−0 14 01	33.174
520	520	20	54.174	0.032609	53.169	−0.557	−0 36 00	53.172
500	500	20	74.174	0.06113				
T_2	499.175	20.826	75	0.0625	74.971	−1.562	−1 11 38	74.987

12.5.9 Set out of Common Tangents, T, T_1, T_2 and U:

Calculating a set out for these 4 points will allow a check of the data before commitment to a field trip. Combined with the IP, a traverse can be checked for a misclose.

The Point of Intersection of the two straights at IP has coordinates:

E_I 789.14
N_I 863.72

The bearing of straight from T to IP is:

θ_{TI} 48°16′25″

12.5.9.1 Calculate the bearings of the tangents and chords:

See Figure 12-26.

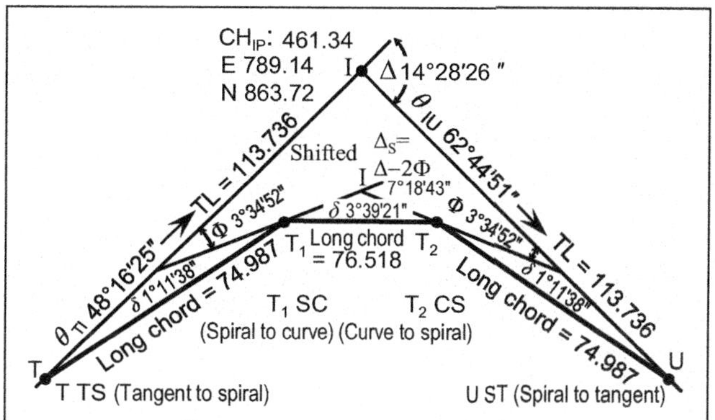

Figure 12-26 Summary of angles.

1. Bearing TI	48°16′25″	
+ δ spiral 1	1°11′37″	
2. Brg T – T_1	**49°28′02″**	
+ (Φ – δ_{spiral}) (3°34′52″ – 1°11′37″)	2°23′15″	
3. Brg common tangent at T_1	**51°51′17″**	
Check Brg TI + Φ = 48°16′25″ + 3°34′52″	= 51°51′17″	
+ δ circular arc	3°39′21″	
4. Brg T_1 – T_2	**55°30′38″**	
+ δ circular arc	3°39′22″	
5. Brg common tangent at T_2	**59°10′00″**	
Brg T_1 + (Δ – 2Φ) = 51°51′17″ + 7°18′43″	= 59°10′00″	
+ (Φ– δ_{spiral}) (3°34′52″ – 1°11′37″)	2°23′15″	
6. Brg T_2 – U	**61°33′15″**	
+ δ spiral 2	1°11′36″	
7. Brg I – U	**62°44′51″**	
CHECKS:		
Brg TI	48°16′25″	
+ Δ IP	14°28′26″	
1. Brg I – U	**62°44′51″**	Checks as reciprocal of Brg UI
BRG TI	48°16′25″	
+ Φ spiral angle 1	3°34′52″	
2. Brg common tangent T_1	**51°51′17″**	Checks as Brg CT at T_1
Brg TI	48°16′25″	
+ Δ IP	14°28′26″	
3. Brg I – U	**62°44′51″**	
– Φ spiral angle 2	3°34′52″	
4. Brg common tangent T_2	**59°09′59″**	Checks as Brg CT at T_2 (1″)

Calculate the bearing of the shifted crown secant from the IP

Intercept angle, ∠IP 180 – Δ = 180 – 14°28′26″ = 165°31′34″

Brg IP - T	228°16′25″
– ∠IP/2	82°45′47″
Brg IP - E	**145°30′38″**

Exsec distance (crown distance) to shifted crown, (R+S)sec(Δ/2) – R = 600.391/cos((14°28′26″)/2) – 600

Distance Shifted IP to E **5.212**

12.5.9.2 Calculate a traverse sheet: I, T, T1, T2, U, I

Table 12-6 Traverse sheet closure.

Stn	Brg ° ′ ″	Dist	dE	dN	E	N
IP					789.140	863.720
T	228 16 25	113.736	–84.885	–75.700	704.255	788.020
T1	49 28 02	74.987	56.993	48.733	761.248	836.753
T2	55 30 38	76.518	63.068	43.329	824.316	880.082
U	61 33 15	74.987	65.934	35.718	890.250	915.800
IP	242 44 51	113.736	–101.111	–52.081	789.139	863.719

The traverse (Table 12-6) closes within 0.002m. If distance to 3d.p. is used the close is exact.

If it's not checked it's wrong.

12.5.10 Calculation of Chainage Station Coordinates:

Now that the traverse has been proved, the intermediate chainages can be introduced to a table to produce coordinated for each point.

One technique is to use a join from each of the transition points.

12.5.10.1 For the first spiral, T to T_1

Table 12-7 Coordinates T to T1. Radiations from T.

a) Use the coordinates of T calculated in the traverse.
b) From the TS at T, to each chainage, add the deflections to the initial bearing, T to IP.
c) Using the tangential distance, calculate the coordinates of each chainage, adding the dEs and dNs to the initial coordinates of T. Coordinates are radiated from T.
d) Check calculated coordinates of T_1 agree with the traverse values.

Deflection angle and distance come from Table 12-3 (deflection and chord length).

Table 12-7 Coordinates T to T1.

Stn	Defl'n ° ′ ″	Brg ° ′ ″	Dist	dE	dN	E	N
T	Brg to I	48 16 25	0.00			704.255	788.020
360	0 01 56	48 18 22	12.395	9.256	8.245	713.511	796.265
380	0 13 22	48 29 47	32.395	24.261	21.467	728.517	809.488
400	0 34 58	48 51 22	52.393	39.455	34.472	743.711	822.493
420	1 06 45	49 23 09	72.385	54.948	47.120	759.203	835.140
T_1	1 11 37	49 28 02	74.987	56.993	48.733	761.248	836.753

12.5.10.2 For the circular arc, T_1 to T_2

Table 12-8 Coordinates T1 to T2. Radiations from T1.

a) Use the coordinates of T_1 calculated in the traverse.
b) From the SC at T_1, to each chainage, add the deflections to the initial bearing of the common tangent.
c) Using the chord distance, calculate the coordinates of each chainage, adding the dEs and dNs to the initial coordinates of T_1.
d) Check calculated coordinates of T_2 agree with the traverse values.

Deflection angle and distance come from Table 12-4 (deflection and chord length).

(i) The shifted crown (E) can be set out from the IP

Table 12-8 Coordinates T1 to T2.

Stn	Defl'n ° ′ ″	Brg ° ′ ″	Dist	dE	dN	E	N
T_1	Brg CT	51 51 17	0.000			761.248	836.753
440	0 52 22	52 43 39	17.395	13.834	10.545	775.083	847.298
460	1 47 10	53 38 27	37.389	30.110	22.166	791.358	858.920
480	2 44 26	54 35 43	57.374	46.764	33.239	808.012	869.993
T_2	3 39 21	55 30 38	76.518	63.069	43.329	824.317	880.082

12.5.10.3 For the second spiral, T_2 to U

Table 12-9 Coordinates T2 to U. Rearranged in chainage order after calculation.

Note that you work backwards from U to T_2.

a) Use the coordinates of U calculated in the traverse. Radiations from **U**.
b) From the ST at U, to each chainage, **add** the (negative) deflections to the initial bearing, U to IP.
c) Using the tangential distance, calculate the coordinates of each chainage, adding the dEs and dNs to the initial coordinates of U.
d) Check calculated coordinates of T_2 agree with the traverse values.

Table 12-9 Coordinates T2 to U. Rearranged in chainage order after calculation.

Stn	Defl'n ° ′ ″	Brg ° ′ ″	Dist	dE	dN	E	N
T_2	–1 11 37	241 33 14	74.987	65.933	35.719	824.317	880.082
500	–1 10 03	241 36 40	74.162	65.225	35.296	825.026	880.505
520	–0 37 22	242 07 29	54.172	47.886	25.328	842.364	890.473
540	–0 14 52	242 29 59	34.175	30.313	15.780	859.937	900.021
560	–0 02 33	242 42 18	14.175	12.596	6.500	877.654	909.301
U	Brg to I	242 44 51	0.000	0.000	0.000	890.250	915.801

Deflection angle and distance come from Table 12-5 (deflection and chord length).

12.5.11 Tangent runout $T_{ro.}$

The transition curve has been calculated from the transition to spiral (TS) point to the spiral to curve (SC) point, around the shifted arc distance, S, to the curve to spiral (CS) point and ending at the spiral to transition (ST) point. The start and end of the spiral are defined as the point at which the "outside" crossfall has risen from the normal tangent straight crossfall, 2 – 3%, to 0% (level) at the start of the spiral. (Figure 13-6 and Figure 13-7, page 207).

This spiral length is variously called spiral distance L_p or spiral runout, S_{ro} as calculated in, e.g., § 12.5.1 as a rounded value of 75m.

To maintain a constant rate of rotation the length of the tangent runout should be proportional to the length of the transition spiral and so we use the normal and superelevated cross falls. V and ω are not involved.

Since we have already designed L_p, or S_{ro} then T_{ro}, proportional to $L_p \Rightarrow T_{ro} = L_p\left(\frac{-e_1}{e_2}\right)$

12.5.12 Length of Superelevation Development, L_e

The Length of Superelevation Development, L_e, is the sum of the tangent runout and the spiral length and is the distance from SSD to CS and SC to ESD. ($L_e = T_{ro} + S_{ro}$).

For example, taking the design parameters for the preceding curve, $e_1 = -3\%$, $e_2 = +6\%$ and $L_p = 75$m.

$T_{ro} = L_p\left(\frac{-e_1}{e_2}\right) = 75\left(\frac{3}{6}\right) = 37.5$. This is logical as the straight has to rise from –3% before the TS and rise another 6% between TS and SC.

$L_e = 75 + 37.5 = 112.5$m. This could be rounded as required.

From chainage SSD = $CH_T - T_{ro}$ = 347.604 – 37.5 = 310.104, ESD = 573.174 + 37.5 = 611.674

From T: E_{SSD} = 704.225 – 27.987 = 676.268, N_{SSD} = 788.020 – 24.959 = 763.061.

From G: Brg 201°26′09″, distance 140.857

From U: E_{ESD} = 890.250 +33.337 = 923.587, N_{ESD} = 915.801 + 17.172 = 932.973.

From G: Brg 72°48′18″, distance 199.848

12.5.13 Set out of Chainage Coordinates from Other Control Points

Figure 12-27 represents the composite curve. Its associated control points for set-out are:

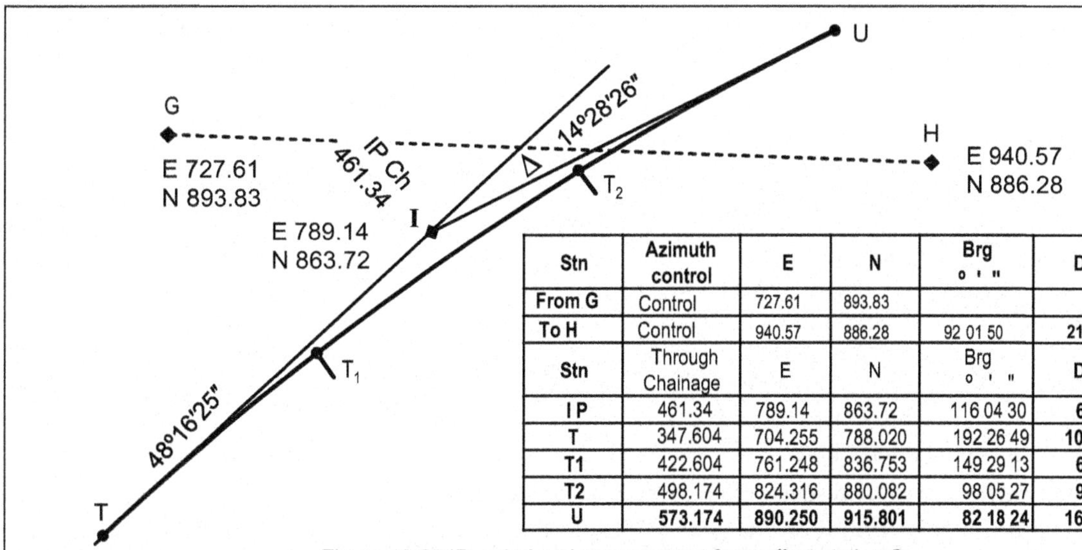

Stn	Azimuth control	E	N	Brg ° ′ ″	Dist
From G	Control	727.61	893.83		
To H	Control	940.57	886.28	92 01 50	**213.09**
Stn	Through Chainage	E	N	Brg ° ′ ″	**Dist**
I P	461.34	789.14	863.72	116 04 30	**68.500**
T	347.604	704.255	788.020	192 26 49	**108.356**
T1	422.604	761.248	836.753	149 29 13	**66.252**
T2	498.174	824.316	880.082	98 05 27	**97.679**
U	**573.174**	**890.250**	**915.801**	**82 18 24**	**164.118**

Figure 12-27 IP, spiral and curve set-out from offset station G.

G E727.61 N893.83

H E940.57 N886.28

The centre line is to be set-out from Control Point G, orientation by the Bearing G-H.

Brg GH = 92°01′50″ Dist GH = 213.09

Table 12-10 shows the set-out bearings and distances from G.

Table 12-10 Set out of chainage stations from control point G.

Stn	Azimuth	E	N	Brg ° ′ ″	Dist
From G	Control	727.61	893.83		
To H	Control	940.57	886.28	92 01 50	213.090
Stn	Chainage (CH)	E	N	Brg ° ′ ″	Dist
IP	461.34	789.140	863.720	116 04 30	68.500
T (TS)	347.605	704.256	788.021	192 26 49	108.356
360	360	713.511	796.265	188 13 21	98.578
380	380	728.517	809.488	179 23 02	84.347
400	400	743.711	822.493	167 16 53	73.132
420	420	759.203	835.140	151 42 21	66.653
T1 (SC)	422.605	761.248	836.753	149 29 13	66.252
440	440	775.083	847.298	134 25 36	66.475
460	460	791.358	858.920	118 42 22	72.681
480	480	808.012	869.993	106 30 49	83.861
T2 (CS)	499.175	824.317	880.082	98 05 27	97.679
500	500	825.026	880.505	97 47 21	98.323
520	520	842.364	890.473	91 40 33	114.803
540	540	859.937	900.021	87 19 17	132.472
560	560	877.654	909.301	84 06 47	150.839
U (ST)	574.175	890.250	915.801	82 18 24	164.118

12.6 Equations for the Clothoid Spiral Curve

12.6.1 General Formula

The following formulæ relate to a compound curve generated by the Clothoid Spiral curves between two straights of deflection angle, Δ, joined by a circular curve of radius, **R**.

Length of transition spiral $L = \dfrac{V^3}{3.6^3 cR}$ (0.3 > c < 0.9, R in metres, V in Km.h^{-1}) § 12.3.1

Spiral Angle $\Phi = L/2R$ (radians). § 12.2.4

Projected tangent distance $X = L\left(1 - \dfrac{\Phi^2}{10} + \dfrac{\Phi^4}{216} - \ldots\right)$ § 12.2.5

Projected tangent offset $Y = L\left(\dfrac{\Phi}{3} - \dfrac{\Phi^3}{42} + \ldots\right)$ § 12.2.5

Shift Distance $S = \dfrac{L^2}{24R} \cdot \left(1 - \dfrac{\Phi^2}{28} + \dfrac{\Phi^4}{1320} - \ldots\right) = L^2/24R$ (approx) § 12.3.1.4

Shift Distance	$S = Y - R(1 - \cos\Phi)$	§ 12.3.1.5
Tangent Distance	$TD = (R+S)\tan(\Delta/2) + (X - R\sin\Phi)$	§ 12.3.1.6
Shifted Arc length	$s = R(\Delta - 2\Phi)^r$	§ 12.3.1.7
Shifted Crown Secant	$= (R+S)\sec(\Delta/2) - R$	§ 12.3.2.2
Tangential angle	$\delta = \arctan((Y/X)$	
Long Chord	$= (X^2 + Y^2)^{1/2}$	
Long Tangent	$= X - Y\times\cot\Phi$	
Short Tangent	$= Y\times\mathrm{cosec}\Phi$	

12.6.2 Spiral Deflections and Chords

Spiral deflection angles and chord distances to a point, p, at length, ℓ, on the spiral:

Tangential angle $= \phi^r = \ell^2/2LR$ or $\phi = (\ell/L)^2\times\Phi$ for $\ell_1, \ell_2, \ldots, \ell_n$

Where $x = \ell\left(1 - \dfrac{\phi^2}{10} + \dfrac{\varphi^4}{216} - \ldots\right)$ for $\ell_1, \ell_2, \ldots, \ell_n$

$y = \ell\left(\dfrac{\phi}{3} - \dfrac{\phi^3}{42} + \ldots\right)$ for $\ell_1, \ell_2, \ldots, \ell_n$

Deflection angle $\delta^\circ = \arctan(y/x)$ for $\ell_1, \ell_2, \ldots, \ell_n$

Chord distance $d = (y^2 + x^2)^{1/2}$

12.6.3 Circular Deflections and Chords

The original design of the simple curve is defined in terms of the design radius, R, and the deflection angle, Δ, at the intersection point, the IP or PI. See Figure 12-28.

The tangent distance, TD, is the distance preceding the PI from the point of curvature, PC, to the following point of tangency, PT.

$TD = R \times \tan(\Delta°/2)$	PC to IP, IP to PT
Arc distance $S = R\Delta^r$	Arc PC to PT
Long chord $= 2R \times \sin(\Delta°/2)$	Chord PC to PT

Circular deflection angles and chord distances to a point, p, at arc length, s, on the centre, O:

Deflection angle $\Delta^r = s/R$ (radians) for $s_1, s_2, , s_n$ see § 12.3.2.1.2

Chord length $\ell_s = 2 \times R\sin(\delta°/2)$ for $s_1, s_2, , s_n$ § 8.6.5, page 116

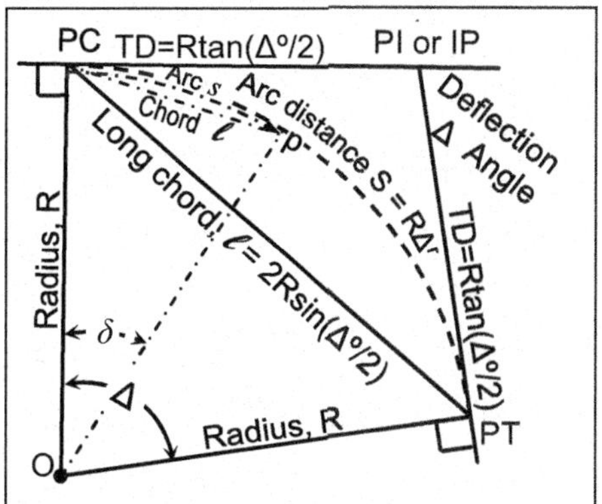

Figure 12-28 Simple circular curve

12.7 Chapter Exercises 12.1

A simple curve in Figure 12-29 is to be improved by the introduction of Transition Spirals at each end.

Given:

Operating speed	$= 85\text{Kmh}^{-1}$
Radius of circular curve	= 400m
Rate of radial acceleration (c)	= 0.3
Coordinates of PI:	E 1220.12, N 1118.15
Chainage at PI	= 2048.94
Bearing of initial tangent	= 61°40′
Bearing of final tangent	= 94°00′

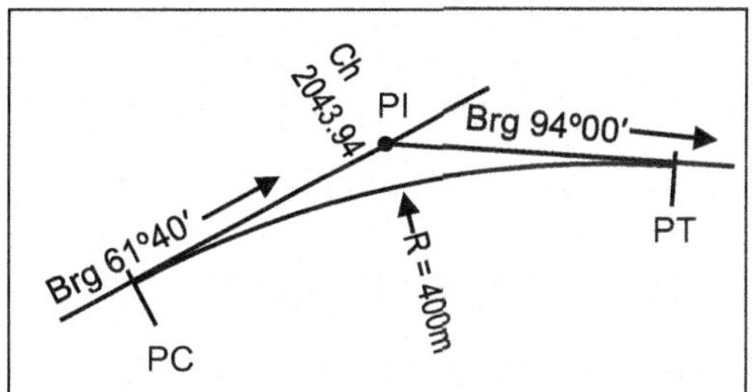

Figure 12-29 Chapter Exercise 12.1

Calculate:

a) Transition length, do not round to integer or suitable length
b) Chainage of TS, SC, CS and ST spiral control points
c) Show closure of tangents and spiral control points

The centre line is to be pegged at even 20m intervals.

Calculate:

d) Data to set out centreline pegs of spirals from TS and ST
e) Data to set out centreline pegs of circular curve from SC
f) Coordinates of all centre pegs.

12.7.1 Answers to Chapter Exercise 12.1.

Calculations:

Tangent deflection, $\Delta = 94° - 61°40' = 32°20'$, $\Delta/2 = 16°10'$

Transition length (spiral, L) = 109.69 (normally increased to, say, 110m) — Eqn 12-14

Spiral angle $\Phi = L/2R = 0.137^r = 7°51'22''$ — Eqn 12-6

Shift $S = L^2/24R = 1.253$ — Eqn 12-15

Run to SC using power series = X = 109.484 — § 12.3.1.3, Eqn 12-8

Offset to SC using power series = Y = 5.007 — Eqn 12-9

a) Tangent Length, TL = 116.322 + 54.811 = 171.133 — Eqn 12-17

b) Chainages TS = 2048.94 − 171.13 = 1877.81

SC = 1877.81 + 109.69 = 1987.50 Long chord TS − SC = 109.60

CS = 1987.50 + 116.04 = 2103.54 Shifted chord SC − CS = 115.63

ST = 2103.54 + 109.69 = 2213.23 Long chord CS − ST = 109.60

c) Show closure of tangents and spiral control points (closes at 3d.p.) (See Table 12-11)

Table 12-11 Control point closure

Stn	Brg	Dist	dE	dN	E	N
PI					1220.12	1118.15
TS	241 40 00	171.13	−150.629	-81.218	1068.491	1036.932
SC	64 17 06	109.60	98.746	47.555	1167.237	1084.487
CS	77 50 00	115.63	113.033	24.370	1280.270	1108.857
ST	91 22 54	109.60	109.568	−2.643	1389.838	1106.214
PI	274 00 00	171.13	−170.713	11.937	1220.125	1118.151
		Miscloses	0.005	0.000		

If it's not checked it's wrong.

The centre line is to be pegged at even 20m intervals.

d) 1) Data to set out centreline pegs of spirals from TS (Ch 1877.81) to SC (Ch 1987.50). (See Table 12-12)

Table 12-12 Setout centreline TS to SC

Stn	Defl'n	Brg	Dist	dE	dN	E	N
TS	Brg to IP	61 40 00	0.00			1068.491	1036.932
1880	0 00 04	61 40 04	2.19	1.928	1.039	1070.419	1037.971
1900	0 06 26	61 46 26	22.19	20.551	10.495	1088.042	1047.427
1920	0 23 15	62 03 15	42.19	37.270	20.772	1105.761	1056.704
1940	0 50 31	62 30 31	62.19	55.167	28.708	1123.658	1065.640
1960	1 28 13	63 08 13	82.17	73.303	37.129	1141.794	1074.061
1980	2 16 21	63 56 21	102.13	91.746	44.868	1160.237	1081.800
SC	2 37 06	64 17 06	109.60	98.746	47.555	1167.237	1084.487

Data to set out centreline pegs of spirals **from ST back to CS.** (See Table 12-13)

Table 12-13 Setout reverse centreline ST to CS

Stn	Defl'n	Brg	Dist	dE	dN	E	N
CS	Brg to IP						
1880	357 22 54	271 22 54	109.60	-109.568	2.643	1280.27	1108.857
1900	358 06 31	272 06 31	93.19	-93.127	3.429	1296.711	1109.643
1920	358 49 59	272 49 59	73.21	-73.121	3.618	1316.717	1109.832
1940	359 23 00	273 23 00	53.22	-53.127	3.141	1336.711	1109.355
1960	359 45 35	273 45 35	33.23	-33.158	2.179	1356.680	1108.393
1980	359 57 43	273 57 43	13.23	-13.198	0.914	1376.640	1107.128
ST	360 00 00	274 00 00	0.00	0.000	0.000	1389.838	1106.214

e) Data to set out centreline pegs of circular curve from SC to CS. (See Table 12-14).
Brg Common Tangent at SC = Brg TS to IP + Spiral angle = 61°40′00″ + 7°51′22″ = 69°31′22″.

Table 12-14 Setout centreline of arc SC to CS

Stn	Defl'n	Brg	Dist	dE	dN	E	N
SC	Brg CT	69 31 22	0.00			1167.237	1084.487
2000	0 53 43	70 25 05	12.50	11.777	4.189	1179.014	1088.676
2020	2 19 38	71 51 00	32.50	30.874	10.121	1198.111	1094.608
2040	3 45 10	73 16 32	52.50	50.243	15.090	1217.480	1099.577
2060	5 11 07	74 42 29	72.50	69.839	20.086	1237.076	1103.573
2080	6 37 30	76 08 52	92.50	89.616	22.099	1256.853	1106.586
2100	8 03 27	77 34 48	112.50	109.506	24.116	1276.743	1108.603
CS		77 50 00	116.04	113.033	24.370	1280.270	1108.857

12.8 References for Chapter 12

Elfick, M.H, J.G Fryer, R.C. Brinker and P.R Wolf. Elementary Surveying, Eighth edition. Longman, 1994. ISBN 0-00-499001-3

Austroads Ltd, Sydney, 2021. Edition 3.4 AGRD03-16: Guide to Road Design Part 3: Geometric Design. ISBN 978-1-922382-15-3

Perco, P., Desirable Length of Spiral Curves for Two-Lane Rural Roads Transportation Research Record 2006 1961:1, 1-8. https://doi.org/10.1177/0361198106196100101

Chapter 13 CALCULATIONS OF SUPERELEVATION ON A COMPOSITE CURVE

In this chapter, we extend the concept of Chapter 12 by examining superelevation on composite curves.

13.1 Superelevation. Road Profile on a Composite Curve

A **composite** curve is a **transitional** curve incorporating **superelevation** in the road design. The purpose of introducing superelevation into the transverse road profile (the road cross-section) is to provide for a comfortable, stable and safe vehicular passage around a curve.

In determining the generation of superelevation, we will consider only the carriageway of the entire road profile. The carriageway consists of the traffic lane(s) and the road shoulders and is carried on the road pavement. Figure 13-1 illustrates a typical two lane two-way rural road.

Urban roads will share the same characteristics in relation to the carriageway. Single carriageway two-way rural road design is based on a 3.1m/3.5m traffic lanes and 1.5m shoulders for a total carriageway of 10m for up to 1000 annual average daily traffic (AADT). Austroads 2021, Table 4.5. The traffic lane width determines whether one or both passing vehicles have to use the shoulder. Figure 13-2 et seq illustrations show the left hand lane as the elevating lane for a right hand drive vehicle turning to the right as it proceeds around the curve.

In the design of two-lane rural roads, the criteria we must assess are:

1. Design speed – topography, cadastral (boundary) restraints and nature of terrain
2. Minimum circular radius – design speed
3. Superelevation cross fall – nature of terrain, flat, undulating, hilly

The process is iterative between the mathematical solution of equations and road design recommendations.

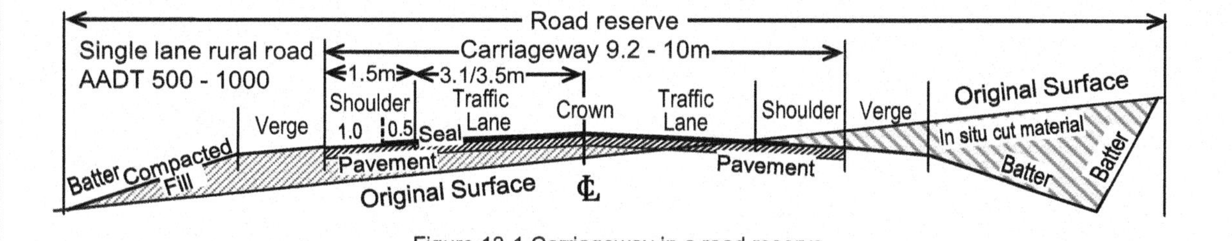

Figure 13-1 Carriageway in a road reserve

The crown of the road, the centreline of the carriageway, is the point from which the slope of the road cross section is developed.

The slope, or cross fall, or camber, is expressed either as a percentage or as the equivalent number. i.e. 3% = 0.03. It is also the tangent of the slope, so arctan(0.03) = 1.7°. The purpose of cross fall is to provide water drainage from the pavement surface.

By convention the **sign** of the slope of indicates the relationship between the centreline and endpoint of the slope. A straight section of road has a, say, −3% slope. This shows that the carriageway edges are below the centreline. (see Figure 13-2).

In keeping with § 6.15, page 82, we use the same sign convention for cross fall as for ground slope, k.

The minimum drainage cross fall depends on the pavement surface material.

Zero crossfall, k_0, (see Figure 13-3) occurs at the start of the spiral, rising to the road centreline.

Elevated crossfall (see Figure 13-4) can thus be considered as a **positive** slope, rising above the road centreline, at the normal road cross fall, k_0.

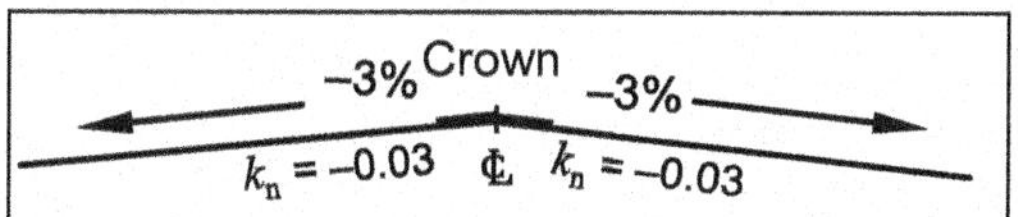

Figure 13-2 Normal crossfall at SSD

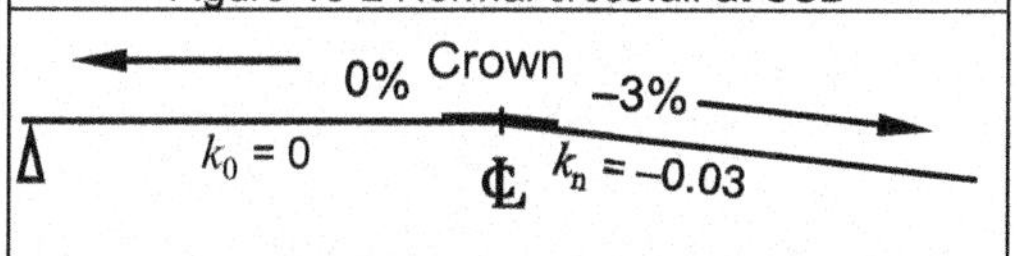

Figure 13-3 Zero crossfall at TS

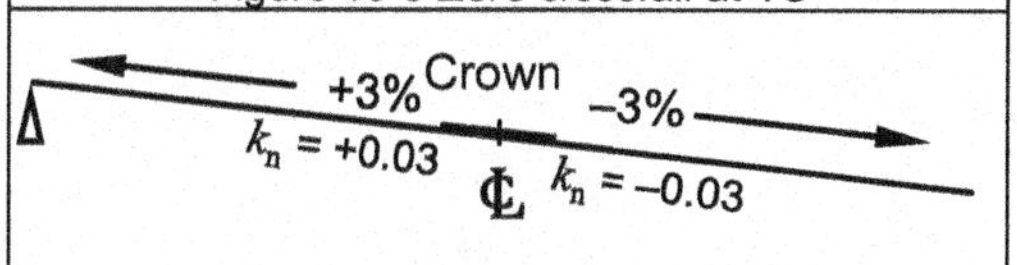

Figure 13-4 Elevated crossfall at X

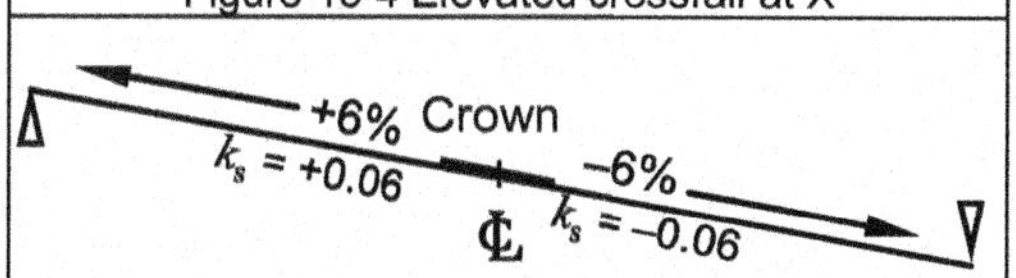

Figure 13-5 Superelevated crossfall at SC

Superelevation crossfall, $k_{s,}$, (see Figure 13-5) is increased above the normal cross slope, k_n. It is needed to overcome the side forces acting on a vehicle following the circular arc portion of composite curve, it can increase to a recommended maximum of 10%. (k_s = ±10% or ±0.1)

J. Walker and J. Awange, *Surveying Computations*,
https://doi.org/10.1007/978-3-031-94382-9_13

13.2 Developing superelevation between the straight and circular curve

Figure 13-6 shows that **prior** to the Tangent to Spiral (TS) point, and **after** the Spiral to Tangent (ST) point, there is a portion of the straight called the **Tangent Runout, T_{ro}**. This is the length over which the normal cross fall is elevated to, or reduced from, the level cross fall section at the beginning and end of the spiral. (TS and ST)

The formula for length of superelevation development, L_k, between SSD and SC provides the value at a chosen (maximum recommended) rate of rotation, ω (omega). However, the main requirement of superelevation development is to provide a **constant rate** of rotation (ω_{max}) around the spiral. This rate should then be extended back along the Tangent Runout, T_{ro}. Figure 13-6 shows the relationship between L_e, T_{ro} and S_{ro} for development from the normal crossfall, $-k_n$ to superelevated $+k_s$.

13.2.1 Length of development of superelevation based on rate of rotation

The criterion used in developing a road profile from normal cross fall on a straight to a fully superelevated cross fall on a transition curve is the **rate of rotation** of the cross fall.

The minimum time needed by a driver to make a change of transition path (Austroads (2021) Spiral Length, page 172) is between 2 and 3 seconds. (Perco, 2006). Generally, the transition path change should occur within about 4 seconds, otherwise perceptions about the curve will wander, and change driving behaviour.

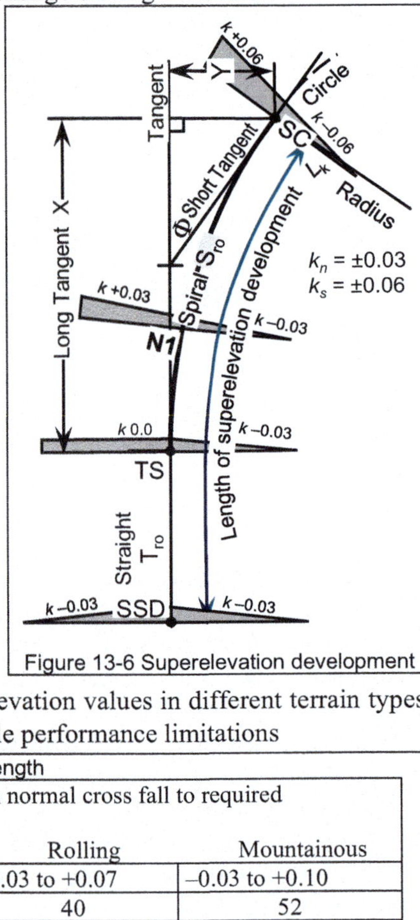

Figure 13-6 Superelevation development

The maximum rate of rotation, ω, recommended is:
speed < 80Kmh^{-1} $\omega = 3.5\%\text{sec}^{-1} = 0.035\text{radians.sec}^{-1}$ and
speed > 80Kmh^{-1} $\omega = 2.5\%\text{sec}^{-1} = 0.025\text{radians.sec}^{-1}$.

It follows that the length required to develop the superelevation depends on:
operating speed, change of relative grade from the straight to the curve.

v = operating speed (m.sec^{-1})
k_n = normal cross fall (negative value)
k_s = superelevated cross fall (positive value)
ω = rate of rotation

Ensure that k_n, k_s and ω operate in either radians or percentage.

Length of superelevation development:	$L_k = \dfrac{v(k_s - k_n)}{\omega}$	Eqn 13-1
Converting v m·sec^{-1} to Kmh^{-1}:	$L_k = \dfrac{V(k_s - k_n)}{3.6\omega}$	Eqn 13-2

For $\omega = 0.035\text{radians.sec}^{-1}$, $V < 80\text{Kmh}^{-1}$:
$L_{k(min)} = V(k_n - k_n)/0.126$
For $\omega = 0.025\text{radians.sec}^{-1}$, $V > 80\text{Kmh}^{-1}$:
$L_{k(min)} = V(k_s - k_n)/0.09$

Table 13-1 shows the length of superelevation development based on rate of rotation, ω_v, from the above equations. Note the different superelevation values in different terrain types. Rolling and mountainous terrain will have decreased speeds due to vehicle performance limitations

Table 13-1 Superelevation development length

Operating Speed (Kmh^{-1})	Rate of rotation R.sec^{-1}	Length of super elevation development from normal cross fall to required superelevation for a single lane. Flat	Undulating	Rolling	Mountainous
	ω	−0.03 to +0.03	−0.03 to +0.05	−0.03 to +0.07	−0.03 to +0.10
50	0.035	24	32	40	52
60	0.035	29	38	48	62
70	0.035	33	44	56	72
80	0.025	53	71	89	116
90	0.025	60	80	100	130
100	0.025	67	89	111	--

13.2.1.1 *Caution.* Use of Table 13-1

*The tables have been incorporated into this book to provide a means of checking your calculations of the derived formulae. They are **not** to be used for other than course work within units provided by the educational institution.*

13.2.1.2 The length of superelevation development:

From Eqn 13-1 the length of superelevation development, L_k, takes place between the SSD and the ESD/SC, before the circular curve. (See Figure 13-6)

$$L_k = \frac{v(k_s - k_n)}{\omega} \text{ from } L = vt \text{ where } t = \frac{(k_s - k_n)}{\omega}.$$

Use the same values for k and ω. Speed $v = (V/3.6)$ m.sec^{-1}.

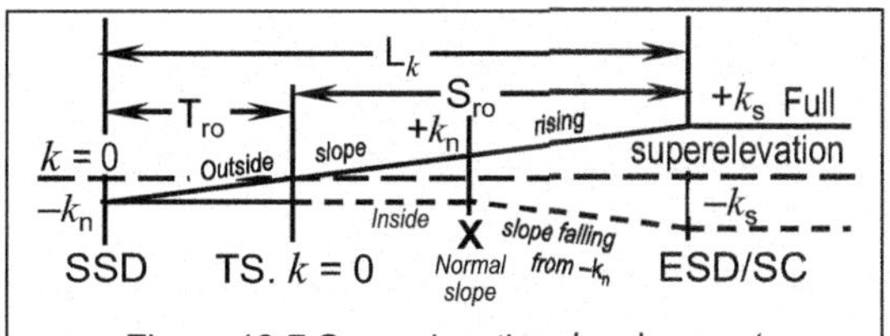

Figure 13-7 Superelevation development

Thus, for a given length, L_k, the rate of rotation, ω, can be found: $\omega = \dfrac{v(k_s - k_n)}{L_k}$.

At a constant rotation rate, ω, the length of superelevation development, L_k, is split proportionally by the ratio of the normal cross slope and the superelevated cross slope into the tangent runout, T_{ro} and superelevation runoff, S_{ro} at the tangent to spiral point, TS. See Figure 13-7. $L_k = T_{ro} + S_{ro}$. Remember that S_{ro} is the spiral length, L

13.2.1.2.1 Spiral runoff, S_{ro}

From L_k being the length of the superelevation development length (see Eqn 13-2), the Spiral Runoff:

$$S_{ro} = L_k \frac{k_s}{(k_s - k_n)}$$ is calculated proportionally where:

k_n = normal crossfall at SSD, remembering that normal cross fall is negative

k_s = superelevation cross fall at SC

13.2.1.2.2 Tangent runout, T_{ro}

An expression for the Tangent Runout, T_{ro}, can be developed from:

the normal cross fall, k_n, at the SSD

the superelevation cross fall, k_s, at SC and

the length of the superelevation development, L_k.

$$T_{ro} = L_k \frac{(-k_n)}{(k_s - k_n)}$$ (note that k_n is a negative value at SSD)

Alternately, if Spiral Runoff, or spiral length, S_{ro}, is known then Tangent Runout, T_{ro}, can be found. It provides a constant rate of rotation from SSD to SC.

The T_{ro} should always be less than S_{ro} because $|k_n|$ will be less than k_s.

13.3 Worked Example 15.1 Superelevation on a Transition Curve

The pavement profile of the road designed in Worked Example 12.4 (§ 12.4, p195) has the following cross falls:

k_n = –0.03, (–3%) normal cross fall at SSD

k_s = +0.06, (+6%) superelevation cross fall at SC

Using the data from Worked Example 12.4, determine the chainage of SSD and ESD and show that the rate of rotation of the development of the superelevation, ω, does not exceed the design maximum, 0.025radians.sec^{-1}.

Prepare a table of cross fall values and RLs either side of the centre line at the even 20m chainages. Carriageway width, W, for set out = 10m.

1) Previous design data. (page 195)

Length of spiral	L	=	73.13m. **75**m (S_{ro}) Recall § 12.5.1
Chainage at T		=	347.605. TS
Chainage at U		=	574.175. ST
Design speed	v	=	85Kmh^{-1}

2) Calculate tangent runout, T_{ro} (See Figure 13-8)

$$T_{ro} = L\frac{(-k_n)}{k_s} = 75\frac{(-(-0.03))}{0.06} = 37.5$$

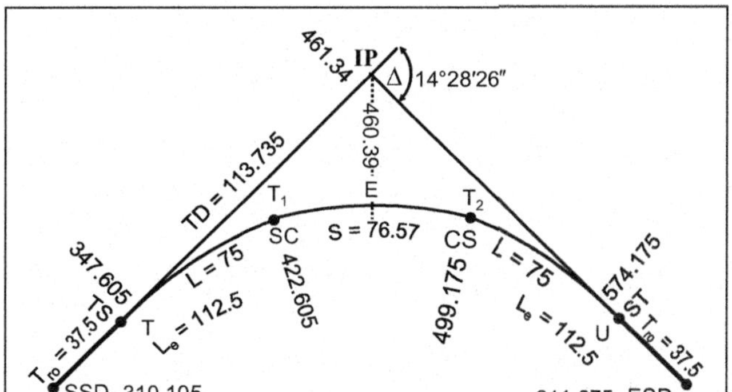

Figure 13-8 Incorporating tangent runout

3) Calculate chainages

SSD $= TS - T_{ro} = 347.605 - 37.5$

SSD = 310.105

ESD $= ST + T_{ro} = 574.175 + 37.5$

ESD = 611.675

4) Calculate rate of rotation of superelevation development.

$$\omega = \frac{v(k_s - k_n)}{L} \qquad \text{For } L \text{ use either}$$

(i) the change on the T_{ro},

(ii) the spiral length, L, or

(iii) the length of superelevation development, $L_k = L + T_{ro}$.

Choosing method (iii).

$L_k = 75 + 37.5 = 112.5$

$k_n = -0.03$

$k_s = +0.06$

$v = 85\text{Kmh}^{-1} = 85/3.6 = 23.6\text{msec}^{-1}$

$$\omega = \frac{v(k_s - k_n)}{L_k} = 23.6\frac{(0.06-(-0.03))}{112.5} = 0.019$$

$\omega < 0.025\text{radians.sec}^{-1}$. Design criteria met.

5) Calculate the linear rate of slope change per metre of curve length. $\delta k = \frac{\omega}{v} = \frac{0.019}{23.6} = 0.0008$

Super elevation starts after the normal cross fall of the outer carriageway exceeds $-k_n$, i.e. when the outer cross fall, k_n, =+0.03.

This distance is either:

a) $(0.03-(-0.03))/\delta k$ (rotate from -0.03 to $+0.03 = 0.06$)
$= 0.06/0.0008 = 75$m from SSD, or

b) $(0.03-0)/\delta k$ (rotate from -0.03 to level $= 0.03$)
$= 0.03/0.0008 = 37.5$m from T (TS point).

Chainage elevation rising above normal cross fall, N1

= SSD + 75.0 = 386.04 or

= T + 37.5 = 386.04.

Complete the curve, see Figure 13-9.

Similarly, the superelevation falls to normal crossfall, N2 after:

= ESD − 75 = 535.70 or

= U − 37.5 = 535.70.

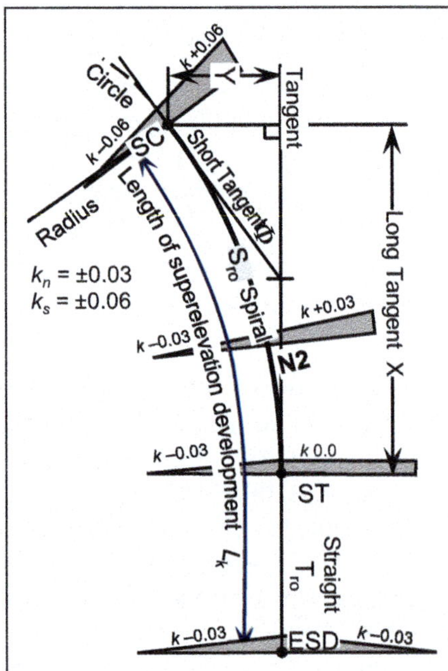

Figure 13-9 Completion of curve

6) Calculate cross fall and RL at each chainage. Add N1 and N2 to Table 13-2:

a) Tabulate slope at each chainage.
The change in slope for the superelevation is linear:
$\delta k = 0.0008$radians/m

b) Tabulate RLs at each chainage. Table 13-2.
Centreline is elevation = 0.00
Carriageway, W = 10 (Shoulders and carriageway. 1.5 + 3.5 + 3.5 +1.5 = 10)
$k_{CH} = k_n + \delta k \times S$ (through distance) to $k_s = 0.06$. ($k_n = -0.03$)
RL = $k \times W/2$
$k_s = +0.06$ and -0.06 from T1 to T2, superelevated section.

Table 13-2 Chainages and RLs reference ℄ level.

Stn	Through Chainage	Incremental distance, d	Through distance, S.	k_{Left} outside CL	k_{Right} inside CL	RL_{Left}	RL_{Right}
SSD	310.105	0	0	–0.03	–0.03	–0.15	–0.15
320	320	9.895	9.895	–0.022	–0.03	–0.12	–0.15
340	340	29.895	29.895	–0.006	–0.03	–0.04	–0.15
T	347.605	37.5	37.5	+0.000	–0.03	0.00	–0.15
360	360	12.395	49.895	+0.010	–0.03	+0.05	–0.15
380	380	32.395	69.895	+0.026	–0.03	+0.13	–0.15
N1	385.105	37.5	75	+0.030	**–0.03**	**+0.15**	**–0.15**
400	400	52.395	89.895	+0.042	–0.042	+0.21	–0.21
420	420	109.895	109.895	+0.058	–0.058	+0.29	–0.29
T1	422.605	112.5	112.5	+0.060	–0.06	+0.30	–0.30
440	440	17.395	129.895	+0.06	–0.06	+0.30	–0.30
460	460	37.395	149.895	+0.06	–0.06	+0.30	–0.30
480	480	57.395	169.895	+0.06	–0.06	+0.30	–0.30
500	500	77.395	189.895	+0.06	–0.06	+0.30	–0.30
T2	499.175	76.57	189.07	+0.06	–0.06	+0.30	–0.30
520	520	20.825	209.895	+0.043	–0.043	+0.22	–0.22
N2	536.675	37.5	226.57	+0.030	**–0.03**	**+0.15**	**–0.15**
540	540	40.825	229.895	+0.027	–0.03	+0.14	–0.15
560	560	60.825	249.895	+0.011	–0.03	+0.05	–0.15
U	574.175	75	264.07	0.000	–0.03	+0.00	–0.15
580	580	5.825	269.895	–0.005	–0.03	–0.03	–0.15
600	600	25.825	289.895	–0.021	–0.03	–0.11	–0.15
ESD	611.675	37.5	301.57	–0.030	–0.03	–0.15	–0.15

The following diagrams (Figure 13-10) show the pavement cross section at selected chainages. They illustrate the development of the superelevation from the start of superelevation development, SSD, through to the fully developed superelevation at the start of the circular curve, SC.

Note that the decrease in elevation on the inside of the turn does not occur until the outside elevation, $k_{Left} > +0.03$. In this case, where normal cross fall, $k_n = \frac{1}{2} k_s$, it occurs near chainage 385m, which is half way around the spiral from TS. (TS + L/2 = 348.54 + 73.13/2 = 385.10).

At the end of the circular curve the process is reversed back to the ESD. Refer to the previous Figure 13-9.

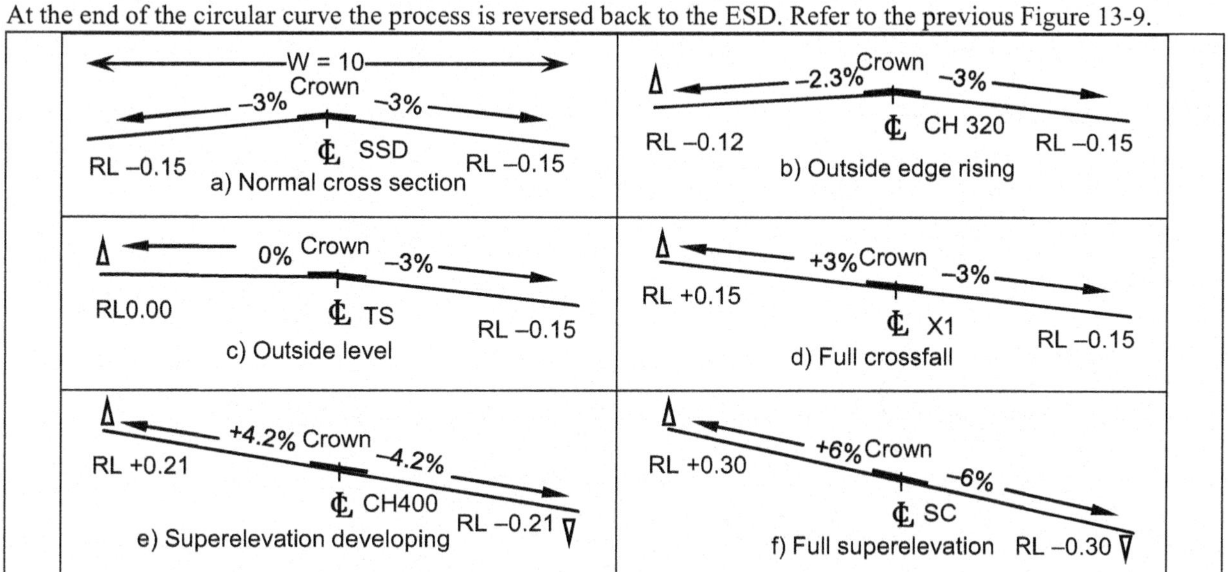

Figure 13-10 Development of superelevation from SSD – SC.

13.4 Profiles of single and two way rural roads.

The carriageway width of rural roads is constrained by costs which are determined by the Annual Average Daily Traffic (AADT),

Figure 13-11 shows a sealed low volume country road. Passing opposite direction traffic requires one or both drivers to use the unsealed shoulder.

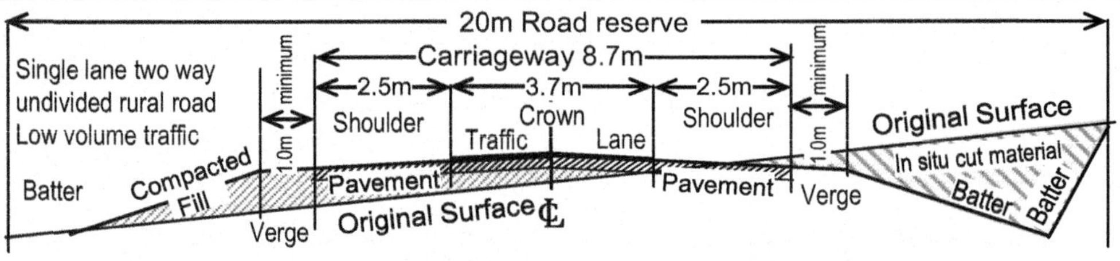

Figure 13-11 Single lane two way undivided low volume road design

Figure 13-12 is designed for higher traffic volumes, but with a minimum carriageway of 9.2m to accommodate two traffic lanes 3.1m wide. In general, the passing vehicles will be able to remain on the sealed surface

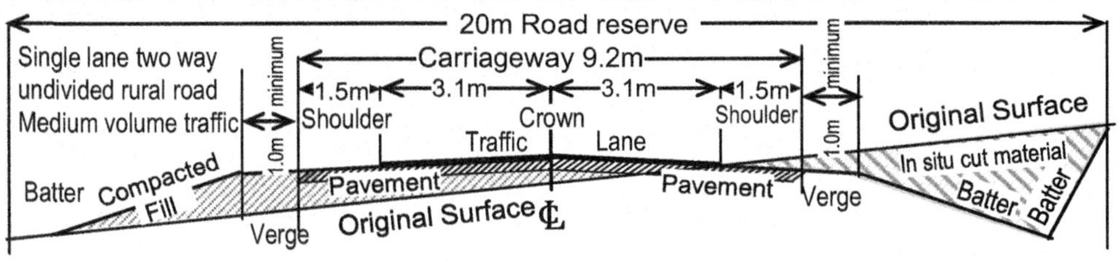

Figure 13-12 Single lane two way undivided higher volume road design

Figure 13-13 is designed for higher traffic volumes, having dual traffic lanes in a two way undivided rural road. The road can handle quite high traffic volumes and heavier vehicles and articulated trucks.

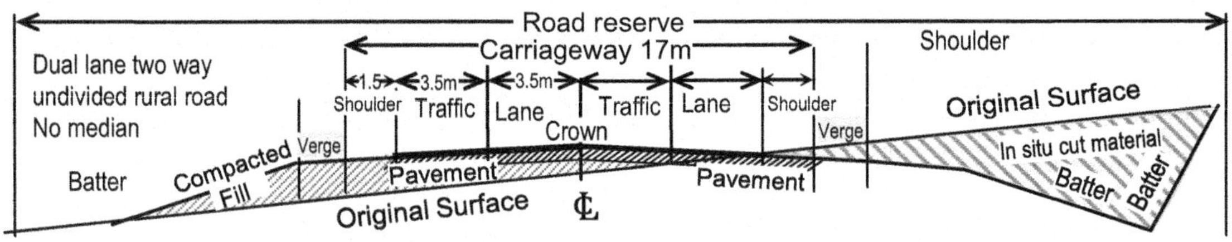

Figure 13-13 Dual lane two way undivided rural road

A rural road median strip between opposite directions may be provided. They range from dual unbroken lines to about a 1m painted median about the centre line.

13.5 Chapter exercise 13.1

With the answers to Chapter Exercise 12.1, page 203, now calculate road profile from the start of superelevation development (SSD) to the end of superelevation development (ESD).

Carriageway width is 10m.

The road formation has a normal cross fall of –0.03 (–3%). Superelevation around the curve is to be a maximum of +0.07 (+7%).

Calculate:

a) Tangent runout
b) Chainage at SSD and ESD
c) Rate of rotation of superelevation development at operating speed
d) Cross fall and RL of each even 20m chainage from SSD to ESD,
   include spiral control points.
e) Chainages where inside cross fall deviates from normal cross fall.

13.5.1 Answers to Chapter Exercise 13.1

Calculate:

a) Tangent runout = 47.01
b) Chainage SSD = 1877.81 – 47.01 = 1830.80
   ESD = 2213.23 + 47.01 = 2260.24
c) $\omega_{85} = 0.014$ radians.sec^{-1} ($\omega_{85} < \omega_{80(max)} = 0.025$)
d) Cross fall and RL of each even 20m chainage from SSD to ESD, include spiral control points (See Table 13-3).

Table 13-3 Crossfall calculations at 20m intervals

Stn	Through Chainage	Incremental distance, d	k_{Left} outside CL	k_{Right} inside CL	RL_{Left}	RL_{Right}
SSD	1830.80	0	-0.030	-0.030	-0.15	-0.15
1840	1840.00	9.2	-0.024	-0.030	-0.12	-0.15
1860	1860.00	29.2	-0.011	-0.030	-0.06	-0.15
TS	1877.81	47.01	0.000	-0.030	0.00	-0.15
1880	1880.00	49.2	0.001	-0.030	0.01	-0.15
1900	1900.00	69.2	0.014	-0.030	0.07	-0.15
1920	1920.00	89.2	0.027	-0.030	0.13	-0.15
1940	1940.00	109.2	0.040	-0.040	0.20	-0.20
1960	1960.00	129.2	0.052	-0.052	0.26	-0.26
1980	1980.00	149.2	0.065	-0.065	0.33	-0.33
SC	1987.50	156.7	0.070	-0.070	0.35	-0.35
2000	2000.00	12.5	0.070	-0.070	0.35	-0.35
2020	2020.00	32.5	0.070	-0.070	0.35	-0.35
2040	2040.00	52.5	0.070	-0.070	0.35	-0.35
2060	2060.00	72.5	0.070	-0.070	0.35	-0.35
2080	2080.00	92.5	0.070	-0.070	0.35	-0.35
2100	2100.00	112.5	0.070	-0.070	0.35	-0.35
CS	2103.54	116.04	0.070	-0.070	0.35	-0.35
2120	2120.00	16.46	0.059	-0.059	0.30	-0.30
2140	2140.00	36.46	0.047	-0.047	0.23	-0.23
2160	2160.00	56.46	0.034	-0.034	0.17	-0.17
2180	2180.00	76.46	0.021	-0.030	0.11	-0.15
2200	2200.00	96.46	0.008	-0.030	0.04	-0.15
ST	2213.23	109.69	0.000	-0.030	0.00	-0.15
2220	2220.00	116.46	-0.004	-0.030	-0.02	-0.15
2240	2240.00	136.46	-0.017	-0.030	-0.09	-0.15
2260	2260.00	156.46	-0.030	-0.030	-0.15	-0.15
ESD	2260.24	156.7	-0.030	-0.030	-0.15	-0.15

e) Chainages where inside cross fall sags below normal cross fall.
   $= TS + T_{ro} = 1877.81 + 47.01 = 1924.82$
   $= ST - T_{ro} = 2213.23 - 47.01 = 2166.22$

Other values as checks:

spiral angle, $\Phi = 7°51'21.5''$
tangent length, L = 109.69
shifted deflection angle $\Delta - 2\Phi = 16°37'17''$
circular arc, s = 116.04

13.6 Worked Example 13.2 Deconstruction of a Transition Curve Calculation

A transition curve set out can be part of a surveying fieldwork exercise. Unless the curve has been fully specified in terms of design parameters there are often inconsistences in the design that can cause confusion.

The interplay between design speed, circular radius, terrain type and superelevation determine the design "greenfield" roads but can impose limitations on existing designs. The limitation may be caused by road boundaries, or existing road formations being upgraded.

The purpose of both the transition curve and superelevation calculations is to not only design the roads with suitable parameters, but to show how those parameters are incorporated into a set of field layout coordinates.

The design of a road is properly the task of a road traffic engineer.

13.6.1 Defined existing transition curve.

A computer-generated transition curve was used as a field set out exercise for surveying students. An examination of the coordinates showed inconsistencies in the straight and transition portions of the curve that did not meet AGRD03-16 recommendations.

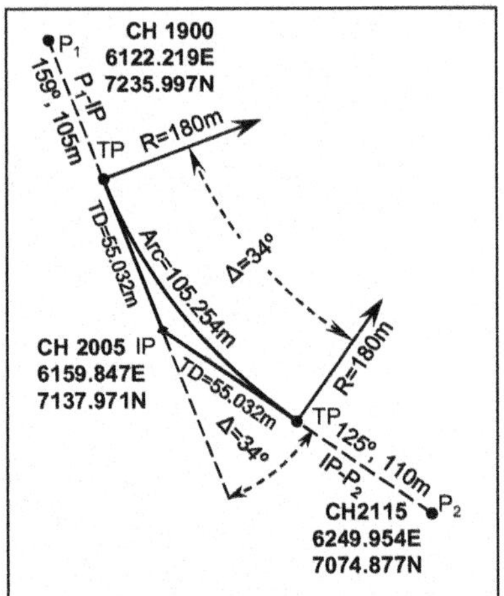

Figure 13-14 Redesign circular curve

13.6.1.1 Basic parameters

Curve radius R = 180m, tangent deflection angle Δ = 34°. Bounding circular curve data is shown in Figure 13-14.
Design speed V = 70Kmh^{-1}, spiral curve length, S_{ro} or L_p = 25m derived from the exercise curve data.

The road was described as a two-lane rural road. This implies one lane each way, with a 3.5m carriage way (Figure 13-12). No crossfalls were specified, however normal value $k_n = -3\%$ and superelevated $k_s = +5\%$ were used for an urban/rural road. The speed limit of 70Kmh would imply an urban road.

Chainages: SSD 1900
TS 1937.426, T_{ro} = 37.426m.
SC 1962.426, S_{ro} = 25.0m. Spiral angle, $\Phi = 0.0694^r = 3°58'44''$. Eqn 12-6
CS 2044.240, Shifted arc S_S = 81.814m. Thus, shifted deflection $\delta_S = 0.4545^r = 26°02'32''$.
Long chord SC - CS= 81.112m.

These exercise values of T_{ro} and S_{ro} are incorrect, showing that they were incorrectly calculated from the formula used in Austroads 2021 §7.7.6, equations 11 and 12, page 182. (At 70kmh^{-1} the rotation rate, ω, changes at the TS from ω_{Tro} = 0.016rad.sec^{-1} to ω_{Sro} 0.04rad.sec^{-1}. That's a big jolt.). See the values in the next § 13.6.3.

Length of superelevation development L_e = 62.426m (from chainage). This figure agrees with Austroads 2021 Table 7.11, page 186, but has been applied to a 2-lane curve. Recommended L_e = 51m for single lane curve. Crossfalls $k_n = -3\%$ and $k_s = +5\%$

For L_e = 62, values of T_{ro} = 62×3/8 = 23.3m, thus S_{ro} = 38.7m. Valid because $ks > |k_n|$.
Similarly, for L_e = 51, values of T_{ro} = 51×3/8 = 19.1m, thus S_{ro} = 31.9m.

13.6.2 Through bearings, fitting the deflection angle Δ.

The sum of $2\times\Phi + \delta = 34°00'00''$ equals tangent deflection angle Δ = 34°.

In the transition curves studied so far, the final deflection angle around the curve = 4×Φ = Δ.
Fitting the curve will require the shifted deflection angle to be found. $\delta_S = \Delta - 2\Phi$. Thus, case $\delta_S = 26°02'32''$.

13.6.3 Transition spiral components.

Transition rectangular X, Y.

L_e 62m (AGRD03 Table 7.11)		62m (2 lane)	51m (Single lane)	
T_{ro}	37.4m(ω = 0.016)	23m(ω = 0.025)	19m(ω = 0.03)	
S_{ro}	25m (ω = 0.040)	39m(ω = 0.025)	32m(ω = 0.03)	
Φ	3°58'44"	6°12'25"	5°05'35"	Eqn 12-6
X =	24.988	38.954	31.975	Eqn 12-9, § 12.3.1.3
Y =	0.579	1.407	0.948	Eqn 12-10
ϕ =	1°19'35"	2°04'08"	1°41'51"	
Tangent length	67.574	74.631	71.100	Eqn 12-17
Long chord	24.995	38.980	31.989	
Long tangent	16.671	26.016	21.342	
Deflection X	8.317	12.938	10.633	
Shift S	0.145	0.352	0.237	
Δ_S	26°02'32"	21°35'09"	23°48'51"	Eqn 12-7
Shifted arc	81.814	67.814	74.814	
Shifted chord	81.112	67.414	74.277	
Shifted secant	0.151	0.368	0.248	

13.6.4 Chainages around the curve.

The coordinates of the IP provide a vector to the start at CH1900. The distance is 105m, thus the chainage of the IP (CHIP) = 2005. From the IP the chainage to the SSD and TS can be calculated.

From the SSD the chainages of the TS, SC, CS, ST and ESD are available.

CH IP	2005.000	2005.000	2005.000
Tangent length	−67.574	−74.631	−71.100
CH TS	1937.426	1930.369	1933.900
T_{ro}	−37.426	−23.000	−19.000
CH SSD	1900.000	1907.369	1914.900
T_{ro}	+37.426	+23.000	+19.000
CH TS	1937.426	1930.369	1933.900
S_{ro}	+25	+39.000	+32.000
CH SC	1962.426	1969.369	1965.900
Shifted arc	+81.814	+67.814	+74.814
CH CS	2044.240	2037.183	2040.714
S_{ro}	+25	+39.000	+32.000
CH ST	2069.240	2076.183	2072.714
T_{ro}	+37.426	+23.000	+19.000
CH ESD	2106.66	2099.183	2091.714

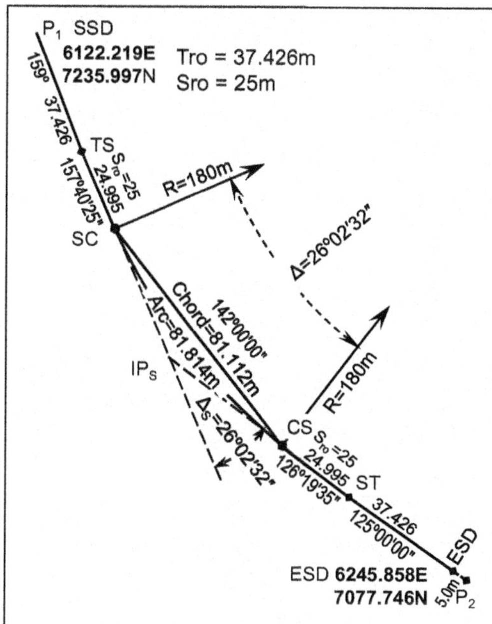

Figure 13-15 Transition traverse

13.6.5 Circular traverse SSD – IP - ESD. (See Figure 13-15)

Find the coordinates of the SSD/ESD from the circular IP.

IP E 6159.847 BRG SSD to IP = 159°, Dist = 105m

IP N 7137.971

CH IP	2005.000	2005.000	2005.000	
CH SSD	1900.000	1907.369	1914.900	
SSD E	6122.218	6124.859	6127.558	
SSD N	7235.997	7229.188	7222.086	
Δ IP = −34°	105.000m	97.631m	90.100m	BRG/Distance to IP to ESD = 125°
ESD E	6245.858	6239.822	6233.652	
ESD N	7077.746	7081.972	7086.292	

13.6.6 Traverse through the straights

The traverse of the straights produces coordinates of each point for initial set out. Check SSD/ESD with § 13.6.5 circular traverse.

SSD E	6122.218	6124.804	6127.540	
SSD N	7235.997	7229.258	7222.133	
θ SSD to TS	159°	159°	159°	
Dist to TS, T_{ro}, m	37.426	23.000	19.000	
TS E	6135.631	6133.048	6134.349	
TS N	7201.057	7207.786	7204.395	
θ TS to SC	157°40′25″	156°55′52″	157°18′09″	ϕ = 1°19′35″ 2°04′08″ 1°41′51 Initial BRG – ϕ
Long chord, m	24.995	39.980	31.989	
SC E	6145.126	6148.312	6146.690	
SC N	7177.936	7171.918	7174.882	
θ SC to CS	142°00′00″	142°00′00″	142°00′00″	Φ = 3°58′44″ 6°12′25″ 5°05′35″Init. BRG – Φ – $\Delta_S/2$
Δ_S	26°02′32″	21°40′52″	23°58′51″	
Shifted chord, m	81.112	67.414	74.277	
CS E	6195.063	6189.879	6192.440	
CS N	7114.019	7118.660	7116.306	
θ SC to ST	126°19′35″	127°04′08″	126°41′32″	
ϕ =	1°19′35″	2°04′08″	1°41′51″	θ SC to CS – Φ – $\Delta_S/2$+ ϕ
Δ_S	26°02′32″	21°40′52″	23°58′51″	
Long chord, m	24.995	38.980	31.989	A mirror of the inbound curve
ST E	6215.200	6220.997	6218.089	
ST N	7099.212	7095.164	7097.190	
θ ST to ESD	125°00′00″	125°00′00″	125°00′00″	θ TS to SSD – Φ – $\Delta_S/2$
T_{ro}	37.426	23.000	19.000	
ESD E	6245.858	6239.822	6233.652	
ESD N	7077.746	7081.971	7086.292	P_2 is 125°, 5.001m from ESD

If it's not checked it's wrong.

13.6.7 Traverse via chainages around the curve.

With the basic traverse framework coordinated the coordinates of the chosen chainages on the straights, spirals and curve can be calculated.

All deflections have to be converted to bearings to calculate coordinates of the points.

Recalling spiral (§ 12.6.2, page 202) and circular formula:

Tangential angle $= \phi^r = \ell^2/2LR$ or $\phi = (\ell/L)^2 \times \Phi$ for $\ell_1, \ell_2, \ldots, \ell_n$

$$x = \ell\left(1 - \frac{\phi^2}{10} + \frac{\varphi^4}{216} - \ldots\right) \text{ for } \ell_1, \ell_2, \ldots, \ell_n \qquad y = \ell\left(\frac{\phi}{3} - \frac{\phi^3}{42} + \ldots\right) \text{ for } \ell_1, \ell_2, \ldots, \ell_n$$

Spiral deflection angle $\delta_S{}^o = \arctan(y/x)$ for $\ell_1, \ell_2, \ldots, \ell_n$ Spiral chord distance $d_S = (y^2 + x^2)^{1/2}$

Review formula: s_{CH} is the arc distance and Δ_{CH} is the deflection angle from SC to the chainage for radius R.

$\Delta_{CH} = s_{CH}/R$, chord $\ell_{CH} = 2R(\sin\Delta_{CH}/2)$, $x_{CH} = \ell_{CH}(\sin(\Delta_{CH}/2))$, $y_{CH} = \ell_{CH}(\cos(\Delta_{CH}/2))$.

The following diagrams are greatly exaggerated to show how the deflection angles are applied to the initial bearing of the ingoing tangent.

13.6.7.1 SSD to TS, the T_{ro}.

The distance from the SSD at CH1900 to the TS is the tangent runout, T_{ro}. The coordinates of the TS are on a constant BRG of 159° for a distance of 37.426m from CH1900.

13.6.7.2 TS to SC, S_{ro}

The 25m transition curve is coordinated at CH 1950, about halfway around the curve. Notice that the coordinates will be generated by radiation from the TS (Figure 13-16), which is on a constant BRG 159°, 37.426m from the SSD, CH 1900.

	CH	ℓ	x	y	δ	chord	BRG from TS.
TS	1937.426	0					159°00′00″
	1950	12.574	12.573	0.074	0°20′08″	12.574	158°39′52″
SC	1962.426	25.0	24.988	0.578	1°19′34″	24.995	157°40′25″

The check is that the SC coordinates match the traverse in § 13.6.6.

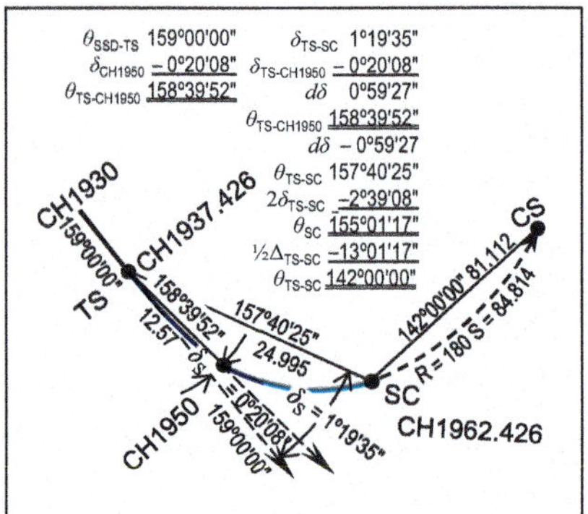

Figure 13-16 TS to SC

13.6.7.3 SC to CS, the shifted arc, S_{shift}.

This is the normal part of the circular curve. The tricky bit is calculating the bearing to the chainages from the SC. The coordinates of the curve chainages are calculated using radiated chords from the SC. (Figure 13-17).

	CH	Arc s	δ	$½\delta$	chord	BRG from TS.
SC	1962.426	0				155°01′17″
	1965.0	2.574	0°49′09″	0°24′35″	2.574	154°36′42″
	1980.0	17.574	5°35′38″	2°47′49″	17.567	152°13′28″
	1995.0	32.574	10°22′07″	5°11′03″	32.529	149°50′13″
	2010 – 2040 in like manner.					
CS	2044.240	81.814	26°02′32″	13°01′16″	81.112	142°00′00″

The check is that the CS coordinates match the traverse in § 13.6.6.

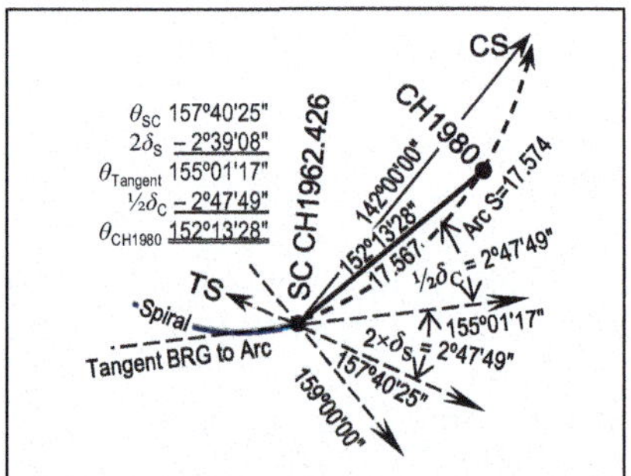

Figure 13-17 SC to CS

13.6.7.4 CS to ST, the S_{ro}.

Because the radius from the spiral at CS increases from radius, R, to infinity at the ST the easy way to calculate the deflections and chords is to work from the ST back to the CS. Thus ℓ = ST 2069.240 – 2055 = 14.2m.

(See Figure 13-18).

ST 2069.240 calculating back to the CS.

	CH	ℓ	x	y	δ	chord	BRG from TS.
	2055	14.24	14.759	0.119	0°27′44″	14.759	126°19′35″
CS	2044.240	25.0	24.988	0.578	1°19′34″	24.995	125°27′44″

The ST coordinates can be calculated from the CS forward, but the CH2055 coordinates must be calculated by the reciprocal BRG from ST.

The check is that the ST coordinates match the traverse in § 13.6.6.

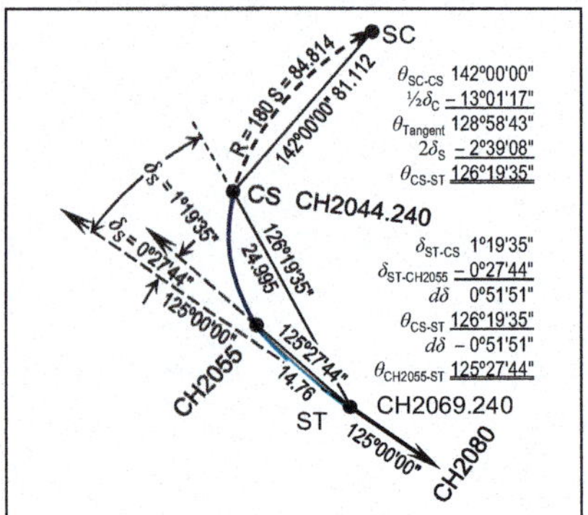

Figure 13-18 CS to ST

If it's not checked it's wrong.

13.6.7.5 ST to ESD, the T_{ro}.

From The ST to the ESD, the traverse is a straight at a BRG 125° for distance, T_{ro}, of 37.426m to the ESD. The traverse continues for another 5.001m to end at CH2111.667.

The check is that the ESD coordinates match the traverse in § 13.6.6.

13.6.8 Chainages and coordinates around the curve.

Table 13-4 is a summary of the coordinates generated for the three different curves. The original curve has an incorrect superelevation runout, S_{ro} = 25m. The rate of rotation from k_n to k_s is not constant.

The second curve basically maintains the superelevation development length, L_e, of 62m but with rounded, but sensible, T_{ro} = 23m and S_{ro} = 39m which maintain a constant rate of rotation from k_n to k_s.

The third curve aligns with AGRD03-16 table 7.11 for a single lane (3.5m) road at 70Kmh^{-1} with an L_e = 51m where T_{ro} = 19m and S_{ro} = 32m. Again, maintaining the rotation rate.

The coordinates are truncated MGA2020 Zone 50.

Table 13-4 Three scenarios

L_e = 62.43m, S_{ro} = 25m, T_{ro} = 37.43m				L_e = 62, S_{ro} = 39m, T_{ro} = 23m				L_e = 51m, S_{ro} = 32m, T_{ro} = 19m			
Point	Chainage	390000+	6450000+	Point	Chainage	390000+	6450000+	Point	Chainage	390000+	6450000+
SSD	1900	6122.218	7235.997	START	1900	6122.218	7235.997	Start	1900.000	6122.218	7235.997
	1920	6129.386	7217.325	SSD	1907.369	6124.859	7229.118	SSD	1914.900	6127.558	7222.086
	1930	6132.969	7207.990		1920	6129.386	7217.325		1920.000	6129.386	7217.325
TS	1937.426	6135.631	7201.057		1930	6132.969	7207.989		1930.000	6132.969	7207.989
	1950	6140.205	7189.345	TS	1930.369	6133.102	7207.645	TS	1933.900	6134.367	7204.348
SC	1962.426	6145.126	7177.936		1950	6140.304	7189.383		1950.000	6140.249	7189.362
	1965	6146.229	7175.610	SC	1969.369	6148.375	7171.782	SC	1965.900	6146.711	7174.837
	1980	6153.312	7162.393		1985	6156.116	7158.208		1980.000	6153.401	7162.430
	1995	6161.470	7149.811		2000	6164.625	7145.860		1995.000	6161.562	7149.849
	2010	6170.648	7137.951		2015	6174.132	7134.263		2010.000	6170.741	7137.991
	2025	6180.781	7126.897		2030	6184.571	7123.498		2025.000	6180.875	7126.938
	2040	6191.798	7116.725	CS	2037.183	6189.879	7118.659	CS	2040.714	6192.440	7116.306
CS	2044.24	6195.063	7114.019		2055	6203.776	7107.497		2055.000	6203.671	7107.481
	2055	6203.571	7107.467	ST	2076.183	6220.982	7095.164	ST	2072.714	6218.088	7097.189
ST	2069.24	6215.200	7099.212		2080	6224.124	7092.974		2080.000	6224.056	7093.010
	2080	6224.014	7093.041		2090	6232.315	7087.238		2090.000	6232.248	7087.275
	2090	6232.205	7087.305	ESD	2099.183	6239.822	7081.971	ESD	2091.714	6233.652	7086.291
	2100	6240.397	7081.569	END	2111.551	6249.954	7074.878	END	2111.615	6249.954	7074.878
ESD	2106.666	6245.858	7077.745								
END	2111.667	6249.954	7074.877								

13.7 Critique of the curve

The exercise curve was designed as a set out task. Trying to decipher the published parameters can lead to confusion if standard transition formulæ from Chapter 12 are attempted as the various parameters don't align.

Using the basic parameters from § 13.6.1.1 where:

L_e = 62m, V = 70Kmh^{-1}, R = 180, c = 0.3, k_n = –0.03 and k_s = +0.05 then, depending on the parameter chosen:

Note the value of c = 0.31 is the recommended absolute maximum for V = 70 (Austroads 2021, Table 7.5). Choosing any 3 of the 4 design parameters, with (k_s–k_n) = 0.08.

Set V, R, c. ⇒ L_e = 135m t = 7 seconds, ω = 0.012radians. sec^{-1} = 1.2%. sec^{-1}

Set L_e, V, R. ⇒ c = 0.66 t =3.2 seconds. ω = 0.025radians. sec^{-1} = 2.5%. sec^{-1} Table 13-1

Set L_e, c, R. ⇒ V = 54Kmh^{-1} t = 4.1seconds ω = 0.019radians.sec^{-1} = 2%. sec^{-1}

Set V, L_e, c. ⇒ R = 392m t = 3.2seconds, ω = 0.025radians.sec^{-1} = 2.5%. sec^{-1}

The circular radius used in the calculations is R = 180m which is slightly greater than the desirable minimum radius calculated using V and f parameters in Table 13-5.

13.7.1 Minimum radii recommendations

Combing recommendations from Austroads 2021 generally agree with Table 13-5

Table 7.5 f Recommended side friction, cars

Table 7.6 R Minimum radii,

Table 7.8 k_s = e_2, = 5% urban superelevation

Table 13-5 Minimum radii,

Speed	$f_{(des)}$	$f_{(max)}$	$Rmin_{(des)}$	$Rmin_{(abs)}$
70	0.2	0.3	160	110

13.8 References for Chapter 13

Elfick, M.H, J.G Fryer, R.C. Brinker and P.R Wolf. Elementary Surveying, Eighth edition. Longman, 1994. ISBN 0-00-499001-3

Austroads Ltd, Sydney, 2021. Edition 3.4 AGRD03-16: Guide to Road Design Part 3: Geometric Design. ISBN 978-1-922382-15-3

Perco, P., Desirable Length of Spiral Curves for Two-Lane Rural Roads Transportation Research Record Volume 1961, Issue 1, 2006, pp 1-8. https://doi.org/10.1177/0361198106196100101

Hickerson, T.F. Route Survey and Design. Fourth edition. McGraw-Hill Book Company, Inc. New York. 1959

Chapter 14 VERTICAL CURVES

Vertical curves are parabolic geometrical figures used to join one portion of a rising road section (rising gradient) to another portion of the road where the gradient changes. This chapter uses a parabolic equation of order 2, i.e., polynomial equation of order two to model the parameters that enable the setting out of a vertical curve. First, we show how the elements of the parabola are computed from partial differential equations (first and second derivatives). Next, we show how the high (summit) and low (sag) portion of the curves are computed and finally, we give examples based on the rural roads of Australia.

14.1 Introductory Remarks

At the end of this Chapter, you should:

- ✓ Understand and appreciate the geometric properties of a vertical curve.
- ✓ Understand what gradients are and the limitations that are imposed on their values.
- ✓ Appreciate the role of vertical curves in improving safety and comfort of passengers travelling from one intersecting gradient to another.
- ✓ Be able to design a vertical curve.

Worked Example 14.5 involves calculating the volume occupied by a series of vertical curves through modelled terrain.

14.1.1 Definition

The vertical curve, Figure 14-1, is a curve joining two different grades (or gradients) in a vertical profile. The purpose of the curve is to smooth the passage from one gradient to another.

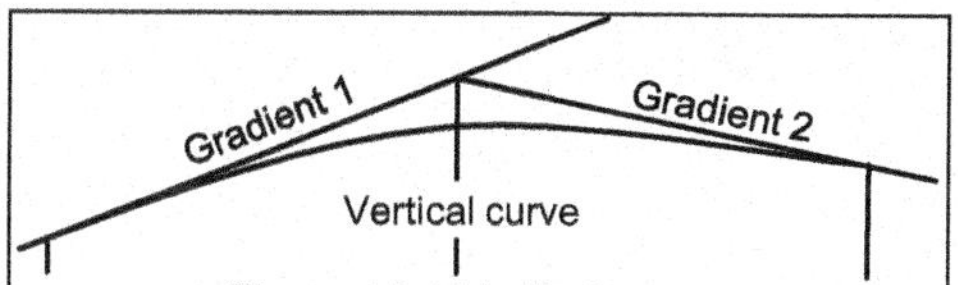

Figure 14-1 Vertical curve.

14.1.2 Uses

- ➢ Introduced between two intersecting gradients to smooth the passage from one gradient to another in the vertical plane.
- ➢ A vehicle travelling along a vertical curve experiences a radial force that tries to force the vehicle from the centre of curvature of the vertical curve. This forces the vehicle to leave the road in case of a crest design, while the vehicles could come into contact with a road surface for a sag design. Both result in discomfort and danger to passengers and are minimized through restricting the gradients and choosing a suitable type and length of curve.
- ➢ Adequate visibility: Vertical curves enable vehicles travelling at a given design speed to stop or overtake safely.
- ➢ Considerations include safety, riding comfort and sight distance.

14.1.3 Components of a Vertical Curve

The **gradient** or grade of a slope is expressed as a **percentage.** A slope is expressed as a ratio of **vertical** to **horizontal**, generally 1: n.

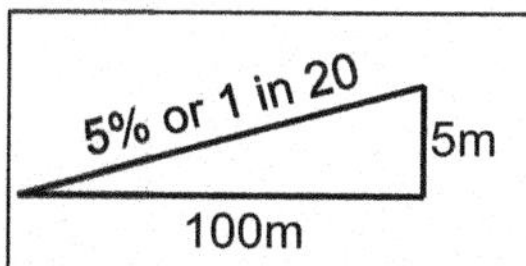

Figure 14-2 Gradient.

The gradient of a slope of 1 vertical in 20 horizontal (1:20)
= 1/20 = 0.05 = 5%, i.e.,
Gradient = 5m vertical for every 100m horizontal (see Figure 14-2).
The slope **angle** = arctangent (slope). 1:20 = 0.05 = arctan(0.05) = 2.862° = 2°52′.

A rising gradient in the direction of the running chainage is positive gradient $+g_1\%$ while a falling gradient in the direction of the running chainage is negative gradient $-g_2\%$. See Figure 14-3.

In road design a vertical curve is generally a **parabolic** curve in the longitudinal profile of the carriageway to provide a change of grade at a specified vertical acceleration. In developing examples of the vertical curve, the design criteria used is that published in Austroads, 2021.

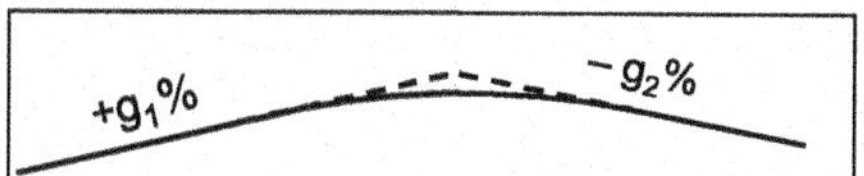

Figure 14-3 Positive/negative gradients

14.2 Elements of the Vertical Curve

Two gradients intersect at a point V, the Point of Intersection (PI). The gradients are **tangential** to a parabolic section running from the initial Point of Curvature (PC) to the final Point of Tangency (PT).

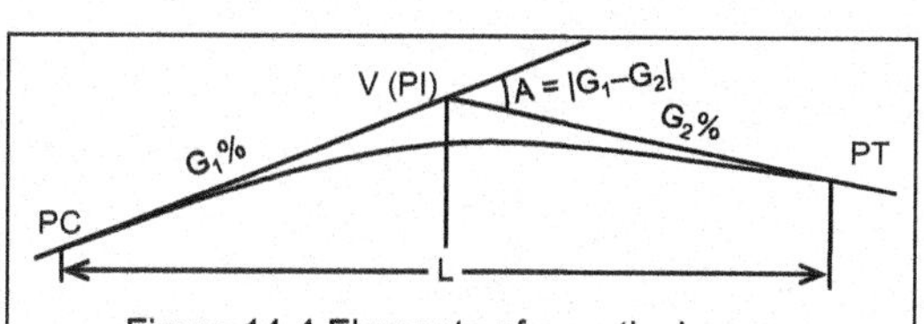

Figure 14-4 Elements of a vertical curve.

Notations: (see Figure 14-4)

G_1 Initial gradient, expressed as a percentage
G_2 succeeding gradient, expressed as a percentage

J. Walker and J. Awange, *Surveying Computations*,
https://doi.org/10.1007/978-3-031-94382-9_14

V or PI Vertex or Point of Intersection
PC Point of Curvature, the start of the curve
PT Point of Tangency, the end of the curve
L Length of the curve. It is measured **horizontally** in units of 100m.
A Gradient difference between initial and subsequent gradient.

A vertical curve with a horizontal distance of 350m has **L=3.5**. The arc length of the parabola can be found by summing the chord lengths between chainage stations.

14.2.1 Two Properties of a Parabola:

Figure 14-5 shows the **first** of two properties of a parabola. The horizontal projections of the tangents at PC and PT are equal. They are equal to half the horizontal projection of the curve, i.e., ½*L*.

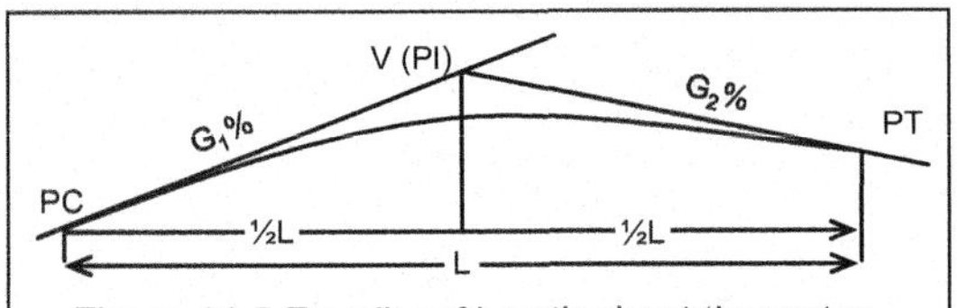

Figure 14-5 Equality of length about the vertex.

The **second** property is that the rate of change of grade of the curve is constant. The gradient at any point along the parabola can be expressed using the horizontal length of the curve.

Rate of change of grade	$r\%\% = \dfrac{G_2 - G_1}{L}$	Eqn 14-1

where **r%%** can be read as **r "percent per 100"**

Figure 14-6 shows a curve between two grades, $G_1 = +5\%$, $G_2 = -4\%$, joining a parabolic curve of length = 300m. L = 3. Find r.

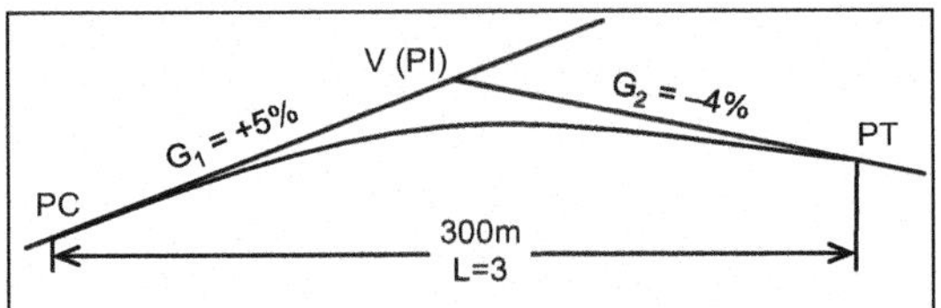

Figure 14-6 Change of gradient.

From Eqn 14-1:

$$r\%\% = \frac{(-4)-(+5)}{3} = \frac{-9}{3} = -3\%\%$$

To find the gradient of the curve at PC + 100 where: (see Figure 14-7)

final gradient = Initial gradient + gradient change.

At PC + 100: L = 1, r = –3

$G_{100} = 5 + (-3\times1)$
$= 2\%$

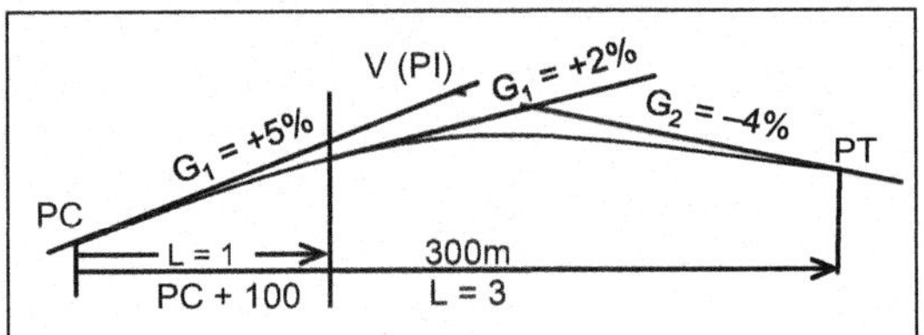

Figure 14-7 Gradient varies with *L*.

14.2.2 Sign of **r**

If r > 0 (positive) = sag curve
If r < 0 (negative) = crest curve as shown in Figure 14-8.

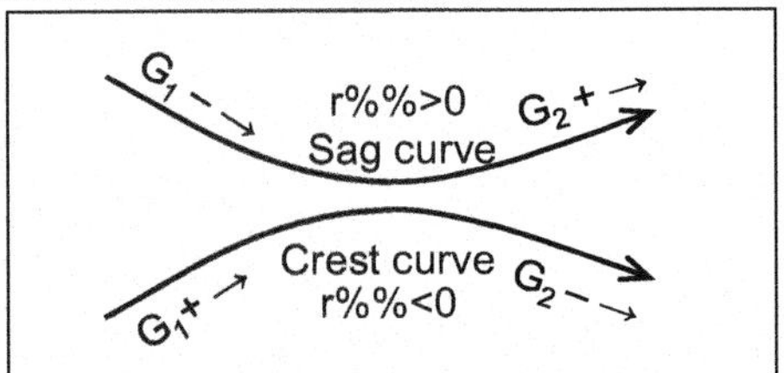

Figure 14-8 Sag and crest curves.

If G_1 and G_2 are numerically equal, but opposite sign, then the high or low point will be at the midpoint. Otherwise it must be calculated.

Crest curves, see Figure 14-9, occur when the succeeding gradient is less than the initial gradient.

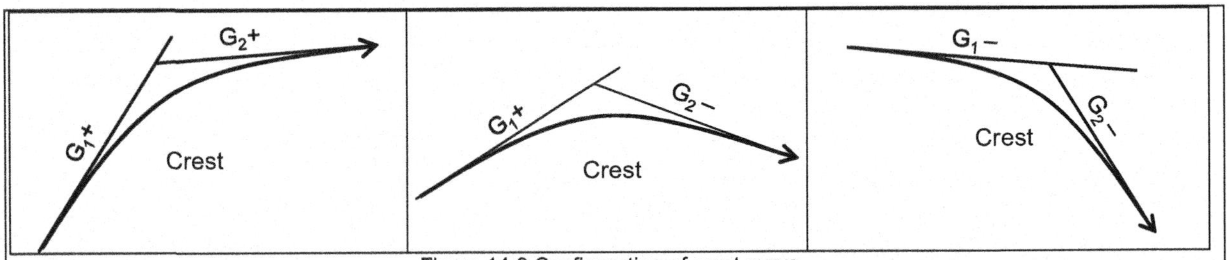

Figure 14-9 Configuration of crest curve.

Sag curves, see Figure 14-10, occur when the succeeding gradient is greater than the initial gradient.

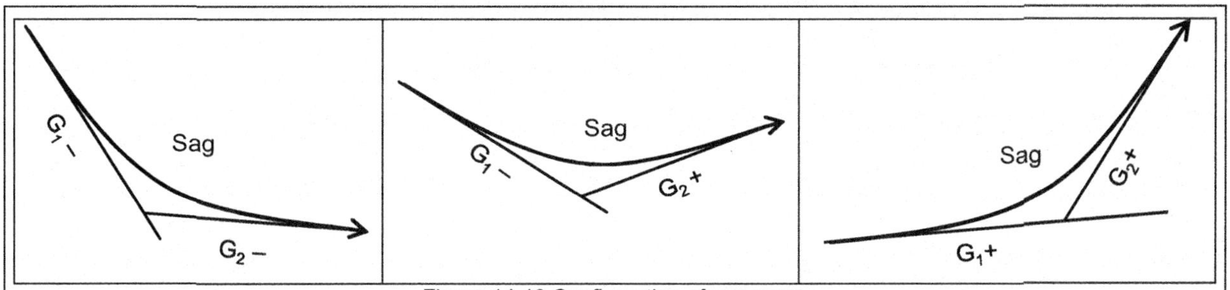

Figure 14-10 Configuration of sag curve.

14.3 Calculus of the Parabola

The parabola is a conic curve of a second-degree equation. If there is no xy term in the expression, then the principal axis is either horizontal, $y = \pm kx^2$ or vertical, $x = \pm ky^2$. The shape, or rate of change of gradient, is given by the constant, ***k***.

The general form of the parabola, with its principal axis parallel to the vertical axis is

General parabola	$y = ax^2 + bx + c$	Eqn 14-2

This equation will be developed to enable calculation of the vertical coordinates, the RLs, of points on the curve.

Reference Figure 14-11, ***y*** will be the RL of any point on the curve at a horizontal distance, ***x***, from the PC.

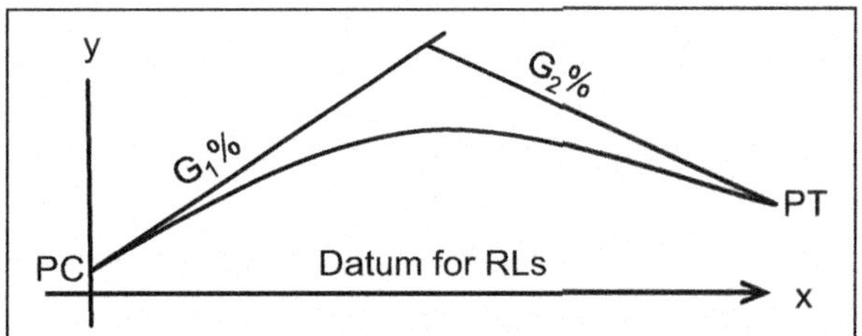

Figure 14-11 Parabola axes and tangent gradients

14.3.1 Developing the General Equation

Convert the general equation: $y = ax^2 + bx + c$ into an expression covered by the elements of the vertical curve.

1. Find the value of y at $x = 0$

At $x = 0$	$y = c =$ RL of PC	Eqn 14-3

2. The first derivative of Eqn 14-2 provides the gradient:

$$\frac{dy}{dx} = 2ax + b \text{, and again}$$

When $x = 0$, $\frac{dy}{dx} = b$ i.e.:

$b =$ gradient of curve when $x = 0$

$\boxed{b = G_1}$

3. The second derivative of Eqn 14-2 of a curve provides the rate of change of grade:

$\frac{d^2y}{d^2x} = 2a$. We have defined ***r*** as the rate of change of gradient. Thus:

Rate of change of gradient	$r = 2a$ or $a = \frac{1}{2}r$	Eqn 14-4

4. Substitute derived constants into Eqn 14-2 gives the expression:

Value of y	$y = \frac{1}{2}rx^2 + G_1x + RL_{PC}$	Eqn 14-5

14.3.2 Deriving Data from the General Equation

The general parabolic Eqn 14-2, for the vertical curve is expressed in terms related to the design of the curve (i.e., Eqn 14-5), is defined as:

$y = \frac{1}{2}rx^2 + G_1x + RL_{PC}$ where

r = rate of change of gradient (as percent per 100)

G_1 = initial gradient of slope (as a percentage)

RL_{PC} = RL of Point of Curvature, at the start of vertical curve.

14.3.3 High or Low Point of the Vertical Curve

The high/low point of the curve is the point at which the gradient of the curve is zero.

Find the value of x where slope = 0

Take first derivative, using Eqn 14-5, $\frac{dy}{dx} = rx + G_1$.

High/low point of parabola	$x = \frac{-G_1}{r}$	Eqn 14-6

Gradient is zero when $\frac{dy}{dx} = 0$, i.e., $rx + G_1 = 0$

where x = distance from PC in units of 100.

Chainage high/low point	Chainage high/low point $=$ Chainage PC $+ 100 \times \frac{-G_1}{r}$	Eqn 14-7

Note: If the high/low point is on the parabola between the PC and PT then $\frac{-G_1}{r}$ will be positive because G_1 and r will be opposite signs.

Taking the vertical curve in Figure 14-6 as an example, remember G_1 is negative in the equation: Eqn 14-1

$r = -3$

$G_1 = 5$

$x = \frac{-5}{-3} = 1.67$, the chainage of the high point is thus:

$CH_{PC} + 100x = CH_{PC} + 167$.

14.3.4 Level of High or Low Point of the Vertical Curve

The height (Δh) of the high/low point of the curve is the product of the distance at which the gradient of the curve is zero, x and gradient of the line AP. See Figure 14-12.

Line AP is parallel to the tangent through the mean point of the **parabola** AP. The gradient of the curve at the mean point will be the mean of the gradients at the end points, A and, in this case, P, where the gradient is zero.

Examine the gradients:

Gradient at A $= G_1$

Gradient at P $= 0$

Gradient AP $= \frac{1}{2}(G_1 + 0) = \frac{1}{2}G_1$

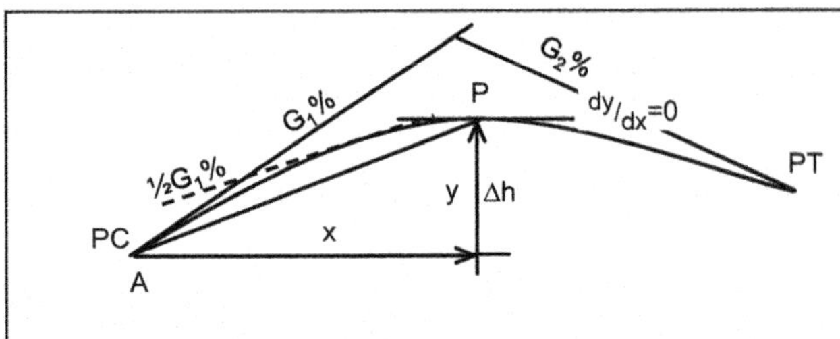

Figure 14-12 High/low point. Gradient zero

Height of high/low point	$\Delta h = \frac{1}{2} G_1 x$	Eqn 14-9

The RL of the low point will use the same formula because G_1 is negative.

RL high/low point	$RL_P = RL_{PC} + \Delta h = RL_{PC} + \frac{1}{2} G_1 x$	Eqn 14-8

It can also be seen that, substituting x from :

Alternate Δh	$\Delta h = \frac{-\frac{1}{2} G_1^2}{r}$	Eqn 14-10

14.3.4.1 Worked Example 14.1

Two intersecting profile gradients consist of a +5% gradient meeting a – 3% gradient.

They are to be joined by a parabolic vertical curve 200m long. $L = 2$. See Figure 14-13.

Chainage of the PC = 1400m,

RL PC = 299m.

Calculate the chainage and RL of the highest point.

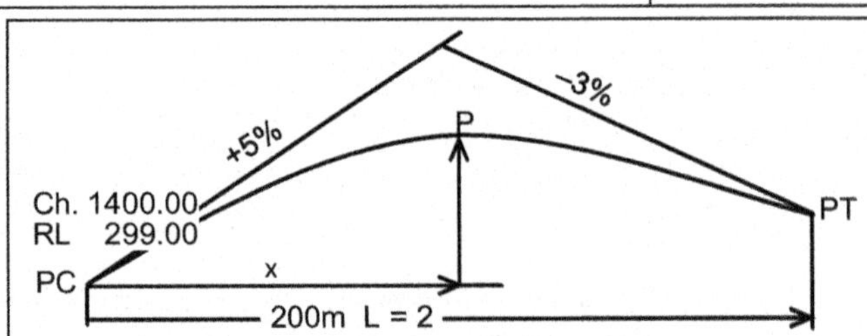

Figure 14-13 Worked Examples 14.1 and 14.2.

1. Calculate r%% using Eqn 14-1:
$$r=\frac{G_2-G_1}{L}=\frac{-3-(+5)}{2}=-4$$
2. Chainage to high point, P: (Eqn 14-6):
$$x=\frac{-G_1}{r}=\frac{-5}{-4}=1.25\times 100=125m$$
Chainage P = $CH_{PC}+x$ = 1400 + 125 = **1525**m.
3. Δh of high point, P: (Eqn 14-9):
$$\Delta h=\frac{1}{2}G_1x=\frac{1}{2}\times 5\times 1.25=3.125$$
RL of high point, P = $RL_P + \Delta h$ = 299 + 3.125 = **302.125**m.

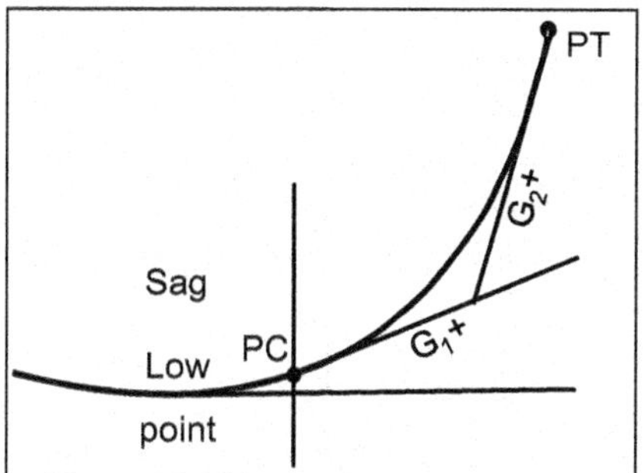

Figure 14-14 Low point outside curve.

Note that the high/low point of the parabola $\left(\frac{dy}{dx}=0\right)$ may not occur on the section of parabola under investigation (see Figure 14-14).

14.3.5 The Midpoint of the Vertical Curve

The "external" distance, ***e***, is the vertical distance from the vertex, or point of intersection, of the two gradients to the midpoint of the curve. In Figure 14-15, the gradient G_1 has been produced to point C vertically above the PT.

Horizontal distance PC to PT = L (units of 100).

Angle A is the algebraic change of grade. $A=|G_2-G_1|$

N is the midpoint of the **chord** PC to PT, and is vertically below V.

Note that, if $|G_1|=|G_2|$ then M is the high point, P, (Figure 14-12).

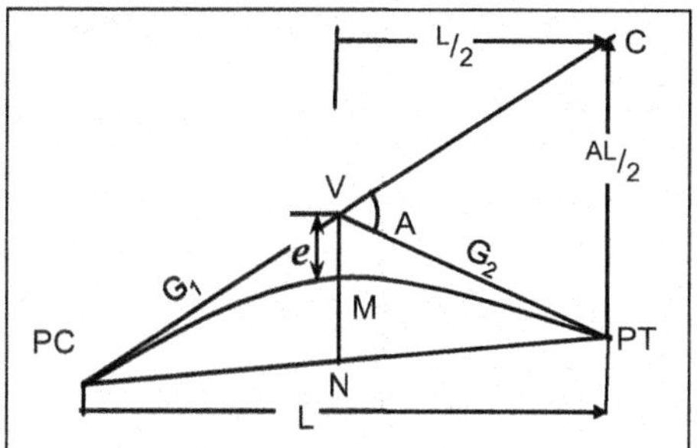

Figure 14-15 Mid point of vertical curve.

1. The Δh of the PT with reference to PC
$\Delta h_{PT}=\Delta h(\text{PC to V})-\Delta h(\text{V to PT})$
remember that G_2 is negative
$\Delta h_{PC\text{-}PT}=½(G_1L)+½(G_2L)=½(G_1+G_2)L.$
2. It follows that Δh between C and PT
$=\Delta h(\text{PC to C})-\Delta h(\text{PC to PT})$
$=G_1L-½(G_1+G_2)L$ $\quad=½(G_1-G_2)L$, but $G_1-G_2=|G_2-G_1|=A$
$=½AL$ or $\frac{AL}{2}$
3. Similarly: $VN=\Delta h(\text{PC to V})-\Delta h(\text{PC to N})$
$=½G_1L-½(½(G_1+G_2)L)$
$=¼(G_1-G_2)L$
$=¼AL$ or $\frac{AL}{4}$.

One of the properties of the parabola is:

Offsets from the tangent vary with the square of the distance from the PC.

In general terms: $e=\left(\frac{n}{m}\right)^2\times\frac{AL}{2}$ and $\frac{AL}{2}$ is a constant for the curve.

The offset to V from the tangent at M (this is ***e***), which is ½L from the PC
$$h_{VM}=e=(½)^2L\times\frac{A}{2}=¼L\times\frac{A}{2}=\frac{AL}{8}$$
This shows that e = ½ the difference between the RL of V and the RL of N.

Using the squared offset method, a number of other "e" values can be found such that, if L is the length of the curve:

for ¼L: $e=\frac{1}{16}\cdot\frac{AL}{2}$ for ¾L: $e=\frac{9}{16}\cdot\frac{AL}{2}$.

14.3.5.1 Worked Example 14.2

Two intersecting profile gradients consist of a +5% gradient meeting a –3% gradient. They are to be joined by a parabolic vertical curve 200m long. (See Figure 14-16).

Find the offset, ***e***, from the initial gradient G_1 at the midpoint and also at the ¼ and ¾ point of length of the curve.

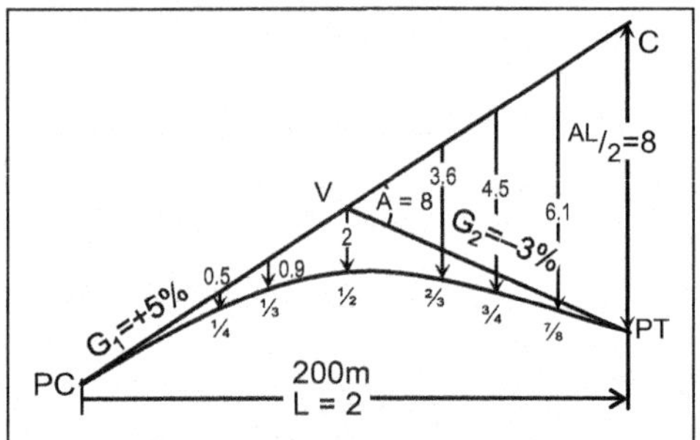

Figure 14-16 Worked Example 14.2.

From § 14.3.4.1, Worked example 4.1

Worked Example 14.1 $L = 2$

$G_1 = +5$

$G_2 = -3$

$A = |G_2 - G_1| = 8$

$$\frac{AL}{2} = \frac{2\times 8}{2} = 8$$

$$e_{\text{midpoint}} = \frac{1}{4}\times\frac{AL}{2} = \frac{AL}{8} = \frac{2\times 8}{8} = 2\text{m}$$

$$e_{\frac{1}{4}} = \frac{1}{16}\times\frac{AL}{2} = \frac{AL}{32} = \frac{1\times 2\times 8}{32} = 0.5\text{m}$$

$$e_{\frac{3}{4}} = \frac{9}{16}\times\frac{AL}{2} = \frac{9\times AL}{32} = \frac{9\times 8\times 2}{32} = 4.5\text{m}$$

Similarly, the offset from the initial gradient can be found at any portion of the length of the parabola once $\frac{AL}{2}$ has been evaluated.

e.g. $e\frac{1}{3} = \frac{1}{9}\times 8 = 0.89\text{m}$

$e\frac{2}{3} = \frac{4}{9}\times 8 = 3.56\text{m}.$

The technique has limited application compared with modern calculation methods, but the midpoint value of $e = \frac{AL}{8}$ is a handy check of the calculation.

14.3.6 Pass a Curve Through a Fixed Point

It is often necessary to have a curve pass through a fixed point when two gradients are being joined. A practical example is the requirement to provide a certain amount of "clearance" below an overpass, or a bridge.

The task may be to find the **length** of the parabola joining two designated grades.

Or it may be necessary to find the **gradient** of the succeeding gradient if the length of the parabola is fixed.

The governing variable in these cases is the value ***r***, the rate of change of gradient (r%%), used in the parabolic formula:

$$y = \tfrac{1}{2}rx^2 + G_1x + RL_{PC},\ \text{(Eqn 14-5)},\qquad \text{where } r = \frac{G_2 - G_1}{L}$$

and x is units of 100.

14.3.6.1 Worked Example 14.3

A road is to be redesigned using a parabolic vertical curve to provide clearance beneath a bridge (see Figure 14-17).

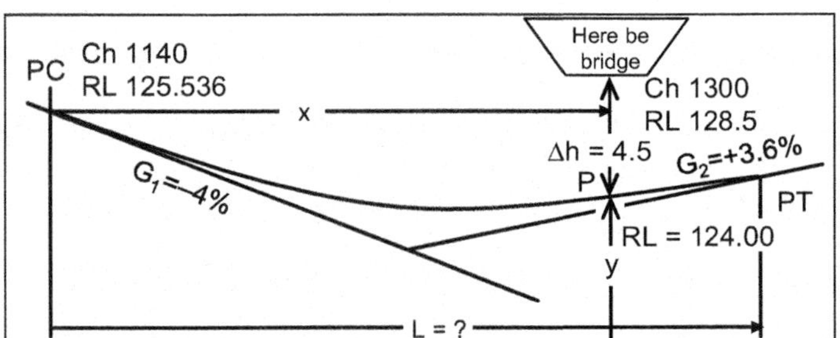

Figure 14-17 Worked Example 14.3.

The PC has chainage 1140.00m, with an RL of 125.604m.

The initial gradient, $G_1 = -4\%$.

A bridge at chainage 1300.00 has an RL = 128.5m at its lowest point.

You must provide **4.5m** clearance from the underside of the bridge. The parabola will finish with a succeeding gradient, $G_2 = +3.6\%$.

What is the length of the curve?

Formula of vertical curve: (Eqn 14-5),

$y = \frac{1}{2}rx^2 + G_1x + RL_{PC}$,

Data: x = (1300 – 1140)/100 = 1.6 (units of 100 to allow for grade %)

y = 128.5 – 4.5 = 124.0 (RL of clearance)

RL_{PC} = 125.536

G_1 = – 4%

G_2 = +3.6%

Using the formula of vertical curve above, find **r%%**: to allow calculation of L.

$$124.0 = \tfrac{1}{2}r\times(1.6)^2 + (-4\times1.6) + 125.536$$

$$r = \frac{2(y - Gx - RL_{PC})}{x^2} = \frac{2\times(124-(-4\times1.6)-125.536)}{1.6^2} = 3.8$$

But: $r = \dfrac{G_2 - G_1}{L}$, thus $L = \dfrac{G_2 - G_1}{r} = \dfrac{3.6-(-4)}{3.8} = 2.0$ in units of 100

Length of curve = $L \times 100$ = **200**m

14.4 Calculate the RLs of Points Along the Vertical Curve

Once a vertical curve has been designed it is necessary to generate a tabulated list of RLs at intermediate stations. As with the horizontal curve, they will most likely be at regular whole number chainages. Remember that for the vertical curve, we measure the chainages horizontally from the PC.

A constant pegging interval allows a check of the calculations to be performed using the second difference between RLs.

The first difference (Δ') is the difference between successive RLs. The second difference (Δ") is the difference between successive first differences, the Δ's. As the parabola is a second-degree equation the Δ" must be constant for a constant interval.

The first and second differences can be tabulated in the pegging data.

Chord length can be tabulated to find approximate arc length.

14.4.1 Worked Example 14.4. Single Vertical Curve

Figure 14-18 shows the design of a vertical curve to be pegged at 20m intervals from the PC to the PT over a length of 200m.

Using the vertical curve in Worked Example 21.3, and Figure 14-18 for visualisation:

Tabulate the RLs, show the second difference similarities.

Find the chainage and RL for the low point. Use chords to find arc length.

Ch PC = 1140.00m,

RL PC = 125.536m.

Ch PT = 1340.00m.

G_1 = –4% , G_2 = +3.6%.

L = 2.0,

r = $(G_2 – G_1)/L$

= (3.6 – (–4))/2

r = +3.8 (sag curve).

Create the formula for the curve in terms of x from:

x = 0, 0.2, 0.4, …, 1.8, 2:

$RL_{Ch} = \frac{1}{2}rx^2+G_1x+RL_{PC}$

$= 1.9x^2 - 4x + 125.536$.

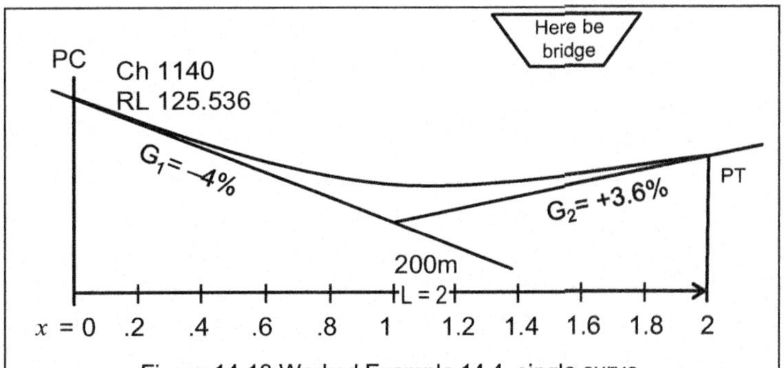

Figure 14-18 Worked Example 14.4, single curve.

Table 14-1 shows the results of the calculations for RL at each chainage as well as the first and second RL differences between the chainages. The arc length is the sum of the chords between each chainage pair.

Table 14-1 Worked Example 14.4. Table of chainages, RL, first & second differences, chords. Checks.

Chainage	x	RL	Δ′	Δ″	Chord
1140	0.0	125.536			
			–0.724		20.013
1160	0.2	124.812		0.152	
			–0.572		20.008
1180	0.4	124.240		0.152	
			–0.420		20.004
1200	0.6	123.82 0		0.152	
			–0.268		20.002
1220	0.8	123.552		0.152	
			–0.116		20.000
1240	1.0	123.436		0.152	
			0.036		20.000
1260	1.2	123.472		0.152	
			+0.188		20.001
1280	1.4	123.660		0.152	
			+0.340		20.003
1300	1.6	124.000		0.152	
			+0.492		20.006
1320	1.8	124.492		0.152	
			+0.644		20.010
1340	2.0	125.136			
Arc Length =					**200.047**

First difference, $\Delta' = RL_{n+1} - RL_n$

Second difference, $\Delta'' = r(dx)^2$

where: dx = peg interval/100

$= 20/100$

$= 0.2$

i.e., $\Delta'' = r(dx)^2$

$= 3.8\times(0.2)^2$

$= \mathbf{0.152}$

Check $\Delta'' = \Delta'_{n+1} - \Delta'_n$

e.g.,

$\Delta'_{1160} - \Delta'_{1140} = -0.572-(-0.724)$

$= 0.152$

The **chord** of each interval (dx) is calculated using the first difference Δ′.

$$\text{Chord}_{(n,\ n+1)} = \sqrt{(100dx)^2 + (\Delta')^2_{(n,n+1)}}$$

$$\text{e.g., Chord}_{(1140,\ 1160)} = \sqrt{(100\times 0.2)^2 + (-0.724)^2}$$

$= 20.013.$

Arc length = Σ(Chords)

Find the low point:

Distance to low point Eqn 14-6: $x = \dfrac{-G_1}{r} = \dfrac{-(-4)}{3.8} = 1.053$ (distance = $100x$ = 105.3)

Chainage low point = 1140 + 105.3 = **1245.30**, checks by inspection against pegging sheet.

RL of low point at distance × (Eqn 14-8):

$RL = \frac{1}{2}G_1x + RL_{PC} = -2\times1.053 + 125.536 = -2.105 + 125.536$

RL low point = **123.431**, checks against pegging sheet between Ch1240 & 1260.

14.4.2 Worked Example 14.5. Multiple Vertical Curves

Figure 14-19 shows the design of two vertical curves, joined by straight sections, to be pegged at 15m intervals over a length of 105m.

The RL at CH 00 is 10.000

Vertical Curve design:

1. The path rises at a slope of +2% between CH 00 and CH 15.
2. The path increases, through a **sag** curve, to a slope of +5% over a length of 15m.
3. The path continues a straight slope (5%) to CH75.
4. The path flattens, via a **crest** curve, to a slope, 0%, between CH 75 and CH105.

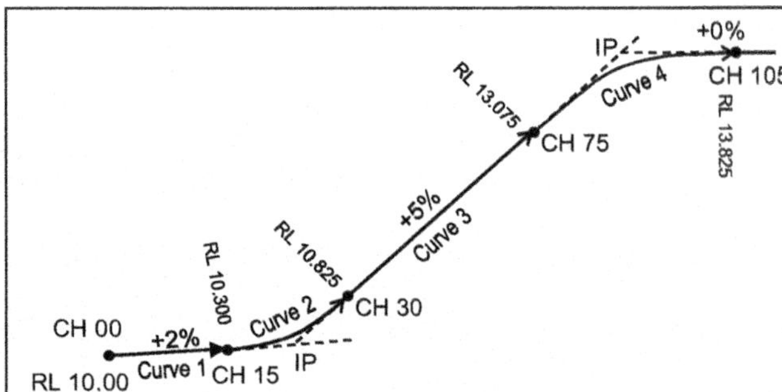

Figure 14-19 Joining multiple vertical curves.

The calculation is broken into four separate parts, the main task being to calculate the RLs at each change of grade. The formula for calculating the RLs over multiple curves replaces the term for the level at the "point of curvature", RL_{PC}, with the expression, "beginning of vertical curve", $\mathbf{RL_{BVC}}$.

The RL at the **end** of one vertical curve is the RL at the **beginning** of the next vertical curve.

Only one formula is uses: Eqn 14-5. $RL_{CH} = \frac{1}{2}rx^2 + G_1x + RL_{BVC}$.

By definition, the rate of change on a straight slope is, $r = 0$, because $G_1 = G_2$.

1. Straight gradient: Ch00 to CH15, L = 15m = 0.15
   CH 00, RL_{BVC} = 10.000, G_1 = +2%, G_2 = +2% This RL_{BVC} is the start of curve 1.
   CH 15, x = 0.15 (15m)
   $r = (G_2 - G_1)/L = (2 - 2)/0.15 = 0$
   $RL_{CH15} = \frac{1}{2}rx^2 + G_1x + RL_{CH00} = 0+2\times0.15+10 = 10.300.$ This is RL_{BVC} curve 2.
2. Sag curve, CH15 to CH30, L =15m = 0.15
   CH 15, RL_{BVC} = 10.300, G_1 = +2%, G_2 = +5%. $r = (G_2 - G_1)/L = (5 - 2)/0.15 = 20$

CH 30, x = 0.15 ($CH30$ – CH15 = 15m)
RL_{CH30} = $\frac{1}{2}rx^2 + G_1x + RL_{CH15}$
= ½(20×0.15²)+2×0.15+10.300 = 0.225+0.3+10.3 = 10.825. RL_{BVC} curve 3.

3. Straight gradient, CH30 to CH75, L =45m = 0.45
CH 30, RL_{BVC} = 10.825, G_1 = +5%, G_2 = +5%,
$r = (G_2 - G_1)/L = (5 - 5)/0.45 = 0$
CH 45, x = 0.15 (CH45 – CH30 = 15m)
RL_{CH45} = $\frac{1}{2}rx^2 + G_1x + RL_{CH30}$ = 0+5×0.15+10.825 = 0.75+10.825 = 11.575
CH 60, x = 0.3 (CH60 – CH30 = 30m)
RL_{CH60} = $\frac{1}{2}rx^2 + G_1x + RL_{CH30}$ = 0+5×0.30+10.825 = 1.50+10.825 = 12.325
CH 75, x = 0.45 (CH75 – CH30 = 45m)
RL_{CH75} = $\frac{1}{2}rx^2 + G_1x + RL_{CH30}$ = 0+5×0.45+10.825 = 2.25+10.825 = 13.075 RL_{BVC} curve 4.

4. Crest curve, CH75 to CH105, L =30m = 0.30
CH 75, RL_{BVC} = 13.075, G_1 = +5%, G_2 = +0%,
$r = (G_2 - G_1)/L = (0 - +5)/0.3 = -16.67$
CH 90, x = 0.15 (CH90 - CH75 = 15m)
RL_{CH90} = $\frac{1}{2}rx^2 + G_1x + RL_{CH75}$
= ½(–16.67×0.15²)+5×0.15+13.075 = –0.188+0.75+13.075 = 13.638
CH 105, x = 0.30 (CH105 – CH75 = 30m)
RL_{CH105} = $\frac{1}{2}rx^2 + G_1x + RL_{CH75}$
= ½(–16.67×0.30²)+5×0.30+13.075 = –0.750+1.50+13.075 = 13.825. RL **end** of curve.

Table 14-2 shows the results of solving the above problem (§ 14.4.2), including first and second differences, and chord lengths.

Table 14-2 Vertical curve profile with multiple slopes.

Chainage	x	RL	Δ′	Δ″	Chord
00	0.0	10.000			
			+0.300		15.003
15	0.15	10.300		0.225	
			+0.525		15.009
30	0.15	10.825		0.225	
			+0.750		15.019
45	0.15	11.575		0.000	
			+0.750		15.019
60	0.3	12.325		0.000	
			+0.750		15.019
75	0.45	13.075		-0.187	
			+0.563		15.010
90	0.15	13.638		-0.376	
			+0.187		15.001
105	0.3	13.825			
Arc Length =					**105.080**

Note the two sets of formula.
One for the curve:
$y = \frac{1}{2}rx^2 + G_1x + RL_{PC}$
and one for the straight:
$y = G_1x + RL_{PC}$
The curve formula can be used for the straight because, as $G_2 = G_1$.
and $r = (G_2 - G_1)/L$,
then $r = 0$ and thus
$\frac{1}{2}rx^2 = 0$.
This may be of help in generalizing a spreadsheet application.

14.5 Calculate End Areas and Excavation Volume

Chapter 6, Areas of regular and rectilinear figures, page 73 introduced the calculation of cross section areas in route excavations. The methods are detailed in § 6.15, page 82. Combining longitudinal and cross section profiles imposed on a vertical curve allows a volume of, in the following case, to be calculated in accordance with § 6.16.

14.5.1 Worked Example 14.6.

To calculate the formation level, batters, end areas and volume of a vertical curve between CH 00 and CH 105 from a previously conducted area levelling exercise.

1. You are given the parameters of a vertical curve designed to link the path at CH 00 through to the path at CH 105 traversing the pine grove to the North of BLD 204.

You are provided with the Ground Level (GL) and cross fall on the centre line (Table 14-3) at:
CH 00, CH 15, CH 30, CH 45, CH 60, CH 75, CH 90 and CH 105.

3. Your design is for a path of formation width 2.0m, excavated to formation level defined by the vertical curve design parameters. The side batters have a slope, m, of 1:3 to the surface.

4. Calculate all the parameters of the path at each chainage point:
   the RL of the path
   depth of cut, h, at the centre line between GL and path RL.
5. Path design: formation width; b = 2.0m, batter slope, m = 3.
6. Figure 14-20 shows formation design.

Table 14-3 Ground RL and cross slopes

CH	GL	K left	K right
0	10.0	15	10000
15	11.42	14	−15
30	12.44	12	−10
45	13.17	13	−17
60	13.56	15	−25
75	13.76	13	−18
90	13.98	11	−16
105	13.83	10000	10000

Calculate:

- the batter width, WL and WR, at the intersection with the ground slope, k,
- RL of batter slope intersection with ground slope k (RL_L, RL_R)
- cross section area by formula or coordinate cross multiplication
- volume by end area method and prismoidal method.

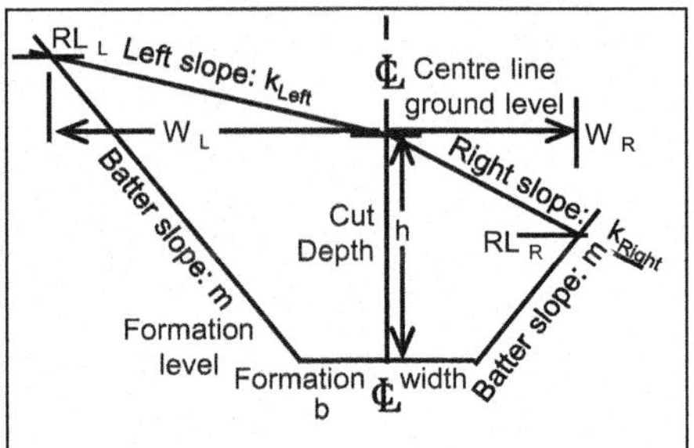

Figure 14-20 Formation design

Path Vertical Curve design parameters:

1. The path rises at a slope of +2.0% between CH 00 and CH 15.
2. The path increases, through a sag curve to a slope of 5% over a length of 15m to CH30.
3. The path is straight to CH75.
4. The path flattens, via a crest curve, to a flat slope, 0%, between CH 75 and CH105.

Provide a summary of values.

1. Calculated path RL at designated chainages.
2. Calculated h at each chainage.
3. Calculated batter intercepts, distance from CL and RL.
4. Calculated end areas
5. Calculated volume of excavation
6. Draw profile of centre line GL and path formation level to scale horizontal 1:400, Vertical 1:40

14.5.1.1 Computations for Worked Example 14.6

The levels of the formation have been calculated in § 14.4.2 in Worked Example 14.5. A table of chainages cut depth, batter width, batter RL and cross section area is produced.

Batter widths and RLs at batter intercept with the surface are shown in Table 14-4.

Figure 14-21 show the ground level and excavation level profile.

14.5.1.1.1 Area from Eqn 6-7, page 86.

$$Area = \frac{1}{2m}\left(\left(\frac{b}{2}+mh\right)\left(W_1+W_2\right)-\left(\frac{b^2}{2}\right)\right)$$

Table 14-4 Worked Example 14.6 calculations

CH	h	W_{left}	RL_{left}	W_{right}	RL_{right}	Area
0	0.000	1.250	10.083	1.000	10.000	0.375
15	1.120	5.549	11.816	3.633	11.178	26.200
30	1.615	7.793	13.089	4.496	11.990	45.206
45	1.595	7.521	13.749	4.917	12.881	45.414
60	1.235	5.881	13.952	4.201	13.392	30.800
75	0.685	3.972	14.066	2.619	13.615	13.656
90	0.342	2.788	14.233	1.707	13.873	6.052
105	0.005	1.015	13.830	1.015	13.830	0.061

14.5.1.1.2 Volumes from § 6.16, page 88.

Volume by end areas: Eqn 6-10.

L = 15m.

$$Volume = \frac{L}{2}\left(A_1 + A_n + 2\left(A_2 + A_3 + \ldots A_{n-1}\right)\right)$$

$V = 2513m^3$

Prismoidal volume: Eqn 6-11

L = 15m.

$$Volume = \frac{L}{3}\left(A_1 + A_n + 4\left(\Sigma A_{even}\right) + 2\left(\Sigma A_{odd}\right)\right)$$

$V = 2523m^3$

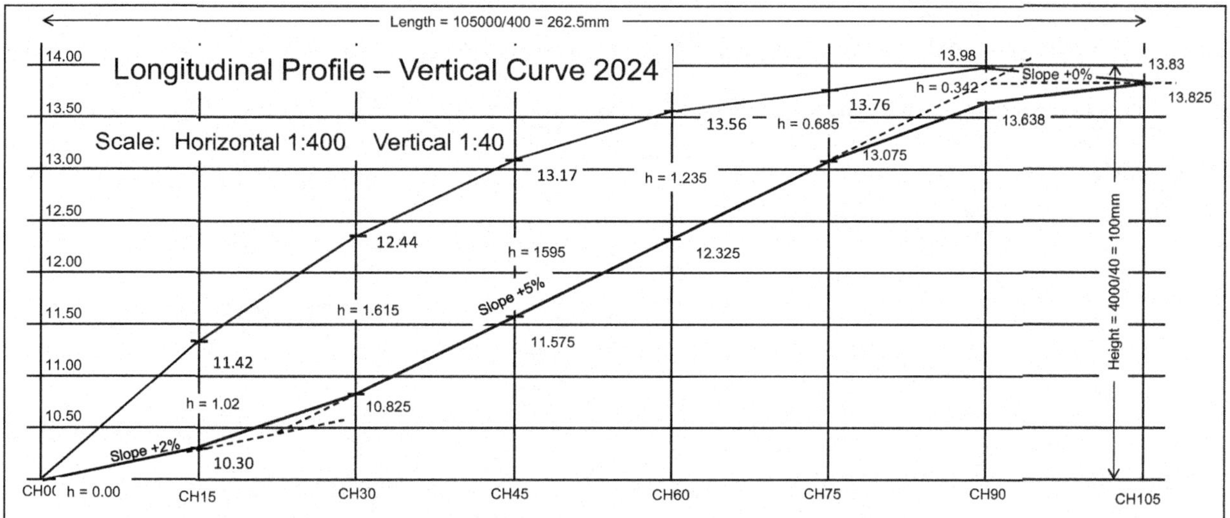

Figure 14-21 Longitudinal profiles and depth of cut (not to scale).

14.6 Vertical Curves in Rural Road Design

To the surveyor, the common use of the vertical curve is in rural road design and construction.

Refer to Austroads, 2021, Chapter 5. Sight Distance (page 129), and commentaries C11 – C 17 (pp 359 – 366) comprehensive treatment of vertical alignment.

The vertical profile of a road is a series of grades joined by parabolic curves.

Grades should be as flat as possible taking into consideration:

terrain,
longitudinal drainage,
economy.

Steeper grades produce greater variations in vehicle speed between vehicles with varying power to weight ratios. These variations lead to higher relative speeds between vehicles, increasing risk of rear end accident rates. They also lead to traffic queuing and overtaking requirements, which increase traffic safety problems, especially at higher traffic densities. There is an additional freight cost due to increased power requirements and lower speeds.

14.6.1 Effects of Grade on Vehicle Type

Austroads, 2021, Table 8.2, page 213 describes the effects of various grades on climbing and descending traffic speeds. Generally maximum grades are described as flat, rolling or mountainous between 3% and 10%, limited by operating speed as described in subsequent Austroads, 2021, Table 8.3.

14.6.2 Grades in Rural Road Design

There are two limiting grades in road design.

The first is the minimum grade. It can be zero except in cut where it should be a minimum of 0.5% to provide longitudinal road drainage. An absolute minimum of 0.33% (1 in 300) may be used as long as the table drains maintain a minimum of 0.5% to ensure drainage from the cutting.

Maximum grades affect vehicle performance, and therefore safety. The length of the maximum grade also affects the performance, especially heavy vehicles. A short length of steep grade may be tolerated before performance degradation becomes a problem.

Terrain also affects maximum grades, with the ability to use higher grades over short distances. This may have an effect on construction costs.

14.6.3 Length of Vertical Curves in Rural Road Design

Lengths of vertical curve design has been heavily enhanced in Austroads 2021 in Chapter 8.

Vertical lengths have now been divided into crest and sag curve categories using sight distance criteria. The length of a vertical curve should be as great as possible, within economic constraints.

The minimum length of vertical curves is governed by:

1. Crest or sag curve.
2. Sight distance:
   a requirement in all situations for driver safety.

3. Appearance:
   generally required in situations of low embankment and flat terrain.
4. Riding comfort:
   A general requirement to limit increased vertical acceleration through a sag curve to 0.05g. Through approaches to floodways, low standard roads and intersections the increase in vertical acceleration may rise to 0.1g.

14.6.4 "K" Value in Vertical Curve Design

The minimum vertical curvature allowed in a vertical road profile and can be expressed as a single value, the **K** value, to allow computerised road design.

$K = L/\mathrm{A}$ where:

K = length of vertical curve for 1% change of grade (metres)
L = length of vertical curve (metres)
A = algebraic change of grade (%).

14.6.4.1 Worked Example 14.6. Sag curve

Worked Example 14.4 (§14.4.1) used a sag curve of length $L = 200$m. From the example:

A = 7.6%
$K = L/\mathrm{A} = 200/7.6 = 26$

From Austroads (2021), Figure 8.9, page 226, this K value represents an operating speed of 100Km/h for a 0.05g sag curve on a lighted rural road, but 75Km/h for an unlit rural road.

14.7 Sight Distance in Vertical Curve Design

Austroads (2021), Chapters 5 and 8, have a comprehensive treatment of sight distances to which you are referred.

Sight distance (see Figure 14-22) is the distance necessary for a driver to see, react to and avoid an object on a road. A number of parameters are incorporated in various scenarios to calculate sight distance.

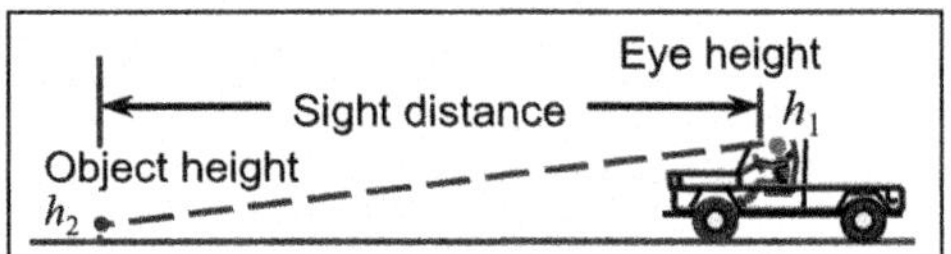

Figure 14-22 Sight distance

The 3 main parameters are:

Reaction time, seconds t_r: normal 2.5; alert, 2.0; extra alert, 1.5 (≤90Km/h).
Driver eye height, h_1: car, 1.05m; truck, 2.4m; bus, 1.8m
Object height, h_2: intersection line, 0.0m; object, 0.2m; taillights, 0.6m.

14.7.1 Stopping distance.

A vehicle's stopping distance at a certain velocity, $\boldsymbol{v}$, comprises the distance travelled during the reaction time, $\boldsymbol{t_r}$, and the deceleration distance due to the frictional coefficient between the tyres and the pavement, $\boldsymbol{f}$, and the pavement gradient, $\boldsymbol{a}$.

14.7.2 Various types of sight distances can be determined:

1. Stopping sight distance (**SSD**). The distance necessary for an alert driver, travelling on a **wet** pavement at design speed, to perceive, react and brake to a stop before reaching a hazard on the road ahead.
2. Headlight sight distance (**HSD**). The distance that can safely be assumed an object is visible on an unlit road at night. This distance is between 120m and 150m. This distance corresponds to a SSD speed range between 80Kmh^{-1} and 90Kmh^{-1}, or a manoeuvring sight distance speed of 100Kmh^{-1}. HSD is predicated on headlight high beam with a beam elevation of about 1°.

Active visibility objects such as retro-reflective targets (reflectors), car headlights, tail lights and large or light coloured objects may allow SSD at design speed at night.

Increase in light output has little effect on visibility. A fivefold output increase will only allow a 15Km/h increase in speed. A tenfold increase in light output will allow a reduction of only 50% in object size. Beam intensity is limited by the requirements of driving visibility and glare to oncoming vehicles.

3. Overtaking sight distance. The distance necessary for a driver to overtake safely a slower vehicle without interfering with the speed of an oncoming vehicle.
4. Manoeuvring sight distance. The distance needed for an alert driver to perceive, react, brake and manoeuvre to avoid an object. It was generally about 6% less than the SSD. Now omitted from the Guide.

14.7.2.1 Horizontal curve perception distance.

This is the distance needed for a driver to perceive and react to the presence of a horizontal curve that may require speed adjustment.

A driver needs to see enough of the curve to judge its curvature. The minimums are:

- 5 degrees of curve, (radius × 5° in radians = R × 0.09, 9% radius.)
- about 80m of curve length (3 – 4 seconds travel)
- the whole curve.
- Horizontal curves "hidden" over a crest curve can cause particular perception problems.

These various sight distances produce a recommended range of "K" values that are needed to be incorporated into the design of the vertical profile curve. The designer's task is to ensure that the design exceeds the minimum values for the design operating speed in various profile configurations.

The surveyor's task is to calculate the RLs of the vertical stations to the parabolic curve design parameters.

14.7.3 Crest vertical curve design

Crest curve design, see Figure 14-23, is normally associated with finding the minimum length of the vertical curve to allow a normally alert driver to perceive an object on the road, react to the perception and decelerate to a stop. This is the stopping sight distance (SSD).

Previously available tables of deceleration were produced for each speed.

Examples of the 4 main parameters are:

- Reaction time (r_t): normal 2.5 secs; alert, 2.0 sec, 1.5 sec in controlled low speed conditions. (Table 5.2).
- Driver eye height (h_1): car, 1.1m; truck, 2.4m; bus, 1.8m (Table 5.1)
- Object height (h_2): intersection line, 0.0m; object, 0.2m; taillights, 0.6m (Table 5.1)
- Deceleration rate (braking on sealed roads); f = 0.61 (dry); 0.46 (wet, with hazard), 0.36 (wet, minimum SSD, rail level crossings). f = 0.27 (unsealed, dependent on surface) (Table 5.3).

14.7.3.1 Car Stopping Sight Distance, level road grade.

SSD has two components: (Eqn 14-11)

- reaction distance associated with speed, v, and reaction time, t_r. $D_R = vt_r$.
- deceleration distance; speed, v, deceleration coefficient (longitudinal friction factor), d or f and longitudinal grade, a%.

f = is a constrained design value depending on vehicle type and pavement conditions.

The deceleration distance is the distance needed to dissipate the vehicle's kinetic energy at speed, v, to zero by using the brakes of friction coefficient f at weight w.

Work done for braking distance. WD = Fd = fwd where F is the maximum developed frictional force.

The kinetic energy of the vehicle at initial speed v.

Equating WD with KE. $fwd = \frac{wv^2}{2g}$, then $\frac{\not{w}v^2}{2g} \times \frac{1}{\not{w}f} = \frac{v^2}{2gf} = \frac{V^2 Kmh^{-1}}{2 \times 9.8 \times 3.6^2 \times f}$.

Deceleration distance is independent of vehicle weight, it relies only on reaction time, speed and braking coefficient. Austroads 2021 Table 5.5 lists various braking coefficients, f, for cars, trucks and buses. Car values are specified for wet sealed roads. In calculating stopping distance for trucks, a lower f of 0.29 is used. For buses, f = 0.15 for passenger comfort and safety. Recall Figure 12-7, page 185 and the discussion of static/kinetic friction.

Stopping sight distance	$SSD = vt_r + \frac{v^2}{2gf} \Rightarrow \frac{t_r V}{3.6} + \frac{V^2}{254(f + 0.01a)}$	Eqn 14-11

Table 14-5 Stopping Sight Distance (Eqn 14-11).

Operating Speed (Kmh^{-1})	Desirable min. f = 0.4	
	t_r = 2	t_r = 2.5
50	52	59
60	69	77
70	87	97
80	107	119
90	130	142
100	164	168

Braking coefficient is modified by the grade of the pavement. A constant of 0.01a% is added to the f value in the denominator. For a level road, a = 0%.

As an example, Table 14-5 tabulates SSD from Eqn 14-11 for a flat wet asphalt road using f = 0.4 for two reaction times, t_r = 2 or 2.5 seconds. They assume a driver eye height of 1.1m and an object height of 0.2m. Studies in several countries have set eye heights between 1.0m and 1.2m. Australia reduced from 1.15m to 1.1m after about 2002.

14.7.3.2 Car Stopping Sight Distance, on grade.

Distance corrections due to grade (see Table 14-6) can be included in the SSD calculation by the application of the slope grade, a%, or G_1. The example deceleration used is $f = 0.4$.

It is recommended that stopping sight distances be rounded up to the next 5m.

(see Figure Table 14-6 Grade distance corrections. $f = 0.4$

Grade, a% speed ↓	–7	–5	–3	–1	1	3	5	7
50	5	4	2	1	–1	–2	–3	–4
60	8	5	3	1	–1	–2	–4	–5
70	10	7	4	1	–1	–3	–5	–7
80	13	9	5	2	–2	–4	–7	–9
90	17	11	6	2	–2	–6	–9	–12
100	21	14	8	3	–2	–7	–11	–15

14.7.3.3 Length of a crest curve incorporating SSD.

A crest curve (see Figure 14-23) introduces the problem of the driver being able to perceive an object over the crest of the curve. In general, the SSD will be less than the length, L, of the curve. (SSD < L).

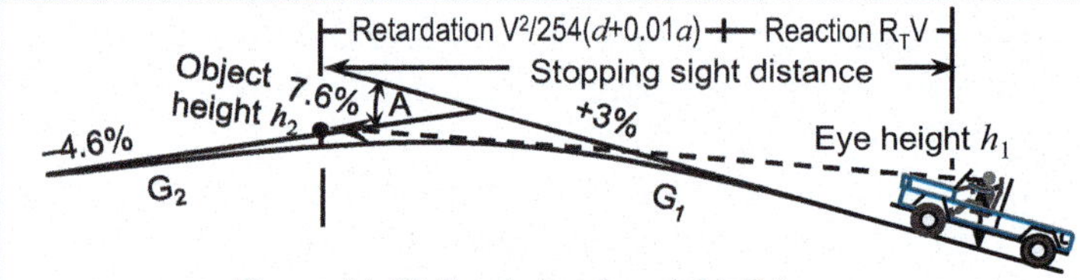

Figure 14-23 Crest stopping sight distance

Apart from the heights, h_1 and h_2, the other components are the percentage algebraic change of grade, A, and the SSD. For cars the accepted values of h_1 and h_2 are 1.1m and 0.2m. See Eqn 14-12, where the approximation matches the above h_1 and h_2.

$L_{crest\ curve}$ for SSD	$L = \dfrac{SSD^2}{200\left(\sqrt{h_1}+\sqrt{h_2}\right)^2} \times A\%$ when $SSD < L$ $L \cong \dfrac{SSD^2}{450} \times A\%$	Eqn 14-12

14.7.3.3.1 Worked Example 4.7. Crest curve.

For example, where $h_1 = 4$ feet (1.2m) and $h_2 = 6$ inches (0.15m) and, using Figure 14-23 at $V = 70\text{Kmh}^{-1}$.

For reaction time $t_r = 2$ seconds SSD = 87m (Table 14-5).

Grade distance correction, ($a = G_1 = +3\%$) = –3m (Eqn 14-6)

SSD = 84m and rounded to 85m

Minimum length crest curve, $L = 122$m (Eqn 14-12)

14.7.4 Sag curve design

14.7.4.1 Parabolic constant, K

K is defined as a parabolic vertical curve constant and is the length of the curve required for a change of grade of 1%. It is the reciprocal of the 100×r%% (grade rate of change per 100, see 14.2.1, page 218). It can be thought of as a "pitch change rate" value through a sag curve. The lower the number, the more uncomfortable (higher a value of g) the change.

K constant	$K = \dfrac{L}{A} \Rightarrow L = KA.\ \ L =$ curve length, $A = \%$ algebraic grade change $\lvert G_2 - G_1 \rvert$	Eqn 14-13

Using Figure 14-18 Worked Example 14.4, single curve. page 223, $L = 200$m, $A = 7.6\% \Rightarrow$ r%% = 3.8.

$$K = 200/7.6 \ = \ 26 \equiv \frac{1}{r\%\%} \times 100 = \frac{100}{3.8} = 26$$

Sag curve design incorporates the comfort criterion as a minimum. A maximum of 0.05g vertical acceleration is considered the comfort limit for urban and rural lighted roads. On low standard roads $a = 0.1$g.

K sag curve comfort	$K = \dfrac{V^2}{1296a}$ where $a_{max} = 0.05\,(\text{m/sec}^2)$	Eqn 14-14

On unlighted urban and rural roads, the K value is dependent on the HSD (120 – 150m).

Austroads 2021 Figure 8.9, page 226, graphs minimum K values versus speed on various classifications of road.

As for crest curves, appearance in a sag curve is an important consideration for small and larger grade changes. They should be designed for headlight sight distance (HSD) criteria on unlit rural and urban roads. Highways and freeways will normally be designed with the sag curve length exceeding the HSD.

Drainage of the surface and subgrade is also an important consideration on sag curves. Long sag curves (greater than 750m) should be avoided on two lane rural roads.

14.7.4.2 Headlight Sight Distance (HSD), sag curve.

HSD in sag curves is not limited during daylight hours, or on well-lit roads at night.

On unlit roads at night the limiting factor for HSD (see Figure 14-24) is the limitations of the vehicle's headlights. High beam HSD is between 120m and 150m in sag curves. This HSD allows an adequate stopping distance up to 90Kmh^{-1} (Eqn 14-11, $t_r = 2, f = 0.4$) and about a 5 second manoeuvring time from 100Km/h.

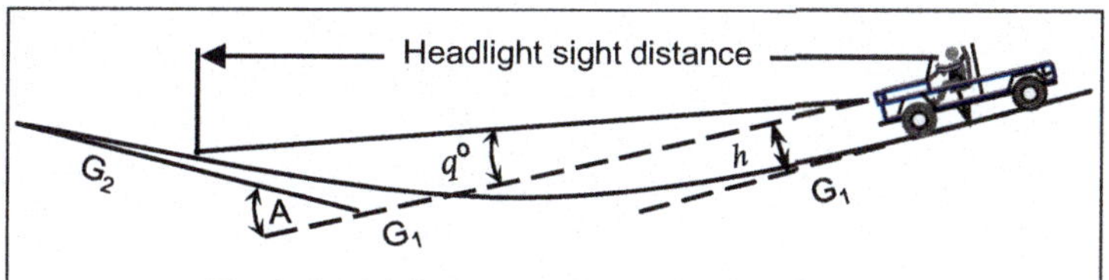

Figure 14-24 Sag curve headlight sight distance

With $h = 0.7$m and $q = 1°$ (high beam) then K = 26 for HSD = 120m and K = 34 for HSD = 150m. These limits to K give maximum speeds of about 80 – 90Kmh^{-1} on unlit roads. (See Eqn 14-15).

In all cases of night driving involving sag curves the emphasis must be on retro-reflective materials on guideposts, centrelines and depth gauges in sag floodways.

K HSD	$K = \dfrac{HSD^2}{200(h + HSD\sin q)}$ where $HSD < L$	Eqn 14-15

14.8 The Arc Length of the Vertical Curve

The task may be to find the **length** of the parabola joining two designated grades.

This is mostly done by summing the chord lengths between each pegging station, if indeed that level of accuracy is required. In general, the **length** of the vertical curve, L, will suffice.

Worked Example 14.4 produced an answer of 200.0480m at 20m intervals (10 steps).
Iterating at a $dx = 1$m (200 steps) produced an answer of 200.0485m

14.9 References for Chapter 14

Allan, A.L., Hollwey, J.R., and Maynes, J.H.B. Practical Field Surveying and Computations. William Heinemann Ltd, London. 1971

Uren, J and B. Price. (2010) Surveying for engineers. Fourth edition, Palgrave Macmillan, Chap. 14.

Austroads Ltd, Sydney, 2021. Edition 3.4 AGRD03-16: Guide to Road Design Part 3: Geometric Design. ISBN 978-1-922382-15-3.

American Association of State Highway and Transportation Officials. (AASHTO) A Policy on Geometric Design of Highways and Streets, 7th Edition, 2018 ISBN: 978-1-56051-676-7

Hickerson, T.F. Route Survey and Design. Fourth edition. McGraw-Hill Book Company, Inc. New York. 1959

Chapter 15 Strike And Dip to an Embedded Plane

Mining engineers are involved in the extraction of minerals from ore bodies that often incorporate deposition seams.

Civil engineers may also need an understanding of slopes associated with earthworks on sloping terrain. Typical use could be on bulk earth structures such as water reservoir dam walls, tank bunds and the like.

The ability to calculate depths and directions of drill holes, the slopes of drives and the calculation of inclined angles is not well covered in most surveying texts. This chapter attempts to show some of the simple calculation techniques available for use in the field, or as an adjunct to normal mine planning software. Field measurement of strike and dip of an exposed seam can be made using Total Stations, automatic levels or geologists' compasses, e.g., the Brunton type. There are few references to strike and dip. Schofield and Breach (2007). Shepherd (1968, 1981) covers the subject in detail. Most field exploration by geologists will be with compass and inclinometer, the Brunton and like instruments. Thus, dip and strike directions will be measured in degrees magnetic which is converted to true azimuth by the application of magnetic declination (from maps or on-line resources).

15.1 Strike and Dip

Geologists describe the orientation of a seam of mineralisation in terms of the direction of the horizontal plane intersecting the seam, the **strike**, and the maximum inclination of the plane, the **dip**. Exploratory drilling will intersect different geological structures, some of which may warrant development into a mine. Having located the seam, we now need to describe its location and orientation so that a mine may be planned to expose the seam.

For simplicity, the examples refer to vertical drill holes. Structural geologists of ten drill inclines holes for better intercept interpretation. § 15.5.1 examines inclines drill holes.

15.2 Strike and Dip to a Plane

The relationship between the ground surface and the inferred seam, defined by three drill holes, DH0, DH1, and DH2, is illustrated in Figure 15-1.

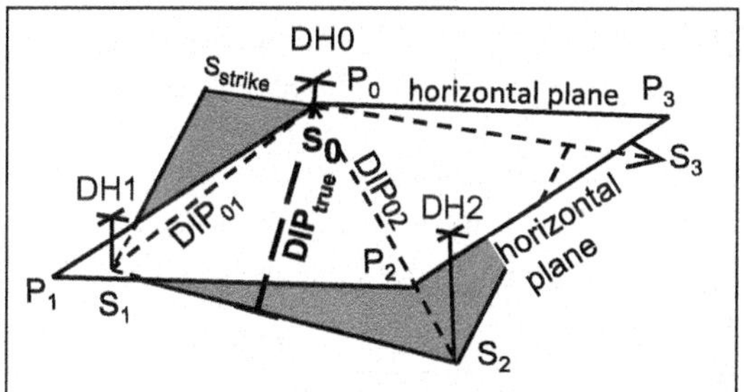

Figure 15-1 Dipping plane.

Plane P_0, P_1, P_2, P_3 is the **horizontal** plane surface through S_0, (below DH0) which is the "highest" intersect point. S_0 and P_0 can be considered to be a hinge point.

The **dipping** plane, the intersected seam, is defined below the surface of points DH0, DH1 and DH2 at S_0, S_1, and S_2.

The line S_{strike} to S_3 defines the direction, as yet unknown, of the **strike line**.

The lines $S_0 - S_1$, and $S_0 - S_2$ are the **apparent** strike lines from S_0. Apparent dip angle, α.

The **true, or maximum,** dip, θ, from S_0 (DIP_{true}) is perpendicular to the strike line. So, if you find the direction of maximum dip; then you have defined the strike direction.

15.3 Deriving Strike and Dip of a Plane

The dipping plane is represented by the plane *apbdra* containing S_0, S_1, and S_2 in Figure 15-2. *rcqbdr* is the plane containing S_1 and S_2 at a depth **ac** below S_0. Angles at *c* and *q* are right angles.

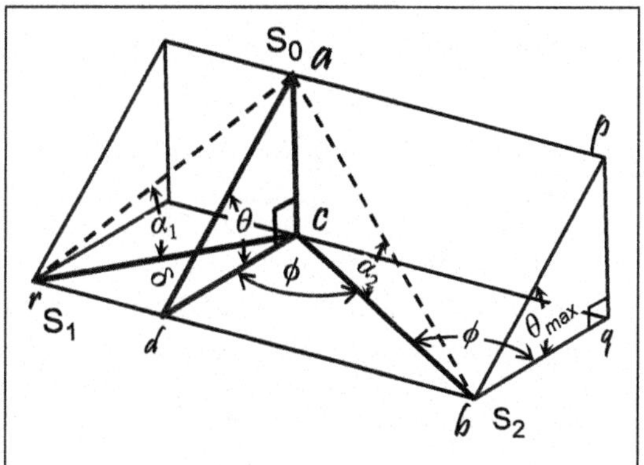

Figure 15-2 Dip and strike.

There are 4 angles to consider:

1. The **true** dip, θ_{max} (theta), is the slope *ad* where *ac* = rise and *cd* = run. Slopes (dips) are expressed as a value of 1:*k*, *k* = rise/run.
2. The apparent dip from a (S_0) to *b* at S_2 (α_2).
3. The apparent dip from *a* to *r* at S_1 (α_1) ("apparent" means as observed or measured).
4. The included angle, ϕ (phi), is between the apparent dip to *b*, S_2, and the true dip to *d*. ϕ is the angle *dcb*.

The line $bp = ad\cos\phi = \dfrac{cd}{cb}$

From Figure 15-2 we will re-label θ_{max} as θ for the development of the true dip formula.

J. Walker and J. Awange, *Surveying Computations*,
https://doi.org/10.1007/978-3-031-94382-9_15

If it's not checked it's wrong.

1. $\tan\theta = \dfrac{ac}{cd} = \dfrac{pq}{qb}$ maximum dip, and, using α $(1/k)$ to denote apparent dip (k = slope):

2. $\tan\alpha_2 = \dfrac{ac}{cb} = \dfrac{pq}{cb}$ because $ac = pq$. [$\tan\alpha_2 = 1/k_2$]

3. $\dfrac{pq}{cb}$ is the cross product of the apparent and true dips at **b** (S_2).

	$\dfrac{pq}{cb} = \dfrac{pq}{bq} \times \dfrac{bq}{cb} = \tan\theta\cos\phi$	Eqn 15-1

4. $\tan\alpha_1 = \dfrac{ac}{cr} = \dfrac{pq}{cr}$, [$\tan\alpha_1 = 1/k_1$]

and again, is the cross product of the apparent and true dips at **r** (S_1).

	$\dfrac{pq}{cr} = \dfrac{pq}{bq} \times \dfrac{bq}{cr} = \tan\theta\cos\delta$	Eqn 15-2

There are now three dips defined: true dip, $\tan\theta$; apparent dip, $\tan\alpha_1$; and apparent dip, $\tan\alpha_2$. They are related by the common side, ac.

The two apparent dips are defined by their respective **cross** products.

$\tan\alpha_2 = \dfrac{pq}{cb}$, $\tan\theta = \dfrac{pq}{qb}$, $\cos\phi = \dfrac{bq}{cb}$. Thus, by cross product:

	$\tan\alpha_2 = \tan\theta\cos\delta$	Eqn 15-3

The included angle between the two apparent dips can be designated **Δ** (capital delta), the angle rcb. This is the measured angle at DH0 between DH1 and DH2.

ϕ is the included angle between the true dip and apparent dip to S_2, (see Figure 15-3).

Label the angle between the true dip and the apparent dip to S_1, as $\boldsymbol{\delta}$ (delta).

So $\boldsymbol{\delta} = \boldsymbol{\Delta} - \boldsymbol{\phi}$, thus $\phi = \Delta - \delta$.

$\tan\alpha_2 = \tan\theta\cos\phi \Rightarrow \tan\theta\cos(\Delta - \delta)$

Expanding: $\cos(\Delta \pm \delta) \Leftrightarrow (\cos\Delta\cos\delta \mp \sin\Delta\sin\delta)$

$\tan\alpha_2 = \tan\theta\cos\phi \Rightarrow \tan\theta\cos(\Delta - \delta)$	Eqn 15-4
$\tan\theta = \dfrac{\tan\alpha_2}{(\cos\Delta\cos\delta + \sin\Delta\sin\delta)}$ and, as $\alpha_2 = \dfrac{1}{k_2}$ $\tan\theta = \dfrac{1}{k_2(\cos\Delta\cos\delta + \sin\Delta\sin\delta)}$	Eqn 15-5

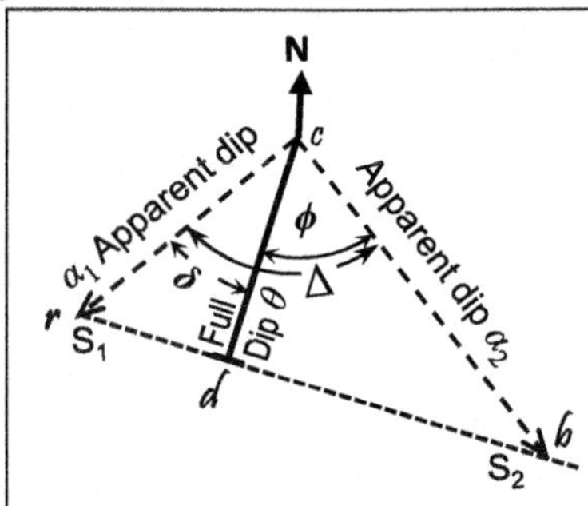

Figure 15-3 Resolution of dip angles.

Apparent dip $\tan\alpha_1$ is defined by its cross product such that:

$\tan\theta_1 = \dfrac{pq}{qr}$, $\tan\theta = \dfrac{pq}{qb}$, $\cos\delta = \dfrac{bq}{cr} \Rightarrow \tan\theta_1 = \tan\theta\cos\delta$	Eqn 15-6
$\tan\theta = \dfrac{\tan\alpha_1}{\cos\delta}$ and, similarly, $\tan\alpha_1 = \dfrac{1}{k_1} \Rightarrow \tan\theta = \dfrac{1}{k_1\cos\delta}$	Eqn 15-7

As **Δ** is known, there are two expressions with **tan*θ*** common, linked by **sin*δ*** and **cos*δ***, which will allow the calculation of the angle $\boldsymbol{\delta}$.

Remember, the preceding equation to calculate the deflection angle, δ, which is measured from the true dip, $\tan\theta$, to the apparent dip at S_1, $\tan\alpha_1$.

Thus, the true dip, (following Eqn 15-6), $\tan\theta = \dfrac{\tan\alpha_1}{\cos\delta}$.

Dip angle:	$\theta = \arctan\left(\dfrac{\tan\alpha_1}{\cos\delta}\right)$	Eqn 15-8

Knowing the orientation (azimuth) of either of the apparent dips and the included angle between the two apparent dip lines, then the azimuth of the true dip can be calculated.

The **strike line** has **no** dip and is perpendicular to the true dip. If the plane is dipping down, then the strike azimuth is the dip azimuth – 90º.

Solve for the angle δ since it is common in both Eqn 15-7 and Eqn 15-5.

Solution for angle δ	$\tan\theta = \dfrac{1}{k_1\cos\delta}$, and $\tan\theta = \dfrac{1}{k_2\left(\cos\Delta\cos\delta + \sin\Delta\sin\delta\right)}$, inverting and equating $k_1\cos\delta = k_2\left(\cos\Delta\cos\delta + \sin\Delta\sin\delta\right) \Rightarrow k_1\cos\delta = k_2\cos\Delta\cos\delta + k_2\sin\Delta\sin\delta$ $k_1\cos\delta - k_2\cos\Delta\cos\delta = k_2\sin\Delta\sin\delta$. Ans since Δ, k_1 and k_2 are known: $\dfrac{k_2\sin\Delta\sin\delta}{k_1\cos\delta - k_2\cos\Delta\cos\delta} \Rightarrow \dfrac{\sin\delta}{\cos\delta} = \tan\delta$	Eqn 15-9

15.3.1 Strike and Dip of a Plane, Worked Example 15.1 by Calculation

Geologist field trip notes provide the data on three vertical drill holes: See Figure 15-4.

DH0, RL100, intercepted a seam, S_0, at a depth of 10m

DH1, RL95, 72m from DH0, azimuth 063º intercepted the seam, S_1, at a depth of 22m

DH2, RL90, 111m from DH0, azimuth 130º intercepted the seam, S_2, at a depth of 32m

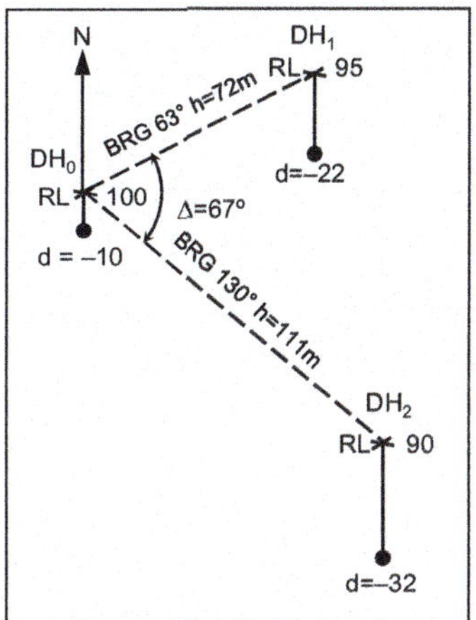

Figure 15-4 Field sketch.

15.3.1.1 Calculate the strike and dip of the seam

Figure 15-5 shows the two planes.

Invert level (IL) of seam intersects and depth (dIL) below S_0 datum.

IL $S_0 = 100 - 10 = 90$ (the highest)

IL $S_1 = 95 - 22 = 73$

$dIL_{S0-S1} = 90 - 73 = 17$

IL $S_2 = 90 - 32 = 58$

$dIL_{S0-S2} = 90 - 58 = 32$

Horizontal distances and apparent dips ($\tan\alpha$=rise/run).

α is the apparent dip angle.

d_{S1} = 72m, dip_{S1} ($\tan\alpha_1$) =17/72 = 0.236 = 1:4.24 (k_1)

d_{S2} =111m, dip_{S2} ($\tan\alpha_2$) =32/111=0.288 = 1:3.47 (k_2)

Included angle between apparent dip lines

$\Delta = Az_{S2} - Az_{S1} = 130 - 63 = 67°$

See the dip plan in Figure 15-6.

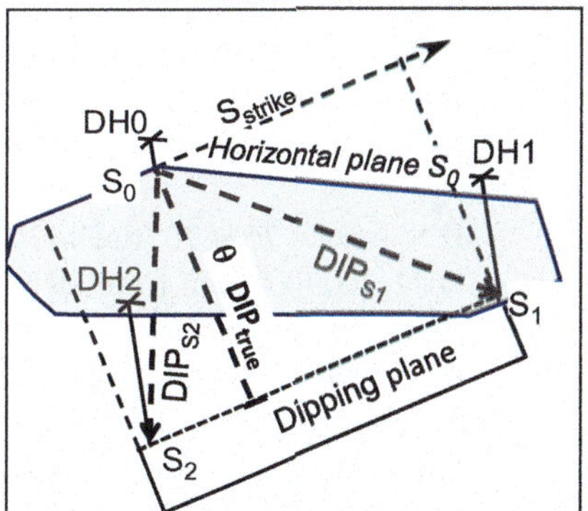

Figure 15-5 Dip and horizontal planes

15.3.1.2 Calculate the strike and dip directions

Evaluate dip to S_1 (α_1) _Eqn 15-4:

$$\tan\alpha_1 = \left(\cos\Delta\cos\delta - \sin\Delta\sin\delta\right)$$

$$\frac{1}{4.24} = \tan\theta\left(\cos 67°\cos\delta + \sin 67°\sin\delta\right)$$

$$\tan\theta = \frac{1}{4.24\left(\cos 67°\cos\delta + \sin 67°\sin\delta\right)}$$

Evaluate dip to S_2 (α_2) Eqn 15-3: $\tan\alpha_2 = \tan\theta\cos\delta$

$$\frac{1}{3.47} = \tan\theta\cos\delta$$

$$\tan\theta = \frac{1}{3.47\cos\delta}$$

We now have two equal expressions for tan θ. Invert each one and express the equality:

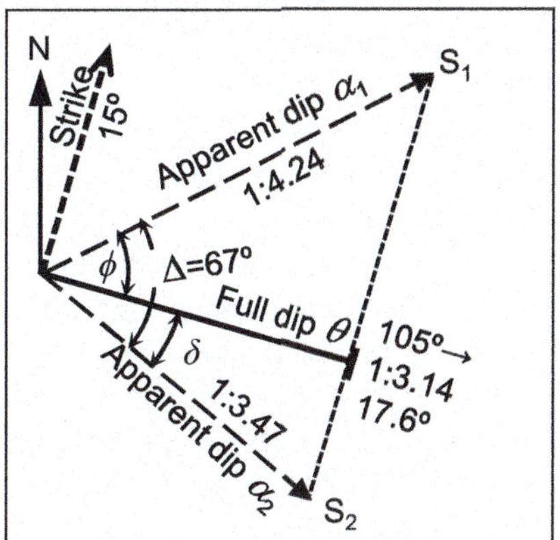

Figure 15-6 Dip plan diagram.

$$4.24\left(\cos 67°\cos\delta + \sin 67°\sin\delta\right) = 3.47\cos\delta$$

$$4.24\left(0.39\times\cos\delta + 0.62\times\sin\delta\right) = 3.47\cos\delta \Rightarrow 1.65\cos\delta + 3.90\sin\delta = 3.47\cos\delta$$

Gather terms: $3.90\sin\delta = 3.47\cos\delta - 1.65\cos\delta \Rightarrow 3.90\sin\delta = 1.82\cos\delta$

Now, as tan δ can be expressed in terms of sinδ and cosδ, solve for $\boldsymbol{\delta}$, Eqn 15-9

$$\tan\delta = \frac{\sin\delta}{\cos\delta} = \frac{1.82}{3.90} = 0.467$$
$$\delta = \arctan(0.467) = 23.03° = 23°01'$$

Azimuth to full dip = $Az_{S2} - \delta$ = 130° – 25° = **105°**
Strike azimuth = dip direction – 90° =105° – 90° = **15°**

15.3.1.3 Calculate the full dip angle

See Figure 15-6

Now evaluate full dip **tan*θ*** using Eqn 15-8, Dip S_2, where tan α_2 = 1:3.47. From Eqn 15-3:

$$\tan\alpha_2 = \tan\theta\cos\delta$$

Measured $S_2(\tan\alpha_2) = 1:3.47$ and $\delta = 25°$. Now full dip, $\tan\theta = 1:x$

$$\frac{1}{3.47} = \frac{1}{x}\cos(25°) \Rightarrow x = 3.47 \times \cos(25°) = 3.14$$

$$\text{Rate of full dip}, \theta = 1:3.14 = \arctan\left(\frac{1}{3.14}\right) = 17.6°$$

Having solved for the orientation of the strike and true dip you are now able to calculate depths to seam from any point in the exploration area.

Volume of overburden and all the other tasks associated with mine development can now be calculated.

15.3.2 Strike and Dip of a Plane, the Graphical Solution

The previous complex calculations can be simplified by using a graphical solution derived from the observed data. It also allows visualisation of the observations and can be done in the field. Referring to the plan in it can be seen that the true dip, θ, is less than any apparent dip, α_1. The slope is at its steepest point at θ_{max}.

A geologist returns from a field trip with the following information on three different vertical drill holes: Figure 15-7 is the field sketch.

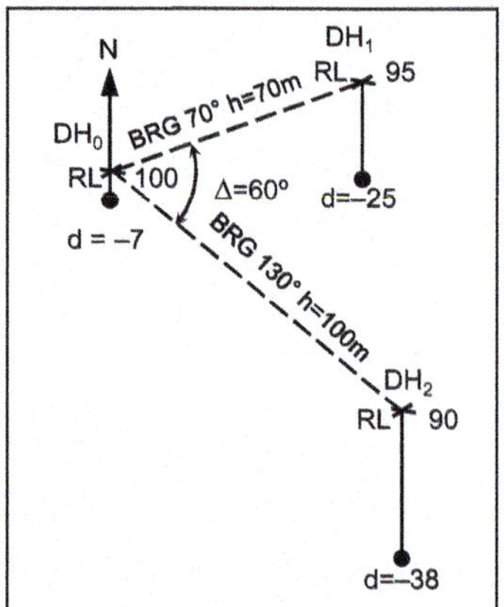

Figure 15-7 Field sketch.

DH0, RL100, intercepted a seam, S_0 at a depth of 7m.

ILS_0 = 100 – 7 = 93

DH1, RL95, 70m from DH0, azimuth 070° intercepted seam, S_1, at a depth of 25m.

ILS_1 = 95 – 25 = 70, rise to RLS_0 = 93 – 70 = 23, run = 70

DH2, RL90, 100m from DH0, azimuth 130° intercepted the seam, S_2, at a depth of 38m.

ILS_2 = 90 – 38 = 52, rise to RLS_0 = 93 – 52 = 41, run = 100

From this information, where tan = rise/run:

dip_1 (tan a_1) = 23/70 = 0.33 = 1:3.03.
Azimuth, Az_1 = 070°
dip_2 (tan a_2) = 41/100 = 0.41 = 1:2.44.
Azimuth, Az_2 = 130°.

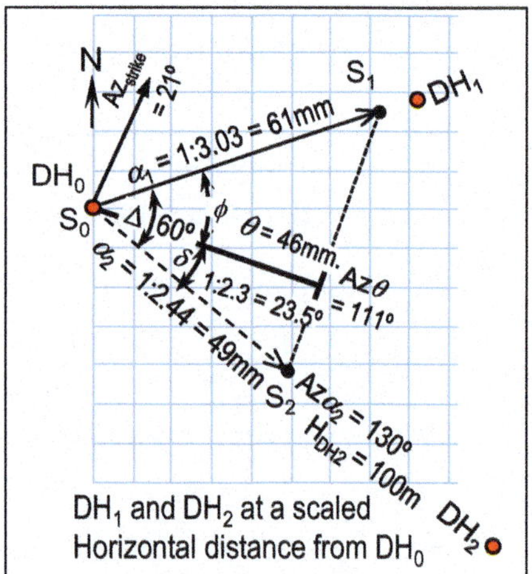

Figure 15-8 Graphic dip solution.

Draw a plan closely to scale. You are drawing dip angles, not horizontal distances. They are designed to fit in an A5 field book. See Figure 15-8. Using a scale of 20mm per dip unit:

1) From S_0,
   a) draw a line on a bearing of 70° towards S_1
      for a distance 3.03 units (say 3.03 × 20 = 60.6mm)
   b) draw a line on a bearing of 130° towards S_2
      for a distance of 2.44 units (2.44 × 20 = 48.8mm).
2) Join S_1 *to* S_2. From this base line, draw perpendicular through S_0.
   Measure the distance from the base to S_0 (46mm) and convert to dip units 46/20 = 2.30.
   The **true** dip,*θ*, is 1:2.30 = arctan (1/2.30) = 23.5°
   measure the bearing of the true dip from S_0 = 111°
   Calculate bearing of strike = 111 – 90 = **21°**

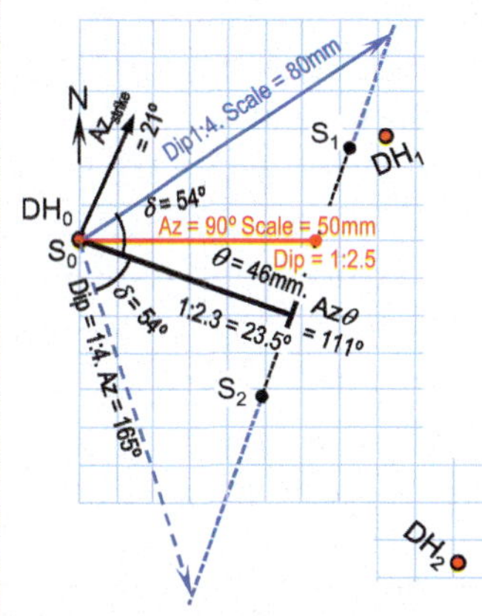

Figure 15-9 Draw solutions.

With a carefully drawn diagram it is then possible to find the apparent dip in any direction by scaling from S_0 to the strike line between S_1 and S_2. Figure 15-9 illustrates the next two problems.

1) What is the apparent dip on a bearing of 90° from S_0?

Draw a line at a bearing of 90° to intercept the strike line between S_1 and S_2. It measures 50mm and at a scale of 1:20. The dip = 1:2.5 (22°).

2) What is the azimuth of the 1:4 dip line?

The scale factor requires an 80mm line be drawn to intercept the strike line (extended past S_1). Measure its azimuth as 57°.

The angle between apparent and full dip, recall Figure 15-9 , $\delta = 111 - 57 = 54°$. Because of symmetry to the dipping plane, there is another azimuth to the 1:4 apparent dip.

It is true dip + δ, = 111 + 54 = 165°

15.3.3 Outlier Problems

Blythe application of the above formula will sometimes result in "funny" answers. Two problems arise:

1. The angle Δ, the angle between the directions to the two drill holes, is greater than 90°. In this case cosΔ in Eqn 15-4 will produce a negative value (cos>90 is negative). The resulting value of δ will be wrong because, in this case, $\Delta = 90 - \delta$.
2. Tanδ is negative, thus δ is negative. This means that the maximum dip, θ, direction is "behind" the dip to the second drill hole. What we must do is accept the negative answer and move forward with the calculation of the direction of the true dip. From Eqn 15-4 dip direction, *cd*, = (direction *cr* (to S_1) – δ). This holds true in the graphical solutions illustrated in § 15.3.3.2 (Figure 15-11).

   Az to S_2 = 80, $\delta = -25$. Az dip = 80 – (–25) = 105°

15.3.3.1 The case of included angle Δ >90°

By using a graphic solution check, the mathematical solution to a calculation arising from Field Practical 4 can be shown to be incorrect as illustrated in Figure 15-10.

From DH0, the directions to DH2 and DH1 are 78.6° and 178.2°. Thus Δ = 99.6°.

Referring to § 15.3, and re-assigning the *k* subscripts:

$k_1 = \alpha_1 = 5.1$, $k_2 = \alpha_2 = 9.2$. (Eqn 15-9)

$$\tan\delta = \frac{k_1 \sin\Delta \sin\delta}{k_2 \cos\delta - k_1 \cos\Delta \cos\delta}$$

$$= \frac{5.1 \sin(99.6°) \sin\delta}{9.2\cos\delta - 5.1\cos(99.6°)\cos\delta}$$

$$= \frac{5.03 \sin\delta}{10.05 \cos\delta} \Rightarrow \tan\delta = \frac{5.03}{10.05} = 0.05$$

$$\delta = 26.58°$$

$$\tan\theta = \frac{1}{k_2 \cos\delta} = \frac{1}{9.2\cos(26.58°)} \Rightarrow \frac{1}{9.2 \times 0.89}$$

$$= \frac{1}{8.23} \Rightarrow \theta = 6.93° \text{ (Dip angle)}$$

True dip, $\theta = 1:8.23$

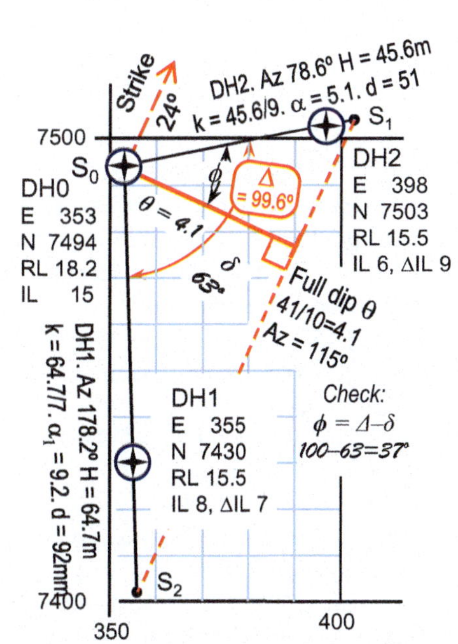

Figure 15-10 Solution: Field Practical 4.

Check by diagram:

By measurement we see that the angle Δ ≈ 99°. OK.

We measure δ, between DH1 and the full dip, as 63°.

By scaling we measure dip, θ, as 4.1.

Calculation by Eqn 15-9 gives results of $\delta = 26.6°$ and $\theta = 8.2$.

What has gone wrong?

The problem is that, as Δ is greater than 90°, it is in quadrant 2 where cos(Δ) is negative.

So $\delta = 90 - \delta = 90 - 26.6 = 63.4°$. (This agrees with our diagram.)

If it's not checked it's wrong.

Recalculating with $\delta = 63.4°$

True dip $\theta = 9.2 \times \cos(63.4) = \mathbf{4.1} = 13.8°$. (Agrees!)

Dip azimuth $= AZ_{DH1} - \delta = 178.2 - 63.4 = 114.8°$.

Strike azimuth $= 114.8 - 90 = 24.8° = \mathbf{25°}$.

Check: $\phi = \Delta - \phi$. We see that $\phi = 99.6 - 63.4 = 36.2°$.

$\theta = k_1\cos\phi = 5.1 \times \cos(36.2) = 4.1$. (Agrees!)

15.3.3.2 Maximum dip "outside" apparent dip lines

Some drill hole data may lead to some "funny" results. *For illustration* ***change*** *DH2 values.*

It can be seen that if, from Worked Example in § 15.3.1, Figure 15-5, the field data for DH2 had been, e.g., DH2, RL90, 111m from DH0, azimuth **80°** intercepted the seam, S_2, at a depth of 32m, then a scale drawing as shown in Figure 15-11 would have revealed the direction of full dip as being further south of DH2. The S_2 distance can intercept the strike line at S_2 or S_2', as shown.

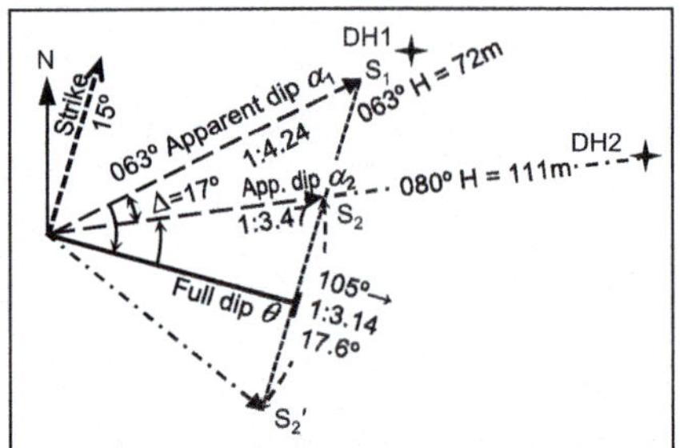

Figure 15-11 Ambiguous dip.

Calculation may have caused concern with a negative answer because the angle $\Delta = 17°$, and δ would result in a value of $\delta = -25°$.

The full dip azimuth, Az_θ, is still;

$Az_\theta = Az_2 - \delta = 80 - (-25) = 105°$.

Just as with the DH2 azimuth being 130°, the formula provides for Az = 80°, where $\Delta = 17°$ (Figure 15-11):

Azimuth to full dip $= Az_{S2} - \delta = 80 - (-25) = \mathbf{105°}$

The maximum dip azimuth is equidistant between two equal apparent dips.

15.3.4 Using a strike oriented grid

The previous examples have worked on a chosen **azimuth** orientation, defining the strike as azimuth. In practical field surveying it may be easier, once the strike line is known, to give **it** an azimuth of zero. The resultant grid makes the area much easier for field mapping. The depth to the plane at any point can be found as a $\Delta E/\theta$ (E offset/true dip) below the Total Station as shown in Figure 15-12. The TS will read BRG and H distance. From this, use the [REC(] function to find x (displayed) and [RCL] [tan^F] for y, (ΔE). Then F= 73÷3.14 ANS = 22.3.

Set the TS up on a known point with depth to seam, RL and true dip, θ, available. Orient on the strike line and set 0°. With the prism pole (HT) set to instrument height (HI), you can now map any point on the surface and deduce the depth to seam below the surface point.

From § 15.3.1.2 and Figure 15-6 the true dip is $\theta = 3.14$

DH_0 RL = 100, Depth to seam = –10, RL S_0 = 90.

Field book entry:

Brg 73°00'00", ZA 99°20'00" S = 74

H = 73, V = –12.0

RL DH_P = 100–12 = 88.0

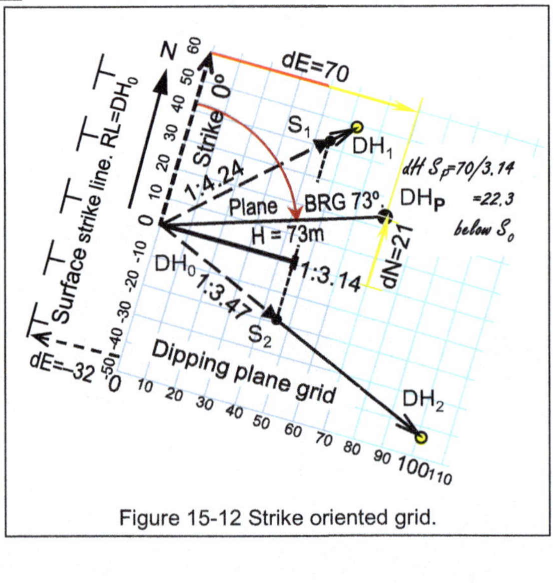

Figure 15-12 Strike oriented grid.

Find ΔE using REC(73,73°00'00")RCL tan^F ⇒ F=70.0

Depth to seam = $\Delta E/\theta$ = 70/3.14 = 22.3

RL seam at DH_P = S_P = RL S_0 – depth = 90–22.3 = 67.7

Depth to S_P =RL DH_P – S_P = 88.0–67.7 = 28.6m.

With confidence in your calculator, the RL of the drill hole and the depth to seam can be derived in the field and marked on the locating post.

15.4 Depth to Seam

Once the strike line azimuth and the true dip, θ, have been determined, it is possible to calculate the depth to the seam at any point. The reference level is initially the level of S_0. The depth to the seam is then dependent on the departure (distance east or west) from the strike line.

In Worked Example 1 in § 15.3.1, assuming the ground to the west of DH0 was flat (RL100) then the seam, 10m below DH_0, penetrates the surface some 31m west on the dip line through of DH_0 (10 × true dip, $\theta = 31.4$).

Figure 15-12 is a plan of the drilling programme.

Bearings and distances to DH_1 and DH_2 from DH_0 are to scale. A scaled 10m grid has overlaid the seam, oriented in the direction of the strike.

With S_0 as datum, the depth to any point on the seam can be calculated by scaling in the direction of the dip. By inspection, scaled S_2 appears to be about 101m east of S_0.

Depth to S_2 below S_0 = 101/3.14 = 32.2.

This agrees with original data of 32m.

The easiest way to determine position on the rotated grid is to set up at a known point (say DH_0) and zero-set the Total Station to the strike azimuth.

Grid Brg to $DH_1 = Az_{DH1} - Az\ Strike = 63 - 15 = 48°$.

Sight to DH1, note the direction, subtract 48°, turn to the new direction. Then Zero Set the TS [0 SET]. E.g., HAR to DH_1 = 105°, turn to HAR (105-48) = 57°. Zero Set to strike azimuth.

Thus, any bearing and horizontal distance (H) can be converted to local dE and local dN by the standard rectangular formulae, dE = H×sin(brg), dN = H×cos(brg).

The depth (h) to the seam at any point (reference RL S_0) can be calculated. (In this case RLS_0 is RLDH0–10):

$dH = dE/\theta$

15.4.1 Worked Example 15.2

Regard point DH_P in Figure 15-12.

At DH_0 a sight was taken to P,

Az (HAR) = 73, ZA = 99°20′,

slope distance = 74m.

Grid Brg from strike line = 73°,

Dist H = 73m. dE = 70, dN = 21.

$h_P = dE_{S0}$/true dip

h_P = 70/3.14 = 22.3m below S_0

IL S_P = RL $S_0 - h_P$ = 90 – 22.3 = 67.7

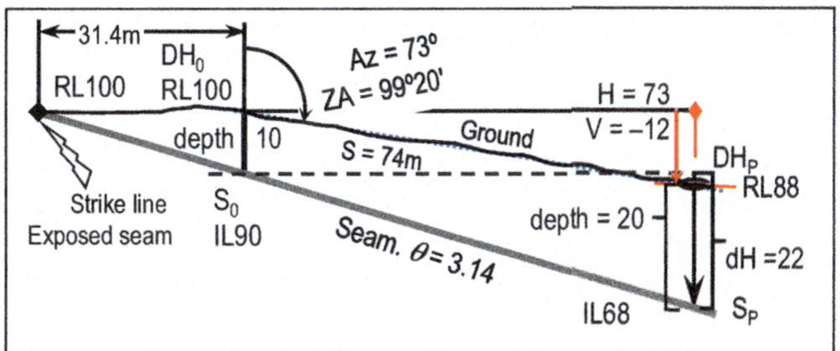

Figure 15-13 Drilling in Worked Example 15.2.

The profile between DH_0 and DH_P is shown in Figure 15-13.

IL S_P = 68m, the RL of the seam at P.

From DH_1, RL = 100, the difference in height to P, dH = 74 × cos(99.33°) = –12m

RL DH_P = RL DH_0 – dH = 100 – 12 = 88m

Depth to seam at DH_P = 88 – 68 = 20m.

15.5 Seam Thickness

Having calculated the true dip, θ, the **thickness** of the seam can be calculated.

From the Worked Example in §15.3.1, the seam was intercepted 10m vertically below DH0. Continued vertical drilling resulted in an **apparent** seam thickness (d_a) of 16m. What is the **true** thickness (d_t) of the seam?

In Figure 15-14,

Depth to seam, h = 10m

True dip, θ= 1:3.*14* = 17.6°

true thickness, d_t = apparent $(d_a)\cos\theta$

seam $(d_t) = 16\times\cos(17.6°)$ = **15.2m.**

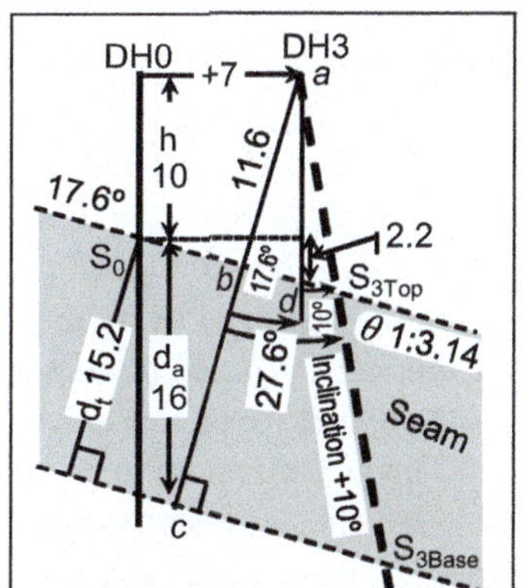

Figure 15-14 Inclined drill.

15.5.1 Worked Example 15.3

Consider an inclined drill hole, DH3 in Figure 15-15, which is inclined away from the strike line. Drilling starts at +7m down the dip line through DH0 at an inclination of +10° from the vertical away "down slope" (Inclination 27.6° perpendicular to seam).

At DH3 (referring to Figure 15-14):

Inclination angle to seam normal,	I = 27.6°
Advance from DH0,	s = 7m
Vertical depth to seam	h + s/θ = 10 + (7/3.14) = 12.2m
Incline depth to seam from DH3:	ab = 12.2 ×cos(17.6) = 11.6m
Seam thickness,	bc = d_t = 15.2m
Distance to top seam (S_{3T})	ab/cosI = 11.6/cos(27.6) = 13.1m
Distance to base seam (S_{3B})	(ab+bc)/cosI = (27.8)/cos(27.6) = 30.2m
Apparent core interval	$S_{3B} - S_{3T}$ = 30.2 – 13.1 = 17.1m
Or, core length	d_t/cosI = 15.2/cos(27.6) = 17.1m.

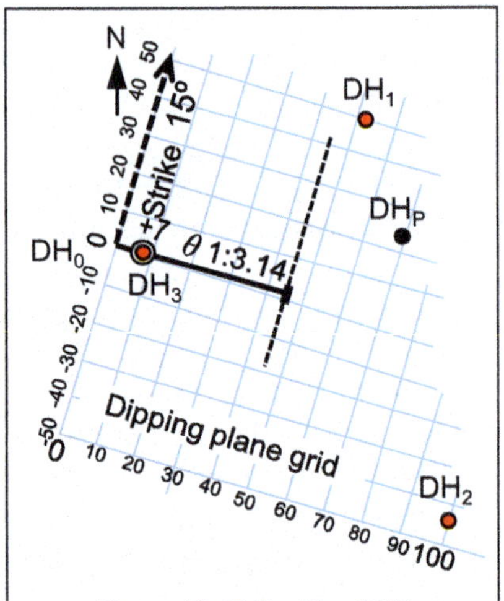

Figure 15-15 Position DH3

15.6 Direction of Any Slope Over the Dipping Surface

Investigate the **direction** of desired apparent dip. **Draw** a design apparent dip of 1:6.

The direction of **any** apparent dip can be realised graphically by scaling the dip until it meets the strike line perpendicular to the true dip. There will be two intercepts as illustrated.

Use a compass to draw the dip radius of 6 units, α_6, to the same scale used for the true dip, θ, to intercept the extended strike line through S_1, and S_2, Using Worked Example 15.1 in §15.3.1

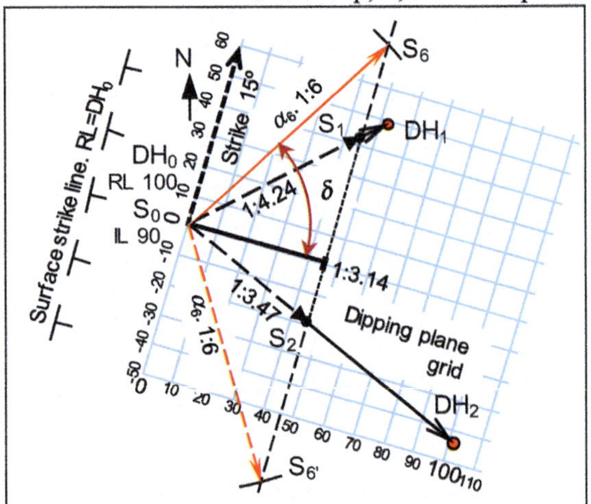

Figure 15-16 Direction of fixed slope.

Figure 15-16 shows an apparent dip of 1:6 (α = 6, 9.5°). A protractor shows an azimuth of about 47° or 163°.

Calculate the dip deflection angle, δ, between true dip, θ, and apparent dip, α_6, as a check.

From the expression (Eqn 15-3) from §15.3, page 234:

$$\tan\alpha_6 = \cos\theta\cos\delta \Rightarrow \cos\delta = \frac{\tan\alpha_6}{\tan\theta}$$

$\tan\alpha_6$ is the apparent dip, 1:6. 1/6 = 0.6667
$\tan\theta$ is the true dip, 1:3.14. 1/3.14 = 0.3185

$$\cos\delta = \frac{0.6667}{0.3185} = 0.5233 \qquad \delta = 58.4°$$

Azimuth of apparent dip 1:6 = true dip ± δ
= 105 + 58 = 163° or
= 105 – 58 = 47°.

This agrees with the graphical solution, and it is up to the geologist or engineer to pick the required direction. Any point on a line of bearing of 47° (or 163°) will have an apparent dip of 1:6 on this seam plane.

At a distance (H) of 100m, the depth (h) to the plane, below DH0 at S_0, will be:

$h = H_{dist}.\tan\alpha_6$.

H_{dist} = 100, h = 100/6 = 16.7m. And from the , RL S_0 = 90m.

IL S_6 = 90 – 17 = 73m.

We still require the RL at the surface at the chosen point, DH6, to establish the drill depth to the seam. This would be achieved using the Zenith Angle (ZA) and slope distance when setting the position of DH6 from DH0 as shown in Worked Example 15.1. (§15.3.1, p235).

Note that we have used azimuth to define the direction to S_6.

The grid direction to S_6 is azimuth – strike = 47 – 15 = 32° or 163 – 15 = 148°.

As a check of depth to S_6: BRG 32°, H_{dist} = 100. dE = 53, h = 53/3.14 = 16.8

15.7 Horizontal Angles Projected on to an Inclined Plane

The deflection angles that have been developed for dipping plane have been **horizontal** angles. The question arises as to the value of the included angle on the inclined plane. Theodolites measure horizontal angles and inclinations (vertical angles). Only a **sextant,** sighting to two targets on an inclined plane, can read the inclined angle. The astronomic "method of lunars" solved longitude (time) problems by using a sextant to measure the inclined angle between a bright star or planet and the moon.

Inclined angles are needed for the manufacture of pipework or ducting elbows along any inclined surface. The angular difference between horizontal and inclined may not be great, but it may be significant on steep inclinations.

In Figure 15-17:
AC_1 and BC_1 are inclined to horizontal plane ACB.
inclinations at A and B are $\boldsymbol{\alpha}$ and $\boldsymbol{\beta}$ respectively.
horizontal angle ACB = θ
incline angle AC_1B = λ
height difference CC_1 = h

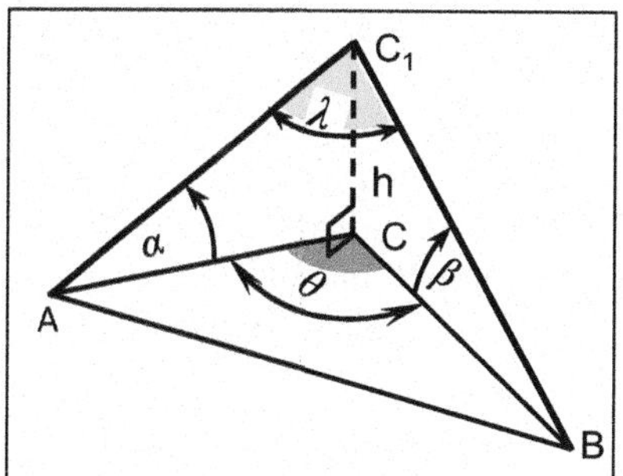

Figure 15-17 Horizontal and inclined planes to an intersection.

$$AC = h\cot\alpha, AC_1 = h\csc\alpha \text{ where } \cot = \frac{1}{\tan}, \csc(\text{cosec}) = \frac{1}{\sin}$$

$$BC = h\cot\beta, BC_1 = h\csc\beta$$

In triangle ACB, by cos rule for sides, in Figure 15-17,

$$AB^2 = AC^2 + BC^2 - 2AC \times BC\cos\theta$$

$$AB^2 = h^2\cot^2\alpha + h^2\cot^2\beta - 2h\cot\alpha\cot\beta\cos\theta$$

Similarly, in triangle AC_1B, by cos rule for sides:

$$AB^2 = h^2\csc^2\alpha + h^2\csc^2\beta - 2h\csc\alpha\csc\beta\cos\theta$$

By similarity, gathering terms and cancelling h^2:

$$h^2\cot^2\alpha + h^2\cot^2\beta - 2h^2\cot\alpha\cot\beta\cos\theta = h^2\csc^2\alpha + h^2\csc^2\beta - 2h^2\csc\alpha\csc\beta\cos\lambda$$

$$2h^2\csc\alpha\csc\beta\cos\lambda = h^2\csc^2\alpha + h^2\csc^2\beta - \left(h^2\cot^2\alpha + h^2\cot^2\beta - 2h^2\cot\alpha\cot\beta\cos\theta\right)$$

$$2\csc\alpha\csc\beta\cos\lambda = \csc^2\alpha + \csc^2\beta - \left(\cot^2\alpha + \cot^2\beta - 2\cot\alpha\cot\beta\cos\theta\right)$$

Thus, expressing the similarity in terms of the inclined angle λ:

$$\cos\lambda = \frac{\left(\csc^2\alpha - \cot^2\alpha\right) + \left(\csc^2\beta - \cot^2\beta\right) + 2\cot\alpha\cot\beta\cos\theta}{2\csc\alpha\csc\beta}$$

Because, from general trigonometric proof:

As $1 + \cot^2 A = \csc^2 A$, then $\csc^2 A - \cot^2 A$, and then:

$$\csc^2\alpha - \cot^2\alpha = 1 = \csc^2\beta - \cot^2\beta$$

Thrashing valiantly through the enlightening jungle of trigonometric obfuscation to the sunlit up-lands,

$$\cos\lambda = \frac{1 + 1 + 2\cot\alpha\cot\beta\cos\theta}{2\csc\alpha\csc\beta} \Rightarrow \frac{2\left(1 + \cot\alpha\cot\beta\cos\theta\right)}{2\csc\alpha\csc\beta} \Rightarrow \frac{\frac{1}{1} + \frac{1}{\tan\alpha}\cdot\frac{1}{\tan\beta}\cdot\frac{\cos\theta}{1}}{\frac{1}{\sin\alpha}\cdot\frac{1}{\sin\beta}}$$

Extract the **inclined** angle, λ, in terms of inclination and **horizontal** angle, θ.

Inclined angle, λ	$\cos\lambda = \sin\alpha\sin\beta + \cos\alpha\cos\beta\cos\theta$	Eqn 15-10

Extract the **horizontal** angle, θ, in terms of inclination and **inclined** angle, λ.

Horizontal angle θ:	$\cos\theta = \dfrac{\cos\lambda - \sin\alpha\sin\beta}{\cos\alpha\cos\beta}$	Eqn 15-11

The formula is sensitive to the sign of the inclination of the two lines **at** the intersection (the bend). A continuing falling decline will have one positive inclination to, and one negative inclination away from, the intersection.

15.7.1 Worked Example 15.4

A pipe is to be installed along two intersecting declines. The coordinates of the pipe supports provide the following information, see Figure 15-18:

The horizontal angle at the bend is 63°
one decline, A-C_1, rises at a gradient of 1:3.
the second decline, B-C_1, rises on a gradient of 1:4

Both declines are rising **to** the intersection, so each gradient is positive to the high point.

Determine the inclined angle at the intersection, C_1:

Data:

$C = \theta = 63°$

$A = \alpha = \arctan(1/3) = 18.43°$

$B = \beta = \arctan(1/4) = 14.03°$

Using Eqn 15-10:

$\cos\lambda = \sin\alpha\sin\beta + \cos\alpha\cos\beta\cos\theta$

Solution:

$\cos\lambda = \sin(18.43)\sin(14.03) + \cos(18.43)\cos(14.03)\cos(63)$

$= 0.316 \times 0.242 + 0.949 \times 0.97 \times 0.454 = 0.077 + 0.417$

$\cos\lambda = 0.494$

$\lambda = \arccos(0.494) = 60.36°$

Inclined angle, $C_1 = 60°21'$. Horizontal angle $C = \theta = 63°$

Note nearly 3° difference between the inclines and horizontal angle.

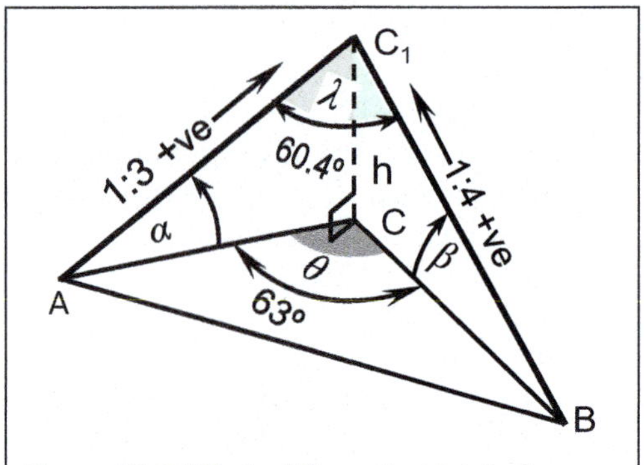

Figure 15-18 Worked Example 15.4, both grades positive, "rising" to the intersection

15.7.2 Worked Example 15.5

A pipe is to be installed along two intersecting declines. The coordinates of the pipe supports provides the following information: See Figure 15-19

The horizontal angle at the bend is 63°

one decline, A - C_1, rises at a gradient of 1:3.

the second decline, C_1-B, continues a rising gradient of 1:4

One decline rises **to** the intersection, C_1,

the other continues rising **from** C_1

so one gradient is positive: A-C_1

the other is negative: B-C_1. (It falls **to** the intersection)

Determine the inclined angle at intersection, C_1:

See Figure 15-19.

Data:

$C = \theta = 63°$

$A = \alpha = \arctan(1/3) = 18.43°$

$B = \beta = \arctan(1/{-4}) = -14.03°$

Using Eqn 15-10:

$\cos\lambda = \sin\alpha\sin\beta + \cos\alpha\cos\beta\cos\theta$

Solution:

$\cos\lambda = \sin(18.43)\sin(-14.03) + \cos(18.43)\cos(-14.03)\cos(63)$

$= 0.316 \cdot -0.242 + 0.949 \cdot 0.97 \cdot 0.454 = -0.077 + 0.417$

$\cos\lambda = 0.340$

$\lambda = \arccos(0.340)\lambda = 70.14°$

Inclined angle, $C_1 = 70°08'$. Horizontal angle $C = \theta = 63°$

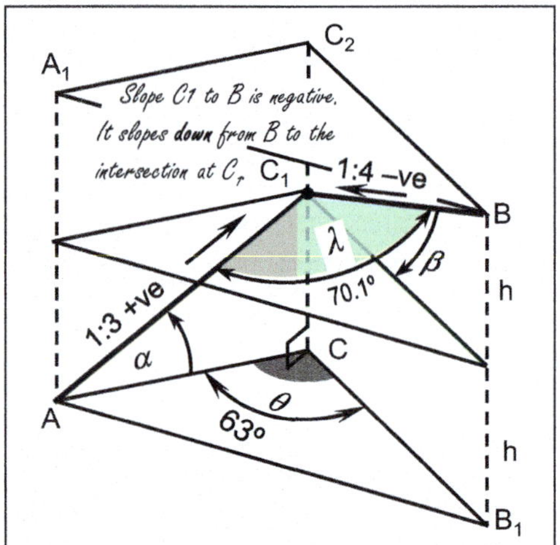

Figure 15-19 Worked Example 15.5, Grades rising and falling to the intersection.

Note that only the **sine** changes sign as the **sign** of the inclination changes. If both inclinations were negative, it wouldn't matter as the product of the two negative sines is a positive, as in Worked Example 1.

15.8 Concluding Remarks

This chapter is a demonstration of elements used to calculate on the dipping plane. We hope we have shown that the process can, in simple cases, be solved in the field, either by simple mathematics or by graphical means.

Courses in Structural Geology introduce the subject of a dipping plane. They have not been investigated because the references to this chapter cover the subject more fully from the surveying and computational aspect.

15.9 Reference for Chapter 15

Shepherd, F.A, Surveying Problems and Solutions, Edward Arnold, 1968

Shepherd, F.A, Engineering Surveying, Problems and Solutions, Edward Arnold, 2nd ed, 1983

Schofield, W and M Breach. Engineering Surveying, Sixth edition. Elsevier. 2007

Chapter 16 UTM AND GDA2020 REDUCTIONS AND COMPUTATIONS

For surveyors and engineers in Australia, more often than not, they will be working on Geocentric Datum of Australia (2020) ellipsoid (see Chapter 1 on ellipsoid). This chapter presents reductions that are undertaken on GDA2020 in addition to the UTM (Universal Transverse Mercator) Southern hemisphere projection, Figure 16-1.

16.1 Check for scale factor in horizontal distances.

Total stations display slope, **S**, horizontal, **H** and vertical, **V** distances. H and V are generated from the slope distance and zenith angle. H distance may, or may not, incorporate scale factor. It depends on the TS software. And whether the TS is in basic (display only) or program (data collection) mode.

16.1.1 Distances

Defining measured horizontal distance: *Geodetic Datum of Australia 2020 (GDA2020) Technical Manual v 1.6.*

1. s, Ellipsoidal distance; the distance on the ellipsoid (spheroid). From reduced distance observations.
2. S, Grid distance; the length measured on the grid, along the arc of the projected ellipsoid distance.

If your reference coordinates are modelled on, say the Universal Transverse Mercator (UTM) projection (e.g. MGA2020) then there will be a point or line scale factor (S.F., *k* or *K*) in the horizontal distances in your work area. Your observed horizontal distances must incorporate the scale factor to maintain the projection coordinates. The scale factor may be introduced via:

1. Adjusting the scale factor in the JOB parameters in the total station program. The displayed H distance then incorporates the scale factor,
2. Applying the scale factor to the horizontal distance calculated from ZA and slope distance. The H distance then has S.F. = 1.

Use S.F. = 1 in the field and reduce the H distance to the projection distance by incorporating the S.F. in the office calculations.

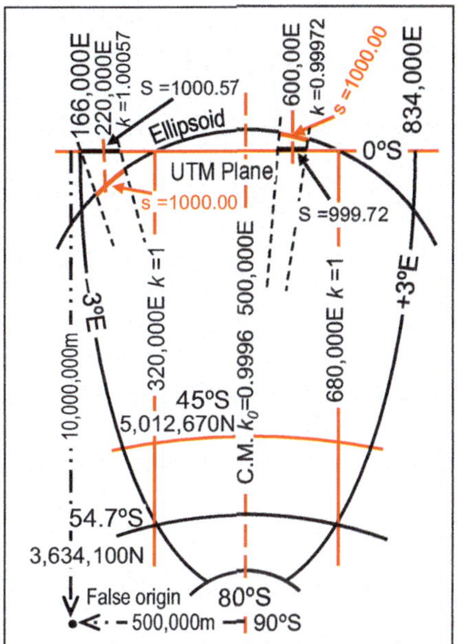

Figure 16-1 UTM projection Southern hemisphere

The opening and closing distance observations in a closed link traverse allow for checking the presence of horizontal scale factor (S.F.) in the coordinate frame. The calculated S.F. should agree with the area's declared scale factor. Disagreements should be resolved before proceeding with the adjustment.

Observations over the base line should incorporate zenith angle (ZA) and slope distance (S) measurements. This will allow verification of the horizontal distance with the (H) distance displayed on the instrument. The vertical distance (V) will allow rudimentary checking of station RLs.

16.1.2 What's this UTM and scale factor?

All our raw measurements are on the surface of the earth. Its generalized shape is ellipsoidal, so that one of the first tasks of position generation is to reduce the distance and angle measurements to the ellipsoid. The next step is to reduce to the mapping projection plane. The projection used in Australian surveying is the Universal Transverse Mercator projection defined by the Map Grid of Australia (MGA).

Aviation uses the World Aeronautic Chart (WAC) 1:1,000,000 Lambert Conformal Projection with Two Standard Parallels.

Standard surveying practice for Australia is to use the MGA2020 projection on a national or regional basis. Local projections grids are used over areas of limited extent to provide, basically, a plane coordinate system coincident between the ellipsoid and the mapping plane. This implies a very nearly 1:1 correspondence, i.e., the scale factor between the plane and the ellipsoid is very nearly k = 1. An example is the Perth Coastal Grid (PCG2020) where k_0 = 0.9999906, and the zone width is only 40′.

The UTM is defined in terms of the extent of the zone width (width of the "map"), latitude limits and central meridian scale factor, k_0. You are referred to the ISCM publication, Geodetic Datum of Australia 2020 (GDA2020) Technical Manual v 1.6. illustrates the characteristics Southern hemisphere of a UTM Zone and some of the limitations imposed on the survey area. The limit of the UTM zone is 80°S and 84°N. The GDA is limited to between 8°S and 45°S. Figure 16-2 illustrates the changing scale factor between the C.M. and the ±3° boundary. Of special note is the fact that the unitary scale factor, k = 1, is approximately ±180,000m from the C.M. Near the

J. Walker and J. Awange, *Surveying Computations*,
https://doi.org/10.1007/978-3-031-94382-9_16

zone boundary an overlap of about 0.5° allows cross zone coordinates to be calculated without significant scale factor errors.

There are 36 zones around the Earth, each zone width is 6° wide about a Central Meridian (C.M.). The first C.M. is at 177°W and the zones proceed Eastwards. Perth is in Zone 50 with a C.M. at 117°E.

To avoid negative numbers in coordinate calculation, the central meridian is offset by a false origin of 500,000m. The central scale factor, $k_0 = 0.9996$. These Easting coordinates are termed false Eastings (E) and are calculated by adding 500,000 to the Easting distance, E′, from the central meridian.

The GDA extends from Zone 49, 111°E at its western boundary, to Zone 57 with an eastern boundary of 162° (to include Lord Howe Island). The latitude extents are generally 8°S to 45°S to cover continental Australia and its offshore islands.

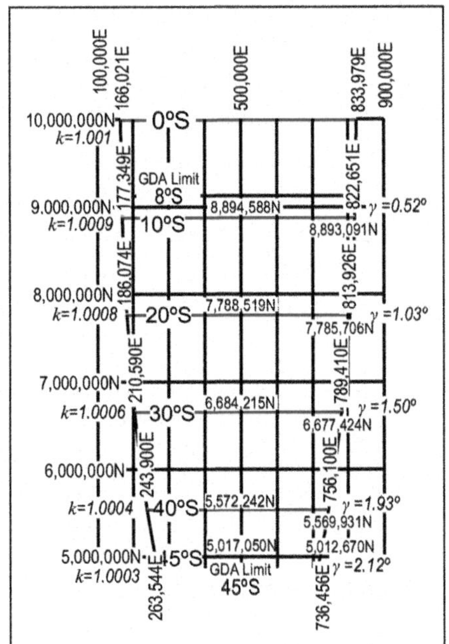

Figure 16-2 GDA2020 UTM zone limits

16.1.2.1 Point scale factor, k

The problem of projecting an ellipsoid onto a plane is one of reducing the distortion over the extent of the map area. Scale refers to the correspondence between the projection distance on the plane and the "size" of the resultant map. Scale factor defines the acceptable minor change in scale over the map plane.

The UTM operates between latitudes 84°N and 80°S. On choosing a central scale factor, the task was to decide on a value such that the point scale factor at the edge of the zone balanced the central scale factor. Now, allowing that the half zone width is 3° and the mid latitude is about 40°. The "rule" is that ($2\text{-}k_0$) at the zone boundary balances k_0 at the C.M. such that the mean is $k = 1$ between $k_0 = 0.9996$, $2\text{-}k_0 = 1.0004$. At a distance of 3° from the C.M., at a latitude of 40° the value for the point scale factor is $k = 1.0004$. Point scale factor is dependent on the distance from the C.M and can be calculated by $k = k_0[1+(dE^2/2rm^2)]$ to 2 parts in 10^7. dE is distance from C.M. and rm is mean radius of ellipsoid at the point. A mean radius, rm = 6,367,445m, would not impact the accuracy.

Figure 16-2 shows the point scale factor at the edge of the UTM zone at a number of latitudes. Figure 16-3 is a nomograph of point scale factor versus Easting (E′) either side of the C.M. As latitude increases away from the equator the point scale factor decreases within the UTM restraints of $\omega = \pm 3.5°$ (zone boundary ±30′) as E′ decreases.

16.1.2.2 Line scale factor, K

All plane surveying is now undertaken over a very limited area. This means that any point scale factor will serve the survey area computations without resort to calculating an "average" scale factor for a measured line.

Only long geodetic lines need to consider line scale factor, a subject left to the geodetic surveyor. If a line scale factor is needed, then one acceptable method is to calculate (or read from Figure 16-3) at the mid Easting, E′.

At 8°S a 20,000m ellipsoid distance, at the zone boundary, will have a line scale factor of 1.00089139 at E 177,000. The grid distance is 20,017.827m. If the mean of the end points (E 167,000 and 187,000) is used the grid distance is 20,017.844, a difference of 0.017m.

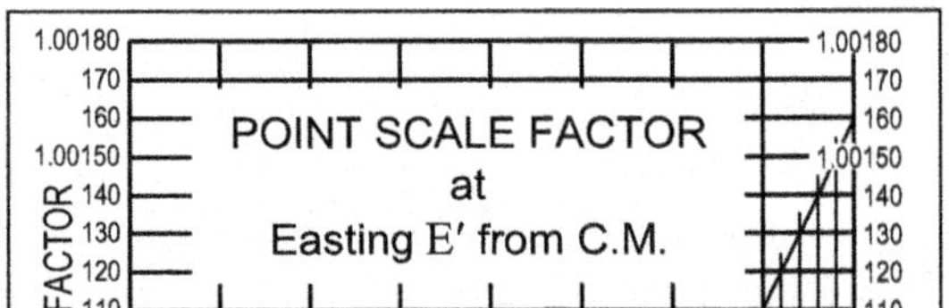

Figure 16-3 Point scale factor at E′ from C.M.

16.1.2.3 Grid distance calculation from ellipsoid distance

The observed plane or ellipsoidal horizontal distance, s, is reduced to the ellipsoidal grid distance, S, by the application of the scale factor, ***k*** or ***K***.

$S = Ks$. This calculation is demonstrated in Figure 16-1.

16.1.3 Scale factor in coordinate calculations

Referring back to Figure 5-16, page 58: H(calc) relates to the calculated distances derived from the station coordinates. H(obs) is the observed horizontal (H) distance calculated from the slope distance (S) and the zenith angle (ZA) as recorded in the field book, where the S.F. = 1 for the new intermediate distances, H_1 to H_4. The figure has been adjusted accordingly.

From the given coordinates it is impossible to know where the survey area is located. Thus, location data is needed. The station data should also indicate the projection and the scale factor.

16.1.3.1 Local plane coordinates

The observations of a closed connecting traverse (§ 5.13.2, page 58) are repeated in Table 16-1 and Figure 16-4. On the face of the example an assumption has been made that a plane coordinate system is being used. But examination of the field book showed the following: (Table 16-1).

H18-20(calc) = 478.410, H18-20(obs) = 478.601.

S.F. = 478.410/487.601 = 0.9996

H35-52(calc) = 600.000, H35-52(obs) = 600.240.

S.F. = 600.000/600.240 = 0.9996

Observed distances with S.F. = 1:

H_1 = 465.656, H_2 = 438.605, H_3 = 278.671, H_4 = 740.016

Table 16-1 Field book. Observed angles, plane distances (S.F. = 1)

Line/angle	Adj angle	H obs	H reduced
18 - 20		*478.601*	478.410
∠20	70 00 45		
20 – A, H_1		*465.656*	465.47
∠A	241 31 25		
A – B, H_2		*438.605*	438.43
∠B	257 34 05		
B – C, H_3		*278.671*	278.65
∠C	57 53 20		
C – 35, H_4		*740.016*	739.72
∠35	274 27 50		
35 - 52		*600.240*	600.000

The projection S.F for the area = 0.9996 has been confirmed by observations and has been used to reduce observed horizontal distances to the projection plane. The observed and reduced H distances are shown and have been reduced to projection distances using S.F. 0.9996 as shown in Figure 16-4.

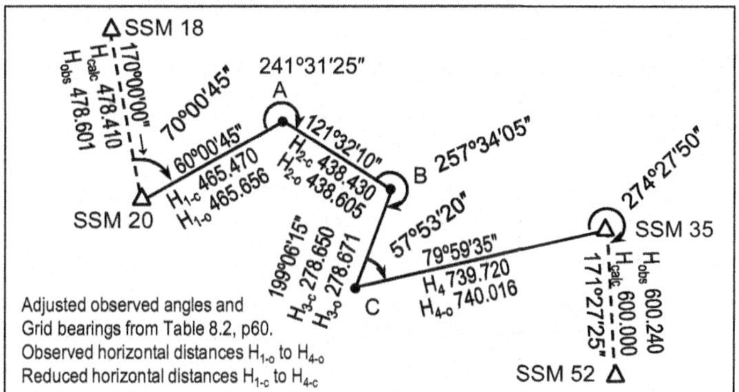

Figure 16-4 Unadjusted linked traverse with reduced adjusted observations

The surveyor would suspect that the coordinates used in the station description are truncated MGA or GDA coordinates and would have to find the location to determine where it relates to. "It's just north of the Beverley-Mawson road". Beverley, Western Australia, is a town about 117°E, so it is in zone 50 near the central meridian.

Since the SF is 0.9996 it must be close to the zone 50 C.M. which is 500,000E. Thus, one could assume that, for SSM20 there has been a truncation of 500,000, giving SSM20 an E coordinate of 501,000. The Beverley-Mawson road has a UTM N coordinate of 6,454,500N. Similar reasoning gives SSM20 a coordinate of 6,455,000N. The surveyor could then conclude that the given coordinates have been truncated by 500,00 in E and 6,450,000 in N. However, note that this truncation in North can occur every 10,000m above or below the current position. A station diagram and geographic coordinates are required to confirm the location of the truncated coordinates.

Truncation in Easting coordinates could be more problematic around the C.M because scale factor, k, changes slowly. At 491,000E k = 0.999601 and k = 0.999604 at 481,000E One possible cause for the confusion is the name of the control stations. The abbreviation SSM generally stands for "State Survey Mark" and implies backing by a regulatory authority whose coordinate system is UTM. Figure 16-5 shows the survey area with reference to the UTM Zone 50 coordinates.

Some notes on the Google Earth presentation (see Figure 16-5).

1. The image date format is MM/DD/YYY.
2. The false North UTM coordinate (N) shows as 6455621.62 m S which is their way of showing southern hemisphere rectangular coordinates.

   The Northing (N') coordinate, measured from the equator is N – 10^7 = – 3545150.27m
3. G.E switches zones AT the zone boundary. It does not do overlap. The Mt Holland position flips to Zone 51H at about 783,401E, 6,445,000N.

16.1.3.2 Testing the zone limits

The following examples Table 16-2 show the limits of Zone 50 in Western Australia which incorporate the truncated SSM20 coordinates, SSM20 1,000E, 5,000N. The point scale factor, k, has been calculated at the various SSM20 coordinates. The centre easting of the figure SSM18 – SSM52 is about 743m East of SSM20, at about STN B. The line scale factor, *K*, has been calculated for that easting. The biggest difference in ellipsoidal distance between SSM20 and SSM35 is 0.005m in 1031.6m due to using *k* instead of *K*. This is an error of 1:206,000.

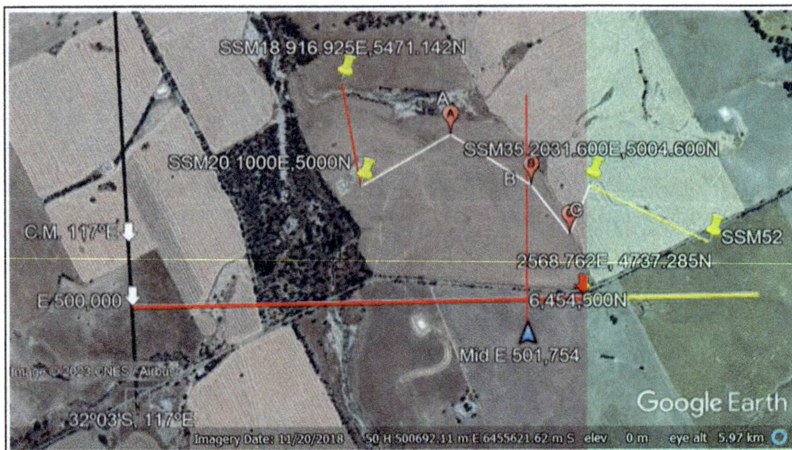

Figure 16-5 Survey area, Zone 50

Table 16-2 Some limits in Zone 50

Location	UTM E	UTM N	Latitude	Longitude	Grid conv., λ	Scale factor *k*	Line scale factor *K*
Kalbarri	231,000	6,925,000	27.77°S	114.27°E	–1.27°	1.000493	1.000488
4Km E Yallingup	231,000	6,275,000	33.63°S	114.06E	–1.61°	1.0004892	1.000487
47Km ENE Mt Holland	781,000	6,455,000	32.01S	120.97°E	1.57°	1.000574	1.000579
Beverley	501,000	6,455,000	32.04°S	117.01°E	0.005°	0.999600	0.999600

16.1.4 Calculating UTM parameters

UTM calculations, based on the GDA Technical Manual, are posted on the Geoscience Australia website.
Spreadsheet: (https://github.com/icsm-au/DatumSpreadsheets)
Access to solutions: www.ga.gov.au/scientific-topics/positioning-navigation/geodesy/transform-convert

16.1.4.1 Coordinate computations using Vincenty's Direct and Inverse solutions:

Vincenty's Direct: Given the latitude and longitude of a point (1) and the forward geodetic azimuth (1-2) and ellipsoidal distance to a second point (2), calculate the latitude and longitude of the second point and the reverse azimuth (2-1).

Vincenty's direct: (http://www.ga.gov.au/geodesy/datums/vincenty_direct.jsp)

Vincenty's Inverse: Given the latitude and longitude of two points compute the ellipsoidal distance and forward and reverse azimuths between the points.

Vincenty's inverse: (http://www.ga.gov.au/geodesy/datums/vincenty_inverse.jsp)

16.1.4.2 Coordinate conversions between geographic and grid using Krueger n-series or Krueger λ-series

Two methods are presented to convert Geographic to/from Grid coordinates: Krueger n-series equations and Krueger λ-series equations (Redfearn's equations). The Krueger λ-series equations are accurate to better than 1 mm within any zone of the Map Grid of Australia 2020 and can still be used for many purposes. However, for applications where users are working across multiple UTM/MGA zones, Krueger n-series equations are recommended. The Krueger n-series equations are particularly beneficial in software to avoid error build up when conversions are done back and forth between geographic and grid coordinates.

Grid to Geographic: (http://www.ga.gov.au/geodesy/datums/redfearn_geo_to_grid.jsp)
Grid to Geographic: (http://www.ga.gov.au/geodesy/datums/redfearn_grid_to_geo.jsp)

16.1.5 Grid convergence and bearings

Grid convergence is the angular quantity to be added algebraically to an azimuth to obtain a grid bearing. In determining bearing between two stations, we are calculating the plane bearing defined by the rectangular coordinate frame. The actual orientation of the figure with reference to the True North meridian cannot be realised until we have some information on the projection being used. Because the UTM projection is projecting the Earth, as an ellipsoid, on to a rectangular plane the meridians of longitude curve away from the central meridian with reference to the rectangular grid of the projection. This curving away is called **grid convergence** and, at any point, is the horizontal angle between true north and grid north. It is proportional to the longitude difference between the place and the central meridian and is latitude dependent.

Spherical grid convergence, γ, can be calculated from latitude, ϕ, and ω, the longitude difference between the point and the central meridian, by: $g = \arctan(-\sin\phi \times \tan\omega)$. Elliptical γ is a power series.

The MGA grid convergence is a maximum at the zone boundary, ±25′ at 8°S to ± 2°07′ at 45°S.

16.1.5.1 Converting between true azimuth and grid azimuth

In many texts the term bearing is applied loosely. Unless it is more explicitly defined in relation to a reference direction it is a confusing term. The term azimuth is used to denote a horizontal direction in relation to a defined meridian. This may be True North, Grid North or Magnetic North.

Convert from Grid Azimuth to True Azimuth	Convert from True Azimuth to Grid Azimuth
grid azimuth, β = true azimuth, α + grid convergence, γ	true azimuth, α = grid azimuth, β − grid convergence, γ
$\beta = \alpha + \gamma$	$\alpha = \beta - \gamma$

Selected definitions from:

Geodetic Datum of Australia 2020 (GDA2020) Technical Manual v 1.6. Terms and Definitions, page 7. (GDA TM) are illustrated in Figure 16-6.

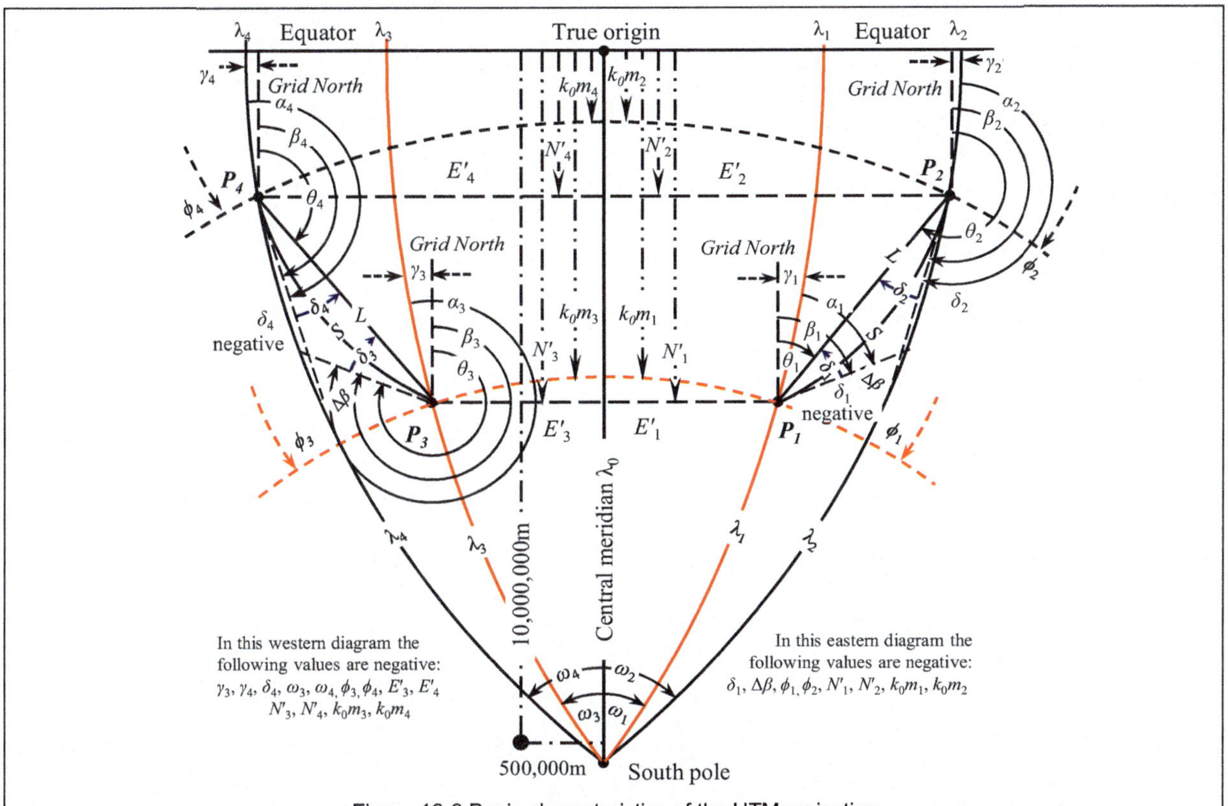

Figure 16-6 Basic characteristics of the UTM projection

- ϕ Geodetic latitude; negative south of the equator.
- λ Geodetic longitude; positive measured eastwards from the prime meridian.
- ω Geodetic longitude difference.
- E' Easting; positive measured eastwards from a central meridian.
- E False Easting; measured eastwards from a false origin, 500,00m west of the central meridian. 500,000 + E′.
- N' Northing; negative measured southwards from the equator.
- N Northing; measured northwards from a false origin 10,000,000m south of the equator. 10,000,000 + N′.
- α Azimuth; the horizontal angle measured from the meridian clockwise from true north.
- γ Grid convergence; the angular quantity to be added algebraically to an azimuth to obtain a grid bearing. In the southern hemisphere, grid convergence is positive for points east of the central meridian (grid north is west of true north) and negative for points west of the central meridian (grid north is east of true north). $\beta = \alpha + \gamma$
- β Grid bearing; the angle between grid north and the tangent to the arc at the point. It is measured from grid north clockwise through 360°.
- δ Arc-to-chord correction; the angular quantity to be added algebraically to a grid bearing to obtain a plane bearing. $\theta = \beta + \delta = \alpha + \gamma + \delta$

- θ Plane bearing; the angle measured clockwise through 360°, between grid north and the straight line on the grid between the ends of the arc formed by the projection of the ellipsoidal distance.
- k_0 Central scale factor; the scale factor on the central meridian.
- k Point scale factor; the scale factor at any point on the projection.
- K Line scale factor; the ratio of a plane distance, L, to the corresponding ellipsoidal distance s. K=L/s ≈ S/s.
- $\Delta\beta$ Line curvature; the change in grid bearing between two points on the arc.
- m Meridian distance is the geodesic distance from the equator along the meridian.
- s Ellipsoidal distance; the distance on the ellipsoid.
- S Grid distance; the length measured on the grid, along the arc of the projected ellipsoid distance.
- L Plane distance is the straight-line distance on the grid between the ends of the arc of the projected ellipsoidal distance. The difference in length between the plane distance (L) and the grid distance (S) is nearly always negligible. Using plane bearings and plane distances, the formulae of plane trigonometry hold rigorously. $L{=}Ks$ $\tan\theta{=}\Delta E/\Delta N$, $\Delta E{=}L\sin\theta$, $\Delta N{=}L\cos\theta$

16.1.6 True bearings (azimuths) and spheroidal distances across the zone

Figure 16-7 shows the grid bearings and distances of our survey. What are the azimuths and elliptical distances of the three lines at Kalbarri, Beverley, and Mt Holland?

Refer to Table 16-2 for convergence and scale factor:

1. applying the grid to azimuth formula:
   True azimuth = grid azimuth – grid convergence
2. applying the scale factor, k/K, over the each line of grid distance S (and S ≈ L) the elliptical horizontal distance, s, (the observed distance) is:
   $s = S/\mathrm{K}$ where $K \approx k$ over short distances:
   K for the mid E′ each line has been calculated

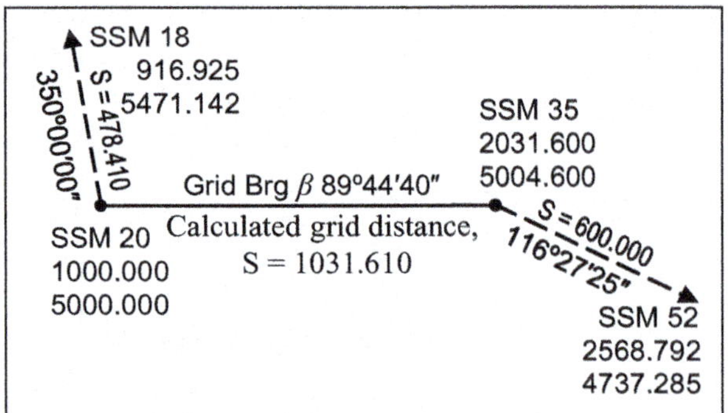

Figure 16-7 Polar vector SSM 20 to SSM 35

Table 16-3 Change of azimuth and elliptical distance across the zone.

Line		Kalbarri		Beverley		Mt Holland	
At SSM20	Grid S	Azimuth β-γ. γ= −1°16′22″	Elliptical s K=1.000493	Azimuth β-γ. γ 0°00′20″	Elliptical s K=0.9996	Azimuth β-γ. γ = 1°34′39″	Elliptical s K=1.000574
20 - 18	478.410	351°16′22″	478.174	349°59′40	478.601	348°25′21″	478.136
20 - 35	1031.610	91°01′02″	1031.102	89°44′20″	1032.023	87°10′01″	1031.018
At SSM35		Azimuth γ= −1°16′04″	Elliptical s K=1.000486	Azimuth γ= 0°00′41″	Elliptical s K=0.9996	Azimuth γ=1°35′00″	Elliptical s K=1.000581
35 - 20	1031.610	271°00′44″	1031.109	269°43′59″	1032.023	268°09′40″	1031.011
At SSM52		Azimuth γ= −1°15′55″	Elliptical s K=1.000486	Azimuth γ= 0°00′52″	Elliptical s K=0.9996	Azimuth γ= 1°35′11″	Elliptical s K=1.000584
35 - 52	600.000	117°43′29″	599.708	116°26′44″	600.240	294°52′14″	599.650

Note that the forward azimuth from SSM20 applies the grid convergence at SSM20 to the lines SSM 20 - SSM 18 and SSM 20 - SSM 35 However the grid convergence changes by about 20″ over the line SSM 20 - SSM 35 and the new convergence value at SSM35 must be used for the azimuth SSM 35 - SSM 52.

The reverse azimuth, SSM35 - SSM20 is reverse grid – grid convergence = 279°44′20″ – γ, as shown in Table 16-3. Similarly, setting out azimuths from SSM 18 – SSM 20 and SSM 52 – SSM 35 would require recalculation of convergence at those points.

Figure 16-8 illustrates the values in Table 16-2 depending on SSM 35 being on the western, central or eastern edge of the zone.

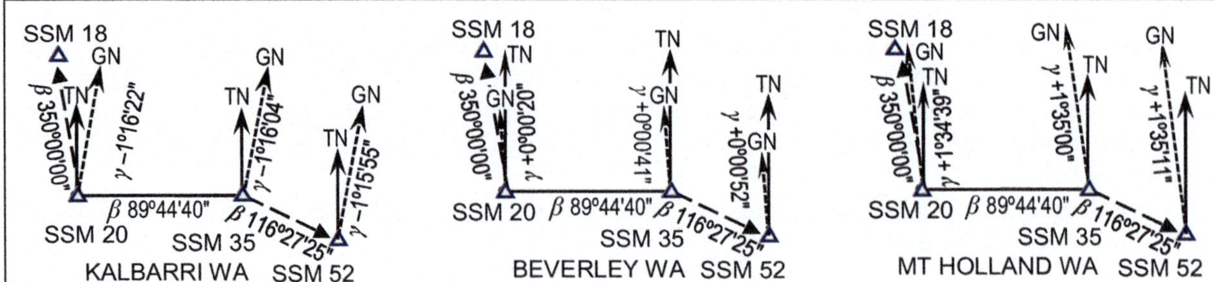

Figure 16-8 Grid convergence across the zone

16.1.7 Beware the long line in construction

Many construction projects use UTM coordinates for documenting structural elements of the project. Long physical features using off-site structural elements can run into problems if there is a dimensioning calculated on the UTM. Any constructed physical device must be measured on the terrain, not on the ellipsoid.

Cadastral and Geodetic Services of Land and Spatial Information Group, Department of Natural Resources and Mines. © State of Queensland, 2020. Cadastral Survey Requirements Version 7.1 July 2020, Reprint 2

"8.6.2 Use of terrain distances Section 48B of the Land Title Act 1994 requires that distances shown on cadastral plans must be corrected so that they are a horizontal distance between the marks on the ground. This is interpreted as being at the mean terrain height of the line (referred to as "level terrain distances" in this section).

When using a total station, the measurements are made at the height of the terrain and level terrain distances are typically achieved by simply applying a slope correction to the measured distances. However, distances resulting from GNSS measurements are computed on the ellipsoid. Surveyors should note that ellipsoidal distances can be significantly different to [sic] terrain distances depending on the height of the terrain in the area. A change in height of 6.5m causes approximately a 1ppm effect on distances. Therefore, at 650m above the ellipsoid, the difference between terrain and ellipsoidal distances is approximately 100ppm. Similarly, variations in height across a cadastral survey covering a large area may also be significant." (https://www.dnrm.qld.gov.au/?a=105601)

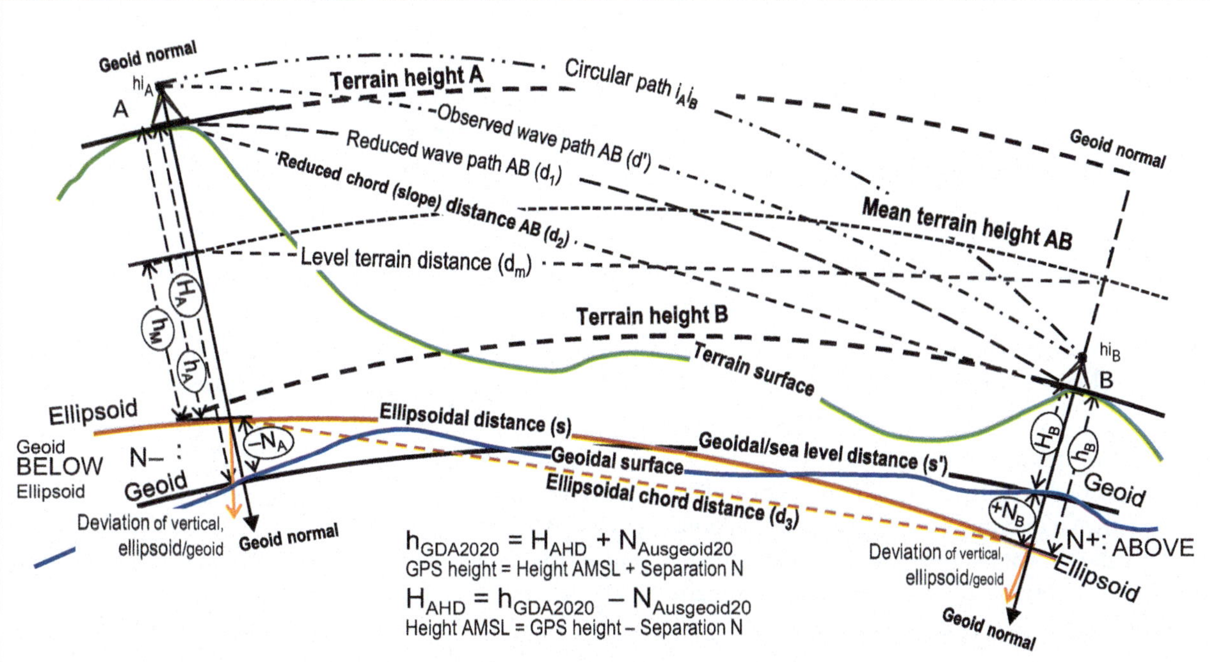

Figure 16-9 EDM data reductions to the ellipsoid

16.1.7.1 Ellipsoidal computation

Because of atmospheric refraction, the light waves or microwaves used by Electronic Distance Measurement (EDM) devices follow a curved path. Before this curved wave path distance can be used for geodetic computations, it should be reduced to the surface of the ellipsoid by the application of both physical and geometric corrections. Figure 16-9 shows the various EDM wave paths and the terrain, ellipsoid and geoid profiles.

16.1.7.2 Atmospheric velocity corrections.

The *observed* wave path d' establishes the *reduced* wave path d_1 by the application of first and second velocity corrections.

The **first** velocity correction K', determined by the manufacturer, is a combination of refractive constant (light frequency) and air density (meteorological) parameters, entered into the total station at the time of observation. (MET or EDM corrections). See Figure 16-10.

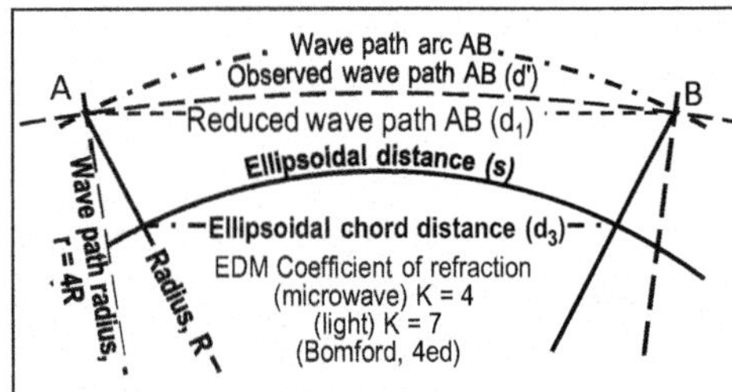

Figure 16-10 EDM wave path reductions

The **second** velocity correction, K'', is the difference between the reduced wave path *length* d_1 and the wave path *chord* (slope distance) d_2. The radius of the wave path, r, is not the same as the radius of the spheroid, R. Wave path radius is a function of the EDM (light waves or microwaves) used and also of the topographic and meteorological conditions prevailing along the wave path at the time of measurement. The relationship between R and r is determined by the coefficient of refraction, k, for the line. $k = R/r$. A general value for light emitting total station EDMs is $k = 1/7$ (0.14). Previous microwave EDMs used in Australia, Tellurometers (MRA101, CA1000), used $k = 1/4$ (0.25).

"This difference can often be ignored for distance measurements of up to 15 kilometres, using either light waves or microwaves. These physical corrections, which involve the application of certain velocity corrections to the measured wave path distance, are not discussed in this manual." (GDA TM)

Both velocity corrections are discussed extensively in Bomford 4th edition and Rueger (2nd ed 1984. § 3.11)

16.1.7.2.1 First velocity correction K'

First velocity correction K' is from EDM wavelength and atmospheric modelling determined by the manufacturer.

Sokkia SET530 $K' = d' \times 10^{-6}\left(282.59 - \dfrac{0.2942 \times p}{1 + 0.003661 \times t}\right)$ λ_{EDM}=690nm, 1013hPa, 15°C, 0%RH [§28.1]

Leica TS15 $K' = d' \times 10^{-6}\left(286.34 - \dfrac{0.29525 \times p}{1 + 0.003661 \times t}\right)$ λ_{EDM}=658nm, 1013.25hPa, 12°C, 60%RH

p = atmospheric pressure (hPa) and t = temperature (°C).

High precision long distance measurements in hot, high humidity conditions may justify humidity corrections.

16.1.7.3 Second velocity correction

Second velocity correction, K'', uses values empirically obtained from long line observations over various topography.

The computations rely on the refractive index of different "layers" of air in the wave path. (Rueger, § 3.12)

Suppose the wave path d' = 15,000m, R ≅ a ≅ 6,380,000. For light $k_{(LED)} \cong 1/7 = 0.14$

Second velocity correction: $$K'' = -\left(k - k^2\right)\frac{d'^3}{12R^2} = -\left(0.14 - 0.14^2\right)\frac{15000^3}{12 \times 6380000^3} = 0.0008m$$ (Rueger 1984, § 3.12(3.15))

Similarly for microwaves, $k_{(MWM)} \cong 1/4 = 0.25$ at d'=15,000. $K'' = -0.0013$.

The second velocity correction for either EDM is around –1mm at 15Km.

At 50Km a microwave Tellurometer would have a $K''= -0.047$m.

This shows that the reduced wave path d_1 is the same as the observed wave path d' after the application of the first velocity corrections, K', from the atmospheric observations that have been entered as EDM corrections.

16.1.7.4 Geometric corrections, combined formula

Wave path distance d_1 is influenced by the prevailing atmospheric conditions and is corrected by the inclusion of meteorological parameters (temperature and station level pressure) in the total station. We can take the wave path chord d_2 as being the same as d_1.

The reduction of the wave path chord distance d_2, to the ellipsoidal chord distance d_3, can be given as a single rigorous formula (Clark, 1966).

a. The wave path chord, which may be a slope distance, is reduced to the horizontal ellipsoidal chord distance by:

$$d_3 = \sqrt{\left[\left(d_2^2 - \left(h_A - h_B\right)^2\right) / \left(1 + h_A / R_\alpha\right)\left(1 + h_B / R_\alpha\right)\right]}$$

Note that ellipsoidal heights h_A and h_B are used. R_α is the radius of curvature in the azimuth of the line. Latitude dependent $R_m = \sqrt{(\rho\nu)}$ is the mean radius of curvature of the ellipsoid. (GDA TM § 5.1.1.4)

GRS80 R_m of 6,368,740m at latitude –32° is an example.

b. The ellipsoidal chord distance d_3 is then reduced to the ellipsoidal arc distance s.

$$s = d_3\left[1+\left(\frac{d_3^{\,2}}{24R\alpha^2}+\frac{3d_3^{\,4}}{640R\alpha^4}+\ldots\right)\right]$$

where R_α is the radius of curvature in the azimuth of the line. Again, R_m may be used

"For a distance of 30 kilometres in the Australian region the chord-to-arc correction is 0.028 m. For a distance of 50 km, the correction reaches about 0.13m, and it is more than 1m at 100km. The second term in the chord-to-arc correction is less than 1 mm for lines up to 100km anywhere in Australia and can usually be ignored." (GDA TM 1.6).

16.1.7.4.1 Separate formulæ

"The combined formula above includes the slope and ellipsoid level corrections. The slope correction reduces the wave path chord d_2 to a horizontal distance at the mean elevation of the terminals of the line (terrain distance) and the ellipsoid level correction reduces the horizontal distance to the ellipsoid chord distance d_3. The chord-to-arc correction is then applied to the ellipsoid chord distance, as with the combined formula, to give the ellipsoidal arc distance s." (GDA TM).

Slope correction	$\sqrt{\left(d_2^{\,2}-\Delta h^2\right)}-d_2$	$d_2 \Rightarrow d_m$
Ellipsoidal correction	$\left(h_m / R\alpha\right)\sqrt{\left(d_2^{\,2}-\Delta h^2\right)}$	$d_2 \Rightarrow d_3$
Chord to arc correction	$+d_3^{\,3} / 24R_\alpha^{\,2}\left\{+d_3^{\,5} / 640R_\alpha^{\,2}+\cdots\right\}$	$d_3 \Rightarrow s$

The rise of GNSS surveying techniques has revolutionized geodetic surveying. Modern surveying with EDM is limited to short distances. A total station is generally limited to about 5,000m and is unlikely to exceed 1,000m in most applications. Thus, many of the cautions in the preceding reductions need not apply.

The main check is to ensure that the horizontal observed distances and the projection distances agree so that mapping takes place on the correct grid. This means confirming the project scale factor.

16.2 Local grids on the UTM

The preceding section has focussed on the Australian use of the Universal Transverse Mercator (UTM) projection for defining position in terms of geocentric (GDA2020) and rectangular E, N (MGA2020) coordinates. As has been shown, there is a need, at the local level, to be able to map orthogonally at a scale factor very close to unity ($k = 1$). This is a posh way of saying that we use plane distances and angles.

As students, we are introduced to the "flat earth" concept in our initial survey practical sessions. And this book concentrated on calculations the plane.

16.2.1 Introduction

Over the past several years building works on the Curtin University T.L. Robertson Library restricted survey exercises to the southern portion of Henderson Court Upper. This work has restricted CVEN2000 exercises to about 72 metres from the northern pathway of Building 204. With the completion of Library works, a return to the previous exercise area requires a re-establishment of survey control to the new exercise area.

The works also impacted other areas of Henderson Court, including the demolition of Building 599 and the renewal of access roads through the site. Consequently, survey marks have been destroyed, including some placed by a student's project in 2022.

Another requirement is to re-establish a grid coordinate system for use over the site. The most appropriate is a truncated Perth Coastal Grid, epoch 2020, to coincide with the Geodetic Datum of Australia 2020 (GDA2020) and the Map Grid of Australia 2020 (MGA2020) projection. Grid coordinates allow plane trigonometric solutions.

To satisfy the CVEN2000 unit requirements the resurvey has been based on two existing survey marks, CSM200 and CSM201 that are located adjacent Spatial Science Building 207. GDA2020 and MGA2020 coordinates are detailed on the Blackboard LMS under Curtin Survey Marks published as part of a student project.

16.2.2 Rationale.

For the initial survey exercises a rectangular grid of internally consistent measurements is the main requirement of control coordinates. Relative consistency is the important point.

However, many other Survey practical units now utilize the area, so a consistent set of coordinates in both local grid and national grid are needed. To this end the GDA2020 and its MGA2020 projection are transformed to Perth Costal Grid, 2020 (PCG2020) using Landgate published transformation parameters.

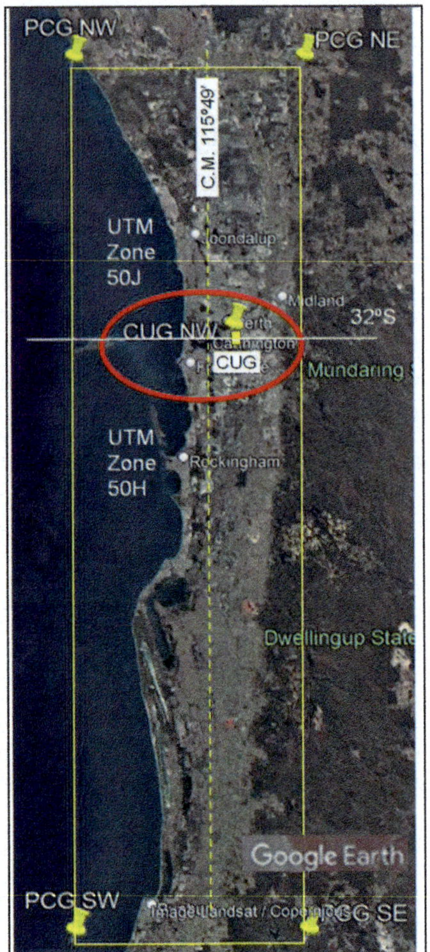

Figure 16-11 Perth Coastal Grid

16.2.3 Proposed Curtin University Grid 2020

The main problem for students using rectangular grid coordinates is the plethora of digits involved in each coordinate pair, leading to transcription errors and mathematical mistakes. To this end it is proposed that a truncated PCG2020 set of coordinates be initiated by using an offset datum of 56,000m in Easting and 350,000m in Northing. This allows a common set of truncated PCG2020 coordinates across the campus, including Edinburgh Oval and the Landgate pillars.

For example, CSM200 has PCG2020 coordinates of 57393.970E, 357399.343N. The corresponding CUG2020 coordinates are 1393.970E, 7399.343N. The numbers are not likely to be misinterpreted for orientation between a 4.3 digit E and N coordinate pairs.

16.2.4 Transformation from GDA2020 to PCG2020 coordinates

Projection parameters: Perth Coastal Grid 2020 (See Figure 16-11).

False Easting (E′)	=50,000
False Northing (N′)	= 3,900,000
Central scale factor k_0	= 0.99999906
Central meridian λ_0	= 115°49′00″
SE Limits	= 33°25′00″S, 115°26′30″E
NW Limits	= 31°20′00″S, 116°05′00″E

Curtin University Grid 2020 (See Figure 16-12).

False Easting (E′)	-6,000	
False Northing (N′)	3,550,000	
Central scale factor k_0	= 0.99999906	
Central meridian λ_0	= 115°49′00″	(PCG2020)
SE Limits	=32°01′30″S, 115°54′25″E	2528E, 5375N
NW Limits	=31°59′30″S, 115°52′50″E	38E, 9073N
AHD limit	= 165m	
Datum	GDA2020	

However, data held by Spatial Sciences for Curtin Survey Marks, whilst updated to GDA2020 and MGA2020 coordinates, has not been transformed to PCG2020.

Local points in vicinity of Henderson Court Upper are transformed by use of Redfearn's (1948) geodetic to/from grid formulæ using the Krueger (1912) λ-series equations. The conversion is sufficient, to the millimetre, anywhere in any zone to a width of 6°.

Redfearn's conversions are available on-line, but only between the various national systems (GDA/MGA). The transformation has been written in Excel, which allows conversion of any local UTM grid between geographic and grid coordinates by user defined parameters (e.g., CM, λ_0, central scale factor, k_0).

In contrast to the MGA Zone 50 covering 32°S, 114° to 40°S; 120°E (Zone 50J), both the PCG and proposed CUG straddle the Zone 50H and 50J boundary of 32°S. Extent, the standard $\Delta\lambda$=8° × $\Delta\omega$=6° of the UTM grid Limits Zone 50J Central meridian. λ_0 = 117°,

central scale factor k_0 = 0.9996.

Extent, 32°S, 114°E to 40°S, 120°E.

False Easting, 500,000,

False Northing, 10,000,000.

16.2.4.1 Extent of proposed survey projection area

PCG2020 C.M. λ_0 = 115°49′, central scale factor k_0 = 0.9999906.

Limit, 31°20′S, 115°26′E to 33°25′S, 116°05′E.

Extent, custom $\Delta\lambda$=2°05′ × $\Delta\omega$=0°39′ of the UTM grid

False Easting, 50,000, false Northing, 3,900,000. (Figure 16-11).

CUG2020 C.M. λ_0 = 115°49′, central scale factor k_0 = 0.9999906.

Limit, 31°59′30″S, 115°52′50″E to 32°01′30″S, 115°54′25″E.

Custom $\Delta\lambda$=0°02′ × $\Delta\omega$=0°01′35″of the UTM grid

False Easting, –6,000, false Northing, 3,550,000

Figure 16-12 shows the limits of the proposed coordinate area, it covers both the campus and Centennial Park where some field work practical sessions are conducted. It is a very small subset of PCG2020.

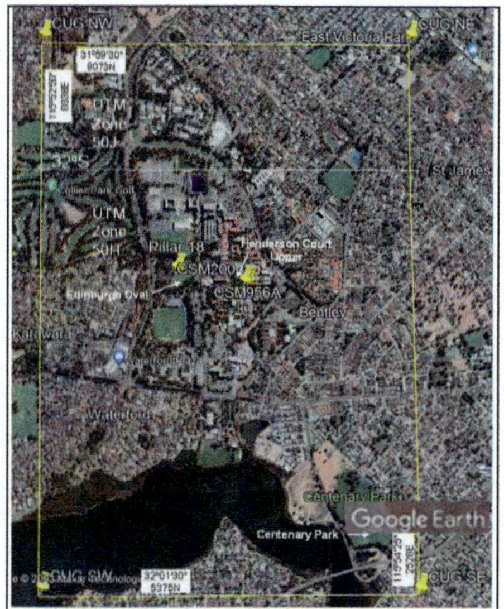

Figure 16-12 Proposed limits of CUG2020

16.2.5 Local (project) grids on the UTM in Western Australia.

The basis of local conversion is to modify Redfearn's forward and reverse conversion spreadsheets so that they can utilize local grid parameters. This is best done using published components of the conversion: i.e., Central Meridian (CM), central scale factor (k_0), false Easting measured from the CM and false Northing measured from the equator. Other parameters may include latitude and longitude boundaries, and height limitations.

Landgate, the Western Australian mapping authority https://www.landgate.wa.gov.au/location-data-and-services/surveying/geodesy/ has published some 57 project grids involving AGD84, GDA94 and MGA2020 horizontal datums. https://www.landgate.wa.gov.au/siteassets/documents/location-data-and-services/surveying/waprojectgrids.xlsx

Main Roads WA maintain a number of project grids related to GDA94 and GDA2020 in Esri PRJ files and 12d Carto 4d files. https://www.mainroads.wa.gov.au/technical-commercial/technical-library/

16.3 UTM Conversion between Grid and Ellipsoid

16.3.1 Using MGA local grids.

Geographic/grid coordinate transformations used by Geoscience Australia use the Karney (2011) - Krueger (1912) direct and reverse methods in the GDA2020 Technical Manual. For illustration Redfearn's formulæ (Krueger λ) is used as a simpler (?) implementation, avoiding iteration (and accuracy) used in the GDA2020 Krueger n method.

> "Two methods are presented to convert geographic to / from grid coordinates: Krueger n series equations and Krueger λ-series equations (Krueger, 1912). The Krueger λ-series equations are also known as Redfearn's formulae (Redfearn, 1948) and were used in the GDA94 Technical Manual. These equations are accurate to better than 1 mm within any zone of the Map Grid of Australia 1994 and Map Grid of Australia 2020 and can still be used for many purposes. However, for applications where users are working across multiple UTM/MGA zones, Krueger n-series equations are recommended and are explained in this Technical Manual. The Krueger n-series equations are particularly beneficial in software to avoid error build up when conversions are done back and forth between geographic and grid coordinates."

16.3.2 Redfearn's Grid to Geographic formula

This formula transforms UTM grid coordinates to geographic latitude and longitude. The arguments used are:

Project ellipsoid, e.g., GDA2020 major semi-axis, a = 6,378,137, flattening (GRS80), f = (1/297.25722101).

UTM zone, based on 6° zone widths from 180°W (anti-prime meridian) Eastwards. Z = INT(180+λ)/6+1.

UTM zone Central Meridian, λ_0, defined by zone number table, or λ_0 = (ZONE×6)–183

Central Meridian scale factor, k_0 = 0.9996

This example uses the Technical Manual position of **BUNINGYONG** Control Point to confirm the result.

Table 16-4 Grid to ellipsoid conversion

Transformation between UTM grid and ellipsoid.	
Redfearn's Formula	
UTM COORDINATES	**MGA2020**
Input East **E**	**758173.797**
Input North **N**	**5828674.34**
Input Zone	**54**
Summary ELLIPSOIDAL COORDINATES GRS80	
Latitude	**S37° 39' 10.1561"**
Longitude	**E143° 55' 35.3839"**
Start Calculations	
(Input E – false origin) **E'**	258173.797
E'/k_0	258277.1078
(Input N – false origin) **N'**	-4171325.66
$m = N'/k_0$	-4172994.858
UTM Constants for GDA (GRS80) false origins	
Central Scale Factor. **k_0**	0.9996
E = E'+500000	500000
N = N'+10000000	10000000
Find Central Meridian from zone input	
Central meridian. λ_0	141
Define ellipsoid (ANS, WGS84, **GRS80)**	
	GRS80
Major semi-axis (m), **a**	6378137
Inverse flattening, **1/f**	298.257222101
Flattening (a-b)/a, **f**	0.0033528106812
Derived values for spheroid	
Minor semi-axis **$b = a.(1-f)$**	6356752.314
Eccentricity² **$e^2 = 2f - f^2$**	0.00669438
Second eccentricity² ε (epsilon) or **$e'^2 = e^2/(1-e^2)$**	0.006739497
Foot Point Latitude **ϕ'** (theta prime)	
$n = f/(2-f)$	0.00167922
$G = a(1-n)(1-n^2)(1+9/4n^2+225/64n^4).\pi/180$	111132.9525
$\sigma = m\pi/(G.180)$	-0.6553636727
$\phi' = \sigma$	-0.6553636727
$+(3n/2-27n^3/32).\sin(2\sigma)$	-0.0024341241
$+(21n^2/16-55n^4/32).\sin(4\sigma)$	-1.83937E-06
$+(151n^3/96).\sin(6\sigma)$	5.29367E-09
$+(1097n^4/512).\sin(8\sigma)$	1.46943E-11
Sum power series $\phi' = \Sigma$	-0.6577996309
ϕ'°	-37.68914261
Sec ϕ'	1.2636802847
$t' = \tan(\phi')$	-0.772585181

Derived values for foot point latitude ϕ'	
ature ρ' (rho) $= a(1-e^2)/(1-e^2\sin^2\phi')^{3/2}$	6359293.091
curvature ν' (nu) $= a/(1-e^2\sin^2\phi')^{1/2}$	6386131.817
ψ' (psi) $= \nu'/\rho'$	1.004220395
$x = E'/k_0\nu'$	0.040443435
For Scale Factor use **$x' = E'^2/k_0^2\rho'\nu'$**	0.001642575
$t'/k_0\rho'$	-1.21538E-07
Latitude (radians) from foot point latitude ϕ'	
$\phi = \phi'$	-0.6577996309
$-(t'/k_0\rho')\,xE'/2$	0.0006345144
$+(t'/k_0\rho')\,(x^3E'/24)[-4y'^2+9y'(1-t'^2)+12t'^2]$	-5.8571E-07
$-(t'/k_0\rho')\,(x^5E'/720)[8\psi'^4(11-24t'^2)-12\psi'^3(21-71t'^2)+15\psi'^2(15-98t'^2+15t'^4)+180\psi'(5t'^2-3t'^4)+360t'^4]$	6.13854E-10
$+(t'/k_0\rho')\,(x^7E'/40320)[(385+3633t'^2+4095t'^4+1575t'^6)$	-7.36512E-13
Sum power series ϕ (radians) $= \Sigma$	-0.6571657016
Latitude ϕ°	-37.65282114
	S37° 39' 10.1561"
Longitude (radians) ω (omega) **from CM**	
$\omega = \sec\phi'\,x$	0.0511075716
$-\sec\phi'\,x^3/6.(\psi'+2t'^2)$	-3.06237E-05
$+\sec\phi'\,x^5/120.[-4\psi'^3(1-6t'^2)+\psi'^2(9-68t'^2)+72\psi't'^2+24t'^4]$	3.45356E-08
$-\sec\phi'\,x^7/5040.(61+662t'^2+1320t'^4+720t'^6)$	-4.79049E-11
Sum power series ω (radians) $= \Sigma$	0.051076982
Longitude difference from CM ω	2.926495525
λ° = Central Meridian $\lambda_0 + \omega$	143.9264955
	E143° 55' 35.3839"
Grid convergence (radians) γ (gamma)	
$\gamma = -t'\,x$	0.0312459986
$+t'\,x^3/3.(-2\psi'^2+3\psi'+t'^2)$	-2.71322E-05
$-t'\,x^5/15.[\psi'^4(11-24t'^2)-3\psi'^3(8-23t'^2)+5\psi'^2(3-14t'^2)+30\psi't'^2+3t'^4]$	3.37969E-08
$+t'\,x^7/315.(17+77t'^2+105t'^4+45t'^6)$	-4.77225E-11
Sum power series γ (radians) $= \Sigma$	0.0312189002
γ°	1.788711224
	+1° 47' 19.3604"
Point Scale Factor k	
$4\psi'(1-6t'^2)-3(1-16t'^2)-24t'^2/\psi']+x'^3/720\}$	**1.0004210730**

The author has developed spreadsheets that enable differing datums and central scale factors to be called for projection on to the UTM grid. This includes local projection grids (cf § 16.2).

The algorithms to produce the longitude difference (ω) combine the longitude of the C.M (λ_0), which is derived from the zone number, to produce the longitude, $\lambda = \omega + \lambda_0$ where ω is a maximum of ±3°. If you want the Easting, E, in an adjacent zone then λ_0 has to be amended in the algorithm, limited to keeping λ_0 to a maximum overlap of ±3.5°.

16.3.3 Redfearn's Geodetic to Grid formula

Redfearn's formula converts geodetic latitude and longitude to UTM grid coordinates. The arguments used are:

Project ellipsoid, e.g., GDA2020 major semi-axis, a = 6,378,137, flattening (GRS80), f = (1/297.25722101). This example uses the Technical Manual position of **BUNINGYONG** Control Point to confirm the result.

If it's not checked it's wrong.

Table 16-5 Geodetic to GRID **Conversion between ellipsoid and UTM grid**	
	GDA2020
Display check Latitude °, ′, ″ ϕ	**S37° 39′ 10.1561″**
Display check Longitude ° ′ ″ λ	**E143° 55′ 35.3839″**
Input Latitude, dec. degrees ϕ	−37.652821139
Input Longitude, dec. degrees λ	143.926495528
Height (above ellipsoid) **h**	737.000
Summary of results: UTM	**MGA2020**
East **E**	**758173.797**
North **N**	**5828674.340**
Zone	**54**
Define ellipsoid	**GRS80**
Major semi–axis (m), **a**	6378137
Inverse flattening, **1/f**	298.257222101
Flattening (a–b)/a, **f**	0.0033528107
Derived values	
Minor semi-axis **b** = a.(1–f)	6356752.314
Eccentricity² $e^2 = 2f - f^2$	0.00669438
Second eccentricity² ε (epsilon) or $e'^2 = e^2/(1-e^2)$	0.006739497
Radius of meridian curvature ρ (rho) $= a(1-e^2)/(1-e^2\sin^2\phi)^{3/2}$	6359253.829
Radius of prime vertical curvature ν (nu) $= a/(1-e^2\sin^2\phi)^{1/2}$	6386118.674
Another method of calculating ν is from $W^2 = (1-e^2\sin^2\phi)$	0.997501867
Normal radius ν (nu) = a/W	6386118.674
ψ (psi) = ν/ρ	1.004224528
Geographic to Cartesian Coordinates	
X = $(\nu+h).\cos\phi.\cos\lambda$	−4087095.231
Y = $(\nu+h).\cos\phi.\sin\lambda$	2977467.051
$Z = (n(1-e^2)+h).\sin\phi$ or $Z = [(b^2 \cdot n/a^2)+h].\sin\phi$	−3875456.899
UTM Constants	
Central Scale Factor. k_0	0.9996
E = E′+500000	500000
N = N′+10000000	10000000
Zone	54
Central Meridian λ_0	141
$\omega = \lambda - \lambda_0$	2.9264955278
ω (radians)	0.051076983
ϕ (radians)	−0.657165702
t = $\tan\phi$	−0.771573362
$\cos\phi$	0.791726812
$\sin\phi$	−0.610875319
Find Meridian Distance, using the power terms	
Series expansion for meridian distance: $m = a(A_0\phi - A_2\sin 2\phi + A_4\sin 4\phi - A_6\sin 6\phi)$	

$A_0 = 1-e^2/4-3e^4/64-5e^6/256$, 1st term = $a.A_0.\phi$	−4184469.185
$A_2 = (3/8)(e^2+e^4/4+15e^6/128)$, 2nd term = $-a.A_2.\sin 2\phi$	15513.9327
$A_4 = (15/256)(e^4+3e^6/4)$, 3rd term = $a.A_4.\sin 4\phi$	−8.260102238
$A_6 = 35e^6/3072$, 4th term = $-a.A_6.\sin 6\phi$	−0.015660185
m = $\Sigma(A_0 - A_6)$	−4168963.528
Find Easting, using power functions	
1st term, $\nu.\omega.\cos\phi$	258248.3587
2nd term, $\nu.\omega^3/6.\cos^3\phi.(\psi-t^2)$	28.78085009
3rd term, $\nu.\omega^5/120.\cos^5\phi.[4\psi^3(1-6t^2)+\psi^2(1+8t^2)-\psi(2t^2)+t^4]$	−0.031357767
4th term, $\nu.\omega^7/5040.\cos^7\phi.(61-479t^2+179t^4-t^6)$	−3.60623E−05
Σ_{terms}	258277.108123
E′ = $\Sigma.k_0$	258173.7973
E	**758173.7973**
Find Northing, using power functions	
Meridian distance, m	−4168963.5280
1st term, $\nu\sin\phi\,\omega^2/2\cos\phi$	−4028.889769
2nd term, $\nu\sin\phi\,\omega^4/24\cos^3\phi.(4\psi^2+\psi-t^2)$	−2.439264054
3d term, $\nu\sin\phi\,\omega^6/720\cos^5\phi.[8\psi^4(11-24t^2)-28\psi^3(1-6t^2)+\psi^2(1-32t^2)-\psi(2t^2)+t^4]$	−0.000812168
4th term, $\nu\sin\phi\,\omega^8/40320\cos^7\phi.(1385-3111t^2+543t^4-t^6)$	2.40188E−07
Σ	−4172994.858
N′ = $\Sigma.k_0$	−4171325.66
N	**5828674.3401**
Find Grid Convergence	
1st term, $-\sin\phi\,\omega$	0.031201668
2nd term, $-\sin\phi\,\omega^3/3\cos^2\phi(2\psi^2-\psi)$	1.72243E−05
3rd term, $-\sin\phi\,\omega^5/15\cos^4\phi[\psi^4(11-24t^2)-\psi^3(11-36t^2)+2\psi^2(1-7t^2)+\psi t^2]$	7.95701E−09
4th term, $-\sin\phi\,\omega^7/315\cos^6\phi(17-26t^2+2t^4)$	9.66154E−13
γ (radians) = Σ	0.0312189
γ (degrees)	1.788711226
	+1° 47′ 20.3604″
Find Point Scale Factor	
1st term, $1+\omega^2/2\cos^2\phi\psi$	1.000821111
2nd term, $\omega^4/24\cos^4\phi[4\psi^3(1-6t^2)+\psi^2(1+24t^2)-4\psi t^2]$	2.90507E−07
3rd term, $\omega^6/720\cos^6\phi(61-148t^2+16t^4)$	−1.30211E−10
Σ	1.000821402
k = $\Sigma.k_0$	**1.0004210731**

The algorithms produce the longitude difference (ω) from the longitude of the C.M (λ_0), which is derived from the zone number (where ω is a maximum of ±3°), to produce the Easting, E', and Northing, N'. If you want the grid coordinates, E N, in an adjacent zone then λ_0 has to be amended in the algorithm, limited to keeping λ_0 to a maximum overlap of ±3.5°. This produces adjacent zone grid coordinates from the geographic coordinates.

16.4 Geodetic to Cartesian conversion formulæ

The process of converting geocentric coordinates from one reference epoch to another, say GDA94 to GDA2020, requires the cartesian (X, Y, Z) coordinates and a 7-parameter transformation matrix to supply a conformal transformation. The 7 parameters are: 3 shift, 3 rotation and a scale.

The formulæ in supply Cartesian (X, Y, Z) coordinates based on the Australian Terrestrial Reference Frame 2014 (ATRF2014), coincident with the ITRF2014 at epoch 2020.

https://www.spatial.nsw.gov.au/__data/assets/pdf_file/0018/217224/2017_Janssen_Position90_GDA2020_AUSGeoid2020_and_ATRF_explained.pdf
https://www.spatial.nsw.gov.au/__data/assets/pdf_file/0020/216614/2017_Janssen_APAS2017_GDA2020-AUSGeoid2020-ATRF_introduction.pdf

This is a direct conversion process with h being the height above the reference ellipsoid. It may be necessary to convert geoidal height, H, (from levelling runs) by applying the geoid-ellipse separation, N.

$h = H + N$ where N is positive when the geoid is **above** the ellipsoid and negative when the geoid is **below** the ellipsoid. Australia "tilts" from N = –32m at Cape Leeuwin (W.A.) to N = + 72m at Cape York (Qld).

Conversion from one to the other height datum on the latest Ausgeoid09 is available through:
https://geodesyapps.ga.gov.au/ausgeoid09

Otherwise the Earth Gravitational Model 2008 (EGM08), maintained by the US NGA Office of Geomatics https://earth-info.nga.mil/index.php?dir=wgs84&action=egm96-geoid-calc proves the geoid/ellipsoid separation on a global scale as part of the WGS84 model.

Ausgeoid09 and EGM08 differ by about 0.5m

16.4.1 Ellipsoidal constants

GRS80 Ellipsoidal constants	a) $a =$ major semi-axis b) $f = \dfrac{a-b}{a}$, flattening, where $\dfrac{1}{f} =$ c) $b = a(1-f)$, minor semi-axis d) $e^2 = \dfrac{a^2-b^2}{a^2} = f(2-f)$, first eccentricity squared e) $e'^2 = \dfrac{a^2-b^2}{b^2} = \dfrac{e^2}{1-e^2}$, second eccentricity squared	6378137 298.257222101 6356752.314 0.00669438 0.006739497	Eqn 16-1

16.4.1.1 Geocentric to Cartesian formula

The forward conversion from geocentric to Cartesian coordinates is a direct solution using Eqn 16-2.

Direct Cartesian coordinate conversions	$X = (\nu + h)\cos\phi\cos\lambda$ $Y = (\nu + h)\cos\phi\sin\lambda$ $Z = (\nu(1-e^2)+h)\sin\phi$ or $Z = \left(\left(\dfrac{\nu b^2}{a^2}\right)+h\right)\sin\phi$, where $\nu = \dfrac{a}{\sqrt{1-e^2\sin^2\phi}}$	Eqn 16-2

16.4.1.2 Cartesian to Geocentric formula

This inverse X, Y, Z ⇒ ϕ, λ, h is a complex problem since there is no simple link between the latitude, ϕ, height, h, and the X, Y, Z coordinates. That's because we are dealing with an elliptical meridian on which the latitude, ϕ, is calculated. Figure 16-13 details the components of the meridian plane.

The longitude of the meridian λ, is simply obtained from the X and Y coordinates, i.e., $\tan\lambda = Y/X$.

However, calculation of geodetic latitude, ϕ, and thus the ellipsoidal height, h, on the elliptical plane of the meridian is an iterative process as shown in Eqn 16-3c. An initial value of ϕ_0 from Eqn 16-3b is introduced to the Eqn 16-3c and iterated until a suitable convergence is achieved, four iterations are shown in Table 16-6 if ϕ_0 is within 5′ of ϕ.

Geodetic coordinate conversion by iteration.	a. $\lambda = \arctan\left(\frac{X}{Y}\right)$ b. $\phi_0 = \arctan\left(\frac{aZ}{bp}\right)$ where $p = \sqrt{X^2+Y^2}$ (perpendicular distance from rotational axis) c. $\phi = \arctan\left(\dfrac{Z+\nu e^2\sin\phi}{p}\right)$, $\nu = \dfrac{a}{\sqrt{1-e^2\sin^2\phi}}$, and iterate to a suitable convergence d. $h = \dfrac{p}{\cos\phi} - \nu$	Eqn 16-3

16.4.1.2.1 A good enough direct method for ϕ.

Gerdan and Deakin[1], RMIT Melbourne, investigated several methods of an approximately suitable "direct" method for daily use and, apart from the iterative process being the fastest, found that Bowring's method was the quickest of the "direct" solutions.

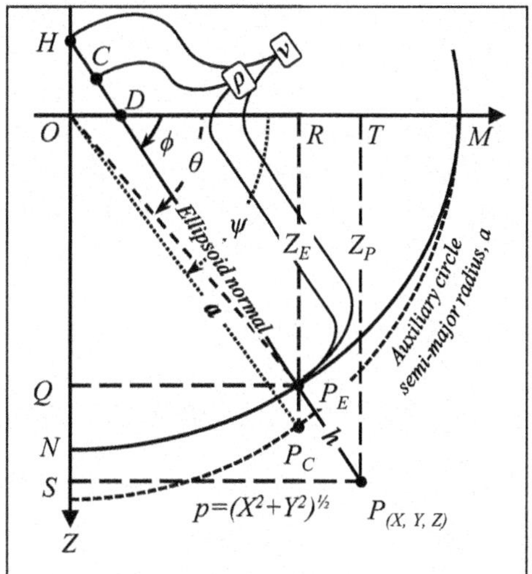

Figure 16-13 The ellipsoidal meridian plane

In Figure 16-13 the following latitude relationships exist:

OM = a, ON = b: ellipsoid semi-major, semi-minor radii

$\angle$PDM = ϕ (geodetic latitude)

$\angle P_C$OM = ψ (parametric latitude)

$\angle P_E$OM = θ (geocentric latitude)

$$\psi = \arctan\left(\frac{b}{a}\tan\phi\right)$$

$$\theta = \arctan\left(\frac{b}{a}\tan\psi\right) = \frac{b^2}{a^2}\tan\phi$$

BOWRING'S METHOD

This method (Bowring 1976) is iterative, but owing to the nature of the equation used can be regarded as exact, since for all practical purposes, second or third iterations are not required (Bowring 1976, p.326).

From the ellipsoid relationships, perpendicular distances from C (the centre of curvature of the meridian ellipse) to the Z- and OM-axes are $e^2a\cos^3\psi$ and $-e'^2b\sin^3\psi$ respectively, thus an expression for tan ϕ in terms of the parametric latitude ψ ($\angle P_C$OM) is:

$$\tan\phi = \frac{Z + be'^2\sin^3\psi}{p + ae'^2\cos^3\psi} \qquad \text{(Bowring 6.1)}$$

This equation, Eqn 16-3c, with ν having functions of ϕ on both sides of the equation, clearly requires an iterative solution for tanϕ. An initial value ψ_0 is obtained from tanθ' = Z/p, θ' being a good approximation of the geocentric latitude θ, and since tanψ=(a/b)tanθ $\approx$ (a/b)tanθ', then:

$$\tan\psi_0 = \frac{aZ}{bp} \qquad \text{(Bowring 6.2)}$$

Bowring (1976) shows that for all Earth-bound points ($-5{,}000\text{m} \le h \le 10{,}000\text{m}$) the maximum error in ϕ, induced by using only a single iteration, is 0.000 000 030″. Thus, for all practical purposes, the evaluation of tanϕ by (6.1), with a first approximation of ψ from (6.2), can be regarded as exact.

Cartesian coordinates converted to geodetic latitude using Bowring's method are shown in Table 16-7.

Bowring's geodetic latitude coordinate conversions	$\lambda = \arctan\left(\frac{Y}{X}\right)$ $\phi = \arctan\left(\frac{Z + be'^2\sin^3\psi}{p - ae^2\cos^3\psi}\right)$, where $\psi = \psi_0 = \arctan\left(\frac{aZ}{bp}\right)$ Bowring (1976) $p = \sqrt{X^2 + Y^2}$ (perpendicular distance from rotational axis) $\nu = \frac{a}{\sqrt{1 - e^2\sin^2\phi}}$ (where e^2 and e'^2 are the first and second eccentricities squared) $h = \frac{p}{\cos\phi} - \nu$	Eqn 16-4

[1] Gerdan and Deakin, RMIT Melbourne, (George P. Gerdan and Rodney E. Deakin. Transforming Cartesian Coordinates *X, Y, Z* to Geographical coordinates to $\boldsymbol{\phi}, \boldsymbol{\lambda}, \boldsymbol{h}$. *The Australian Surveyor*, Vol 44, No 1, pp 53-66, June 1999)

16.4.1.3 Geographic coordinates from the geodetic solution

The geocentric latitude coordinate is calculated from the mass centre of the Earth. That is, from the reference polar axis and OM, the plane of the **equator**, shown in Figure 16-14. The difference between geodetic and geocentric latitude is nearly 12′ at $\phi = 45°$. Geocentric latitude also varies with height, h, from the ellipsoid surface.

Using the Geoscience Australia test data as validation, Table 16-7 and Table 16-8 show an 11′ difference in latitude values between the parametric latitude ($\tan \psi_0$ = (aZ/bp), Bowring (6.2)) and the final (iterative) value.

16.4.1.4 Latitude change for varying h.

Using the two formulæ in Eqn 16-5, from Table 16-8 geocentric latitude at Buninyong changes by 12″ for h = 737m between P_E and P.

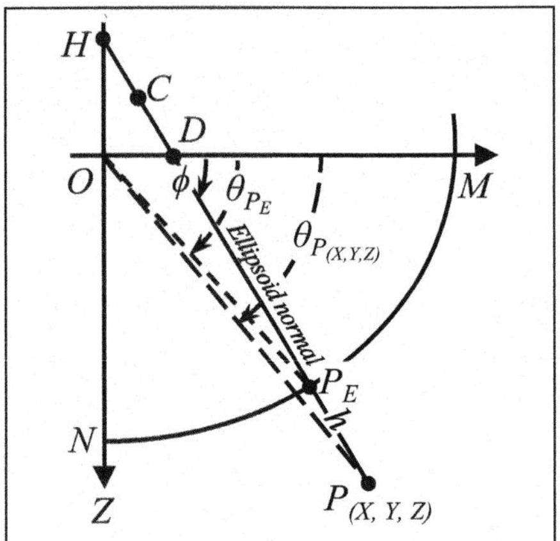

Figure 16-14 Geocentric latitude

Geocentric coordinate conversions for height, h	$\lambda = \arctan\left(\frac{Y}{X}\right)$ $\theta_{P_E} = \arctan\left(\left(1-e^2\right)\tan\phi\right)$ P projected on the ellipse. $\theta_{P_{X,Y,Z}} = \arctan\left(\frac{\nu\left(1-e^2\right)+h}{\nu+h}\tan\phi\right)$ $P_{(X,Y,Z)}$ above ellipse.	Eqn 16-5

16.4.1.5 Iterative compared with Bowring latitude solutions.

This test data is modelled on the Geoscience Australia control point **BUNINYONG** near Ballarat in Victoria.

Table 16-6 Iterative solution

Cartesian Coordinates to Geographic	**GRS80**
Iterative Reverse Transformation	GA test data
X	-4,087,095.384
Y	2,977,467.494
Z	-3,875,457.340
Define spheroid	**GRS80**
Major semi-axis (m), **a**	6378137
Inverse flattening, **1/f**	298.257222101
Flattening (a-b)/a, **f**	0.00335281068
Derived values	
Minor semi-axis **b** = a.(1-f)	6356752.3141
Eccentricity² **e²** = 2f-f²	0.0066943800
Second eccentricity² ε (epsilon) or **e'²** = e²/(1-e²)	0.0067394968
Cartesian to Geographic coordinates	
Calculate longitude directly tanλ = (Y/X)	-0.728504528
λ° = degrees(atan2(X,Y)) [ATAN2(X,Y) for correct quadrant]	143.9264925
λ	**E143° 55' 35.3730"**
Zone CM. λ_0 TRUNC((180+λ)/6)*6-180+3	141
$\omega = \lambda - \lambda_0$	26.92649249
Zone	54
Calculate approx. latitude using (radius) p = (X²+Y²)^½	5056645.267
tanu = (Z/p)·(a/b) = [aZ/bp]	-0.76898705
Initial u	-37.55981802
Initial ν = a/(1-e²sin²u)^1/2	6386085.043
Iterate with **u** to find approximate **ϕ'**	S37° 33' 35.3449"
tanϕ' = (Z+νe²sinu)/p	-0.771562492
ϕ'	-37.65243072
First ν' = a/(1-e²sin²ϕ')^1/2	6386118.533
Second approximation: Iterate with ϕ' to find approximate ϕ"	S37° 39' 08.7506"
tanϕ"= (Z+ν'e²sinϕ')/p	-0.771573345
ϕ"	-37.65282052
Second ν" = a/(1-e²sin²ϕ")^1/2	6386118.674
Third approximation: Iterate with ϕ" to find approximate ϕ"'	S37° 39' 10.1539"
tanϕ"'= (Z+ν"e²sinϕ")/p	-0.771573391
ϕ"'	-37.65282216
Third ν"' = a/(1-e²sin²ϕ"')^1/2	6386118.675
Fourth approximation: Iterate with ϕ"' to find ϕ(+1)	***S37° 39' 10.1598"***
tanϕ(+1)= (Z+νe²sinu)/p	-0.771573391
ϕ(+1)	-37.65282217
Next ν(+1) = a/(1-e²sin²ϕ(+1))^1/2	6386118.675
ϕ	***S37° 39' 10.1598"***
Now find ν(nu) **from** ϕ, **(just solved)**	
Radius of prime vertical ν (nu) = a/(1-e²sin²ϕ)^1/2	6386118.675
Ellipsoidal height, h = (p×secϕ) – ν	737.5737

Table 16-7 Bowring's solution

Cartesian Coordinates to Geographic	**GRS80**
Bowring's Reverse Transformation	GA test data
X	-4,087,095.384
Y	2,977,467.494
Z	-3,875,457.340
Define spheroid	**GRS80**
Major semi-axis (m), **a**	6378137
Inverse flattening, **1/f**	298.257222101
Flattening (a-b)/a, **f**	0.00335281068
Derived values	
Minor semi-axis **b** = a.(1-f)	6356752.3141
Eccentricity² **e²** = 2f-f²	0.0066943800
Second eccentricity² ε (epsilon) or **e'²** = e²/(1-e²)	0.0067394968
Cartesian to Geographic coordinates	
Calculate longitude directly tanλ = (Y/X)	-0.728504528
λ° = degrees(ATAN2(X,Y)) [ATAN2(X,Y) for correct quadrant]	143.9264925
λ	**E143° 55' 35.3730"**
Zone CM. λ_0 TRUNC((180+λ)/6)*6-180+3	141
$\omega = \lambda - \lambda_0$	26.92649249
Zone	54
Parametric latitude using (radius) p = (X²+Y²)^½	5056645.267
tan ψ_0 = (aZ/bp)	-0.76898705
Parametric latitude ψ_0 =	-37.55981802
Parametric Latitude *ψ* °, ', "	**S37° 33' 35.3449"**
tanϕ = (Z+be'²sin³ψ)/(p–ae²cos³ψ)	
Numerator a) Z+be'²sin³ψ	-3885161.88
Denominator b) p–ae²cos³ψ	5035375.669
tan ϕ = a/b	-0.771573391
ϕ ° =	-37.65282217
Latitude *ϕ* °, ', "	**S37° 39' 10.1598"**
elliptical radius =(p×secϕ) =	6386856.248
ν (nu) = a/(1–e²sin²ϕ)^1/2	6386118.675
Ellipsoidal height, h = (p×secϕ) – ν	737.5737

Table 16-8 Geocentric latitude

Geocentric latitude reference ellipsoid height	
tan θ, (h=0) = (1-e²)×tanϕ	-0.766408186
θ° =	-37.46685097
Latitude *θ* °, ', "	**S37° 28' 00.6635"**
tanθ,(at h = 737.6) = (ν(1-e²)+h/(ν+h))×tanϕ	-0.766497392
θ° =	-37.4700707
Latitude *θ* °, ', "	**S37° 28' 12.2545"**

16.5 Datum transformation using the 7-parameter matrix

Having expresses the cartesian (X, Y, Z) coordinates of a point defined in a particular datum then those coordinates can be expressed in a different datum. This is particularly used in updating old coordinate values to a new datum. In Australia the most common transformations would be between AGD94 ⇒ GDA94 ⇒ GDA2020. Earlier transformations are generally by published grid point data from State authorities.

For lower precision transformations the 3-parameter transformation, basically a translation from one origin to another, is suitable to about the ±5m level. It is used in handheld GPS receivers to covert from WGS84 to the local datum.

16.5.1 7-parameter transformation matrix

GDA2020 Technical Manual V1.7 para 3.1, p26

$$\begin{bmatrix} X'_{GDA2020} \\ Y'_{GDA2020} \\ Z'_{GDA2020} \end{bmatrix} = \begin{bmatrix} t_x \\ t_y \\ t_z \end{bmatrix} + (1+s_c)\begin{bmatrix} 1 & r_z & -r_y \\ -r_z & 1 & r_x \\ r_y & -r_x & 1 \end{bmatrix}\begin{bmatrix} X_{GDA94} \\ Y_{GDA94} \\ Z_{GDA94} \end{bmatrix}$$

16.5.2 Transformation parameters GDA94 to GDA2020

https://www.icsm.gov.au/sites/default/files/InterimReleaseNoteV1.0.pdf

Table 16-9 Transformation parameters for AGD84 to GDA2020

	Translation			Scale	Rotation		
	t_x m	t_y m	t_z m	s_c ppm	r_x arcseconds	r_y arcseconds	r_z arcseconds
***t* parameters**	+0.06155	−0.01087	−0.04019	0.0099940	−0.039424	−0.0327221	−0.0328979
uncertainty±	0.0007	0.0006	0.0007	0.0001	0.000011	0.00001	0.000011

16.5.3 Example transformation

Applecross 97A (Landgate Pillar 22)
GDA94
Latitude: S 32° 00′ 11.64766″ X: −2,363,822.244
Longitude: E 115° 53′ 20.46132″ Y: 4,870,535.714
Ellipsoidal Height: −21.125 Z: −3,360,724.487
Orthometric Height (AHD): 11.077

16.5.3.1 Generating the transformation matrices.

The transformation matrices can be calculated in Excel using the matrix functions MMULT(Array1,Array2)
The matrix transformation moves from RIGHT to LEFT as shown in § 16.5.1.

1. Rotate the original X, Y, Z frame

$$\text{Array 1} = \begin{bmatrix} 1.00 & -1.595^{-7} & 1.586^{-7} \\ 1.595^{-7} & 1.00 & -1.915^{-7} \\ -1.589^{-7} & 1.915^{-7} & 1.00 \end{bmatrix} \quad \text{Array 2} = \begin{bmatrix} -2363822.244 \\ 4870535.714 \\ -3360724.487 \end{bmatrix}^{GDA94}$$

$$\text{Rotate GDA94 ECEF frame} = \text{MMULT(Array1,Array2)} = \begin{bmatrix} -2363823.554 \\ 4870535.980 \\ -3360723.180 \end{bmatrix}$$

2. Scale the rotated X, Y, Z frame

$$1+s_c = 0.999999990006\times \text{ Rotated frame } = \begin{bmatrix} -2363823.554 \\ 4870535.980 \\ -3360723.180 \end{bmatrix} = \begin{bmatrix} -2363823.531 \\ 4870535.932 \\ -3360723.146 \end{bmatrix}$$

3. Add the shifts to the scaled, rotated X, Y, Z frame

$$\text{Shift + scale} \times \text{MMULT(Array 1, Array2)} = \begin{bmatrix} 0.062 \\ -0.011 \\ -0.040 \end{bmatrix} + \begin{bmatrix} -2363823.531 \\ 4870535.932 \\ -3360723.146 \end{bmatrix} = \begin{bmatrix} -2363823.469 \\ 4870535.921 \\ -3360723.186 \end{bmatrix}^{GDA2020}$$

GDA2020
X: −2363823.469 Latitude: S 32° 00′ 21.49367″
Y: 4870535.921 Longitude: E 115° 53′ 24.49696″
Z: −3360723.186 Ellipsoidal Height: −21.144m
Orthometric Height (AHD): 11.077m

The published coordinates may differ slightly from the above figures due to different adjustment processes used.

16.6 Horizontal Time-Dependent Positioning (HTDP)

A dynamic datum. 14-parameter transformation

https://geodesy.noaa.gov/TOOLS/Htdp/Htdp.shtml

16.6.1 Transformations before GDA2020

Between GDA1994 and GDA2020, as fundamental control positioning was refined, there were a lot of time dependent transformations where annual changes were introduced to the 7 static parameters and 7 rate change parameters. With the coincidence of the GDA2020, ITRF2014 and ATRF2014 a new set of parameters, based on 109 GNSS stations in the Australian Fiducial Network (AFN) was mandated.

ITRF to GDA94 coordinate transformation, John Dawson and Alex Wood, Journal of Applied Geodesy 4 (2010), 189–199. Table 2: page 193.

This publication showed 5 sets of transformation parameters to GDA94 from ITRF1996 to ITRF2008.

https://www.ga.gov.au/__data/assets/pdf_file/0008/61100/ITRF-to-GDA94-coordinate-transformations.pdf

16.6.2 Propagating between ATRF2014 and GDA2020

A key requirement to ensure the successful implementation of a time-dependent reference frame is that software can transform/propagate data from the epoch in which it is natively captured or stored in, to a user-defined epoch. For most use cases, propagating between GDA2020 and ATRF2014 will be achieved by using the Australian plate motion model (PMM).

The Australian PMM accounts for the horizontal motion of the Australian tectonic plate, see Figure 16-15. This is about ~7 cm / year movement to the northeast. The Australian PMM can be used to propagate GDA2020 coordinates to ATRF2014 coordinates. For example, to align real-time vehicle positions with published road maps.

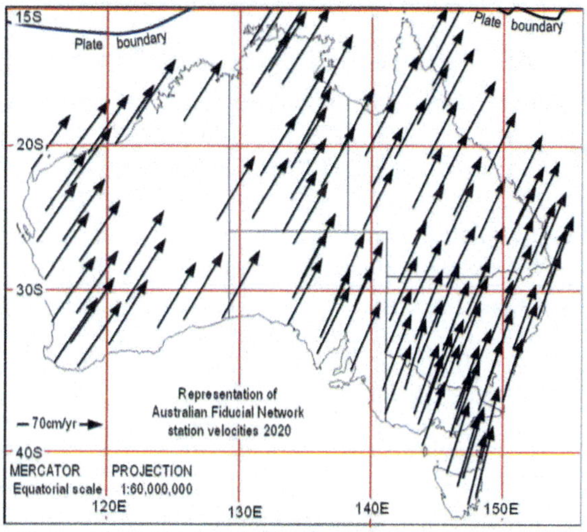

Figure 16-15 Continental Australian plate motion model (PMM)

16.6.3 WGS84 and ITRF.

WGS84(G1762) and ITRF2014 positions can be considered the same at the same epoch. Thus, any GNSS receiver using WGS84 would detect tectonic movements of the Australian plate resulting in changed WGS84 ⁄ ITRF2014 coordinates for Australian stations but does not affect the GDA2020 coordinates.

16.6.4 National Measurements Act 1960 RVS Determination

https://www.legislation.gov.au/F2017L01352/latest/text

National Measurement (Recognized-Value Standard of Measurement of Position) Determination 2017

Reference Ellipsoid means the Geodetic Reference System 1980 (GRS80) ellipsoid with a semi-major axis (a) of 6 378 137 metres exactly and an inverse flattening (1/f) of 298.257 222 101.

Reference Epoch means 2020.0.

Reference Frame means the Geocentric Datum of Australia 2020 (GDA2020) realised by the coordinates of the Australian Fiducial Network (AFN) geodetic stations, referred to in Schedule 1 of this instrument, in the International Terrestrial Reference Frame 2014 (ITRF2014) at the Reference Epoch

16.6.4.1 Calculation of global Cartesian coordinates at an epoch *t* years.

Global Cartesian coordinates of the AFN can be expressed at an epoch t (years) through the application of the following linear model using the coordinates (X, Y, Z) and velocities (VX, VY, VZ) listed in Schedule 1:

This model is valid for 15 years either side of the Reference Epoch: $|t - 2020| \leq 15$

Table 16-10 : Transformation parameters for ATRF2014 / ITRF2014 to GDA2020 along with their one sigma uncertainties (1σ). Units are in metres (m) and m/yr for the translations and their rates, respectively, parts-per-million (ppm) and ppm/yr for scale and its rate, respectively, and arcseconds and arcseconds/yr for rotations and their rates, respectively. The reference epoch t_0 is 2020.0.

NOTE: This refers to coordinates that are in the same reference frame (i.e. GDA2020, ITRF2014, ATRF2014), not coordinates from different reference frames. Transformation coordinates between different frames will require a transformation to GDA2020 (or ITRF2014) using the appropriate parameters (see GDA2020 Technical Manual).

Table 16-10 Transformation parameters for ATRF2014 / ITRF2014 to GDA2020							
	$t_x, \dot{t}_x$ m	$t_y, \dot{t}_t$ m	$t_z, \dot{t}_z$ m	s_c, $\dot{s}_c$ ppm	r_x, $\dot{r}_x$ arcseconds	r_y, $\dot{r}_y$ arcseconds	r_z, $\dot{r}_z$ arcseconds
t parameters	0.00	0.00	0.00	0.00	0.00	0.00	0.00
uncertainty	0.00	0.00	0.00	0.00	0.00	0.00	0.00
$\dot{t}$ rates	0.00	0.00	0.00	0.00	0.00150379	0.00118346	0.00120716
uncertainty	0.00	0.00	0.00	0.00	0.00000417	0.00000401	0.00000370

16.6.4.2 Transformation matrix.

In this transformation we are dealing only with the 3 rotational rates $\dot{r}_x$, $\dot{r}_y$, $\dot{r}_z$ (Table 16-10) over the time period (*dt*) years over 15 years either side of 2020. (2005-2035).

Remember this is transforming ITRF2014 **to** GDA2020, sending the current position GNSS WGS84(G1762), see § 16.6.3, **back** to the MGA2020 origin point. Note there are no shifts and no scale parameters.

$$\begin{bmatrix} X_{GDA2020} \\ Y_{GDA2020} \\ Z_{GDA2020} \end{bmatrix} = \begin{bmatrix} 0 \\ 0 \\ 0 \end{bmatrix} + [1] \begin{bmatrix} 1 & \dot{r}_z(dt) & -\dot{r}_y(dt) \\ -\dot{r}_z(dt) & 1 & \dot{r}_x(dt) \\ \dot{r}_y(dt) & -\dot{r}_x(dt) & 1 \end{bmatrix} \begin{bmatrix} X^t{}_{ITRF2014} \\ Y^t{}_{ITRF2014} \\ Z^t{}_{ITRF2014} \end{bmatrix} \text{where } dt = (t_{y.yy} - 2020)$$

16.6.4.3 Example transformation IFRF2014 to GDA2020

A set of WGS84(G1762) coordinates was observed on 21/02/2024 with ECEF coordinates:

X: −2363822.244 Latitude: S 32° 00′ 11.64756″
Y: 4870535.714 Longitude: E 115° 53′ 20.46131″
Z: −3360724.487 Ellipsoidal Height: −21.125

Set the rotation parameters for *dt* = 4.14 years (21/01/2024-01/01/2020 = 1512/365.25).

We are using the parameters in Table 16-10 in the transformation.

$\dot{r}_z$ arcseconds = 0.00120716 × 4.14 = RADIANS(4.998^{-3}) or 4.998^{-3}/206265 = 2.424^{-8}.

$$\text{Array 1} = \begin{bmatrix} 1.00 & 2.424^{-8} & -2.376^{-8} \\ -2.424^{-8} & 1.00 & 3.019^{-8} \\ 2.376^{-8} & -3.019^{-8} & 1.00 \end{bmatrix} \quad \text{Array 2} = \begin{bmatrix} -2363822.244 \\ 4870535.714 \\ -3360724.487 \end{bmatrix}^{ITRF2014}$$

$$\text{Rotate ITRF2014 ECEF frame} = \text{MMULT(Array1,Array2)} = \begin{bmatrix} -2363822.046 \\ 4870535.670 \\ -3360724.690 \end{bmatrix}^{GDA2020}$$

A comparison of the two ECEF coordinates shows a 3D shift of 0.287m in 4.14 years = 0.069m.y^{-1}. About the 7cm annual shift quoted.

GDA2020

X: −2363822.046 Latitude: S 32° 00′ 11.65542″
Y: 4870535.670 Longitude: E 115° 53′ 20.45526″
Z: −3360724.690 Ellipsoidal Height: −21.125

Caution: Many documents quote translations in millimetres (mm), scale in parts per billion (ppb) and rotations in milliarcseconds (mas). This saves in table column width. $\dot{r}_z$ = 1.20716mas. (See Dawson and Wood Table 2, § 16.6.1). Be careful.

16.7 Join between two geographic points.

The REVERSE PROBLEM: ROBBIN'S FORMULÆ.

AUSTRALIAN GEODETIC DATUM TECHNICAL MANUAL. Special edition no 10.

National Mapping Council of Australia, 1986. ISBN 0 642 515476.

16.7.1 Robbins versus Vincenty

The use of Robbin's formulæ in the AGD TM is used because of its simplicity instead of the later Vincenty's inverse and direct formulæ used in the current GDA2020 Technical Manual V1.7. Vincenty requires iteration, and the Geoscience Australia implementations are recommended.

Vincenty's inverse: (http://www.ga.gov.au/geodesy/datums/vincenty_inverse.jsp)
Vincenty's direct: (http://www.ga.gov.au/geodesy/datums/vincenty_direct.jsp)
Spreadsheet (https://github.com/icsm-au/DatumSpreadsheets).

Over the 55Km test line between Flinders Peak and Buninyong the disagreement is about 0.002m length and 0.01″ in azimuth between the old Robbins and Vincenty.

For shorter lines the Robbins implementation should suffice. Use the GA provided spreadsheets.

The author's spreadsheets allow for any ellipsoid and projection to be used by using lookup tables holding the appropriate parameters. This includes the various WA local grids, and any other defined UTM projection.

https://www.landgate.wa.gov.au/siteassets/documents/location-data-and-services/surveying/waprojectgrids.xlsx.

shows Robbin's reverse formula to determine the grid bearing and distance between two GDA stations. They can be compared with Vincenty.

Table 16-11 Join between two geographic points.

AUSTRALIAN GEODETIC DATUM TECHNICAL MANUAL.
Special edition no 10.
National Mapping Council of Australia, 1986. ISBN 0 642 515476
John D. Walker, April 1999
Join between two geographic points. The REVERSE PROBLEM: ROBBIN'S FORMULAE

		GDA2020	**GDA2020**
Latitude	ϕ	**S37°57'03.7203"**	**S37°39'10.1561"**
Longitude	λ	**E144°25'29.5244"**	**E143°55'35.3839"**
		Station 1	**Station 2**
Latitude	ϕ	-37.95103342	-37.65282114
Longitude	λ	144.42486789	143.92649553
ΔLongitude	$\Delta\lambda$	-0.49837236	0.498372361
Summary of results:			
Forward/Reverse BRG Dist		**GDA2020**	
Brg 1- 2	α_{12}	**306.8681578**	**306° 52' 5.36"**
Brg 2 - 1	α_{21}	**127.1736292**	**127° 10' 25.06"**
Distance	s	**54972.273**	
Define spheroid		**GDA2020**	
Major semi-axis (m),	a	6378137	6378137
Inverse flattening,	1/f	298.2572221	298.2572221
Flattening (a-b)/a,	f	0.0033528107	0.0033528107
Derived values			
Minor semi-axis **b** = a.(1-f)		6356752.314	6356752.314
Eccentricity² **e²** = 2f-f²		0.006694380	0.006694380
Second eccentricity² ε (epsilon) or **e'²** = e²/(1-e²)		0.006739497	0.006739497
Radius of meridian curvature ρ (rho) = a(1-e²)/(1-e²sin²ϕ)^{3/2}		6359576.573	6359253.829
Radius of prime vertical curvature ν (nu) = a/(1-e²sin²ϕ)^{1/2}		6386226.708	6386118.674
Another method of calculating ν is from **W²** = (1-e²sin²ϕ)		0.997468119	0.997501867
Normal radius ν (nu) = a/W		6386226.708	6386118.674
ψ (psi) = ν/ρ		1.004190552	1.004224528
Find Azimuth		Suffix 1	Suffix 2
$\tan\zeta_2 = (1-e^2)\tan\phi_2 + e^2\nu_1\sin\phi_1/(\nu_2\cos\phi_2)$		-0.779875244	-0.771608223
ζ_2 radians		-0.662348724	-0.657187553
$\tau_1=\cos\phi_1\tan\zeta_2-\sin\phi_1\cos(\lambda_2-\lambda_1)$		0.006523184	-0.006595931
$\tan\alpha_{12} = \sin(\lambda_2-\lambda_1)/\tau_1$		-1.333417688	-1.318711227
	α	-53.13184223	-52.82637081
		306.8681578	127.1736292
Find DISTANCE		**GDA2020**	
$\chi=\tau_1/\cos\alpha_{12}$		0.010872414	
$\sin\sigma' = \chi\cos\zeta_2$		0.008607836	
σ' radians		0.008607942	
$g=e'\sin\phi_1$		-0.050487077	
$h'=e'\cos\phi_1\cos\alpha_{12}$		0.03883911	
σ'^2 term = $(\sigma'^2 h'^2(1-h'^2))/6$		1.86007E-08	
σ'^3 term = $\sigma'^3 gh'(1-2h'^2)/8$		-1.561E-10	
σ'^4 term = $\sigma'^4(h'^2(4-7h'^2) - 3g^2(1-7h'^2))/120$		-7.08302E-14	
σ'^5 term = $\sigma'^5 gh'/48$		-1.93066E-15	
Sum = (0 - s' 2 term + s' 3 term +s' 4 term - s'5 term)		-1.87569E-08	
Distance s = $\nu_1\sigma'[1-\text{Sum}]$		54972.2732	

16.8 Polar vector from one geographic point.

The DIRECT PROBLEM: ROBBIN'S FORMULÆ.

AUSTRALIAN GEODETIC DATUM TECHNICAL MANUAL. Special edition no 10. National Mapping Council of Australia, 1986. ISBN 0 642 515476.

Table 16-12 Terminal station from a geographical point given grid bearing and distance

RIGOROUS FORMULAE BETWEEN SPHEROID AND GRID
AUSTRALIAN GEODETIC DATUM TECHNICAL MANUAL. Special edition no 10.
National Mapping Council of Australia, 1986. ISBN 0 642 515476 John D. Walker, April 1999
Join between two geographic points. The REVERSE PROBLEM: ROBBIN'S FORMULAE

Ellipsoid (ANS, GDA, WGS, CLA choice)		**GDA**	
Confirm input Latitude D M S	ϕ_1	**S37° 57' 03.7203"**	
Confirm input Longitude D M S	λ_1	**E144° 25' 29.5244"**	
		Station 1	
Input Latitude in negative D.dd	ϕ_1	-37.95103342	
Input Longitude in D.dd	λ_1	144.4248679	
ΔLongitude	$\Delta\lambda$	-0.49837240	
Calculated $\Delta\lambda$ D M S		**– 0° 29' 54.1406"**	
Summary of results: Forward/Reverse BRG Dist		**306° 52' 05.36"**	
Input Brg 1- 2 D.dd	α_{12}	306.868155	
Distance metres	s	**54972.273**	
		Stn 2 Result D M S	
Confirm result Latitude D M S	ϕ_2	**S37° 39' 10.1561"**	
Confirm result Longitude D M S	λ_2	**E143° 55' 35.3838"**	
		Result Station 2	
Result Latitude in D.dd	ϕ_2	-37.65282115	
Result Longitude in D.dd	λ_2	143.92649549	
Check Δlongitude	$\lambda_2 - \lambda_1 = -\Delta\lambda$	0.498372396	
Calculated reverse $\Delta\lambda$ D M S		**+ 0° 29' 54.1406"**	
Resultant reverse Brg 2 - 1 D.dd	α_{21}	**127.173627**	
Reverse BRG D M S		**127° 10' 25.06"**	
Define spheroid	**from table**	**GDA**	
Major semi-axis (m),	**a**	6378137	
Inverse flattening,	**1/f**	298.2572221	
Flattening (a-b)/a,	**f**	0.0033528107	
Derived values			
Minor semi-axis **b** = a.(1-f)		6356752.314	
Eccentricity² **e²** = 2f-f²		0.006694380023	
Second eccentricity² ε (epsilon) or **e'²** = e²/(1-e²)		0.006739496775	
Radius of meridian curvature ρ (rho) = $a(1-e^2)/(1-e^2\sin^2\phi)^{3/2}$		6359576.573	
Radius of prime vertical curvature ν (nu) = $a/(1-e^2\sin^2\phi)^{1/2}$		6386226.708	
Another method of calculating ν is from $W^2 = (1-e^2\sin^2\phi)$		0.997468119	
Normal radius ν (nu) = a/W		6386226.708	
ψ (psi) = ν/ρ		1.004190552	
Parameters			
$h' = e'\cos\phi_1\cos\alpha_{12}$		0.038839108	
$g = e'\sin\phi_1$		-0.050487077	
$\eta = s/\nu_1$		0.008607943	
$\sigma' = \eta\{1+\eta^2h'^2(1-h'^2)/6-\eta^3gh'(1-2h'^2)/8-(\eta^4/120)[h'^2(4-17h'^2)-3g^2(1-7h'^2)]+\eta^5gh'/48]\}$			
$+\eta^2h'^2(1-h'^2)/6$		1.86007E-08	
$-\eta^3gh'(1-2h'^2)/8$		1.55864E-10	
$-(\eta^4/120)[h'^2(4-17h'^2)-3g^2(1-7h'^2)]$		-7.18714E-14	
$+\eta^5gh'/48]\}$		-1.93066E-15	
$\Sigma(\eta^2 \ldots \eta^5) =$		1.87565E-08	
$\eta(1+\Sigma\eta) = \sigma' =$		0.008607943	
D.dd		0.493198791	+0° 29' 35.52"
5b. $\sin\zeta_2 = \sin\phi_1\cos\sigma'+\cos\phi_1\cos\alpha_{12}\sin\sigma'$		-0.6108926189	
6b. $\sin\Delta\lambda = \sin\sigma'\sin\alpha_{12}/\cos\zeta_2$		-0.00869813	
$\Delta\lambda =$		-0.498372396	-0° 29' 54.14"
7b. $\sin\alpha'_{21} = -\cos\phi_1\sin\alpha_{12}/\cos\zeta_2$		0.796808088	
α'_{21} = D.dd		52.82636924	+52° 49' 34.93"
Find LATITUDE and REV BRG			
$\mu = 1+1/2e'^2(\sin\zeta_2-\sin\phi_1)^2$		1.000000057	
$\tan\phi_2 = \tan\zeta_2(1+e'^2)(1-e^2\mu\sin\phi_1/\sin\zeta_2)$		-0.77157336266	
$\phi_2 =$		-37.65282115	– 37° 39' 10.16"
$\alpha_{21} = \alpha'_{21}-(\phi_2-\zeta_2)\sin\alpha'_{21}\tan\sigma'/2)$ D.dd in reverse quadrant		52.82637254	
Quadrant test for α_{21} using SIGN($\delta\phi$) and SIGN($\Delta\lambda$) α_{21} =		127.173627463676	+127° 10' 25.06"

If it's not checked it's wrong.

Use the GA provided spreadsheets.

The method can use any defined ellipsoid, not just the GDA2020 (GRS80).

uses the grid vector from one GDA station to provide the geographic coordinates of the terminal.

https://www.ga.gov.au/scientific-topics/positioning-navigation/australian-geospatial-reference-system/agrstoolsandmodels

16.9 Long line ellipsoidal vector between two grid points.

The difference in UTM grid and plane calculations over short distances is negligible apart from the length of the grid line, S, and the plane line, L. The difference is due to the line scale factor, K = S/L. And for short distances $K \approx k$ (point scale factor, § 16.1, page 243). In reality, changing plane distance (K = 1) to grid distance, S, is just S = L/k. Working locally, any defined UTM grid point will have a corresponding given point scale factor, k.

Over long lines the solution of the grid bearing, $\theta_{1\text{-}2}$, and distance, L, between two points on plane and the ellipsoidal grid bearing, β_1, β_2, and distance, s, between two points on the UTM grid involves solving the problems of a line scale factor, K, and arc-to-chord corrections, δ''_1, δ''_2, between the grid and the plane.

The line scale factor, K, is the ratio of the plane distance, L, to the corresponding ellipsoidal distance, s. K = L/s ≅ S/s. Where, by definition, S = grid distance and L = plane distance.

Point scale factor, k, varies along the line. K can be calculated using the ellipsoidal radius, R_m, at the mid-latitude of the line.

The DIRECT GRID PROBLEM.

AUSTRALIAN GEODETIC DATUM TECHNICAL MANUAL. Special edition no 10.

National Mapping Council of Australia, 1986. ISBN 0 642 515476.

16.9.1 Mid latitude function,

Mid-latitude, ϕ_m, of a line can be obtained from:

1. Calculated the latitudes of the Northings, N_1', N_2', of the line terminals.
2. Scaled from a chart of the area.
3. Using a truncated formula of the foot-point latitude. Uses the grid coordinates.

By definition, the radius, r, at any latitude point is the mean of the radius in the meridian, ρ, and the prime vertical, ν, at the central meridian.

At any latitude, ϕ, $R^2 = \rho\nu \Rightarrow r^2 = R^2 k_0^2 = \rho\nu k_0^2$ and $r_m^2 = \rho\nu k_0^2$ at ϕ_m.

Thus $r_m^2 = \rho\nu k_0^2 = \dfrac{k_0^2 a^2 (1-e^2)}{\left(1-e^2\sin^2\phi_m\right)^2}$ and, calculated for $\phi_m = -32°$, $r_m^2 = 4.05285^{+13}$

Thus $1/6r_m^2$ at $\phi = -32°$, gives 4.11233^{-15}.

An acceptable method, again using $\phi = -32°$, is from the formula:

$1/6r_m^2 = \left[4.07227 + \dfrac{\sin^2(90-\phi_m)}{18}\right] \times 10^{-15}$ $= 4.11222^{-15}$ The difference is about 1.1^{-19}, not much.

16.9.1.1 Mid latitude approximation from N_m'

The truncated formula $\phi_m = N'_m \times 9^{-6}$ can be used anywhere in the GDA area (8°S – 45°S) and will suffice to the fourth decimal place of $1/6r_m^2$ compared with a rigorously computed ϕ_m. $N'_m = (N_1'+N_2')/2$.

$$1/6r_m^2 = \left[4.07227 + \frac{\sin^2(90 - N'_m \times 9^{-6})}{18}\right] \times 10^{-15}$$

16.9.1.2 Line scale factor.

The solution of the ellipsoidal distance **s =L/K** where line scale factor:

$$K = k_0\left[1+\left(E_1'^2 + E_1'E_2' + E_2'^2\right)/6r_m^2\right]\left[1+\left(E_1'^2 + E_1'E_2' + E_2'^2\right)/36r_m^2\right]$$

It can be truncated, with little penalty to:

$K = k_0\left[1+\left(E_1'^2 + E_1'E_2' + E_2'^2\right)/6r_m^2\right]$ $K = k_0\left[1+\left(E_1'^2 + E_1'E_2' + E_2'^2\right)/6r_m^2\right]$

16.9.2 Arc-to-chord corrections.

The arc-to-chord corrections allow the plane bearing to be converted to a grid bearing between the two points. The grid bearing is the bearing along the geodesic to the other station. And the geodesic, as is the great circle on a sphere, is the shortest distance over the geoid.

Compared with the line bearing, θ, the grid bearing, β, bends away from the central meridian by the arc-to-chord correction, δ. In computations, δ will be determined by the algorithm and applied such that $\beta = \theta - \delta$. This is illustrated in Figure 16-16.

Caution must be exercised using this general diagram rule when the line **crosses** the CM and you are directed to § 5.12.4 of the AGD TM, SE 10 or Appendix C1 of the GDA2020 TM V1.7.

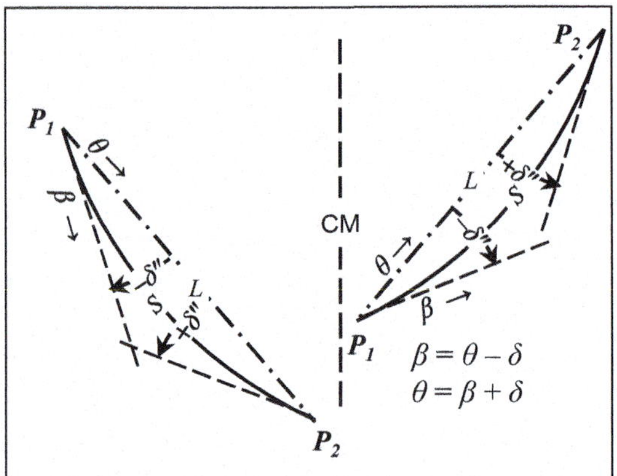

Figure 16-16 Arc-to-chord corrections

16.9.2.1 Computing the arc-to chord correction, δ.

Computation of δ is made in the directional sense as the plane bearing, θ, as:

$$\delta''_{12} = -(N_2 - N_1)\left(E_2' + 2E_1'\right)\left[1 - \left(E_2' + 2E_1'\right)^2 / 27r_m^{\ 2}\right] / 6r_m^{\ 2} \sin 1''$$

$$\delta''_{21} = (N_2 - N_1)\left(2E_2' + E_1'\right)\left[1 - \left(2E_2' + E_1'\right)^2 / 27r_m^{\ 2}\right] / 6r_m^{\ 2} \sin 1''$$

Note that the true Northings, N have been used.

Without significant error, the formulas can be reduced to:

$$\delta''_{12} = -(N_2 - N_1)\left(E_2' + 2E_1'\right) / 6r_m^{\ 2} \sin 1''$$

$$\delta''_{21} = (N_2 - N_1)\left(2E_2' + E_1'\right) / 6r_m^{\ 2} \sin 1''$$

where $1/6r_m^2 \sin 1'' = 1/6r_m^2 \times 206265$

These formulæ have been omitted in the GDA2020 Technical Manual V 1.7.

16.9.2.2 Magnitude of δ illustrated.

Figure 16-17 has been illustrated in the AGD Technical Manual, S.P 10 as Figure 5.3. It shows δ'' on the macro scale of long lines on the UTM with their displacement from the central meridian in Km.

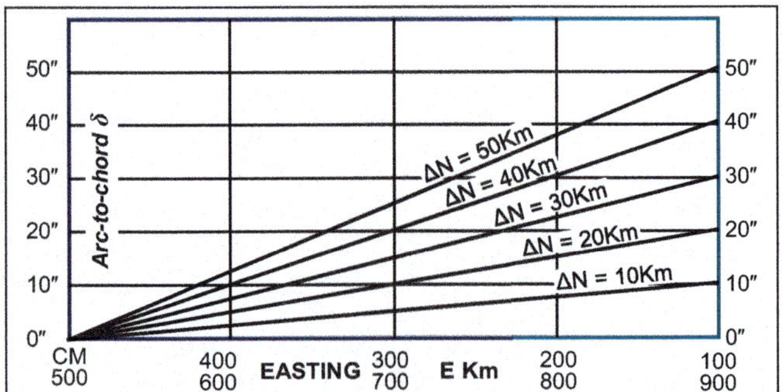

Figure 16-17 UTM zone long-line arc-to-chord corrections (AGD TM SP 10 Fig 5.3)

16.9.3 Effects of simplified formulæ.

The simplified formulæ in this section only start to differ by a few centimetres or millimetres over lines now not contemplated in normal surveying. The days of the Geodimeter® and Tellurometer® and long precise traverse lines (up to 60Km) have been passed to GNSS surveying. On short lines, and in all cases within the MGA2020 area, the accuracies formulæ are within a few parts per million.

Table 16-13 show the result of calculations using simplified formulæ on a 55Km line.

16.10 Coordinates from a long line ellipsoidal vector from one grid point.

The process of finding the terminal coordinates from an initial grid bearing and ellipsoidal distance uses many of the formulæ derived in § 16.9.

Since we know the grid bearing, β, and ellipsoidal distance, s, we have to convert them to a plane bearing, θ, and plane distance, L, through the arc-to-chord correction and the line scale factor, K.

The main requirement is to find the mid-latitude, N_m and subsequently $1/6r_m^2$.

16.10.1 Approximate coordinates of the terminal point.

$E_2' = E_1' + k_1 \times s \times \sin\beta_1$ where $E_1' = E_1 - 500\,000$

$N_2 = N_1 + k_1 \times s \times \cos\beta_1$

$k_1 \approx 0.9996 + 1.23E'^2 \times 10^{-14}$ or from published station data.

$N_m = (N_1 + N_2)/2$ or $N_m = N_1 + \frac{1}{2}(k_1 \times s \times \cos\beta_1)$

From § 16.9.1.1, find $1/6r_m^2$:

$$1/6r_m^{\,2} = \left[4.07227 + \frac{\sin^2(90 - N_m' \times 9^{-6})}{18}\right] \times 10^{-15} \text{ and,}$$

from § 16.9.1.2, find line scale factor, K:

$$\mathrm{K} = k_0\left[1 + \left(E_1'^2 + E_1'E_2' + E_2'^2\right)/6r_m^{\,2}\right]$$

16.10.1.1 Solving the terminal coordinates.

$L = s\mathrm{K}$ Plane distance

$\delta_{12}'' = -(N_2 - N_1)\left(E_2' + 2E_1'\right)/6r_m^{\,2}\sin 1''$

$\theta = \beta_1 + \delta_{12}$ Plane bearing

$\delta_{21}'' = (N_2 - N_1)\left(2E_2' + E_1'\right)/6r_m^{\,2}\sin 1''$

$\beta_{21} = \theta \pm 180 - \delta_{21}$ Reverse grid bearing

$\Delta E = L \times \sin\theta$ $\quad E_2 = E_1 + \Delta E$

$\Delta N = L \times \cos\theta$ $\quad N_2 = N_1 + \Delta N$

16.11 Test ellipsoid to grid algorithms.

16.11.1 Grid bearing and distance between two stations.

The following Table 16-13 shows the join between BUNINYONG and FLINDERS PEAK using the Australian Geodetic Datum coordinates.

Table 16-13 Join between two geoid points

RIGOROUS FORMULAE BETWEEN SPHEROID AND GRID
AUSTRALIAN GEODETIC DATUM TECHNICAL MANUAL. Special edition no 10.
National Mapping Council of Australia,
1986. ISBN 0 642 515476
John D. Walker, April 1999
Join between two geographic points.

Ellipsoid (ANS, GDA, WGS)		**AGD**	
Input Easting UTM E	E_1	**228742.0770**	
	E'_1	-271257.9230	
Input Northing UTM N	N_1	**5828074.2080**	
	N'_1	-4171925.7920	
	$m = N_m'/k_0$	**-4189486.0724**	
Input Easting UTM E	E_2	**273629.436**	
	E'_2	-226370.5640	
Input Northing UTM N	N_2	**5796305.236**	
	N'_2	-4203694.7640	
	N_m'	**-4187810.2780**	
	$m = N_m'/k_0$	**-4189486.0724**	
		Find mid latitude ϕ_m	
ϕ approx $=(N'/k_0)/G = m/111133 =$		-37.69794816	**– 37° 41' 52.6134"**
$\phi=(N_m'/k_0+16039\sin 2\phi)/G =$ $(m+16039\sin 2\phi)/111133 =$		-37.8376078	**– 37° 50' 15.3882"**
Use $\phi=N_m'\times 9\times 10^{-6}$		**-37.6902925**	**– 37° 41' 25.0530"**
$1/6r^2$ $= [4.07227+(\sin^2(90-\phi_m))/18]\times E^{-15}$		**4.10706E-15**	
$1/6r^2\sin 1'' =$		**8.47E-10**	
dE' =		**44887.359**	
dN' =		**-31768.972**	
$\tan\theta = dE'/dN'$		**-1.412931**	
$\theta =$ in second quadrant		**-54.7111**	**– 54° 42' 39.9475"**
$\theta =$		**125.2889**	**125° 17' 20.0525"**
$L = dE'/\sin\theta =$		**54992.2047**	
$K = k_0[1+(E_1'^2+E_1'E_2'+E_2'^2)/6r_m^2]$		1.00036455	
Ellipsoidal distance **s = L/K** =		**54972.1647**	
		Arc-to-chord δ	
$\delta_1'' = -(N_2-N_1)(E_2'+2E_1')/6r_m^2\sin 1''$		-20.69	**– 20.69"**
$\delta_2'' = (N_2-N_1)(2E_2'+E_1')/6r_m^2\sin 1''$		20.48	**+ 20.48"**
Grid BRG $\beta_{12} = \theta_{12} - \delta =$		**125.294652**	**125° 17' 40.75"**
Grid BRG $\beta_{21} = \theta_{12} \pm 180 - \delta =$		**305.2835**	**305° 17' 00.57"**

16.11.2 Calculating the terminal coordinates of a station given a grid bearing and distance from one station.

The following Table 16-14 show the grid coordinates of BUNINYONG using a grid bearing and distance from FLINDERS PEAK, Australian Geodetic Datum coordinates.

Table 16-14 Grid vector to a terminal station.

RIGOROUS FORMULAE BETWEEN SPHEROID AND GRID
AUSTRALIAN GEODETIC DATUM TECHNICAL MANUAL. Special Edition no 10.
National Mapping Council of Australia, 1986. ISBN 0 642 515476
John D. Walker, April 1999
Polar to a grid point.

Ellipsoid (ANS, GDA, WGS, CLA choice)		**AGD**	
Input Easting UTM E	E_1	**228742.077**	
	E'_1	-271257.923	
Input Northing UTM N	N_1	**5828074.208**	
	N'_1	-4171925.792	
Find k_1 at $E'_1 = k_0+1.23E'^2_1\times10^{-14}$ =		**1.00051**	
Input: Forward BRG Dist		**125° 17' 40.72"**	
Input GRID Brg 1- 2 D.dd	β_{12}	125.2946444	
GRID Distance metres	s	**54972.161**	
$E'_2 = E'_1+k_1\times s\times\sin\beta_1$		-226367.4467	
$N'_2 = N'_1+k_1\times s\times\cos\beta_1$		5796296.278	
$\mathbf{N_m} = (N1+(k_1\times s\times\cos\beta_1)/2$		**5812185.2431**	
$\mathbf{N_m'} = (N_m-10{,}000{,}000)$		**-4187814.7569**	
		Find mid latitude ϕ_m	
Use $\phi=N_m'\times9\times10^{-6}$		**-37.69033281**	**– 37° 41' 25.20"**
$\mathbf{1/6r^2} = [4.07227+(\sin^2(90-\phi_m))/18]\times E^{-15}$		**4.10706E-15**	
1/6r²sin1" =		**8.47E-10**	
$K = k_0[1+(E'^2_1+E'_1E'_2+E'^2_2)/6r_m^2]$		**1.00036454**	
GRID distance **L =s×K** =		**54992.200**	
		Arc-to-chord δ	
$\delta_1'' = -(N_2-N_1)(E_2'+2E_1')/6r_m^2\sin1''$		-20.7	
$\delta_2'' = (N_2-N_1)(2E_2'+E_1')/6r_m^2\sin1''$		20.5	
PLANE BRG $\theta_{12} = \beta_{12} + \delta$ =		**125.288895**	**125° 17' 20.02"**
Reverse Grid BRG $\beta_{21} = \theta_{12} \pm180 - \delta$ =		**305.2835**	**305° 17' 00.53"**
ΔE =		**44887.360**	
ΔN =		**-31768.963**	
E_2 =		**273629.437**	
N_2 =		**5796305.245**	

16.12 Zone to zone transformations.

UTM calculations are based on a 6° zone width. However, there is sufficient conformality to allow a zone overlap of ±½° for calculations near the zone boundary where it may be easier, using grid coordinates, to operate in the one zone. This is illustrated in the MGA TM V1.7. Smeaton - Buninyong are on the Eastern edge of Zone 54 while the rest of the traverse stations to Bass are in Zone 55.

16.12.1 Transformation methods.

1. Indirect solution.
   Convert the grid coordinates to geodetic using the Krueger n-series or Krueger λ series (Redfearn's formula) and recalculate them in the adjacent zone. This is the normal method.
2. Direct solution.
   Use formulæ from Jordan & Eggert, 1941, Grossman, 1964 as described in the MGA2020 TM, V1.7.

This is a bit of a trap as the example is in the second edition National Mapping Council publication, The Australian Map Grid Technical Manual, Special Publication 7 (Canberra, 1972). It used lookup tables (Annex E) for the various parameters (calculators just coming on the market). The tables are based on the Australian National

Spheroid (ANS) and require arcane manipulation to function. The formula remains in later editions, but without the worksheet. The direct solution is discussed in Chapter 22, page 329.

16.12.1.1 Krueger's n-series or Krueger's λ (Redfearn's) indirect grid conversion.

There is no modern direct solution to the zone-to-zone grid coordinate transformation. A grid point is converted to geographic coordinates and the grid coordinates in the adjacent zone calculated.

The author's own Redfearn's forward and reverse routines are used as illustrations.

P1 $_{Z50}$ has GDA coordinates of 32°S 119°36′E with MGA Zone 50 coords 745621.129E, 6456609.836N.

It is within 30′ of the Zone 51 boundary. The zone geodetic longitude difference, ω_{z}, = $\lambda_{P1} - \lambda_{CM}$.

Zone 50 CM = 117°. ω_{z}, = 119.6–117 = 2.6°.

Zone 51 CM = 123°. ω_{z}, = 119.6–123 = –3.4°.

P1 $_{Z51}$ has GDA coordinates of 32°S 119°36′E with MGA Zone 51 coords 178768.580E, 6454509.925N.

Conversion of grid coordinates of P1 in Zone 51 (within the overlap limit) are shown in Figure 16-18:

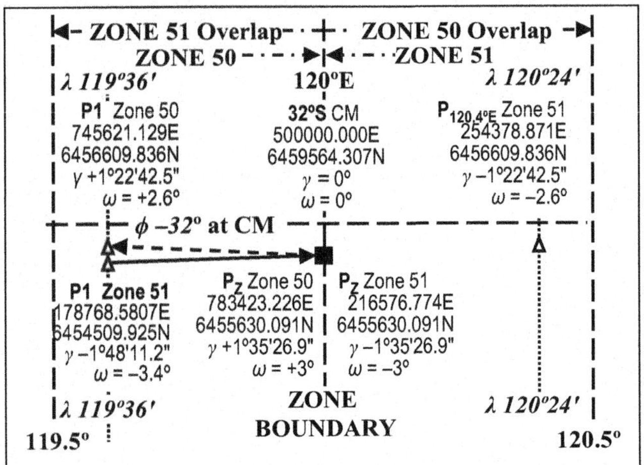

Figure 16-18 Zone to zone transformation

If MGA zone grid coordinates are known, then they must be converted to GDA latitude and longitude before being converted to the adjacent zone. The online calculator at:

https://geodesyapps.ga.gov.au/grid-to-geographic is the convenient method. Very slight discrepancies may exist between Geoscience's Krueger's n-series and Redfearn's method. Batch processing is available.

By way of illustration, converting P2 MGA coordinates from Zone 50 to Zone 51:

P2 $_{Z50}$ MGA Zone 50 coordinates 745630.000E, 6456600.000N convert to GDA 32°00′00.32″S, 119°36′00.35″E

P2 $_{Z51}$ GDA 32°00′00.32″S, 119°36′00.35″E become MGA Zone 51 coordinates 178777.988E, 6454500.592N.

Note the 100m difference in Northing due to the 1.8° difference in ω_{z}.

16.13 References for Chapter 16

Bomford, G. Geodesy 4th ed. Oxford University Press 1980. ISBN 0-19-851946-X

National Mapping Council of Australia. The Australian Map Grid Technical Manual, Special Publication 7. AGPS, Canberra 1972. ISBN 0 642 00057 3

National Mapping Council of Australia. The Australian Geodetic Datum Technical Manual, Special Publication 10. Canberra 1972. ISBN 0 642 515476

Intergovernmental Committee on Surveying and Mapping (ICSM). Geocentric Datum of Australia 2020 Technical Manual, Version 1.7, Geodesy Working Group, 26 October 2021

Rueger, J.M. Introduction to Electronic Distance Measurement, Second Edition June 1984. School of Surveying, The University of New South Wales. ISBN 0 85837 027 2

Chapter 17 COORDINATE TRANSFORMATIONS.

Since most engineering works are undertaken on planes (2-dimesnional x and y systems), it is essential to know how to transform coordinates from one system to another. For example, coordinates (positions) captured from a moving aircraft would normally be in pixel formats that need to be transformed into plane x and y systems that subsequently need to be transformed to more usable ground- based system. This chapter presents the techniques of obtaining transformation parameters (scales and translations). Methods presented range from simple affine transformation to least squares that are programmable.

17.1 Transformation of plane coordinates.

There are a number of different coordinate systems that can be used for an exploration or mining project. Generally, an existing mine site will be based on a Local Mine Grid (Flat Earth system) but should be connected into the Geodetic Datum of Australian (GDA). It is quite probable that an initial exploration program will be based on MGA94 coordinates. In both situations it then becomes necessary to be able to transform a set of coordinates from one system to another.

Generally, a Local Mine Grid (see Figure 17-1) is based on a selected datum point located in a safe area on site and is given an arbitrary set of coordinates. In addition, the coordinate system will be orientated by selecting a second datum point in a particular direction. Common Local Mine Grid coordinate grid orientations can be based on; the direction of the ore body strike, Magnetic North, True North or any other direction that seemed appropriate at the time.

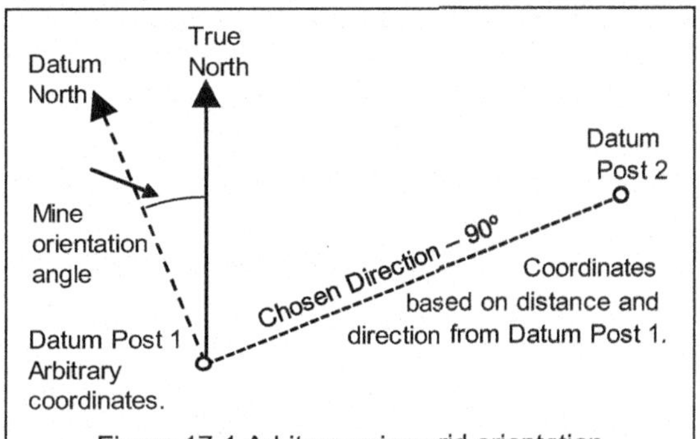

Figure 17-1 Arbitrary mine grid orientation

The most common Coordinate systems in use today within the mining industry today tend to be either:

1. The Map Grid of Australia 2020 (MGA2020), based on the 2020 Geodetic Datum of Australia from January 2020 onwards or,
2. Local Mine Grid, a plane grid system based on arbitrary coordinates and orientation.

However, if you are using data or information that precedes January 2000 then there are a few other coordinate systems that could have been used apart from a Local Mine Grid. These systems are:

1. The Australian Map Grid 84 (AMG84) Coordinates, based on the 1984 Australian Geodetic Datum. This coordinate system was the standard System from 1984 to until 1st January 2000 in Western Australia, South Australia and Queensland (Kirby 2010). This system was replaced with the MGA94 system after this date. Note that coordinates in this system will vary from an MGA94 coordinate of the same point by approximately 200m in a South Westerly direction.
2. Australian Map Grid 66 (AMG66) Coordinates. This system of coordinates replaced the individual State based systems in 1966 and was used up until the introduction of the AMG84 system in 1984, except NSW, Victoria and Tasmania. These states continued to use the AGD66 system up until 1st January 2000.
3. Coordinate systems predating 1966 were generally State or regionally based systems.

Given the number of coordinate systems in existence it is often a common requirement to be able to transforms a set of points coordinates from one system to another. Invariable for mining or exploration areas set up after 1st January 2000 this will involve the transformation between MGA94 and the Local mine Grid system.

17.2 Two-dimensional plane coordinate transformation.

For small areas it is possible to use a 2-Dimensional Transformation based on 2 points having coordinates in either system. This allows the following parameters to be determined that can then be applied to convert coordinates from one system to the other.

A) A rotation shift (**θ**) between the two systems.
B) A datum shift, **Eo, No**, from one system to the other.
C) A scale change (**s**) between the two systems.

These parameters (see Figure 17-2) can be determined by writing the relationship for two common coordinated points in terms of four equations.

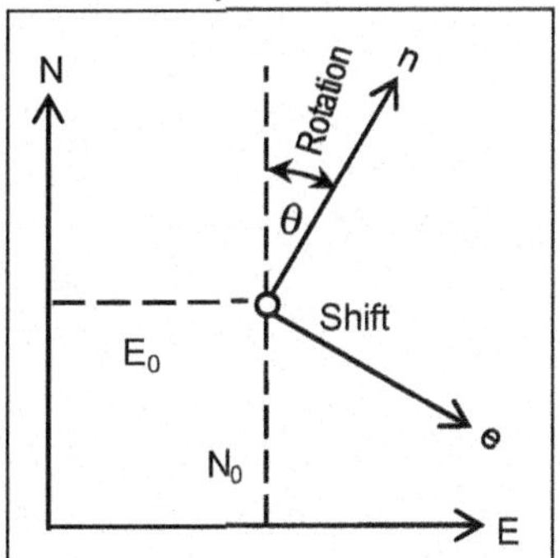

Figure 17-2 2-D transformation

If it's not checked it's wrong.

J. Walker and J. Awange, *Surveying Computations*,
https://doi.org/10.1007/978-3-031-94382-9_17

This process gives rise to four transformation parameters (Uren et al, 2006) written as (Eqn 17-1):

Conformal 2-D Transformation parameters	$a = s\cos(\theta)$ $b = s\sin(\theta)$ $c = E_0$ $d = N_0$	Eqn 17-1

The four equations can be used to solve for the four transformation parameters ***a, b, E_0*** and ***N_0***. Once these four parameters are known they can then be used to convert any point in the secondary coordinate system to the primary coordinate system. These parameters can also be expressed as ***a, b, c***, and ***d***.

Any transformation parameters calculated will always be based on data from actual survey work so that any errors within either system will manifest themselves in the calculated parameters. This means that the transformation parameters are always going to be a "best fit" only. Any small errors in the transformation parameters resulting from the two survey systems will increase as you progress out from the datum points used.

It is good practice to calculate the transformation parameters based on points outside of the subject area. Always work from without to within. A minimum of 2 points is required to calculate the transformation parameters but it is always good practice to have 1 or 2 other dual coordinated points located outside the area to check the transformation parameter accuracies.

A two-dimensional (2D) transformation described above can only be used over small areas of approximately 5 kilometres or less when converted from MGA2020 coordinates to Local Mine Grid Coordinates or vice versa. Over large areas this type of transformation will not accurately model the scale variations of the MGA projection onto a "Flat Earth" system such as a Local Mine Grid system (see Figure 17-3). In addition, these errors will increase in size as you move out from the central meridian of your MGA zone towards the zone boundary. Another source of error will be the height above the ellipsoid and variation of heights across the transformation area.

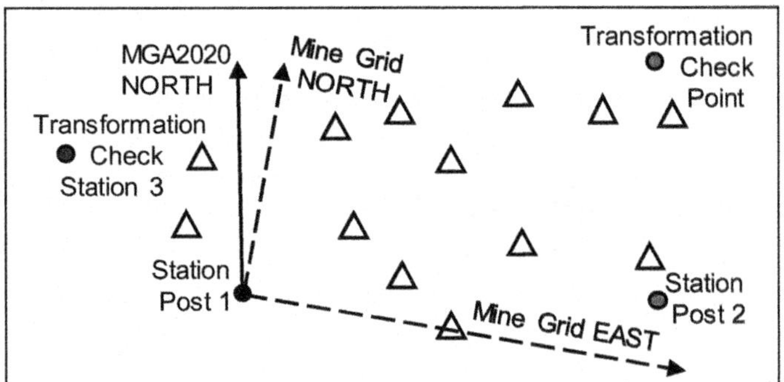

Figure 17-3 Typical mine grid layout (W,E, 17.1)

In Table 17-1 a transformation area having a uniform height over the area less than 370m above the ellipsoid and within 40km of the central meridian will have an error of 0ppm due to the transformation parameters used. In the case where the area in question ranges in height between 145m and 595m the errors associated with the transformation will be as much as 25ppm.

Table 17-1. Approximate maximum relative errors in parts per million (ppm) of a 2D Transformation based on the distance from the central meridian and height (ellipsoid) range over the area.

	0 ppm	5 ppm	10 ppm	15 ppm	20 ppm	25 ppm
0 km	170	125 - 215	80 - 260	35 - 305	10 - 350	55 - 395
10 km	180	135 - 225	90 - 270	45 - 315	0 - 360	45 - 405
20 km	200	155 - 245	110 - 290	65 - 335	20 - 380	25 - 425
30 km	245	200 - 290	155 - 335	110 - 380	65 - 425	20 - 470
40 km	300	255 - 345	210 - 390	165 - 435	120 - 480	75 - 525
50 km	370	325 - 415	280 - 460	235 - 505	190 - 550	145 - 595

Table courtesy of Dr Michael Kuhn (Curtin University), 2015.

The 2D transformation is only suitable over small areas reasonably close to the central meridian. If areas are larger or more accuracy is required, then other types of coordinate transformations need to be used. The most rigorous and accurate transformation is the grid to file transformation that can deliver accuracies of between 0.02 to 0.2m. Information regarding this type of transformation, software and data grid files can be obtained from the Geoscience Australian website. (www.ga.gov.au)

17.2.1 Worked Example 17.1. Conformal transformation.

A typical problem experienced is in an existing mining area where a local mine coordinate system has been used and you are surveying with GPS to locate or set out points in terms of MGA2020 coordinates. Define a set of transformation parameters that will allow coordinate transformation between either system.

As discussed previously, a simple two-dimensional transformation requires at least two (2) points with known coordinates in both systems. It is always good survey practice to have at least one more known point with coordinates in both systems to check the accuracy of your transformation parameters.

The given Worked Example is the determination of the transformation parameters required to change GNSS MGA2020 coordinates in a local mine grid coordinate system. Station 1 and station 2 are the original mine datum stations. The surveyor of the day set out his baseline approximately east and defined the bearing of the line as 90°. Station 3 is another point on site having both MGA2020 coordinates and Mine Grid coordinates.

Calculate the transformation parameters needed to transform other mine grid coordinates to MGA2020.

Table 17-2 Mine and MGA2020 common coordinates

Station	e_{mine}	n_{mine}	E_{MGA}	N_{MGA}
Post 1	50000.000	125000.000	377990.614	6458835.443
Post 2	51235.300	125000.000	379218.354	6458701.729
3	50684.385	125224.717	378695.153	6458984.707

17.2.1.1 Step 1: Set equations

Using the coordinates for Station 1 and 2, (Table 17-2), set up the 4 transformation equations.

$e_1 = E_1a - N_1b + E_0$ $\quad 50000.000 = 377990.614a - 6458835.443b + E_0$ (1)

$n_1 = N_1a + E_1b + N_0$ $\quad 125000.000 = 6458835.443a + 377990.614b + N_0$ (2)

$e_2 = E_2a - N_2b + E_0$ $\quad 51235.300 = 379218.354a - 6458701.729b + E_0$ (3)

$n_2 = N_2a + E_2b + N_0$ $\quad 125000.000 = 6458701.729a + 379218.354b + N_0$ (4)

17.2.1.2 Step 2: Subtract equations

Subtract equation (3) from (1) and equation (4) from (2) to give:

$-1235.300 = -1227.740a - 133.714b$ (5)

$0.000 = 133.714a + (-1227.740)b$ (6)

17.2.1.3 Step 3: Multiply equations

Multiply equation (6) by a real number to make the **a** value or the **b** value equal to the corresponding value in equation (5). In this case we have chosen the **b** to be the same in both equations (5) and (6)

$$\text{Multiplication factor} = \frac{133.714}{1227.740} = 0.108910681$$

The value above is now multiplied to both sides of equation (6) to give equation (7)

$0 \times 0.108911 = 133.714 \times 0.108911a - 1227.740 \times 0.108911b$

$0.000 = 14.562883a - 133.714b$ (7)

17.2.1.4 Step 4: Solve for ***a***

Subtract equation (7) from equation (5) and solve for **a**.

$-1235.300 = -1227.740a - 133.714b$ (5)

Subtract $0.000 = 14.562883a - 133.714b$ (7)

$-1235.300 = -1242.302883\boldsymbol{a}$

$$a = \frac{-1235.300}{-1242.302883} = 0.99432983$$

17.2.1.5 Step 5: Solve ***b*** using value of ***a***

Using the value for ***a*** use equation 5, 6 or 7 to solve for ***b***.

Now $0.000 = 133.714a - 1227.740b$ (6)

Substitute ***a***: $0.000 = 133.714 \times 0.994362983 - 1227.740b$

$-133.714 = -1227.740\boldsymbol{b}$

$$b = \frac{-133.714}{-1227.740} = 0.10829675$$

17.2.1.6 Step 6: Solve for ***E*₀**, (*c*), and ***N*₀**, (*d*)

Using the calculated values of ***a*** and ***b*** you can use either equations (1) and (2) for point one or equations (3) and (4) for point 2. To check, solve for both sets of equations and **check** to see that the E_0 and N_0 are the same for both points.

$e_1 = E_1a - N_1b + E_0$

$50000.000 = 377990.614 \times 0.994362983 - 6458835.443 \times 0.10829675 + E_0$

$\boldsymbol{E_0 = 373{,}611.013}$

$n_1 = N_1a + E_1b + N_0$

$125000.000 = 6458835.443 \times 0.994362983 + 377990.614 \times 0.10829675 + N_0$

$\boldsymbol{N_0 = -6{,}338{,}362.031}$ (note this is a **minus** shift).

17.2.1.7 Step 7: Checks

The final step of the process should be a check of your transformation parameters using a third common point. (Check station 3, Figure 17-3).

$\boldsymbol{a = 0.994363}$ $\boldsymbol{b = 0.108297}$ $\boldsymbol{E_0 = 373{,}611.013}$ $\boldsymbol{N_0 = -6{,}338{,}362.031}$

$$e_3 = E_3a - N_3b + E_0$$

$$= 378695.153 \times 0.994362983 - 6458984.707 \times 0.10829675 + 373611.013$$

$\boldsymbol{e_3 = 50{,}684.403}$

$$n_3 = N_3a + E_3b + N_0$$

$$= 6458984.707 \times 0.994362983 + 378695.153 \times 0.10829675 + (-6338362.031)$$

$\boldsymbol{n_3 = 125{,}224.722}$

Note: The check mine grid coordinates will always be a little different from the given surveyed coordinates. This is because of the small survey errors that exist in both coordinate networks.

The **rotation** of the mine grid from MGA94 can be determined from:

$$\theta = \arctan\left(\frac{b}{a}\right) = \arctan\left(\frac{0.108297}{0.994363}\right) = \arctan(0.108911) = 6.215624°$$

$\boldsymbol{\theta = 6°12'56''}$

The **scale factor, *S***, between MGA2020 and the mine grid can be determined from:

$$S = \left(\frac{a}{\cos(\theta)}\right) = \left(\frac{0.994363}{\cos(6°12'56'')}\right) = \left(\frac{0.994363}{0.994121}\right) \quad or$$

$$S = \sqrt{a^2 + b^2} = \sqrt{0.994363^2 + 0.108297^2}$$

$\boldsymbol{S} = 1.000243$

To determine the transformation parameters for the opposite direction, Local Mine Grid to MGA2020, the transformation equation need to be rewritten and solved as in the above example.

$$E_1 = e_1a - n_1b + e0$$
$$N_1 = n_1a + e_1b + n_0$$
$$E_2 = e_2a - n_2b + e0$$
$$N_2 = n_2a + e_2b + n_0$$

The parameters of the transformation are:

$\boldsymbol{a = 0.993880}$ $\boldsymbol{b = -0.108244}$ $\boldsymbol{e_0 = 314{,}766.094}$ $\boldsymbol{n_0 = 6{,}340{,}012.647}$

$\boldsymbol{\theta = -6°12'56''}$ (same rotation as previous, opposite sign)

$\boldsymbol{S = 0.999757}$ (reciprocal of previous scale)

Mine grid point **3** (e_3 = 50 684.385, n_3 = 125 224.717) has the corresponding MGA2020 coordinates:

E_3 = 378,695.135

N_3 = 6,458,984.704

Note the small survey errors that exist in both coordinate networks.

17.3 Coordinate Transformations – the matrix method of a unique solution.

The given Worked Example is the determination of the transformation parameters required to change GNSS MGA2020 coordinates into local mine grid coordinates. Station 1 and station 2 are the original mine datum stations. The surveyor of the day set out his baseline approximately east and defined the bearing of the line as 90°. Station 3 is another point on site having both MGA2020 coordinates and Mine Grid coordinates.

Calculate the transformation parameters needed to transform other mine grid coordinates to MGA2020.

Table 17-3 Mine and MGA2020 common coordinates, for conformal transformation

Station	e_{mine}	n_{mine}	E_{MGA}	N_{MGA}
Post 1	50000.000	125000.000	377990.614	6458835.443
Post 2	51235.300	125000.000	379218.354	6458701.729
3	50684.385	125224.717	378695.153	6458984.707

17.3.1 Conformal transformation by matrices.

Whilst the preceding was a direct solution of a unique conformal transformation, a matrix method is introduced to solve the unique solution. It will form the basis of a Least Squares solution of over-determined transformations, from the simple conformal to the Affine and beyond.

General observations in surveying can be reduced to the matrix form:

$\mathbf{A}\cdot\mathbf{x} = \mathbf{b} + \mathbf{v}$

- **x** a vector of terms to be computed, including coordinates
- **b** a vector relating to the observations
- **A** a **matrix** of coefficients
- **v** a vector of residuals, to be a minimum.

Taking the original conformal transformation, and recognizing that there ARE the minimum four (4) transformation parameters,

The formulae look like

$$\begin{matrix} e_1 \\ n_1 \\ e_2 \\ n_2 \end{matrix} = \begin{bmatrix} E_1 a - N_1 b + E_0 \\ N_1 a + E_1 b + N_0 \\ E_2 a - N_2 b + E_0 \\ N_2 a + E_2 b + N_0 \end{bmatrix}$$

and the **A** matrix of coefficients is:

$$\mathbf{A} = \begin{matrix} \text{Point} \\ 1 \\ 1 \\ 2 \\ 2 \end{matrix} \begin{matrix} \begin{matrix} a & b & E_0 & N_0 \end{matrix} \\ \begin{bmatrix} E_1 & -N_1 & E_0 & 0 \\ N_1 & E_1 & 0 & N_0 \\ E_2 & -N_2 & E_0 & 0 \\ N_2 & E_2 & 0 & N_0 \end{bmatrix} \end{matrix}$$

The previous solution is unique, using two points, so there will be no residuals. And thus no **v** matrix of residuals.

17.3.2 Worked Example 17.2. Matrix based Conformal transformation.

Using Table 17-3 for the common mine coordinates.

17.3.2.1 Transform MGA2020 to mine grid

1. Identify the **x** matrix of the parameters of the transformed points X:

$$\mathbf{x} = \begin{bmatrix} a \\ b \\ E_0 \\ N_0 \end{bmatrix} \text{ is a } 4 \times 1 \text{ matrix}$$

2. Form the **A** (4 x 4) matrix of the observation coefficients for E and N.
The solution comprises the control points 1 and 2, considered **fixed.**

$$\mathbf{A} = \begin{matrix} \begin{matrix} a & b & E_0 & N_0 \end{matrix} \\ \begin{bmatrix} 377990.614 & -6458835.443 & 1 & 0 \\ 6458835.443 & 377990.614 & 0 & 1 \\ 379218.354 & -6458701.729 & 1 & 0 \\ 6458701.729 & 379218.354 & 0 & 1 \end{bmatrix} \end{matrix} \text{ is a } 4 \times 4 \text{ matrix}$$

3. Form the **b** matrix (4 × 1); the matrix of the observed coordinates in the **Mine** system.

$$\mathbf{b} = \begin{matrix} e_1 \\ n_1 \\ e_2 \\ n_2 \end{matrix} \begin{bmatrix} 50000.000 \\ 125000.000 \\ 51235.300 \\ 125000.000 \end{bmatrix} \text{ is a } 4 \times 1 \text{ matrix}$$

4. Because there are no residuals (unique solution) we have the matrix:

A·x = **b**. then:

x = **A**^{1}·**b**

The A matrix is a 4 × 4 and is a major exercise to invert by hand. Using **Matlab** or Microsoft **Excel** the inversion routine {=MINVERSE(A11:A44)} produces (displayed to 6DP).

$$\mathbf{A}^{-1}=\begin{bmatrix} -0.000805 & 8.766838^{-5} & 0.000805 & -8.766838^{-5} \\ -8.766838^{-5} & -0.000805 & 8.766838^{-5} & 0.000805 \\ -260.969573 & -5232.220538 & 261.969573 & 5232.220538 \\ 5232.220538 & -260.969573 & -5232.220538 & -261.969573 \end{bmatrix}$$

5. Now find the **x** (4 × 1) matrix: **x** = **A**$^{-1}$·**b** (4 × 4) × (4 × 1) = (4 × 1)

$$\mathbf{A}^{-1}=\begin{bmatrix} -0.000805 & 8.766838^{-5} & 0.000805 & -8.766838^{-5} \\ -8.766838^{-5} & -0.000805 & 8.766838^{-5} & 0.000805 \\ -260.969573 & -5232.220538 & 261.969573 & 5232.220538 \\ 5232.220538 & -260.969573 & -5232.220538 & -261.969573 \end{bmatrix}\times\mathbf{b}=\begin{bmatrix} 50000.000 \\ 125000.000 \\ 51235.300 \\ 125000.000 \end{bmatrix}$$

{X11:X41} = {MMULT(MINVERSE(A11:A44),b11:b41)}

$$\mathbf{x}=\begin{matrix} a \\ b \\ E_0 \\ N_0 \end{matrix}\begin{bmatrix} 0.994363 \\ 0.108297 \\ 373611.013 \\ 6338362.031 \end{bmatrix}$$

6. Find the Scale Factor from:

$$S=\sqrt{a^2+b^2}=\sqrt{0.994363^2+0.108297^2}$$

***S* = 1.000243**

7. And the Rotation:

$$\theta=\arctan\left(\frac{b}{a}\right)=\arctan\left(\frac{0.108297}{0.994363}\right)=\arctan(0.108911)=6.215624°$$

***θ* = 6°12′56″**

8. Calculating the coordinates of point **3**, using coefficients of observations (MGA2020 coordinates of Station 3):

$$\begin{matrix} e_3 \\ n_3 \end{matrix}=\begin{bmatrix} E_3a-N_3b+E_0-0N_0 \\ N_3a+E_3b+0E_0+N_0 \end{bmatrix} \qquad \mathbf{A}_3=\begin{matrix} & a & b & E_0 & N_0 \\ & \end{matrix}\begin{bmatrix} 378695.153 & -6458984.707 & 1 & 0 \\ 6458984.707 & 378695.153 & 0 & 1 \end{bmatrix}$$

Result Matrix = A3·x {a11:a21} = {MMULT(A311:A324,X11:X41)}

e3 = 50,684.403

n3 = 125,224.722

17.3.2.2 Transform mine grid to MGA2020 coordinates

To determine the transformation parameters for the opposite direction, Local Mine Grid to MGA2020, the transformation equation need to be rewritten and solved as in the above example.

The formulae look like and the **A** matrix of coefficients is:

$$\begin{matrix} E_1 \\ N_1 \\ E_2 \\ N_2 \end{matrix}=\begin{bmatrix} e_1a-n_1b+e_0 \\ n_1a+n_1b+n_0 \\ e_2a-n_2b+e_0 \\ n_2a+e_2b+n_0 \end{bmatrix} \qquad \mathbf{A}=\begin{matrix} \text{Point} \\ 1 \\ 1 \\ 2 \\ 2 \end{matrix}\begin{matrix} a & b & e_0 & n_0 \\ e_1 & -n_1 & e_0 & 0 \\ n_1 & e_1 & 0 & n_0 \\ e_2 & -n_2 & e_0 & 0 \\ n_2 & e_2 & 0 & n_0 \end{matrix}$$

1. Form the **A** and **A** inverse (**A**$^{-1}$) (4 × 4) matrix of the observation coefficients for e and n in the mine system. The solution comprises the control points 1 and 2, considered **fixed**.

$$\mathbf{A}=\overset{\begin{array}{cccc}\quad a\quad & \quad b\quad & e_0 & n_0\end{array}}{\begin{bmatrix}50000 & -125000 & 1 & 0\\125000 & 50000 & 0 & 1\\51235.3 & -125000 & 1 & 0\\125000 & 51235.3 & 0 & 1\end{bmatrix}}\qquad \mathbf{A}^{-1}=\begin{bmatrix}-0.00081 & 0 & 0.00081 & 0\\0 & -0.00081 & 0 & 0.00081\\41.4760 & -101.1900 & -40.4760 & 101.1900\\101.1900 & 41.4760 & -101.1900 & -40.4760\end{bmatrix}$$

2. Form the **b** matrix (4 × 1); the matrix of the observed coordinates in the **MGA** system.

$$\mathbf{b}=\begin{array}{c}E_1\\N_1\\E_2\\N_2\end{array}\begin{bmatrix}377990.614\\6458835.443\\378215.354\\6458701.729\end{bmatrix}$$

3. Because there are no residuals (unique solution) we have the matrix:
   A·x = **b**. then:
   x = **A^{-1}·b**

4. Now find the **x** (4 × 1) matrix: **x = A^{-1}·b** (4 × 4) x (4 × 1) = (4 × 1)
   $\{x_{11}:x_{41}\} = \{MMULT(MINVERSE(A_{11}:A_{44}),b_{11}:b_{41})\}$

$$\mathbf{x}=\begin{array}{c}a\\b\\e_0\\n_0\end{array}\begin{bmatrix}0.993880\\-0.108244\\314766.094\\6340012.647\end{bmatrix}$$

The parameters of the transformation are:

a = 0.993880 **b = – 0.108244** **e$_0$ = 314,766.094** **n$_0$ = 6,340,012.647**

θ = – 6°12′56″ (same rotation as previous, opposite sign)

S = 0.999757 (reciprocal of previous scale)

5. Calculating the coordinates of point **3**, using coefficients of observations (MINE coordinates of Station 3):

$$\begin{array}{c}E_3=\\N_3=\end{array}\begin{bmatrix}e_3a & -n_3b & e_0 & 0n_0\\n_3a & e_3b & 0e_0 & n_0\end{bmatrix}\qquad A_3=\overset{\begin{array}{ccccc}\text{Point} & a & b & E_0 & N_0\end{array}}{\begin{bmatrix}50684.385 & -125224.717 & 1 & 0\\125224.717 & 50684.385 & 0 & 1\end{bmatrix}}$$

Result Matrix = A$_3$·x $\{a_{11}:a_{21}\} = \{MMULT(A3_{11}:A3_{24},x_{11}:x_{41})\}$

E$_3$ = 378,695.135

N$_3$ = 6,458,984.704

Note the small survey errors that exist in both coordinate networks.

17.4 Coordinate Transformations – a Least Squares solution.

Taking the original conformal transformation, and introducing the third point, Station 3, the transformation becomes over-determined because there are more coefficients, six (6), than the minimum four (4) transformation parameters required. We saw a small transformation error between our forward and backward solutions in § 17.2.1, page 272. Now we will distribute that error to provide a better estimate of the transformation parameters (a, b, E$_0$, N$_0$) to allow a "better" solution of unknown transformation points.

The formulae look like: and the **A** matrix of coefficients is:

$$\begin{array}{c}e_1=\\n_1=\\e_2=\\n_2=\\e_3=\\n_3=\end{array}\begin{bmatrix}E_1a & -N_1b & +E_0 & +0N_0\\N_1a & +E_1b & +0E_0 & +N_0\\E_2a & -N_2b & +E_0 & +0N_0\\N_2a & +E_2b & +0E_0 & +N_0\\E_3a & -N_3b & +E_0 & +0N_0\\N_3a & +E_3b & +0E_0 & +N_0\end{bmatrix}\qquad A=\overset{\begin{array}{ccccc}\text{Point} & a & b & E_0 & N_0\end{array}}{\begin{array}{c}1\\1\\2\\2\\3\\3\end{array}\begin{bmatrix}E_1 & -N_1 & E_0 & 0\\N_1 & E_1 & 0 & N_0\\E_2 & -N_2 & E_0 & 0\\N_2 & E_2 & 0 & N_0\\E_3 & -N_3 & E_0 & 0\\N_3 & E_3 & 0 & N_0\end{bmatrix}}$$

This solution is non-unique, using three points, so there will be residual errors.

There will be a **v** matrix, the **residual** matrix.

17.4.1 Worked Example 17.3. Over-determined Least Squares Conformal transformation.

The use of Microsoft Excel™ to solve these problems.

1. Form the **x** matrix of the parameters of the transformed points X:

$$\mathbf{x} = \begin{bmatrix} a \\ b \\ E_0 \\ N_0 \end{bmatrix} \text{ is a } 4 \times 1 \text{ matrix.}$$

2. Form the **A** (6 × 4) matrix of the observation coefficients for E and N.
   The solution comprises the control points 1, 2 and 3, considered **fixed**.

$$\mathbf{A} = \begin{array}{c} \begin{array}{cccc} \text{a} & \text{b} & \text{E}_0 & \text{N}_0 \end{array} \\ \begin{bmatrix} 377990.614 & -6458835.443 & 1 & 0 \\ 6458835.443 & 377990.614 & 0 & 1 \\ 379218.354 & -6458701.729 & 1 & 0 \\ 6458701.729 & 379218.354 & 0 & 1 \\ 378695.153 & -6458984.707 & 1 & 0 \\ 6458984.707 & 378695.153 & 0 & 1 \end{bmatrix} \end{array}$$

   The $\mathbf{A}^{-1}$ (inverse) matrix is not shown as it is generated directly by the MINVERSE(:) function.

3. Form the **b** matrix (6 × 1); the matrix of the observed coordinates in the **Mine** system.

$$\mathbf{b} = \begin{array}{c} e_1 \\ n_1 \\ e_2 \\ n_2 \\ e_3 \\ n_3 \end{array} \begin{bmatrix} 5000.000 \\ 125000.000 \\ 51235.300 \\ 125000.000 \\ 50684.385 \\ 125224.717 \end{bmatrix}$$

4. Because there are now residuals (it's an over-determined solution) we can generate the **v** matrix (6 × 1) of residuals, *ν*, (Greek letter, nu) for each coordinate.

$$\mathbf{v} = \begin{array}{c} e_1 \\ n_1 \\ e_2 \\ n_2 \\ e_3 \\ n_3 \end{array} \begin{bmatrix} ve_1 \\ vn_1 \\ ve_2 \\ vn_2 \\ ve_3 \\ vn_3 \end{bmatrix}$$

The later iterations of the Excel matrix functions now operate with automatic result array generation. Make sure you put the **A**, **b**, arrays in their proper format. Then put the matrix function (MMULT(), TRANSPOSE(), MINVERSE() etc in its desired cell (with room for the resultant array dimension available).

Use CTRL+ENTER to generate the array.

Solve the matrix $\mathbf{A}\cdot\mathbf{x} = \mathbf{b} + \mathbf{v}$.

The matrix must first be **normalised** to the form:

$\mathbf{A}^{\mathrm{T}}\cdot\mathbf{A}\cdot\mathbf{x} \quad = \mathbf{A}^{\mathrm{T}}\cdot\mathbf{b} + \mathbf{v}$, then

$\qquad\qquad \mathbf{x} \quad = (\mathbf{A}^{\mathrm{T}}\mathbf{A})^{-1}\cdot\mathbf{A}^{\mathrm{T}}\mathbf{b} + \mathbf{v}$

The generation of the normalised $\mathbf{A}^{\mathrm{T}}\cdot\mathbf{A}$ matrix (4 × 6) × (6 × 4) is a 4 × 4 and is a major exercise to calculate, then invert, by hand. Using **Matlab** or Microsoft **Excel** the matrix transpose, multiplication and inversion routines produces:

$$\mathsf{A}^{\mathsf{T}}\mathsf{A} = \begin{bmatrix} 1.2558^{14} & 0 & 1135904.121 & 19376521.88 \\ 6548835.443 & 1.2588^{14} & -19376521.88 & 1135904.121 \\ 1135904.121 & -19376521.88 & 3 & 0 \\ 19376521.88 & 1135904.121 & 0 & 3 \end{bmatrix}$$

1. Find the N^{-1}·(inverse) matrix: note the nesting of the Excel matrix functions.
{N-1 11:N-1 41} = {MINVERSE(MMULT(TRANSPOSE(A11:A64), A11:A64)}

$$\left(A^T A\right)^{-1} = N^{-1} = \begin{bmatrix} 1.2512013^{-6} & 1.1016075^{-15} & -0.4737482 & -8.0813100 \\ -2.5361499^{-15} & 1.2512013^{-6} & 8.0813101 & -0.4737482 \\ -0.4737483 & 8.0813101 & 52375271.894 & 0.07613824 \\ -8.0813100 & -0.4737483 & -0.0761382 & 52375271.380 \end{bmatrix}$$

2. Now find the $\mathbf{A^T \cdot b}$ (4 × 1) matrix:)

$$A^T b = \begin{bmatrix} 2.48104^{12} \\ -8.39152^{11} \\ 151919.685 \\ 375224.717 \end{bmatrix}$$

3. Now find the **x** (4 × 1) matrix:
$\mathbf{x} = \mathbf{N^{-1} \cdot (A^T \cdot b)}$ (4 × 4) × (4 x 1) = (4 × 1)
{x11:x41} = {MMULT(N-1 11:N-1 44),ATb11:ATb41)}

$$x = \begin{matrix} a \\ b \\ c \\ d \end{matrix} \begin{bmatrix} 0.9943608 \\ 0.1082996 \\ 373630.072 \\ -6338348.973 \end{bmatrix}$$

4. Find the Scale Factor from a and b in the **x** matrix:

$$s = \sqrt{a^2 + b^2} = \sqrt{0.9943608^2 + 0.1082996^2}$$

***S* = 1.000241**

5. And the Rotation:

$$\theta = \arctan\left(\frac{b}{a}\right) = \arctan\left(\frac{0.108300}{0.994361}\right) = \arctan(0.108914) = 6.215798°$$

***θ* = 6°12′57″**

6. Calculating the adjusted coordinates of the three points, using coefficients of observations. (MGA2020 coordinates of the stations). The residuals are the differences between the observed and transformed coordinates in the mine system:

Result Matrix: $\mathbf{b_{calc} = A \cdot x}$ **$\mathbf{v = b_{obs} - b_{calc}}$ (residuals)**

$$\underset{\mathbf{A.x}}{b_{calc}} = \begin{matrix} e_1 \\ n_1 \\ e_2 \\ n_2 \\ e_3 \\ n_3 \end{matrix} \begin{bmatrix} 49999.996 \\ 124999.997 \\ 51235.293 \\ 125000.001 \\ 50684.396 \\ 125224.720 \end{bmatrix} \quad b_{obs} = \begin{matrix} e_1 \\ n_1 \\ e_2 \\ n_2 \\ e_3 \\ n_3 \end{matrix} \begin{bmatrix} 5000.000 \\ 125000.000 \\ 51235.300 \\ 125000.000 \\ 50684.385 \\ 125224.717 \end{bmatrix} \quad v_{obs-calc} = \begin{matrix} e_1 \\ n_1 \\ e_2 \\ n_2 \\ e_3 \\ n_3 \end{matrix} \begin{bmatrix} 0.004 \\ 0.003 \\ 0.007 \\ -0.001 \\ -0.011 \\ -0.003 \end{bmatrix} \quad \Sigma v^2 = 0.000215$$

17.4.2 The use of MATLAB to solve these problems.

Your course may introduce you to the MATLAB environment as a programming and problem solving tool. There are many excellent books on the subject; there is plenty of expertise around the teaching department in both the GNSS and remote sensing using both airborne and terrestrial laser scanners (ALS and TLS). The line header **EDU>>** shows an academic licence.

The following solution is presented to allow the two methods to be contrasted on facing pages.

This section is not a lesson in either Excel or MATLAB. It assumes a working knowledge of matrix arithmetic. The section shows the author's use of the programs.

The aim is to present the spreadsheet solution on one page and the MATLAB solution in parallel on the other.

Using MATLAB as a matrix calculator (Worked Example 25.3, data from Table 17-3, page 275):

We will use the comma (,) as the variable separator, although a space or a tab will do the same.

Entry of a multi-dimensional matrix requires the use of the Enter key at the end of each line; using the symbol ↵ to denote the [Enter ↵] key.

Refer to the matrix formation on the opposite page.

1. Form the **A** (6 × 4) matrix of the observation coefficients for E and N.
   The solution comprises the control points 1, 2 and 3, considered **fixed**.

Enter the a, b, E_0, N_0 coefficients for each row:

```
EDU>> A = zeros [ 377990.614, -6458835.443, 1, 0 ↵
6458835.443, 377990.614, 0, 1 ↵
379218.354, -6458701.729,1, 0 ↵
6458701.729, 379218.354, 0, 1 ↵
378695.153, -6458984.707, 1, 0 ↵
6458984.707, 378695.153, 0, 1]; ↵
```

Note a) the **zeros** statement ensures the matrix will be formed correctly
b) the matrix is initiated with an opening square bracket **[**
c) the matrix is finalized with an closing square bracket **]**
d) the semi-colon, **;**, after the closing bracket suppresses the display of the **A** matrix.

2. Form the **b** matrix (6 × 1); the matrix of the observed coordinates in the **Mine** system. ($\mathbf{b_{obs}}$)

```
EDU>> b = [ 50000 ↵
125000 ↵
51235.3 ↵
125000 ↵
50684.385 ↵
125224.717]; ↵
```

3. Solve the matrix $\mathbf{A \cdot x = b + v}$.
   The matrix must first be **normalised** to the form:
   $\mathbf{A^TA \cdot x = A^Tb + v}$, then
   $\mathbf{x = (A^TA)^{-1} \cdot A^Tb + v}$
   The **x** and the **v** matrix can be handled as normal calculations.

```
EDU>>x = (inv(A'*A))*(A'*b) ↵
```

Note a) the `inv(A'*A)` function finds the inverse of the bracketed expression,
b) `A'` (`A` prime) is the **transpose** of the `A` matrix,
c) no semi-colon, the `x` vector (matrix) is displayed to 4 dp with scaling (`1.0e+006 *` $=1\times10^6$),
d) extra (unnecessary) brackets have been included for clarity.

```
x =
   1.0e+006 *
     0.0000
     0.0000
     0.3736
    -6.3383
```

The answers to 4 or 3dp derived from **format long**

x1	a	0.9944
x2	b	0.1083
x3	E_0	373630.072
x4	N_0	-6338348.973

4. Find the Scale Factor, $SF = \sqrt{a^2+b^2}$, using the **norm (vector)** function.

```
EDU>>SF = norm(x(1:2)) ↵
SF =
     1.0002
```

`x(1:2)` represents the first two elements, **a** and **b**, of **x** vector.

SF = 1.000241

5. And the Rotation in radians: $\theta = \arctan(b/a)$

```
EDU>>Rotation = atan(x(2)/x(1)) ↵
Rotation =
     0.1085
```

`x(2)` $= b$, `x(1)` $= a$

to convert to degrees.

```
EDU>>R_degrees = Rotation * 180/pi ↵
R_degrees =
     6.2158
```

$\theta = 6°12'57''$

6. Calculate the adjusted coordinates of the three points, $\mathbf{b_{calc}}$ and check residuals, **v**.

Result Matrix: $\mathbf{b_{calc} = A \cdot x}$ $\mathbf{v = b - b_{calc}}$ (residuals)

```
EDU» bcalc = A*x ↵
bcalc =
   1.0e+005 *
     0.5000
     1.2500
```

Calculate adjusted values of transformed coordinates.

These are the values to 3dp. scaling (`1.0e+005 *` $=1\times10^5$)

e_1 49999.996
n_1 124999.997

If it's not checked it's wrong.

```
      0.5124            e2   51235.293
      1.2500            n2  125000.000
      0.5068            e3   50684.396
      1.2522            n3  125224.720
EDU» v = b-bcalc ↵      Find residuals between observed and calculated coordinates.
v =
      0.0045
      0.0034
      0.0068
     -0.0003
     -0.0113
     -0.0031            Check that the sum of the residuals vector is zero.
EDU» sumv = sum(v) ↵
sumv =
      0.0000            Sum residuals.
```

17.5 Coordinate Translations – a VERY HELPFUL technique.

You will have noted that working in MGA coordinates, and even in the Local Mine Grid, that you have some serious numbers to manipulate. Used directly in the program, these numbers are well outside the capability of even a 12-digit calculator. In fact, Microsoft Excel introduces errors even with its 16 digit computations.

What to do? Translate the axes of one or both coordinate systems. That's what!

Why use numbers like E378,000, N6,500,000 (MGA) when the work area is only covering an area of dE < 1,300 and dN < 300? Why not move (translate) the ORIGIN up close to the work site?

1. This is the original data:

Station	e_{mine}	n_{mine}	E_{MGA}	N_{MGA}
Post 1	50000.000	125000.000	377990.614	6458835.443
Post 2	51235.300	125000.000	379218.354	6458701.729
3	50684.385	125224.717	378695.153	6458984.707

By examination we can choose some minimum numbers for both systems that will leave the system in the first quadrant:

False origin mine		False origin MGA	
e_{mine}	50000	E_{MGA}	377,900
n_{mine}	125000	N_{MGA}	6,458,700

2. Subtracting BOTH sets of false origins, the TRANSLATED coordinates are:

Station	e_{mine}	n_{mine}	E_{MGA}	N_{MGA}
Post 1	0.000	0.000	390.614	135.443
Post 2	1235.300	0.000	1318.354	11.729
3	684.385	224.717	795.153	284.707

Interesting things now happen:

The $\mathbf{x}_{orig}$ matrix was: becomes translated $\mathbf{x}_{shift}$:

$$x = \begin{matrix} a \\ b \\ E_o \\ N_o \end{matrix} \begin{bmatrix} 0.9943608 \\ 0.1082996 \\ 373630.072 \\ -6338348.973 \end{bmatrix} \qquad x_{shift} = \begin{matrix} a \\ b \\ E_o \\ N_o \end{matrix} \begin{bmatrix} 0.9943608 \\ 0.1082996 \\ -75.439 \\ -144.496 \end{bmatrix}$$

There is almost no change for scale factor, **s**, or the rotation, **θ**. However the shifts, $\mathbf{E_0}$ and $\mathbf{N_0}$, are much nicer to deal with. But it is the residuals that provide food for thought. The transformed translated coordinates and

$$b_{original} = \begin{matrix} e_1 \\ n_1 \\ e_2 \\ n_2 \\ e_3 \\ n_3 \end{matrix} \begin{bmatrix} 5000.000 \\ 125000.000 \\ 51235.300 \\ 125000.000 \\ 50684.385 \\ 125224.717 \end{bmatrix} \quad v_{original} = \begin{matrix} e_1 \\ n_1 \\ e_2 \\ n_2 \\ e_3 \\ n_3 \end{matrix} \begin{bmatrix} 0.004 \\ 0.003 \\ 0.007 \\ -0.001 \\ -0.011 \\ -0.003 \end{bmatrix} \quad \Sigma v^2 = 2.15^{-4}$$

$$b_{shifted} = \begin{matrix} e_1 \\ n_1 \\ e_2 \\ n_2 \\ e_3 \\ n_3 \end{matrix} \begin{bmatrix} -0.004 \\ -0.003 \\ 1235.293 \\ 0.000 \\ 684.396 \\ 224.720 \end{bmatrix} \quad v_{shifted} = \begin{matrix} e_1 \\ n_1 \\ e_2 \\ n_2 \\ e_3 \\ n_3 \end{matrix} \begin{bmatrix} 0.004 \\ 0.003 \\ 0.007 \\ -0.000 \\ -0.011 \\ -0.003 \end{bmatrix} \quad \Sigma v^2 = 2.15^{-4}$$

This shows that some errors have crept into the transformations because of computational limitations of the Excel program.

Using a 12-digit calculator on the original MGA problem would have provided meaningless answers, even those programmed to do least squares adjustments.

3. Finally ADD the false origins to the transformed points. This returns you to the original reference frame.

Note that computational limits did not arise in MATLAB.

17.6 Affine Coordinate Transformation with Weights.

The conformal transformation merely translates one set of coordinates to another by a scaling, shift and rotation. The basic SHAPE of the figure remains; the transformation is CONFORMAL.

The next transformation is designed to work where there are different scales in X and Y. It is popularly used in photogrammetry to account for differential shrinkage in photo film images, or in non-conformal maps.

An affine transformation is often used in photogrammetry to compensate for uneven film shrinkage derived from the camera fiducial system and arbitrarily measured fiducial marks on the film. It makes allowances for two different scale factors; one in the X, the other in the Y direction. The camera has four fiducial marks, this provides an overdetermined solution as only 3 fiducials are needed to generate the 6 parameters of the affine transformation; 2 translation, 1 rotation, 1 skew angle and 2 scaling. The transformation allows other points measured on the film to be coordinated in the camera fiducial system.

17.6.1 Derivation of the affine transformation.

The derivation of the affine transformation is detailed in § 17.7, page 289 and is summarized here.

The mathematical model is of the form:, $\begin{matrix} ax+by+c=X \\ dx+ey+f=Y \end{matrix}$ in the matrix form $\begin{bmatrix} X \\ Y \end{bmatrix} = \begin{bmatrix} a & b \\ d & e \end{bmatrix} \begin{bmatrix} x \\ y \end{bmatrix} + \begin{bmatrix} e \\ f \end{bmatrix}$ and,

because it's overdetermined, $\begin{bmatrix} X \\ Y \end{bmatrix} = \begin{bmatrix} X \\ Y \end{bmatrix} + \begin{bmatrix} Vx \\ Vy \end{bmatrix}$

It can be expressed in the matrix form **Ax=b + v**.

In the **affine** transformation there are three (3) common points required for a unique solution. This provides 6 parameters, a, b, c, d, e and f.

The observation equations for one point are:

ax + by + c = X + Vx

dx + ey + f = Y + Vy.

expanding the equations to include all parameters in each line, they become:

ax + by + 1c + 0d + 0e + 0f = X + Vx

0a + 0b + 0c + dx + ey + 1f = Y + Vy.

The standard deviations of the observations, if known, can be included as a vector of precisions (**p**), generating a **weight** matrix of the observations.

Noting that naming conventions can change from author to author, in general, observations in surveying can be reduced to the matrix form:

AWx = Wb + v

x	a vector of terms to be computed, including coordinates;
b	a vector relating to the observations;
A	a **matrix** of coefficients;
p	a **vector** of observation precisions, the standard deviation;
v	a vector of residuals, to be a minimum;
W	a diagonal weight matrix of the inverse of the observation variances ($W = {}^{1}/_{p^2}$).

Taking the original affine transformation, and recognizing that there ARE the minimum six (6) transformation parameters,

the formulae are the **A** matrix of coefficients, and the **p** vector:

$$\textbf{Observations} = \begin{bmatrix} a_{x1} + b_{y1} + c = X_1 + v_{X1} \\ d_{x1} + e_{y1} + f = Y_1 + v_{Y1} \\ a_{x2} + b_{y2} + c = X_2 + v_{X2} \\ d_{x2} + e_{y2} + f = Y_2 + v_{Y2} \\ a_{x3} + b_{y3} + c = X_3 + v_{X3} \\ d_{x3} + e_{y3} + f = Y_3 + v_{Y3} \end{bmatrix} \quad \mathbf{A} = \begin{bmatrix} x_1 & y_1 & 1 & 0 & 0 & 0 \\ 0 & 0 & 0 & x_1 & y_1 & 1 \\ x_2 & y_2 & 1 & 0 & 0 & 0 \\ 0 & 0 & 0 & x_2 & y_2 & 1 \\ x_3 & y_3 & 1 & 0 & 0 & 0 \\ 0 & 0 & 0 & x_3 & y_3 & 1 \end{bmatrix} \quad \mathbf{p} = \begin{bmatrix} p_{X1} \\ p_{Y1} \\ p_{X2} \\ p_{Y2} \\ p_{X3} \\ p_{Y3} \end{bmatrix}$$

The form of the least squares solution is as shown. Because this solution is unique, using three points shown here, there will be no residuals.
And thus the **v** vector = 0.

$$\mathbf{A} = \begin{bmatrix} x_1 & y_1 & 1 & 0 & 0 & 0 \\ 0 & 0 & 0 & x_1 & y_1 & 1 \\ x_2 & y_2 & 1 & 0 & 0 & 0 \\ 0 & 0 & 0 & x_2 & y_2 & 1 \\ x_3 & y_3 & 1 & 0 & 0 & 0 \\ 0 & 0 & 0 & x_3 & y_3 & 1 \end{bmatrix} \mathbf{W} = \begin{bmatrix} W_{x1} & & & & & \\ & W_{y1} & & & & \\ & & W_{x2} & & & \\ & & & W_{y2} & & \\ & & & & W_{x3} & \\ & & & & & W_{y3} \end{bmatrix} \mathbf{x} = \begin{bmatrix} a \\ b \\ c \\ d \\ e \\ f \end{bmatrix} \mathbf{b} = \begin{bmatrix} X_1 \\ Y_1 \\ X_2 \\ Y_2 \\ X_3 \\ Y_3 \end{bmatrix} \mathbf{v} = \begin{bmatrix} v_{X1} \\ v_{Y1} \\ v_{X2} \\ v_{Y2} \\ v_{X3} \\ v_{Y3} \end{bmatrix}$$

The following example will use 4 common points, thus over-determining the transformation. It will result in a least squares solution with the introduction of the **v** vector, a set of residuals. The standard deviation of each observation has been estimated, and this will allow the introduction of **weighted** observations in the solution.

17.6.2 Worked Example 17.4. Over-determined Least Squares Affine transformation.

The affine transformation was useful in rectifying aerial photographs (traditionally 9″×9″ (230×230mm) format).

The coordinates of the four camera fiducial marks, X and Y, are supplied by the manufacturer. They are expressed in mm (0.001m). The image being measured is a copy of the original film image. It may subject to differential shrinkage due to copying, heat, humidity etc.

The film (image) fiducial marks and two pass points have been read in a coordinating digitiser with estimates for the observations of the fiducials. The measurements are in **inches** down to 1/1000" (one thou inch = 40μm [microns]) Calculate the adjusted coordinates of the fiducials and the two pass points.

Standard deviations of the fiducial observations, σx, σy, generate a weight matrix, **W**. This results in a weighted matrix, **WAx** = **Wb** + **v**

The solution for the adjustment is shown on the right-hand side.

Point	X orig	Y orig	x obs	y obs	σx obs	σy obs	**X adj**	**Y adj**	**±σX adj**	**±σy adj**
1	-113.000	0.003	0.764	5.960	0.104	0.112	-112.899	0.052	0.132	0.141
2	0.001	112.993	5.062	10.541	0.096	0.120	-0.085	112.936	0.125	0.147
3	112.998	0.003	9.663	6.243	0.112	0.088	113.115	0.033	0.139	0.118
4	0.001	-112.999	5.350	1.654	0.096	0.104	-0.085	-113.042	0.125	0.134
10			1.746	9.354			-85.193	85.470	0.134	0.154
11			5.329	9.463			5.803	85.337	0.107	0.123

The Figure 17-4 Camera fiducial marks shows the calibrated coordinates of the manufacturer's camera register frame (mm) and the resulting observed coordinate reader values (inches).

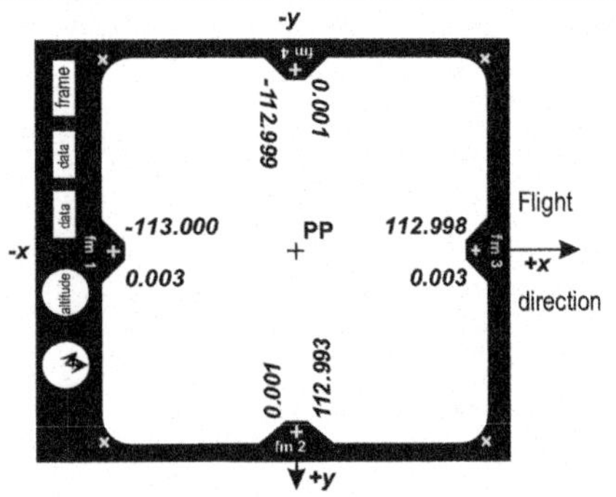

Figure 17-4 Camera fiducial marks

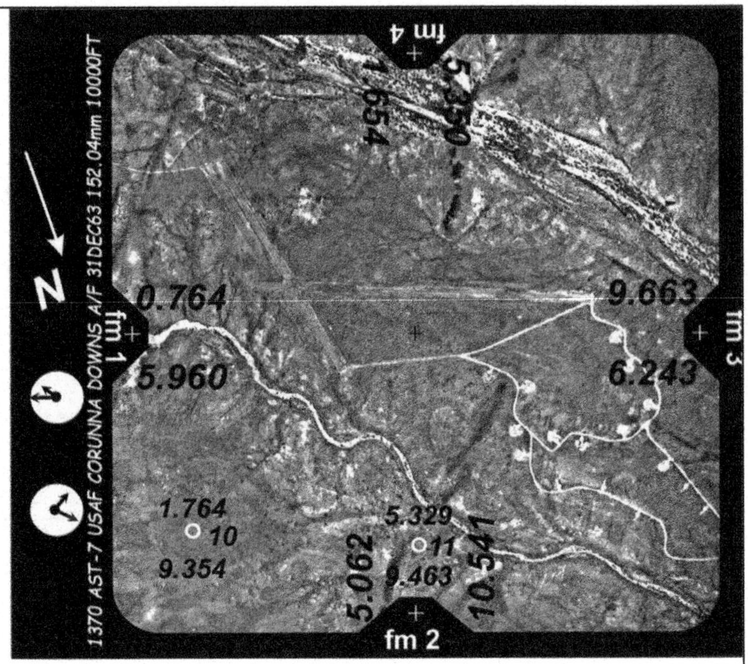

The film mask of a 230mm (9″) aerial camera. It shows the camera's calibrated fiducial marks (fm) and principal point (PP) of the lens in mm. Opposite are the coordinator observations of the Corunna Downs image, the fiducials and two pass points (in inches). (Author's collection)
Corunna Downs A/F, 1963.
(-21°26',119°46'. 2165m runway)
Elevation 250m (830ft).
Photo scale 1:4,600

The Affine transformation will be carried out using Matlab. Excel is a valuable tool but at this level it starts to get unwieldy.

1. Form the **x** matrix of the parameters of the transformed points X:

$$\mathbf{x} = \begin{bmatrix} a \\ b \\ c \\ d \\ e \\ f \end{bmatrix}$$

2. Form the **A** (8 x 6) matrix of the observation coefficients for x and y. These are the measurements taken by the digitizer. The standard deviation, σx_i, σy_i, of each observation is included here as the **p** vector, which will allow generation of the **weight** matrix. The solution comprises the x y control points 1, 2, 3 and 4, which are **fixed**.

$$\mathbf{A} = \begin{matrix} x_1 \\ y_1 \\ x_2 \\ y_2 \\ x_3 \\ y_3 \\ x_4 \\ y_4 \end{matrix} \begin{bmatrix} 0.764 & 5.960 & 1 & 0 & 0 & 0 \\ 0 & 0 & 0 & 0.764 & 5.960 & 1 \\ 5.062 & 10.451 & 1 & 0 & 0 & 0 \\ 0 & 0 & 0 & 5.062 & 10.451 & 1 \\ 9.663 & 6.243 & 1 & 0 & 0 & 0 \\ 0 & 0 & 0 & 9.663 & 6.243 & 1 \\ 3.530 & 1.654 & 1 & 0 & 0 & 0 \\ 0 & 0 & 0 & 5.350 & 1.654 & 1 \end{bmatrix} \quad \mathbf{p} = \begin{bmatrix} 0.104 \\ 0.112 \\ 0.096 \\ 0.120 \\ 0.112 \\ 0.088 \\ 0.096 \\ 0.104 \end{bmatrix}$$

3. Using the p vector, generate the diagonal W matrix where $W_{ii} = \frac{1}{p^2}$

$$\mathbf{W} = \begin{bmatrix} 92.456 &&&&&&& \\ & 79.719 &&&&&& \\ && 108.507 &&&&& \\ &&& 69.444 &&&& \\ &&&& 71.719 &&& \\ &&&&& 129.132 && \\ &&&&&& 108.507 & \\ &&&&&&& 92.456 \end{bmatrix}$$

4. Form the **b** matrix (8 x 1); the matrix of the fixed

Because there are now residuals (over-determined solution)

coordinates in the **Camera** system (the fiducial marks).

$$
\mathbf{b} = \begin{matrix} x_1 \\ y_1 \\ x_2 \\ y_2 \\ x_3 \\ y_3 \\ x_4 \\ y_4 \end{matrix}
\begin{bmatrix} -113.000 \\ 0.003 \\ 0.001 \\ 112.993 \\ 112.998 \\ 0.003 \\ 0.001 \\ -112.999 \end{bmatrix}
$$

we have to generate the **v** vector (8×1) of residuals, **v** (nu).

$$
\mathbf{v} = \begin{bmatrix} v_{X1} \\ v_{Y1} \\ v_{X2} \\ v_{Y2} \\ v_{X3} \\ v_{Y3} \\ v_{X4} \\ v_{Y4} \end{bmatrix}
$$

17.6.3 Affine transformation using MATLAB.

The manipulation of the preceding matrices becomes unwieldy in Excel. Continue using MATLAB.

a. Pre-dimension the **x** matrix of the transformation parameters ***a*** to ***f***; it is a column vector of dimension 6. This is not strictly necessary as the **x** vector will be generated as the result of matrix manipulation. Notice that the output has been suppressed with the semicolon operator, **;**

```
EDU» x = [zeros(1,6)]'; 
```
the ' command creates a column matrix.

Form the **A** matrix of observations for the six (6) unknown **x** elements.

The elements are separated by a comma for clarity. A space will do.

```
EDU» A = [0.764, 5.96, 1, 0, 0, 0
0, 0, 0, 0.764, 5.96, 1
5.062, 10.541, 1, 0, 0, 0
0, 0, 0, 5.062, 10.541, 1
9.663, 6.243, 1, 0, 0, 0
0, 0, 0, 9.663, 6.243, 1
5.35, 1.654, 1, 0, 0, 0
0, 0, 0, 5.35, 1.654, 1];
```

b. Form the **b** matrix of the fiducials. The input is shown as a transposed row vector to make it a column vector, using the transpose operator, `'`.

```
EDU» b = [-113, 0.003, 0.001, 112.996, 112.998, 0.003, .001, -112.999]';
```

c. Form the **p** matrix of the observation errors. It is a transposed row vector.

```
EDU» p = [0.104, 0.112, 0.096, 0.12, 0.112, 0.088, 0.096, 0.104]';
```

d. Form the **W** (weight) matrix of observations errors.
This is a diagonal matrix where the weight of an observation, $\mathbf{W}_{ii}$, is generated from $\mathbf{W} = \mathbf{1}/\sigma_i^2$.
The function `diag(p)` turns the **p** vector column into a square matrix with **p** on the diagonal.
`diag(p)^2` squares the error on the diagonal and
`inv(diag(p)^2)` produces the required weight matrix **W** directly.

```
EDU» W = inv(diag(p)^2)
```
Output format edited for this page.

```
W =
 92.4556         0         0         0         0         0         0         0
       0   79.7194         0         0         0         0         0         0
       0         0  108.5069         0         0         0         0         0
       0         0         0   69.4444         0         0         0         0
       0         0         0         0   79.7194         0         0         0
       0         0         0         0         0  129.1322         0         0
       0         0         0         0         0         0  108.5069         0
       0         0         0         0         0         0         0   92.4556
```

e. Weighted observations in surveying can be reduced to the matrix form:

```
WAx = Wb + v
```

x	a vector of terms to be computed;
b	a vector relating to the observations;
A	a **matrix** of coefficients;
v	a vector of residuals, to be a minimum and
W	a **matrix** of **uncorrelated** observation precisions.

f. To solve the parameters vector, **x**, the matrix form is first normalized:
$\mathbf{A}^T\mathbf{W}\mathbf{A}\mathbf{x} = \quad \mathbf{A}^T\mathbf{W}\mathbf{b} + \mathbf{v}$
Calculate the six unknown transformation parameters, $\mathbf{x}_1$ to $\mathbf{x}_6$ being elements **a** to **f**.

If it's not checked it's wrong.

The left-division operator, \, can solve **Ax** = **b** for a unique solution **(x=A\b)**. It is slightly more accurate, using the LU factorisation approach, and less computationally intensive, than **x=inv(A)*b** which is the standard way of solving for **x**.

Consider now our normalised observation equations, using the \ operator.

For this exercise we will form matrices by solving the normalized **A** and **b** matrices and assign labels ("names") to the result.

```
EDU» N=A'*W*A;       | N = A^TWA   Output suppressed
EDU» Cb=A'*W*b;      | Cb = A^TWb  Output suppressed
```

g. We now have the solution in the form **Nx** = **Cb** + **v** which leads to the solution of **x** by the left division operator

```
EDU>> x=N\Cb         | Equivalent to     EDU» x=(A'*W*A)\(A'*W*b).
x =
    25.3715          | a
     0.8222          | b
  -137.1827          | c
    -0.8100          | d
    25.4020          | e
  -150.7236          | f
```

h. Now it is time to calculate:

a) **b_adj** = **Ax**; the adjusted coordinates of the fiducials;

b) **v** = **b_adj-b;** residuals;

c) **Σv**; the sum of the residuals.

Following these calculations an analysis of the results will be made.

```
EDU» b_adj=A*x       | % Adjusted fiducial coordinates
b_adj =
  -112.8986          | X1
     0.0524          | Y1
    -0.0853          | X2
   112.9363          | Y2
   113.1152          | X3
     0.0334          | Y3
    -0.0852          | X4
  -113.0415          | Y4
EDU» v=b_adj-b       | % Residual = adjusted - true
v =
     0.1014          | vX1
     0.0494          | vY1
    -0.0863          | vX2
    -0.0567          | vY2
     0.1172          | vX3
     0.0304          | vY3
    -0.0862          | vX4
    -0.0425          | vY4
EDU» sum_v=sum(v)    | % Sum residuals
sum_v =
     0.0267
```

i. Analyze the standard deviation of the transformation parameters, the **x** vector. Standard error Wolf p212 (12.15).

1) Find **reference standard deviation of observations:**

a) **SE** = $\mathbf{v^TWv}$ the standard error of the observations

b) degrees of freedom (dof, o) = (no. observations – no. parameters) of observation. matrix

c) Reference standard variance of observations, σ^2o,= SE/dof

d) Reference standard deviation, σo = $\sqrt{\sigma^2 o}$,

```
EDU» SE=v'*W*v        % Standard error Wolf p212 (12.15)
```

```
SE =
    4.3656
EDU» [m,n]=size(A);   % Size of A matrix; m rows, n columns
EDU» dof=m-n          % Degrees of freedom
dof =
     2
EDU» Ref_SV=(SE/dof)       % Reference standard variance of observations
Ref_SV =               | σ²o,  Used in the formation of the standard deviation of
    2.1828             |       the calculated coordinates.
EDU» Ref_SD=sqrt(SV)       % Reference standard deviation of observations
Ref_SD =
    1.4774             | σo
```

Broadly speaking the reference standard deviation, $\sigma_O = 1.48$, shows that there is no evidence of gross error in the observations. Schofield & Breach, p251 (7.9.2)

2) Find **variance-covariance matrix** of the observations (**VCV$_{Obs}$**) from the square root of the elements of the **inverted** weighted set of OBSERVATIONS.

VCV$_{Obs}$ $= \sqrt{(\text{abs}(\mathbf{A^TWA})^{-1})}$. Remember $\mathbf{N} = \mathbf{A^TWA}$ Wolf p224, (13.21)

```
EDU» format short
EDU» Sqrt_VCV_obs_parameters=sqrt(abs(inv(N))) % VCV matrix
Sqrt_VCV_obs_parameters =
    0.0171    0.0015    0.0387         0         0         0
    0.0015    0.0153    0.0378         0         0         0
    0.0387    0.0378    0.1375         0         0         0
         0         0         0    0.0158    0.0026    0.0376
         0         0         0    0.0026    0.0177    0.0424
         0         0         0    0.0376    0.0424    0.1464
```

3) Find the estimated standard deviation of the elements of the observed parameters through the expression:

$S_i = S_o\sqrt{q_{xx}}$ Wolf p224, (13.24)

```
EDU»SD_obs_parameters=diag(Sqrt_VCV_obs_parameters*Ref_SD)
SD_obs_parameters =
    0.0253     | ±σa
    0.0226     | ±σb
    0.2031     | ±σc
    0.0234     | ±σd
    0.0262     | ±σe
    0.2163     | ±σf
```

j. Analyze the standard deviation of the adjusted coordinates of the fiducial points, the **b_adj** vector.

1) The **reference standard deviation of observations** has been determined

a) Display it again, σ_O,

```
EDU» Ref_SD          % Call Reference standard deviation of observations
Ref_SD =
    1.4774           | σo
```

2) **Find** variance-covariance matrix of the weighted set of **calculations**

$\boldsymbol{\sigma^2_{calc}} = \mathbf{AN^{-1}A^T}$

```
EDU» VCV_b_adj=A*inv(N)*A'; % Variance-covariance matrix of the 8 adjusted
                              fiducials
EDU» Sqrt_VCV_b_adj=sqrt(abs(VCV_b_adj)) % Sqrt of VCV of adjusted fiducials
Sqrt_VCV_b_adj =
  0.0895        0   0.0489        0   0.0570        0   0.0489        0
       0   0.0953        0   0.0631        0   0.0462        0   0.0546
  0.0489        0   0.0847        0   0.0526        0   0.0451        0
       0   0.0631        0   0.0992        0   0.0495        0   0.0585
  0.0570        0   0.0526        0   0.0938        0   0.0525        0
       0   0.0462        0   0.0495        0   0.0802        0   0.0428
  0.0489        0   0.0451        0   0.0525        0   0.0848        0
       0   0.0546        0   0.0585        0   0.0428        0   0.0908
```

3) Find the estimated standard deviation of the elements of the adjusted coordinates through the expression $S_i = S_o\sqrt{q_{xx}}$ Wolf p224, (13.24)

```
EDU» SD_b_adj=diag(Sqrt_VCV_b_adj*Ref_SD)% Standard deviation of adjusted fiducials
SD_b_adj =          Standard deviation of adjusted fiducials
    0.1322        | ±σX1
    0.1408        | ±σY1
    0.1252        | ±σX2
    0.1465        | ±σY2
    0.1385        | ±σX3
    0.1185        | ±σY3
    0.1252        | ±σX4
    0.1342        | ±σY4
```

k. To calculate the adjusted coordinates of the pass points 10 & 11. We to form the observation equations and use the transformation parameters, **x**.

1) Form the observation equations for 10 & 11.

```
EDU» App=[1.746, 9.354, 1, 0, 0, 0   % Observation equation of points 10 & 11
0, 0, 0, 1.746, 9.354, 1
5.329, 9.463, 1, 0, 0, 0
0, 0, 0, 5.329, 9.463, 1];
```

2) Calculate the pass point coordinates in the fiducial system.

```
EDU» b_adj_pp=App*x  % Adjusted pass point coordinates of 10 & 11
b_adj_pp =
  -85.1932        | X10
   85.4703        | Y10
    5.8025        | X11
   85.3371        | Y11
```

l. Analyze the standard deviation of the adjusted coordinates of the pass points, the **b_adj_pp** vector.

1) The **reference standard deviation of observations** has been determined

a) Display it again, σo,

```
EDU» Ref_SD          % Call Reference standard deviation of observations
Ref_SD =
    1.4774        | σo
```

2) Find variance-covariance matrix of the weighted set of **calculations**

$$\sigma^2_{calc} = \mathbf{AN^{-1}A^T}$$

```
EDU» VCV_b_adj=A*inv(N)*A';  % Variance-covariance matrix of the 8 adjusted
                               fiducials
EDU» Sqrt_VCV_b_adj_pp=sqrt(abs(VCV_b_adj_pp))  % Sqrt of VCV of adjusted
                                                  pass points
Sqrt_VCV_b_adj_pp =
   0.0180        0   0.0106        0
        0   0.0238        0   0.0159
   0.0106        0   0.0114        0
        0   0.0159        0   0.0151
```

3) Find the estimated standard deviation of the elements of the adjusted coordinates through the expression

$S_i = S_o\sqrt{q_{xx}}$ Wolf p224, (13.24)

```
EDU» SD_b_adj=diag(Sqrt_VCV_b_adj_pp*Ref_SD)  % Standard deviation of adjusted
                                                fiducials
SD_b_adj_pp =                                   Standard deviation of adjusted fiducials
    0.1341        | ±σX10
    0.1543        | ±σY10
    0.1069        | ±σX11
    0.1228        | ±σY11
```

m. Present the results in a table.

The solution for the adjustment is shown on the right-hand side.

Point	X orig	Y orig	x obs	y obs	σx obs	σy obs	**X adj**	**Y adj**	**±σx adj**	**±σy adj**
1	-113.000	0.003	0.764	5.960	0.104	0.112	**-112.899**	**0.052**	**0.132**	**0.141**
2	0.001	112.993	5.062	10.541	0.096	0.120	**-0.085**	**112.936**	**0.125**	**0.147**
3	112.998	0.003	9.663	6.243	0.112	0.088	**113.115**	**0.033**	**0.139**	**0.118**
4	0.001	-112.999	5.350	1.654	0.096	0.104	**-0.085**	**-113.042**	**0.125**	**0.134**
10			1.746	9.354			**-85.193**	**85.470**	**0.134**	**0.154**
11			5.329	9.463			**5.803**	**85.337**	**0.107**	**0.123**

17.7 Affine Coordinate Transformation with Weights

The conformal transformation merely translates one set of coordinates to another by a scaling, shift and rotation. The basic SHAPE of the figure remains; the transformation is CONFORMAL.

The next transformation is designed to work where there are different scales in X and Y. It is popularly used in photogrammetry the interior orientation to account for differential shrinkage in photo film images, or in non-conformal maps.

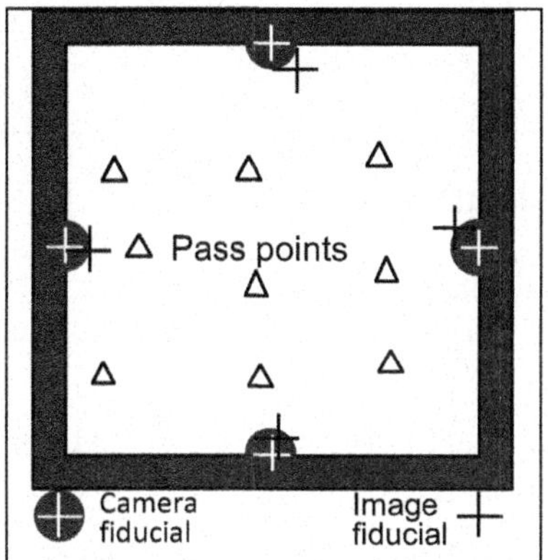

Figure 17-5 Camera/image marks.

17.7.1 Developing the Affine Transformation

An affine transformation is used in photogrammetry to compensate for uneven film shrinkage derived from the camera fiducial mark system and arbitrarily measured fiducial marks on the film. (See Figure 17-5). It makes allowances for two different scale factors; one in the X, the other in the Y direction.

The camera has four (or 8) fiducial marks that provides an overdetermined solution as only 3 fiducials are needed to generate the 6 parameters of the affine transformation: 2 scaling and 1 skew angle, 1 rotation, and 2 translations. The transformation allows other points (pass points, joining adjacent overlapping photos) measured on the film to be coordinated in the camera fiducial system.

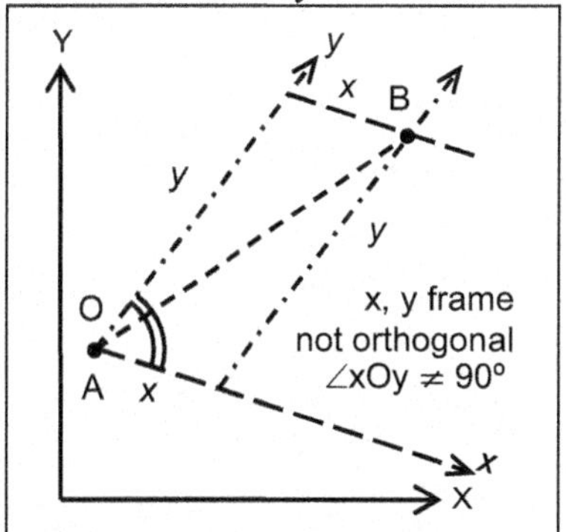

Figure 17-6 Non-orthogonal image

The 6-parameter affine transformation equation is:

Affine transform parameters	$E = ax + by + c$ $N = dx + ey + f$	Eqn 17-2

In matrix form:

	$\begin{bmatrix} X \\ Y \end{bmatrix} = \begin{bmatrix} a & b \\ d & e \end{bmatrix} \begin{bmatrix} x \\ y \end{bmatrix} + \begin{bmatrix} c \\ f \end{bmatrix}$	Eqn 17-3

The coordinate reference frame x, y of points on the image is **non-orthogonal** (not at right angles), see Figure 17-6.

Start by assuming the points are measured **along** the x and y (non-orthogonal) axes.

1. Make the figure orthogonal by *skewing* the Oy axis around to Oy' to be perpendicular to the Ox axis. Call the skew angle α, positive anticlockwise. The frame is now $x'Oy'$. Project the coordinates of B in the orthogonal frame, including s_x and s_y as scale factors for the coordinates. See Figure 17-7.

   $x'_B = s_x x + s_y(y \sin \alpha)$

   $y'_B = s_x x + s_y(y \cos \alpha)$ but, starting at O, $x = 0$.

   $y'_B = s_x 0 + s_y(y \cos \alpha)$

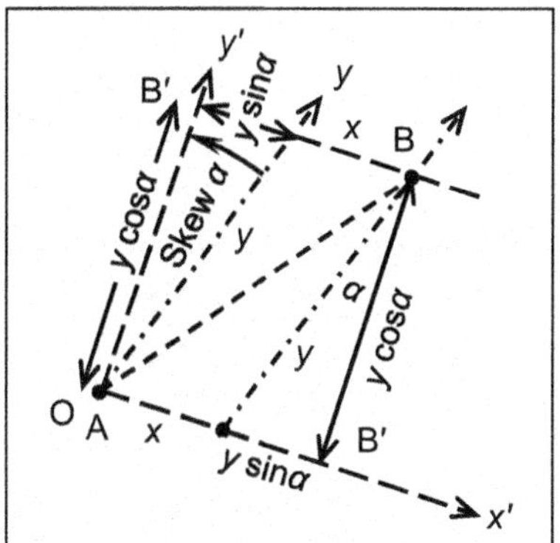

Figure 17-7 De-skew the y axis.

Matrix form:	$\begin{bmatrix} x' \\ y' \end{bmatrix} = \begin{bmatrix} s_x & s_y \sin\alpha \\ 0 & s_y \cos\alpha \end{bmatrix} \begin{bmatrix} x \\ y \end{bmatrix}$	Eqn 17-4

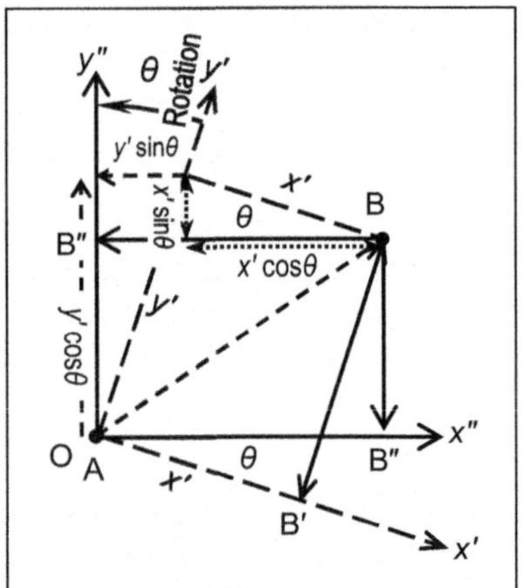

Figure 17-8 Rotate axes

2. Now rotate the figure in the $x'Oy'$ frame into the $x''Oy''$ system. This is parallel the X, Y system but shifted (translated) from its origin. See Figure 17-8.
   $x''_B = x'\cos\theta + y'\sin\theta$
   $y''_B = -x'\sin\theta + y'\cos\theta$

Matrix form:	$\begin{bmatrix} x'' \\ y'' \end{bmatrix} = \begin{bmatrix} \cos\theta & \sin\theta \\ -\sin\theta & \cos\theta \end{bmatrix} \begin{bmatrix} x' \\ y' \end{bmatrix}$	Eqn 17-5

   This is the normal rotation process used in conformal transformations. We have turned the skewed projection into a conformal projection via two matrix operation, scaling and de-skewing.
3. Next shift the figure in the $x''Oy''$ frame to the XY system. The two systems are parallel. Call the shifts X_o and Y_o. See Figure 17-9.
   $X_B = x''_B + X_0$
   $Y_B = y''_B + Y_0$

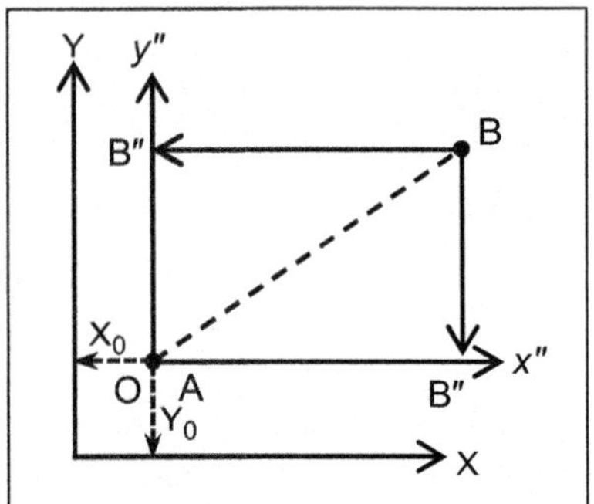

Figure 17-9 Shift to origin.

Matrix form:	$\begin{bmatrix} X \\ Y \end{bmatrix} = \begin{bmatrix} x'' \\ y'' \end{bmatrix} + \begin{bmatrix} X_0 \\ Y_0 \end{bmatrix}$	Eqn 17-6

4. Combining Eqn 17-3, Eqn 17-4 and Eqn 17-5, and renaming X_0 and Y_0 as c and f: (partitioned for clarity), results in the matrix:

	$\begin{bmatrix} X \\ Y \end{bmatrix} = \begin{bmatrix} \cos\theta & \sin\theta \\ -\sin\theta & \cos\theta \end{bmatrix} \begin{bmatrix} s_x & s_y\sin\alpha \\ 0 & s_y\cos\alpha \end{bmatrix} \begin{bmatrix} x \\ y \end{bmatrix} + \begin{bmatrix} c \\ f \end{bmatrix}$ $= \begin{bmatrix} s_x\cos\theta & s_y\cos\theta\sin\alpha + s_y\sin\theta\cos\alpha \\ -s_x\sin\theta & -s_y\sin\theta\sin\alpha + s_y\cos\theta\cos\alpha \end{bmatrix} \begin{bmatrix} x \\ y \end{bmatrix} + \begin{bmatrix} c \\ f \end{bmatrix}$	Eqn 17-7

5. Substituting elements of Eqn 17-7 into Eqn 17-2 gives the coefficients:

$$a = s_x \cos\theta \quad (17.1)$$

$$b = s_y(\cos\theta\sin\alpha + \sin\theta\cos\alpha) = s_y\sin(\theta+\alpha) \quad (17.2)$$

$$d = -s_x\sin\theta \quad (17.3)$$

$$e = s_y(\cos\theta\sin\alpha - \sin\theta\cos\alpha) = s_y\cos(\theta+\alpha) \quad (17.4)$$

From these elements we deduce: 2 scales, 1 rotation angle and 1 skew angle.

$s_x = \sqrt{a^2+b^2}$, scale in x,

$s_y = \sqrt{b^2+e^2}$, scale in y,

$\tan\theta = \frac{-d}{a}$, or $\cos\theta = \frac{a}{s_x}$, rotation angle, θ

$\tan(\theta+\alpha) = \frac{b}{e}$, or $\cos(\theta+\alpha) = \frac{e}{s_y}$, sum of rotation and skew, derive skew angle, α.

17.7.2 Dealing with the Skew Angle

The derivation started with the x and y coordinates being measured along their non-orthogonal axes. Coordinate measuring systems are **orthogonal** in x and y so the skew angle $\alpha = 0$.

Eqn 17-7 reduces to (because $\sin\alpha = 0$, $\cos\alpha = 1$):

Reduced Eqn 17-8.	$\begin{bmatrix} X \\ Y \end{bmatrix} = \begin{bmatrix} s_x\cos\theta & s_y\sin\theta \\ -s_x\sin\theta & s_y\cos\theta \end{bmatrix} \begin{bmatrix} x \\ y \end{bmatrix} + \begin{bmatrix} c \\ f \end{bmatrix}$	Eqn 17-8

6. Substituting the elements of Eqn 17-8 into Eqn 17-2 gives the coefficients:
   $a = s_x\cos\theta$ (17.5)
   $b = s_y\sin\theta$ (17.6)
   $d = -s_x\sin\theta$ (17.7)
   $e = s_y\cos\theta$ (17.8)
7. From these elements we deduce: 2 scales, 1 rotation angle, skew angle = 0.

$$s_x = \sqrt{a^2 + d^2},$$

$$s_y = \sqrt{b^2 + e^2},$$

$$\tan\theta = \frac{b}{a}$$

The shift elements X_0 and Y_0 are represented by ***c*** and ***f***.

The mathematical model is of the form: $\begin{matrix} ax + by + c = X \\ dx + ey + f = Y \end{matrix}$

in the matrix form: $\begin{bmatrix} X \\ Y \end{bmatrix} = \begin{bmatrix} a & b \\ d & e \end{bmatrix} \begin{bmatrix} x \\ y \end{bmatrix} + \begin{bmatrix} c \\ f \end{bmatrix}$

and, because it's overdetermined: $\begin{bmatrix} X \\ Y \end{bmatrix} = \begin{bmatrix} X \\ Y \end{bmatrix} + \begin{bmatrix} V_x \\ V_y \end{bmatrix}$

It can be expressed in the matrix form **Ax=b + v**.

In the **affine** transformation there are three (3) common points required for a unique solution. This provides the 6 parameters, *a*, *b*, *c*, *d*, *e* and *f*.

The observation equations for one point are:

$ax + by + c = X + Vx$

$dx + ey + f = Y + Vy.$

Expanding the equations to include all parameters in each line, they become:

	$ax + by + 1c + 0d + 0e + 0f = X + V_x$ $0a + 0b + 0c + dx + ey + 1f = Y + Vy$	Eqn 17-9

The standard deviations of the observations, if known, can be included as a vector of precisions (**p**), generating a **weight** matrix of the observations.

Noting that naming conventions can change from author to author, in general, observations in surveying can be reduced to the matrix form: **AWx = Wb + v**

- **A** a **matrix** of coefficients;
- **p** a **vector** of observation precisions, the standard deviation;
- **W** a diagonal weight matrix of the inverse of the observation variances ($W = {}^1/_{p^2}$).
- **x** a vector of terms to be computed, including coordinates;
- **b** a vector relating to the observations;
- **v** a vector of residuals, to be a minimum.

Taking the original affine transformation and recognizing there ARE a minimum six (6) transformation parameters, the equations are: the **A** matrix of coefficients and the **p** vector.

$Observations = \begin{bmatrix} a_{x1} + b_{y1} + c = X_1 + v_{X1} \\ d_{x1} + e_{y1} + f = Y_1 + v_{Y1} \\ a_{x2} + b_{y2} + c = X_2 + v_{X2} \\ d_{x2} + e_{y2} + f = Y_2 + v_{Y2} \\ a_{x3} + b_{y3} + c = X_3 + v_{X3} \\ d_{x3} + e_{y3} + f = Y_3 + v_{Y3} \end{bmatrix} \quad \mathbf{A} = \begin{bmatrix} x_1 & y_1 & 1 & 0 & 0 & 0 \\ 0 & 0 & 0 & x_1 & y_1 & 1 \\ x_2 & y_2 & 1 & 0 & 0 & 0 \\ 0 & 0 & 0 & x_2 & y_2 & 1 \\ x_3 & y_3 & 1 & 0 & 0 & 0 \\ 0 & 0 & 0 & x_3 & y_3 & 1 \end{bmatrix} \quad \mathbf{p} = \begin{bmatrix} p_{X1} \\ p_{Y1} \\ p_{X2} \\ p_{Y2} \\ p_{X3} \\ p_{Y3} \end{bmatrix}$	Eqn 17-10

The form of the least squares solution is as shown. Because this solution is unique, using three points shown here, there will be no residuals.

And thus the **v** vector = 0.

$$\mathbf{A}=\begin{bmatrix} x_1 & y_1 & 1 & 0 & 0 & 0\\ 0 & 0 & 0 & x_1 & y_1 & 1\\ x_2 & y_2 & 1 & 0 & 0 & 0\\ 0 & 0 & 0 & x_2 & y_2 & 1\\ x_3 & y_3 & 1 & 0 & 0 & 0\\ 0 & 0 & 0 & x_3 & y_3 & 1\end{bmatrix}\quad \mathbf{W}=\begin{bmatrix} W_{x1} & & & & & \\ & W_{y1} & & & & \\ & & W_{x2} & & & \\ & & & W_{y2} & & \\ & & & & W_{x3} & \\ & & & & & W_{y3}\end{bmatrix}\quad \mathbf{x}=\begin{bmatrix} a\\ b\\ c\\ d\\ e\\ f\end{bmatrix}\quad \mathbf{b}=\begin{bmatrix} X_1\\ Y_1\\ X_2\\ Y_2\\ X_3\\ Y_3\end{bmatrix} \qquad \text{Eqn 17-11}$$

This then concludes the derivation of the affine transformation used in the Worked Example 19-4 in § 17.6, page 282.

17.8 MATLAB script used for this Affine adjustment.

```
% Affine transformation of 4 fiducial points and two pass points on a photograph
%
% Form of Least Squares solution
%       Ax = b + v
% For weighted matrix:
%       WAx = Wb + v
% Solution of over-determined matrix requires normalising:
%       A'WAx = A'Wb + v
%
% Initialise and form observation equations
x=[zeros(1,6)]';                    % Initialise transformation parameters vector
A=[.764 5.96 1 0 0 0                           % Observation equations of fiducials
0 0 0 .764 5.96 1
5.062 10.541 1 0 0 0
0 0 0 5.062 10.541 1
9.663 6.243 1 0 0 0
0 0 0 9.663 6.243 1
5.35 1.654 1 0 0 0
0 0 0 5.35 1.654 1];
b=[-113 .003 .001 112.993 112.998 .003 .001 -112.999]'; % Fiducial coordinates
p=[.104 .112 .096 .12 .112 .088 .096 .104]';            % Sdev of observations
%
% Form weight matrix
W=inv(diag(p)^2);                              % Weight matrix W = 1/p^2 on diagonal
N=A'*W*A;                                        % Normalized weighted observations
Cb=A'*W*b;                                          % Normalized weighted fiducials
%
% Solve for transformation parameters
x=N\Cb                     % Transformation vector using left-division operator
%
% Calculate transformed fiducial coordinates
b_adj=A*x                                           % Adjusted fiducial coordinates
%
% Calculate residuals: adjusted - true
v=b_adj-b                                              % Residual = adjusted - true
sum_v=sum(v)                                                       % Sum residuals
%
% Check transformation, is the Reference Standard Deviation acceptable?
% Reference Standard deviation
SE=v'*W*v                                                         % Standard error
[m,n]=size(A);                      % Dimension of A matrix; m rows, n columns
dof=m-n                                                       % Degrees of freedom
Ref_SV=SE/dof                                          % Reference Standard variance
Ref_SD=sqrt(Ref_SV)                                   % Reference Standard deviation

% Analysis of transformation parameters
VCV_obs_parameters=inv(N)                              % Variance-covariance matrix
                                         of 6 observed parameters. (N = A'*W*A)
Sqrt_VCV_obs_parameters=sqrt(abs(VCV_obs_parameters))        % VCV matrix of obs;
                                                      avoid complex number results
SD_obs_parameters=diag(Sqrt_VCV_obs_parameters*Ref_SD)  % Sdev of the observed
                                                         transformation parameters

% Analysis of adjusted fiducials (8 points)
VCV_b_adj=A*inv(N)*A';                     % VCV matrix of the 8 adjusted fiducials
Sqrt_VCV_b_adj=sqrt(abs(VCV_b_adj))                         % VCV adjusted fiducials
SD_b_adj=diag(Sqrt_VCV_b_adj*Ref_SD)                    % Sdev of adjusted fiducials
```

```
% Form observation equation for the pass points 10 & 11
App=[1.746 9.354 1 0 0 0                    % Observation equation of points 10 & 11
0 0 0 1.746 9.354 1
5.329 9.463 1 0 0 0
0 0 0 5.329 9.463 1];
%
% and transform them into the fiducial system
b_pp_adj=App*x                                   % Adjusted coordinates of 10 & 11
%
% Analysis of adjusted passpoints
VCV_obs_pp=App*inv(N)*App';   % VCV matrix of the adjusted passpoints 10 & 11
Variance_obs_pp=VCV_obs_pp*Ref_SV                          %Variance of 10 & 11
SD_obs_pp=diag(sqrt(abs(Variance_obs_pp)))  % Sdev adjusted passpoints 10 & 11
% END
```

17.9 References for Chapter 17.

Ghilani, C.D., Adjustment computations. Spatial data analysis. Fifth edition, John Wiley & Sons, Inc. 2010. ISBN 978-0-4701-46491-5

Hahn Brian H. and D.T Valentine. Essential MATLAB for Engineers and Scientists, Fourth Edition. Elsevier.

Kirby J. F., 2010. Coordinate and mapping Systems 282/582. Department of Spatial Sciences, Curtin University.

Palm William J.III. A Concise Introduction to MATLAB. McGraw-Hill Higher Education.

Schofield, W and M Breach. Engineering Surveying, Sixth edition. Elsevier. 2007. ISBN 978-0-7506-6949-8

Uren J. and Price W. F., 2006. Surveying for Engineers, 4th edition. Palgrave MacMillan.

Chapter 18 SURVEY FORMULÆ AND APPLICATION OF CALCULATORS

An assembly of the plane trigonometric formulæ used to this portion of the book. Metric imperial measure conversions are included. There is also an expansion of the published functions of algebraic calculators pertinent to normal survey calculations.

18.1 SIN Rules

$$\frac{a}{\sin A} = \frac{b}{\sin B} = \frac{c}{\sin C}$$

$$\frac{\sin A}{a} = \frac{\sin B}{b} = \frac{\sin C}{c}$$

$$\frac{a+b+c}{\sin A + \sin B + \sin C}$$

18.2 COS Rules

$$\cos A = \frac{b^2 + c^2 - a^2}{2bc}$$

$$a = \sqrt{b^2 + c^2 - 2bc\cos A}$$

$$\cos(B-C) = \left(\frac{4\,\text{Area}\sin A}{a^2}\right) - \cos A$$

18.3 TAN Rule

$$\left(\frac{A-B}{2}\right) = \arctan\left(\left(\frac{a-b}{a+b}\right)\cot\left(\frac{C}{2}\right)\right)$$

$$\left(\frac{A+B}{2}\right) = 90 - \left(\frac{C}{2}\right)$$

18.4 Perimeter Rules for angles

$$S = \frac{a+b+c}{2} \text{ semi-perimeter}$$

$$\sin\frac{A}{2} = \sqrt{\frac{(S-b)(S-c)}{bc}}$$

$$\cos\frac{A}{2} = \sqrt{\frac{S(S-a)}{bc}}$$

$$\tan\frac{A}{2} = \sqrt{\frac{(S-b)(S-c)}{(S(S-a)}}$$

$$\text{Area} = \sqrt{S(S-a)(S-b)(S-c)}$$

18.5 Area Rules

$$\text{Area} = \tfrac{1}{2}bc\sin A$$

$$\text{Area} = \frac{a^2\sin B\sin C}{2\sin A}$$

$$\text{Area} = \frac{a^2}{2\,(\cot B + \cot C)}$$

$$\text{Side } a = \sin A\sqrt{\frac{2\text{Area}}{\sin A\sin B\sin C}}$$

18.6 Elements of part of a circle. Chapter 8

Δ = deflection angle at centre

Δ^r = deflection angle in RADIANS

R = Radius of circle

Tangent distance (TD) = R·Tan($^{\Delta}/_{2}$)

Radius (R) = TD/Tan($^{\Delta}/_{2}$)

Arc length (S) = R·Δ^r

Crown Secant = R·Sec($^{\Delta}/_{2}$) – R

Long Chord= 2R·Sin($^{\Delta}/_{2}$)

Chord = 2R·Sin(δ)

δ = deflection, tangent to arc

$\delta = {}^{\Delta}/_{2}$

$$R_2 = \frac{S^2 - R_1^2}{2(S\cos\alpha - R_1)}$$ **§ 14.9.3**

18.7 Area of parts of a circle. Chapter 6

Sector = $\tfrac{1}{2}\cdot R^2\cdot\Delta^r$

Segment = $\tfrac{1}{2}\cdot R^2\cdot(\Delta^r - \text{Sin}(\Delta))$

$$\text{Truncation} = R^2\left(\tan\left(\frac{\Delta}{2}\right) - \frac{\Delta^r}{2}\right)$$ **W.Ex 8.5**

18.8 Systematic Errors

§ 11.6.1, page 174

$$u = ax + by, \Delta u = a\Delta x + b\Delta y$$

$$u = F(x,y), \Delta u = \frac{\partial F}{\partial x}\Delta x + \frac{\partial F}{\partial y}\Delta y$$

18.9 Random Errors

§ 11.6.2, page 174.

$$u = ax + by,\ \sigma_u{}^2 = a^2\sigma_x{}^2 + b^2\sigma_y{}^2$$

$$u = F(x,y),\ \sigma_u{}^2 = \left(\frac{\partial F}{\partial x}\right)^2\sigma_x{}^2 + \left(\frac{\partial F}{\partial y}\right)^2\sigma_y{}^2$$

J. Walker and J. Awange, *Surveying Computations*,
https://doi.org/10.1007/978-3-031-94382-9_18

If it's not checked it's wrong.

18.10 Intersections

Chapter 7, page 99

Intersection by Bearings B_2 & B_3

Clockwise cyclic order of points 1, 2 & 3.

$$N_1 = \frac{E_3 - E_2 + N_2 \tan B_2 - N_3 \tan B_3}{\tan B_2 - \tan B_3}$$

$$E_1 = (N_1 - N_2)\tan B_2 + E_2$$

Intersection by Angles α & β

Point 1 is the intersection

$$N_1 = \frac{E_2 - E_3 + N_2 \cot\beta + N_3 \cot\alpha}{\cot\alpha + \cot\beta}$$

$$E_1 = \frac{N_3 - N_2 + E_2 \cot\beta + E_3 \cot\alpha}{\cot\alpha + \cot\beta}$$

Intersection by Distances H_1 baseline distance, H_2, H_3 measured to Pt 1. This is the standard CAD distance/distance intersection computation. Position of point 1 with respect to the line 2 to 3 decides the intersect coordinate. Refer Chapter 7 and Figure 18-1.

$E_1 = E_2 + p\sin\theta_{23} \pm h\cos\theta_{23}$ Point 1 left/right line 2-3: use $\pm$ (plus to left/ minus to right)

$N_1 = N_2 + p\cos\theta_{23} \mp h\sin\theta_{23}$ Point 1 left/right line 2-3: use $\mp$ (minus to left/ plus to right)

where H_1 =

$$H_1 = \sqrt{(E_3 - E_2)^2 + (N_3 - N_2)^2},\ p = \frac{H_3^2 + H_1^2 - H_2^2}{2H_1},\ h = \sqrt{h_3^2 - p^2}$$

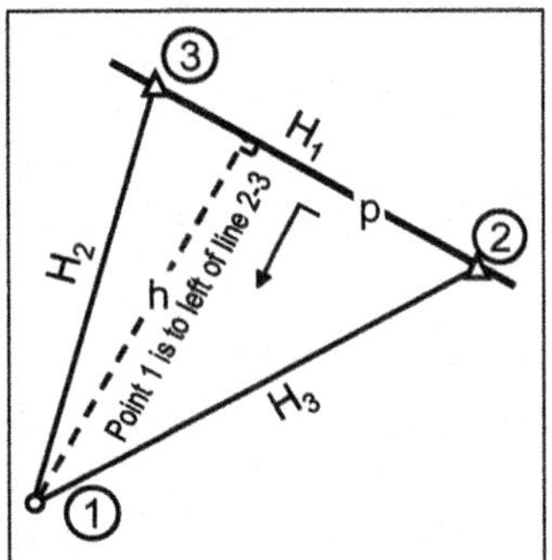

Figure 18-1 Left or right?

Alternately, recognising that:

$$\sin\theta_{23} = \left(\frac{E_3 - E_2}{H_1}\right) = a \text{ and } \cos\theta_{23} = \left(\frac{N_3 - N_2}{H_1}\right) = b$$

then the formulæ can be simplified to:

$E_1 = E_2 + \mathbf{a}p \pm \mathbf{b}h$ Point 1 left/right line 2-3: use $\pm$

$N_1 = N_2 + \mathbf{b}p \mp \mathbf{a}h$ Point 1 left/right line 2-3: use $\mp$

18.11 Trigonometric functions.

SOH CAH TOA

With reference to Figure 18-2, 7 functions and 3 inverse functions can be defined as follows:

Sine	$\sin A$	$= \frac{a}{c}$	$\frac{opposite}{hypotenuse}$	
Cosine	$\cos A$	$= \frac{b}{c}$	$\frac{adjacent}{hypotenuse}$	
Tangent	$\tan A$	$= \frac{a}{b}$	$\frac{opposite}{adjacent}$	$= \frac{\sin A}{\cos A}$
Cotangent	$\cot A$	$= \frac{b}{a}$	$\frac{adjacent}{opposite}$	$= \frac{1}{\tan A}$
Secant	$\sec A$	$= \frac{c}{b}$	$\frac{hypotenuse}{adjacent}$	$= \frac{1}{\cos A}$
Cosecant	$\csc A$	$= \frac{c}{a}$	$\frac{hypotenuse}{opposite}$	$= \frac{1}{\sin A}$

(also written as cosec$A$)

Versine	vers$A$	$= (1 - \cos A)$	$= 1 - \frac{b}{c}$

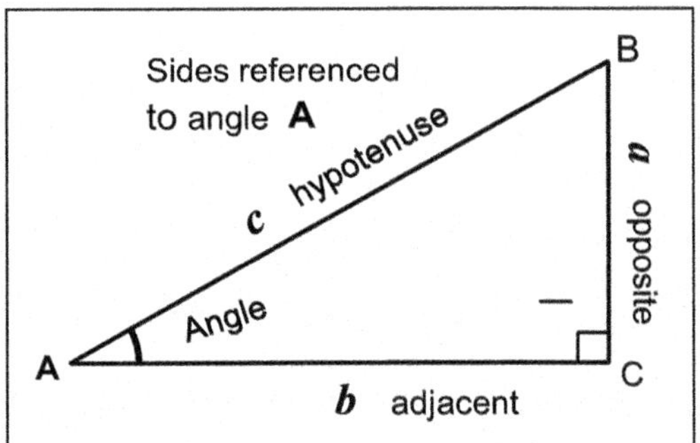

Figure 18-2 The right triangle. SOH CAH TOA.

The inverse of a function is generally noted as, e.g., ASIN, ACOS, ATAN, or SIN^{-1}, COS^{-1}, TAN^{-1}.

Arc Sine	$\arcsin\left(\frac{a}{c}\right) = A$	the angle whose Sine is $\frac{a}{c}$
Arc Cosine	$\arccos\left(\frac{b}{c}\right) = A$	the angle whose Cosine is $\frac{b}{c}$
Arc Tangent	$\arctan\left(\frac{a}{b}\right) = A$	the angle whose Tangent is $\frac{a}{b}$

18.12 Exact linear conversions

12 inches = 1 foot
3 feet = 1 yard
66 feet = 22 yards = 1 chain
100 links = 1 chain
1 link = 7.92 inches
1 inch = 25.4mm
1 U.S. survey foot = 1200/3937 = 0.304800609601…m
Links × 0.201168 = metres
Metres ÷ 0.201168 = links
Feet × 0.3048 = metres
Metres ÷ 0.3048 = feet
U.S. survey feet × (1200/3937) = metres (US legislation)

18.13 Conversions used in cadastral surveying.

5 links = 1.00584m = 1.006m
1 chain = 20.1168 m = 20.117m
80 chain = 1 mile (statute) = 1760 yards = 5280 feet = 1609.344m

1 acre = 10 sq chains = 4840 sq yards = 43,560 sq feet
1 acre = 4 roods = 160 perches
1 rood = 40 perches
1 acre = 4046.8564224m^2 = 0.4046856Ha (0.4047Ha to m^2 level)

1 hectare (Ha) = 10,000m^2
1 Ha = 107,639.1042 sq feet = 11,959.9005 sq yards
1 Ha = 2.4710538acres = 2 acres, 1 rood, 35.37perch

18.14 Circular measure. (Using the superscript ^{r} for radians)

$\pi^r = 180°\pi = 3.141\ 592\ 653\ 589\ 79$ approx(355/113)

$1° =$ 0.017 453 292 519^r
$1' =$ 0.000 290 888 209^r
$1'' =$ 0.000 004 848 137^r 4.8481E^{-6}

$1^r =$ 57.295 7759°
$1^r =$ 3437.7467′
$1^r =$ 206264.806 2471″ (use **206 265″**)

$1° =$ 60′
$1' =$ 60″
$1° =$ 3600″

18.15 The BIG 3 numbers

These are the only numbers you NEED to remember:

0.3048 feet to/from metres
0.201168 links to/from metres
206265 seconds arc/radians

18.16 Survey Calculations on HP 10s+, HP300s+, Casio, Sharp

18.17 Introduction

The HP 10s+ Scientific Calculator (Figure 18-3) is one of the calculators specified for use in engineering examinations. Consult your user manual, not particularly informative, on the basic functions of the calculator.

A new model, the HP300s+, has been introduced. It features a 3-line display. The screen may not provide enough contrast for easy viewing. The use of mid blue printing on a shiny black case makes for difficult viewing. The comma key, [,], is now SHIFT [)]. Access to the Pol and Rec functions is now via SHIFT [+] and SHIFT [−]. The y, x and r, θ components are stored in the X and Y memory registers.

Two other calculators used are those manufactured by Sharp and Casio. The function in the same way as the HP 10s+ but they store the rectangular and polar components is different storage registers (memory locations). These different locations apply to § 18.18. The differences are explained in § 18.20.3.

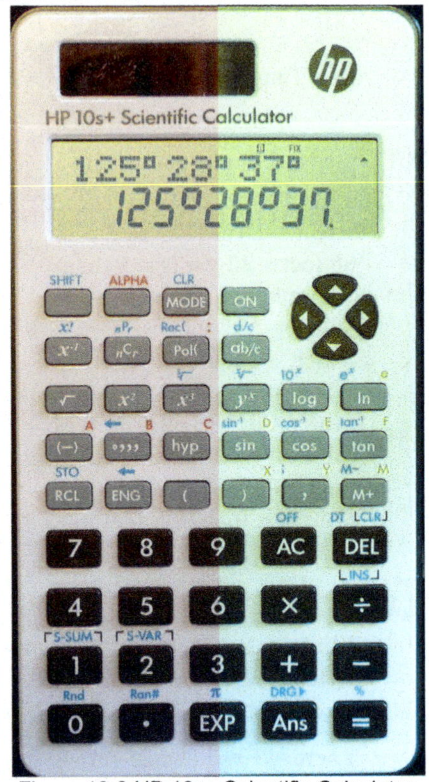

Figure 18-3 HP 10s+ Scientific Calculator.

18.17.1 HP10s+. Set-up for surveying.

Getting up the calculator ready for survey computations involves the [MODE] key for setting base parameters.

First press [MODE] 1 COMP for computations,
Next press [MODE] 1 Deg for degrees mode,
Third press [MODE] 1 Fix, press 1 for
Fix 0~9?

choose 6 to maintain precision if you are not using memory locations for Ans storage.

18.17.2 Time/angle Calculations:

You can use the [°’”] key to manipulate angles. The display uses the degree ° symbol as the unit (minutes, seconds) separator. (125°28°37°).

125°28°37° is input as 125 [°’”] 28 [°’”] 37 [°’”] The top line, the display **input** line, will show 125°28°37°. Pressing [=] puts 125°28°37° in the second line as the **output**, or result line. Importantly, it is the line that fills the Ans memory, after the [=] key is used. Using the [°’”] key will toggle the Ans line between 125°28°37° and 125.476944, but the Ans memory only holds 125°28°37°. This means you can’t convert DMS to decimal degrees to use for following calculations (converting to radians, say).

18.17.3 DMS to Decimal Degrees: on Ans line.

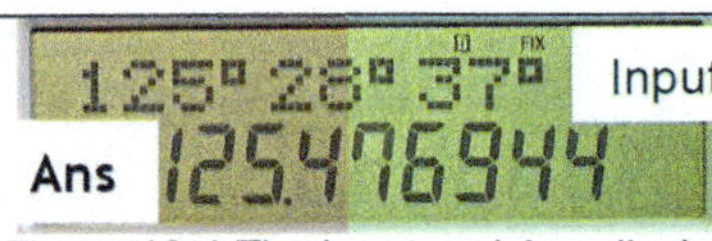

Figure 18-4 The Input and Ans display

The [°’”] key toggles between Degrees, Minutes, Seconds (DMS) and decimal degrees display on the Ans line:

125°28°37° [°’”] 125.476944.

The degrees display expressed as degrees, minutes, seconds (*125°28°37.*) returns an Ans to any problem in the same format: degrees, minutes, seconds.

If you input **decimal** degress, you can convert them to DMS after pressing [=] to put them in the Ans line, then use [°’”] to toggle to DMS.

125.4769 [=] 125.4769 [°’”] 125°28°36.8°,

but the Ans memory still holds the decimal input, 125.4769.

If you want to convert DMS to decimal and use the decimal value in chain calculations you must do it on the input line: 125+28÷60+37÷3600 [=] 125.476944, which will be in the Ans buffer.

18.17.4 Degrees to Radians:

A nasty, tricky conversion with this calculator, and there are several ways to handle it:

1. The old faithful: Deg × π ÷180 [=]
unfortunately it provides a funny looking answer if you use the [°’’’] input:
125°28°37°×π÷180 [=] 2°11°23.95 This is the correct answer to the problem of dividing 125°28′37″ by 57.29578 (180/π). It is expressed in degrees, minutes, seconds! However, it is incorrect in this case because radians are decimals. You have to use the [°’’’] key to get 2.189986, the correct answer in radians.

2. The new faithful, using a memory [STO] (SHIFT RCL) in a memory location A - D, to store the number of degrees in a radian. 57.29578 (π/180).

180÷π [=] 57.29578 [STO] (SHIFT RCL) A

125°28°37°×A [=] 2°11°23.95 then use the [°’’’] key to get 2.189986.

3. Frogging around using the DRG key (SHIFT Ans), probably forget it.

4. Work in decimal degrees and use either 1 or 2 above. But you must convert DMS to decimal degrees to decimals first;

125+28÷60+37÷3600 [=] 125.476944, then [Ans]÷A [=] 2.189986,

5. Consult: *h20331.www2.hp.com/Hpsub/downloads/10s_03_operating_modes.pdf*.

18.17.5 Why Radians to Degrees?

A little bit easier, and where is it needed? With circular curves (Figure 18-5), that’s where.

Curve or arc length is an expression of the distance around the arc of radius generated by an angle between the end radii of the arc.

Arc distance, S = R θ (radians) where θ is the arc deflection angle (at the centre).

If we are given radius, R, and **arc** length, S, then:

arc deflections, θ, = S/R expressed in **radians**.

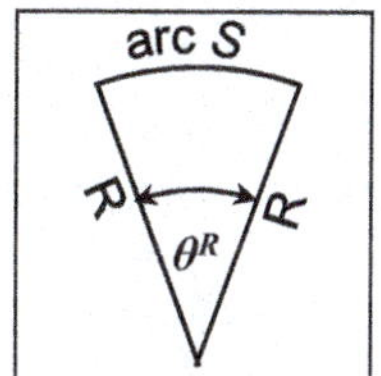

Figure 18-5 Radian measure

In this book, concerning circular curves, we tend to use the symbol Δ (delta) because it denotes the **deflection** angle between two tangents joined by a circular curve. It is the same as the **arc** deflection angle at the centre between the two radii joining the PC and the PT.

18.17.5.1 Radians to Degrees:

The radian answer will come from a computations like, for example:

328.498÷150 [=] 2.189986.

The usual conversion with this calculator :

1. The old faithful: Radians × 180 ÷ π [=]

[Ans]×180÷π [=] 125.460556 is the answer in decimal degrees.

It can be converted to DMS using the [°’’’] 125°28°37

2. The new faithful, using a memory [STO] (SHIFT RCL) in a memory location A - D, to store the number of degrees in a radian. 57.29578 (π/180).

180÷π [=] 57.29578 [STO] A. then, from the previous answer 2.189986 (scroll to it):

[Ans]×A [=] 125.460556, then use the [°’’’] key to get 125°28°37°. But the value in Ans will still be in decimal degrees.

18.18 Vector Conversion and Axes Rotation

Normal polar vectors are represented in a mathematically oriented (anticlockwise) frame. The zero-degree axis is along the + x axis, pointing horizontally to the right. Figure 18-7 shows the vector rotates anticlockwise to 90° through the + y axis, 180° through to – x, 270° through to – y and back to 360° at + x.

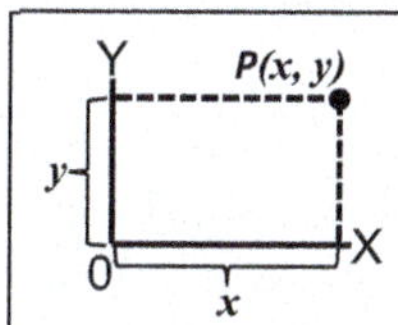

Figure 18-6 Rectangular coordinates

Surveying polar vectors are represented in an azimuthally (**clockwise**) oriented frame. Zero degrees points vertically up to North (N, +y). It then rotates **clockwise** to 90° on the East axis (E, +x), 180° on the South axis (–N, –y), 270° on the West axis (–E, –x) and returns to 360° on the N axis.

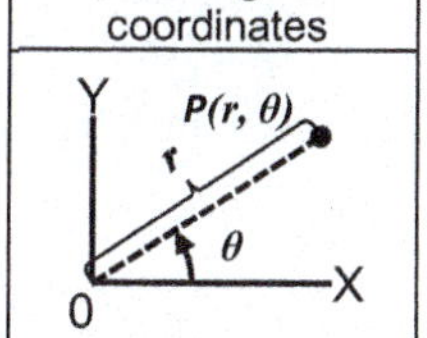

Figure 18-7 Polar coordinates

To use the SHIFT [Pol(] and [Rec(] functions we have to swap our rectangular coordinate orientation (Figure 18-6) to make the calculator think it’s working in math orientation. The polar coordinates don’t change, it’s only the rectangular coordinates that need understanding. This is amplified in § 18.20.

18.18.1 Them (the US) and Us (Australia)

A word on the difference between rectangular/polar coordinate order using the SHIFT [Pol(] and [Rec(] functions in azimuth vector mode.

The HP10S+ (and all its ilk) conform to the American order of:

dN (Latitude difference distance, y), dE (Departure difference distance, x) for rectangular coordinates and:

r (Distance), **θ** (whole circle bearing) for polar coordinates.

18.18.2 The Importance of the SIGN of the Rectangular Coordinates

It is imperative that you maintain the SIGN of the dN and dE in the rectangular coordinates. The sign is necessary to produce the required ATN2() angle, and vital to produce the ATAN() angle. Especially for the ATAN, because the answer; positive in two, negative in two quadrants, is the angle referenced the North/South axis. Combined with the SIGN of the latitude difference (ΔN or Δy), the BEARING of the vector can be ascertained. Consult Chapter 7 Manipulating Vectors for revision.

ATAN() function marked as $\tan^{-1}$ (SHIFT [tan^F]): § 18.20.1.1, Figure 18-8

In using the ATAN() function examine the SIGN of the δN.

If δN is POSITIVE the ANGLE is in quadrants I and IV. If necessary, ADD 360°.

If δN is NEGATIVE, the angle is in quadrants II or III, so ADD 180°.

ATAN2() function. (Used by the Pol(function key).

The returned angle (stored in memory F (RCL [tan^F]) is either positive, (0 clockwise to 180°)

or negative (0 anticlockwise to −180°) § 18.20.1.1

if positive it is the BEARING or,

if negative, the BEARING is 360° + RCL [tan^F] .

18.19 Converting between Polar and Rectangular Coordinates using Function Keys

18.19.1 Trigonometric Conversion to Rectangular Coordinates

The examples in this manual, converting bearing/distance to dE, dN via the standard method:

dE = distance × SIN(brg)

dN = distance × COS(brg)

are tedious and error prone from having to key in the distance and bearing twice, and record each answer.

18.19.2 Rectangular Trigonometric Conversion using the [Rec(] Function Key

To convert from the polar BRG/DIST coordinates to dE, dN you should use the [Rec(] function; (SHIFT [Pol(]). However, there are a couple of gotchas with these two functions:

3. the input is in the sequence **r,θ**, i.e. distance, bearing. Note the comma separator.
4. the answers, dE and dN, are held in two memory locations, E and F.

[cos^E] holds the dN and [tan^F] holds dE. The display show the dN value (memory E)

dE is accessed by recalling memory F (RCL [tan^F])

dN is accessed by recalling memory E (RCL [cos^E])

Example: Convert the following BRG/DIST to dE, dN:

BRG 125°15′ 20″, distance 102.536m.

Conventionally: 102.536×sin 125°15°20° [=] 83.729420, dE, then re-keying for dN

102.536×cos 125°15°20° [=] -59.186278, dN.

Note the requirement to repeatedly key the same data.

Using [Rec(] : input the distance, the comma, then the BRG in either DMS, using the [°’”] key for each value of D, M and S; or use decimal degrees.

There is no need to close the parenthesis [)] key before enter, [=] :

Rec(102.536,125°15°20° [=] -59.186278, dN, then, to retrieve dE

RCL [tan^F] F= 83.729420, dE.

dN and dE will be retained in [cos^E] and [tan^F] until overwritten.

18.20 Converting between Rectangular and Polar Coordinates Using Function Keys

18.20.1 Trigonometric Conversion of Rectangular to Polar Coordinates

The examples in this manual, converting dE, dN to bearing/distance via the standard,

θ = Atan(tan⁻¹)(dE/dN), converted to a whole circle bearing in the correct quadrant (Figure 18-8).

r = Sqrt(√)(dE^2+dN^2), Pythagoras,

is also tedious and error prone from having to key in the dE and dN twice, and record each answer.

Conventionally we calculate and record the dE and dN between two points, then go ahead through a number of steps to get the result.

Let's try an example and attack it three ways: twice conventionally, then using the [Pol(] function key.

E_A	=	135.872	N_A	=	7450.124
E_B	=	<u>125.126</u>	N_B	=	<u>7540.067</u>
dE_{AB}	=	–10.746	dN_{AB}	=	89.943.

18.20.1.1 Finding the tangent

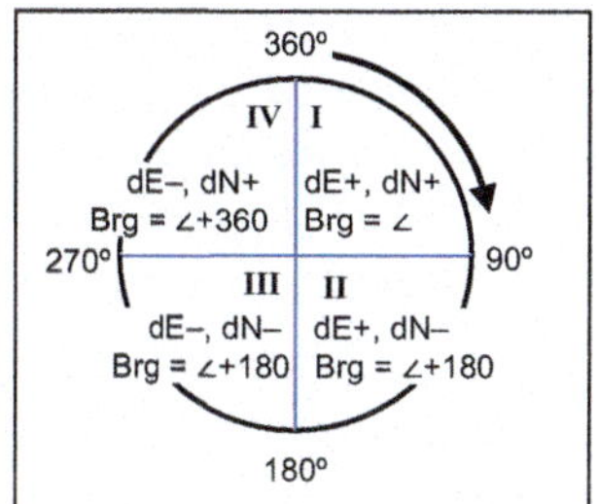

Figure 18-8 ATAN() quadrants

Using `-10.746 ÷ 89.943` [=] -0.119476, (note –ve sign from division)

find ATAN(), `tan⁻¹Ans` [=] -6.813156, (note –ve sign)

it is in quadrant 4. (dE –ve, dN+ve), so add 360:

`360+Ans` [=] 353.186844 [°’”] 353°11°12.6

Then, using Pythagoras for distance, using the [x^2] key and the [√] keys

`(-10.746)²+89.943²` [=] 8205.219765. (Note the compulsory use of parentheses for the negative dE_{AB}).

`√Ans` [=] 90.5827

(Note: $-10.746^2+89.943^2$ [=] 7974.2667; and √Ans [=] 89.2987). This is BIMDAS at work. See HP 10s+ manual page 8 and pp 34-35).

18.20.1.2 Shortcutting it a bit

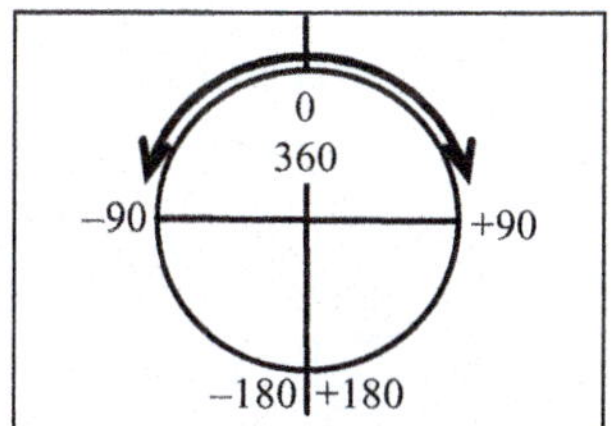

Figure 18-9 ATAN2() hemispheres

and knowing the angle is in quadrant 4 (+360)

`360+tan⁻¹(-10.746÷89.943)` [=] 353.186844 [°’”] 353°11°12.6

Straight in to Pythagoras: `√((-10.746)²+89.943²)` [=] 90.5827

or, shorter (?) by entering the E and N values directly

`360+tan⁻¹((125.126-135.872)÷(7540.067-7450.124))` [=]

353.186844 [°’”] 353°11°12.6

`√((125.126-135.872)²+(7540.067-7450.124)²)` [=] 90.5827

18.20.2 Polar Trigonometric Conversion using the [Pol(] Function Key

Using the [Pol(] function key: – converting rectangular(dN, dE) to polar (**r**,**θ**) coordinates

Note: we have to input dN, then dE,

`Pol(89.943` [,] `-10.746)` [=] 90.5827, distance, **r**,

then, to retrieve bearing, **θ**,

RCL [tan^F] F= -6.813156, bearing, **θ**, is negative,

this means **θ** is in the western hemisphere, add 360;

360+ RCL [tan^F] [=] 353.186844 [°’”] 353°11°12.6

The [Pol(] function uses the ATN2(dN,dE) function, providing a bearing between (see Figure 18-9):

0 and +180 clockwise for quadrants 1 and 2 and

0 and –180 anti-clockwise for quadrants 4 and 3.

Thus a negative result requires the addition of 360°.

r and **θ** will be retained in [cos^E] and [tan^F] until overwritten.

18.20.3 Coordinate conversion in different calculators.

Apart from the HP 10S+ features in this chapter, there are two other popular calculators, the Casio fx-82AU and the Sharp D.A.L. EL-531 used in the course. They have similar functions to the HP 10S+, but their keyboards and memory locations are slightly different. The comma separator between the two-input field is hard to find, especially when non-contrasting fore/background colours are used (Casio).

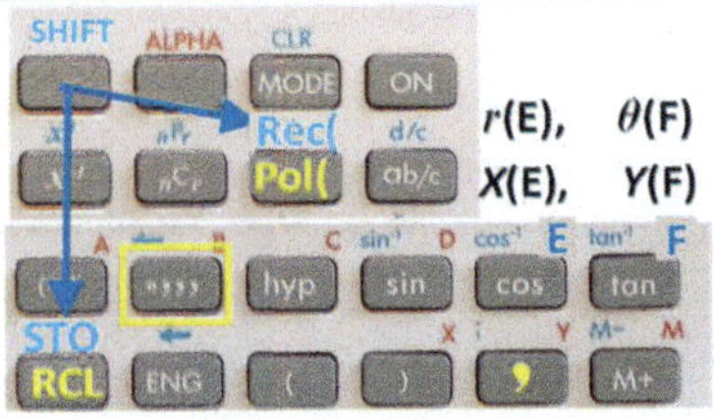

Figure 18-10 HP 10S+ memory locations

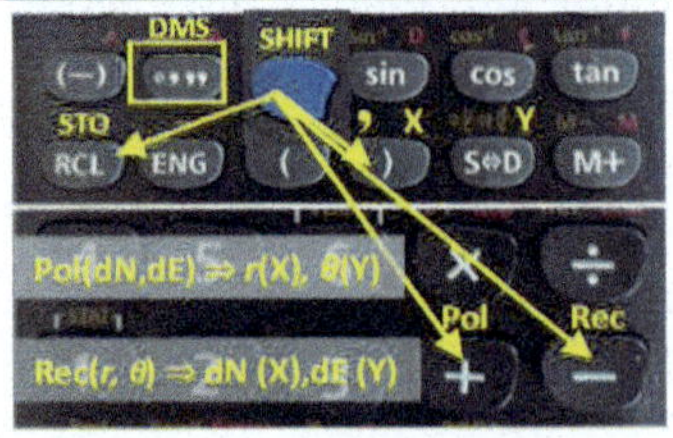

Figure 18-11 Casio fx-82AU memory locations

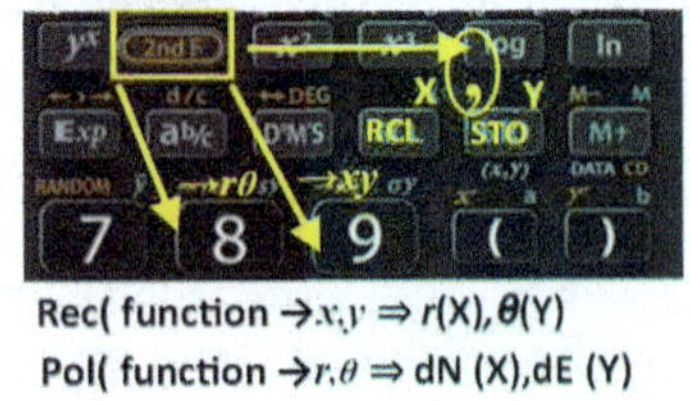

Figure 18-12 Sharp EL-531 memory locations

Coordinate conversion between rectangular and polar coordinates is a basic skill in surveying that is your constant companion in the field. You MUST be comfortable with your calculator.

Following conversion, the values are STORED in the appropriate registers. You have to use the RCL F on the HP10S+ to see the second result. The Casio displays the X, Y or r, θ registers on the screen. It's a bit easier.

The **Pol(** function provides polar coordinates (r, θ) from rectangular input. This is the one to use when finding distances between pairs of coordinates or coordinate differences. (See § 18.20.2).

The **Rec(** function provides rectangular (dN, dE) coordinates from polar input. This function can be used to convert zenith angle (ZA) and slope distance (S) to the vertical (V) and horizontal (H) components. (See § 18.19.2).

18.20.4 The ATAN2() Function in Spreadsheets.

ATN2(dN,dE) is a useful function in Excel, note the order: dN, dE. The best way to get decimal degrees in the correct bearing is to use the expression:

=MOD(DEGREES(ATAN2(dN, dE)),360).

18.21 Memory Locations and Calculations

18.21.1 Variable memory manipulation

There are nine storage locations (names) for variables, A to F, X, Y and M. (Figure 18-13). See the HP10s+ User Guide pp 14, 15.

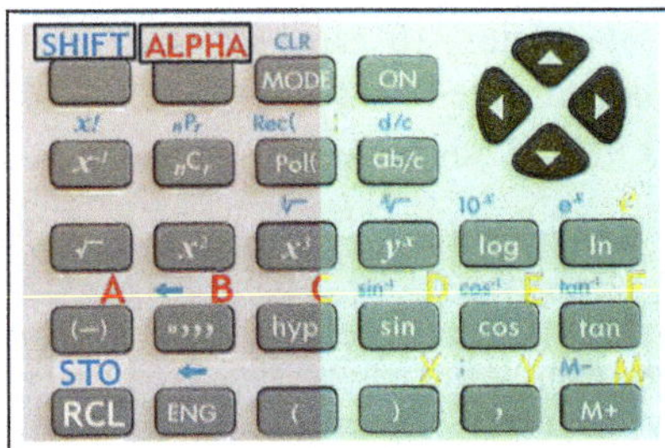

Figure 18-13 Variable memory locations

18.21.1.1 STO a variable

A variable is stored using keystrokes [SHIFT] [STO] and the variable name, e.g., 10.125 [SHIFT] [STO] [(–)] stores 10.125 in memory location A. (Figure 18-14).

The [(–)] key itself allows input of a negative number (a preceding sign key).

18.21.1.2 RCL a variable

A variable is recalled from the memory location using the [RCL] key and the variable name. Use the [ALPHA] key to access the name. e.g., [RCL] [ALPHA] [(–)] recalls the variable in location A. It places the variable name (A) in the top line where it is now available for manipulation (Figure 18-15). The [=] key will place the value in the second line. (And in the ANS location).

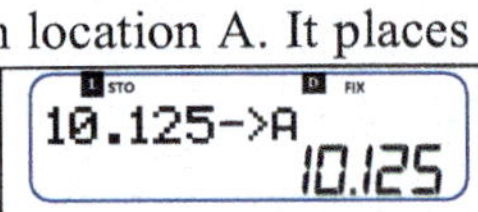

Figure 18-14 STO var A

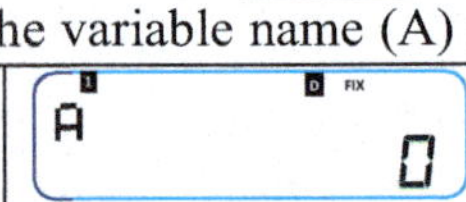

Figure 18-15 RCL var A

However, you may chain calculations by continuing with the normal math operators. e.g.,

[RCL] [ALPHA] [(–)] [x^2] [=] → 102.516. (Figure 18-16).

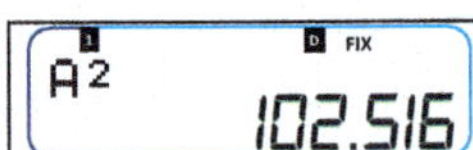

Figure 18-16 RCL A^2 =

Euler's number, ***e***, the base of natural logarithms, is a constant recalled by [RCL] [ALPHA] [LN] [=] → 2.718

Not to be confused with Euler's constant, $\gamma = 0.55721\ldots$

18.21.1.3 Clear variables

To clear any variable, store a zero (0) in it. To clear ALL variables, use [SHIFT] [STO] [1] [=].

18.21.2 The [Ans] key

The Ans memory location is updated with a result value every time the [=] key is pressed. It is also updated when a value is stored in any of the memory locations A to F, X, Y or M.

18.21.3 Using the M+ Key for Accumulate Operations

Using the example on Table 18-1

$RL_{BM} = 10.125$

Load the memory with RL_{BM}:

10.125 STO M+

this puts 10.125 in M+

Get the first rise/fall:

1.742 − 1.628 = 0.114 (note rise)

M+ puts result in M+ (don't use the = key).

Use RCL M+ M= 10.239 to see RL

Next rise/fall:

1.628 − 1.205 = 0.423 (note rise)

M+ Use RCL M+ M= 10.662 to see RL.

Next rise/fall:

1.205 − 1.1.585 = -0.380 (note **fall**)

M+ puts result in M+ (don't use the = key).Use RCL M+ M= 10.282 to see RL.

Table 18-1 Accumulate operations with M+

BS	IS	FS	Rise	Fall	RL
1.742				BM	10.125
	1.628		0.114		10.239
	1.205		0.423		10.662
	1.585			0.380	10.282
	1.625			0.040	10.242
	1.550		0.075		10.317
1.580		1.522	0.028		10.345
	1.635			0.055	10.290
	1.640			0.005	10.285
	1.825			0.185	10.100
	1.745		0.080		10.180
		1.892		0.147	10.033
Sum		Sum	Sum	Sum	
3.322		**3.414**	**0.720**	**0.812**	
Σ BS − Σ FS			**Σ Rises − Σ Falls**		
	-0.092		**Diff =**	**-0.092**	**-0.092**

The M+ key is a tricky beast to get accustomed to but may be useful for chain calculations.

18.22 Statistics, Mean and Standard Deviations

During the field observation sessions there are occasions when the mean result of a set of observations is required. A good check is to evaluate the standard deviation (SD) of the results to form an opinion of the acceptability of the observation, a form of blunder detection. SD is generally denoted with the small Greek symbol sigma, $\boldsymbol{\sigma}$.

Establishing control points, using FL/FR observations in a field traverse exercise, will produces four horizontal, and four vertical distances for each leg of the control traverse. Check the mean (average or $\bar{x}$) and the SD ($\sigma_{\bar{x}}$) of the mean to guard against mistakes.

The easiest process for using the statistics functions require that you calculate the horizontal and vertical distances of each observation using the Rec(function to convert from zenith angle (ZA) and slope distance. If both FL and FR observations are taken, then the absolute value need to be assigned to the horizontal distance, otherwise the FR horizontal distance will be negative. (sin (ZA = 90) = +1, sin (ZA = 270) = −1).

Using Rec(125.257,89°30°45°= 1.066, RCL tan^F F= 125.252466 and making a table for the four observations, the SD mode can be then be initiated.

Set SD mode: MODE 2 = .

Clear registers (Clr):

CLR (SHIFT MODE 1 = .

Stat clear_ = 0.0000.

The M+ key is the DT (DaTa entry) key.

n = input count.

Table 18-2 Input for SD mode

Obs	ZA	S dist	|H dist|	V dist
A − B FL	89° 30′ 45″	125.257	125.2525	+1.066
A − B FR	270° 29′ 30″	125.259	125.2544	+1.075
B − A FL	90° 29′ 30″	125.256	125.2514	-1.075
B − A FR	269° 30′ 15″	125.255	125.2503	-1.084

125.2525 M+ n=1.00)

125.2544 M+ n =2.00)

125.2514 M+ n =3.00)

125.2503 M+ n =4.00)

Find mean: S-VAR (SHIFT 2) 1 = $\bar{x}$ 125.2521

Find standard deviation of mean: S-VAR (SHIFT 2) 2 = $x\sigma n$ 0.0015

This method will catch number transposition blunders, the booker must be alert for anomalies in the data recording. Similarly, for V distance: mean = $\bar{x}$ 1.075, SD_{mean} = $x\sigma n$ 0.0064.

Note the sensitivity of the SD of the V height versus the H (and slope) distance; 6.4mm against 1.5mm. It shows that accurate sighting of the vertical angle is important in determining height differences. In the above Table

18-2 the spread of 15″ between FL/FR is too much. If H distance is the only result needed, then the observations are adequate.

Statistically we are finding the standard deviation of the entire sample (the 4 observations) $x\sigma n$, not the SD of the mean of the entire population (where we could consider the 4 observations as randomly selected from, say, 12 observations).

That's where we would use $x\sigma_{n-1}$ and in this case H $x\sigma_{n-1}$ = 0.0017 and V $x\sigma_{n-1}$ = 0.0073.

After about 30 observations the difference in the SD of the entire population and the sample population $x\sigma_{n-1}$ and $x\sigma_n$ is insignificant. But for small populations (all 4 observations) you should be aware of the difference.

Sometimes the two are defined as Sx for the SD of the sample mean of the population and $\sigma\bar{x}$ for the SD of the entire population mean.

Return to COMP mode: MODE 1 = to continue normal calculator mode.

18.23 Concluding remarks

It is hoped that this expansion on the capabilities of the HP 10S+ calculator and similar Sharp and Casio models, mandated for some courses, will be of use in allowing you to explore its vector capabilities.

18.24 References for Chapter 18

Uren, J. and B. Price. Surveying for engineers. Fifth edition. Palgrave Macmillan, London 2010. ISBN978-0-230-22157-4

Shepherd, F.A, Surveying Problems and Solutions, Edward Arnold, 1968

Bradley, J.M. Mathematical formulae and statistical tables for tertiary institutions. Murdoch University, 2014. ISBN 0-86905-786-3

Abbott, P and H. Neill. Trigonometry. A complete introduction. John Murray Learning, London 2018. ISBN 978 1 473 67849 1

Hewlett-Packard, HP 10s+ Scientific Calculator User Guide, Hewlett-Packard Development Company, 2012.

Sharp Scientific Calculator Operation Guide, EL-531 series, Sharp Corporation 2017.

Casio fx-82AU User's Guide, CASIO Computer Co., Ltd

PART 3: A miscellany of survey topics

If it's not checked it's wrong.

Chapter 19 TARGET DESIGN.

The precision of readings taken for direction are highly influenced by the design of the target plate and the design is examined in many undergraduate Surveying units.

19.1 Pointing accuracy

19.1.1 Pointing to a "natural" target.

Natural targets are generally characterised as building edges, towers, masts and geodetic station targets. The common problem encountered with these is the interplay between the telescope cross hairs and the illumination of the target. The sighting to a mast or tower should be such that it is withing the paired vertical wires. There is a tendency to be influenced (biased) by the more illuminated side of the target.

19.1.2 Probable error of a single pointing.

The point being examined is how well a single pointing can be made to a target, from which the direction can be measured. A classic reference published by Washer and Williams (1946) was arrived at over indoors and outdoors observations to targets over a range between 100m and 13,600m.

Initial indoor testing by two observers over 2m to a collimator produced a probable error (PE_S) of 0.25″ from 60 sets of 10 observation "rounds" by each observer.

Outdoors testings to various targets such as buildings, masts and clock towers produced a PE_s =0.62″ from 4,700 observations split between two observers. Four different observation groups, characterised by distances of 100m – 455m, 700m – 1,400m, 2,700m – 4,500m and 4,800m – 13,600m covered a range of observation conditions such as weather and visibility.

The instrument used was a level telescope of 39mm aperture and 37 power, quite a high magnification compared with modern instruments. Rather than turning off line and re-bisecting the target the pointing deviation was measured by use of a weak prism (optical wedge) deviating the target from the vertical hair. The target was then brought back to the line-of-sight wire and the deviation noted.

Washer and Willaims (1946) determined a least squares solution for probable error of a single pointing to an outdoors target as $PE_s = 0.064d^{1/4} + 0.19$.

19.1.3 Probable error due to atmosphere.

Washer and Scott (1947) conducted further experiments to evaluate the effects of atmosphere and magnification on PE_S, the atmosphere effect being mainly due to distance to the target (Figure 19-1). Of interest to surveyors is the PE_S over the range of theodolite and total station magnifications, most of which are about 30 times.

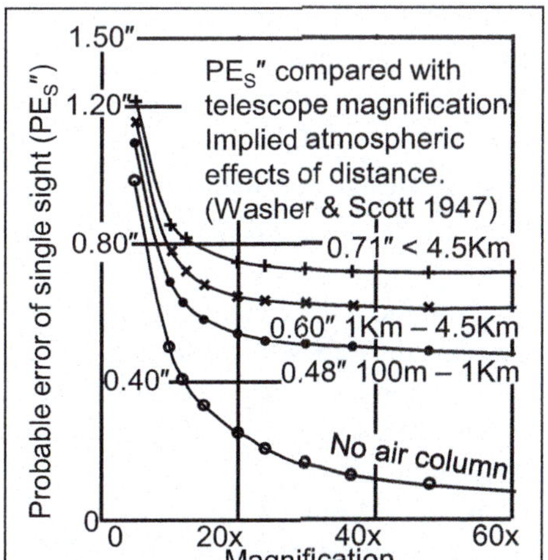

Figure 19-1 PE_S v distance/magnification

For outdoor targets the value $PE_S = \sqrt{\left(\frac{4.96}{M}\right)^2 + 0.60^2}$ yields a result of 0.62″ (in line with earlier work). This then equates to a standard deviation (1σ), $\sigma_{30x} = \frac{0.62''}{0.6745} = 0.92''$

Does this then mean that we should expect a standard deviation of about one second ($\sigma_{30X} = 1''$) from a single pointing using a good instrument in good atmospheric condition, and that no further amount of observations will significantly reduce this value?

19.2 The Y target.

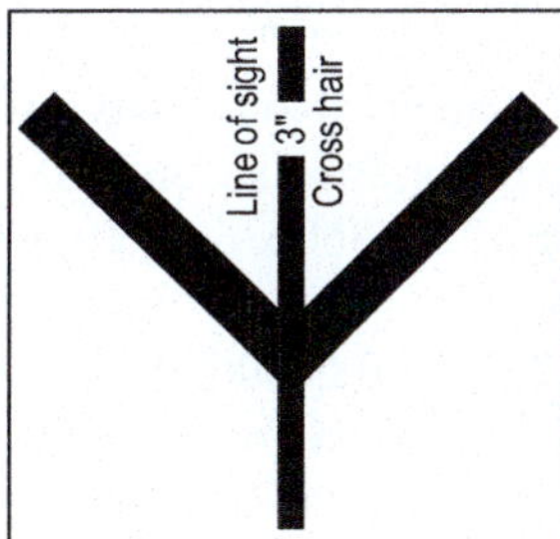

Figure 19-2 The **Y** target

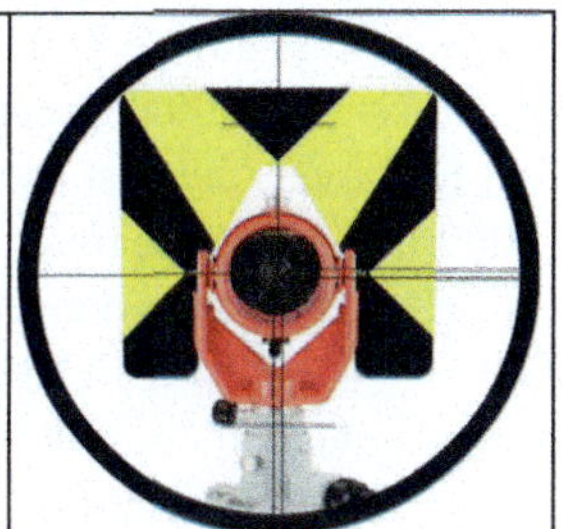
Figure 19-3
Prism plate with **V** targets

Target plates associated with EDM prisms visually fall into the designation of a **Y** type target (Figure 19-2). The cross hair (line of sight) intersects the apex of a **V**. The EDM plate surrounds the prism carrier and the intersection of the **V** is the plumbline axis of the prism tribrach. The sighting should be taken to the prism plate, not to the centre of the (perhaps off axis) reflector. The plate is coincident with the

J. Walker and J. Awange, *Surveying Computations*,
https://doi.org/10.1007/978-3-031-94382-9_19

rotational axis of the plumbline and the trunnion of the reflector carrier (see Figure 19-3). At short range quite precise directions can be observed.

19.3 The paired line target.

Empirical experiments have produced the graphic table shown in Figure 19-4 (no attribute found) which shows generic target dimensions in terms of angular measure compared with pointing accuracy. This topic deals with the calculations of the design of the target to conform with the figure.

The diagram shows that with about a five (5) second gap either side of the theodolite crosshair, a maximum pointing error of less than 0.25″ can be expected. As the gap decreases or increases from this value, so the accuracy of the pointing deteriorates.

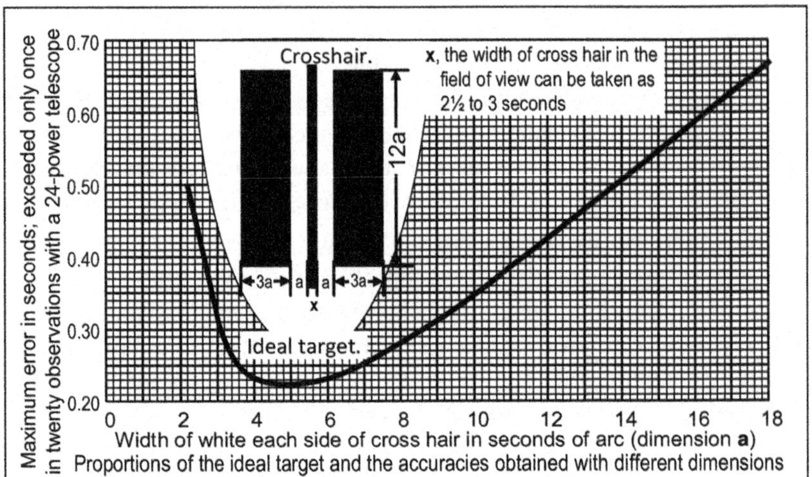

Figure 19-4 Ideal target design

In the first instance the design of the target plate will be examined for a single distance. Then the design will be examined to determine a range of distances at which a certain maximum error can be tolerated.

Basically, the target consists of two rectangles of a 1:4 aspect ratio (3a:12a) separated by a white background gap which is bisected by the theodolite crosshair.

Remember that the width of the crosshair, embedded in the reticule, is about a constant 3″. Its visual "size" does not change. However, the "gap" is fixed, so its optical size changes with observation distance. At a sufficiently long distance the width of the crosshair can fill the gap as in Figure 19-5.

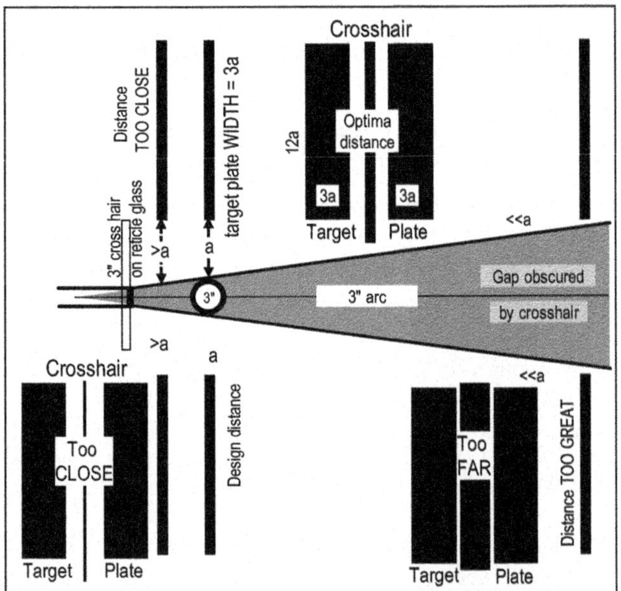

Figure 19-5 Aperture degradation away from idea distance

19.4 Design for distance.

What are the dimensions of the target to be observed at a distance of 100m?

Examining the diagram in Figure 19-6 it can be seen that the **gap** between the target rectangles is:

a + x + a = **2a** + **x** seconds of arc.

Given that a = 5, and x = 3 the fixed width of the crosshair.

Then 1) gap = 5 + 3 + 5 = 13″

2) target width = 3a = 15″

3) target height = 12a = 60″

Remember, from circular measure, that:

arc **s** (gap) = R(distance) × $\theta^{radians}$.

So 13″ = 13/3600 degrees, converted to radians = 0.00006303

However, the best conversion from seconds of arc to radians is the famous number, 206265. This is the inverse of 2π/(360 x 60 x60), which is 1″ arc converted to radians. 1″ = 0.00000485 radians.

Now, 13″/**206265** = 0.00006303radians

Gap **s** = 0.000063 x 100m = 0.0063m = 6.3mm.

Target width = 3a = 15″ = 15/206265 = 0.000073 x 100 = 7.3mm

Target height = 12a = 60″ = 60/206265 = 0.000291 x 100 = 29mm

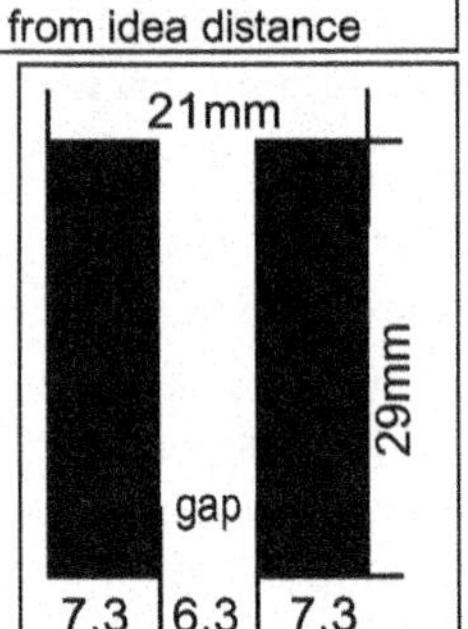

Figure 19-6 100m target

So our target drawing (see) is 7.3 + 6.3 + 7.3 = 20.9
= 21mm wide with a 6.3mm gap and
= 29mm high

19.5 Distance range of a target.

While we can achieve a maximum pointing accuracy with a 5″ gap, it may be useful to use the target over a range of distances. If we are prepared to accept, say, 0.4″ maximum error, what is the limit of the distances?

Examining Figure 19-4, we can see that the gap, the dimension **a**, can vary between a = 2.6″ and 11.4″ for an error < 0.4″ when 2.6″ > a < 11.4″.

Assuming our newly designed target with a gap of 6.3mm (0.0063m).

1. Maximum observation distance occurs when:
   gap = 2.6 + 3 + 2.6 = 8.2″ = 8.2/206264 = 0.00004 radians
   S = Rθ, R = s/θ = 0.0063/0.00004
   R = 158m
2. Minimum observation distance occurs when:
   gap = 11.4 + 3 + 11.4 = 25.8″ = 25.8/206264 = 0.000125 radians
   R = s/ θ = 0.0063/0.000125
   R = 50m

Thus, our target can be used over a range from 50m to 160m, with a maximum pointing error of 0.4″.

19.5.1 Design exercise 1. Geodetic survey exercise.

Examining the case of the geodetic survey exercise, design a target plate to work between 1600m and 2200m, the long sights over the extent of the survey area.

Calculate over, say, 1800m, optimum gap = 13″/206265 = 0.00006303radians

Gap **s** = 0.000063 x 1800m = 0.113m
= 113mm

Target width = 3a = 15″ = 15/206265 = 0.000073 x 1800
= 131mm

Target height = 12a = 60″ = 60/206265 = 0.000291 x 1800
= 525mm

So our target drawing is 131 + 113 + 131 = 375
375mm wide and 525mm high.

This target will have to be drawn on an A2 sheet (420mm x 594mm)

Accepting the gap, 2a + 3″ = 0.113m, show the value for **a:**
at 1600m = 5.8″
at 2200m = 3.8″

These two values still show a pointing error of less than 0.25″.

At a maximum pointing error of 0.4″ the minimum and maximum distances are about 900m and 2800m.

19.5.2 ISO12857 (DIN18723) How many arcs to get a decent angle?

Given that the accuracy standards for Total Stations incorporated in DIN18723 (ISO 12857) can be expressed by:

$\sigma_{\alpha_{pr}} = \dfrac{2\sigma_{DIN}}{\sqrt{n}}$ How many arcs do we need to observe to achieve the **minimum** pointing error of our target?

Minimum pointing error = 0.25″ per target, the angular reading error could be σ_α = SQRT($0.25^2+0.25^2$) = 0.35″.

Using a 1 second (DIN18723) TS, and equation (1) the number of observations,

$n = (2\ \sigma_{DIN} / \sigma_\alpha)^2$
$= (2 \text{ x } 1/0.35)^2$
= 40 pointings.

Thus 20 pairs of pointings on FL/FR are needed to maximize the pointing accuracy offered by the target.

Allowing a maximum pointing error of 0.4″ to the target:

σ_α = SQRT($0.4^2 + 0.4^2$) = 0.57
$n = (2 \text{ x } 1/0.57)^2$
= 12 pointing.

Thus six (6) pairs of pointings on FL/FR are needed to achieve a maximum error of 0.4″ on the target.

19.5.2.1 Reference

DIN 18723, (1990): Feldverfahren zur Genauigkeitsuntersuchung geodätischer Instrumente. (Field procedures for precision testing of surveying instruments).

ISO 12857, (1997): Optics and optical instruments – Field procedures for determining accuracy– Part 2 – 4

http://www.leicaus.com/support/TPS1200/faqs/TPS/tpsinfos/TPS37_00.pdf

http://www.fig.net/pub/proceedings/korea/full-papers/pdf/ws_com5_3/zeiske.pdf

Surveying 481 (Course manual), Curtin University, 1987.

19.5.3 Design exercise 2. Salter Point – Volta Park.

Examining the case of a survey exercise between two stations that required the observation of vertical angles. Design a target plate to work over 2700m, the sight distance over the extent of the survey area.

We achieve a maximum pointing accuracy (0.25″, a = 5″) with a 13″ gap between rectangular targets, see Figure 19-4.

Over, say, 2700m, optimum gap = 13″/206265 = 0.00006303radians.

Target gap, **s** = 0.000063 x 2700m = 0.170m = 170mm.

Target width = 3a = 15″ = 15/206265 = 0.000073 x 2700 = 196mm.

Target height = 12a = 60″ = 60/206265 = 0.000291 x 1800 = 785mm.

So our target drawing is 196 + 170 + 196 = 562

Tilted sideways, it is = 560mm high and = 785mm wide.

This target will have to be drawn on A1 sized paper (841 x 594) or painted on a 900 x 600 composite material (MDF) sheet.

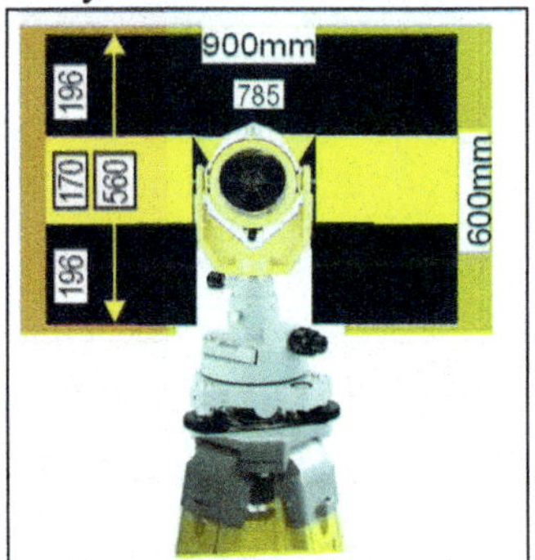

Figure 19-7 Vertical angle target designed for 2700m

Observing Figure 19-4, we can see that the gap, the dimension **a**, can vary between a = 2.6″ and a = 11.4″ for a maximum error < 0.4″.

Accepting the optimum designed gap for 2700m = 0.170m, show the minimum and maximum distances over which the target can perform at a maximum pointing error of 0.4″.

Gap **s** = 0.170m = 170mm

Minimum gap = (a″ + 3″ + a″) = 2.6 + 3 + 2.6 = 8.2″ = 0.000040radians

Maximum distance = 0.17/.000038 = 4280m.

Maximum gap = (a″ + 3″ + a″) = 11.4 + 3 + 11.4 = 25.8″ = 0.000125radians

Minimum distance = 0.17/.000125 = 1360m.

Thus, our target can be used over a range from 1400m to 4200m, maximum pointing error less than 0.4″.

We can thus pick three numbers to represent the minimum, optimum and maximum gap, in radians, to assist in target design.

Optimum gap (a = 5.0):	13″
= 0.000063 (6.3^{E-5})	Optimum distance
Minimum gap (a = 2.6):	8.2″
= 0.000040 (4.0^{E-5})	Maximum distance
Maximum gap (a = 11.4):	25.8″
= 0.000125 (12.5^{E-5})	Minimum distance.

Design for an observation distance to find the optimum gap between the two target rectangles.

1. gap between target rectangles
   = observation distance x 0.000063 (13″)
2. target rectangle width (3a)
   = distance x 0.000073 (15″)
3. target rectangle height (12a) = width x 4.

From the optimum gap size, we then find:

4. Minimum observation distance = gap/0.000125
5. maximum observation distance = gap/0.00004.

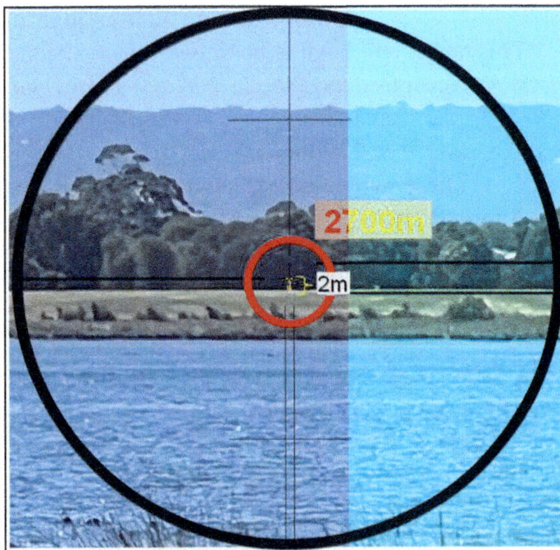

Figure 19-8 Target at 2700m

Figure 19-8 show the target, Figure 19-7, as it appears over the 2700m sighting range from Sandon Park. The top of the target is about 2m above the ground. Obviously the target will require good "seeing" conditions; clear and stable atmosphere, contrasting background, good illumination and minimum shimmer.

19.6 References for Chapter 19

Hoar, G.J. The Analysis, precision and optimization of Control Surveys. School of Surveying, UNSW, Kensington, NSW. July 1974.

Washer, F.E. and H.B. Williams. Precision of Pointing for Outdoor Targets. Journal of Research, National Bureau of Standards (U.S) Paper No RP1717, 36. 1946

Washer, F.E. and L.W. Scott. Influence of Atmosphere on the Precision of Telescope Pointing. Journal of Research, National Bureau of Standards (U.S), Paper No RP1829, Vol 39. 1947.

Munsie, S.W. Surveying 481. Department of Surveying and Mapping, Curtin University 1990

Chapter 20 Resection to a Point by EDM – a Least Squares solution.

Modern positioning methods using Total Stations (TS) use the resection from known points to the unknown observation point occupied by the TS.

Resection is a function found in all TS software and this chapter is intended as an introduction to the method of resection – including a least squares solution to calculate the most likely unknown station.

A Total Station measures distances and directions directly to targets set over known points. However, the TS also shows directions. The **directions** to the known points thus form the **angle** between the two known points at the observation point. Whilst we may consider this angle to be **redundant** information (it isn't used in the solution by distances); it can still be used as a check of our calculations. By calculating the directions from the solved unknown point to the two known points we can get the included angle at the unknown point.

But, more than a check, it can now be included in the solution of the unknown coordinates so that the coordinate values we calculate are as close to the true value as we can make, and prove, them from ALL in information available. The extra observations above the minimum required for a unique solution are called redundant observations

20.1 Resection using a Least Squares minimisation.

Refer to Chapter 7 Intersections, page 99.

Recalling § 7.5, page 103, it can be seen that a Total Station will give horizontal distances to two known points. The unknown point can be calculated directly.

However, the TS also shows directions. The **directions** to the known points thus form the **angle** between the two known points at the observation point.

Whilst we may consider this angle to be **redundant** information (it isn't used in the solution by distances); it can still be used as a check of our calculations. By calculating the directions from the solved unknown point to the two known points we can get the included angle at the unknown point.

But, more than a check, it can now be included in the solution of the unknown coordinates so that the coordinate values we calculate are as close to the true value as we can make, and prove, them from ALL in information available.

The method we use is the method of **Least Squares** where the sum of the squares of the differences between the calculated and the true values, the **residuals**, is a minimum. From this information we can further derive information about the "accuracy" of our observations, the standard deviations.

Most least squares solutions are beyond the scope of hand calculation. You will be introduced to Least Squares adjustment software in later years of your course. This will do the work for you. However this section, with a simple problem, will be solved manually using matrix manipulation.

The ability to understand and manipulate matrices is, along with partial differentiation and linearization of functions, vital to the understanding of Least Squares adjustment. **Matlab** will be the software of instruction for matrix work. Microsoft **Excel** an also manipulate matrices, as long as you know the rules, and aren't too ambitious.

This section will use either hand calculation of matrices (up to 3 x 3) or Excel for larger systems. The calculations will be performed in 2D space (X, Y or E, N), however 3D space (Z or height) can be relatively easily incorporated.

All the calculations have been made in Microsoft Excel to full precision and then displayed as shown. The numbers have not been truncated.

The basic structure of the solutions presented comes from: "Engineering Surveying", W Schofield and M. Breach, Elsevier Ltd, Sixth edition, 2007.

The Least Squares method of adjustment is discussed in chapter 7, Rigorous methods of control.

Calculus of Observations 281/282, WAIT, 1982.

20.2 Resection to a Point by EDM

20.2.1 Worked Example 20.1 Resection by observation of two known points.

The following least squares solution follows Schofield and Breach (2007) chapter 7, Rigorous methods of control, section 7.7, Design matrix and observations vector (page 235).

In Figure 20-1 the adjusted coordinates of X are to be calculated from the following observations:

Control	A	E1000.000, N5000.000, fixed
points	B	E1160.000, N5190.000, fixed

J. Walker and J. Awange, *Surveying Computations*,
https://doi.org/10.1007/978-3-031-94382-9_20

At station X the following **observations** were made:

XA $= 295.055 \pm 0.01$m (H_3 distance)

XB $= 304.864 \pm 0.01$m (H_2 distance)

$\angle AXB = 48°53'15'' \pm 10''$

Initial coordinates for **X** have been calculated from the method described in § 7.5, page 103, intersection by distances.

The diagram has been annotated to help you confirm the coordinates of X (point 1).

X E1281, N4910 (rounded initial values for calculation).

The most likely values of the X coordinates are:

$E_X = e_X + \delta E_X$, the initial values + small error

$N_X = n_X + \delta N_X$.

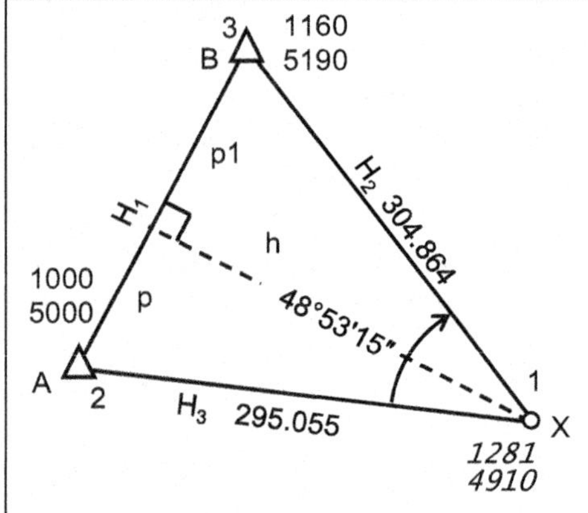

Figure 20-1 Resection to a point.

All surveying equations involving observations can be reduced to the form:

A·**x** = **b** + **v**
- **x** a vector of the terms to be computed, including coordinates
- **b** a vector relating to the observations
- **A** a **matrix** of coefficients
- **v** a vector of residuals, to be a minimum.

1. Form the **x** matrix of the errors in Easting and Northing of the point X:

$$\mathbf{x} = \begin{bmatrix} \delta E_X \\ \delta N_X \end{bmatrix} \text{ is a } 2 \times 1 \text{ matrix}$$

2. The bearings and distances from the initial coordinates of X to A and B are **calculated.**

$\theta_{XA} = 287°45'34''$ $s_{XA} = 295.061$

$\theta_{XB} = 336°37'44''$ $s_{XB} = 305.026$

$\angle AXB = 48°52'10''$ (calculated angle, to be used later)

3. Form the **A** (3 × 2) matrix of the observation coefficients for δE and δE.
In general the solution should include the uncertainties of the control points A and B. However, they are considered **fixed** and so we only consider the uncertainties in the measurements from **X, vis;**
angle **AXB**
distance **XA** s_{XA}
distance **XB** s_{XB}

4. Points A and B are considered fixed (no errors) and thus have not appeared in the **x** matrix, and also do not appear in the **A** matrix.
From the **calculated** bearings and distances from X to A and B, the angle AXB can be found as the difference between the two calculated bearings.

$$\mathbf{A} = \begin{bmatrix} \dfrac{\cos\theta_{XA}}{s_{XA}\sin 1''} - \dfrac{\cos\theta_{XB}}{s_{XB}\sin 1''} & \dfrac{\sin\theta_{XA}}{s_{XB}\sin 1''} - \dfrac{\sin\theta_{XB}}{s_{XA}\sin 1''} \\ -\sin\theta_{XA} & -\cos\theta_{XA} \\ -\sin\theta_{XB} & -\cos\theta_{XB} \end{bmatrix} = \begin{bmatrix} -407.5 & 397.5 \\ 0.9523 & -0.3050 \\ 0.3976 & -0.9180 \end{bmatrix}$$

(columns: δE, δN)

Evaluating A_{11} as an example
note that $\sin 1'' = 0.000004848137 = 1/206265$

$$\frac{\cos\theta_{XA}}{s_{XA}\sin 1''} - \frac{\cos\theta_{XB}}{s_{XB}\sin 1''}$$

$$= \frac{\cos(287°45'34'')}{295.061/206265} - \frac{\cos(336°37'44'')}{305.026/206265} = \frac{0.3050}{0.00143} - \frac{0.9179}{0.00148}$$

$= 213.228 - 620.739$

$= -407.511$

5. The **W** (weight) matrix is a square 3 × 3 matrix with the weights of the observations on the diagonal. Since they are assumed uncorrelated, the rest of the matrix is filled with zeros.

Form the **W** matrix from the standard deviations of the observations. The weight is $1/\sigma^2$:

$\sigma_\theta = \pm 10''$ $(\sigma_\theta)^2 = 100$ $W_\theta = 0.01$

$\sigma_s = \pm 0.01\text{m}$ $(\sigma_s)^2 = 0.0001$ $W_s = 10{,}000$

$$\mathbf{W} = \begin{bmatrix} 0.01 & 0 & 0 \\ 0 & 10000 & 0 \\ 0 & 0 & 10000 \end{bmatrix} \begin{matrix} \text{refers to } \angle AXB \\ \text{refers to } s_{XA} \\ \text{refers to } s_{XB} \end{matrix}$$

6. Form the **b** matrix (3 × 1); the matrix of the difference between observed and calculated observations. The sense is **obs – calc** in the same order as the **v** and **W** matrices. Units are converted to seconds and metres.

observed ∠AXB 48°53′ 15″

calculated ∠AXB <u>48°52′ 10″</u>

obs – calc = 65″

$H_{XA} - s_{XA} = 295.055 - 295.061 = -0.006$

$H_{XB} - s_{XB} = 304.864 - 305.026 = -0.162$

$$\mathbf{b} = \begin{bmatrix} 65'' \\ -0.006 \\ -0.162 \end{bmatrix} \begin{matrix} \text{refers to difference in } \angle AXB \\ \text{refers to difference in } s_{XA} \\ \text{refers to difference in } s_{XB} \end{matrix}$$

7. the matrix of unknown corrections in dE and dN, the **x** matrix (2 × 1), results from manipulation of the **A**, **W** and **b** matrices. This discussion is NOT a lesson in matrix manipulation.

x matrix = $(\mathbf{A}^T\cdot\mathbf{W}\cdot\mathbf{A})^{-1}\cdot(\mathbf{A}^T\cdot\mathbf{W}\cdot\mathbf{b})$

8. Start by calculating the product $\mathbf{A}^T\cdot\mathbf{W}$ as it is used in both matrices; (2 × 3)·× (3 × 3) = (2 × 3)

$$\mathbf{A}^T = \begin{bmatrix} -4017.5 & 0.9523 & 0.3967 \\ 397.5 & -0.3050 & 0.9180 \end{bmatrix} \quad \mathbf{W} = \begin{bmatrix} 0.01 & 0 & 0 \\ 0 & 10000 & 0 \\ 0 & 0 & 10000 \end{bmatrix}$$

$$\mathbf{A}^T\mathbf{W} = \begin{bmatrix} -4.075 & 9523.5 & 3966.9 \\ 3.975 & 3050.2 & 9197.5 \end{bmatrix}$$

9. Now calculate the product $(\mathbf{A}^T\cdot\mathbf{W})\cdot\mathbf{A}$; (2 × 3) × (3 × 2) = (2 × 2)

$$\mathbf{A}^T\mathbf{W} = \begin{bmatrix} -4.075 & 9523.5 & 3966.9 \\ 3.975 & 3050.2 & 9197.5 \end{bmatrix} \quad \mathbf{A} = \begin{bmatrix} -407.5 & 397.5 \\ 0.9523 & -0.3050 \\ 0.3967 & -0.9180 \end{bmatrix}$$

$$\mathbf{A}^T\mathbf{W}\mathbf{A} = \begin{bmatrix} 12306.5 & -8166.9 \\ -8166.9 & 10935.8 \end{bmatrix}$$

10. We have normalised the equations so that we will have $(\mathbf{A}^T\mathbf{W}\mathbf{A})\mathbf{x} = \mathbf{A}^T\mathbf{W}\mathbf{b}$.

This means that $\mathbf{x} = \dfrac{A^TWb}{A^TWA} = \left(A^TWA\right)^{-1} \times \left(A^TWb\right)$

Calculate and check the inverse $(\mathbf{A}^T\mathbf{W}\mathbf{A})^{-1}$, a (2 × 2) matrix. Previously an example using the adjugate method was detailed. However, anything over a (2 × 2) matrix is too hard. Use the Excel MINVERSE{} function.

Let $(\mathbf{A}^T\mathbf{W}\mathbf{A})$ be called **matrix M**

$$\mathbf{M} = \left(A^TWA\right) = \begin{bmatrix} 12306.5 & -8166.9 \\ -8166.9 & 10935.8 \end{bmatrix} \quad \text{Function:MINVERSE}\{\text{M}\} = \text{M}^{-1} = \left(A^TWA\right)^{-1} = \begin{bmatrix} 1.611^{-4} & 1.203^{-4} \\ 1.203^{-4} & 1.813^{-4} \end{bmatrix}$$

$$\text{CHECK}: \; M^{-1}M = I = \begin{bmatrix} 12306.5 & -8166.9 \\ -8166.9 & 10935.8 \end{bmatrix} \begin{bmatrix} 1.611^{-4} & 1.203^{-4} \\ 1.203^{-4} & 1.813^{-4} \end{bmatrix} = \begin{bmatrix} 1 & 0 \\ 0 & 1 \end{bmatrix} \text{ The identity matrix}$$

11. Now, for the second time, use $(\mathbf{A}^T\mathbf{W})$ to calculate the product $(\mathbf{A}^T\mathbf{W})\cdot\mathbf{b}$; (2 × 3)·× (3 × 1) = (2 × 1)

$$\mathbf{A}^T\mathbf{W} = \begin{bmatrix} -4.075 & 9523.5 & 3966.9 \\ 3.975 & 3050.2 & 9197.5 \end{bmatrix} \quad \mathbf{b} = \begin{bmatrix} 65 \\ -0.006 \\ -0.162 \end{bmatrix}$$

$$\mathbf{A}^T\mathbf{W}\mathbf{b} = \begin{bmatrix} -963.13 \\ 1798.00 \end{bmatrix}$$

12. The **x** matrix is the product of $(\mathbf{A^TWA})^{-1}$ and $(\mathbf{A^TWb})$; $(2 \times 2) \times\cdot(2 \times 1) = (2 \times 1)$

$$\left(\mathbf{A^TWA}\right)^{-1} = \begin{bmatrix} 1.611^{-4} & 1.203^{-4} \\ 1.203^{-4} & 1.813^{-4} \end{bmatrix} \quad \mathbf{A^TWb} = \begin{bmatrix} -963.13 \\ 1798.00 \end{bmatrix}$$

$$\mathbf{x} = \begin{bmatrix} 0.057 \\ 0.204 \end{bmatrix} \begin{matrix} = \delta Ex \\ = \delta Ex \end{matrix}$$

We started by saying that the most likely values of the X coordinates are:

$E_X = e_X + \delta E_X$

$N_X = n_X + \delta N_X$

From the initial estimated coordinates, add the errors to find the adjusted coordinates:

$\mathbf{E_X = 1281 + 0.057 = 1281.057}$

$\mathbf{N_X = 4910 + 0.204 = 4910.204}$

13. The **v** matrix of **residuals** is defined by: $\mathbf{v = Ax - b}$; $(3 \times 2)\cdot\times (2 \times 1) = (3 \times 1) - (3 \times 1)$

$$\mathbf{A} = \begin{bmatrix} -407.5 & 397.5 \\ 0.9523 & -0.3050 \\ 0.3967 & -0.9018 \end{bmatrix} \quad \mathbf{x} = \begin{bmatrix} 0.057 \\ 0.204 \end{bmatrix}$$

$$\mathbf{Ax} = \begin{bmatrix} 58.0 \\ -0.0083 \\ -0.1648 \end{bmatrix} - \mathbf{b} = \begin{bmatrix} 65.6 \\ -0.006 \\ -0.162 \end{bmatrix} = \mathbf{v} = \begin{bmatrix} -7.6'' \\ -0.0022 \\ -0.0026 \end{bmatrix} \begin{matrix} \text{residual error in } \angle AXB \\ \text{residual error in } H_{XA} \\ \text{residual error in } H_{XB} \end{matrix}$$

14. Now we must look at the quality of the adjustment by examining the **standard** error of a unit weight. This allows a check against blunders and is also an assessment of whether the weights assigned to the adjustment are appropriate.

The **variance** factor $\boldsymbol{\sigma_0^2}$ is the sum of the weighted squares of the residuals $(\Sigma(w_iV_i^2))$ divided by the degrees of freedom $(m - n)$. $\boldsymbol{\sigma_0^2}$ can also be defined: $\boldsymbol{\sigma_0^2} = \mathbf{v^TWv}/(m\text{-}n)$,

m = number observations,

n = number of parameters

$$\mathbf{v^T} = \begin{bmatrix} -7.6 & -0.0022 & -0.0026 \end{bmatrix} \quad \mathbf{W} = \begin{bmatrix} 0.01 & 0 & 0 \\ 0 & 10000 & 0 \\ 0 & 0 & 10000 \end{bmatrix}$$

$$\mathbf{v^TW} = \begin{bmatrix} -0.076 & -21.838 & -25.592 \end{bmatrix} \quad \mathbf{v} = \begin{bmatrix} -7.6 \\ -0.0022 \\ -0.0026 \end{bmatrix} \qquad \boldsymbol{\sigma_0^2} = [0.689]$$

Degrees of freedom

m = number of observations = 3,

n = number of parameters = 2

m - n = 1 degree of freedom

Thus $\boldsymbol{\sigma_0^2} = 0.689$ and at the 5% level of the is acceptable on the $\boldsymbol{\chi^2}$ test.

$\boldsymbol{\sigma_0} = 0.830$, which is an acceptable figure for the **standard error**.

15. The **variance-covariance** matrix is an indication as to whether the various off diagonal elements of the adjustment show any correlation between each other. By examining the upper diagonal of the inverse of the **normal** equation matrix, $(\mathbf{A^TWA})^{-1}$, we can find the **variances** of the parameters **on** the diagonal, and the **covariances** off the diagonal.

$$\left(\mathbf{A^TWA}\right)^{-1} = \begin{bmatrix} \sigma_1^2 & \sigma_{12} \\ \sigma_{21} & \sigma_2^2 \end{bmatrix} = \begin{bmatrix} 1.609^{-4} & 1.202^{-4} \\ 1.202^{-4} & 1.813^{-4} \end{bmatrix}$$

The correlation coefficient between any two parameters can be found from:

$r_{ij} = \sigma_{ij}/(\sigma_{ii}\sigma_{jj})$

$$\mathbf{r_{ij}} = \begin{bmatrix} \sigma_1 & \sigma_{12} \\ & \sigma_2 \end{bmatrix} = \begin{bmatrix} \sqrt{\sigma_1^2} & \sigma_{12} / \left(\sqrt{\sigma_1^2}\sqrt{\sigma_2^2}\right) \\ & \sqrt{\sigma_2^2} \end{bmatrix} = \begin{bmatrix} 0.0127 & 0.7038 \\ & 0.0135 \end{bmatrix}$$

16. Once the initial coordinates have been adjusted, we have a better estimate of the coordinates. By **iterating**, feeding the adjusted coordinates back in at the start of the exercise, a new set of adjusted coordinates can be found. The iterations continue until the adjustment **converges** so that the "improvement" in coordinates stabilises.
Use $E_X = 1281.057$, $N_X = 4910.204$ as the new coordinates and recalculate the adjustment

$$\left(\mathbf{A^TWA}\right)^{-1} = \begin{bmatrix} 1.609^{-4} & 1.202^{-4} \\ 1.202^{-4} & 1.813^{-4} \end{bmatrix} \quad \mathbf{A^TWb} = \begin{bmatrix} -1.0575 \\ 0.6753 \end{bmatrix} \quad \mathbf{x} = \begin{bmatrix} 0.000 \\ 0.000 \end{bmatrix} \begin{matrix} = \delta Ex \\ = \delta Nx \end{matrix}$$

E_X = 1281.057, N_X = 4910.204, which is no change because the estimates for $\hat{\mathbf{x}}$ are zero.

This result is to be expected given accurate measurements and a good initial estimate. The other factor is that there is only one degree of freedom, so there is hardly any over-determination.

Recalculating the join from the new X coordinates produced standard deviations of:

$\delta S_{XA} = 0.002$, $\delta S_{XB} = 0.003$ and $\delta\angle AXB = 7.6''$.

Compare the least squares result $E_X = 1281.057$, $N_X = 4910.204$
with the distance intersection result $E_X = 1281.058$, $N_X = 4910.202$.
Remember we started the exercise with approximate coordinates: 1281, 4910.

20.2.1.1 Visualising the point's errors using a standard error ellipse.

An absolute error ellipse can show the uncertainty of a point's position. The square roots of the variances on the diagonal of the variance-covariance matrix give the standard errors in Easting and Northing (σE and σN). An imaginary box of dimensions 2σE by 2σN could be constructed around the point to demonstrate the error in the computed coordinate. However, an ellipse is a better description of a bivariate error distribution. The box axes are also not necessarily oriented N-S and E-W. They are aligned by way of the covariance. Schofield and Breach (2007), topic 7.11, provides the derivation of the orientation and magnitude of the major and minor axes of the ellipse.

17. Ellipse orientation:

From the previous section 15, the variance-covariance matrix:

$$\left(\mathbf{A^TWA}\right)^{-1} = \sigma^2 = \begin{bmatrix} \sigma_E{}^2 & \sigma_{EN} \\ \sigma_{NE} & \sigma_N{}^2 \end{bmatrix} = \begin{bmatrix} 1.609^{-4} & 1.202^{-4} \\ 1.202^{-4} & 1.813^{-4} \end{bmatrix} \quad \sigma = \begin{bmatrix} 0.0127 & 0.0110 \\ 0.0110 & 0.0135 \end{bmatrix}$$

$$\tan 2\theta = \frac{2\sigma_{EN}}{\sigma_N{}^2 - \sigma_E{}^2} = \frac{2(1.202^{-4})}{1.813^{-4} - 1.609^{-4}} = 1.486.$$

$$2\theta = 85.137^\circ.$$

$$\theta = 42.57^\circ \text{ Major axis orientation.}$$

If $\sigma_{EN} < 0$ (*negative*) then $\theta = \theta + 90^\circ$

18. Magnitude of major and minor axes:

$$\text{Major axis, } a: \quad \sigma_{\max}{}^2 = \frac{1}{2}\left[\sigma_N{}^2 + \sigma_E{}^2 + \left\{\left(\sigma_N{}^2 - \sigma_E{}^2\right)^2 + 4\sigma_{EN}{}^2\right\}^{\frac{1}{2}}\right] \quad \text{Note } \sigma_{EN} \text{ is squared.}$$

$$= \frac{1}{2}\left[1.81^{-4} + 1.61^{-4} + \left\{\left(1.81^{-4} - 1.61^{-4}\right)^2 + 4\times\left(1.20^{-4}\right)^2\right\}^{\frac{1}{2}}\right] = 2.92^{-4}$$

$$a = \sigma_{\max} = 0.0171. \text{ Major axis}$$

$$\text{Minor axis, } b: \quad \sigma_{\min}{}^2 = \frac{1}{2}\left[\sigma_N{}^2 + \sigma_E{}^2 - \left\{\left(\sigma_N{}^2 - \sigma_E{}^2\right)^2 + 4\sigma_{EN}{}^2\right\}^{\frac{1}{2}}\right] = 5.05^{-5}$$

$$b = \sigma_{\min} = 0.0071. \text{ Minor axis}$$

20.2.1.2 Drawing the error ellipse and the pedal curve.

The ellipse is dimensionless; that means you choose a size that suits the presentation.

This ellipse is quite eccentric, suggesting that the weights of the distances and the angle are not suitable. A highly eccentric ellipse may suggest a problem with network weighting, or, in a network, an error in an observation.

The ellipse has shown the orientation and magnitude of the major and minor standard deviations.

What is the magnitude of the variance in any other direction? If it is known then a curve of that magnitude can show the area of the actual error, not just the error ellipse of the major and minor errors.

A curve, , can be formed on the error ellipse and is called a pedal curve. It is an orthogonal projection on the tangents to a curve. It has many famous forms. The curve perimeter shows the magnitude of the variance in any direction. See Figure 20-2.

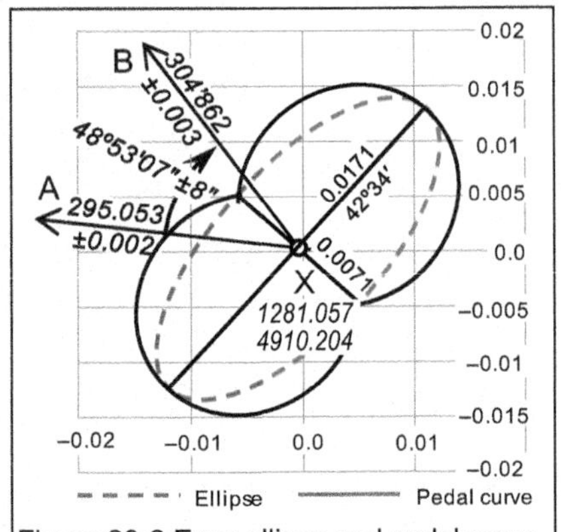

Figure 20-2 Error ellipse and pedal curve

The variance σ_u^2 in any direction, ϕ, is:

$$\sigma_u^2 = \sigma_E^2 \sin^2 \phi + \sigma_N^2 \cos^2 \phi + 2\sigma_{EN} \sin\phi \cos\phi .$$

This will delineate the pedal curve, which is the true error figure. Thanks to Dr David Belton, Curtin University, for preparing it for the diagram.

20.3 An over-determined Least Squares solution using trilateration.

Plane coordinates from distance observations. "Intersection by distances." Ghilani, C.D. "Adjustment Computations" Wiley 5th ed Section 14.2 p242.

Observation of two distances from the unknown control point to **two** known points produces a unique solution to the point's plane position. It only requires two observations, the distances, to produce the two unknowns, the coordinates in E and N. If more distance observations are made, say to a third known point, then there are three (3) observations for the two unknowns. The system is **over-determined** (i.e., more observations than unknowns).

The third observation (distance), in conjunction with one of the other distances, could be used to check the first answer. In all we could arrive at three sets of coordinates for the same point using the three combinatorial pairs of observed distances, - this could be the "easy" way, given the simplicity of the calculations.

20.3.1 Worked Example 20.2. Over-determined by distances.

By way of example, observations to 3 control points produced the following observations: (See Figure 20-3).

Refer to § 7.5.4, page 105:

Point Fixed coordinates

Measured horizontal distance, ℓ, from **U**.

	E	N	ℓ
A	1188.113	5333.960	152.420m
B	1069.120	5290.789	138.580m
C	1046.174	5348.466	84.110m

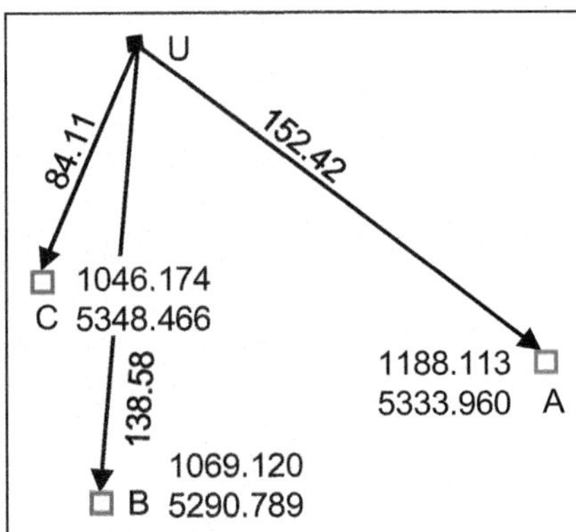

Figure 20-3 Worked Example 20.2.

The three sets of coordinates calculated for the unknown point **U**, using the distance pairs UA, UB; UA, UC and UB, UC are:

U_{AB}		U_{AC}		U_{BC}	
U_{E1}	U_{N1}	U_{E2}	U_{N2}	U_{E3}	U_{N3}
1069.248	5429.369	1069.235	5429.353	1069.178	5429.369

The average of these observations from U is:

U_E	U_N		S_E	S_N
1069.220	5429.364	with a standard deviation of:	0.037	0.009.

Is there a better way of doing this, allowing us to find the "most likely" values of the unknown point, **U**?

Again, we tackle the problem with a least squares solution by using an observation equation with an observation equation that minimises the sum of the squares of the residuals.

It is the normal matrix solution that we are after:

A·x = b + v, it looks similar and is solved using the same matrix manipulation techniques used previously, but with non-linear functions.

Since we are using **distances** (trilateration) to find the coordinates of U the errors in the distances have to be minimised.

For any line, *IJ*, there is the observed distance, ℓ_{ij},
plus its associated residual v_{ij} (Figure 20-4).

By plane trigonometry it stands that (Figure 20-3):

Line *IJ*	$IJ = \ell_{ij} + v_{ij}\left[\left(x_j - x_i\right)^2 + \left(y_j - y_i\right)^2\right]^{½}$	Eqn 20-1

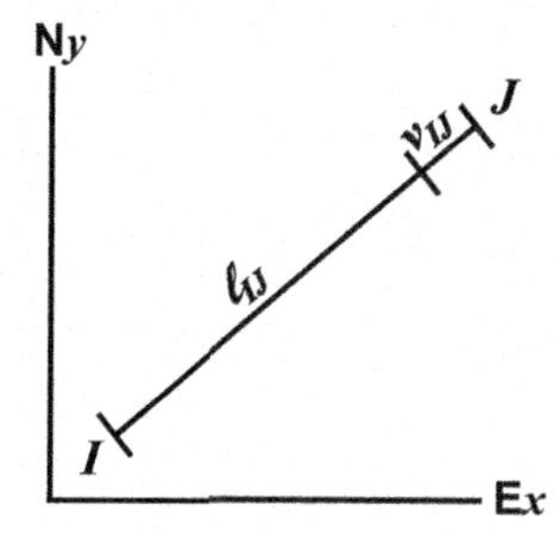

Figure 20-4 Length plus error of a line.

Eqn 20-1 is a non-linear function with unknown parameters:

x_i, x_j, y_i, y_j and the function can be expressed as:

Non-linear function *F(x,y)*	$F\left(x_i, y_i, x_j, y_j\right) = \sqrt{\left(x_j - x_i\right)^2 + \left(y_j - y_i\right)^2}$	Eqn 20-2

Eqn 20-2 can be linearized and solved by the first-order Taylor series expansion by finding the partial derivatives of the elements of the function:

in essence, the final function = initial function + corrections

F() = *F*()0 + *d*(*F*)

$F(x_i, y_i, x_j, y_j) = F(x_{i0}, y_{i0}, x_{j0}, y_{j0})$

Taylor expansion	$F\left(x_{i0}, y_{i0}, x_{j0}, y_{j0}\right) = \left(\frac{\partial F}{\partial x_i}\right)_0 dx_i + \left(\frac{\partial F}{\partial y_i}\right)_0 dy_i + \left(\frac{\partial F}{\partial x_j}\right)_0 dx_j + \left(\frac{\partial F}{\partial y_j}\right)_0 dy_j$	Eqn 20-3

Thus, for the start coordinates, *i*: $x_i = x_{i0} + dx_i$ $y_i = y_{i0} + dy_i$

and, for the end coordinates, *j*: $x_j = x_{j0} + dx_j$ $y_j = y_{j0} + dy_j$

Evaluating the partial derivatives using $\partial F/\partial x_i$, with Eqn 20-2,

Evaluate	$F\left(x_i, y_i, x_j, y_j\right) = \left[\left(x_j - x_i\right)^2 + \left(y_j - y_i\right)^2\right]^{\frac{1}{2}}$	Eqn 20-4

The partial derivative with respect to x_i is:

Partial derivative of x_i	$\left(\frac{\partial F}{\partial x_i}\right) = \frac{1}{2}\left[\left(x_j - x_i\right)^2 + \left(y_j - y_i\right)^2\right]^{-\frac{1}{2}}\left[-2\left(x_j - x_i\right)\right]$	Eqn 20-5

Reduce Eqn 20-5	$\frac{\partial F}{\partial x_i} = \frac{-2(x_j - x_i)}{2\left[(x_j - x_i)^2 + (y_j - y_i)^2\right]^{½}} = \frac{-(x_j - x_i)}{\sqrt{\left(\left(x_j - x_i\right)^2 + \left(y_j - y_i\right)^2\right)}} = \frac{x_i - x_j}{IJ}.$	Eqn 20-6

Similarly, with respect to y_i, the partial derivative is:

Partial derivative of y_i	$\frac{\partial F}{\partial y_i} = \frac{y_i - y_j}{IJ}.$	Eqn 20-7

But, with respect to the ***j*** subscript:

$$\left(\frac{\partial F}{\partial x_j}\right) = \frac{1}{2}\left[\left(x_j - x_i\right)^2 + \left(y_j - y_i\right)^2\right]^{-\frac{1}{2}}\left[2(1)\left(x_j - x_i\right)\right] = \frac{(x_j - x_i)}{\sqrt{\left(\left(x_j - x_i\right)^2 + \left(y_j - y_i\right)^2\right)}} = \frac{(x_j - x_i)}{IJ}$$

Note the change of subscript order from Eqn 20-6.

Partial derivative of x_j	$\frac{\partial F}{\partial x_j} = \frac{x_j - x_i}{IJ}.$	Eqn 20-8
Partial derivative of y_j	$\frac{\partial F}{\partial y_j} = \frac{y_j - y_i}{IJ}.$	Eqn 20-9

Create equation	$\left(\frac{x_i - x_j}{IJ}\right)_0 dx_i + \left(\frac{y_i - y_j}{IJ}\right)_0 dy_i + \left(\frac{x_j - x_i}{IJ}\right)_0 dx_j + \left(\frac{y_j - x_i}{IJ}\right)_0 dy_j = b_{l_{ij}} + v_{l_{ij}}$	Eqn 20-10

With respect to point x_j and y_j the partial derivates are:

The initial parameters are calculated using the approximate (initial) values, $b_{l_{ij}} = l_{ij} - IJ_0$.

This is the **b** matrix. $v_{l_{ij}}$ is the **v** residuals matrix. And:

Initial length IJ	$IJ_0 = F\left(x_{i0}, y_{i0}, y_{j0}, y_{j0}\right) = \sqrt{\left(x_{j0} - x_{i0}\right)^2 + \left(y_{j0} - y_{i0}\right)^2}$	Eqn 20-11

20.3.1.1 First Iteration, i_0:

Start with the estimates of **A·x = b + v**.

The initial estimated coordinates of **U₀**, from the unique solution using two distances, U-A and U-B, are: $\mathbf{U_{e0} = 1069.248}$, $\mathbf{U_{n0} = 5429.369}$. See the start of the Worked Example 20.2 for coordinates $\mathbf{U_{E0}}$ and $\mathbf{U_{N0}}$

In the first iteration, since we have used two measured distances to define approximate coordinates of U, we can say that $b_{l_{ij}} = l_{ij} - IJ_0 = 0$ with, in this case, $b_{UC} = l_{UC} - UC_0 =$ the difference between the measured and calculated distance to the third (over-determined) point. Point U is determined by distances AU and BU. CU is the over-determined distance.

And because we have turned a non-linear function into a linear function by the Taylor's expansion we can use the normal linear matrix methods for the solution of the least squares adjustment. The **A** matrix of observations:

$$\mathbf{A_0} = \begin{bmatrix} \left(\frac{x_{AU} - x_A}{AU}\right) & \left(\frac{y_U - y_A}{AU}\right) \\ \left(\frac{x_U - x_B}{BU}\right) & \left(\frac{y_U - y_B}{BU}\right) \\ \left(\frac{x_U - x_C}{CU}\right) & \left(\frac{y_U - y_C}{CU}\right) \end{bmatrix} = \begin{bmatrix} \left(\frac{1069.248 - 1188.113}{152.42}\right) & \left(\frac{5429.369 - 5333.960}{152.42}\right) \\ \left(\frac{1069.248 - 1069.120}{138.580}\right) & \left(\frac{5429.369 - 5290.783}{138.580}\right) \\ \left(\frac{1069.248 - 1046.174}{84.110}\right) & \left(\frac{5429.369 - 5348.466}{84.110}\right) \end{bmatrix} = \begin{bmatrix} -0.78985 & 0.62596 \\ 0.00072 & 1.00000 \\ 0.27433 & 0.96187 \end{bmatrix}$$

The **b** matrix elements are the residual distances from U to A, B and C. Distances AU and BU were used to calculate the position of U, so initial error is assumed to be 0.000, and $\ell_{CU} - C_{U0}$ is the over-determined distance with an error of (measured – calculated) = –0.019.

$$\mathbf{b_0} = \begin{bmatrix} l_{AU} - AU_0 \\ l_{BU} - BU_0 \\ l_{CU} - CU_0 \end{bmatrix} = \begin{bmatrix} 154.420 - 154.420 \\ 138.580 - 138.580 \\ 84.110 - 84.129 \end{bmatrix} = \begin{bmatrix} 0.000 \\ 0.000 \\ -0.019 \end{bmatrix}$$

Normalised matrix: $\mathbf{A^TAx = A^Tb}$

Solve for x $\mathbf{x_0 = ((A^TA)^{-1})(A^Tb)}$, break the problem down:

$$\mathbf{A^TA} = \begin{bmatrix} 0.683 & -0.223 \\ -0.223 & 2.317 \end{bmatrix} \quad \mathbf{(A^TA)^{-1}} = \begin{bmatrix} 1.511 & 0.146 \\ 0.146 & 0.446 \end{bmatrix} \quad \mathbf{A^Tb} = \begin{bmatrix} -0.006 \\ -0.182 \end{bmatrix} \quad \mathbf{x_0} = \begin{bmatrix} -0.011 \\ -0.009 \end{bmatrix}$$

At the end of the first iteration the coordinates of U become,

as $U_1 = E_0 + x_{E1}, N_0 + x_{N1}$):

1069.248 + (–0.011) = 1069.237

5429.369 + (–0.009) = 5429.360

20.3.1.2 Second Iteration. i_1:

Run the second iteration, where now, using the U1:

$$\mathbf{A_1} = \begin{bmatrix} \frac{1069.237 - 1188.113}{152.42} & \frac{5429.360 - 5333.960}{152.42} \\ \frac{1069.237 - 1069.120}{138.580} & \frac{5429.360 - 5290.783}{138.580} \\ \frac{1069.237 - 1046.174}{84.110} & \frac{5429.360 - 5348.466}{84.110} \end{bmatrix} = \begin{bmatrix} -0.779923 & 0.625903 \\ 0.000845 & 0.999936 \\ 0.274201 & 0.961765 \end{bmatrix}$$

Using the new coordinates for U, the distances AU_1, BU_1 and CU_1 are calculated, and shown to 4DP to create the **b** matrix:

$AU_1 = 152.4226$
$BU_1 = 138.5711$
$CU_1 = 84.1175$

$$\mathbf{b_1} = \begin{bmatrix} l_{AU} - AU_1 \\ l_{BU} - BU_1 \\ l_{CU} - CU_1 \end{bmatrix} = \begin{bmatrix} 154.420 - 154.423 \\ 138.580 - 138.571 \\ 84.110 - 84.118 \end{bmatrix} = \begin{bmatrix} -0.003 \\ 0.009 \\ -0.008 \end{bmatrix}$$

Normalise matrix: $\mathbf{A^T A x} = \mathbf{A^T b}$

Solve for x $\mathbf{x_1} = ((\mathbf{A^T A})^{-1})\mathbf{A^T b}$, break the problem down:

$$\mathbf{A^T A} = \begin{bmatrix} 0.6835 & -0.2236 \\ -0.2236 & 2.3166 \end{bmatrix} \quad (\mathbf{A^T A})^{-1} = \begin{bmatrix} 1.511 & 0.146 \\ 0.146 & 0.446 \end{bmatrix} \quad \mathbf{A^T b} = \begin{bmatrix} 1.29^{-7} \\ -3.35^{-6} \end{bmatrix}$$

$$\mathbf{x_1} = ((\mathbf{A^T A})^{-1})(\mathbf{A^T b}) = \begin{bmatrix} -2.9^{-7} \\ -1.5^{-6} \end{bmatrix} = \begin{bmatrix} 0.000 \\ 0.000 \end{bmatrix}$$

The dE (x_{E1}) and dN (x_{N1}) corrections are now effectively zero.

At the end of the second iteration the coordinates of U become, as $U_2 = (E_1 + x_{E1}, N_1 + x_{N1})$:

1069.237 + (–0.000) = 1069.237
5429.360 + (–0.000) = 5429.360

20.3.1.3 Third Iteration, i_2:

There is no need to run a third iteration as the **x** matrix of corrections has converged to almost zero.

Final least squares derived coordinates. The best estimate for U is thus:

E_U 1069.237
N_U 5429.360.

Since there is no change to the coordinates of **U** the distances remain the same as previously calculated.

We started by saying we would have residuals, **v**, from the estimates, where **Ax=b + v**.

Thus, at the end of the iterations, $\mathbf{v} = \mathbf{A_1 x_1} - \mathbf{b_1}$.

$$\mathbf{A_1 x_1} = \begin{bmatrix} -6.97^{-7} \\ -1.47^{-6} \\ -1.50^{-6} \end{bmatrix} \quad \mathbf{b_1} = \begin{bmatrix} -0.003 \\ 0.009 \\ -0.008 \end{bmatrix} \quad \mathbf{v} = \begin{bmatrix} 0.003 \\ -0.009 \\ 0.008 \end{bmatrix}$$

The variance, and hence the standard deviation, is found from the sum of the residuals squared, Σv^2, divided by the number of redundant observations ($m - n = 1$).

$$v^2 = \begin{bmatrix} 6.93^{-6} \\ 7.88^{-5} \\ 5.65^{-5} \end{bmatrix} \quad \Sigma v^2 = 1.42^{-4} \quad \sigma^2 = \frac{1.42^{-4}}{(3-2)}$$

$\Sigma v^2 = \sigma^2 = 0.000142$, thus

$\boldsymbol{\sigma_0 = \pm 0.012}$.

Finally, compare the **adjusted** coordinates E 1069.237 N 5429.360
with the averaged coordinates E 1069.220 N 5429.364.

20.3.1.4 The absolute error ellipse.

Carrying on from § 20.2.1.1 (Schofield and Breach, section 7.11, page 255) we can form the error ellipse for point U from the inverted normalised unweighted observation matrix.

$$\left(\mathbf{A^T A}\right)^{-1} = \sigma^2 = \begin{bmatrix} \sigma_E{}^2 & \sigma_{EN} \\ \sigma_{NE} & \sigma_N{}^2 \end{bmatrix} = \begin{bmatrix} 1.511 & 0.146 \\ 0.146 & 0.446 \end{bmatrix} \quad \sigma = \begin{bmatrix} 1.229 & 0.382 \\ 0.382 & 0.668 \end{bmatrix}$$

$$\tan 2\theta = \frac{2\sigma_{EN}}{\sigma_N{}^2 - \sigma E^2} = \frac{2(0.146)}{0.446-1.511} = -0.2676$$

$2\theta = -15.332°$. as $\tan 2\theta$ is negative, add 90°

$\theta = -7.666° + 90° = 82°20'$. Major axis orientation.

Major axis, a: $\sigma_{max}{}^2 = \frac{1}{2}\left[\sigma_N{}^2 + \sigma_E{}^2 + \left\{\left(\sigma_N{}^2 - \sigma_E{}^2\right)^2 + 4\sigma_{EN}{}^2\right\}^{\frac{1}{2}}\right]$ Note σ_{EN} is squared

$$= \frac{1}{2}\left[0.446 + 1.511 + \left\{(0.446-1.511)^2 + 4\times(0.146)^2\right\}^{\frac{1}{2}}\right] = 1.531$$

$a = \sigma_{max} = 1.237$. Major axis

Minor axis, b: $\sigma_{min}{}^2 = \frac{1}{2}\left[\sigma_N{}^2 + \sigma_E{}^2 - \left\{\left(\sigma_N{}^2 - \sigma_E{}^2\right) + 4\sigma_{EN}{}^2\right\}^{\frac{1}{2}}\right] = 0.426$

$b = \sigma_{min} = 0.653$. Minor axis

20.3.1.5 The error ellipse and associated pedal curve.

Referring to § 20.2.1.2, again the ellipse is dimensionless; that means you choose a size that suits the presentation. This ellipse is quite eccentric, suggesting that one of the observed distances may be in error. The residuals of two observation:

UB = –0.007 and

UC = +0.008

show a suspect set of distances.

The narrow intersection angle between them (∠CUB = 17°) shows poor geometry.

The ellipse, Figure 20-5, shows the orientation and magnitude of the major and minor standard deviations.

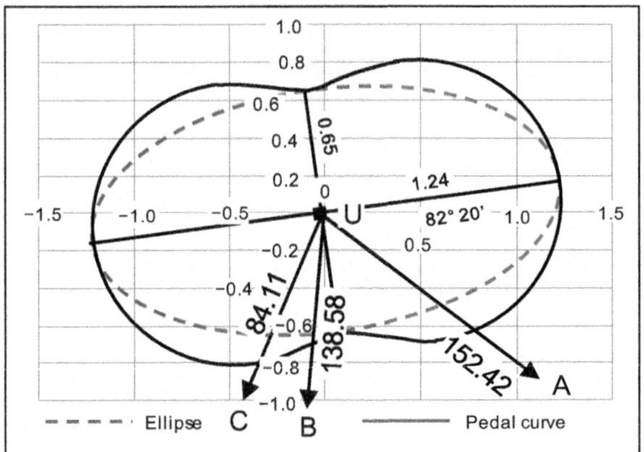

Figure 20-5

20.3.2 A note of caution.

This article on least squares has not been peer reviewed by your lecturers. There may be errors and mistakes, especially in partial differentiation. The problems have yet to be run through least squares software for validation.

20.4 References for Chapter 20

Richardus, P., Project Surveying, North-Holland Publishing Company, Amsterdam, 1966

Schofield, W and M Breach. Engineering Surveying, Sixth edition. Elsevier. 2007. ISBN 978-0-7506-6949-8

Chapter 21 TRIAL LINES, OR PARALLEL OFFSETS.

Cadastral boundaries, defined by on line and corner pegs, often rely on reestablishment to verify property entitlement or to incorporate subdivisions. This chapter investigated direct methods of resurveying to allow replacement of missing physical boundary points. Corner pegs are normally guarded by witness or reference marks (RMs) which are recorded in cadastral field books. These must be used to check or replace a boundary marker to eliminate disturbance. In some cases there may be only a witness marks to a missing boundary peg.

21.1 Trial lines for boundaries

This topic is examined in some detail because boundary offsets, and more particularly the observational techniques, are not usually covered in textbooks. It is of particular interest to licensed survey work related to property boundaries as well as engineering layout.

The boundaries of reserves for roads, railways, pipelines, drainage channels, of easements (R.O.W's), etc. are usually formed by a series of pairs of parallel lines at some fixed distance apart. Refer to Chapter 9, page 139, Road Intersections.

The solution to the problem of establishing a smaller **parallel** offset, generally 2 to 5 metres away from a boundary, generates the **trial line.** This line is then used to define the offset corners of intersecting boundaries.

The dimensions of a figure for which the trial lines will be established are known. Of particular interest is the problem of (re)establishing property boundaries when the boundary line is inaccessible, especially when survey corner pegs or witness marks are missing. To confirm the dimensions of the existing lot, it may be necessary to girt the property with 4 trial lines.

In Figure 21-1 a property ABCD is to be subdivided. An inaccessible boundary line means trial lines are needed for the setout. The parallel trial lines **intersect** on the diagonal from the boundary. The intersect angle equals the original lot angle.

It is usually required to calculate:

1. the lengths of the trial lines boundaries, A′B′, B′C′, C′D′, D′A′.
   length of original boundary plus extension to the intersect.
2. the lengths of the diagonals from the lot corners, A, B, C, D,
   to the trial line intersect corners, A′, B′, C′, D′.

To calculate these lengths the trial line offset must be decided. Generally, the smaller the offset the better, but 2m is about the minimum offset imposed by instrument set-up limitations. **Assuming equal offsets.**

Solution of:

1. requires distances A′a = A′a′,
   B′b = B′b′, C′c = C′c′ and D′d = D′d′,
   <u>the cotangent distances</u>. (The boundary extension distances).
2. requires distances AA′, BB′, CC′ and DD′,
   <u>the cosecant distances</u>, (the intersection diagonal distances).

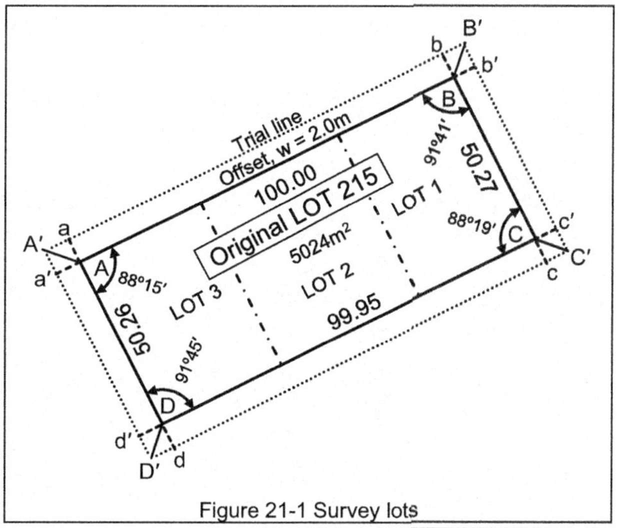

Figure 21-1 Survey lots

Boundary lengths AB, BC, CD and DA and Angles A, B, C and D are known from field books or plans.

Offset distance is planned.

Subdivision of Lot 215 into 3 lots of equal area.

Lot 215 closes to 1:21,000. Marks found at all corners.

21.1.1 Calculation of boundary extension and cosecant (diagonal) distances.

21.1.1.1 Equal offsets

In Figure 21-2: Bb = Bb′ = the equal offsets.

∠B′bB = ∠B′b′B = 90° (perpendicular distances).

The triangles B′bB and B′b′B are congruent, (2 sides, 1 angle) and therefore B′b = B′b′, and ∠BB′b = ∠BB′b′

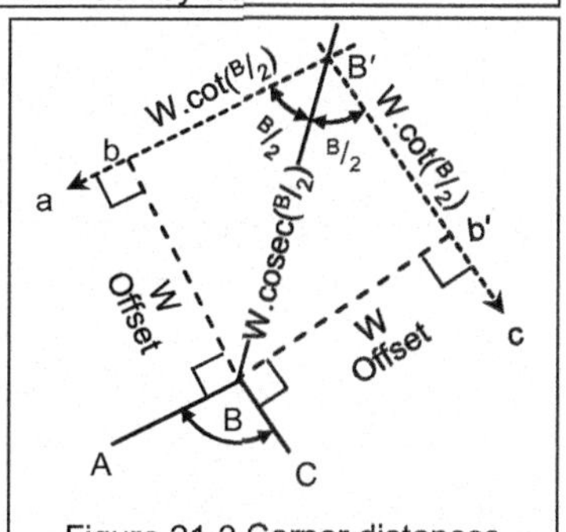

Figure 21-2 Corner distances

If it's not checked it's wrong.

J. Walker and J. Awange, *Surveying Computations*,
https://doi.org/10.1007/978-3-031-94382-9_21

When ∠ABC is known, ∠BB′b. = ½∠ABC = ∠BB′b′

∠ABB′ = ∠CBB′ = ½(360° - ∠ABC)

B′B = Bb×cosec(∠BB′b) = Bb×cosec(½∠ABC) = $Bb \div \sin(\tfrac{1}{2}\angle ABC)$ = <u>Diagonal distance</u>

bB′ = B′b′ = Bb×cot(∠BB′b) = Bb×cot(½∠ABC) = $Bb \div \tan(\tfrac{1}{2}\angle ABC)$ = <u>Extension distance</u>

21.1.2 Worked Example 21.1 Constant (equal) offset distance.

Field book data of boundaries and measurements of the subdivision at corner B in Figure 21-3 are as follows:

∠B = ∠ABC = 91°41′ offset w = 2.000

Calculate the lengths of the trial lines extensions and the intersection diagonal distance.

∠B = 91°41′, ∠½B = 45°50′30″

cot(½B) = 0.971044 = 1/ tan(½B)

cosec(½B) = 1.393889.

bB′ = B′b′ = Bb·Cot(∠BB′b) = Bb·Cot(½∠B)

= 2.000×cot(45°50′30″) = 2.000 x 0.97104

= 1.9421

BB′ = Bb·Cosec(∠BB′b) = Bb·Cosec(½∠B)

= 2.000×cosec(45°50′30″) = 2.000 x 1.3934

= 2.7878

See Figure 21-3 for the results.

So there is no mystery in equal parallel offsets.

Extension = Offset×cot(½Angle) or Offset/tan(½Angle),

Diagonal = Offset×cosec(½Angle) or Offset/sin(½Angle).

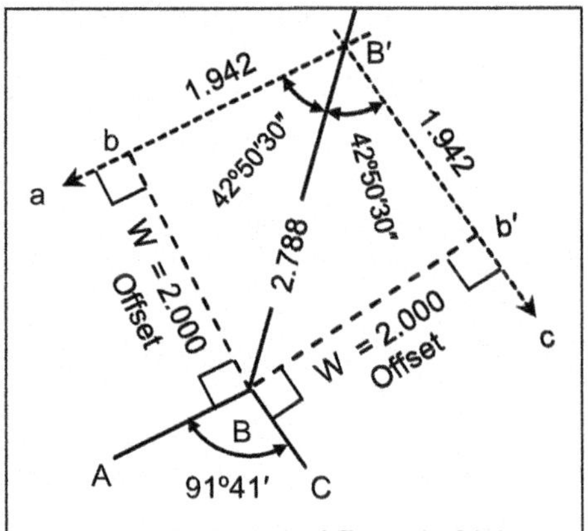

Figure 21-3 Worked Example 21.1

The mystery is in the precise setout of the offset points and the intersect point. More anon.

21.1.3 Calculating boundaries and cosecant (intersection) distances for offsets of different (unequal) values.

In Figure 21-4 let ABC be the lot boundaries intersecting at points B. YB′X is portion of the unequal parallel offset boundaries.

X and Y are the intersections of the prolongation of the boundaries AB and BC respectively with the parallel offsets.

Bb(W_1) and Bb′(W_2) are the perpendicular offsets from boundaries AB and BC respectively.

Many texts use the supplementary angle, ∠α = 180 – ∠B, for calculations.

This approach is not necessary.

Using only the internal angle, ∠B, the cosecant (1/sin) is positive in QI and QII.

The cotangent (1/tan) supplies:

- the sign of the offset
- correction between the offset distance and the side adjacent the prolongated intersect (phew).

bY and b′X are the sides referred, see Figure 21-4.

If ∠B > 90° then bY and b′X will be negative.

Thus: noting W_1 and W_2 order,

B′**b** = B′Y + bY = W_2/sin(B) + W_1/tan(B)

B′b′= B′X + b′Y = W_1/sin(B) + W_2/tan(B).

This avoids the Cosec and Cot functions.

These two values, coupled with their respective offsets, provide the rectangular coordinates from which the polar distance (the diagonal) across the corner can be calculated (and checked).

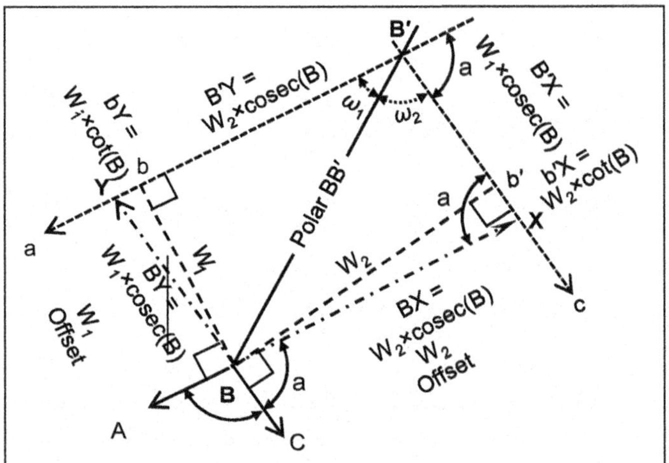

Figure 21-4 Differing offsets

The two angles, ∠BB′Y and ∠BB′X can be calculated by the polar method, confirming their sum equals ∠B.

∠BB′Y = ω_1 = arctan(W_1/B′**b**), BB′ = B′b/sin(ω_1)

∠BB′X = ω_2 = arctan(W_2/B′b′), BB′ = B′b′/sin(ω_2)

Check ∠B = ω_1 + ω_2.

If it's not checked it's wrong.

21.1.4 Worked Example 21.2. Different offsets.

Taking the same corner as Worked Example 21.1 but using 2m and 3m offsets. (See Figure 21-5)

$\angle B = 91°41'$

$\cot(B) = -0.029388 = 1/\tan(B)$

$\mathrm{cosec}(B) = 1.000432 = 1/\sin(B)$

offset $W_1 = 2.000$, $W_2 = 3.000$

Instead, taking the same corner as Worked Example 21.1 and using 2m and 3m offsets. (See Figure 21-5)

$\angle B = 91°41'$

$\cot(B) = -0.029388 = 1/\tan(B)$

$\mathrm{cosec}(B) = 1.000432 = 1/\sin(B)$

offset $W_1 = 2.000$, $W_2 = 3.000$

$B'Y = W_2 \times \mathrm{cosec}(B) = 3 \times 1.0004 = 3.0013$

$bY = W_1 \cdot \cot(B) = 2 \times -1.0294 = -0.0588$

$B'b = 3.0013 + (-0.0588) = 2.9425$

Similarly:

$B'X = W_1 \cdot \mathrm{cosec}(B) = 2.0009$

$b'X = W_2 \cdot \cot(B) = -0.0881$

$B'b' = 1.9127$

$BB' = 3.5579$ from both triangles.

$\angle BB'Y = \arctan(W_1/B'b)$

$= \arctan(2/2.9425) = 34°\ 12'\ 13''$

$\angle BB'X = \arctan(w_2/B'b')$

$= \arctan(3/1.9127) = 57°\ 28'\ 47''$.

Check: Sum $= 91°\ 41' = \angle B$

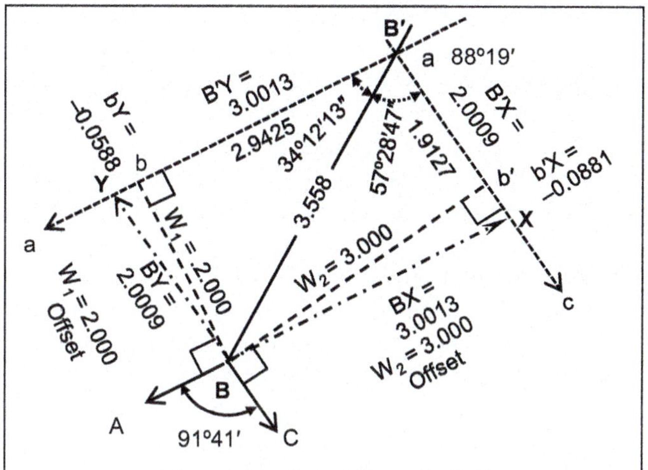

Figure 21-5 Worked Example 21.2

21.2 Practical aspects of trial line set out.

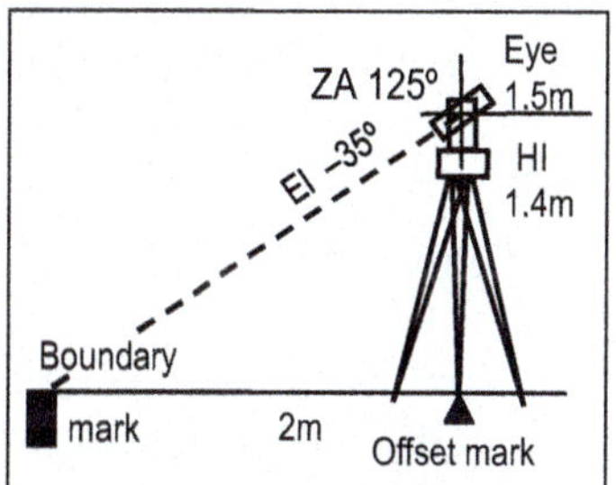

Figure 21-6 Over the offset point

After reconnaissance and positive identifications of survey marks the trial line involves an approximate, but close, set out.

The first order of business is to set the offset points perpendicular to the boundary line at the marks. It is easily achieved with an optical square. Measure each offset mark carefully with an offset tape and use a GI nail at each offset.

Keep the offsets as short as possible. This will involve setting the total station up about 100mm lower than usual because the tilt of the telescope lifts the eyepiece about 90mm above the trunnion axis at a depression angle of 35°. (See Figure 21-6). This is about the physical limit of telescope depression.

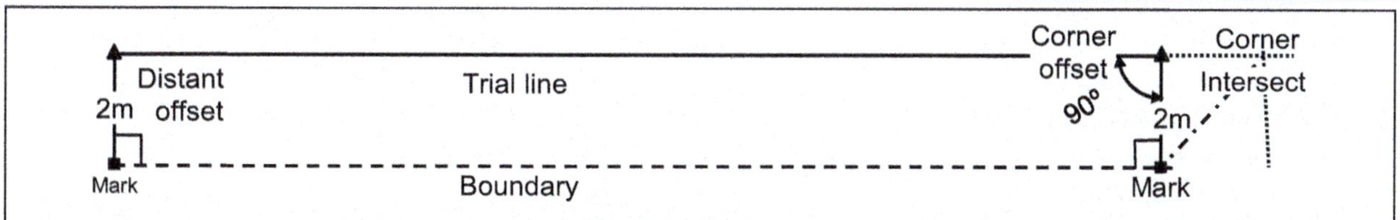

Figure 21-7 Trial line setout

A 2m offset, with depression 35° (ZA 125°) has an eye height of about 1.5m.

Set a prism over the distant offset mark and plumb the total station over the corner offset. (See Figure 21-7).

Back sight to the distant offset, turn 90° and observe the mark. With an offset tape at the mark, measure the misclose between the cross hair and the station mark. (See Figure 21-8).

Setup as above, the 2.5m slope distance field of view is about 65mm.

The offset mark is 25mm beyond the boundary.

The mark has to be moved. But how?

What does the problem look like? (See Figure 21-9)

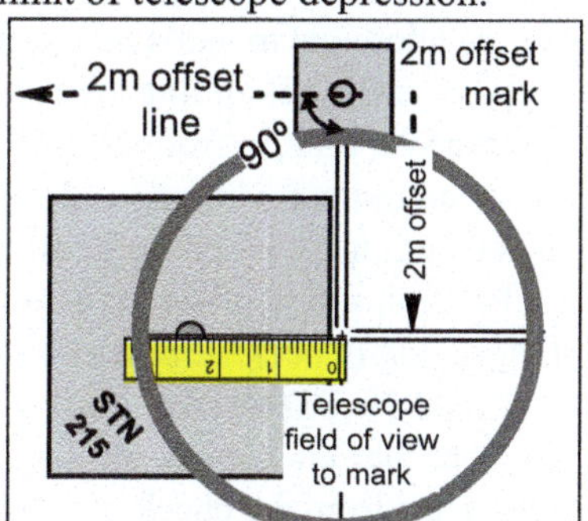

Figure 21-8 Offset to station

The only problem is the "along offset" **misclose** distance. The 25mm error does not affect the offset line, the diagonal is still 2000mm, with 43′ misalignment.

From simple trig: varying misclose, offset shift

65mm; 2000mm diagonal; 1° 51′ skew, 1999mm offset

90mm; 2000mm diagonal; 2° 35′ skew, 1998mm offset

110mm; 2000mm diagonal; 3° 09′ skew, 1997mm offset.

Even a badly skewed 2000mm set-out line, with say 100mm misclose, still keeps the offset distance on the head of the GI nail. You have to only move along the line.

It means that a reasonable offset at the far end of the line will have NO effect on the measured misclose at the mark.

How so?

Well, let's accept a 3mm offset error, say over a 100m line. (See Figure 21-7).

Angular error = 0.003/100 radians = 6″.

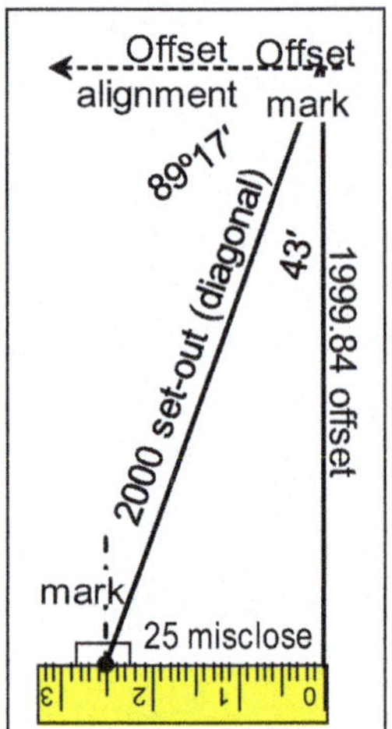

Figure 21-9
25mm offset angle

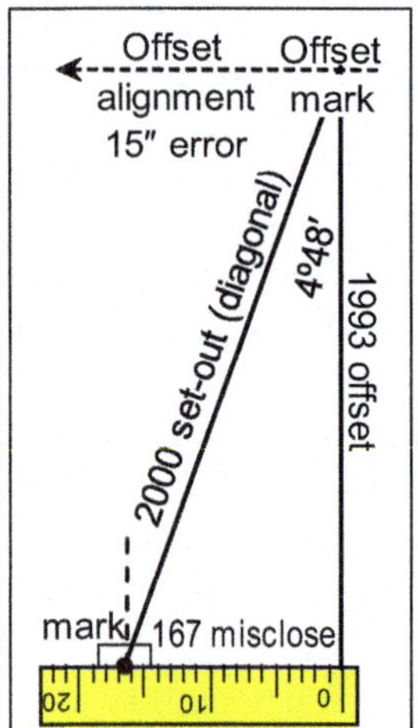

Figure 21-10
15″ alignment error

That's not bad, but how about we say 15″ over the 100m line? (See Figure 21-10)

15″ at 100m = 0.007 (7mm); measured 2000mm diagonal;

1993mm offset, 167mm misclose; 4° 48′ skew.

Turning 90° from the back sight to the mark (survey from the whole to the part rule)

Error 15″ at 2m = 0.00014m = 0.14mm.

Thus even a badly skewed back sight, and a poorly turned right angle, do not affect the misclose distance.

1′50″ angular error gives an "along alignment" error of 1mm. (See Figure 21-11)

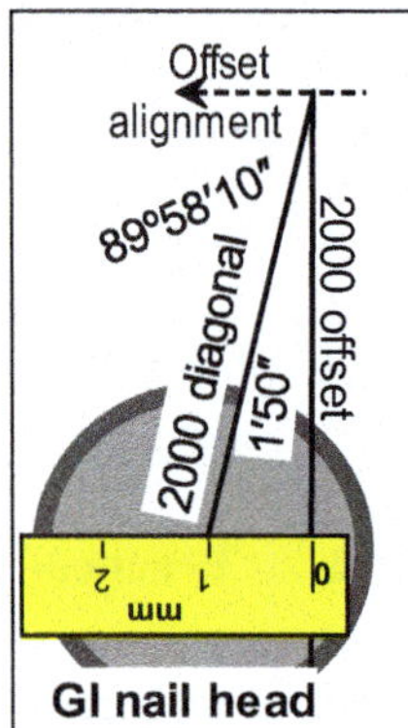

Figure 21-11
1mm offset

21.3 Setting out the intersection point.

First, check the length of the boundary from the currently positioned set-up.

Field Book boundary length:	100.000m
H. dist. to distant offset (backsight prism):	100.030
Observed misclose (towards intersection – subtract):	–0.025
Offset boundary distance	100.005
Misclose from Field Book:	+0.005.

0.005m in 100m = 1:20,000 error. If it was an error, it would have NO influence on the offset distance from the alignment corner peg, A.

The offset alignment is acceptable.

From the pre-calculations for the intersect (§ 21.1.2 Worked Example 21.1):

Intersect distance from offset:	1.942
Offset misclose (towards intersection – subtract):	–0.025
Distance to corner intersection:	1.917.

Back sight trial line to offset mark (prism),(See Figure 21-12).

Plunge telescope towards intersection:

place a GI nail **on line** at about 2m (1.92m is a bit steep at 36°)

measure 1.917m from offset mark to intersection,

set alignment with edge of offset tape, check with telescope.

Check alignment by plunging from BS on the second face.

Position a new offset mark at the 1.942m distance from the intersection (Check: 25mm away from the current mark.)

Look! No EDM, no angles!

Note that all this set-out has been accomplished with an offset tape, apart from the check measure to the BS prism.

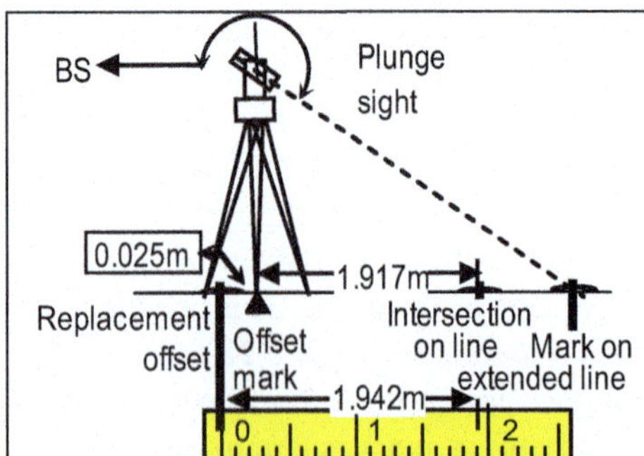

Figure 21-12 Setting offset mark

And no angles need reading, just the back sight and 90° turn for the offset, or back sight and plunge for the projection of the offset line intersect.

Figure 21-13 shows the plan measurement to the offset intersect.

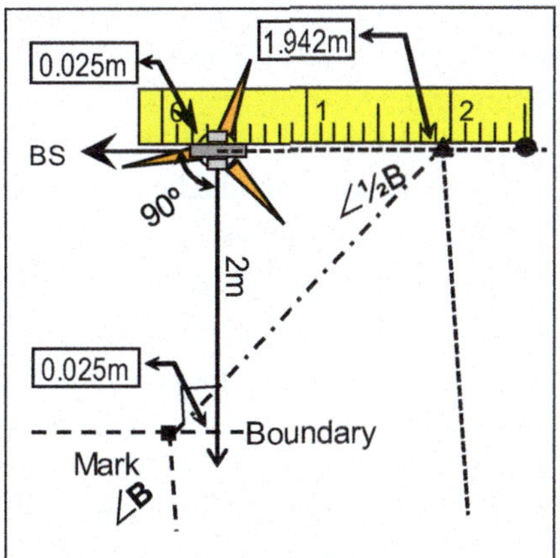

Figure 21-13 Plan of offset

21.3.1 Reprise of equal offsets.

Calculating:

boundary extension to intersect point B′.

bB′ = W×cot(½B)

diagonal distance.

BB′ = W×cosec((½B)

(See Figure 21-14 and example in Figure 21-3, page 324)

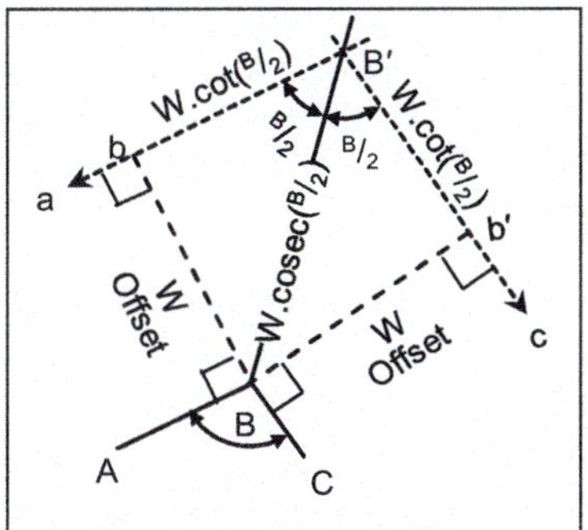

Figure 21-14 Equal offsets

21.4 Middling in – getting "online" before the internet.

What to do when there's a hill in the way? Make a "waddy" or two, just a cardboard red and white 40mm square target (R&W) nailed hammered in with a non-geodetic plumb bob on top of a 1.2m timber stake.

Sometimes the two offsets are not inter-visible and you have to set an intermediate station, and it has to be "on line". Set up and plumb a waddy at each end of the line over your offsets. Set it as shown in Figure 21-15 so that you align with the plumb bob string, which bisects the R&W, rather than having the R&W vertical and trying to intersect the colour boundary.

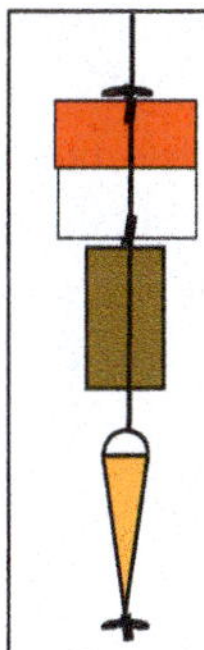
Figure 21-15 R&W

You can, of course, set up prisms over your over your offset marks and do the same process.

Set up at your middling in point. Use an optical plumb to the waddies, and guess the distance ratio (half way, 1/3 way etc).

Backsight and plunge the TS. Measure (or guess) the offset from the waddy.

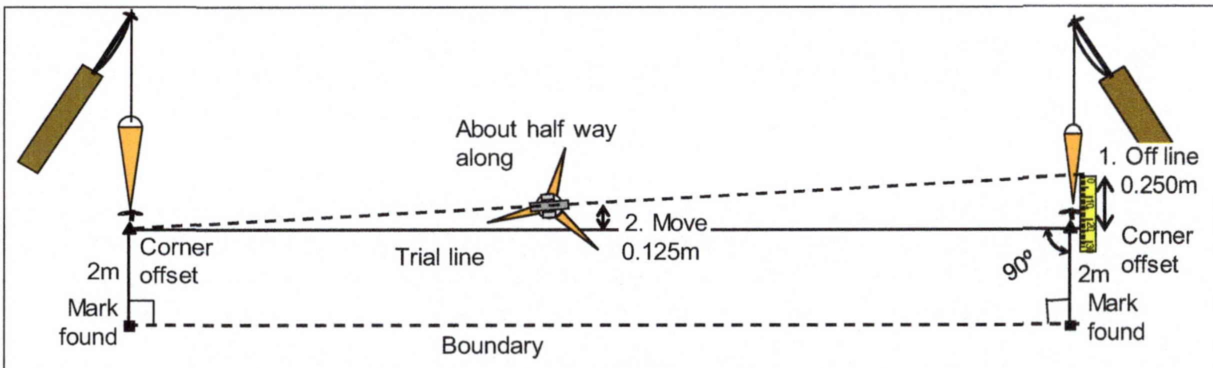

Figure 21-16 Getting on line

In Figure 21-16 the tape shows a misalignment of 0.250m (250mm). Half way along? Align the end of your offset tape with the TS plummet and put a GI nail at 125mm towards the offset line. ⅓ along? Move over 80mm.

Sighting on, or near, the R&W move the TS tribrach across the tripod head by the proportional error. Move by sighting through the optical plummet to the tape on the ground. (At halfway I just missed the R&W which is 50mm wide, so 25mm looks OK as a refinement. Figure 21-17).

BS, plunge, check and adjust. If it's acceptable, check on the other face.

Set the alignment mark under the plummet, checking plummet concentricity. (™).

You can now leave the tripod and tribrach in place which will allow you, now, to check your corner offsets and continue with the trial line survey.

Hey, look Mum. No EDM, no calculator. Just a plumb bob and tape.

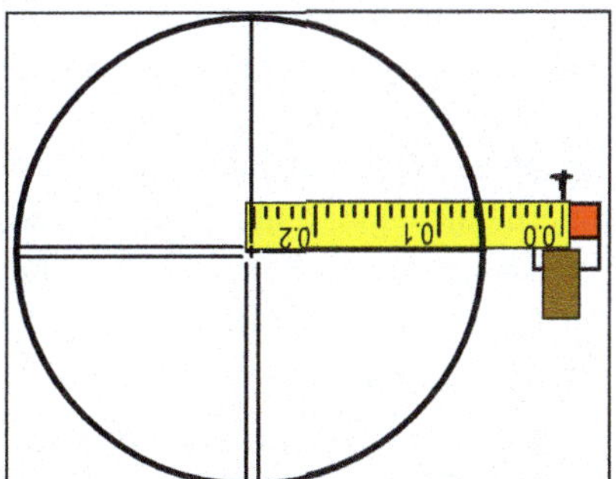

Figure 21-17 After the first adjustment

If it's not checked it's wrong.

If you do have the prisms set up then, before moving from the middling point, read the distances to get an uncorrected boundary length between the offset marks. Then you can proceed to set out the corner points as previously discussed, using the now "on line" target for alignment.

Remember that, although our offset mark may be slightly inaccurately set along the line, the offset distance should be reasonably good, and generally on the top of the GI nail. A couple of mm off line (Figure 21-5, page 325) only puts the halfway line station off by half that amount. Happy middling in.

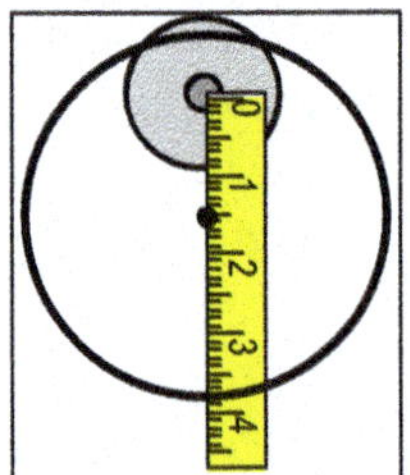

Figure 21-18
Adjust the head

21.5 Is this boundary line straight?

Long lines connecting lot boundaries are meant to be straight between the corners. A long boundary line is required to have intermediate marks on line to help define the position. If all these marks are found undisturbed a careful examination may show that line wobbles about the straight line between the corners. Why is this so? And does it matter in the definition of the boundary?

The line marks are there for intermediate distance determination, providing a check on the total boundary. Any distance error introduced by the observed "wobble" has no effect on the measured distance. The distance between the boundary corners is used to establish the location of each lot, one to another, and to assist in finding and re-establishing the corner. Total distance, and its intermediates, are guides only. The position of the peg is the important thing that must be verified by witness marks and comparison with adjacent boundaries.

Is there a wobble?

21.5.1 Survey and Plan Practice Manual for Western Australia- Edition 9.0 June 2013 Landgate

"4.18 Visibility of Marks

Intermediate spikes should be placed on every boundary where necessary such that (in addition to the requirement of General Regulation 39) from each end mark or intermediate mark on that boundary another mark should be visible on line in both directions (allowing that a pole up to 4 metres high may be used). This need applies in both rural and urban surveys and regardless of the length of the boundary. It is a practical requirement to allow fencing of the boundary or its identification by the public."

LICENSED SURVEYORS (GENERAL SURVEYING PRACTICE) REGULATIONS 1961 - REG 39

"39 Intermediate marks

(1) On all permanent boundaries that exceed 250 m in length a mark consisting of an iron spike at least 0.01 m in diameter and 0.4 m long driven flush, shall be placed exactly on the alignment at intervals not exceeding 250 m such that from each mark at least one other mark is visible forward and backward.

(2) Where such marks are not reasonably available or conditions are unsuitable for their use, alternative marks of other materials of equivalent durability and stability may be used.

(3) A full description of alternative marks used must be shown in the field record."

[Regulation 39 inserted in Gazette 9 Nov 1979 p. 3520; amended in Gazette 26 Jul 2013 p. 3470.]

It comes down to how the original surveyor established the line and placed the intermediates.

A lot of rural boundaries have line marks every 10 chain (200m). From a corner the surveyor establishes the direction of the line and sends the "chainies" out (with axe and chain) to cut a line and put the intermediate mark on the line as required. The surveyor advances to the intermediate, backsights to the corner and then produces the line forward to the next intermediate by plunging the telescope and measuring the distance. Advance to the next mark, backsight, plunge and mark forward. And so on. Any collimation error on plunging is taken out with FL/FR observations at each intermediate mark. Early instruments may have had small collimation errors. If FL/FR was not used, then a small direction error may creep in. However, by using alternate FL and FR plunges at each intermediate mark, the general direction of the line would be maintained.

The main requirement of the boundary intermediate distance spikes is to provide the distance between the line terminal posts. The minor intermediate misalignments will not affect the boundary distance.

21.6 References for Chapter 21

Uren, J. and B. Price. Surveying for engineers. Fifth edition. Palgrave Macmillan, London 2010. ISBN978-0-230-22157- 4

Chapter 22 UTM ZONE TO ZONE COORDINATE DIRECT TRANSFORMATION

22.1 An examination of direct transformation.

22.1.1 National Mapping direct coordinate zone-to-zone transformation using method 2 with tables

The inclusion of this method in current modern text is puzzling, and the author has tried to appreciate the rationale. The most interesting point is the assumptions of the method, and the accuracy of the results.

The method advocated is to convert UTM coordinates to geographic, then back to UTM in the adjacent zone, as per § 16.12.1.1. Prior to computers this was computationally intensive.

To allow direct rectangular coordinate conversion Natmap's AMG TM SE 7 (TM) section 5.12.3 (page 30) used a set of tables (Annexe E) and a form 5.13. This produced an accuracy to 10mm anywhere within half a degree of the zone boundary.

Section 6.6 (page 55) of the TM introduced simplified formula for the zone-to-zone transformation but with accuracies of only 1 metre within 10Km and 10m within 40Km of the boundary.

You have the coordinates of a known station, E_1, N_1 and the similar coordinates of E'_Z, N_Z on the zone boundary (ω = 3°) *chosen closest to* N_1. The table gives the convergence, γ, on the boundary in the form of the $\sin 2\gamma_z$, $(1-\cos 2\gamma_z)\times 10^4$, J_1 and $H_i \times 10^8$.

In the following formulae the definitions are:

Z	= a point on the zone boundary (CM±3°) depending on direction of transform.
E_1, N_1	= coordinates of known point to be transformed. $E'_1 = 500000 - E_1$. (Result negative in Eastern Zone)
E_2, N_2	= coordinates of resultant transformed point. $E'_2 = 500000 - E_2$.
E'_Z, N_Z	= coordinates of latitude on the zone boundary at $\omega_z = \pm 3°$. Tabulated in 15′ intervals. (Annexe E).
θ_z	= plane bearing from Z to the point to be transformed, E_1, N_1. $\tan\theta_z = (E'_1 - E'_z)/(N_1 - N_z)$
L	= $(E'_1 - E'_z)/\sin\theta_z$, $(N'_1 - N'_z)/\cos\theta_z$. Plane distance P_Z to P_1.
ω_z	= ±3° for AMG in radians (0.052359878).
ρ_Z	= radius of curvature in the meridian is latitude dependant. $\rho = a(1-e^2)/(1-e^2\sin^2\phi)^{3/2}$.
γ_Z	= convergence at θ_z at ω_z

Zone to zone transformation	$E_2 = 500000 - E'_z + (E'_1 - E'_z)\cos 2\gamma_z - (N_1 - N_z)\sin 2\gamma_z + H_1 L^2 \sin(2\theta_z + J_1)$ $N_2 = N_z + (N_1 - N_z)\cos 2\gamma_z + (E'_1 - E'_z)\sin 2\gamma_z + H_1 L^2 \cos(2\theta_z + J_1)$	Eqn 22-1
J_1 parameter	$\tan J_1 = \dfrac{\left[\omega_z^2 \cos^2\phi_z \left(1 + 31\tan^2\phi_z\right) - 6\left(1 + e'^2\cos^2\phi_z\right)\right]}{\left[18\omega_z \sin\phi_z \left(1 + e'^2\cos^2\phi_z\right)\right]}$	Eqn 22-2
H_1 parameter	$H_1 = \dfrac{-3\omega_z^2 \sin\phi_z \cos\phi_z}{\left(\rho_z \cos J_1\right)}$	Eqn 22-3
Convergence	$\gamma = -\sin\phi\omega - \sin\phi\dfrac{\omega}{3}\cos^2\phi\left(2\psi^2 - \psi\right) - \cdots$	Eqn 22-4

Table 22-1 is a calculated excerpt of the AMG zone to zone transformation parameters at the zone boundary. using the values of J_1 (Eqn 22-2) and $H_i.10^8$ (Eqn 22-3). Convergence, γ, loses less than 4″ by using the first term of Eqn 22-4 only. The second power contributes 1.85^{-5} radians, the 4″ error

The values of E'_Z (distance to zone boundary at ϕ_Z) and $\sin 2\gamma_Z$ (based on convergence) are unsigned in the table and are used in accordance of the transformation direction, East to West or vice versa.

Table 22-1 Example zone to zone parameters

LATITUDE	NORTHING N_z	E'_z	$\sin 2\gamma_Z$	$(1-\cos 2\gamma_Z)10^4$	J_1	$Hi.10^8$
32 0	6 455 617.802	283 424.255	0.055 5015	15.414	85 13	0.6983
15	6 427 889.918	282 631.506	0.055 8879	15.630	85 11	0.6964
30	6 400 161.236	281 873.372	0.056 2733	15.846	85 9	0.6944
45	6 372 431.754	281 089.867	0.056 6575	16.063	85 7	0.6925
37 0	5 900 905.064	266 963.091	0.063 0174	19.876	84 34	0.6570
15	5 873 160.663	266 085.501	0.063 3809	20.106	84 32	0.6548
30	5 845 415.408	265 202.834	0.063 7433	20.337	84 30	0.6526
37 45	5 817 669.298	264 315.106	0.064 1044	20.568	84 29	0.6504

J_1 is approximately $90° - 3\gamma$ at ϕ_Z on the CM.

J. Walker and J. Awange, *Surveying Computations*,
https://doi.org/10.1007/978-3-031-94382-9_22

If it's not checked it's wrong.

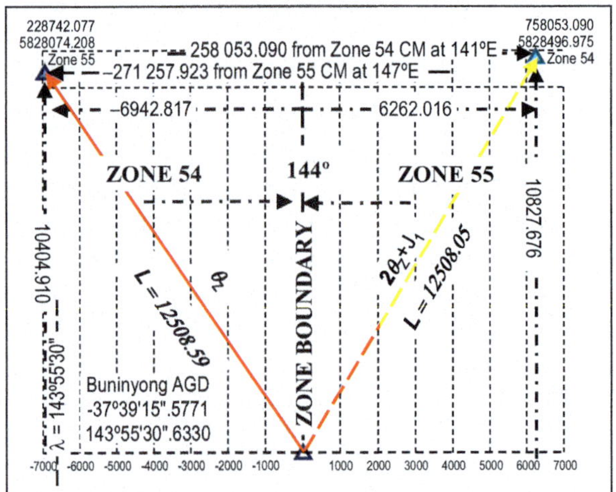

Figure 22-1 Zone transformation

22.1.1.1 Mimicking the AMG example

Buninyong: –37°39′15″.5571, 143°55′30″.6330

AMG Zone 54: 758053.090E, 5828496.974N

AMG Zone 55: 228742.077E, 5828074.208N

Convert from East zone 55 to West zone 54.

[15] E_1 = 228,742.077

[26] N_1 = 5,828,074.208

[16] False easting = 500,000.000

The SIGN convention of the various calculated values depends on:

a) Zone conversion direction (E to W / W to E)

b) Bearing θ_Z

c) Bearing $2\theta_Z+J_1$

The signs (Figure 22-2) are inserted manually in the calculation table. West zone to East zone requires the signs be swapped. Some of the variables are scales to help with logarithm solutions (1969)

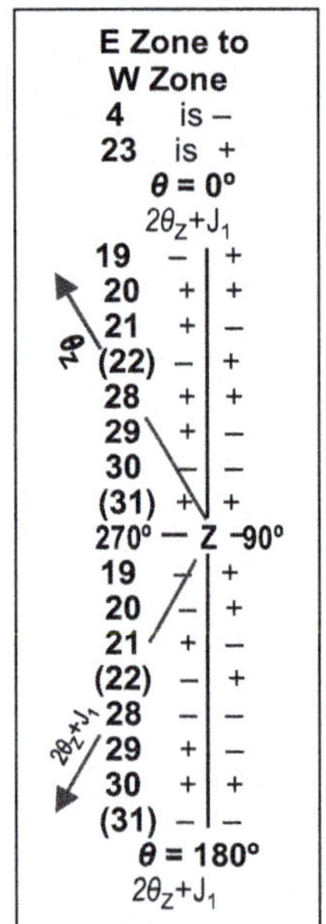

Figure 22-2 Signs

1. Find Latitude of nearest N_Z for N_1.

[18] E'_Z = 264315.106 (False Easting to CM)

From Table 22-1 Latitude = 37° 45′

[17] E'_1 = [15] – [16] (False Easting of E_1)

= 500000 – 228742.077 = 271257.923

[19] $\Delta E' = E'_Z - E'_1$

= 264315.106 – 271257.923 = –6942.817

[5] [19]×10^{-4} = –0.694282 ($\Delta E'$)

[6] [28]×10^{-4} = 1.04049 (ΔN)

[7] $\tan(\theta_Z) = \Delta E'/\Delta N$ = [5]/[6] = –0.667264.

[8] θ_Z = –33°42′49″+360 = 326°17′11″ (Quadrant IV)

[9] $2\theta_Z$ = 652°34′21″ = 292°34′21″

Evaluate J_1 from Eqn 22-3 where $e'2$ = 0.006739661 (Second eccentricity) ⇒ 84°30.4′

[4] J_1 = –84°29′ from Table 22-1. (Quadrant IV East to West ≅ –(90 – 3λ)

Calculate ρ_Z at 37°45′ = 6359381.381 using Eqn 22-2. J_1 = 84°30.4′.

[10] $2\theta+J_1$ = 208°05′21″

[11] $\sin(\theta_Z)$ = 0.555042

[12] $([19]/[11])^2 \times 10^{-8}$ = 1.56466 (Hypotenuse squared, $L^2\times10^{-8}$)

[13] sin([10]) = –0.470845

[14] cos([10]) = –0.882216

2. Calculate E'_2 in adjacent zone.

Accounts for the difference in ω in the adjacent zone.

[1] $\sin 2\gamma_Z$ = 0.0641044

[2] $(1-\cos\gamma_Z)\times10^4$ = 20.568

[3] $H_1\times10^8$ = 0.6504

[20] [1]×[28] ($\sin2\gamma_Z\times\Delta N$) = 667.000

[21] [2]×[5] ($(1-\cos\gamma_Z)\times10^4\times\Delta' E\times10^{-4}$) = 14.278

[22] [3]×[12]×[13] ($H_i\times10^8\times L^2\times10^{-8}\times\sin(2\theta+J_1)$) = –0.479

[23] [18]+[19]+sum([20]:[22]) E'_2 = 258053.088

[24] [23]+[16] False easting 500000.000

[25] [23] + [24] $\mathbf{E_2}$ = **758053.088**

3. Calculate N_2 in adjacent zone

From Table 22-1.

[26] N_1 = 5828074.208

[27] N_Z = 5817669 298

[28] [26] – [27] = ΔN = 10404.910

[29] [1]×[19] ($\sin2\gamma_Z\times\Delta E'$) = 445.065

[30] [2]×[6] ($1 - \cos2\gamma_Z)\times\Delta N$ = –21.401

[31] [3]×[12]×[14] ⇒

($H_i\times10^8\times L^2\times10^{-8}\times\cos(2\theta+J_1)$) = –0.897

[32] sum([27]:[31]) $\mathbf{N_2}$= **5828496.974**

The calculations, introduced about 1969, were carried out before calculators became available. So all work was done with logarithms. Scaling applied to H_i ($\times10^8$) and L^2 (10^{-8}), and ($1-\cos\gamma_Z$) ($\times10^4$) and $\Delta E'$ ($\times10^{-4}$) allowed sensible numbers to be generated before finding the logarithms. Jordan's earlier (1941) tables supplied the logarithms of the numbers. Adherence to the sign of various arguments relied on the direction of transformation.

Thus the ΔE′, ΔE′$_Z$ and ΔN′ were unsigned differences that had to be signed in the calculation depending on transformation direction (W to E or E to W) and quadrant (north or south of ϕ_Z, the nearest N_Z). J_1 and E′$_2$ are also dependent on transformation direction.

The arrival of the first HP35 calculator in 1972 obviated the use of logs and probably contributed to the format illustrated.

22.1.2 Comparison of methods.

The AMG Technical Manual S.P. 7, 1972 transformation was based on:

Jordan, W and Eggert, O. Handbuch der Vermessungskunde. Dritter Band. Zweiter Halbband. Sphäroidische Berechnungen, konforme Abbildung des Erdellipsoids und Aufgaben der Erdmessung. Metzlersche Verlagsbuchhandlung, Stuttgart, 1941 or the U.S. Army Corps of Engineers Army Map Service 1962 translation titled "Jordan's Handbook of Geodesy, Volume III, second half."

Their work refers back to:

Hristow. Transformation zwischen zwei Gauß-Krügerschen Streifen. Zeitschritt für P Vermessungswesen Heft 17 Band LXVII, 1938.

Ueber die Transformation zwischen zwei Gauß-Krügerschen Streifen mit Anwendung auf Deutschland.
Von Dr. Wl. K. Hristow, Sofia.

Their work was based on the Potsdam Datum of the German Main Triangulation Network (DHDN) Bessel 1841 ellipsoid. Compared with the AMG TM the equivalent variables are:

Equivalents to DHDN: $J_1 = u2$, $L = r$, $H_1 = s_2$, $\theta = v$. There is no equivalent of s_3 or u_3, these being very small numbers. J_1, u_2 are perpendicular to the CM ± 3 times the convergence at the zone boundary (ω = 3° or 1.5°). Hristow and Jordan round u_3 to whole degrees.

Apart from the Bessel 1841 versus the AMG ellipsoid the other main difference is that the DHDN G-K zone width is 3° and $k_0 = 1$ compared with UTM zone with 6° and $k_0 = 0.9996$. The method can be applied using the GDA ellipsoid as an exercise.

The author has yet to successfully calculate Jordan's s_3 and u_3 parameters, but they should be of little consequence if omitted as shown in the extreme zone offsets in § 22.1.3.4.2. They are well withing the tolerance of the AMG method.

22.1.3 Transformation of Coordinates for two Gauss-Kruger Systems. DHDN Bessel 1841 ellipsoid

Jordan's Handbook of Geodesy Volume III, second half, Chapter IV, section 38 page 208, (U.S. Army Corps) provided an example based on the Potsdam Datum, linked to the nationwide "German Main Triangle Network" (Deutsches Hauptdreiecksnetz (DHDN)).

The DHDN Gauss-Krüger (G-K) projection covers 5 zones, each 3° wide. The limits cover 1.5°E to 16.5°E (Paris to Vienna) in 5 indexed bands, with index 1 having a central meridian of 3°E.

The G-K coordinates use a 7-digit Easting coordinate, the first digit being the zone index number. False Easting for each zone is 500,000 + index×10^6. (1,500,000 to 5,500,000). Thus, any G-K coordinate can be converted to geographic because its zone is implicit in the G-K Easting.

The UTM system used worldwide requires the zone of the rectangular coordinates be specified so that the CM of the zone can be generated. The zone is generated from the geographic longitude. However UTM rectangular coordinates do not provide a "zone index" to allow transformation back to geographic coordinates. It must be specified.

22.1.3.1 Handling logarithms

Calculation with logarithms is a lost art. Basically, for surveying, the multiplication/division of numbers is performed by the addition/subtraction of their logarithmic values. The reader is referred to external sources regarding logarithms, or Chamber's Mathematical Tables. Figure 22-3 is an illustrates the zone parameters discussed.

Change from a west system (x, y) to an east system (x', y').

x	φ_0	x_0	y_0	$\log(1-\cos 2\gamma_0)$	$\log \sin 2\gamma_0$	$\log s_2$	u_2	$\log s_3$	u_3
m		m	m						
5 180 000	46° 30′	5 152 233,647	115 129,255	6.858 1339·9	8.579 5038·3	1.4518–10	+ 93° 15,9′	4.20–20	— 89°
5 792 000	52 0	5 763 813,247	103 001,595	6.930 0418·2	8.615 4454·3	1.4027	+ 93 32,8	4.23	— 87
5 847 000	52 30	5 819 438,155	101 849,888	6.935 9101·4	8.618 3774·1	1.3978	+ 93 34,3	4.24	— 87

Figure 22-3 Jordan-Eggert Bessel 1841 zone to zone parameters

The values shown in the table are small numbers where the *characteristic* (preceding the decimal point) is implied as a *bar* number and the remainder (after the decimal) is the *mantissa*. The mantissa is used to find the antilog from the log table. The characteristic then determines the number's decimal place. It also means that, to evaluate a log from an antilog on a calculator, it has to be converted to a 10's complement negative number such that, from , for φ 52 30′ where *log*(sin2γ) = 8.618 3774·1 the 10's complement is −1.381623 and the result is that $10^{\wedge -1.381623} = 0.0415315$. For log tables it is understood (in this context) 8.618 3774 the characteristic is *bar* $\bar{8}$. We find that the *antilog* of the mantissa 618 3774, from 7 figure logarithm tables (Chamber's p 65), falls between 415310 (618 3724) and 415320 (618 3828) which is proportional to 415315. The 10's complement of $\bar{8}$ is −2. The answer is 4.15315×10^{-2} or 0.0415315.

The convergence, γ, can be extracted directly from the logarithm sine tables (Chamber's pp 240 - 285) where *log*sin2γ = 8.618 3774 falls (p 185) between 2°22′ (8.615 8910) and 2°23′ (8.618 9369). By proportion from the differences value (30459 (83774–58910) = 24864/30459 ≅ 0.82′ = 49′. Thus 2γ = 2°22′49″ and γ = 1°11′24″.

Because *log* s_2 and *log* s_3 are very small numbers, the characteristic is evaluated from the 10's compliment before the mantissa is used to find the antilog. *Log* s_2 = 1.4518–10. The mantissa 4518 evaluates to 28301 and the characteristic (1–10 = –9) is the power of the antilog. The value is 2.8301×10^{-9}. Note *log* s_3 has a power 10^{-16}.

Table 22-2 Log solutions of

Log Number for 52° 30′	*log*(1-cos2γ)	*log*(sin2γ)	*log* s_2 (H_1)	u_2 (J_1)	*log* s_3	u_2
Number	$\bar{6}$.9359101	$\bar{8}$.6183774	1.3978–10		4.24–20	
10's complement	-3.0640899	-1.381623	-8.6022		-15.76	
Result: $10^{\wedge logn}$	0.0008628 (8.6280^{-4})	0.0415315 (4.153148^{-2})	2.4992^{-9}	93 34.3′	1.738^{-16}	-87º

22.1.3.2 Hristrow's 1938 calculations

In 1938 Dr. Wl. K. Hristow published a paper setting out the work on which Jordan's later calculations were based. Jordan truncated some of Hristow's tables which leads to slight differences in methods, and requires an understanding of the direction of the transformation, East zone to West zone or vice versa.

Transformation von einem Westsystem in ein Ostsystem

	φ	$x_0 = x_0'$ (m)	y_0 (m)	y_0' (m)	$1-k_{1.1}$	$k_{.2}$	$\varkappa_2$	ω_2 (0 ′)	$\varkappa_3$	ω_3 (0)	x (m)	
I	47° 0′	5207809.275	.614069.260	..385 930.740	+0.00073327	+0.03828834	1.4477–10	+93 17.5	4.20–20	–88	5236000	I
II	47 30	5263389.432	.613000.516	..386 999.484	+0.00074519	+0.03859823	1.4435	+93 19.1	4.20	–88	[illegible]	II

Transformation von einem Ostsystem in ein Westsystem

	φ	$x_0 = x_0'$ (m)	y_0 (m)	y_0' (m)	$1-k_{1.1}$	$k_{1.2}$	$\varkappa_2$	ω_2 (0 ′)	$\varkappa_3$	ω_3 (0)	x (m)	
I	47° 0′	5207809.275	..385 930.740	.614069.260	+0.00073327	–0.03828834	1.4477–10	–93 17.5	4.20–20	+88	5236000	I
II	47 30	5263389.432	..386 999.484	.613000.516	+0.00074519	–0.03859823	1.4435	–93 19.1	4.20	+88	[illegible]	II

Figure 22-4 Hristow transformation between West and East systems

Hristow also provided the G-K y_0 (E_0) coordinates at the zone boundary Figure 22-4 either side of the central meridian ($p = \pm1.5° = p/\varrho = 0.0262$radians) whereas Jordan provided only the E'_0 (y_0 in), the distance from the CM to the boundary. Note also the change of sign of many of the parameters $k_{1.2}$, ω_2 and ω_3. The table shows the values of $1-k_{1.1}$ and $k_{1.2}$ instead of the logarithms of $(1-\cos2\gamma_0)$ and $\sin2\gamma_0$ in Jordan's table. The values ω_2 and ω_3 are equivalent to $90 \cong \pm3\gamma_0$, tabulated as u_2, u_3 in Jordan and J_1 in the AMG TM. Log u_2, u_3 in Jordan, natural χ_2, χ_3 in Hristow and H_1 in TM are equivalent (depending on spheroid).

In all three sets of transformations it is necessary to calculate the plane bearing, either θ, υ or ϑ and the plane distance L, r or ϱ as is the want of the author. Tan(θ) occurs in many places and is generally denoted as t or t_0 as a shortcut.

22.1.3.3 Jordan's example. DHDN Bessel 1841 ellipsoid

υ (Greek upsilon) is the plane bearing from the chosen CM latitude, ϕ_Z, Northing x_0, y_0 to P_1, x, y.

Jordan's is a simplification of Hristow's tables and shows that the computer needed a detailed grasp of the transformation direction to enable the correct variable signs to be applied.

Jordan-Eggert Zone to zone transformation	$x' = x_0 + \Delta x - (1-\cos2\gamma_0)\Delta x - \sin2\gamma_0\Delta y + s_2r^2\cos(2\upsilon+u_2) + s_3r^3\cos(3\upsilon+u_3)$ $y' = -y_0 + \Delta y + \sin2\gamma_0\Delta x - (1-\cos2\gamma_0)\Delta y + s_2r^2\sin(2\upsilon+u_2) + s_3r^3\sin(3\upsilon+u_3)$ $r\sin\upsilon = \Delta y \qquad r\cos\upsilon = \Delta x$	Eqn 22-5

Jordans examples on pages 209-210 have been calculated to show the method in its original form, avoiding the complementary numbers needed for logarithmic calculations. A further set of forward/reverse calculation across the zone boundary were undertaken to examine the effect of 30′ deviation from the boundary.

The first example is a replication of Jordan's transformation of the Grunewald I survey point to validate the author's interpretation. § 22.1.3.4.1 and § 22.1.3.4.2 is a forward/reverse transformation near the Fahrlander church to test errors at the extreme overlap limit (30′) between zones and adjacent latitude (28′) parameters.

22.1.3.3.1 Grunewald I

Bessel ellipsoid 1841: Gauss-Krüger projection

a = 6377397.155 m. f = 1/299.15281285. b = 6356078.962822 m.

Scale factor, k_0 = 1.0. Zone width 3°. False E = 4500000, False N = 0.

Station: Transforming from West zone 4 to East zone 5.

Grunewald I: 4,586,714.070E, 5,814,976.154N, index 4 (CM 12°E). (False E = 4,500,000)

ϕ = 52°27′45.081″, λ = 13°16′33.568″.

The example illustrated in the translated Volume III, on pages 209 - 210, produces adjacent zone transformed G-K coordinates of 5,382,841.496E, 5,815,608.314N in zone 5, CM 15°E.

From x_0 = 5819438.155 y_0 = 101849.888 at 52 30′

$s_2 = 2.4992\times10^{-9}$ $s_3 = 1.738\times10^{-16}$

u_2 = 93°34.3′ u_3 = –87°

i) x (N) = 5814976.154 x_0 = 5819438.155 $\Delta x = x - x_0 =$ –4462.001

ii) y (E) = (4586714.070 – 4500000)y_0 = 101849.888 $\Delta y = y - y_0 =$ –15135.818 (Zone boundary 13.5°E)

iii) υ (θ) = arctan(Δy(dE))/Δx(dN)) = arctan (3.39216) = 73°34′29″ in quad. II = 253°34′29″

$r = \Delta y/\sin\upsilon$ = 15779.802

a) sin2γ = 0.04153148, (1–cos2γ_0)= 0.0008628)	2γ = 2°22′49″
b) u_2 = 90+3γ = 93°34′18″,	u_3 = 3γ–90 = –86°39′31″
c) 2υ + u_2 = 2×253.575°+(90+3γ) = 240°43′19″	3υ + u_3 = 3×253.575°+(3γ–90) = –45.9°

Applying Jordan's (49, page 208) equation.

a)	x_0	5819438.155	$-y_0$	–101849.888
b)	Δx	–4462.001	Δy	–15135.818
c)	$-(1-\cos2\gamma_0)\Delta x$	3.850	$\sin2\gamma_0\Delta x$	–185.315
d)	$-\sin2\gamma_0\Delta y$	628.613	$-(1-\cos2\gamma_0)\Delta y$	13.059
e)	$s_2r^2\cos(2\upsilon + u_2)$	–0.304	$s_2r^2\sin(2\upsilon + u_2)$	–0.542
f)	$s_3r^3\cos(3\upsilon + u_3)$	0.0005	$s_3r^3\sin(3\upsilon + u_3)$	–0.0005
g)	N′ = x'	5,815,608.314	y'	–117158.504
h)			y'+500,000+zone index = E′	5,382,841.496

The calculated values agree with Jordan's example on page 210. His example uses complementary numbers for some of the results.

22.1.3.4 Testing zone extremes in ω_0 and Δ_0

The following example was designed to show the effects of having the transformation point P at about 30′ from the zone boundary and 30′ from the ϕ_0 latitude in the tables. Jordan and Hristow use 30′ latitude intevals.

22.1.3.4.1 Fahrlander Kirche West to East, ϕ_0 = **52 30′**

Bessel ellipsoid 1841: Gauss-Krüger projection

False E = 4,500,000, False N = 0.

Station: Transforming from West zone 4 to East zone 5.

Fahrlander Kirche: 4,569,028.003E (y), 5,815,415.953N (x), index 4 (CM 12°E). (False E = 4,500,000)

ϕ = 52°28′08.383″, λ = 13°00′57.179″.

This example, at 30′ from the zone boundary, produces adjacent zone transformed G-K coordinates of 5,365,186.801E, 5,816,782.079N in zone 5, CM 15°E.

From x_0 = 5819438.155 y_0 = 101849.888 at **52 30′**

$s_2 = 2.4992\times10^{-9}$ $s_3 = 1.738\times10^{-16}$

u_2 = 93°34.3′ u_3 = –87°

i) x (N) = 5815415.953 x_0 = 5819438.155 $\Delta x = x - x_0$ = –1173.765

ii) y (E) = (4569028.003 – 4500000), y_0 = –103001.595, $\Delta y = y - y_0$ = –14156.909 (Zone boundary 13.5°E)

iii) υ (θ) = arctan($\Delta y/\Delta x$) = arctan (12.0611) = 85°15′37″ in quad. II = 265°15′37″ $r = \Delta y/\sin\upsilon$ = 14205.485

a) sin2γ = 0.0412520 (1–cos2γ_0) = 0.000862815)	2γ = 2°22′49″
b) u_2 = 90+3γ = +93°34′14″ (93°34.3 from table),	u_3 = 3γ–90 = –86°27′13″ (–87° from table)
c) 2υ + u_2 = 2×265.5260°–(90+3γ) = 76°58′27″	3υ + u_3 = 3×265.5260°–(3γ–90) = –9.9°

If it's not checked it's wrong.

Applying Jordan's (49, page 208) equation.

i)	x_0	5819438.155	$-y_0$	−101849.888
j)	Δx	−4022.202	Δy	−32821.885
k)	$-(1-\cos 2y_0)\Delta x$	3.470	$\sin 2y_0\Delta x$	−167.048
l)	$-\sin 2y_0\Delta y$	1363.142	$-(1-\cos 2y_0)\Delta y$	28.319
m)	$s_2r^2\cos(2\upsilon + u_2)$	−0.493	$s_2r^2\sin(2\upsilon + u_2)$	−2.688
n)	$s_3r^3\cos(3\upsilon + u_3)$	0.006	$s_3r^3\sin(3\upsilon + u_3)$	−0.002
o)	$N' = x'$	5,816,782.079	y'	−134813.192
p)			y'+500000+zone index (5) = E'	5,365,186.808

Converting these index 5 (CM 15°) DHDN G-K coordinates results in same geographic coordinates of ϕ = 52°28'08.383", λ = 13°00'57.179".

22.1.3.4.2 Fahrlander Kirche East to West, ϕ_0 = **52 00′**

Bessel ellipsoid 1841: Gauss-Krüger projection

Station: Transforming from East zone 5 to West zone 4.

Fahrlander Kirche: 5,365,186.808E, 5,816,682.079N, index 5 (CM 15°E). (False E = **5**,500,000))

ϕ = 52°28'08.383", λ = 13°00'57.179".

This example, near 30′ from the zone boundary, produces adjacent zone transformed G-K coordinates of 5,815,415.341E, 4,569,028.373N in zone 5, CM 12°E. (u_3 is positive; u_2, y_0, $\sin 2\gamma_0$ are negative E to W zone)

From x_0 = 5763813.247 y_0 = 103001.595 at **52 00′**

$s_2 = 2.5276\times10^{-9}$ $s_3 = 1.698\times10^{-16}$

u_2 = 93°32.8′ u_3 = −87°

reversed for East to West transformation (u_3 is positive; u_2, y_0, $\sin 2\gamma_0$ are negative):

i) x (N) = 5816782.079 x_0 = 5763813.247 $\Delta x = x - x_0$ = 52968.831

ii) y (E) = (5382841.496 − 5500000)y_0 = −103001.595$\Delta y = y - y_0$ = −31811.597 (Zone boundary 13.5°E)

iii) υ (θ) = arctan($\Delta y/\Delta x$) = arctan (−0.600572) = −30°59′16″ quad. III = 149°00′44″ $r = \Delta y/\sin\upsilon$ = 61787.335

a) $\sin 2\gamma$ = 0.0412520 $(1-\cos 2\gamma_0)$ = 0.000862815)	2γ = 2°21′51″
b) $u_2 = 90+3\gamma$ = −93°32′47″ (93°32.8′ from table),	$u_3 = 3\gamma-90$ = +86°27′13″ (+87° from table)
c) $2\upsilon + u_2 = 2\times149.012151°+(90+3\gamma)$ = 204°28′40″	$3\upsilon + u_3 = 3\times149.012151° +(3\gamma-90)$ = 174°02′11″

Applying Jordan's (49, page 208) equation. ,

a)	x_0	5763813.247	$-y_0$	103001.595
b)	Δx	52968.381	Δy	−31811.597
c)	$-(1-\cos 2y_0)\Delta x$	−45.702	$\sin 2y_0\Delta x$	−2185.070
d)	$-\sin 2y_0\Delta y$	−1312.292	$-(1-\cos 2y_0)\Delta y$	27.448
e)	$s_2r^2\cos(2\upsilon + u_2)$	−8.782	$s_2r^2\sin(2\upsilon + u_2)$	−3.998
f)	$s_3r^3\cos(3\upsilon + u_3)$	0.040	$s_3r^3\sin(3\upsilon + u_3)$	−0.0004
g)	$N' = x'$	5,815,415.341	y'	69028.373
h)			y'+500000+zone index (4) = E'	4,569,028.373

Converting these index 4 (CM 12°) DHDN G-K coordinates results in same geographic coordinates of ϕ = 52°28'08.363", λ = 13°00'57.198". There is a misclose of ΔE = −0.370 and ΔN = 0.612. This represents an error of (61787/0.715), 1:85,000 over the length of the line. At the 0.1″ level the geographic coordinates are the same.

This transformation illustrates transforming a point at the edge of the overlap ($\delta\lambda$ = 29′) and the 52°00′ latitude table value ($\delta\phi$ = 28′) still produces a result within 1m of the initial value.

22.1.3.4.3 Decoding g_2, h_2, g_3 and h_3

AGD UTM parameters of J_1 and H_1 in § 22.1.1, page 329, are calculated using Eqn 22-2 and Eqn 22-3, noting the author's comments on computational discrepancies.

Jordan's g_2, h_2, g_3 and h_3 and Hristow's $k_{2.1}$, $k_{2.1}$, $k_{3.1}$ and $k_{3.2}$ parameters are shown in Eqn 22-7 and Eqn 22-6. Radii of curvature, spheroid dependant. a = major semi-axis, f = inverse flattening (1/f).

Bessel 1841: e = 6377397.155m f = 1/299.1528129 = 0.003342773

$e^2 = (2f-f^2) = (a^2-b^2)/a^2$ eccentricity	= 0.006674372
$e'^2 = (a^2-b^2)/b^2 = e^2/(1-e^2)$ second eccentricity	= 0.006719219
$c = a/(1-e^2)^{1/2}$	= 6398786.848

Latitude dependant, ϕ, φ = 52°30′N

$V^2 = 1+e'^2\cos^2\phi = \rho/v = \psi$ 1.002490079

If it's not checked it's wrong.

$V^2 = 1+\eta^2 \Rightarrow \eta^2 = e'^2\cos^2\phi$ from (Jordan 1st half page 62, (b))
$\eta = 0.049900686$
ρ =radius of curvature in meridian = $a(1-e^2)/(1-e^2\sin^2\phi)^{3/2} = c/V^3 = M$ = 6374960.802
ν = radius of curvature in prime vertical = $a/(1-e^2\sin^2\phi)^{1/2} = c/V = N$ = 6390834.955 (Jordan).
R = mean radius of curvature = $(\rho\nu)^{1/2} = c/V^2$. = 6382892.944
γ = grid convergence, $\omega = 1.5° = 0.0262 = -\sin\phi\omega - \sin\phi(\omega^3/3)\cos^2\phi(2\psi^2-\psi)\ldots = 0.020771714 = 1°11'24''$
$2\gamma_0 = 2.3803° = 2°22'49''$
$t = \tan\phi = 1.303225373$

Hristow. Transformation components	*Hristow. Transformation zwischen zwei Gauß – Krügerschen Streifen.* _ *Zeitschritt für P Vermessungswesen* 1938 where $p = 1.5°$ in radians $k_{1.1} = \cos 2c_0'' = \cos c_0'$ $k_{1.2} = \sin 2c_0'' = -\sin c_0'$ $(c_0 = 2\gamma_0$ from Jordan$)$ $k_{2.1} = -3p^2 \frac{\cos^2\varphi}{N} t\left(1+\eta^2\right)$ $k_{2.2} = p\frac{\cos\varphi}{N}\left[\left(1+\eta^2\right) - \frac{p^2}{6}\cos^2\varphi\left(1+31t^2\right)\right]$ $k_{3.1} = \frac{p^2}{3}\frac{\cos^2\varphi}{N^2}\left(-3+4t^2\right)$ $k_{3.2} = -\frac{p}{3}\frac{\cos\varphi}{N^2} t\left(1+5\eta^2\right)$	Eqn 22-6

Note the difference in terms between the two sets of transformation components.

The author is unable to reconcile the calculated variables g_3, h_3, $k_{3.1}$ and $k_{3.2}$ with posted u_3 and χ_3 values. Working with diverse notation systems necessitates close scrutiny. In Eqn 22-7 Jordan uses the *rho* with tail symbol ϱ to denote a radian conversion ($\varrho = 57.2958°$). In Eqn 22-8 Hristow also uses ϱ to denote vector length whereas AGD uses L and Jordan uses r.

Jordan. Transformation components	$g_1 = \cos 2\gamma_0$ $h_1 = \sin 2\gamma_0$ $g_2 = -\frac{3}{N_0}\frac{p^2}{\varrho^2}\cos^2\varphi_0 t_0\left(1+\eta_0^2\right) \Rightarrow -\frac{3}{N_0}\omega_0^2\cos^2\varphi_0\tan\varphi_0\psi = s_2\sin u_2$ $h_2 = \frac{1}{N_0}\frac{p}{\varrho}\cos\varphi_0\left(1+\eta_0^2\right) - \frac{1}{6N_0}\frac{p^3}{\varrho^3}\cos^3\varphi_0\left(1+31t_0^2\right) = s_2\cos u_2$ $g_3 = -\frac{1}{3N_0^2}\frac{p^2}{\varrho^2}\cos^2\varphi_0\left(3-4t_0^2\right) = s_3\sin u_3$ $h_3 = -\frac{1}{3N_0^2}\frac{p}{\varrho}\cos\varphi_0 t_0\left(1+\eta_0^2\right) = s_3\cos u_3$	Eqn 22-7

Hristow. Zone to zone transformation	$\chi_2\sin\omega_2 = k_{2.2}$ $\chi_3\sin\omega_3 = k_{3.2}$ $\chi_2\cos\omega_2 = k_{1.2[sic]}k_{2.1}$ $\chi_3\cos\omega_3 = k_{3.1}$ where $\varrho\sin\vartheta = \Delta y = y - y_0$ where $\vartheta = \arctan(\Delta y/\Delta x)$ $\varrho\cos\vartheta = \Delta x = x - x_0$ $\varrho^2 = \left(\Delta y^2 + \Delta x^2\right)$ $x' = x_0' + \Delta x - (1-k_{1.1})\Delta x - k_{1.2}\Delta y + \chi_2\varrho^2\cos(2\vartheta+\omega_2) + \chi_3\varrho^3\cos(3\vartheta+\omega_3)$ $y' = y_0' + \Delta y - (1-k_{1.1})\Delta y - k_{1.2}\Delta x + \chi_2\varrho^2\sin(2\vartheta+\omega_2) + \chi_3\varrho^3\sin(3\vartheta+\omega_3)$	Eqn 22-8

22.1.4 Comments on zone-to-zone transformations.

This chapter has been an exploration of a redundant technique but has given rise to a conundrum with the still current zone to zone formula published in the GDA Technical Manual. Compare the formula to with those of Hristow and Jordan. Hristow's transformation seems to be the easiest method by of its table (Figure 22-4) with the ability to maintain signs.

Table 22-3 Comparison of the three transformation methods

GDA -	$N_2 = N_z + (N_1 - N_z)\cos 2\gamma_z + (E_1' - E_z')\sin 2\gamma_z + H_1 L^2 \cos(2\theta_z + J_1)$
Hristow -	$x' = x_0' + \Delta x - (1 - k_{1.1})\Delta x - k_{1.2}\Delta y + \chi_2 \varrho^2 \cos(2\vartheta + \omega_2) + \chi_3 \varrho^3 \cos(3\vartheta + \omega_3)$
Jordan -	$(N)x' = x_0 + \Delta x - (1 - \cos 2\gamma_0)\Delta x - \sin 2\gamma_0 \Delta y + s_2 r^2 \cos(2\upsilon + u_2) + s_3 r^3 \cos(3\upsilon + u_3)$
GDA -	$E_2 = 500000 - E_z' + (E_1' - E_z')\cos 2\gamma_z - (N_1 - N_z)\sin 2\gamma_z + H_1 L^2 \sin(2\theta_z + J_1)$
Hristow -	$y' = y_0' + \Delta y - (1 - k_{1.1})\Delta y - k_{1.2}\Delta x + \chi_2 \varrho^2 \sin(2\vartheta + \omega_2) + \chi_3 \varrho^3 \sin(3\vartheta + \omega_3)$
Jordan -	$(E)y' = -y_0 + \Delta y + \sin 2\gamma_0 \Delta x - (1 - \cos 2\gamma_0)\Delta y + s_2 r^2 \sin(2\upsilon + u_2) + s_3 r^3 \sin(3\upsilon + u_3)$

The GDA version has dropped the $(1-\cos 2\gamma_0)$ value and substituted $\sin 2\gamma_0$ as a simplification. However the simplification of $(1-\cos 2\gamma_0)$ is $2\sin^2 \gamma_0$.

As an example, if $2\gamma_0 = 2°22'49''$ then $\sin 2\gamma_0 = 0.0415$, $(1-\cos 2\gamma_0) = 0.00086$ and $2\sin^2 \gamma_0 = 0.00086$

22.1.4.1 Correcting the GDA TM

It is of mere academic interest that this rabbit hole was examined. The Natmap computations are a truncation of Hristow and Jordan. Only latitude constants H_1 and J_1 were calculated (corresponding to u_2 an χ_2), perhaps feeling that the 15′ latitude intervals would compensate for the lack of very minor constant u_3 an χ_3. The three methods are displayed in Table 22-3. Further comment may be forthcoming from geodetic historians.

22.1.4.2 Historical spheroids

Whilst we now use GDA parameters, and the software easily produces corresponding AGD coordinates there are many historical spheroids/ellipsoids in use in Australia. Each tried to fit the local or continental figure of the mapping area. Prior to the Australian National Spheroid (ANS) and the Geodetic Datum of Australia (GDA) the Clarke 1858 ellipsoid was popularly used in Australia.

An extract from the ASPE&RS provides a snippet of history.

1. Mugnier, Clifford J., Grids and Datums, Commonwealth of Australia. Photogrammetric Engineering & Remote Sensing. Vol. 87, No 11, November 2021, pp. 795-797

https://www.asprs.org/wp-content/uploads/2022/01/11-21-GD-Australia.pdf

"Prior to the Australian Geodetic Datum of 1966, the Clarke 1858 ellipsoid as used in Tasmania was a = 6,378,293.645 meters and 1/f = 294.26 and in Australia proper was a = 6,378,339.78 meters and 1/f = 294.26. The difference between the two was the Clarke foot = 0.3047972654 meters versus the British foot of 1926 = 0.30479947 meters. Of the earlier more important datum origins, there were: Sydney Observatory where: Φo = 33° 51′ 41.10″ S and Λo = 151° 12′ 17.85″ E, Perth Observatory 1899 where: Φo = 31° 57′ 09.63″ S and Λo = 115° 50′ 26.10″ E, Darwin Origin Pillar where: Φo = 12° 28′ 08.452″ S and Λo = 130° 50′ 19.802″ E, and Lochmaben Astro Station in Tasmania where: Φo = 41° 38′ 23.389″ S and Λo = 147° 17′ 49.725″ E. The astronomic longitudes differed from geodetic longitudes on either the Sidney [sic] or Perth origins on an average of 10″, which indicated the magnitude of the deflections of the vertical".

2. https://www.ga.gov.au/scientific-topics/positioning-navigation/australian-geospatial-reference-system/agrstoolsandmodels
3. Algorithms for geodesics Charles F. F. Karney. 2012, Springerlink.com. J Geod (2013) 87:43–55. DOI 10.1007/s00190-012-0578-

22.2 References for Chapter 22

Jordan's Handbook of Geodesy Volume III, second half, ~~selected portions~~. English translation by Martha W. Carter, Corps of Engineers, U.S. Army, Army Map Service, Washington 25, D.C.

Jordan, W. und O. Eggert, Handbuch der Vermessungskunde. Dritter Band. Zweiter Halbband. Sphäroidische Berechnungen, konforme Abbildung des Erdellipsoids und Aufgaben der Erdmessung. J.B. Metzlersche Verlagsbuchhandlung, Stuttgart, 1941

Jordan, W and Eggert, O. Handbuch der Vermessungskunde. Dritter Band. Zweiter Halbband. Sphäroidische Berechnungen, konforme Abbildung des Erdellipsoids und Aufgaben der Erdmessung. Metzlersche Verlagsbuchhandlung, Stuttgart, 1941

Hristow. Transformation zwischen zwei Gauß-Krügerschen Streifen. Zeitschritt für P Vermessungswesen Heft 17 Band LXVII, 1938.

Ueber die Transformation *zwischen zwei Gauß-Krügerschen Streifen mit Anwendung auf Deutschland.*

Von Dr. Wl. K. Hristow, Sofia.

Chamber's Seven-Figure Mathematical Tables, James Pryde, ed. Full Edition, W.&R. Chambers Ltd, Edinburgh, 1961.

National Mapping Council of Australia. The Australian Map Grid Technical Manual, Special Publication 7. AGPS, Canberra 1972. ISBN 0 642 00057 3

If it's not checked it's wrong.

The manufacturer's authorised representative in the EU is Springer Nature Customer Service Centre GmbH, Europaplatz 3, 69115 Heidelberg, Germany. If you have any concerns regarding our products, please contact ProductSafety@springernature.com

Printed and bound by CPI Group (UK) Ltd, Croydon, CR0 4YY
18/09/2026
02218914-0001